Student Solutions Manual

College Mathematics for the Managerial, Life, and Social Sciences

SEVENTH EDITION

Soo T. Tan
Stonehill College

Australia • Brazil • Canada • Mexico • Singapore • Spain • United Kingdom • United States

Printed in the United States of America

1 2 3 4 5 6 7 11 10 09 08 07

Printer: Thomson/West

ISBN-13: 978-0-495-11974-6
ISBN-10: 0-495-11974-1

Thomson Higher Education
10 Davis Drive
Belmont, CA 94002-3098
USA

For more information about our products, contact us at:
Thomson Learning Academic Resource Center
1-800-423-0563

COLLEGE MATHEMATICS
For The Managerial, Life, and Social Sciences

CONTENTS

CHAPTER 5 MATHEMATICS OF FINANCE

CHAPTER 6 SETS AND COUNTING

CHAPTER 7 PROBABILITY

CHAPTER 8 PROBABILITY DISTRIBUTIONS AND STATISTICS

CHAPTER 9 MARKOV CHAINS

CHAPTER 10 PRECALCULUS REVIEW

CHAPTER 11 FUNCTIONS, LIMITS, AND THE DERIVATIVE

CHAPTER 12 DIFFERENTIATION

CHAPTER 13 APPLICATIONS OF THE DERIVATIVE

CHAPTER 14 EXPONENTIAL AND LOGARITHMIC FUNCTIONS

CHAPTER 15 INTEGRATION

CHAPTER 16 ADDITIONAL TOPICS IN INTEGRATION

CHAPTER 17 CALCULUS OF SEVERAL VARIABLES

CHAPTER 1

1.1 Problem Solving Tips

Suppose you are asked to determine whether a given statement is true or false, and you are also asked to explain your answer. How would you answer the question?

If you think the statement is true, then prove it. On the other hand, if you think the statement is false, then give an example that disproves the statement. For example, the statement "If a and b are real numbers, then $a - b = b - a$" is false and an example that disproves it may be constructed by taking $a = 3$ and $b = 5$. For these values of a and b, we find $a - b = 3 - 5 = -2$ but $b - a = 5 - 3 = 2$ and this shows that $a - b \neq b - a$. Such an example is called a **counterexample**.

1.1 CONCEPT QUESTIONS, page 7

1. a. $a < 0$ and $b > 0$; b. $a < 0$ and $b < 0$ c. $a > 0$ and $b < 0$

EXERCISES 1.1, page 7

1. The coordinates of A are (3, 3) and it is located in Quadrant I.

3. The coordinates of C are (2, -2) and it is located in Quadrant IV.

5. The coordinates of E are (-4, -6) and it is located in Quadrant III.

7. A 9. E, F, and G 11. F

For Exercises 13-19, refer to the figure that follows.

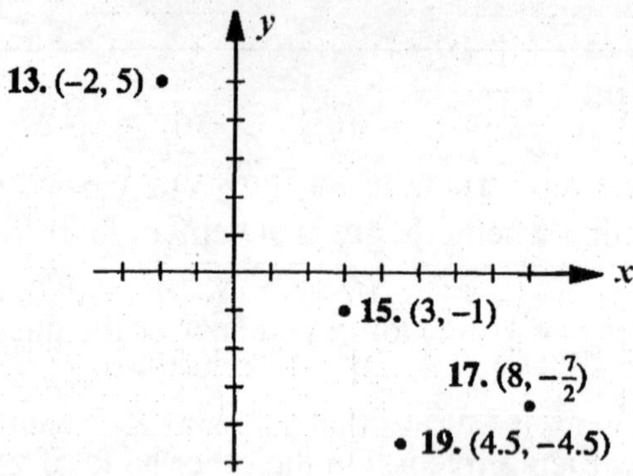

21. Using the distance formula, we find that $\sqrt{(4-1)^2+(7-3)^2}=\sqrt{3^2+4^2}=\sqrt{25}=5$.

23. Using the distance formula, we find that
$$\sqrt{(4-(-1))^2+(9-3)^2}=\sqrt{5^2+6^2}=\sqrt{25+36}=\sqrt{61}.$$

25. The coordinates of the points have the form $(x,-6)$. Since the points are 10 units away from the origin, we have
$$(x-0)^2+(-6-0)^2=10^2$$
$$x^2=64,$$
or $x=\pm 8$. Therefore, the required points are $(-8,-6)$ and $(8,-6)$.

27. The points are shown in the following diagram:

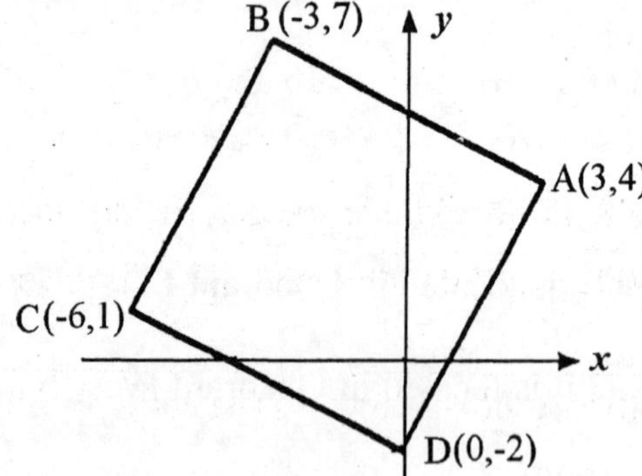

To show that the four sides are equal, we compute the following:

$$d(A,B) = \sqrt{(-3-3)^2 + (7-4)^2} = \sqrt{(-6)^2 + 3^2} = \sqrt{45}$$
$$d(B,C) = \sqrt{[(-6-(-3)]^2 + (1-7)^2} = \sqrt{(-3)^2 + (-6)^2} = \sqrt{45}$$
$$d(C,D) = \sqrt{[0-(-6)]^2 + [(-2)-1]^2} = \sqrt{(6)^2 + (-3)^2} = \sqrt{45}$$
$$d(A,D) = \sqrt{(0-3)^2 + (-2-4)^2} = \sqrt{(3)^2 + (-6)^2} = \sqrt{45}.$$

Next, to show that ΔABC is a right triangle, we show that it satisfies the Pythagorean Theorem. Thus,

$$d(A,C) = \sqrt{(-6-3)^2 + (1-4)^2} = \sqrt{(-9)^2 + (-3)^2} = \sqrt{90} = 3\sqrt{10}$$

and $[d(A,B)]^2 + [d(B,C)]^2 = 90 = [d(A,C)]^2$. Similarly, $d(B,D) = \sqrt{90} = 3\sqrt{10}$, so ΔBAD is a right triangle as well. It follows that $\angle B$ and $\angle D$ are right angles, and we conclude that $ADCB$ is a square.

29. The equation of the circle with radius 5 and center (2, -3) is given by

$$(x-2)^2 + [y-(-3)]^2 = 5^2$$

or $$(x-2)^2 + (y+3)^2 = 25.$$

31. The equation of the circle with radius 5 and center (0, 0) is given by

$$(x-0)^2 + (y-0)^2 = 5^2$$

or $$x^2 + y^2 = 25.$$

33. The distance between the points (5, 2) and (2, -3) is given by

$$d = \sqrt{(5-2)^2 + (2-(-3))^2} = \sqrt{3^2 + 5^2} = \sqrt{34}.$$

Therefore $r = \sqrt{34}$ and the equation of the circle passing through (5, 2) and centered at (2, -3) is

$$(x-2)^2 + [y-(-3)]^2 = 34$$

or $$(x-2)^2 + (y+3)^2 = 34.$$

35. Referring to the diagram on page 8 of the text, we see that the distance from A to B is given by $d(A,B) = \sqrt{400^2 + 300^2} = \sqrt{250{,}000} = 500$. The distance from B to C is given by

$$d(B,C) = \sqrt{(-800-400)^2 + (800-300)^2} = \sqrt{(-1200)^2 + (500)^2}$$
$$= \sqrt{1{,}690{,}000} = 1300.$$

The distance from C to D is given by

$$d(C,D) = \sqrt{[-800-(-800)]^2 + (800-0)^2} = \sqrt{0+800^2} = 800.$$

The distance from D to A is given by

$$d(D,A) = \sqrt{[0-(-800)]^2 + (0-0)} = \sqrt{640000} = 800.$$

Therefore, the total distance covered on the tour, is

$$d(A,B)+d(B,C)+d(C,D)+d(D,A) = 500+1300+800+800$$
$$= 3400, \quad \text{or } 3400 \text{ miles.}$$

37. Referring to the following diagram,

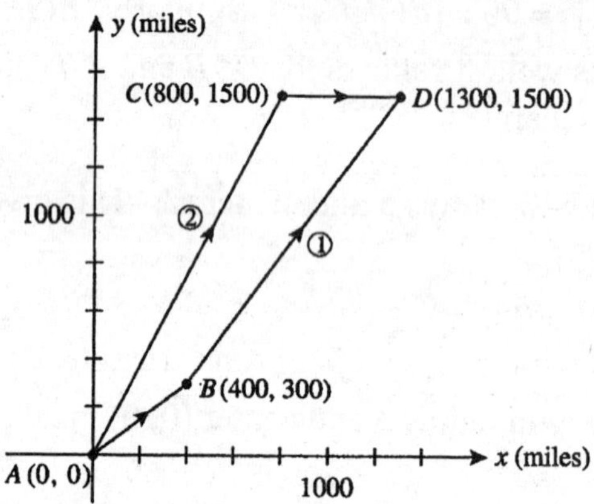

we see that the distance he would cover if he took Route (1) is given by

$$d(A,B)+d(B,D) = \sqrt{400^2+300^2} + \sqrt{(1300-400)^2+(1500-300)^2}$$
$$= \sqrt{250{,}000} + \sqrt{2{,}250{,}000} = 500+1500 = 2000,$$

or 2000 miles. On the other hand, the distance he would cover if he took Route (2) is given by

$$d(A,C)+d(C,D) = \sqrt{800^2+1500^2} + \sqrt{(1300-800)^2}$$
$$= \sqrt{2{,}890{,}000} + \sqrt{250{,}000} = 1700+500 = 2200,$$

or 2200 miles. Comparing these results, we see that he should take Route (1).

39. Calculations to determine VHF requirements:

$$d = \sqrt{25^2+35^2} = \sqrt{625+1225} = \sqrt{1850} \approx 43.01.$$

Models B through D satisfy this requirement.

Calculations to determine UHF requirements:

$$d = \sqrt{20^2+32^2} = \sqrt{400+1024} = \sqrt{1424} \approx 37.74$$

Models C through D satisfy this requirement. Therefore, Model C will allow him to receive both channels at the least cost.

41. a. Let the position of ship A and ship B after t hours be $A(0, y)$ and $B(x, 0)$, respectively. Then $x = 30t$ and $y = 20t$. Therefore, the distance between the two ships is

$$D = \sqrt{(30t)^2 + (20t)^2} = \sqrt{900t^2 + 400t^2} = 10\sqrt{13}t.$$

b. The required distance is obtained by letting $t = 2$ giving $D = 10\sqrt{13}(2)$ or approximately 72.11 miles.

43. False. The distance between $P_1(a,b)$ and $P_3(kc,kd)$ is

$$d = \sqrt{(kc-a)^2 + (kd-b)^2}$$

$$\neq |k|D = |k|\sqrt{(c-a)^2 + (d-b)^2} = \sqrt{k^2(c-a)^2 + k^2(d-b)^2} = \sqrt{[k(c-a)]^2 + [k(d-b)]^2}$$

45. Referring to the figure in the text, we see that the distance between the two points is given by the length of the hypotenuse of the right triangle. That is,

$$d = \sqrt{(x_2 - x_1)^2 + (y_2 - y_1)^2}$$

1.2 Problem Solving Tips

When you solve a problem in the exercises that follow each section, first read the problem. Then, before you start computing or writing out a solution, try to formulate a strategy for solving the problem. Then use your strategy to solve the problem.

Here we summarize some general problem-solving techniques that have been covered in this section.

1. **To show that two lines are parallel,** you need to show that the slopes of the two lines are equal or their slopes are undefined.

2. **To show that two lines L_1 and L_2 are perpendicular,** you need to show that the slope m_1 of L_1 is the negative reciprocal of the slope m_2 of L_2; that is, $m_1 = -1/m_2$.

3. **To find the equation of a line,** you need the slope of the line and a point lying on the line. You can then find the equation of the line by using the point-slope form of the equation of a line: $(y - y_1) = m(x - x_1)$

1.2 CONCEPT QUESTIONS, page 19

1. The slope is $m = \dfrac{y_2 - y_1}{x_2 - x_1}$, where $P(x_1, y_1)$ and $P(x_2, y_2)$ are any two distinct points on the nonvertical line. The slope of a vertical line is undefined.

3. a. $m_1 = m_2$ b. $m_2 = -\dfrac{1}{m_1}$

EXERCISES 1.2, page 19

1. Referring to the figure shown in the text, we see that $m = \dfrac{2-0}{0-(-4)} = \dfrac{1}{2}$.

3. This is a vertical line, and hence its slope is undefined.

5. $m = \dfrac{y_2 - y_1}{x_2 - x_1} = \dfrac{8-3}{5-4} = 5.$

7. $m = \dfrac{y_2 - y_1}{x_2 - x_1} = \dfrac{8-3}{4-(-2)} = \dfrac{5}{6}.$

9. $m = \dfrac{y_2 - y_1}{x_2 - x_1} = \dfrac{d-b}{c-a}.$

11. Since the equation is in the slope-intercept form, we read off the slope $m = 4$.
    a. If x increases by 1 unit, then y increases by 4 units.
    b. If x decreases by 2 units, y decreases by $4(-2) = -8$ units.

13. The slope of the line through A and B is $\frac{-10-(-2)}{-3-1} = \frac{-8}{-4} = 2$.

The slope of the line through C and D is $\frac{1-5}{-1-1} = \frac{-4}{-2} = 2$.

Since the slopes of these two lines are equal, the lines are parallel.

15. The slope of the line through A and B is $\frac{2-5}{4-(-2)} = -\frac{3}{6} = -\frac{1}{2}$.

The slope of the line through C and D is $\frac{6-(-2)}{3-(-1)} = \frac{8}{4} = 2$.

Since the slopes of these two lines are the negative reciprocals of each other, the lines are perpendicular.

17. The slope of the line through the point $(1, a)$ and $(4, -2)$ is $m_1 = \frac{-2-a}{4-1}$, and the slope of the line through $(2, 8)$ and $(-7,\ a+4)$ is $m_2 = \frac{a+4-8}{-7-2}$. Since these two lines are parallel, m_1 is equal to m_2. Therefore,

$$\frac{-2-a}{3} = \frac{a-4}{-9}$$

$$-9(-2-a) = 3(a-4)$$

$$18+9a = 3a-12$$

$$6a = -30 \qquad \text{and } a = -5$$

19. An equation of a horizontal line is of the form $y = b$. In this case $b = -3$, so $y = -3$ is an equation of the line.

21. e 23. a 25. f

27. We use the point-slope form of an equation of a line with the point $(3, -4)$ and slope $m = 2$. Thus $y - y_1 = m(x - x_1)$,

and

$$y - (-4) = 2(x - 3)$$

$$y + 4 = 2x - 6$$

$$y = 2x - 10.$$

29. Since the slope $m = 0$, we know that the line is a horizontal line of the form $y = b$. Since the line passes through $(-3, 2)$, we see that $b = 2$, and an equation of the line is

$y = 2.$

31. We first compute the slope of the line joining the points (2, 4) and (3, 7). Thus,

$$m = \frac{7-4}{3-2} = 3.$$

Using the point-slope form of an equation of a line with the point (2,4) and slope $m = 3$, we find

$$\begin{aligned} y - 4 &= 3(x - 2) \\ y &= 3x - 2. \end{aligned}$$

33. We first compute the slope of the line joining the points (1, 2) and (–3, –2). Thus,

$$m = \frac{-2-2}{-3-1} = \frac{-4}{-4} = 1.$$

Using the point-slope form of an equation of a line with the point (1, 2) and slope $m = 1$, we find

$$\begin{aligned} y - 2 &= x - 1 \\ y &= x + 1. \end{aligned}$$

35. We use the slope-intercept form of an equation of a line: $y = mx + b$. Since $m = 3$, and $b = 4$, the equation is $y = 3x + 4$.

37. We use the slope-intercept form of an equation of a line: $y = mx + b$. Since $m = 0$, and $b = 5$, the equation is $y = 5$.

39. We first write the given equation in the slope-intercept form:

$$\begin{aligned} x - 2y &= 0 \\ -2y &= -x \\ y &= \tfrac{1}{2}x \,. \end{aligned}$$

From this equation, we see that $m = 1/2$ and $b = 0$.

41. We write the equation in slope-intercept form:

$$\begin{aligned} 2x - 3y - 9 &= 0 \\ -3y &= -2x + 9 \\ y &= \tfrac{2}{3}x - 3. \end{aligned}$$

From this equation, we see that $m = 2/3$ and $b = -3$.

43. We write the equation in slope-intercept form:

$$2x + 4y = 14$$

$$4y = -2x + 14$$
$$y = -\tfrac{2}{4}x + \tfrac{14}{4}$$
$$= -\tfrac{1}{2}x + \tfrac{7}{2}.$$

From this equation, we see that $m = -1/2$ and $b = 7/2$.

45. We first write the equation $2x - 4y - 8 = 0$ in slope- intercept form:

$$2x - 4y - 8 = 0$$
$$4y = 2x - 8$$
$$y = \tfrac{1}{2}x - 2$$

Now the required line is parallel to this line, and hence has the same slope. Using the point-slope equation of a line with $m = 1/2$ and the point $(-2, 2)$, we have

$$y - 2 = \tfrac{1}{2}[x - (-2)] = \tfrac{1}{2}x + 1$$
$$y = \tfrac{1}{2}x + 3.$$

47. A line parallel to the x-axis has slope 0 and is of the form $y = b$. Since the line is 6 units below the axis, it passes through $(0,-6)$ and its equation is $y = -6$.

49. We use the point-slope form of an equation of a line to obtain

$$y - b = 0(x - a) \quad \text{or} \quad y = b.$$

51. Since the required line is parallel to the line joining $(-3, 2)$ and $(6, 8)$, it has slope

$$m = \frac{8-2}{6-(-3)} = \frac{6}{9} = \frac{2}{3}.$$

We also know that the required line passes through $(-5, -4)$. Using the point-slope form of an equation of a line, we find

$$y - (-4) = \tfrac{2}{3}[x - (-5)]; \quad y = \tfrac{2}{3}x + \tfrac{10}{3} - 4, \text{ or } y = \tfrac{2}{3}x - \tfrac{2}{3}$$

53. Since the point $(-3, 5)$ lies on the line $kx + 3y + 9 = 0$, it satisfies the equation. Substituting $x = -3$ and $y = 5$ into the equation gives

$$-3k + 15 + 9 = 0 \quad \text{or} \quad k = 8.$$

55. $3x - 2y + 6 = 0$

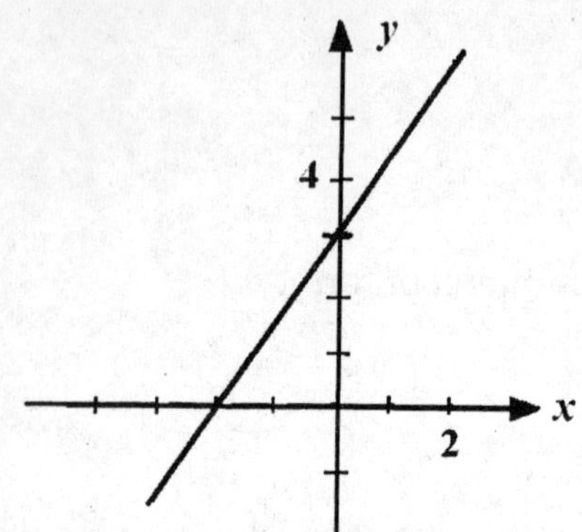

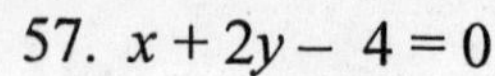

57. $x + 2y - 4 = 0$

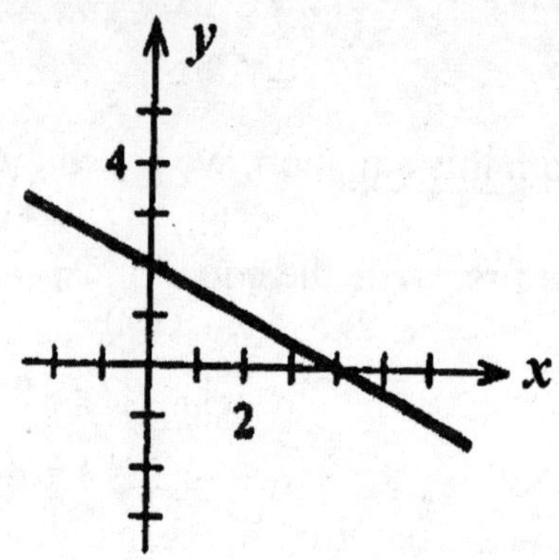

59. $y + 5 = 0$

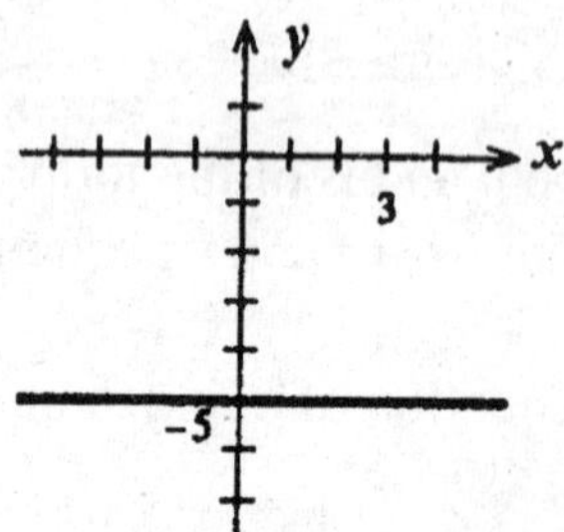

61. Since the line passes through the points $(a, 0)$ and $(0, b)$, its slope is $m = \dfrac{b-0}{0-a} = -\dfrac{b}{a}$. Then, using the point-slope form of an equation of a line with the point $(a, 0)$ we have

$$y - 0 = -\tfrac{b}{a}(x - a)$$

$$y = -\tfrac{b}{a}x + b$$

which may be written in the form $\tfrac{b}{a}x + y = b$. Multiplying this last equation by $1/b$, we have $\dfrac{x}{a} + \dfrac{y}{b} = 1$.

63. Using the equation $\dfrac{x}{a} + \dfrac{y}{b} = 1$ with $a = -2$ and $b = -4$, we have $-\dfrac{x}{2} - \dfrac{y}{4} = 1$.

Then

$$-4x - 2y = 8$$
$$2y = -8 - 4x$$
$$y = -2x - 4.$$

65. Using the equation $\frac{x}{a} + \frac{y}{b} = 1$ with $a = 4$ and $b = -1/2$, we have

$$\frac{x}{4} + \frac{y}{-\frac{1}{2}} = 1$$
$$-\tfrac{1}{4}x + 2y = -1$$
$$2y = \tfrac{1}{4}x - 1, \quad y = \tfrac{1}{8}x - \tfrac{1}{2}.$$

67. The slope of the line passing through A and B is $m = \frac{7-1}{1-(-2)} = \frac{6}{3} = 2$,

and the slope of the line passing through B and C is $m = \frac{13-7}{4-1} = \frac{6}{3} = 2$.

Since the slopes are equal, the points lie on the same line.

69. a.

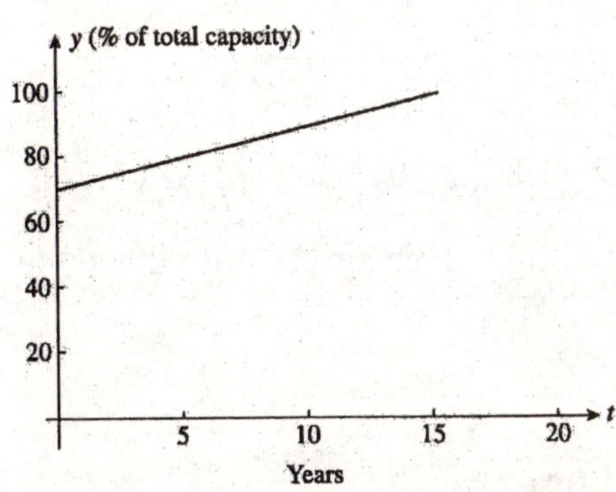

b. The slope is 1.9467 and the y-intercept is 70.082.

c. The output is increasing at the rate of 1.9467%/yr; the output at the beginning of 1990 was 70.082%.

d. We solve the equation $1.9467t + 70.082 = 100$ giving $t = 15.37$. We conclude that the plants will be generating at maximum capacity shortly after 2005.

71. a. $y = 0.55x$

b. Solving the equation $1100 = 0.55x$ for x, we have $x = \frac{1100}{0.55} = 2000$.

73. Using the points (0, 0.68) and (10, 0.80), we see that the slope of the required line is

$$m=\frac{0.80-0.68}{10-0}=\frac{0.12}{10}=.012.$$

Next, using the point-slope form of the equation of a line, we have

$$y-0.68=0.012(t-0)$$

or

$$y=0.012t+0.68.$$

Therefore, when $t=14$, we have

$$y=0.012(14)+0.68$$
$$=.848$$

or 84.8%. That is, in 2004 women's wages are expected to be 84.8% of men's wages.

75. a. – b.

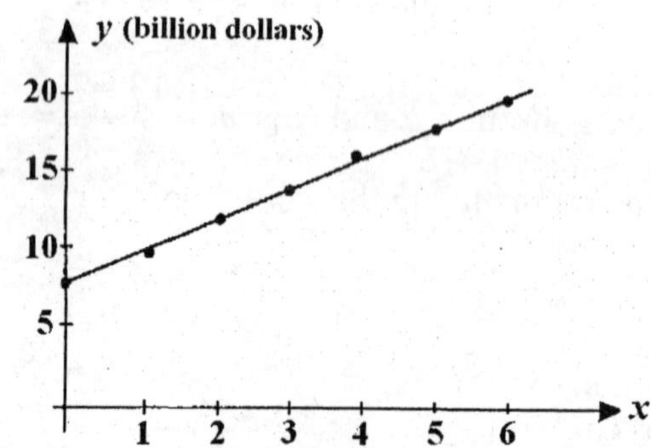

c. $m=\dfrac{18.8-7.9}{6-0}\approx 1.82$, $y-7.9=1.82(x-0)$, or $y=1.82x+7.9$.

d. $y=1.82(5)+7.9\approx 17$ or \$17 billion. This agrees with the actual data for that year.

77. a. – b.

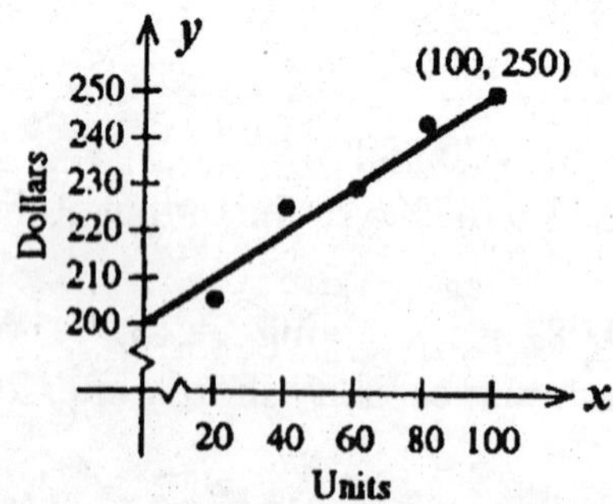

c. Using the points (0, 200) and (100, 250), we see that the slope of the required line is $m=\dfrac{250-200}{100}=\dfrac{1}{2}$. Therefore, the required equation is

$y - 200 = \frac{1}{2}x$ or $y = \frac{1}{2}x + 200$.

d. The approximate cost for producing 54 units of the commodity is
$\frac{1}{2}(54) + 200$, or \$227.

79. a. – b.

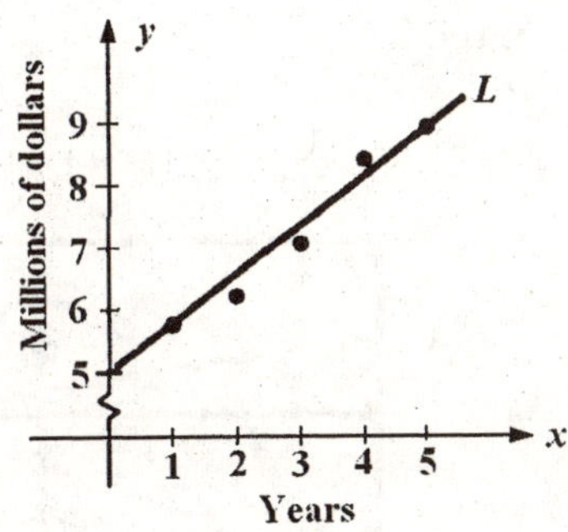

c. The slope of L is $m = \dfrac{9.0 - 5.8}{5 - 1} = \dfrac{3.2}{4} = 0.8$. Using the point-slope form of an equation of a line, we have $y - 5.8 = 0.8(x - 1) = 0.8x - 0.8$, or $y = 0.8x + 5$.

d. Using the equation of part (c) with $x = 9$, we have
$y = 0.8(9) + 5 = 12.2$, or \$12.2 million.

81. a. We obtain a family of parallel lines each having slope m.
b. We obtain a family of straight lines all of which pass through the point $(0,b)$.

83. False. Substituting $x = -1$ and $y = 1$ into the equation gives $3(-1) + 7(1) = 4$, and this is not equal to the right-hand side of the equation. Therefore, the equation is not satisfied and so the given point does not lie on the line.

85. True. The slope of the line $Ax + By + C = 0$ is $-A/B$. (Write it in the slope-intercept form.) Similarly, the slope of the line $ax + by + c = 0$ is $-a/b$. They are parallel if and only if $-\dfrac{A}{B} = -\dfrac{a}{b}$, $Ab = aB$, or $Ab - aB = 0$.

87. True. The slope of the line $ax + by + c_1 = 0$ is $m_1 = -a/b$. The slope of the line $bx - ay + c_2 = 0$ is $m_2 = b/a$. Since $m_1 m_2 = -1$, the straight lines are indeed perpendicular.

89. Writing each equation in the slope-intercept form, we have

$$y = -\frac{a_1}{b_1}x - \frac{c_1}{b_1} \quad (b_1 \neq 0) \quad \text{and} \quad y = -\frac{a_2}{b_2}x - \frac{c_2}{b_2} \quad (b_2 \neq 0)$$

Since two lines are parallel if and only if their slopes are equal, we see that the lines are parallel if and only if $-\frac{a_1}{b_1} = -\frac{a_2}{b_2}$, or $a_1b_2 - b_1a_2 = 0$.

USING TECHNOLOGY EXERCISES 1.2, page 27

1.

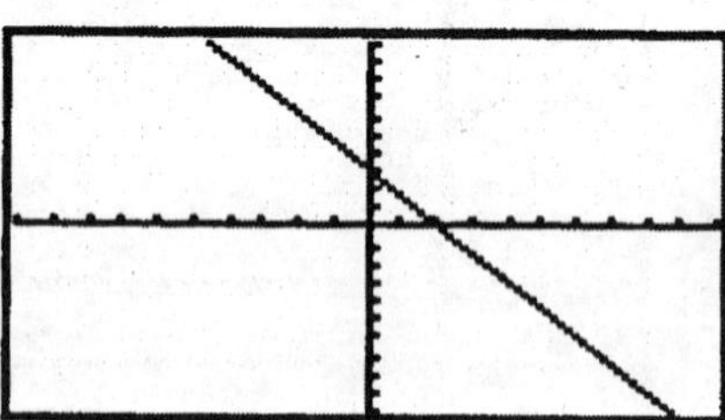

3.

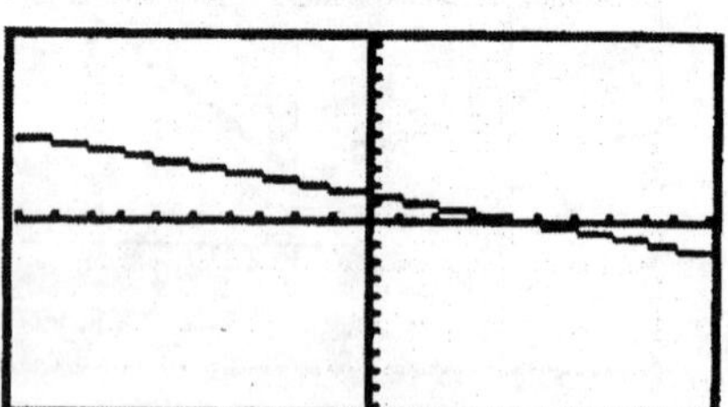

5.

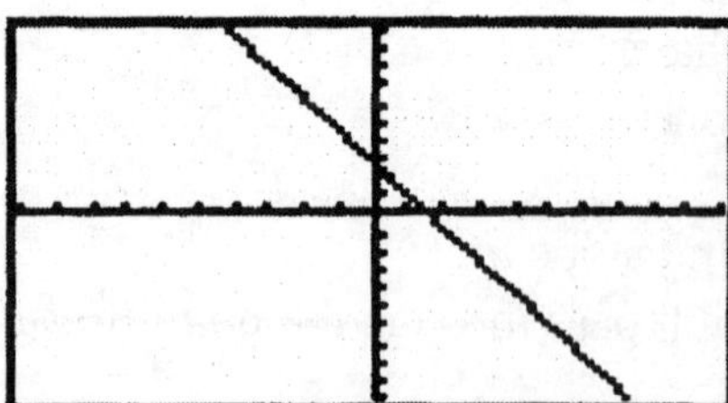

7. a.

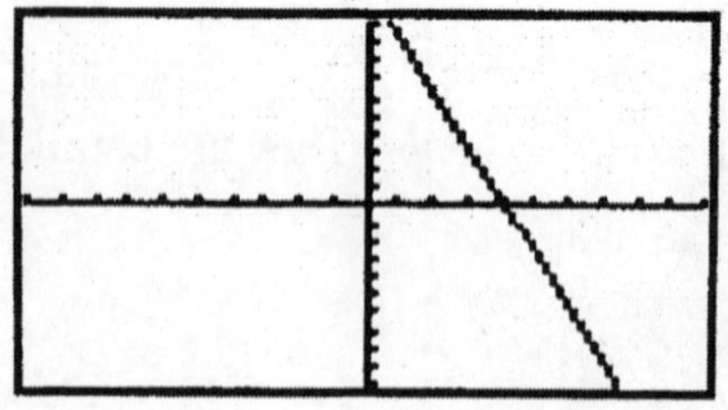

b.

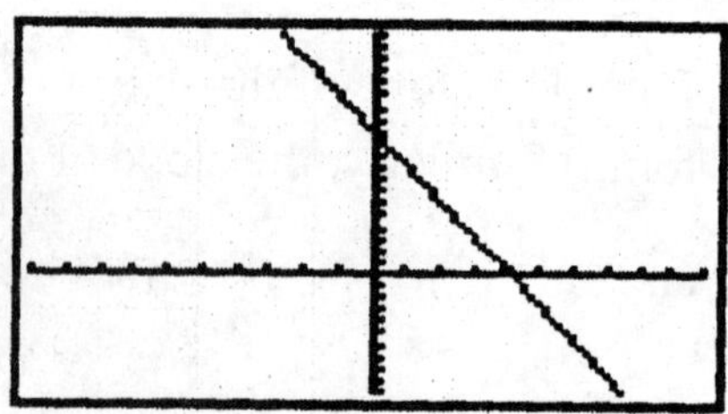

9. a.

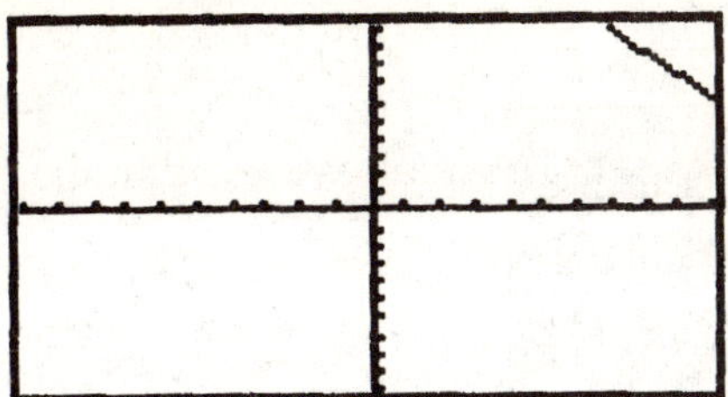

b.

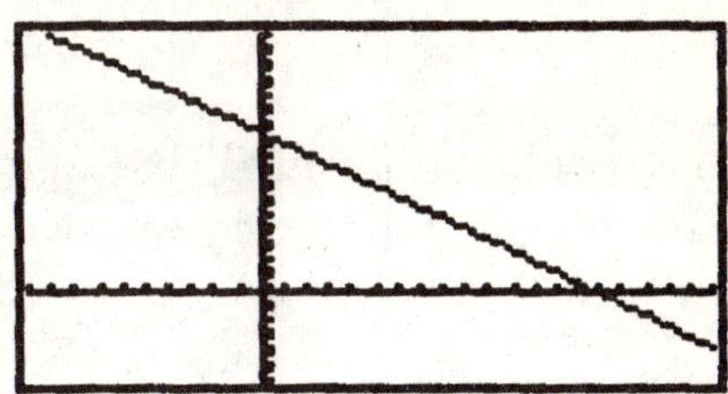

11.

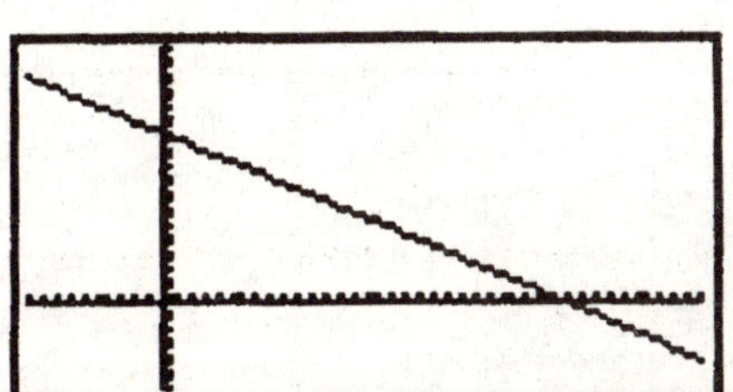

13.

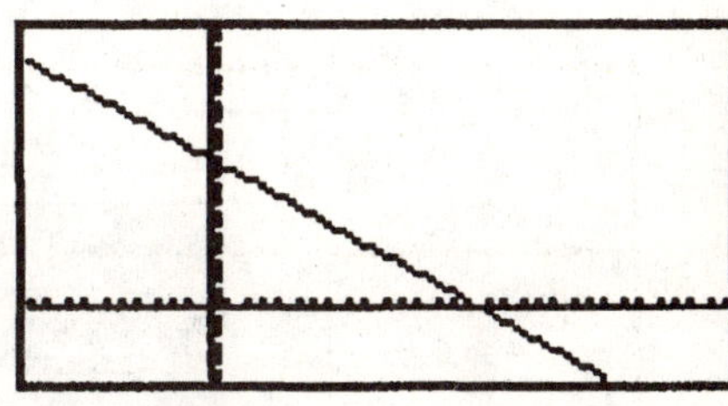

15.

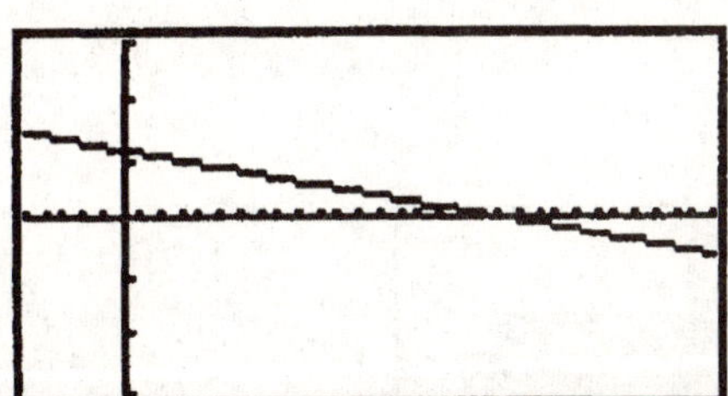

17.

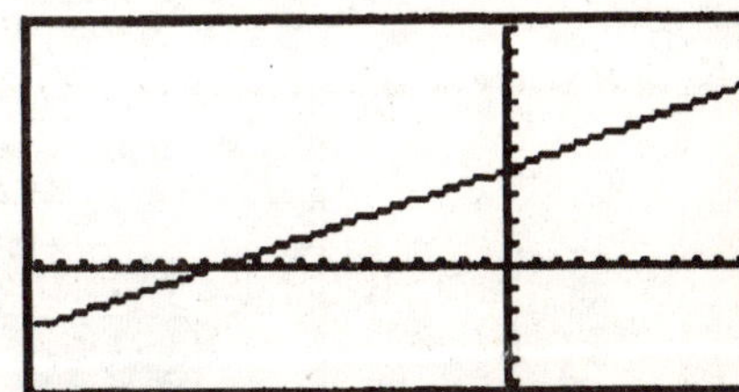

EXCEL

1.

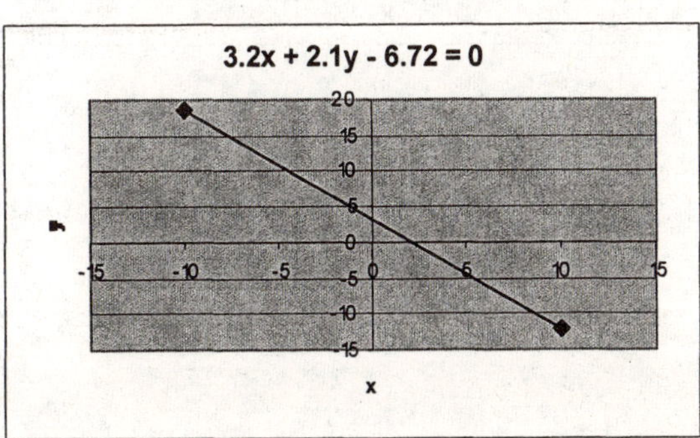

3.

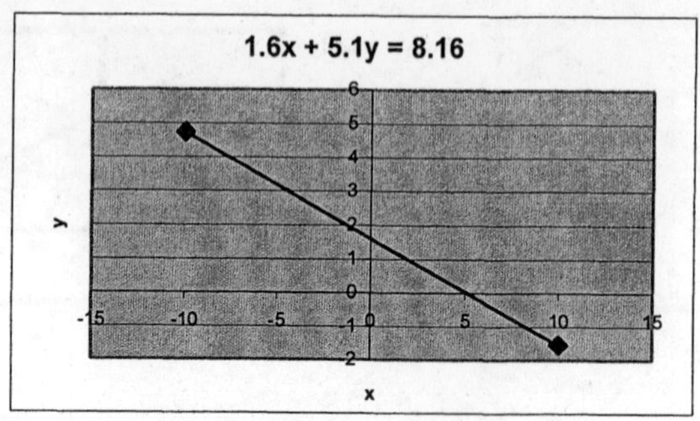
1.6x + 5.1y = 8.16
y
x
-15
-10
-5
0
5
10
15
6
5
4
3
2
1
-1
-2

5.

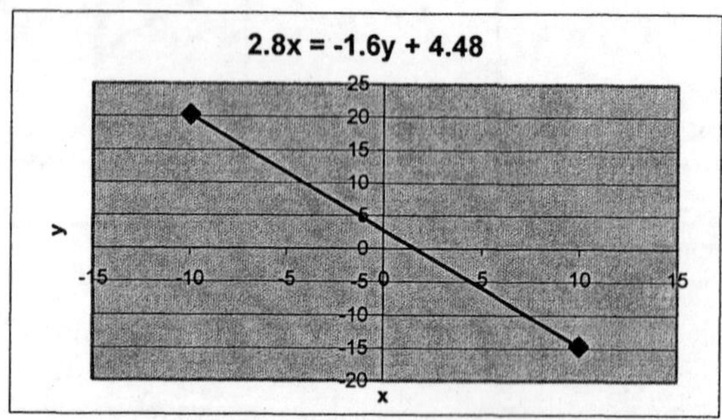
2.8x = -1.6y + 4.48
y
x
-15
-10
-5
0
5
10
15
25
20
15
10
-10
-15
-20

7.

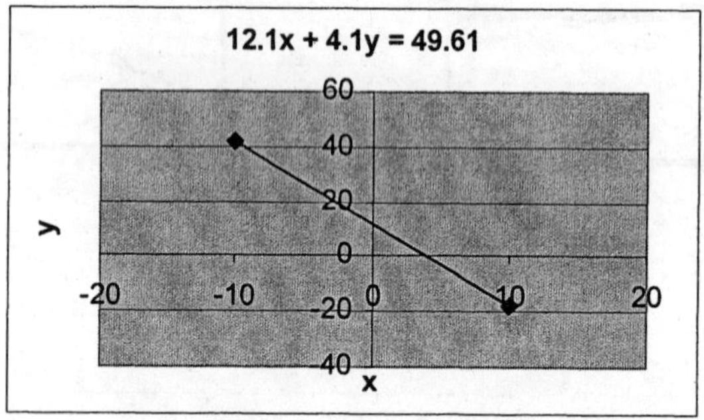
12.1x + 4.1y = 49.61
y
x
-20
-10
0
10
20
60
40
-40

9.

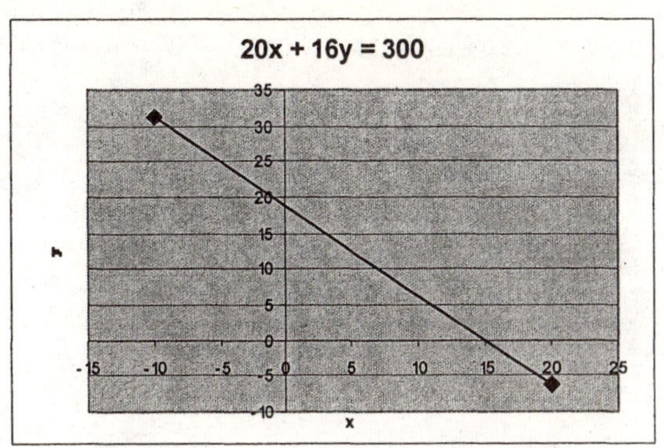

11.

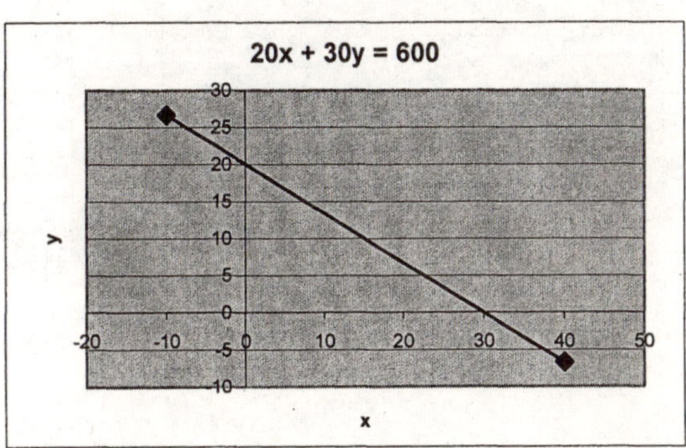

13.

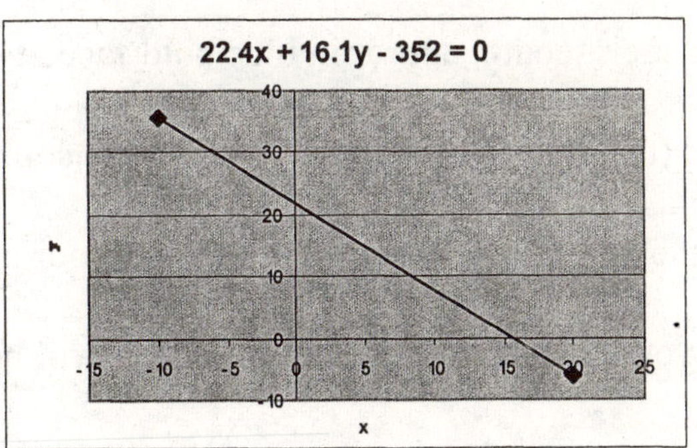

15.

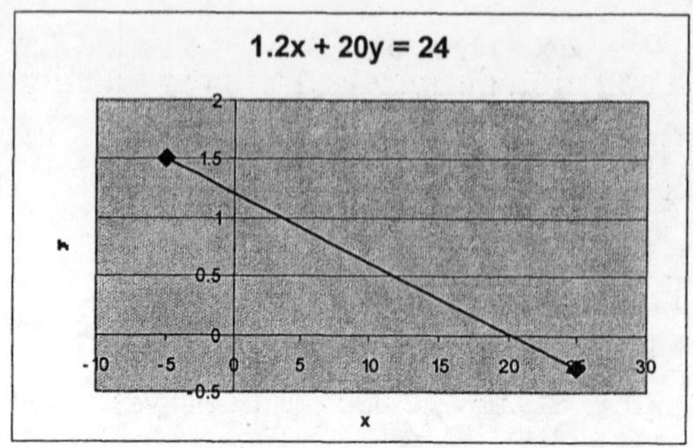

17.

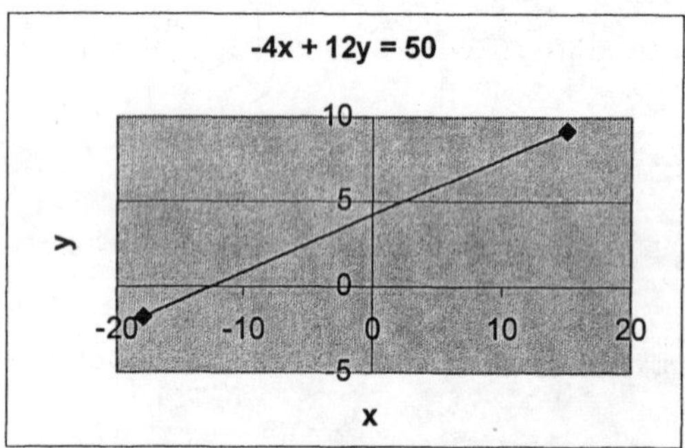

1.3 Problem Solving Tips

New mathematical terms in each section appear in blue bold-faced type along with their definition or they are boxed (the green boxes). Each time you encounter a new term, read through the definition and then try to express the definition in your own words without looking at the book. Once you understand these definitions, it will be easier for you to work the exercise sets that follow each section.

Here are some hints for solving the problems in the exercises that follow:

1. To determine whether a given equation defines y as a linear function of x, check to see that the given equation has the form $Ax + By + C = 0$, where A, B, and C are constants and A and B are not both zero.
2. Since the demand for a commodity decreases as its unit price increases, a demand function is generally a decreasing function So a linear demand function will have a negative slope and its graph will slant downwards as we move from left to right along the x-axis. Similarly, since the supply of a commodity increases as the unit price increases, a supply function is generally an increasing function. So a linear supply function will have positive slope and its graph will slant upwards as we move from left to right along the x-axis.

1.3 CONCEPT QUESTIONS, page 35

1. a. A function is a rule that assigns to each element in a set A one and only one element in a set B.
   b. A linear function is one that has the form $f(x) = mx + b$, (m, b, constants). Example: $f(x) = 2x + 3$.
   c. $(-\infty, \infty)$; $(-\infty, \infty)$ d. A straight line

3. Negative; positive

EXERCISES 1.3, page 35

1. Yes. Solving for y in terms of x, we find $3y = -2x + 6$, or $y = -\frac{2}{3}x + 2$.

3. Yes. Solving for y in terms of x, we find $2y = x + 4$, or $y = \frac{1}{2}x + 2$.

5. Yes. Solving for y in terms of x, we have $4y = 2x + 9$, or $y = \frac{1}{2}x + \frac{9}{4}$.

7. y is not a linear function of x because of the quadratic term $2x^2$.

9. y is not a linear function of x because of the nonlinear term $-3y^2$.

11. a. $C(x) = 8x + 40{,}000$, where x is the number of units produced.
b. $R(x) = 12x$, where x is the number of units sold.
c. $P(x) = R(x) - C(x) = 12x - (8x + 40{,}000) = 4x - 40{,}000$.
d. $P(8{,}000) = 4(8{,}000) - 40{,}000 = -8{,}000$, or a loss of \$8,000.
$P(12{,}000) = 4(12{,}000) - 40{,}000 = 8{,}000$ or a profit of \$8,000.

13. $f(0) = 2$ gives $m(0) + b = 2$, or $b = 2$. So, $f(x) = mx + 2$. Next, $f(3) = -1$ gives $m(3) + 2 = -1$, or $m = -1$.

15. Let V be the book value of the office building after 2005. Since $V = 1{,}000{,}000$ when $t = 0$, the line passes through $(0, 1{,}000{,}000)$. Similarly, when $t = 50$, $V = 0$, so the line passes through $(50, 0)$. Then the slope of the line is given by

$$m = \frac{0 - 1{,}000{,}000}{50 - 0} = -20{,}000.$$

Using the point-slope form of the equation of a line with the point $(0, 1{,}000{,}000)$, we have $V - 1{,}000{,}000 = -20{,}000(t - 0)$, or $V = -20{,}000t + 1{,}000{,}000$.
In 2010, $t = 5$ and $V = -20{,}000(5) + 1{,}000{,}000 = 900{,}000$, or \$900,000.
In 2015, $t = 10$ and $V = -20{,}000(10) + 1{,}000{,}000 = 800{,}000$, or \$800,000.

17. The consumption function is given by $C(x) = 0.75x + 6$. When $x = 0$, we have $C(0) = 0.75(0) + 6 = 6$, or \$6 billion dollars.
If $x = 50$, $C(50) = 0.75(50) + 6 = 43.5$, or \$43.5 billion dollars.
If $x = 100$, $C(100) = 0.75(100) + 6 = 81$, or \$81 billion dollars.

19. a. $y = 1.053x$, where x is the monthly benefit before adjustment, and y is the adjusted monthly benefit.
b. His adjusted monthly benefit will be $(1.053)(1020) = 1074.06$, or \$1074.06.

21. Let the number of tapes produced and sold be x. Then

$$C(x) = 12{,}100 + 0.60x; \qquad R(x) = 1.15x$$

and $\quad P(x) = R(x) - C(x) = 1.15x - (12{,}100 + 0.60x)$
$= 0.55x - 12{,}100.$

23. Let the value of the server after t years be V. When $t = 0$, $V = 60{,}000$ and when $t = 4$, $V = 12{,}000$.

a. Since $\quad m = \dfrac{12{,}000 - 60{,}000}{4} = -\dfrac{48{,}000}{4} = -12{,}000$

the rate of depreciation ($-m$) is \$12,000.

b. Using the point-slope form of the equation of a line with the point (4, 12,000), we have $\quad V - 12{,}000 = -12{,}000(t - 4)$

or $\quad V = -12{,}000t + 60{,}000.$

c.

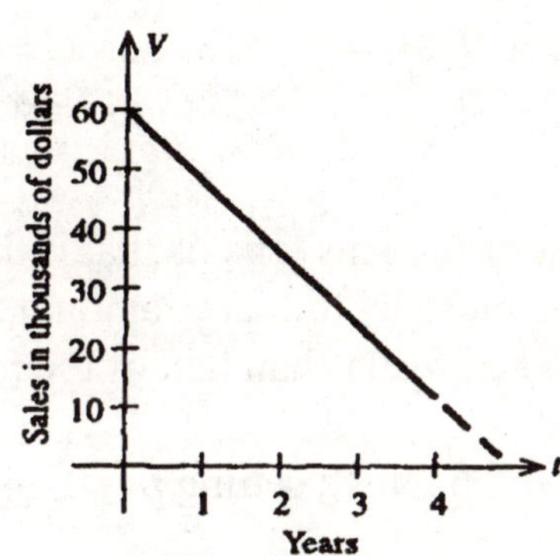

d. When $t = 3$, $V = -12{,}000(3) + 60{,}000 = 24{,}000$, or \$24,000.

25. The formula given in Exercise 24 is $V = C - \dfrac{C - S}{N}t$.

Then, when $C = 1{,}000{,}000$, $N = 50$, and $S = 0$, we have

$$V = 1{,}000{,}000 - \frac{1{,}000{,}000 - 0}{50}t \text{ or } \quad V = 1{,}000{,}000 - 20{,}000t.$$

In 2010, $t = 5$ and $\quad V = 1{,}000{,}000 - 20{,}000(5) = 900{,}000$, or \$900,000.
In 2015, $t = 10$ and $\quad V = 1{,}000{,}000 - 20{,}000(10) = 800{,}000$ or \$800,000.

27. a. $D(S) = \dfrac{Sa}{1.7}$. If we think of D as having the form $D(S) = mS + b$, then

$m = \dfrac{a}{1.7}$, $\quad b = 0$, and D is a linear function of S.

b. $D(0.4) = \dfrac{500(0.4)}{1.7} = 117.647$, or approximately 117.65 mg.

29. a. $f(t) = 7.5t + 20$, where $(0 \le t \le 6)$

b. $f(6) = 7.5(6) + 20 = 65$, or 65 million.

31. a. Since the relationship is linear, we can write $F = mC + b$, where m and b are constants. Using the condition $C = 0$ when $F = 32$, we have $32 = b$, and so $F = mC + 32$. Next, using the condition $C = 100$ when $F = 212$, we have

$$212 = 100m + 32 \quad \text{or } m = \tfrac{9}{5}.$$

Therefore, $F = \frac{9}{5}C + 32$.

b. From (a), we see $F = \frac{9}{5}C + 32$. Next, when $C = 20$, $F = \frac{9}{5}(20) + 32 = 68$ and so the temperature equivalent to 20°C is 68°F.

c. Solving for C in terms of F, we find $\frac{9}{5}C = F - 32$, or $C = \frac{5}{9}F - \frac{160}{9}$. When $F = 70$, $C = \frac{5}{9}(70) - \frac{160}{9} = \frac{190}{9}$, or approximately 21.1°C.

33. The slope of L_2 is greater than that of L_1. This tells us that if the manufacturer lowers the unit price for each model clock radio by the same amount, the additional quantity demanded of model B radios will be greater than that of the model A radios.

35. a. Setting $x = 0$, gives $3p = 18$, or $p = 6$. Next, setting $p = 0$, gives $2x = 18$, or $x = 9$.

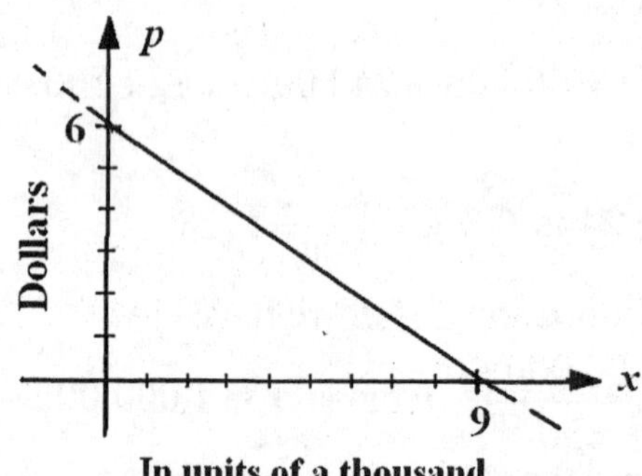

b. If $p = 4$,
$$2x + 3(4) - 18 = 0$$
$$2x = 18 - 12 = 6$$
and $x = 3$. Therefore, the quantity demanded when $p = 4$ is 3000. (Remember x is given in units of a thousand.)

37. a. When $x = 0$, $p = 60$ and when $p = 0$, $-3x = -60$, or $x = 20$.

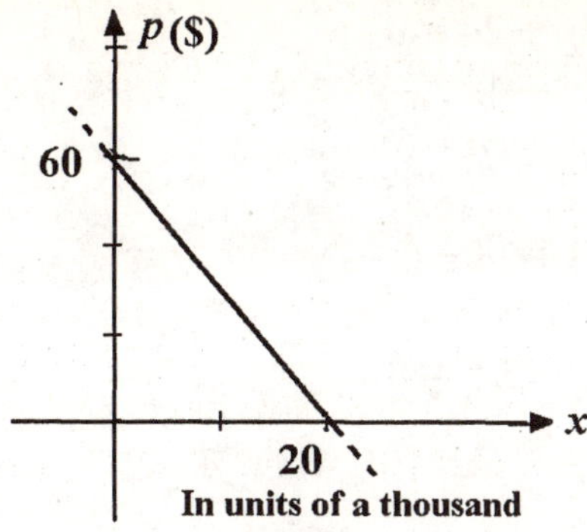

b. When $p = 30$, $\quad 30 = -3x + 60$

$$3x = 30 \qquad \text{and} \quad x = 10.$$

Therefore, the quantity demanded when $p = 30$ is 10,000 units.

39. When $x = 1000$, $p = 55$, and when $x = 600$, $p = 85$. Therefore, the graph of the linear demand equation is the straight line passing through the points (1000, 55) and (600, 85). The slope of the line is

$$\frac{85-55}{600-1000} = -\frac{3}{40}.$$

Using the point (1000, 55) and the slope just found, we find that the required equation is $\quad p - 55 = -\frac{3}{40}(x - 1000)$

$$p = -\frac{3}{40}x + 130 \quad .$$

When $x = 0$, $p = 130$, and this means that there will be no demand above \$130. When $p = 0$, $x = 1733.33$, and this means that 1733 units is the maximum quantity demanded.

41. Since the demand equation is linear, we know that the line passes through the points (1000,9) and (6000,4). Therefore, the slope of the line is given by

$$m = \frac{4-9}{6000-1000} = -\frac{5}{5000} = -0.001\,.$$

Since the equation of the line has the form $p = ax + b$,

$$9 = -0.001(1000) + b, \text{ or } \quad b = 10.$$

Therefore, the equation of the line is $p = -0.001x + 10$.

If $p = 7.50$, $\quad 7.50 = -0.001x + 10$

$$0.001x = 2.50, \text{ or } x = 2500.$$

So, the quantity demanded when the unit price is \$7.50 is 2500 units.

43. a. Setting $x = 0$, we obtain $3(0) - 4p + 24 = 0$

$$-4p = -24$$

or

$$p = 6.$$

Setting $p = 0$, we obtain $3x - 4(0) + 24 = 0$

$$3x = -24$$

or

$$x = -8.$$

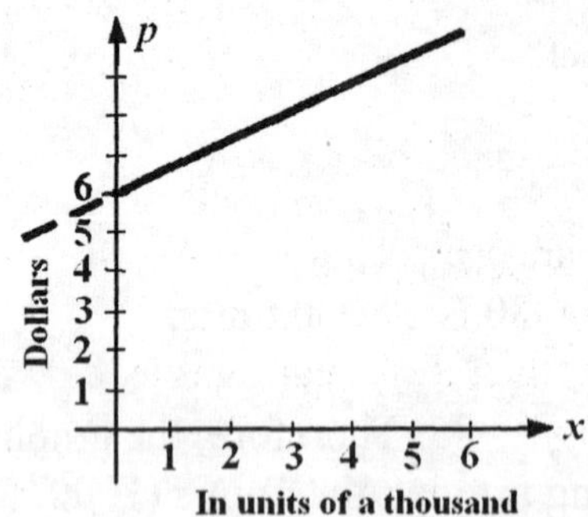

b. When $p = 8$, $3x - 4(8) + 24 = 0$

$$3x = 32 - 24 = 8$$

$$x = 8/3.$$

Therefore, 2667 units of the commodity would be supplied at a unit price of \$8. (Here again x is measured in units of thousands.)

45. a. When $x = 0$, $p = 10$, and when $p = 0$, $x = -5$.

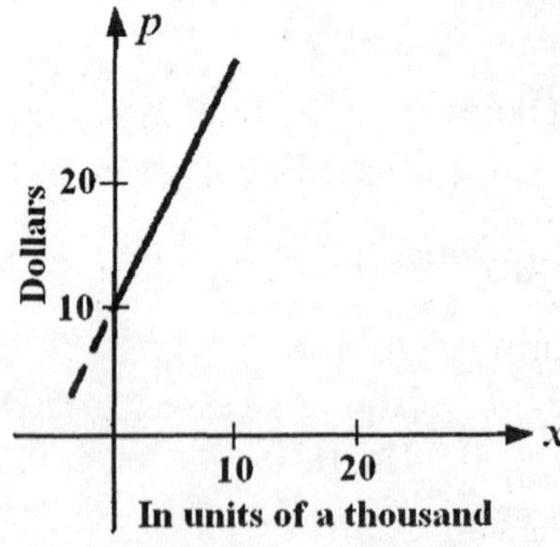

b. $p = 2x + 10$, $14 = 2x + 10$, $2x = 4$, and $x = 2$. Therefore, when $p = 14$, the supplier will make 2000 units of the commodity available.

47. When $x = 10{,}000$, $p = 45$ and when $x = 20{,}000$, $p = 50$. Therefore, the slope of the line passing (10,000, 45) and (20,000, 50) is

$$m = \frac{50 - 45}{20{,}000 - 10{,}000} = \frac{5}{10{,}000} = 0.0005$$

Using the point- slope form of an equation of a line with the point (10,000, 45), we have

$$p - 45 = 0.0005(x - 10{,}000)$$
$$p = 0.0005x - 5 + 45$$

or

$$p = 0.0005x + 40.$$

If $p = 70$, $\quad 70 = 0.0005x + 40$

$$0.0005x = 30, \qquad \text{or } x = \frac{30}{0.0005} = 60{,}000 \quad .$$

(If x is expressed in units of a thousand, then the equation may be written in the form $p = \frac{1}{2}x + 40$.)

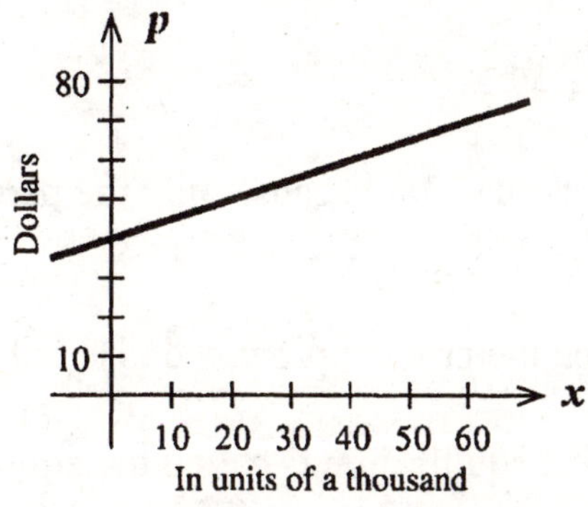

49. False. $P(x) = R(x) - C(x) = sx - (cx + F) = (s - c)x - F$. Therefore, the firm is making a profit if $P(x) = (s - c)x - F > 0$ or $x > \dfrac{F}{s - c}$.

USING TECHNOLOGY EXERCISES 1.3, page 42

1. 2.2875

3. 2.880952381

5. 7.2851648352

7. 2.4680851064

1.4 Problem Solving Tips

Here are some hints for solving the problems in the exercises that follow:

1. **To find the break-even point,** solve the simultaneous equations $p = R(x)$ and $p = C(x)$ for x and p.

2. **To find the market equilibrium for a commodity**, find the point of intersection of the supply and demand equations for the commodity. (Market equilibrium prevails when the quantity produced is equal to the quantity demanded.).

1.4 CONCEPT QUESTIONS, page 49

1. The intersection must lie in the first quadrant because only the parts of the demand and supply curves in the first quadrant are of interest.

3. a. Market equilibrium occurs when the product produced is equal to the quantity demanded.
   b. The quantity produced at market equilibrium is called the equilibrium quantity.
   c. The price of the product produced at market equilibrium is called the equilibrium price.

EXERCISES 1.4, page 49

1. We solve the system $y = 3x + 4$
$$y = -2x + 14.$$
Substituting the first equation into the second yields
$$3x + 4 = -2x + 14$$
$$5x = 10,$$
and $x = 2$. Substituting this value of x into the first equation yields
$$y = 3(2) + 4,$$
or $y = 10$. Thus, the point of intersection is (2, 10).

3. We solve the system $2x - 3y = 6$
$3x + 6y = 16.$
Solving the first equation for y, we obtain

$$3y = 2x - 6$$
$$y = \tfrac{2}{3}x - 2\ .$$

Substituting this value of y into the second equation, we obtain

$$3x + 6(\tfrac{2}{3}x - 2) = 16$$
$$3x + 4x - 12 = 16$$
$$7x = 28$$

and $\quad x = 4.$

Then $\quad y = \tfrac{2}{3}(4) - 2 = \tfrac{2}{3}.$

Therefore, the point of intersection is $(4, \tfrac{2}{3})$.

5. We solve the system $\begin{cases} y = \tfrac{1}{4}x - 5 \\ 2x - \tfrac{3}{2}y = 1 \end{cases}$. Substituting the value of y given in the first equation into the second equation, we obtain

$$2x - \tfrac{3}{2}(\tfrac{1}{4}x - 5) = 1$$
$$2x - \tfrac{3}{8}x + \tfrac{15}{2} = 1$$
$$16x - 3x + 60 = 8$$
$$13x = -52,$$

or $x = -4$. Substituting this value of x in the first equation, we have

$$y = \tfrac{1}{4}(-4) - 5 = -1 - 5,$$

or $y = -6$. Therefore, the point of intersection is $(-4, -6)$.

7. We solve the equation $R(x) = C(x)$, or $15x = 5x + 10{,}000$, obtaining $10x = 10{,}000$, or $x = 1000$. Substituting this value of x into the equation $R(x) = 15x$, we find $R(1000) = 15{,}000$. Therefore, the breakeven point is $(1000, 15{,}000)$.

9. We solve the equation $R(x) = C(x)$, or $0.4x = 0.2x + 120$, obtaining $0.2x = 120$, or $x = 600$. Substituting this value of x into the equation $R(x) = 0.4x$, we find $R(600) = 240$. Therefore, the breakeven point is $(600, 240)$.

11. a.

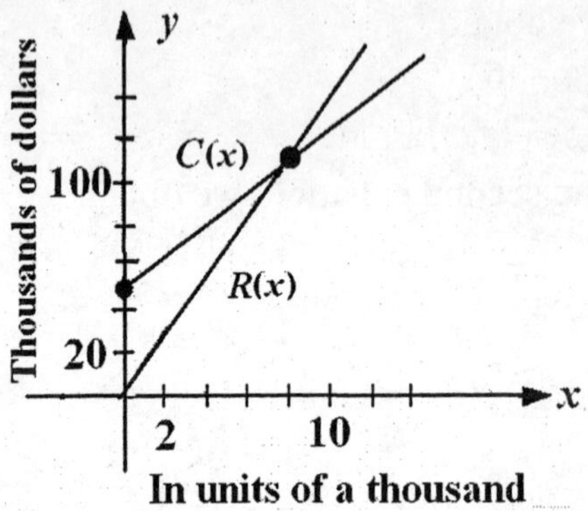

b. We solve the equation $R(x) = C(x)$ or $14x = 8x + 48{,}000$, obtaining $6x = 48{,}000$ or $x = 8000$. Substituting this value of x into the equation $R(x) = 14x$, we find $R(8000) = 14(8000) = 112{,}000$. Therefore, the breakeven point is (8000, 112,000).

c.

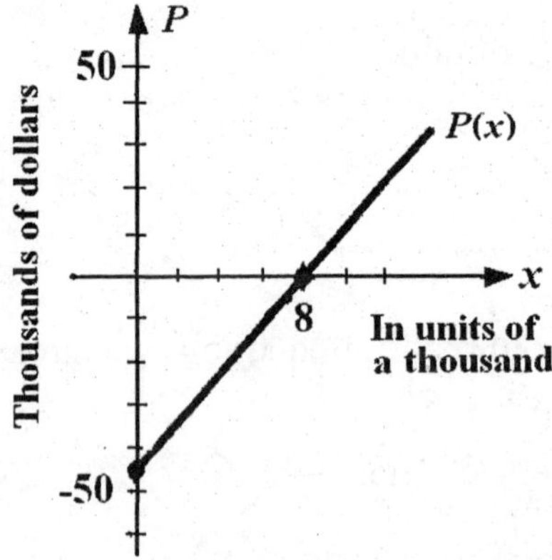

d. $P(x) = R(x) - C(x) = 14x - 8x - 48{,}000 = 6x - 48{,}000$.
The graph of the profit function crosses the x-axis when $P(x) = 0$, or $6x = 48{,}000$ and $x = 8000$. This means that the revenue is equal to the cost when 8000 units are produced and consequently the company breaks even at this point.

13. Let x denote the number of units sold. Then, the revenue function R is given by

$$R(x) = 9x.$$

Since the variable cost is 40 percent of the selling price and the monthly fixed costs are \$50,000, the cost function C is given by

$$C(x) = 0.4(9x) + 50{,}000$$
$$= 3.6x + 50{,}000.$$

To find the breakeven point, we set $R(x) = C(x)$, obtaining

$$9x = 3.6x + 50{,}000$$
$$5.4x = 50{,}000$$
$$x \approx 9259, \text{ or } 9259 \text{ units.}$$

Substituting this value of x into the equation $R(x) = 9x$ gives

$$R(9259) = 9(9259) = 83{,}331.$$

Thus, for a breakeven operation, the firm should manufacture 9259 bicycle pumps resulting in a breakeven revenue of \$83,331.

15. a. The cost function associated with using machine *I* is given by

$$C_1(x) = 18{,}000 + 15x.$$

The cost function associated with using machine II is given by

$$C_2(x) = 15{,}000 + 20x.$$

b.

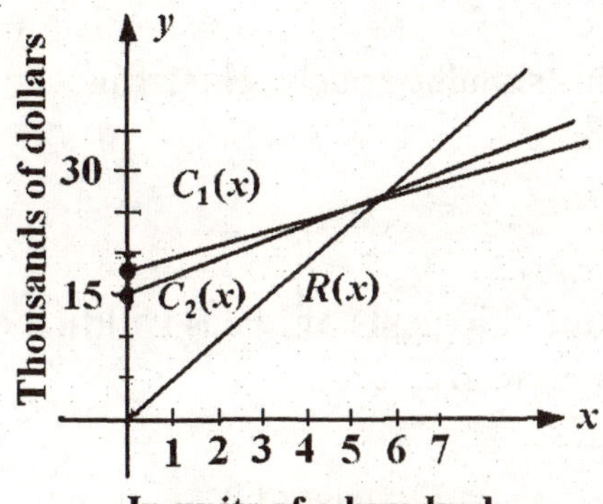

c. Comparing the cost of producing 450 units on each machine, we find

$$C_1(450) = 18{,}000 + 15(450)$$
$$= 24{,}750 \quad \text{or } \$24{,}750 \text{ on machine } I,$$

and

$$C_2(450) = 15{,}000 + 20(450)$$
$$= 24{,}000 \text{ or } \$24{,}000 \text{ on machine } II.$$

Therefore, machine *II* should be used in this case.

Next, comparing the costs of producing 550 units on each machine, we find

$$C_1(550) = 18{,}000 + 15(550)$$
$$= 26{,}250 \text{ or } \$26{,}250 \text{ on machine } I,$$

and

$$C_2(550) = 15{,}000 + 20(550)$$
$$= 26{,}000$$

on machine *II*. Therefore, machine *II* should be used in this instance. Once again, we compare the cost of producing 650 units on each machine and find that

$C_1(650) = 18{,}000 + 15(650)$
$= 27{,}750$, or \$27,750 on machine *I* and

$C_2(650) = 15{,}000 + 20(650)$
$= 28{,}000$,

or \$28,000 on machine *II*. Therefore, machine *I* should be used in this case.

d. We use the equation $P(x) = R(x) - C(x)$ and find

$$P(450) = 50(450) - 24{,}000 = -1500,$$

or a loss of \$1500 when machine *II* is used to produce 450 units. Similarly,

$$P(550) = 50(550) - 26{,}000 = 1500,$$

or a profit of \$1500 when machine *II* is used to produce 550 units.

Finally, $P(650) = 50(650) - 27{,}750 = 4750$,

or a profit of \$4750 when machine *I* is used to produce 650 units.

17. We solve the two equations simultaneously, obtaining

$$18t + 13.4 = -12t + 88$$

$$30t = 74.6$$

$$t \approx 2.486$$

or approximately 2.5. So the shipments of LCDs will first overtake the shipments of CRTs a little before the middle of 2003.

19. a.

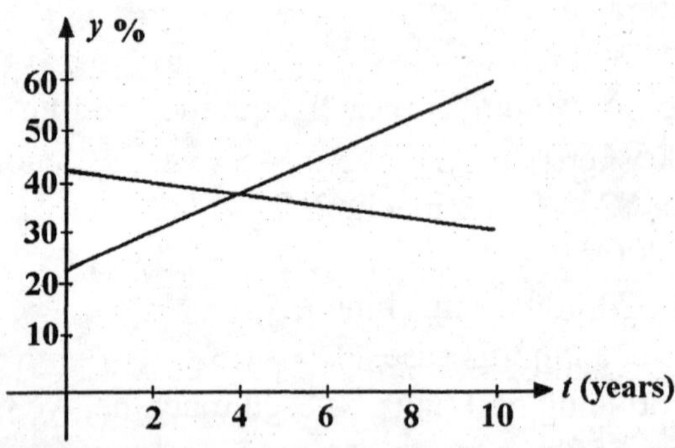

b. We solve the two equations simultaneously. Thus,

$$\tfrac{11}{3}t + 23 = -\tfrac{11}{9} + 43$$

$$\tfrac{44}{9}t = 20$$

$$t = 4.09$$

The transactions done electronically will first exceed those done by check in early 2005.

21. We solve the system

$$\begin{aligned} 4x + 3p &= 59 \\ 5x - 6p &= -14. \end{aligned}$$

Solving the first equation for p, we find $p = -\frac{4}{3}x + \frac{59}{3}$.
Substituting this value of p into the second equation, we have

$$\begin{aligned} 5x - 6(-\tfrac{4}{3}x + \tfrac{59}{3}) &= -14 \\ 5x + 8x - 118 &= -14 \\ 13x &= 104 \\ x &= 8. \end{aligned}$$

Substituting this value of x into the equation

$$p = -\tfrac{4}{3}x + \tfrac{59}{3}$$

we have $p = -\frac{4}{3}(8) + \frac{59}{3} = \frac{27}{3} = 9$

Thus, the equilibrium quantity is 8000 units and the equilibrium price is \$9.

23. We solve the system $p = -2x + 22$
$p = 3x + 12$.

Substituting the first equation into the second, we find

$$\begin{aligned} -2x + 22 &= 3x + 12 \\ 5x &= 10 \end{aligned}$$

and $x = 2.$

Substituting this value of x into the first equation, we obtain

$$p = -2(2) + 22 = 18.$$

Thus, the equilibrium quantity is 2000 units and the equilibrium price is \$18.

25. Let x denote the number of DVD players produced per week, and p denote the price of each DVD player.

a. The slope of the demand curve is given by $\dfrac{\Delta p}{\Delta x} = -\dfrac{20}{250} = -\dfrac{2}{25}$.

Using the point-slope form of the equation of a line with the point (3000, 485), we

have

$$\begin{aligned} p - 485 &= -\tfrac{2}{25}(x - 3000) \\ p &= -\tfrac{2}{25}x + 240 + 485 \end{aligned}$$

or $p = -0.08x + 725.$

b. From the given information, we know that the graph of the supply equation passes through the points (0, 300) and (2500, 525). Therefore, the slope of the supply curve is $$m = \frac{525-300}{2500-0} = \frac{225}{2500} = 0.09.$$

Using the point-slope form of the equation of a line with the point (0, 300), we find that

$$p - 300 = 0.09x$$
$$p = 0.09x + 300.$$

c. Equating the supply and demand equations, we have

$$-0.08x + 725 = 0.09x + 300$$
$$0.17x = 425$$

or $$x = 2500.$$

Then $$p = -0.08(2500) + 725 = 525.$$

We conclude that the equilibrium quantity is 2500 and the equilibrium price is \$525.

27. We solve the system $$3x + p = 1500$$ $$2x - 3p = -1200.$$

Solving the first equation for p, we obtain

$$p = 1500 - 3x.$$

Substituting this value of p into the second equation, we obtain

$$2x - 3(1500 - 3x) = -1200$$
$$11x = 3300$$

or $$x = 300.$$

Next, $$p = 1500 - 3(300) = 600.$$

Thus, the equilibrium quantity is 300 and the equilibrium price is \$600.

29. a. We solve the system of equations $p = cx + d$ and $p = ax + b$. Substituting the first into the second gives

$$cx + d = ax + b$$
$$(c - a)x = b - d$$

or $$x = \frac{b-d}{c-a}.$$

Since $a < 0$ and $c > 0$, and $b > d > 0$, and $c - a \neq 0$, x is well-defined. Substituting this value of x into the second equation, we obtain

$$p = a\left(\frac{b-d}{c-a}\right) + b = \frac{ab - ad + bc - ab}{c-a} = \frac{bc - ad}{c-a} \quad (1)$$

Therefore, the equilibrium quantity is $\dfrac{b-d}{c-a}$ and the equilibrium price is $\dfrac{bc-ad}{c-a}$.

b. If c is increased, the denominator in the expression for x increases and so x gets smaller. At the same time, the first term in equation (1) for p decreases (since a is a negative number) and so p gets larger. This analysis shows that if the unit price for producing the product is increased then the equilibrium quantity decreases while the equilibrium price increases.

c. If b is decreased, the numerator of the expression for x decreases while the denominator stays the same. Therefore x decreases. The expression for p also shows that p decreases. This analysis shows that if the (theoretical) upper bound for the unit price of a commodity is lowered, then both the equilibrium quantity and the equilibrium price drop.

31. True. $P(x) = R(x) - C(x) = sx - (cx + F) = (s-c)x - F$. Therefore, the firm is making a profit if $P(x) = (s-c)x - F > 0$, or $x > \frac{F}{s-c}$ $(s \neq c)$.

33. Solving the two equations simultaneously to find the point(s) of intersection of L_1 and L_2, we obtain

$$m_1x + b_1 = m_2x + b_2$$

$$(m_1 - m_2)x = b_2 - b_1 \qquad (1)$$

a. If $m_1 = m_2$ and $b_2 \neq b_1$, then there is no solution for (1) and in this case L_1 and L_2 do not intersect.
b. If $m_1 \neq m_2$, then Equation (1) can be solved (uniquely) for x and this shows that L_1 and L_2 intersect at precisely one point.
c. If $m_1 = m_2$ and $b_1 = b_2$, then (1) is satisfied for all values of x and this shows that L_1 and L_2 intersect at infinitely many points.

USING TECHNOLOGY EXERCISES 1.4, page 53

1. (0.6, 6.2)   3. (3.8261, 0.1304)   5. (386.9091, 145.3939)

7. a.

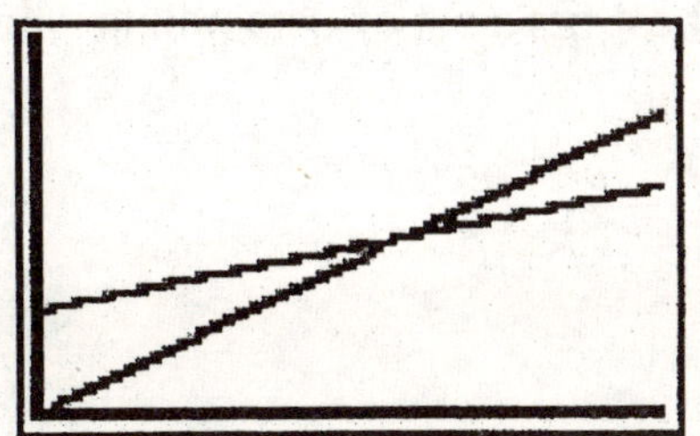

b. (3548, 27,997)

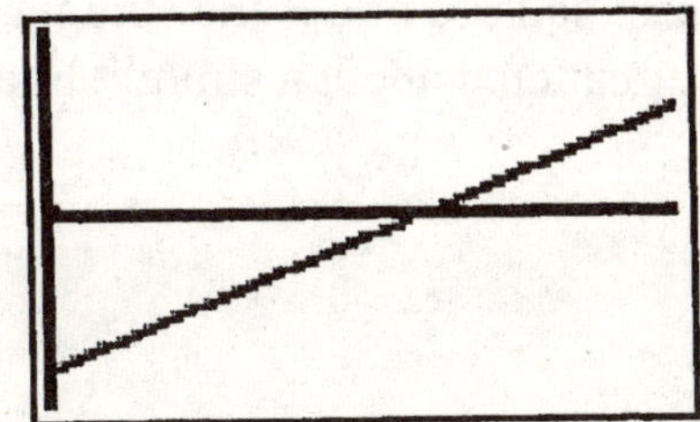

c. x-intercept is approximately 3548

9. a. $C_1(x) = 34 + 0.18x$; $C_2(x) = 28 + 0.22x$

b.

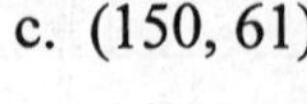
c. (150, 61)

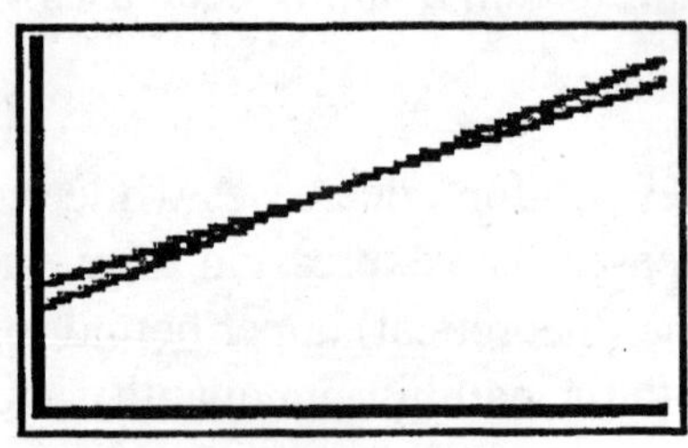

d. If the distance driven is less than or equal to 150 mi, rent from Acme Truck Leasing; if the distance driven is more than 150 mi, rent from Ace Truck Leasing.

11. a. $p = -\frac{1}{10}x + 284$; $p = \frac{1}{60}x + 60$

b. (1920, 92)

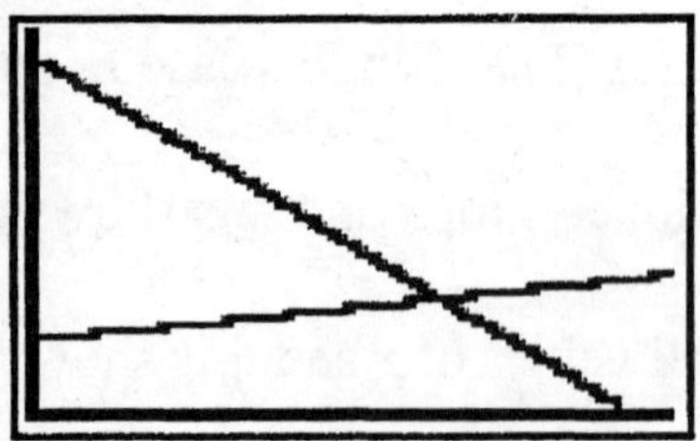

c. 1920/wk; \$92/radio

1.5 CONCEPT QUESTIONS, page 58

1. a. A scatter diagram is a graph of the data points for a problem.

b. The least-squares line is the straight line that best fits a set of data points when the points are scattered about a straight line.

EXERCISES 1.5, page 59

1. a. We first summarize the data:

x	y	x^2	xy
1	4	1	4
2	6	4	12
3	8	9	24
4	11	16	44
10	29	30	84

The normal equations are

$$4b + 10m = 29$$
$$10b + 30m = 84.$$

Solving this system of equations, we obtain $m = 2.3$ and $b = 1.5$. So an equation is $y = 2.3x + 1.5$.

b. The scatter diagram and the least squares line for this data follow:

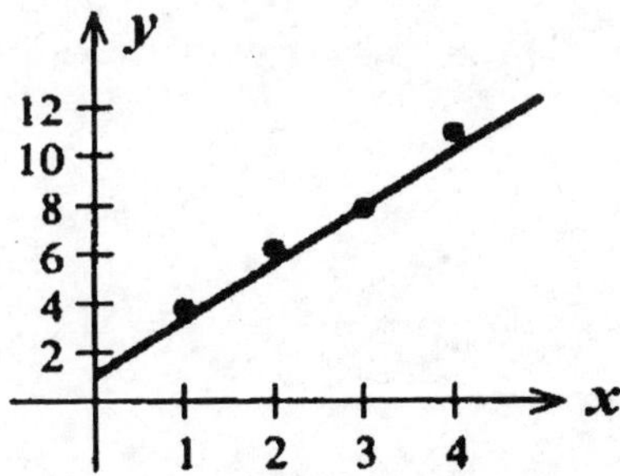

3. a. We first summarize the data:

x	y	x^2	xy
1	4.5	1	4.5
2	5	4	10
3	3	9	9
4	2	16	8
4	3.5	16	14
6	1	36	6
20	19	82	51.5

The normal equations are $6b + 20m = 19$

$$20b + 82m = 51.5.$$

The solutions are $m \approx -0.7717$ and $b \approx 5.7391$ and so a required equation is $y = -0.772x + 5.739$.

b. The scatter diagram and the least-squares line for these data follow.

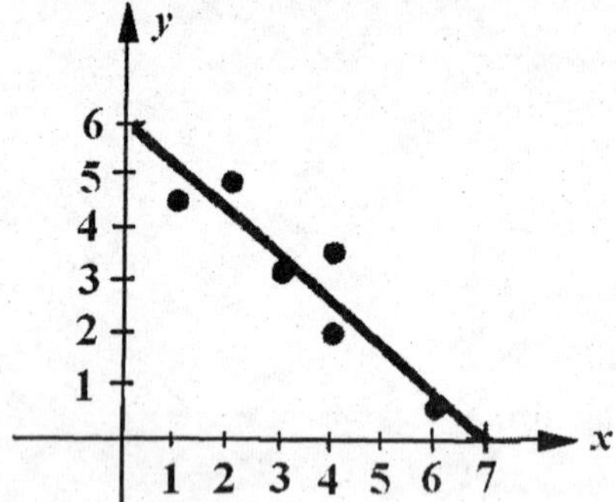

5. a. We first summarize the data:

x	y	x^2	xy
1	3	1	3
2	5	4	10
3	5	9	15
4	7	16	28
5	8	25	40
15	28	55	96

The normal equations are $55m + 15b = 96$

$$15m + 5b = 28.$$

Solving, we find $m = 1.2$ and $b = 2$, so that the required equation is $y = 1.2x + 2$.

b. The scatter diagram and the least-squares line for the given data follow.

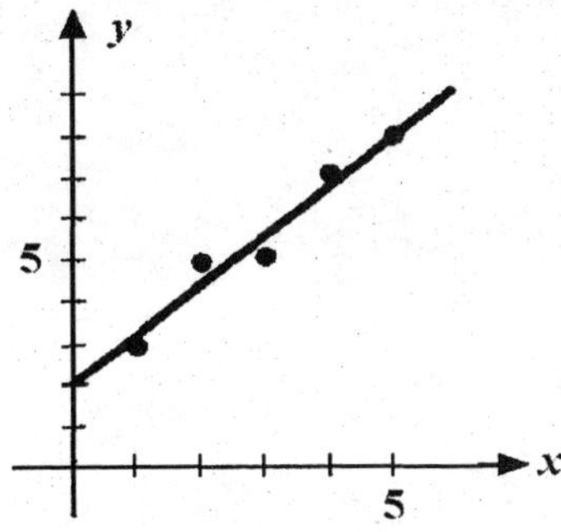

7. a. We first summarize the data:

x	y	x^2	xy
4	0.5	16	2
4.5	0.6	20.25	2.7
5	0.8	25	4
5.5	0.9	30.25	4.95
6	1.2	36	7.2
25	4	127.5	20.85

The normal equations are

$$5b + 25m = 4$$
$$25b + 127.5m = 20.85.$$

The solutions are $m = 0.34$ and $b = -0.9$, and so a required equation is $y = 0.34x - 0.9$.

b. The scatter diagram and the least-squares line for these data follow.

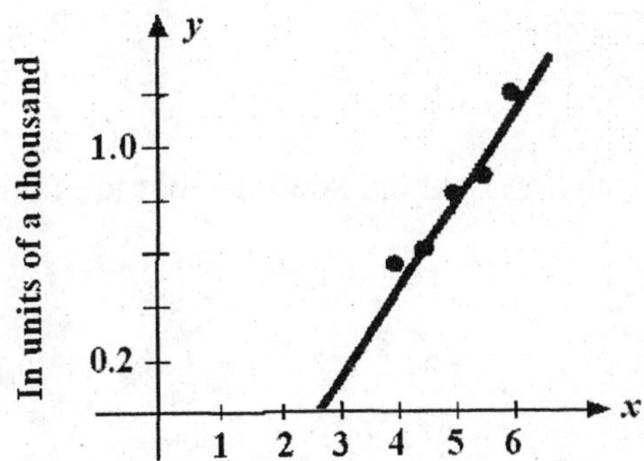

c. If $x = 6.4$, then $y = 0.34(6.4) - 0.9 = 1.276$ and so 1276 completed applications might be expected.

9. a. We first summarize the data:

x	y	x^2	xy
1	436	1	436
2	438	4	876
3	428	9	1284
4	430	16	1720
5	426	25	2130
15	2158	55	6446

The normal equations are

$$5b + 15m = 2158$$
$$15b + 55m = 6446.$$

Solving this system, we find $m = -2.8$ and $b = 440$.
Thus, the equation of the least-squares line is $y = -2.8x + 440$.
b. The scatter diagram and the least-squares line for this data are shown in the figure that follows.

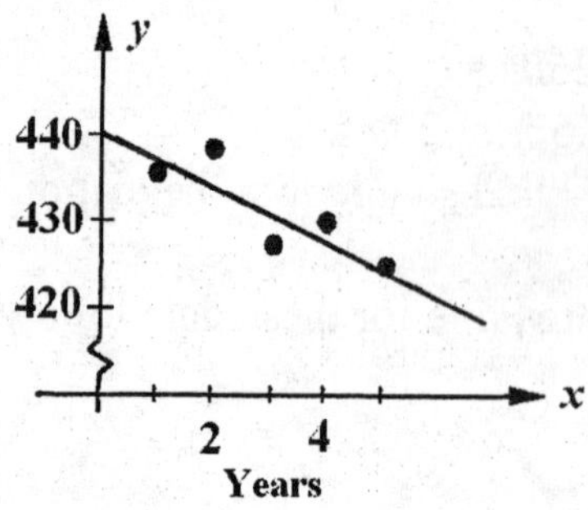

c. Two years from now, the average SAT verbal score in that area will be $y = -2.8(7) + 440 = 420.4$.

11. a. We first summarize the data:

x	y	x^2	xy
1	20	1	20
2	24	4	48
3	26	9	78
4	28	16	112
5	32	25	160
15	130	55	418

The normal equations are $5b + 15m = 130$

$$15b + 55m = 418.$$

The solutions are $m = 2.8$ and $b = 17.6$, and so an equation of the line is

$$y = 2.8x + 17.6.$$

b. When $x = 8$, $y = 2.8(8) + 17.6 = 40$. Hence, the state subsidy is expected to be \$40 million for the eighth year.

13. a.

t	y	t^2	ty
0	126	1	0
1	144	4	144
2	171	9	343
3	191	16	573
4	216	25	864
10	848	30	1923

The normal equations are

$$5b + 10m = 848$$

$$10b + 30m = 1923$$

The solutions are $m \approx 22.7$ and $b \approx 124.2$. Therefore, the required equation is $y = 22.7t + 124.2$.

b. $y = 22.7(6) + 124.2 = 260.4$, or \$260.4 billion.

15. a. We first summarize the data:

x	y	x^2	xy
0	21.7	0	0
1	32.1	1	32.1
2	45.0	4	90
3	58.3	9	174.9
4	69.6	16	278.4
10	226.7	30	575.4

The normal equations are
$$\begin{aligned} 5b + 10m &= 226.7 \\ 10b + 30m &= 575.4 \end{aligned}.$$

The solutions are $m = 12.2$ and $b = 20.9$ and so a required equation is $y = 12.2x + 20.9$.

b. In 2005, $x = 7$ so $y = 12.2(7) + 20.9 = 106.3$, or 106.3 million computers are expected to be connected to the internet in Europe in that year.

17. a. We first summarize the data.

x	y	x^2	xy
0	19.5	0	0
10	20	100	200
20	20.6	400	412
30	21.2	900	636
40	21.8	1600	872
50	22.4	2500	1120
150	125.5	5500	3240

The normal equations are
$$6b + 150m = 125.5$$
$$150b + 5500m = 3240$$

The solutions are $b = 19.45$ and $m = 0.0586$. Therefore, $y = 0.059x + 19.5$.

b. The life expectancy at 65 of a female in 2040 is
$$y = 0.059(40) + 19.5 = 21.86 \quad \text{or } 21.9 \text{ years.}$$

The datum gives a life expectancy of 21.8 years.

c. The life expectancy at 65 of a female in 2030 is
$$y = 0.059(30) + 19.5 = 21.27 \quad \text{or } 21.3 \text{ years.}$$

The datum gives a life expectancy of 21.2 years.

19. a.

x	y	x^2	xy
0	7.9	0	0
1	9.6	1	9.6
2	11.5	4	23
3	13.3	9	39.9
4	15.2	16	60.8
5	16	25	80
6	18.8	36	112.8
21	92.3	30	326.1

The normal equations are

$$7b + 21m = 92.3$$

$$21b + 91m = 326.1$$

The solutions are $m \approx 1.7571$ and $b \approx 7.9143$. Therefore, the required equation is

$$y = 1.7571x + 7.9143$$

b. $y = 1.7571(8) + 7.9143 \approx 21.97$ or \$21.97 billion.

21. a. We first summarize the given data:

x	y	x^2	xy
0	90.4	0	0
1	100.0	1	100
2	110.4	4	220.8
3	120.4	9	361.2
4	130.8	16	523.2
5	140.4	25	702
6	150	36	900
21	842.4	91	2807.2

The normal equations are $\begin{cases} 7b + 21m = 842.4 \\ 21b + 91m = 2807.2 \end{cases}$.

The solutions are $m = 10$ and $b = 90.34$. Therefore, the required equation is $y = 10x + 90.34$.

b. If $x = 6$, then $y = 10(6) + 90.34 = 150.34$, or 150,340,000. This compares well with the actual data for that year-- 150,000,000 subscribers.

23. a.

x	y	x^2	xy
0	501	0	0
1	540	1	540
2	585	4	1170
3	631	9	1893
4	680	16	2720
5	728	25	3640
6	779	36	4674
21	4444	91	14637

The normal equations are

$$7b+21m= 4444$$

$$21b+91m=14637$$

The solutions are $m\approx 46.6071$ and $b\approx 495.0357$. Therefore, the required equation is $y=46.6071x+495.0357$.

b. The rate of change is given by the slope of the least-squares line; that is, approximately \$46.6/buyer/year.

25. a.

x	y	x^2	xy
1	80.4	1	80.4
2	84.9	4	169.8
3	87	9	261
4	87.9	16	351.6
5	90	25	450
6	94.2	36	565.2
21	524.4	91	1878

The normal equations are

$$6b+21m= 524.4$$

$$21b+91m=1878$$

The solutions are $m\approx 2.43$ and $b\approx 78.88$. Therefore the required equation is

$$y=2.43x+78.88$$

b. Then the FICA wage base for the year 2010 is given by

$$y=2.43(10)+78.88 = 103.18, \text{ or } \$103{,}180.$$

27. False. See Example 1, page 55.

29. True. Since there exists one and only one line passing through two distinct points, the two lines must be the same.

USING TECHNOLOGY EXERCISES 1.5, page 65

1. $y = 2.3596x + 3.8639$ 3. $y = -1.1948x + 3.5525$

5. a. $y = 0.55x + 1.17$ b. \$5.57 billion

7. a. $y = 13.321x + 72.57$ b. 192 million tons

9. a. $y = 14.43x + 212.1$ b. 247 trillion cu ft

CHAPTER 1 CONCEPT REVIEW, page 67

1. ordered; abscissa (x-coordinate); ordinate (y-coordinate) 3. $\sqrt{(c-a)^2 + (d-b)^2}$

5. a. $\dfrac{y_2 - y_1}{x_2 - x_1}$ b. undefined c. zero d. positive

7. a. $y - y_1 = m(x - x_1)$; point-slope b. $y = mx + b$; slope-intercept

9. $mx + b$ 11. breakeven

CHAPTER 1 REVIEW EXERCISES, page 68

1. The distance is $d = \sqrt{(6-2)^2 + (4-1)^2} = \sqrt{4^2 + 3^2} = \sqrt{25} = 5$.

3. The distance is $d = \sqrt{[1-(-2)]^2 + [-7-(-3)]^2} = \sqrt{3^2 + (-4)^2} = \sqrt{9+16} = \sqrt{25} = 5$.

5. An equation is $x = -2$.

7. The slope of L is $m = \dfrac{\frac{7}{2}-4}{3-(-2)} = \dfrac{\frac{7-8}{2}}{5} = -\dfrac{1}{10}$ and an equation of L is

$$y-4 = -\tfrac{1}{10}[x-(-2)] = -\tfrac{1}{10}x - \tfrac{1}{5},$$

or

$$y = -\tfrac{1}{10}x + \tfrac{19}{5}$$

The general form of this equation is $x + 10y - 38 = 0$.

9. Writing the given equation in the form $y = \frac{5}{2}x - 3$, we see that the slope of the given line is 5/2. So a required equation is $y - 4 = \frac{5}{2}(x+2)$ or $y = \frac{5}{2}x + 9$
The general form of this equation is $5x - 2y + 18 = 0$.

11. Using the slope-intercept form of the equation of a line, we have $y = -\frac{1}{2}x - 3$.

13. Rewriting the given equation in the slope-intercept form, we have $4y = -3x + 8$
or

$$y = -\tfrac{3}{4}x + 2$$

and conclude that the slope of the required line is $-3/4$. Using the point-slope form of the equation of a line with the point (2,3) and slope $-3/4$, we obtain

$$\begin{aligned} y - 3 &= -\tfrac{3}{4}(x-2) \\ y &= -\tfrac{3}{4}x + \tfrac{6}{4} + 3 \\ &= -\tfrac{3}{4}x + \tfrac{9}{2}. \end{aligned}$$

The general form of this equation is $3x + 4y - 18 = 0$.

15. Rewriting the given equation in the slope-intercept form $y = \frac{2}{3}x - 8$, we see that the slope of the line with this equation is 2/3. The slope of the required line is $-3/2$. Using the point-slope form of the equation of a line with the point $(-2, -4)$ and slope $-3/2$, we have

$$y - (-4) = -\tfrac{3}{2}[x - (-2)]$$

or

$$y = -\tfrac{3}{2}x - 7 .$$

The general form of this equation is $3x + 2y + 14 = 0$.

17. Setting $x = 0$, gives $5y = 15$, or $y = 3$. Setting $y = 0$, gives $-2x = 15$, or $x = -15/2$.
The graph of the equation $-2x + 5y = 15$ follows.

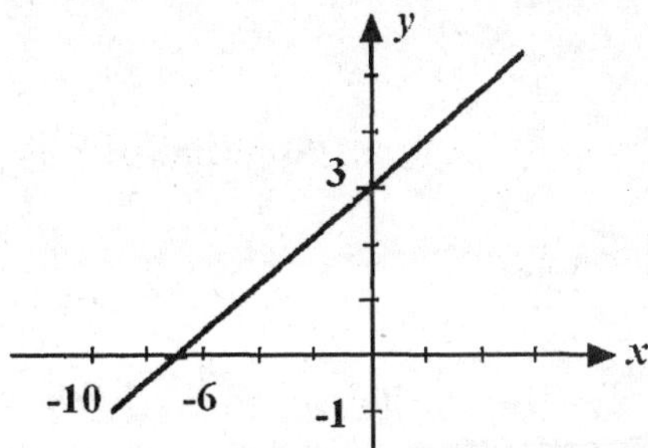

19. Let x denote the time in years. Since the function is linear, we know that it has the form $f(x) = mx + b$.

a. The slope of the line passing through (0, 2.4) and (5, 7.4) is $m = \frac{7.4 - 2.4}{5} = 1$.

Since the line passes through (0, 2.4), we know that the y-intercept is 2.4. Therefore, the required function is $f(x) = x + 2.4$.

b. In 2004 ($x = 3$), the sales were $f(3) = 3 + 2.4 = 5.4$, or \$5.4 million dollars.

21. a. $D(w) = \frac{a}{150}w$. The given equation can be expressed in the form $y = mx + b$,

where $m = \frac{a}{150}$ and $b = 0$.

b. If $a = 500$ and $w = 35$, $D(35) = \frac{500}{150}(35) = 116\frac{2}{3}$, or approximately 117 mg.

23. Let V denote the value of the machine after t years.

a. The rate of depreciation is

$$-\frac{\Delta V}{\Delta t} = \frac{300{,}000 - 30{,}000}{12} = \frac{270{,}000}{12} = 22{,}500, \quad \text{or } \$22{,}500/\text{year}.$$

b. Using the point-slope form of the equation of a line with the point (0, 300,000) and $m = -22{,}500$, we have

$$V - 300{,}000 = -22{,}500(t - 0)$$
$$V = -22{,}500t + 300{,}000.$$

25. The slope of the demand curve is $\frac{\Delta p}{\Delta x} = -\frac{10}{200} = -0.05$.

Using the point-slope form of the equation of a line with the point (0, 200), we have

$$p - 200 = -0.05(x), \text{ or } \quad p = -0.05x + 200.$$

The graph of the demand equation follows.

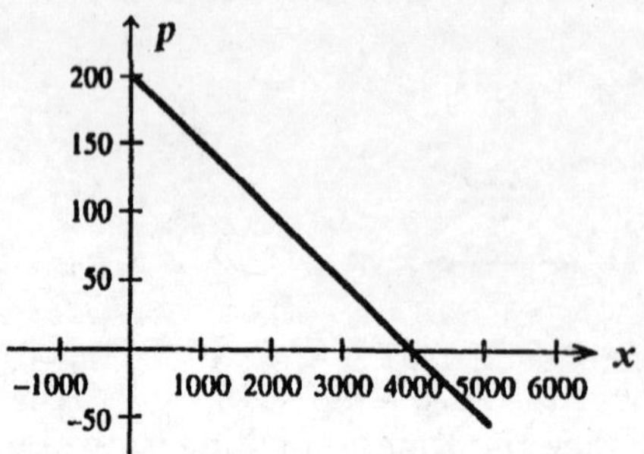

27. We solve the system $3x + 4y = -6$
$$2x + 5y = -11.$$
Solving the first equation for x, we have $3x = -4y - 6$ and $x = -\frac{4}{3}y - 2$.
Substituting this value of x into the second equation yields
$$2(-\tfrac{4}{3}y - 2) + 5y = -11$$
$$-\tfrac{8}{3}y - 4 + 5y = -11$$
$$\tfrac{7}{3}y = -7, \qquad \text{or} \qquad y = -3.$$
Then $\qquad x = -\frac{4}{3}(-3) - 2 = 4 - 2 = 2.$
Therefore, the point of intersection is $(2, -3)$.

29. Setting $C(x) = R(x)$, we have $12x + 20{,}000 = 20x$
$$8x = 20{,}000$$
or $\qquad x = 2500.$
Next, $R(2500) = 20(2500) = 50{,}000$, and we conclude that the breakeven point is $(2500, 50{,}000)$.

31. a. The slope of the line is $m = \dfrac{1 - 0.5}{4 - 2} = 0.25$.
Using the point-slope form of an equation of a line, we have
$$y - 1 = 0.25(x - 4)$$
$$y = 0.25x$$
b. $\qquad y = 0.25(6.4) = 1.6$, or 1600 applications.

CHAPTER 1, BEFORE MOVING ON, page 69

1.

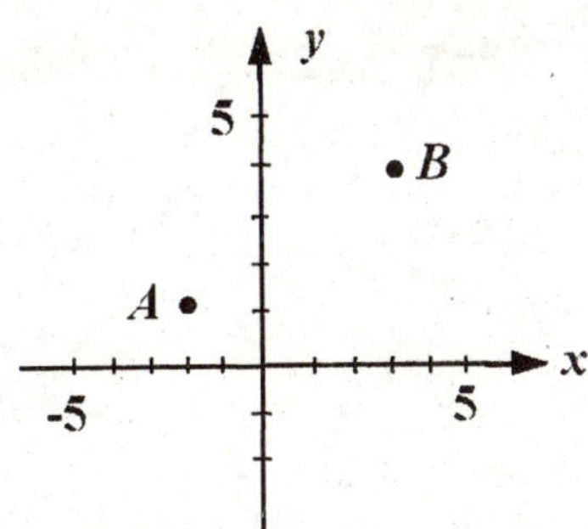

$$d = \sqrt{[3-(-2)]^2 + (4-1)^2} = \sqrt{5^2 + 3^2} = \sqrt{34}$$

2. Solving the equation $3x - y - 4 = 0$ gives $y = 3x - 4$ and this tells us that the slope of the second line is 3. Therefore, the slope of the required line is $m = 3$. It equation is $y - 1 = 3(x - 3)$ or $y = 3x - 8$.

3. The slope of the line passing through (1, 2) and (3, 5) is $m = \dfrac{5-2}{3-1} = \dfrac{3}{2}$. Solving $2x + 3y = 10$ gives $y = -\frac{2}{3}x + \frac{10}{3}$ and the slope of the line with this equation is $m_2 = -\dfrac{2}{3} = -\dfrac{1}{m_1}$ and so the two lines are perpendicular.

4. a. This is given by the coefficient of x in $C(x)$; that is, \$18. b. \$22,000
   c. This is given by the coefficient of x in $R(x)$; that is, \$15.

5. Solving $2x - 3y = -2$ for x gives $x = \frac{3}{2}y - 1$. Substituting into the second equation gives

$$9(\tfrac{3}{2}y - 1) + 12y = 25$$
$$\tfrac{27}{2}y - 9 + 12y = 25$$
$$27y - 18 + 24y = 50$$
$$51y = 68 \text{ or } y = \tfrac{68}{51} = \tfrac{4}{3}$$

Therefore, $x = \frac{3}{2}(\frac{4}{3}) - 1 = 1$. So the point of intersection is $(1, \frac{4}{3})$.

6. We solve the equation $S_1 = S_2$ or

$$4.2 + 0.4t = 2.2 + 0.8t$$

$$2 = 0.4t \quad \text{or} \quad t = \tfrac{2}{0.4} = 5$$

So it will surpass Best's annual sales in 5 years.

CHAPTER 2

2.1 CONCEPT QUESTIONS, page 77

1. a. There may be no solution, a unique solution, or infinitely many solutions.
   b. There is no solution if the two lines represented by the given system of linear equations are parallel and distinct; there is a unique solution if the two lines intersect at precisely one point; there are infinitely many solutions if the two lines are parallel and coincident.

EXERCISES 2.1, page 77

1. Solving the first equation for x, we find $x = 3y - 1$. Substituting this value of x into the second equation yields
$$4(3y-1)+3y=11$$
$$12y-4+3y=11$$
or
$$y=1.$$
Substituting this value of y into the first equation gives $x = 3(1)-1=2$.
Therefore, the unique solution of the system is (2,1).

3. Solving the first equation for x, we have $x = 7 - 4y$. Substituting this value of x into the second equation, we have
$$\tfrac{1}{2}(7-4y)+2y=5$$
$$7-4y+4y=10$$
$$7=10.$$
Clearly, this is impossible and we conclude that the system of equations has no solution.

5. Solving the first equation for x, we obtain $x = 7 - 2y$.
Substituting this value of x into the second equation, we have
$$2(7-2y)-y=4$$
$$14-4y-y=4$$
$$-5y=-10$$
and
$$y=2.$$
Then
$$x=7-2(2)=7-4=3.$$

We conclude that the solution to the system is (3,2).

7. Solving the first equation for x, we have

$$2x = 5y + 10$$

and $$x = \tfrac{5}{2}y + 5.$$

Substituting this value of x into the second equation, we have

$$6(\tfrac{5}{2}y + 5) - 15y = 30$$
$$15y + 30 - 15y = 30$$

or $$0 = 0.$$

This result tells us that the second equation is equivalent to the first. Thus, any ordered pair of numbers (x, y) satisfying the equation

$$2x - 5y = 10 \qquad (\text{or } 6x - 15y = 30)$$

is a solution to the system. In particular, by assigning the value t to x, where t is any real number, we find that $y = -2 + \tfrac{2}{5}t$ so the ordered pair, $(t, \tfrac{2}{5}t - 2)$ is a solution to the system, and we conclude that the system has infinitely many solutions.

9. Solving the first equation for x, we obtain

$$4x - 5y = 14$$
$$4x = 14 + 5y$$
$$x = \tfrac{14}{4} + \tfrac{5}{4}y = \tfrac{7}{2} + \tfrac{5}{4}y.$$

Substituting this value of x into the second equation gives

$$2(\tfrac{7}{2} + \tfrac{5}{4}y) + 3y = -4$$
$$7 + \tfrac{5}{2}y + 3y = -4$$
$$\tfrac{11}{2}y = -11$$

or $$y = -2.$$

Then, $$x = \tfrac{7}{2} + \tfrac{5}{4}(-2) = 1.$$

We conclude that the ordered pair (1, –2) satisfies the given system of equations.

11. Solving the first equation for x, we obtain

$$2x = 3y + 6$$
$$x = \tfrac{3}{2}y + 3$$

Substituting this value of x into the second equation gives

$$6(\tfrac{3}{2}y + 3) - 9y = 12$$
$$9y + 18 - 9y = 12$$
$$18 = 12.$$

which is impossible. We conclude that the system of equations has no solution.

13. Solving the first equation for y, we obtain $y = 2x - 3$. Substituting this value of y into the second equation yields

$$\begin{aligned} 4x + k(2x-3) &= 4, \\ 4x + 2xk - 3k &= 4 \\ 2x(2+k) &= 4 + 3k \\ x &= \frac{4+3k}{2(2+k)}. \end{aligned}$$

Since x is not defined when the denominator of this last expression is zero, we conclude that the system has no solution when $k = -2$.

15. Let x and y denote the number of acres of corn and wheat planted, respectively. Then $x + y = 500$. Since the cost of cultivating corn is \$42/acre and that of wheat \$30/acre and Mr. Johnson has \$18,600 available for cultivation, we have

$$42x + 30y = 18{,}600.$$

Thus, the solution is found by solving the system of equations

$$\begin{aligned} x + \quad y &= \quad 500 \\ 42x + 30y &= 18{,}600 \end{aligned}.$$

17. Let x denote the number of pounds of the \$5.00/lb coffee and y denote the number of pounds of the \$6/lb coffee. Then

$$x + y = 100.$$

Since the blended coffee sells for \$5.60/lb, we know that the blended mixture is worth $(5.60)(100) = \$560$. Therefore,

$$5x + 6y = 560.$$

Thus, the solution is found by solving the system of equations

$$\begin{aligned} x + \quad y &= 100 \\ 5x + 6y &= 560 \end{aligned}$$

19. Let x denote the number of children who rode the bus during the morning shift and y denote the number of adults who rode the bus during the morning shift. Then $x + y = 1000$. Since the total fare collected was \$1300, we have $0.5x + 1.5y = 1300$. Thus, the solution to the problem can be found by solving the system of equations

$$\begin{aligned} x + \quad\quad y &= 1000 \\ 0.5x + 1.5y &= 1300. \end{aligned}$$

21. Let x = the amount of money invested at 6 percent in a savings account
y = the amount of money invested at 8 percent in mutual funds
and z = the amount of money invested at 12 percent in bonds.

Since the total interest was \$21,600, we have

$$0.6x + 0.8y + 0.12z = 21{,}600.$$

Also, since the amount of Mr. Sid's investment in bonds is twice the amount of the investment in the savings account, we have

$$z = 2x.$$

Finally, the interest earned from his investment in bonds was equal to the interest earned on his money mutual funds, so

$$0.08y = 0.12z.$$

Thus, the solution to the problem can be found by solving the system of equations

$$\begin{aligned} 0.06x + 0.08y + 0.12z &= 21{,}600 \\ 2x \qquad\quad - z &= 0 \\ 0.08y - 0.12z &= 0. \end{aligned}$$

23. Let x, y, and z denote the number of 100-lb bags of grade-A, grade-B, and grade-C fertilizers to be produced. The amount of nitrogen required is $18x + 20y + 24z$, and this must be equal to 26,400. So, we have $18x + 20y + 24z = 26{,}400$. Similarly, the constraints on the use of phosphate and potassium lead to the equations $4x + 4y + 3z = 4900$, and $5x + 4y + 6z = 6200$, respectively. Thus we have the problem of finding the solution to the system

$$\begin{aligned} 18x + 20y + 24z &= 26{,}400 \qquad (\textit{Nitrogen}) \\ 4x + 4y + 3z &= 4{,}900 \qquad (\textit{Phosphate})\ . \\ 5x + 4y + 6z &= 6{,}200 \qquad (\textit{Potassium}) \end{aligned}$$

25. Let x, y, and z denote the number of compact, intermediate, and full-size cars, respectively, to be purchased. The cost incurred in buying the specified number of cars is $12000x + 18000y + 24000z$. Since the budget is \$1.5 million, we have the system

$$\begin{aligned} 12{,}000x + 18{,}000y + 24{,}000z &= 1{,}500{,}000 \\ x - \quad 2y \qquad\qquad &= 0 \\ x + \quad y + \quad z &= 100. \end{aligned}$$

27. Let x = the number of ounces of Food I used in the meal
y = the number of ounces of Food II used in the meal
and z = the number of ounces of Food III used in the meal.

Since 100 percent of the daily requirement of proteins, carbohydrates, and iron is to be met by this meal, we have the following system of linear equations:

$$\begin{aligned} 10x + \ 6y + \ 8z &= 100 \\ 10x + 12y + \ 6z &= 100 \\ 5x + \ 4y + 12z &= 100. \end{aligned}$$

29. True. If the three lines coincide, then the system has infinitely many solutions--corresponding to all the points on the (common) line. If at least one line is distinct from the others, then the system has no solution.

2.2 Problem Solving Tips

When you come across new notation, make sure that you understand that notation. If you can't express the notation verbally, you haven't yet grasped its use. For example, in this section we introduced the notation $R_i \leftrightarrow R_j$. This notation tells us to interchange row i with row j.

Here are some hints for solving the problems in the exercises that follow:

1. Make sure you are familiar with the three row operations: (a) $R_i \leftrightarrow R_j$ (b) cR_i (c) $R_i + aR_j$.

2. Before writing the augmented matrix, make sure that the variables in all of the equations are on the left and the constants are on the right side of the equal sign.

3. The last step of the Gauss-Jordan elimination method states that the matrix must be in row-reduced form. This means that
   (a) A row with all zeros lies below any other row that has nonzero entries.
   (b) The first nonzero entry in each row is a 1.
   (c) The leading 1 in any row lies to the right of any leading 1 in a row above that row.
   (d) All columns containing a leading 1 are unit columns.

2.2 CONCEPT QUESTIONS, page 90

1. a. The two systems are equivalent to each other if they have precisely the same

solutions.

b. (i) Interchanging row i with row j (ii) replacing row j by c times row j (iii) replacing row i with the sum of row i and a times row j.

3. a. It lies below any other row having nonzero entries.
   b. It is a 1.
   c. The leading 1 in the lower row lies to the right of the leading 1 in the upper row.
   d. They are all zero.

EXERCISES 2.2, page 91

1. $\left[\begin{array}{cc|c} 2 & -3 & 7 \\ 3 & 1 & 4 \end{array}\right]$

3. $\left[\begin{array}{ccc|c} 0 & -1 & 2 & 6 \\ 2 & 2 & -8 & 7 \\ 0 & 3 & 4 & 0 \end{array}\right]$

5. $\begin{aligned} 3x + 2y &= -4 \\ x - y &= 5 \end{aligned}$

7. $\begin{aligned} x + 3y + 2z &= 4 \\ 2x \qquad\quad &= 5 \\ 3x - 3y + 2z &= 6 \end{aligned}$

9. Yes. Conditions 1-4 are satisfied (see page 84 of the text).

11. No. Condition 3 is violated. The first nonzero entry in the second row does not lie to the right of the first nonzero entry 1 in row 1.

13. Yes. Conditions 1-4 are satisfied.

15. No. Condition 2 and consequently condition 4 are not satisfied. The first nonzero entry in the last row is not a 1 and the column containing that entry does not have zeros elsewhere.

17. No. Condition 1 is violated. The first row consists entirely of zeros and it lies above row 2.

19. $\left[\begin{array}{cc|c} \boxed{2} & 4 & 8 \\ 3 & 1 & 2 \end{array}\right] \xrightarrow{\frac{1}{2}R_1} \left[\begin{array}{cc|c} 1 & 2 & 4 \\ \boxed{3} & 1 & 2 \end{array}\right] \xrightarrow{R_2 - 3R_1} \left[\begin{array}{cc|c} 1 & 2 & 4 \\ 0 & -5 & -10 \end{array}\right]$

21. $\left[\begin{array}{cc|c} \boxed{-1} & 2 & 3 \\ 6 & 4 & 2 \end{array}\right] \xrightarrow{-R_1} \left[\begin{array}{cc|c} 1 & -2 & -3 \\ 6 & 4 & 2 \end{array}\right] \xrightarrow{R_2-6R_1} \left[\begin{array}{cc|c} 1 & -2 & -3 \\ 0 & 16 & 20 \end{array}\right]$

23. $\left[\begin{array}{ccc|c} \boxed{2} & 4 & 6 & 12 \\ 2 & 3 & 1 & 5 \\ 3 & -1 & 2 & 4 \end{array}\right] \xrightarrow{\frac{1}{2}R_1} \left[\begin{array}{ccc|c} 1 & 2 & 3 & 6 \\ 2 & 3 & 1 & 5 \\ 3 & -1 & 2 & 4 \end{array}\right] \xrightarrow[R_3-3R_1]{R_2-2R_1} \left[\begin{array}{ccc|c} 1 & 2 & 3 & 6 \\ 0 & -1 & -5 & -7 \\ 0 & -7 & -7 & -14 \end{array}\right]$

25. $\left[\begin{array}{ccc|c} 0 & 1 & 3 & 4 \\ 2 & 4 & \boxed{1} & 3 \\ 5 & 6 & 2 & -4 \end{array}\right] \xrightarrow[R_3-2R_2]{R_1-3R_2} \left[\begin{array}{ccc|c} -6 & -11 & 0 & -5 \\ 2 & 4 & 1 & 3 \\ 1 & -2 & 0 & -10 \end{array}\right]$

27. $\left[\begin{array}{cc|c} 3 & 9 & 6 \\ 2 & 1 & 4 \end{array}\right] \xrightarrow{\frac{1}{3}R_1} \left[\begin{array}{cc|c} 1 & 3 & 2 \\ 2 & 1 & 4 \end{array}\right] \xrightarrow{R_2-2R_1} \left[\begin{array}{cc|c} 1 & 3 & 2 \\ 0 & -5 & 0 \end{array}\right] \xrightarrow{-\frac{1}{5}R_2}$

$\left[\begin{array}{cc|c} 1 & 3 & 2 \\ 0 & 1 & 0 \end{array}\right] \xrightarrow{R_1-3R_2} \left[\begin{array}{cc|c} 1 & 0 & 2 \\ 0 & 1 & 0 \end{array}\right]$

29. $\left[\begin{array}{ccc|c} 1 & 3 & 1 & 3 \\ 3 & 8 & 3 & 7 \\ 2 & -3 & 1 & -10 \end{array}\right] \xrightarrow[R_3-2R_1]{R_2-3R_1} \left[\begin{array}{ccc|c} 1 & 3 & 1 & 3 \\ 0 & -1 & 0 & -2 \\ 0 & -9 & -1 & -16 \end{array}\right] \xrightarrow{-R_2} \left[\begin{array}{ccc|c} 1 & 3 & 1 & 3 \\ 0 & 1 & 0 & 2 \\ 0 & -9 & -1 & -16 \end{array}\right]$

$\xrightarrow[R_3+9R_2]{R_1-3R_2} \left[\begin{array}{ccc|c} 1 & 0 & 1 & -3 \\ 0 & 1 & 0 & 2 \\ 0 & 0 & -1 & 2 \end{array}\right] \xrightarrow[-R_3]{R_1+R_3} \left[\begin{array}{ccc|c} 1 & 0 & 0 & -1 \\ 0 & 1 & 0 & 2 \\ 0 & 0 & 1 & -2 \end{array}\right]$

31. The augmented matrix is equivalent to the system of linear equations

$$3x+9y=6$$
$$2x+\ \ y=4$$

The ordered pair (2,0) is the solution to the system.

33. The augmented matrix is equivalent to the system of linear equations

$$\begin{aligned} x+3y+\ z &= \ \ 3 \\ 3x+8y+3z &= \ \ 7 \\ 2x-3y+\ z &= -10 \end{aligned}$$

Reading off the solution from the last augmented matrix,

$$\left[\begin{array}{ccc|c} 1 & 0 & 0 & -1 \\ 0 & 1 & 0 & 2 \\ 0 & 0 & 1 & -2 \end{array}\right],$$

which is in row-reduced form, we have $x=-1$, $y=2$, and $z=-2$.

35. Using the Gauss-Jordan method, we have

$$\left[\begin{array}{cc|c} 1 & -2 & 8 \\ 3 & 4 & 4 \end{array}\right] \xrightarrow{R_2-3R_1} \left[\begin{array}{cc|c} 1 & -2 & 8 \\ 0 & 10 & -20 \end{array}\right] \xrightarrow{\frac{1}{10}R_2} \left[\begin{array}{cc|c} 1 & -2 & 8 \\ 0 & 1 & -2 \end{array}\right] \xrightarrow{R_1+2R_2} \left[\begin{array}{cc|c} 1 & 0 & 4 \\ 0 & 1 & -2 \end{array}\right]$$

The solution is (4,–2).

37. Using the Gauss-Jordan method, we have

$$\left[\begin{array}{cc|c} 2 & -3 & -8 \\ 4 & 1 & -2 \end{array}\right] \xrightarrow{\frac{1}{2}R_1} \left[\begin{array}{cc|c} 1 & -\frac{3}{2} & -4 \\ 4 & 1 & -2 \end{array}\right] \xrightarrow{R_2-4R_1} \left[\begin{array}{cc|c} 1 & -\frac{3}{2} & -4 \\ 0 & 7 & 14 \end{array}\right] \xrightarrow{\frac{1}{7}R_2}$$

$$\left[\begin{array}{cc|c} 1 & -\frac{3}{2} & -4 \\ 0 & 1 & 2 \end{array}\right] \xrightarrow{R_1+\frac{3}{2}R_2} \left[\begin{array}{cc|c} 1 & 0 & -1 \\ 0 & 1 & 2 \end{array}\right].$$

The solution is (–1,2).

39. Using the Gauss-Jordan method, we have

$$\left[\begin{array}{ccc|c} 1 & 1 & 1 & 0 \\ 2 & -1 & 1 & 1 \\ 1 & 1 & -2 & 2 \end{array}\right] \xrightarrow[R_3-R_1]{R_2-2R_1} \left[\begin{array}{ccc|c} 1 & 1 & 1 & 0 \\ 0 & -3 & -1 & 1 \\ 0 & 0 & -3 & 2 \end{array}\right] \xrightarrow{-\frac{1}{3}R_2} \left[\begin{array}{ccc|c} 1 & 1 & 1 & 0 \\ 0 & 1 & \frac{1}{3} & -\frac{1}{3} \\ 0 & 0 & -3 & 2 \end{array}\right] \xrightarrow{R_1-R_2}$$

$$\left[\begin{array}{ccc|c} 1 & 0 & \frac{2}{3} & \frac{1}{3} \\ 0 & 1 & \frac{1}{3} & -\frac{1}{3} \\ 0 & 0 & -3 & 2 \end{array}\right] \xrightarrow{-\frac{1}{3}R_3} \left[\begin{array}{ccc|c} 1 & 0 & \frac{2}{3} & \frac{1}{3} \\ 0 & 1 & \frac{1}{3} & -\frac{1}{3} \\ 0 & 0 & 1 & -\frac{2}{3} \end{array}\right] \xrightarrow[R_2-\frac{1}{3}R_3]{R_1-\frac{2}{3}R_3} \left[\begin{array}{ccc|c} 1 & 0 & 0 & \frac{7}{9} \\ 0 & 1 & 0 & -\frac{1}{9} \\ 0 & 0 & 1 & -\frac{2}{3} \end{array}\right].$$

The solution is $(\frac{7}{9}, -\frac{1}{9}, -\frac{2}{3})$.

41. $$\left[\begin{array}{ccc|c} 2 & 2 & 1 & 9 \\ 1 & 0 & 1 & 4 \\ 0 & 4 & -3 & 17 \end{array}\right] \xrightarrow{R_1 \leftrightarrow R_2} \left[\begin{array}{ccc|c} 1 & 0 & 1 & 4 \\ 2 & 2 & 1 & 9 \\ 0 & 4 & -3 & 17 \end{array}\right] \xrightarrow{R_2-2R_1} \left[\begin{array}{ccc|c} 1 & 0 & 1 & 4 \\ 0 & 2 & -1 & 1 \\ 0 & 4 & -3 & 17 \end{array}\right] \xrightarrow{\frac{1}{2}R_2}$$

$$\left[\begin{array}{ccc|c} 1 & 0 & 1 & 4 \\ 0 & 1 & -\frac{1}{2} & \frac{1}{2} \\ 0 & 4 & -3 & 17 \end{array}\right] \xrightarrow{R_3-4R_2} \left[\begin{array}{ccc|c} 1 & 0 & 1 & 4 \\ 0 & 1 & -\frac{1}{2} & \frac{1}{2} \\ 0 & 0 & -1 & 15 \end{array}\right] \xrightarrow{-R_3} \left[\begin{array}{ccc|c} 1 & 0 & 1 & 4 \\ 0 & 1 & -\frac{1}{2} & \frac{1}{2} \\ 0 & 0 & 1 & -15 \end{array}\right] \xrightarrow[R_2+\frac{1}{2}R_3]{R_1-R_3}$$

$$\left[\begin{array}{ccc|c} 1 & 0 & 0 & 19 \\ 0 & 1 & 0 & -7 \\ 0 & 0 & 1 & -15 \end{array}\right].$$ The solution is (19,–7,–15).

43. $$\left[\begin{array}{ccc|c} 0 & -1 & 1 & 2 \\ 4 & -3 & 2 & 16 \\ 3 & 2 & 1 & 11 \end{array}\right] \xrightarrow{R_1 \leftrightarrow R_2} \left[\begin{array}{ccc|c} 4 & -3 & 2 & 16 \\ 0 & -1 & 1 & 2 \\ 3 & 2 & 1 & 11 \end{array}\right] \xrightarrow{R_1-R_3} \left[\begin{array}{ccc|c} 1 & -5 & 1 & 5 \\ 0 & -1 & 1 & 2 \\ 3 & 2 & 1 & 11 \end{array}\right]$$

$$\xrightarrow[R_3-3R_1]{-R_2} \left[\begin{array}{ccc|c} 1 & -5 & 1 & 5 \\ 0 & 1 & -1 & -2 \\ 0 & 17 & -2 & -4 \end{array}\right] \xrightarrow[R_3-17R_2]{R_1+5R_2} \left[\begin{array}{ccc|c} 1 & 0 & -4 & -5 \\ 0 & 1 & -1 & -2 \\ 0 & 0 & 15 & 30 \end{array}\right] \xrightarrow{\frac{1}{15}R_3}$$

$$\left[\begin{array}{ccc|c} 1 & 0 & -4 & -5 \\ 0 & 1 & -1 & -2 \\ 0 & 0 & 1 & 2 \end{array}\right] \xrightarrow[R_2+R_3]{R_1+4R_3} \left[\begin{array}{ccc|c} 1 & 0 & 0 & 3 \\ 0 & 1 & 0 & 0 \\ 0 & 0 & 1 & 2 \end{array}\right].$$

The solution is (3,0,2).

45. Using the Gauss-Jordan method, we have

$$\left[\begin{array}{rrr|r} 1 & -2 & 1 & 6 \\ 2 & 1 & -3 & -3 \\ 1 & -3 & 3 & 10 \end{array}\right] \xrightarrow[R_3-R_1]{R_2-2R_1} \left[\begin{array}{rrr|r} 1 & -2 & 1 & 6 \\ 0 & 5 & -5 & -15 \\ 0 & -1 & 2 & 4 \end{array}\right] \xrightarrow{\frac{1}{5}R_2} \left[\begin{array}{rrr|r} 1 & -2 & 1 & 6 \\ 0 & 1 & -1 & -3 \\ 0 & -1 & 2 & 4 \end{array}\right]$$

$$\xrightarrow[R_3+R_2]{R_1+2R_2} \left[\begin{array}{rrr|r} 1 & 0 & -1 & 0 \\ 0 & 1 & -1 & -3 \\ 0 & 0 & 1 & 1 \end{array}\right] \xrightarrow[R_2+R_3]{R_1+R_3} \left[\begin{array}{rrr|r} 1 & 0 & 0 & 1 \\ 0 & 1 & 0 & -2 \\ 0 & 0 & 1 & 1 \end{array}\right].$$

Therefore, the solution is (1,–2,1).

47. Using the Gauss-Jordan method, we have

$$\left[\begin{array}{rrr|r} 2 & 0 & 3 & -1 \\ 3 & -2 & 1 & 9 \\ 1 & 1 & 4 & 4 \end{array}\right] \xrightarrow{R_1 \leftrightarrow R_3} \left[\begin{array}{rrr|r} 1 & 1 & 4 & 4 \\ 3 & -2 & 1 & 9 \\ 2 & 0 & 3 & -1 \end{array}\right] \xrightarrow[R_3-2R_1]{R_2-3R_1} \left[\begin{array}{rrr|r} 1 & 1 & 4 & 4 \\ 0 & -5 & -11 & -3 \\ 0 & -2 & -5 & -9 \end{array}\right]$$

$$\xrightarrow{-\frac{1}{5}R_2} \left[\begin{array}{rrr|r} 1 & 1 & 4 & 4 \\ 0 & 1 & \frac{11}{5} & \frac{3}{5} \\ 0 & -2 & -5 & -9 \end{array}\right] \xrightarrow[R_3+2R_2]{R_1-R_2} \left[\begin{array}{rrr|r} 1 & 0 & \frac{9}{5} & \frac{17}{5} \\ 0 & 1 & \frac{11}{5} & \frac{3}{5} \\ 0 & 0 & -\frac{3}{5} & -\frac{39}{5} \end{array}\right] \xrightarrow{-\frac{5}{3}R_3}$$

$$\left[\begin{array}{rrr|r} 1 & 0 & \frac{9}{5} & \frac{17}{5} \\ 0 & 1 & \frac{11}{5} & \frac{3}{5} \\ 0 & 0 & 1 & 13 \end{array}\right] \xrightarrow[R_2-\frac{11}{5}R_3]{R_1-\frac{9}{5}R_3} \left[\begin{array}{rrr|r} 1 & 0 & 0 & -20 \\ 0 & 1 & 0 & -28 \\ 0 & 0 & 1 & 13 \end{array}\right].$$

Therefore, the solution is (–20, –28, 13).

49. Using the Gauss-Jordan method, we have

$$\left[\begin{array}{rrr|r} 1 & -1 & 3 & 14 \\ 1 & 1 & 1 & 6 \\ -2 & -1 & 1 & -4 \end{array}\right] \xrightarrow[R_3+2R_1]{R_2-R_1} \left[\begin{array}{rrr|r} 1 & -1 & 3 & 14 \\ 0 & 2 & -2 & -8 \\ 0 & -3 & 7 & 24 \end{array}\right] \xrightarrow{\frac{1}{2}R_2} \left[\begin{array}{rrr|r} 1 & -1 & 3 & 14 \\ 0 & 1 & -1 & -4 \\ 0 & -3 & 7 & 24 \end{array}\right].$$

$$\xrightarrow[R_3+3R_2]{R_1+R_2}\left[\begin{array}{ccc|c}1&0&2&10\\0&1&-1&-4\\0&0&4&12\end{array}\right]\xrightarrow{\frac{1}{4}R_3}\left[\begin{array}{ccc|c}1&0&2&10\\0&1&-1&-4\\0&0&1&3\end{array}\right]\xrightarrow[R_2+R_3]{R_1-2R_3}\left[\begin{array}{ccc|c}1&0&0&4\\0&1&0&-1\\0&0&1&3\end{array}\right]$$

Therefore, the solution is (4, –1, 3).

51. We wish to solve the system of equations

$$\begin{aligned} x + y &= 500 \qquad (x = \text{the number of acres of corn planted})\\ 42x + 30y &= 18{,}600 \qquad (y = \text{the number of acres of wheat planted})\end{aligned}$$

Using the Gauss-Jordan method, we find

$$\left[\begin{array}{cc|c}1&1&500\\42&30&18600\end{array}\right]\xrightarrow{R_2-42R_1}\left[\begin{array}{cc|c}1&1&500\\0&-12&-2400\end{array}\right]\xrightarrow{-\frac{1}{12}R_2}\left[\begin{array}{cc|c}1&1&500\\0&1&200\end{array}\right]$$

$$\xrightarrow{R_1-R_2}\left[\begin{array}{cc|c}1&0&300\\0&1&200\end{array}\right].$$

The solution to this system of equations is $x = 300$ and $y = 200$. We conclude that Jacob should plant 300 acres of corn and 200 acres of wheat.

53. Let x denote the number of pounds of the \$5/lb coffee and y denote the number of pounds of the \$6/lb coffee. Then we are required to solve the system

$$\begin{aligned} x + y &= 100\\ 5x + 6y &= 560\end{aligned}$$

Using the Gauss-Jordan method of elimination, we have

$$\left[\begin{array}{cc|c}1&1&100\\5&6&560\end{array}\right]\xrightarrow{R_2-5R_1}\left[\begin{array}{cc|c}1&1&100\\0&1&60\end{array}\right]\xrightarrow{R_1-R_2}\left[\begin{array}{cc|c}1&0&40\\0&1&60\end{array}\right].$$

Therefore, 40 pounds of the \$5/lb coffee and 60 pounds of the \$6/lb coffee should be used in the 100 lb mixture.

55. Let x and y denote the number of children and adults who rode the bus during the morning shift, respectively. Then the solution to the problem can be found by solving the system of equations

$$\begin{aligned} x + y &= 1000\\ 0.5x + 1.5y &= 1300\end{aligned}$$

Using the Gauss-Jordan elimination method, we have

$$\left[\begin{array}{cc|c} 1 & 1 & 1000 \\ 0.5 & 1.5 & 1300 \end{array}\right] \xrightarrow{R_2-0.5R_1} \left[\begin{array}{cc|c} 1 & 1 & 1000 \\ 0 & 1 & 800 \end{array}\right] \xrightarrow{R_1-R_2} \left[\begin{array}{cc|c} 1 & 0 & 200 \\ 0 & 1 & 800 \end{array}\right].$$

We conclude that 800 adults and 200 children rode the bus during the morning shift.

57. Let x, y, and z, denote the amount of money he should invest in a savings account, in mutual funds, and in bonds, respectively. Then, we are required to solve the system

$$\begin{aligned} 0.06x + 0.08y + 0.12z &= 21{,}600 \\ 2x \qquad\qquad - \quad z &= 0 \\ 0.08y - 0.12z &= 0 \end{aligned}$$

Using the Gauss-Jordan method, we find

$$\left[\begin{array}{ccc|c} 0.06 & 0.08 & 0.12 & 21{,}600 \\ 2 & 0 & -1 & 0 \\ 0 & 0.08 & -0.12 & 0 \end{array}\right] \xrightarrow[\frac{1}{0.08}R_3]{\frac{1}{0.06}R_1} \left[\begin{array}{ccc|c} 1 & \frac{4}{3} & 2 & 360{,}000 \\ 2 & 0 & -1 & 0 \\ 0 & 1 & -\frac{3}{2} & 0 \end{array}\right] \xrightarrow{R_2-2R_1}$$

$$\left[\begin{array}{ccc|c} 1 & \frac{4}{3} & 2 & 360{,}000 \\ 0 & -\frac{8}{3} & -5 & -720{,}000 \\ 0 & 1 & -\frac{3}{2} & 0 \end{array}\right] \xrightarrow{-\frac{3}{8}R_2} \left[\begin{array}{ccc|c} 1 & \frac{4}{3} & 2 & 360{,}000 \\ 0 & 1 & \frac{15}{8} & 270{,}000 \\ 0 & 1 & -\frac{3}{2} & 0 \end{array}\right]$$

$$\xrightarrow[R_3-R_2]{R_1-\frac{4}{3}R_2} \left[\begin{array}{ccc|c} 1 & 0 & -\frac{1}{2} & 0 \\ 0 & 1 & \frac{15}{8} & 270{,}000 \\ 0 & 0 & -\frac{27}{8} & -270{,}000 \end{array}\right] \xrightarrow{-\frac{8}{27}R_3} \left[\begin{array}{ccc|c} 1 & 0 & -\frac{1}{2} & 0 \\ 0 & 1 & \frac{15}{8} & 270{,}000 \\ 0 & 0 & 1 & 80{,}000 \end{array}\right]$$

$$\xrightarrow[R_2-\frac{15}{8}R_3]{R_1+\frac{1}{2}R_3} \left[\begin{array}{ccc|c} 1 & 0 & 0 & 40{,}000 \\ 0 & 1 & 0 & 120{,}000 \\ 0 & 0 & 1 & 80{,}000 \end{array}\right]$$

Therefore, Sid should invest \$40,000 in a savings account, \$120,000 in mutual funds, and \$80,000 in bonds.

59. Refer to Exercise 23, page 78. We obtain the following augmented matrices.

$$\left[\begin{array}{ccc|c} 18 & 20 & 24 & 26400 \\ 4 & 4 & 3 & 4900 \\ 5 & 4 & 6 & 6200 \end{array}\right] \xrightarrow{R_1 \leftrightarrow R_3} \left[\begin{array}{ccc|c} 5 & 4 & 6 & 6200 \\ 4 & 4 & 3 & 4900 \\ 18 & 20 & 24 & 26400 \end{array}\right] \xrightarrow{R_1 - R_2}$$

$$\left[\begin{array}{ccc|c} 1 & 0 & 3 & 1300 \\ 4 & 4 & 3 & 4900 \\ 18 & 20 & 24 & 26400 \end{array}\right] \xrightarrow[R_3 - 18R_1]{R_2 - 4R_1} \left[\begin{array}{ccc|c} 1 & 0 & 3 & 1300 \\ 0 & 4 & -9 & -300 \\ 0 & 20 & -30 & 3000 \end{array}\right] \xrightarrow{\frac{1}{4}R_2}$$

$$\left[\begin{array}{ccc|c} 1 & 0 & 3 & 1300 \\ 0 & 1 & -\frac{9}{4} & -75 \\ 0 & 20 & -30 & 3000 \end{array}\right] \xrightarrow{R_3 - 20R_2} \left[\begin{array}{ccc|c} 1 & 0 & 3 & 1300 \\ 0 & 1 & -\frac{9}{4} & -75 \\ 0 & 0 & 15 & 4500 \end{array}\right] \xrightarrow{\frac{1}{15}R_3}$$

$$\left[\begin{array}{ccc|c} 1 & 0 & 3 & 1300 \\ 0 & 1 & -\frac{9}{4} & -75 \\ 0 & 0 & 1 & 300 \end{array}\right] \xrightarrow[R_2 + \frac{9}{4}R_3]{R_1 - 3R_3} \left[\begin{array}{ccc|c} 1 & 0 & 0 & 400 \\ 0 & 1 & 0 & 600 \\ 0 & 0 & 1 & 300 \end{array}\right]$$

We see that $x = 400$, $y = 600$, and $z = 300$. Therefore, Lawnco should produce 400, 600, and 300 100-lb bags of grade-A, grade-B, and grade-C fertilizer.

61. Let x, y, and z denote the number of compact, intermediate, and full-size cars, respectively, to be purchased. Then the problem can be solved by solving the system

$$\begin{aligned} 12{,}000x + 18{,}000y + 24{,}000z &= 1{,}500{,}000 \\ x - 2y &= 0 \\ x + y + z &= 100 \end{aligned}$$

Using the Gauss-Jordan method, we have

$$\left[\begin{array}{ccc|c} 12{,}000 & 18{,}000 & 24{,}000 & 1{,}500{,}000 \\ 1 & -2 & 0 & 0 \\ 1 & 1 & 1 & 100 \end{array}\right] \xrightarrow{R_1 \leftrightarrow R_3} \left[\begin{array}{ccc|c} 1 & 1 & 1 & 100 \\ 1 & -2 & 0 & 0 \\ 12{,}000 & 18{,}000 & 24{,}000 & 1{,}500{,}000 \end{array}\right]$$

$$\xrightarrow[R_3-12{,}000R_1]{R_2-R_1}\left[\begin{array}{ccc|c}1&1&1&100\\0&-3&-1&-100\\0&6000&12{,}000&300{,}000\end{array}\right]\xrightarrow{-\frac{1}{3}R_2}\left[\begin{array}{ccc|c}1&1&1&100\\0&1&\frac{1}{3}&\frac{100}{3}\\0&6000&12{,}000&300{,}000\end{array}\right]$$

$$\xrightarrow[R_3-6000R_2]{R_1-R_2}\left[\begin{array}{ccc|c}1&0&\frac{2}{3}&\frac{200}{3}\\0&1&\frac{1}{3}&\frac{100}{3}\\0&0&10000&100{,}000\end{array}\right]\xrightarrow{\frac{1}{10{,}000}R_3}\left[\begin{array}{ccc|c}1&0&\frac{2}{3}&\frac{200}{3}\\0&1&\frac{1}{3}&\frac{100}{3}\\0&0&1&10\end{array}\right]\xrightarrow[R_2-\frac{1}{3}R_3]{R_1-\frac{2}{3}R_3}\left[\begin{array}{ccc|c}1&0&0&60\\0&1&0&30\\0&0&1&10\end{array}\right].$$

We conclude that 60 compact cars, 30 intermediate-size cars, and 10 full-size cars will be purchased.

63. Let x, y, and z, represent the number of ounces of Food I, Food II, and Food III used

in the meal, respectively. Then the problem reduces to solving the following system of linear equations:

$$\begin{aligned}10x+\ 6y+\ 8z&=100\\10x+12y+\ 6z&=100\\5x+\ 4y+12z&=100.\end{aligned}$$

Using the Gauss-Jordan method, we obtain

$$\left[\begin{array}{ccc|c}10&6&8&100\\10&12&6&100\\5&4&12&100\end{array}\right]\xrightarrow{\frac{1}{10}R_1}\left[\begin{array}{ccc|c}1&\frac{3}{5}&\frac{4}{5}&10\\10&12&6&100\\5&4&12&100\end{array}\right]\xrightarrow[R_3-5R_1]{R_2-10R_1}$$

$$\left[\begin{array}{ccc|c}1&\frac{3}{5}&\frac{4}{5}&10\\0&6&-2&0\\0&1&8&50\end{array}\right]\xrightarrow{\frac{1}{6}R_2}\left[\begin{array}{ccc|c}1&\frac{3}{5}&\frac{4}{5}&10\\0&1&-\frac{1}{3}&0\\0&1&8&50\end{array}\right]\xrightarrow[R_3-R_2]{R_1-\frac{3}{5}R_2}$$

$$\left[\begin{array}{ccc|c}1&0&1&10\\0&1&-\frac{1}{3}&0\\0&0&\frac{25}{3}&50\end{array}\right]\xrightarrow{\frac{3}{25}R_3}\left[\begin{array}{ccc|c}1&0&1&10\\0&1&-\frac{1}{3}&0\\0&0&1&6\end{array}\right]\xrightarrow[R_2+\frac{1}{3}R_3]{R_1-R_3}\left[\begin{array}{ccc|c}1&0&0&4\\0&1&0&2\\0&0&1&6\end{array}\right].$$

We conclude that 4 oz of Food I, 2 oz of Food II, and 6 oz of Food III should be used to prepare the meal.

65. Let $x =$ the number of front orchestra seats sold
$y =$ the number of rear orchestra seats sold
and $z =$ the number of front balcony seats sold for this performance.
Then, we are required to solve the system

$$\begin{aligned} x + y + z &= 1{,}000 \\ 80x + 60y + 50z &= 62{,}800 \\ x + y - 2z &= 400. \end{aligned}$$

Using the Gauss-Jordan method, we find

$$\left[\begin{array}{ccc|c} 1 & 1 & 1 & 1{,}000 \\ 80 & 60 & 50 & 62{,}800 \\ 1 & 1 & -2 & 400 \end{array}\right] \xrightarrow[R_3 - R_1]{R_2 - 80R_1} \left[\begin{array}{ccc|c} 1 & 1 & 1 & 1{,}000 \\ 0 & -20 & -30 & -17{,}200 \\ 0 & 0 & -3 & -600 \end{array}\right] \xrightarrow[-\frac{1}{3}R_3]{-\frac{1}{20}R_2}$$

$$\left[\begin{array}{ccc|c} 1 & 1 & 1 & 1{,}000 \\ 0 & 1 & \frac{3}{2} & 860 \\ 0 & 0 & 1 & 200 \end{array}\right] \xrightarrow{R_1 - R_2} \left[\begin{array}{ccc|c} 1 & 0 & -\frac{1}{2} & 140 \\ 0 & 1 & \frac{3}{2} & 860 \\ 0 & 0 & 1 & 200 \end{array}\right] \xrightarrow[R_2 - \frac{3}{2}R_3]{R_1 + \frac{1}{2}R_3}$$

$$\left[\begin{array}{ccc|c} 1 & 0 & 0 & 240 \\ 0 & 1 & 0 & 560 \\ 0 & 0 & 1 & 200 \end{array}\right].$$

We conclude that tickets for 240 front orchestra seats, 560 rear orchestra seats, and 200 front balcony seats were sold.

67. Let x, y, and z denote the number of days he spent in London, Paris, and Rome, respectively. We have

$$\begin{aligned} 180x + 230y + 160z &= 2660 \\ 110x + 120y + 90z &= 1520 \\ x - y - z &= 0 \qquad \text{(since } x = y + z\text{)} \end{aligned}$$

Using the Gauss-Jordan method to solve the system, we have

$$\left[\begin{array}{ccc|c} 180 & 230 & 160 & 2660 \\ 110 & 120 & 90 & 1520 \\ 1 & -1 & -1 & 0 \end{array}\right] \xrightarrow{R_1 \leftrightarrow R_3} \left[\begin{array}{ccc|c} 1 & -1 & -1 & 0 \\ 110 & 120 & 90 & 1520 \\ 180 & 230 & 160 & 2660 \end{array}\right] \xrightarrow[R_3 - 180R_1]{R_2 - 110R_1}$$

$$\left[\begin{array}{ccc|c} 1 & -1 & -1 & 0 \\ 0 & 230 & 200 & 1520 \\ 0 & 410 & 340 & 2660 \end{array}\right] \xrightarrow{\frac{1}{230}R_2} \left[\begin{array}{ccc|c} 1 & -1 & -1 & 0 \\ 0 & 1 & \frac{20}{23} & \frac{152}{23} \\ 0 & 410 & 340 & 2660 \end{array}\right] \xrightarrow[R_3-410R_2]{R_1+R_2}$$

$$\left[\begin{array}{ccc|c} 1 & 0 & -\frac{3}{23} & \frac{152}{23} \\ 0 & 1 & \frac{20}{23} & \frac{152}{23} \\ 0 & 0 & -\frac{380}{23} & -\frac{1140}{23} \end{array}\right] \xrightarrow{-\frac{23}{380}R_3} \left[\begin{array}{ccc|c} 1 & 0 & -\frac{3}{23} & \frac{152}{23} \\ 0 & 1 & \frac{20}{23} & \frac{152}{23} \\ 0 & 0 & 1 & 3 \end{array}\right] \xrightarrow[R_2-\frac{20}{23}R_1]{R_1+\frac{3}{23}R_3}$$

$$\left[\begin{array}{ccc|c} 1 & 0 & 0 & 7 \\ 0 & 1 & 0 & 4 \\ 0 & 0 & 1 & 3 \end{array}\right]$$

The solution is $x = 7, y = 4$, and $z = 3$. Therefore, he spent 7 days in London, 4 days in Paris, and 3 days in Rome.

69. False. The constant cannot be zero. The system

$$\begin{aligned} 2x + y &= 1 \\ 3x - y &= 2 \end{aligned}$$

is not equivalent to

$$\begin{aligned} 2x + y &= 1 \\ 0(3x - y) &= 0(2) \end{aligned} \quad \text{or} \quad \begin{aligned} 2x + y &= 1 \\ 0 &= 0 \end{aligned}.$$

USING TECHNOLOGY EXERCISES 2.2, page 96

1. (3, 1, –1, 2) 3. (5, 4, –3, –4) 5. (1,–1, 2, 0, 3)

2.3 CONCEPT QUESTIONS, page 103

1. a. There may be no solution, a unique solution, or infinitely many solutions.
   b. There may be no solution or infinitely many solutions.

EXERCISES 2.3, page 104

1. a. The system has one solution. b. The solution is (3, –1, 2).

3. a. The system has one solution. b. The solution is (2, 4).

5. a. The system has infinitely many solutions.
b. Letting $x_3 = t$, we see that the solutions are given by $(4 - t, -2, t)$, where t is a parameter.

7. a. The system has no solution. The last row contains all zeros to the left of the vertical line and a nonzero number (1) to its right.

9. a. The system has infinitely many solutions.
b. Letting $x_4 = t$, we see that the solutions are given by $(2, -1, 2 - t, t)$, where t is a parameter.

11. a. The system has infinitely many solutions.
b. Letting $x_3 = s$ and $x_4 = t$, the solutions are given by $(2 - 3s, 1 + s, s, t)$, where s and t are parameters.

13. Using the Gauss-Jordan method, we have

$$\left[\begin{array}{cc|c} 2 & -1 & 3 \\ 1 & 2 & 4 \\ 2 & 3 & 7 \end{array}\right] \xrightarrow{R_1 \leftrightarrow R_2} \left[\begin{array}{cc|c} 1 & 2 & 4 \\ 2 & -1 & 3 \\ 2 & 3 & 7 \end{array}\right] \xrightarrow[R_3-2R_1]{R_2-2R_1} \left[\begin{array}{cc|c} 1 & 2 & 4 \\ 0 & -5 & -5 \\ 0 & -1 & -1 \end{array}\right] \xrightarrow{-\frac{1}{5}R_2}$$

$$\left[\begin{array}{cc|c} 1 & 2 & 4 \\ 0 & 1 & 1 \\ 0 & -1 & -1 \end{array}\right] \xrightarrow[R_3+R_2]{R_1-2R_2} \left[\begin{array}{cc|c} 1 & 0 & 2 \\ 0 & 1 & 1 \\ 0 & 0 & 0 \end{array}\right].$$

The solution is (2,1).

15. Using the Gauss-Jordan method, we have

$$\left[\begin{array}{cc|c} 3 & -2 & -3 \\ 2 & 1 & 3 \\ 1 & -2 & -5 \end{array}\right] \xrightarrow{R_1 \leftrightarrow R_3} \left[\begin{array}{cc|c} 1 & -2 & -5 \\ 2 & 1 & 3 \\ 3 & -2 & -3 \end{array}\right] \xrightarrow[R_3-3R_1]{R_2-2R_1} \left[\begin{array}{cc|c} 1 & -2 & -5 \\ 0 & 5 & 13 \\ 0 & 4 & 12 \end{array}\right] \xrightarrow{\frac{1}{5}R_2}$$

$$\left[\begin{array}{cc|c} 1 & -2 & -5 \\ 0 & 1 & \frac{13}{5} \\ 0 & 4 & 12 \end{array}\right] \xrightarrow[R_3-4R_2]{R_1+2R_2} \left[\begin{array}{cc|c} 1 & 0 & \frac{1}{5} \\ 0 & 1 & \frac{13}{5} \\ 0 & 0 & \frac{8}{5} \end{array}\right].$$

Since the last row implies the 0 = 8/5, we conclude that the system of equations is inconsistent and has no solution.

17. $$\left[\begin{array}{cc|c} 3 & -2 & 5 \\ -1 & 3 & -4 \\ 2 & -4 & 6 \end{array}\right] \xrightarrow{R_1 \leftrightarrow R_2} \left[\begin{array}{cc|c} -1 & 3 & -4 \\ 3 & -2 & 5 \\ 2 & -4 & 6 \end{array}\right] \xrightarrow{-R_1} \left[\begin{array}{cc|c} 1 & -3 & 4 \\ 3 & -2 & 5 \\ 2 & -4 & 6 \end{array}\right] \xrightarrow[R_3-2R_1]{R_2-3R_1}$$

$$\left[\begin{array}{cc|c} 1 & -3 & 4 \\ 0 & 7 & -7 \\ 0 & 2 & -2 \end{array}\right] \xrightarrow{\frac{1}{7}R_2} \left[\begin{array}{cc|c} 1 & -3 & 4 \\ 0 & 1 & -1 \\ 0 & 2 & -2 \end{array}\right] \xrightarrow[R_3-2R_2]{R_1+3R_2} \left[\begin{array}{cc|c} 1 & 0 & 1 \\ 0 & 1 & -1 \\ 0 & 0 & 0 \end{array}\right].$$

We conclude that the solution is (1,–1).

19. $$\left[\begin{array}{cc|c} 1 & -2 & 2 \\ 7 & -14 & 14 \\ 3 & -6 & 6 \end{array}\right] \xrightarrow[R_3-3R_1]{R_2-7R_1} \left[\begin{array}{cc|c} 1 & -2 & 2 \\ 0 & 0 & 0 \\ 0 & 0 & 0 \end{array}\right].$$

We conclude that the infinitely many solutions are given by $(2t+2,\ t)$, where t is a parameter.

21. $$\left[\begin{array}{cc|c} 3 & 2 & 4 \\ -\frac{3}{2} & -1 & -2 \\ 6 & 4 & 8 \end{array}\right] \xrightarrow{\frac{1}{3}R_1} \left[\begin{array}{cc|c} 1 & \frac{2}{3} & \frac{4}{3} \\ -\frac{3}{2} & -1 & -2 \\ 6 & 4 & 8 \end{array}\right] \xrightarrow[R_3-6R_1]{R_2+\frac{3}{2}R_1} \left[\begin{array}{cc|c} 1 & \frac{2}{3} & \frac{4}{3} \\ 0 & 0 & 0 \\ 0 & 0 & 0 \end{array}\right].$$

We conclude that the infinitely many solutions are given by $(\frac{4}{3}-\frac{2}{3}t,\ t)$, where t is a parameter.

23. $$\left[\begin{array}{ccc|c} 2 & -1 & 1 & -4 \\ 3 & -\frac{3}{2} & \frac{3}{2} & -6 \\ -6 & 3 & -3 & 12 \end{array}\right] \xrightarrow{\frac{1}{2}R_1} \left[\begin{array}{ccc|c} 1 & -\frac{1}{2} & \frac{1}{2} & -2 \\ 3 & -\frac{3}{2} & \frac{3}{2} & -6 \\ -6 & 3 & -3 & 12 \end{array}\right] \xrightarrow[R_3+6R_1]{R_2-3R_1} \left[\begin{array}{ccc|c} 1 & -\frac{1}{2} & \frac{1}{2} & -2 \\ 0 & 0 & 0 & 0 \\ 0 & 0 & 0 & 0 \end{array}\right].$$

We conclude that the infinitely many solutions are given by $(-2+\frac{1}{2}s-\frac{1}{2}t,\ s\ ,t)$ where s and t are parameters.

25. $\left[\begin{array}{ccc|c} 1 & -2 & 3 & 4 \\ 2 & 3 & -1 & 2 \\ 1 & 2 & -3 & -6 \end{array}\right] \xrightarrow[R_3-R_1]{R_2-2R_1} \left[\begin{array}{ccc|c} 1 & -2 & 3 & 4 \\ 0 & 7 & -7 & -6 \\ 0 & 4 & -6 & -10 \end{array}\right] \xrightarrow{\frac{1}{7}R_2} \left[\begin{array}{ccc|c} 1 & -2 & 3 & 4 \\ 0 & 1 & -1 & -\frac{6}{7} \\ 0 & 4 & -6 & -10 \end{array}\right]$

$\xrightarrow[R_3-4R_2]{R_1+2R_2} \left[\begin{array}{ccc|c} 1 & 0 & 1 & \frac{16}{7} \\ 0 & 1 & -1 & -\frac{6}{7} \\ 0 & 0 & -2 & -\frac{46}{7} \end{array}\right] \xrightarrow{-\frac{1}{2}R_3} \left[\begin{array}{ccc|c} 1 & 0 & 1 & \frac{16}{7} \\ 0 & 1 & -1 & -\frac{6}{7} \\ 0 & 0 & 1 & \frac{23}{7} \end{array}\right] \xrightarrow[R_2+R_3]{R_1-R_3}$

$\left[\begin{array}{ccc|c} 1 & 0 & 0 & -1 \\ 0 & 1 & 0 & \frac{17}{7} \\ 0 & 0 & 1 & \frac{23}{7} \end{array}\right].$

We conclude that the solution is $(-1, \frac{17}{7}, \frac{23}{7})$.

27. $\left[\begin{array}{ccc|c} 4 & 1 & -1 & 4 \\ 8 & 2 & -2 & 8 \end{array}\right] \xrightarrow{\frac{1}{4}R_1} \left[\begin{array}{ccc|c} 1 & \frac{1}{4} & \frac{1}{4} & 1 \\ 8 & 2 & -2 & 8 \end{array}\right] \xrightarrow{R_2-8R_1} \left[\begin{array}{ccc|c} 1 & \frac{1}{4} & -\frac{1}{4} & 1 \\ 0 & 0 & 0 & 0 \end{array}\right]$

We conclude that the infinitely many solutions are given by $(1-\frac{1}{4}s+\frac{1}{4}t, s, t)$, where s and t are parameters.

29. $\left[\begin{array}{ccc|c} 2 & 1 & -3 & 1 \\ 1 & -1 & 2 & 1 \\ 5 & -2 & 3 & 6 \end{array}\right] \xrightarrow{R_1 \leftrightarrow R_2} \left[\begin{array}{ccc|c} 1 & -1 & 2 & 1 \\ 2 & 1 & -3 & 1 \\ 5 & -2 & 3 & 6 \end{array}\right] \xrightarrow[R_3-5R_1]{R_2-2R_1} \left[\begin{array}{ccc|c} 1 & -1 & 2 & 1 \\ 0 & 3 & -7 & -1 \\ 0 & 3 & -7 & 1 \end{array}\right] \xrightarrow{\frac{1}{3}R_2}$

$\left[\begin{array}{ccc|c} 1 & -1 & 2 & 1 \\ 0 & 1 & -\frac{7}{3} & -\frac{1}{3} \\ 0 & 3 & -7 & 1 \end{array}\right] \xrightarrow[R_3-3R_2]{R_1+R_2} \left[\begin{array}{ccc|c} 1 & 0 & -\frac{1}{3} & \frac{2}{3} \\ 0 & 1 & -\frac{7}{3} & -\frac{1}{3} \\ 0 & 0 & 0 & 2 \end{array}\right].$

This last row implies that $0 = 2$, which is impossible. We conclude that the system of equations is inconsistent and has no solution.

31. $\left[\begin{array}{ccc|c} 1 & 2 & -1 & -4 \\ 2 & 1 & 1 & 7 \\ 1 & 3 & 2 & 7 \\ 1 & -3 & 1 & 9 \end{array}\right] \xrightarrow[\substack{R_3-R_1 \\ R_4-R_1}]{R_2-2R_1} \left[\begin{array}{ccc|c} 1 & 2 & -1 & -4 \\ 0 & -3 & 3 & 15 \\ 0 & 1 & 3 & 11 \\ 0 & -5 & 2 & 13 \end{array}\right] \xrightarrow{-\frac{1}{3}R_2} \left[\begin{array}{ccc|c} 1 & 2 & -1 & -4 \\ 0 & 1 & -1 & -5 \\ 0 & 1 & 3 & 11 \\ 0 & -5 & 2 & 13 \end{array}\right]$

$$\xrightarrow[R_4+5R_2]{\substack{R_1-2R_2\\R_3-R_2}}\left[\begin{array}{ccc|c}1&0&1&6\\0&1&-1&-5\\0&0&4&16\\0&0&-3&-12\end{array}\right]\xrightarrow{\frac{1}{4}R_3}\left[\begin{array}{ccc|c}1&0&1&6\\0&1&-1&-5\\0&0&1&4\\0&0&-3&-12\end{array}\right]\xrightarrow[R_4+3R_3]{\substack{R_1-R_3\\R_2+R_3}}\left[\begin{array}{ccc|c}1&0&0&2\\0&1&0&-1\\0&0&1&4\\0&0&0&0\end{array}\right].$$

We conclude that the solution of the system is (2,–1,4).

33. Let x, y, and z represent the number of compact, mid-sized, and full-size cars, respectively, to be purchased. Then the problem can be solved by solving the system

$$\begin{aligned}x+\quad y+\quad z&=60\\12000x+19200y+26400z&=1008000\end{aligned}.$$

Using the Gauss-Jordan method, we have

$$\left[\begin{array}{ccc|c}1&1&1&60\\12000&19200&26400&1008000\end{array}\right]\xrightarrow{R_2-12{,}000R_1}\left[\begin{array}{ccc|c}1&1&1&60\\0&7200&14400&288000\end{array}\right]$$

$$\xrightarrow{\frac{1}{7200}R_2}\left[\begin{array}{ccc|c}1&1&1&60\\0&1&2&40\end{array}\right]\xrightarrow{R_1-R_2}\left[\begin{array}{ccc|c}1&0&-1&20\\0&1&2&40\end{array}\right]$$

and we conclude that the solution is $(20+z, 40-2z, z)$. Letting $z=5$, we see that one possible solution is (25,30,5); that is Hartman should buy 25 compact, 30 mid-sized cars, and 5 full-sized cars. Letting $z=10$, we see that another possible solution is (30,20,10); that is, 30 compact cars, 20 mid-sized cars, and 10 full-sized cars.

35. Let x, y, and z denote the number of ounces of Food I, Food II, and Food III, respectively, that the dietician includes in the meal. Then the problem can be solved by solving the system

$$\begin{aligned}400x+1200y+800z&=8800\\110x+570y+340z&=2160\\90x+30y+60z&=1020.\end{aligned}$$

Using the Gauss-Jordan method, we have

$$\left[\begin{array}{ccc|c}400&1200&800&8800\\110&570&340&2160\\90&30&60&1020\end{array}\right]\xrightarrow{\frac{1}{400}R_1}\left[\begin{array}{ccc|c}1&3&2&22\\110&570&340&2160\\90&30&60&1020\end{array}\right]\xrightarrow[R_3-90R_1]{R_2-110R_1}$$

$$\left[\begin{array}{ccc|c} 1 & 3 & 2 & 22 \\ 0 & 240 & 120 & -260 \\ 0 & -240 & -120 & -960 \end{array}\right] \xrightarrow{\frac{1}{240}R_2} \left[\begin{array}{ccc|c} 1 & 3 & 2 & 22 \\ 0 & 1 & \frac{1}{2} & -\frac{13}{12} \\ 0 & -240 & -120 & -960 \end{array}\right] \xrightarrow[R_3+240R_2]{R_1-3R_2}$$

$$\left[\begin{array}{ccc|c} 1 & 0 & \frac{1}{2} & \frac{101}{4} \\ 0 & 1 & \frac{1}{2} & -\frac{13}{12} \\ 0 & 0 & 0 & -1220 \end{array}\right].$$

This last row implies that $0 = -1220$, which is impossible. We conclude that the system of equations is inconsistent and has no solution--that is, the dietician cannot prepare a meal from these foods and meet the given requirements.

37. a.

$$\begin{aligned} x_1 - x_2 \qquad\qquad\qquad\qquad &= 200 \\ x_1 \qquad\qquad - x_5 \qquad &= 100 \\ - x_2 + x_3 \qquad\qquad + x_6 &= 600 \\ - x_3 + x_4 \qquad\qquad &= 200 \\ x_4 - x_5 + x_6 &= 700. \end{aligned}$$

b.

$$\left[\begin{array}{cccccc|c} 1 & -1 & 0 & 0 & 0 & 0 & 200 \\ 1 & 0 & 0 & 0 & -1 & 0 & 100 \\ 0 & -1 & 1 & 0 & 0 & 1 & 600 \\ 0 & 0 & -1 & 1 & 0 & 0 & 200 \\ 0 & 0 & 0 & 1 & -1 & 1 & 700 \end{array}\right] \xrightarrow{R_2-R_1} \left[\begin{array}{cccccc|c} 1 & -1 & 0 & 0 & 0 & 0 & 200 \\ 0 & 1 & 0 & 0 & -1 & 0 & -100 \\ 0 & -1 & 1 & 0 & 0 & 1 & 600 \\ 0 & 0 & -1 & 1 & 0 & 0 & 200 \\ 0 & 0 & 0 & 1 & -1 & 1 & 700 \end{array}\right] \xrightarrow[R_1+R_2]{R_3+R_2}$$

$$\left[\begin{array}{cccccc|c} 1 & 0 & 0 & 0 & -1 & 0 & 100 \\ 0 & 1 & 0 & 0 & -1 & 0 & -100 \\ 0 & 0 & 1 & 0 & -1 & 1 & 500 \\ 0 & 0 & -1 & 1 & 0 & 0 & 200 \\ 0 & 0 & 0 & 1 & -1 & 1 & 700 \end{array}\right] \xrightarrow{R_4+R_3} \left[\begin{array}{cccccc|c} 1 & 0 & 0 & 0 & -1 & 0 & 100 \\ 0 & 1 & 0 & 0 & -1 & 0 & -100 \\ 0 & 0 & 1 & 0 & -1 & 1 & 500 \\ 0 & 0 & 0 & 1 & -1 & 1 & 700 \\ 0 & 0 & 0 & 1 & -1 & 1 & 700 \end{array}\right] \xrightarrow{R_5-R_4}$$

We conclude that the solution is

$$(s+100,\ s-100,\ s-t+500,\ s-t+700,\ s,\ t).$$

Taking $s = 150$ and $t = 50$, we see that one possible traffic pattern is
(250, 50, 600, 800, 150, 50).

Similarly, taking $s = 200$, and $t = 100$, we see that another possible traffic pattern is
(300, 100, 600, 800, 200, 100).

c. Taking $t = 0$ and $s = 200$, we see that another possible traffic pattern is
(300, 100, 700, 900, 200, 0).

39. We solve the given system by using the Gauss-Jordan method. We have

$$\left[\begin{array}{cc|c} 2 & 3 & 2 \\ 1 & 4 & 6 \\ 5 & k & 2 \end{array}\right] \xrightarrow{R_1 \leftrightarrow R_2} \left[\begin{array}{cc|c} 1 & 4 & 6 \\ 2 & 3 & 2 \\ 5 & k & 2 \end{array}\right] \xrightarrow[R_3 - 5R_1]{R_2 - 2R_1} \left[\begin{array}{cc|c} 1 & 4 & 6 \\ 0 & -5 & -10 \\ 0 & k-20 & -28 \end{array}\right] \xrightarrow{-\frac{1}{5}R_2}$$

$$\left[\begin{array}{cc|c} 1 & 4 & 6 \\ 0 & 1 & 2 \\ 0 & k-20 & -28 \end{array}\right] \xrightarrow[R_3 + aR_2]{R_1 - 4R_2} \left[\begin{array}{cc|c} 1 & 0 & -2 \\ 0 & 1 & 2 \\ 0 & k+a-20 & -28+2a \end{array}\right]$$

From the last matrix, we see that the system has a solution if and only if $x = -2$, $y = 2$, and

$$-28 + 2a = 0, \text{ or } a = 14$$

and $\quad k + a - 20 = k - 6 = 0, \text{ or } k = 6.$

(All the entries in the last row of the matrix must be equal to zero.)

41. False. Such a system cannot have a unique solution.

USING TECHNOLOGY EXERCISES 2.3, page 107

1. $(1 + t, 2 + t, t)$; t, a parameter 3. $(-\frac{17}{7} + \frac{6}{7}t, 3 - t, -\frac{18}{7} + \frac{1}{7}t, t)$ 5. No solution

2.4 Problem Solving Tips

Here are some hints for solving the exercises that follow.

1. If a matrix is of size $m \times n$, then it has m rows and n columns. For example, a matrix of size 4×3 has 4 rows and 3 columns.

2. The sum (difference) of two matrices A and B is only defined if A and B have the same size. To find the sum (difference) of A and B add (subtract) the corresponding entries in the two matrices.

3. To find the scalar product of a real number c and a matrix A, multiply each entry in A by c.

2.4 CONCEPT QUESTIONS, page 114

1. a. A matrix is an ordered rectangular array of real numbers.
   b. A matrix has size (or dimension) $m \times n$ if it has m rows and n columns.
   c. A row matrix is one of size $1 \times n$.
   d. A column matrix is one of size $m \times 1$.
   e. A square matrix is one of size $n \times n$.

3.

$$A = \begin{bmatrix} 1 & 2 & 4 \\ 2 & -2 & 1 \\ 4 & 1 & 3 \end{bmatrix}$$

The entries satisfy $a_{ij} = a_{ji}$, that is A is symmetric with respect to the main diagonal.

EXERCISES 2.4, page 115

1. The size of A is 4×4; the size of B is 4×3; the size of C is 1×5, and the size of D is 4×1.

3. These are entries of the matrix B. The entry b_{13} refers to the entry in the first row and third column and is equal to 2. Similarly, $b_{31} = 3$, and $b_{43} = 8$.

5. The column matrix is the matrix D. The transpose of the matrix D is
$$D^T = [1 \;\; 3 \;\; -2 \;\; 0].$$

7. A is of size 3×2; B is of size 3×2; C and D are of size 3×3.

9. $$A+B=\begin{bmatrix}-1 & 2\\ 3 & -2\\ 4 & 0\end{bmatrix}+\begin{bmatrix}2 & 4\\ 3 & 1\\ -2 & 2\end{bmatrix}=\begin{bmatrix}1 & 6\\ 6 & -1\\ 2 & 2\end{bmatrix}.$$

11. $$C-D=\begin{bmatrix}3 & -1 & 0\\ 2 & -2 & 3\\ 4 & 6 & 2\end{bmatrix}-\begin{bmatrix}2 & -2 & 4\\ 3 & 6 & 2\\ -2 & 3 & 1\end{bmatrix}=\begin{bmatrix}1 & 1 & -4\\ -1 & -8 & 1\\ 6 & 3 & 1\end{bmatrix}.$$

13. $$\begin{bmatrix}6 & 3 & 8\\ 4 & 5 & 6\end{bmatrix}-\begin{bmatrix}3 & -2 & -1\\ 0 & -5 & -7\end{bmatrix}=\begin{bmatrix}3 & 5 & 9\\ 4 & 10 & 13\end{bmatrix}.$$

15. $$\begin{bmatrix}1 & 4 & -5\\ 3 & -8 & 6\end{bmatrix}+\begin{bmatrix}4 & 0 & -2\\ 3 & 6 & 5\end{bmatrix}-\begin{bmatrix}2 & 8 & 9\\ -11 & 2 & -5\end{bmatrix}=\begin{bmatrix}3 & -4 & -16\\ 17 & -4 & 16\end{bmatrix}.$$

17. $$\begin{bmatrix}1.2 & 4.5 & -4.2\\ 8.2 & 6.3 & -3.2\end{bmatrix}-\begin{bmatrix}3.1 & 1.5 & -3.6\\ 2.2 & -3.3 & -4.4\end{bmatrix}=\begin{bmatrix}-1.9 & 3.0 & -0.6\\ 6.0 & 9.6 & 1.2\end{bmatrix}.$$

19. $$\frac{1}{2}\begin{bmatrix}1 & 0 & 0 & -4\\ 3 & 0 & -1 & 6\\ -2 & 1 & -4 & 2\end{bmatrix}+\frac{4}{3}\begin{bmatrix}3 & 0 & -1 & 4\\ -2 & 1 & -6 & 2\\ 8 & 2 & 0 & -2\end{bmatrix}-\frac{1}{3}\begin{bmatrix}3 & -9 & -1 & 0\\ 6 & 2 & 0 & -6\\ 0 & 1 & -3 & 1\end{bmatrix}$$

$$=\begin{bmatrix}\frac{7}{2} & 3 & -1 & \frac{10}{3}\\ -\frac{19}{6} & \frac{2}{3} & -\frac{17}{2} & \frac{23}{3}\\ \frac{29}{3} & \frac{17}{6} & -1 & -2\end{bmatrix}.$$

21. $$\begin{bmatrix}2x-2 & 3 & 2\\ 2 & 4 & y-2\\ 2z & -3 & 2\end{bmatrix}=\begin{bmatrix}3 & u & 2\\ 2 & 4 & 5\\ 4 & -3 & 2\end{bmatrix}.$$

Now, by the definition of equality of matrices,

$u = 3$

$2x - 2 = 3$ and $2x = 5$, or $x = 5/2$,

$y - 2 = 5$, and $y = 7$,

$2z = 4$, and $z = 2$.

23. $\begin{bmatrix} 1 & x \\ 2y & -3 \end{bmatrix} - 4\begin{bmatrix} 2 & -2 \\ 0 & 3 \end{bmatrix} = \begin{bmatrix} 3z & 10 \\ 4 & -u \end{bmatrix}; \begin{bmatrix} -7 & x+8 \\ 2y & -15 \end{bmatrix} = \begin{bmatrix} 3z & 10 \\ 4 & -u \end{bmatrix}.$

Now, by the definition of equality of matrices,

$-u = -15$, so $u = 15$

$x + 8 = 10$, so $x = 2$

$2y = 4$, so $y = 2$

$3z = -7$, so $z = -7/3$.

25. To verify the Commutative Law for matrix addition, let us show that $A + B = B + A$.

Now, $A + B = \begin{bmatrix} 2 & -4 & 3 \\ 4 & 2 & 1 \end{bmatrix} + \begin{bmatrix} 4 & -3 & 2 \\ 1 & 0 & 4 \end{bmatrix} = \begin{bmatrix} 6 & -7 & 5 \\ 5 & 2 & 5 \end{bmatrix}$

$= \begin{bmatrix} 4 & -3 & 2 \\ 1 & 0 & 4 \end{bmatrix} + \begin{bmatrix} 2 & -4 & 3 \\ 4 & 2 & 1 \end{bmatrix} = B + A.$

27. $(3+5)A = 8A = 8\begin{bmatrix} 3 & 1 \\ 2 & 4 \\ -4 & 0 \end{bmatrix} = \begin{bmatrix} 24 & 8 \\ 16 & 32 \\ -32 & 0 \end{bmatrix} = 3\begin{bmatrix} 3 & 1 \\ 2 & 4 \\ -4 & 0 \end{bmatrix} + 5\begin{bmatrix} 3 & 1 \\ 2 & 4 \\ -4 & 0 \end{bmatrix}$

$= 3A + 5A$.

29. $4(A + B) = 4\left[\begin{bmatrix} 3 & 1 \\ 2 & 4 \\ -4 & 0 \end{bmatrix} + \begin{bmatrix} 1 & 2 \\ -1 & 0 \\ 3 & 2 \end{bmatrix}\right] = 4\begin{bmatrix} 4 & 3 \\ 1 & 4 \\ -1 & 2 \end{bmatrix} = \begin{bmatrix} 16 & 12 \\ 4 & 16 \\ -4 & 8 \end{bmatrix}$

$4A + 4B = 4\begin{bmatrix} 3 & 1 \\ 2 & 4 \\ -4 & 0 \end{bmatrix} + 4\begin{bmatrix} 1 & 2 \\ -1 & 0 \\ 3 & 2 \end{bmatrix} = \begin{bmatrix} 16 & 12 \\ 4 & 16 \\ -4 & 8 \end{bmatrix}.$

31. $\begin{bmatrix} 3 & 2 & -1 & 5 \end{bmatrix}^T = \begin{bmatrix} 3 \\ 2 \\ -1 \\ 5 \end{bmatrix}.$

33. $\begin{bmatrix} 1 & -1 & 2 \\ 3 & 4 & 2 \\ 0 & 1 & 0 \end{bmatrix}^T = \begin{bmatrix} 1 & 3 & 0 \\ -1 & 4 & 1 \\ 2 & 2 & 0 \end{bmatrix}.$

35.
$$\begin{array}{r} \\ \textit{Mr. Cross} \\ \textit{Mr. Jones} \\ \textit{Mr. Smith} \end{array}\begin{array}{c} \begin{array}{cccc} 1 & 2 & 3 & 4 \end{array} \\ \begin{bmatrix} 220 & 215 & 210 & 205 \\ 220 & 210 & 200 & 195 \\ 215 & 205 & 195 & 190 \end{bmatrix} \end{array}$$

37. a. $D = A + B - C$

$$= \begin{bmatrix} 2820 & 1470 & 1120 \\ 1030 & 520 & 480 \\ 1170 & 540 & 460 \end{bmatrix} + \begin{bmatrix} 260 & 120 & 110 \\ 140 & 60 & 50 \\ 120 & 70 & 50 \end{bmatrix} - \begin{bmatrix} 120 & 80 & 80 \\ 70 & 30 & 40 \\ 60 & 20 & 40 \end{bmatrix}$$

$$= \begin{bmatrix} 2960 & 1510 & 1150 \\ 1100 & 550 & 490 \\ 1230 & 590 & 470 \end{bmatrix}.$$

b. $E = 1.1D = 1.1\begin{bmatrix} 2960 & 1510 & 1150 \\ 1100 & 550 & 490 \\ 1230 & 590 & 470 \end{bmatrix} = \begin{bmatrix} 3256 & 1661 & 1265 \\ 1210 & 605 & 539 \\ 1353 & 649 & 517 \end{bmatrix}.$

39.

$$A = \begin{array}{r} \text{MA} \\ \text{U.S.} \end{array}\begin{bmatrix} 6.88 & 7.05 & 7.18 \\ 4.13 & 4.09 & 4.06 \end{bmatrix}$$

41.

$$A = \begin{array}{c} \\ W \\ M \end{array}\!\!\begin{array}{c} \begin{array}{ccc} W & B & H \end{array} \\ \begin{bmatrix} 81 & 76.1 & 82.2 \\ 76 & 69.9 & 75.9 \end{bmatrix} \end{array} ; \quad B = \begin{array}{c} \\ W \\ B \\ H \end{array}\!\!\begin{array}{c} \begin{array}{cc} W & M \end{array} \\ \begin{bmatrix} 81 & 76 \\ 76.1 & 69.9 \\ 82.2 & 75.9 \end{bmatrix} \end{array}$$

43. True. Each element in $A+B$ is obtained by adding together the corresponding elements in A and B. Therefore, the matrix $c(A+B)$ is obtained by multiplying each element in $A+B$ by c. On the other hand, cA is obtained by multiplying each element in A by c and cB is obtained by multiplying each element in B by c and $cA + cB$ is obtained by adding the corresponding elements in cA and cB. Thus $c(A+B) = cA + cB$.

45. False. Take $\begin{bmatrix} 1 & 2 \\ 3 & 4 \end{bmatrix}$ and $c = 2$. Then

$$cA = 2\begin{bmatrix} 1 & 2 \\ 3 & 4 \end{bmatrix} = \begin{bmatrix} 2 & 4 \\ 6 & 8 \end{bmatrix} \text{ and } (cA)^T = \begin{bmatrix} 2 & 6 \\ 4 & 8 \end{bmatrix}.$$

On the other hand, $\frac{1}{c}A^T = \frac{1}{2}\begin{bmatrix} 1 & 3 \\ 2 & 4 \end{bmatrix} = \begin{bmatrix} \frac{1}{2} & \frac{3}{2} \\ 1 & 2 \end{bmatrix} \neq (cA)^T$

USING TECHNOLOGY EXERCISES 2.4, page 121

1. $\begin{bmatrix} 15 & 38.75 & -67.5 & 33.75 \\ 51.25 & 40 & 52.5 & -38.75 \\ 21.25 & 35 & -65 & 105 \end{bmatrix}$

3. $\begin{bmatrix} -5 & 6.3 & -6.8 & 3.9 \\ 1 & 0.5 & 5.4 & -4.8 \\ 0.5 & 4.2 & -3.5 & 5.6 \end{bmatrix}$

5. $\begin{bmatrix} 16.44 & -3.65 & -3.66 & 0.63 \\ 12.77 & 10.64 & 2.58 & 0.05 \\ 5.09 & 0.28 & -10.84 & 17.64 \end{bmatrix}$

7. $\begin{bmatrix} 22.2 & -0.3 & -12 & 4.5 \\ 21.6 & 17.7 & 9 & -4.2 \\ 8.7 & 4.2 & -20.7 & 33.6 \end{bmatrix}$

2.5 Problem Solving Tips

1. The *matrix* product of two matrices A and B is only defined if the number of columns in A is equal to the number of rows in B. The *scalar* product of a real number c and a matrix A is always defined.

2. We can write a system of equations in the matrix form $AX = B$ where A is the coefficient matrix, X is the column matrix of unknowns, and B is the column matrix of constants.

2.5 CONCEPT QUESTIONS, page 129

1. Scalar multiplication involves multiplying a matrix A by a scalar c (result: cA); whereas matrix multiplication involves the product of two matrices.
   Example:

$$3\begin{bmatrix} 1 & 2 \\ 2 & 3 \end{bmatrix} = \begin{bmatrix} 3 & 6 \\ 6 & 9 \end{bmatrix} \quad \text{and} \quad \begin{bmatrix} 2 & 1 \\ 3 & 0 \end{bmatrix}\begin{bmatrix} 1 & 3 & 2 \\ 1 & 4 & 3 \end{bmatrix} = \begin{bmatrix} 3 & 10 & 7 \\ 3 & 9 & 6 \end{bmatrix}.$$

EXERCISES 2.5, page 129

1. $(2 \times 3)(3 \times 5)$ so AB has order 2×5.

$$\uparrow\ \uparrow$$
$$=$$

$(3 \times 5)(2 \times 3)$ so BA is not defined.

$$\uparrow\ \uparrow$$
$$\neq$$

3. $(1 \times 7)\ (7 \times 1)$ so AB has order 1×1.

$$\uparrow\ \uparrow$$
$$=$$

$(7 \times 1)\ (1 \times 7)$ so BA has order 7×7.

$$\underset{=}{\uparrow\ \uparrow}$$

5. If AB and BA are defined then $n = s$ and $m = t$.

7. $$\begin{bmatrix} 1 & 2 \\ 3 & 0 \end{bmatrix}\begin{bmatrix} 1 \\ -1 \end{bmatrix} = \begin{bmatrix} -1 \\ 3 \end{bmatrix}$$

9. $$\begin{bmatrix} 3 & 1 & 2 \\ -1 & 2 & 4 \end{bmatrix}\begin{bmatrix} 4 \\ 1 \\ -2 \end{bmatrix} = \begin{bmatrix} 9 \\ -10 \end{bmatrix}$$

11. $$\begin{bmatrix} -1 & 2 \\ 3 & 1 \end{bmatrix}\begin{bmatrix} 2 & 4 \\ 3 & 1 \end{bmatrix} = \begin{bmatrix} 4 & -2 \\ 9 & 13 \end{bmatrix}$$

13. $$\begin{bmatrix} 2 & 1 & 2 \\ 3 & 2 & 4 \end{bmatrix}\begin{bmatrix} -1 & 2 \\ 4 & 3 \\ 0 & 1 \end{bmatrix} = \begin{bmatrix} 2 & 9 \\ 5 & 16 \end{bmatrix}$$

15. $$\begin{bmatrix} 0.1 & 0.9 \\ 0.2 & 0.8 \end{bmatrix}\begin{bmatrix} 1.2 & 0.4 \\ 0.5 & 2.1 \end{bmatrix} = \begin{bmatrix} 0.1(1.2)+0.9(0.5) & 0.1(0.4)+0.9(2.1) \\ 0.2(1.2)+0.8(0.5) & 0.2(0.4)+0.8(2.1) \end{bmatrix} = \begin{bmatrix} 0.57 & 1.93 \\ 0.64 & 1.76 \end{bmatrix}$$

17. $$\begin{bmatrix} 6 & -3 & 0 \\ -2 & 1 & -8 \\ 4 & -4 & 9 \end{bmatrix}\begin{bmatrix} 1 & 0 & 0 \\ 0 & 1 & 0 \\ 0 & 0 & 1 \end{bmatrix} = \begin{bmatrix} 6 & -3 & 0 \\ -2 & 1 & -8 \\ 4 & -4 & 9 \end{bmatrix}.$$

19. $$\begin{bmatrix} 3 & 0 & -2 & 1 \\ 1 & 2 & 0 & -1 \end{bmatrix}\begin{bmatrix} 2 & 1 & -1 \\ -1 & 2 & 0 \\ 0 & 0 & 1 \\ -1 & -2 & 2 \end{bmatrix} = \begin{bmatrix} 5 & 1 & -3 \\ 1 & 7 & -3 \end{bmatrix}.$$

21. $$4\begin{bmatrix} 1 & -2 & 0 \\ 2 & -1 & 1 \\ 3 & 0 & -1 \end{bmatrix}\begin{bmatrix} 1 & 3 & 1 \\ 1 & 4 & 0 \\ 0 & 1 & -2 \end{bmatrix} = \begin{bmatrix} -4 & -20 & 4 \\ 4 & 12 & 0 \\ 12 & 32 & 20 \end{bmatrix}$$

23. $$\begin{bmatrix} 1 & 0 \\ 0 & 1 \end{bmatrix}\begin{bmatrix} 4 & -3 & 2 \\ 7 & 1 & -5 \end{bmatrix}\begin{bmatrix} 1 & 0 & 0 \\ 0 & 1 & 0 \\ 0 & 0 & 1 \end{bmatrix} = \begin{bmatrix} 1 & 0 \\ 0 & 1 \end{bmatrix}\begin{bmatrix} 4 & -3 & 2 \\ 7 & 1 & -5 \end{bmatrix} = \begin{bmatrix} 4 & -3 & 2 \\ 7 & 1 & -5 \end{bmatrix}.$$

25. To verify the associative law for matrix multiplication, we will show that $(AB)C = A(BC)$.

$$AB = \begin{bmatrix} 1 & 0 & -2 \\ 1 & -3 & 2 \\ -2 & 1 & 1 \end{bmatrix}\begin{bmatrix} 3 & 1 & 0 \\ 2 & 2 & 0 \\ 1 & -3 & -1 \end{bmatrix} = \begin{bmatrix} 1 & 7 & 2 \\ -1 & -11 & -2 \\ -3 & -3 & -1 \end{bmatrix}$$

$$(AB)C = \begin{bmatrix} 1 & 7 & 2 \\ -1 & -11 & -2 \\ -3 & -3 & -1 \end{bmatrix}\begin{bmatrix} 2 & -1 & 0 \\ 1 & -1 & 2 \\ 3 & -2 & 1 \end{bmatrix} = \begin{bmatrix} 15 & -12 & 16 \\ -19 & 16 & -24 \\ -12 & 8 & -7 \end{bmatrix}$$

$$BC = \begin{bmatrix} 3 & 1 & 0 \\ 2 & 2 & 0 \\ 1 & -3 & -1 \end{bmatrix}\begin{bmatrix} 2 & -1 & 0 \\ 1 & -1 & 2 \\ 3 & -2 & 1 \end{bmatrix} = \begin{bmatrix} 7 & -4 & 2 \\ 6 & -4 & 4 \\ -4 & 4 & -7 \end{bmatrix}$$

$$A(BC) = \begin{bmatrix} 1 & 0 & -2 \\ 1 & -3 & 2 \\ -2 & 1 & 1 \end{bmatrix}\begin{bmatrix} 7 & -4 & 2 \\ 6 & -4 & 4 \\ -4 & 4 & -7 \end{bmatrix} = \begin{bmatrix} 15 & -12 & 16 \\ -19 & 16 & -24 \\ -12 & 8 & -7 \end{bmatrix}.$$

27. $$AB = \begin{bmatrix} 1 & 2 \\ 3 & 4 \end{bmatrix}\begin{bmatrix} 2 & 1 \\ 4 & 3 \end{bmatrix} = \begin{bmatrix} 10 & 7 \\ 22 & 15 \end{bmatrix}$$

$$BA = \begin{bmatrix} 2 & 1 \\ 4 & 3 \end{bmatrix}\begin{bmatrix} 1 & 2 \\ 3 & 4 \end{bmatrix} = \begin{bmatrix} 5 & 8 \\ 13 & 20 \end{bmatrix}$$

Therefore, $AB \neq BA$ and matrix multiplication is not commutative.

29. $$AB = \begin{bmatrix} 3 & 0 \\ 8 & 0 \end{bmatrix}\begin{bmatrix} 0 & 0 \\ 4 & 5 \end{bmatrix} = \begin{bmatrix} 0 & 0 \\ 0 & 0 \end{bmatrix}$$

$AB = 0$, but neither A nor B is the zero matrix. Therefore, $AB = 0$, does not imply that A or B is the zero matrix.

31. $$\begin{bmatrix} a & b \\ c & d \end{bmatrix}\begin{bmatrix} 1 & 0 \\ -1 & 3 \end{bmatrix} = \begin{bmatrix} a-b & 3b \\ c-d & 3d \end{bmatrix} = \begin{bmatrix} -1 & -3 \\ 3 & 6 \end{bmatrix}$$

Then $3b=-3$, and $b=-1$

$3d=6$, and $d=2$

$a-b=-1$, and $a=b-1=-2$.

$c-d=3$, and $c=d+3=5$

Therefore, $A=\begin{bmatrix}-2 & -1\\ 5 & 2\end{bmatrix}$.

33. a.

$$A^T=\begin{bmatrix}2 & 5\\ 4 & -6\end{bmatrix} \text{ and } (A^T)^T=\begin{bmatrix}2 & 4\\ 5 & -6\end{bmatrix}=A$$

b. $$(A+B)^T=\begin{bmatrix}6 & 12\\ -2 & -3\end{bmatrix}^T=\begin{bmatrix}6 & -2\\ 12 & -3\end{bmatrix}$$

$$A^T+B^T=\begin{bmatrix}2 & 5\\ 4 & -6\end{bmatrix}+\begin{bmatrix}4 & -7\\ 8 & 3\end{bmatrix}=\begin{bmatrix}6 & -2\\ 12 & -3\end{bmatrix}$$

c. $$AB=\begin{bmatrix}2 & 4\\ 5 & -6\end{bmatrix}\begin{bmatrix}4 & 8\\ -7 & 3\end{bmatrix}=\begin{bmatrix}-20 & 28\\ 62 & 22\end{bmatrix}, \quad \text{so} \quad (AB)^T=\begin{bmatrix}-20 & 62\\ 28 & 22\end{bmatrix}.$$

$$B^TA^T=\begin{bmatrix}4 & -7\\ 8 & 3\end{bmatrix}\begin{bmatrix}2 & 5\\ 4 & -6\end{bmatrix}=\begin{bmatrix}-20 & 62\\ 28 & 22\end{bmatrix}=(AB)^T$$

35. The given system of linear equations can be represented by the matrix equation $AX=B$, where

$$A=\begin{bmatrix}2 & -3\\ 3 & -4\end{bmatrix}, \ X=\begin{bmatrix}x\\ y\end{bmatrix}, \text{ and } B=\begin{bmatrix}7\\ 8\end{bmatrix}.$$

37. The given system of linear equations can be represented by the matrix equation $AX=B$, where

$$A = \begin{bmatrix} 2 & -3 & 4 \\ 0 & 2 & -3 \\ 1 & -1 & 2 \end{bmatrix}, \quad X = \begin{bmatrix} x \\ y \\ z \end{bmatrix}, \quad B = \begin{bmatrix} 6 \\ 7 \\ 4 \end{bmatrix}.$$

39. The given system of linear equations can be represented by the matrix equation $AX = B$, where

$$A = \begin{bmatrix} -1 & 1 & 1 \\ 2 & -1 & -1 \\ -3 & 2 & 4 \end{bmatrix}, \quad X = \begin{bmatrix} x_1 \\ x_2 \\ x_3 \end{bmatrix}, \quad B = \begin{bmatrix} 0 \\ 2 \\ 4 \end{bmatrix}.$$

41. a. $$AB = \begin{bmatrix} 200 & 300 & 100 & 200 \\ 100 & 200 & 400 & 0 \end{bmatrix} \begin{bmatrix} 54 \\ 48 \\ 98 \\ 82 \end{bmatrix} = \begin{bmatrix} 51{,}400 \\ 54{,}200 \end{bmatrix}$$

b. The first entry shows that William's total stock holdings are \$51,400, while the second entry shows that Michael's stockholdings are \$54,200.

43. The column vector that represents the profit for each type of house is

$$B = \begin{bmatrix} 20{,}000 \\ 22{,}000 \\ 25{,}000 \\ 30{,}000 \end{bmatrix}.$$

The column vector that gives the total profit for Bond Brothers is

$$AB = \begin{bmatrix} 60 & 80 & 120 & 40 \\ 20 & 30 & 60 & 10 \\ 10 & 15 & 30 & 5 \end{bmatrix} \begin{bmatrix} 20{,}000 \\ 22{,}000 \\ 25{,}000 \\ 30{,}000 \end{bmatrix}$$

$$= \begin{bmatrix} 7{,}160{,}000 \\ 2{,}860{,}000 \\ 1{,}430{,}000 \end{bmatrix}.$$

Therefore, Bond Brothers expects to make \$7,160,000 in New York, \$2,860,000 in Connecticut, and \$1,430,000 in Massachusetts, and the total profit is \$11,450,000.

45.

$$BA = \begin{bmatrix} 30{,}000 & 40{,}000 & 20{,}000 \end{bmatrix} \begin{array}{c} \begin{matrix} D & R & I \end{matrix} \\ \begin{bmatrix} 0.50 & 0.30 & 0.20 \\ 0.45 & 0.40 & 0.15 \\ 0.40 & 0.50 & 0.10 \end{bmatrix} \end{array}$$

$$= \begin{array}{c} \begin{matrix} D & R & I \end{matrix} \\ \begin{bmatrix} 41{,}000 & 35{,}000 & 14{,}000 \end{bmatrix} \end{array}$$

47.

$$AB = \begin{bmatrix} 2700 & 3000 \\ 800 & 700 \\ 500 & 300 \end{bmatrix} \begin{bmatrix} 0.25 & 0.20 & 0.30 & 0.25 \\ 0.30 & 0.35 & 0.25 & 0.10 \end{bmatrix} = \begin{bmatrix} 1575 & 1590 & 1560 & 975 \\ 410 & 405 & 415 & 270 \\ 215 & 205 & 225 & 155 \end{bmatrix}$$

49. $AC = \begin{bmatrix} 80 & 60 & 40 \end{bmatrix} \begin{bmatrix} 0.34 \\ 0.42 \\ 0.48 \end{bmatrix} = 71.6$ $\quad BD = \begin{bmatrix} 300 & 150 & 250 \end{bmatrix} \begin{bmatrix} 0.24 \\ 0.31 \\ 0.35 \end{bmatrix} = 206$

$AC + BD = [277.60]$, or \$277.60. It represents Cindy's long distance bill for phone calls to those 3 cities.

51. a. $MA^T = \begin{bmatrix} 400 & 1200 & 800 \\ 110 & 570 & 340 \\ 90 & 30 & 60 \end{bmatrix} \begin{bmatrix} 7 \\ 1 \\ 6 \end{bmatrix} = \begin{bmatrix} 8800 \\ 3380 \\ 1020 \end{bmatrix}$.

The amounts of vitamin A, vitamin C, and calcium taken by a girl in the first meal are 8800, 3380, and 1020 units respectively.

b.

$$MB^T = \begin{bmatrix} 400 & 1200 & 800 \\ 110 & 570 & 340 \\ 90 & 30 & 60 \end{bmatrix} \begin{bmatrix} 9 \\ 3 \\ 2 \end{bmatrix} = \begin{bmatrix} 8800 \\ 3380 \\ 1020 \end{bmatrix}$$

The amounts of vitamin A, vitamin C, and calcium taken by a girl in the second meal are 8,800, 3380, and 1020 units, respectively.

c. $M(A+B)^T = \begin{bmatrix} 400 & 1200 & 800 \\ 110 & 570 & 340 \\ 90 & 30 & 60 \end{bmatrix} \begin{bmatrix} 16 \\ 4 \\ 8 \end{bmatrix} = \begin{bmatrix} 17{,}600 \\ 6{,}760 \\ 2{,}040 \end{bmatrix}$.

The amounts of vitamin A, vitamin C, and calcium taken by a girl in the two meals are 17,600, 6,760, and 2,040 units respectively.

53. False. Let A be a matrix of order 2×3 and let B be a matrix of order 3×2. Then AB and BA are both defined. But, evidently, neither A nor B is a square matrix.

55. True. In order for the sum $B+C$ to be defined, B and C must have the same size, and in order for the product of A and $(B+C)$ to be defined, the number of columns of A must be equal to the number of rows of $(B+C)$.

USING TECHNOLOGY EXERCISES 2.5, page 136

1. $\begin{bmatrix} 18.66 & 15.2 & -12 \\ 24.48 & 41.88 & 89.82 \\ 15.39 & 7.16 & -1.25 \end{bmatrix}$

3. $\begin{bmatrix} 20.09 & 20.61 & -1.3 \\ 44.42 & 71.6 & 64.89 \\ 20.97 & 7.17 & -60.65 \end{bmatrix}$

5. $\begin{bmatrix} 32.89 & 13.63 & -57.17 \\ -12.85 & -8.37 & 256.92 \\ 13.48 & 14.29 & 181.64 \end{bmatrix}$

7. $\begin{bmatrix} 128.59 & 123.08 & -32.50 \\ 246.73 & 403.12 & 481.52 \\ 125.06 & 47.01 & -264.81 \end{bmatrix}$

9. $\begin{bmatrix} 87 & 68 & 110 & 82 \\ 119 & 176 & 221 & 143 \\ 51 & 128 & 142 & 94 \\ 28 & 174 & 174 & 112 \end{bmatrix}$ $\begin{bmatrix} 113 & 117 & 72 & 101 & 90 \\ 72 & 85 & 36 & 72 & 76 \\ 81 & 69 & 76 & 87 & 30 \\ 133 & 157 & 56 & 121 & 146 \\ 154 & 157 & 94 & 127 & 122 \end{bmatrix}$

11. $\begin{bmatrix} 170 & 18.1 & 133.1 & -106.3 & 341.3 \\ 349 & 226.5 & 324.1 & 164 & 506.4 \\ 245.2 & 157.7 & 231.5 & 125.5 & 312.9 \\ 310 & 245.2 & 291 & 274.3 & 354.2 \end{bmatrix}$

$$\left[\begin{array}{cccccc|c} 1 & 0 & 0 & 0 & -1 & 0 & 100 \\ 0 & 1 & 0 & 0 & 1 & 0 & -100 \\ 0 & 0 & 1 & 0 & -1 & 1 & 500 \\ 0 & 0 & 0 & 1 & -1 & 1 & 700 \\ 0 & 0 & 0 & 0 & 0 & 0 & 0 \end{array}\right].$$

We conclude that the solution is

$(s+100,\ s-100,\ s-t+500,\ s-t+700,\ s,\ t)$.

Taking $s = 150$ and $t = 50$, we see that one possible traffic pattern is

(250, 50, 600, 800, 150, 50).

Similarly, taking $s = 200$, and $t = 100$, we see that another possible traffic pattern is

(300, 100, 600, 800, 200, 100).

c. Taking $t = 0$ and $s = 200$, we see that another possible traffic pattern is

(300, 100, 700, 900, 200, 0).

2.6 Problem Solving Tips

The problem-solving skills that you learned in earlier sections, are building-blocks for the rest of the course. You can't skip a section or a concept and "hope" that you will understand the material in a new section. It just won't work. If you don't build a strong foundation, you won't be able to understand the later concepts. For example, in this section we discussed the process for finding the inverse of a matrix. You need to use the Gauss-Jordan method of elimination to find the inverse of a matrix, so if you don't know how to use that method to solve a system of equations you won't be able to find the inverse of a matrix. If you are having difficulties you may need to go back and review the earlier section before you go on.

Here are some hints for solving the problems in the exercises that follow:

1. Not every square matrix has an inverse. If there is a row to the left of the vertical line in the augmented matrix containing all zeros, then the matrix does not have an inverse.

2. You can use the formula $A^{-1} = \frac{1}{D}\begin{bmatrix} a & b \\ c & d \end{bmatrix}$, where $D = ad - bc \neq 0$ to find the inverse of a 2×2 matrix. Note that the matrix does not have an inverse if $D = 0$.

3. The inverse of a matrix can be used to find the solution of a system of n equations in n unknowns.

2.6 CONCEPT QUESTIONS, page 145

1. The inverse of a square matrix A is the matrix A^{-1} satisfying the conditions $AA^{-1} = A^{-1}A = I$

3. The formula for finding the inverse of a 2×2 matrix are given on page 141 of the text.

EXERCISES 2.6, page 145

1. $\begin{bmatrix} 1 & -3 \\ 1 & -2 \end{bmatrix}\begin{bmatrix} -2 & 3 \\ -1 & 1 \end{bmatrix} = \begin{bmatrix} 1 & 0 \\ 0 & 1 \end{bmatrix}$; $\begin{bmatrix} -2 & 3 \\ -1 & 1 \end{bmatrix}\begin{bmatrix} 1 & -3 \\ 1 & -2 \end{bmatrix} = \begin{bmatrix} 1 & 0 \\ 0 & 1 \end{bmatrix}$

3. $\begin{bmatrix} 3 & 2 & 3 \\ 2 & 2 & 1 \\ 2 & 1 & 1 \end{bmatrix}\begin{bmatrix} -\frac{1}{3} & -\frac{1}{3} & \frac{4}{3} \\ 0 & 1 & -1 \\ \frac{2}{3} & -\frac{1}{3} & -\frac{2}{3} \end{bmatrix} = \begin{bmatrix} 1 & 0 & 0 \\ 0 & 1 & 0 \\ 0 & 0 & 1 \end{bmatrix}$ and

$$\begin{bmatrix} -\frac{1}{3} & -\frac{1}{3} & \frac{4}{3} \\ 0 & 1 & -1 \\ \frac{2}{3} & -\frac{1}{3} & -\frac{2}{3} \end{bmatrix}\begin{bmatrix} 3 & 2 & 3 \\ 2 & 2 & 1 \\ 2 & 1 & 1 \end{bmatrix} = \begin{bmatrix} 1 & 0 & 0 \\ 0 & 1 & 0 \\ 0 & 0 & 1 \end{bmatrix}.$$

5. Using Formula (13), we find $A^{-1} = \frac{1}{(2)(3)-(1)(5)}\begin{bmatrix} 3 & -5 \\ -1 & 2 \end{bmatrix} = \begin{bmatrix} 3 & -5 \\ -1 & 2 \end{bmatrix}$.

7. Since $ad - bc = (3)(2) - (-2)(-3) = 6 - 6 = 0$, the inverse does not exist.

9. $$\left[\begin{array}{ccc|ccc} 2 & -3 & -4 & 1 & 0 & 0 \\ 0 & 0 & -1 & 0 & 1 & 0 \\ 1 & -2 & 1 & 0 & 0 & 1 \end{array}\right] \xrightarrow{R_1 \leftrightarrow R_3} \left[\begin{array}{ccc|ccc} 1 & -2 & 1 & 0 & 0 & 1 \\ 0 & 0 & -1 & 0 & 1 & 0 \\ 2 & -3 & -4 & 1 & 0 & 0 \end{array}\right] \xrightarrow{R_3 - 2R_1}$$

$$\left[\begin{array}{ccc|ccc} 1 & -2 & 1 & 0 & 0 & 1 \\ 0 & 0 & -1 & 0 & 1 & 0 \\ 0 & 1 & -6 & 1 & 0 & -2 \end{array}\right] \xrightarrow{R_2 \leftrightarrow R_3} \left[\begin{array}{ccc|ccc} 1 & -2 & 1 & 0 & 0 & 1 \\ 0 & 1 & -6 & 1 & 0 & -2 \\ 0 & 0 & -1 & 0 & 1 & 0 \end{array}\right] \xrightarrow[-R_3]{R_1 + 2R_2}$$

$$\left[\begin{array}{ccc|ccc} 1 & 0 & -11 & 2 & 0 & -3 \\ 0 & 1 & -6 & 1 & 0 & -2 \\ 0 & 0 & 1 & 0 & -1 & 0 \end{array}\right] \xrightarrow[R_2 + 6R_3]{R_1 + 11R_3} \left[\begin{array}{ccc|ccc} 1 & 0 & 0 & 2 & -11 & -3 \\ 0 & 1 & 0 & 1 & -6 & -2 \\ 0 & 0 & 1 & 0 & -1 & 0 \end{array}\right].$$

Therefore, the required inverse is $\begin{bmatrix} 2 & -11 & -3 \\ 1 & -6 & -2 \\ 0 & -1 & 0 \end{bmatrix}$.

11. $$\left[\begin{array}{ccc|ccc} 4 & 2 & 2 & 1 & 0 & 0 \\ -1 & -3 & 4 & 0 & 1 & 0 \\ 3 & -1 & 6 & 0 & 0 & 1 \end{array}\right] \xrightarrow{R_1 - R_3} \left[\begin{array}{ccc|ccc} 1 & 3 & -4 & 1 & 0 & -1 \\ -1 & -3 & 4 & 0 & 1 & 0 \\ 3 & -1 & 6 & 0 & 0 & 1 \end{array}\right]$$

$$\xrightarrow{R_2 + R_1} \left[\begin{array}{ccc|ccc} 1 & 3 & -4 & 1 & 0 & -1 \\ 0 & 0 & 0 & 1 & 1 & -1 \\ 3 & -1 & 6 & 0 & 0 & 1 \end{array}\right]$$

Because there is a row of zeros to the left of the vertical line, we see that the inverse does not exist.

13. $\left[\begin{array}{rrr|rrr} 1 & 4 & -1 & 1 & 0 & 0 \\ 2 & 3 & -2 & 0 & 1 & 0 \\ -1 & 2 & 3 & 0 & 0 & 1 \end{array}\right] \xrightarrow[R_3+R_1]{R_2-2R_1} \left[\begin{array}{rrr|rrr} 1 & 4 & -1 & 1 & 0 & 0 \\ 0 & -5 & 0 & -2 & 1 & 0 \\ 0 & 6 & 2 & 1 & 0 & 1 \end{array}\right] \xrightarrow{R_2+R_3}$

$\left[\begin{array}{rrr|rrr} 1 & 4 & -1 & 1 & 0 & 0 \\ 0 & 1 & 2 & -1 & 1 & 1 \\ 0 & 6 & 2 & 1 & 0 & 1 \end{array}\right] \xrightarrow[R_3-6R_2]{R_1-4R_2} \left[\begin{array}{rrr|rrr} 1 & 0 & -9 & 5 & -4 & -4 \\ 0 & 1 & 2 & -1 & 1 & 1 \\ 0 & 0 & -10 & 7 & -6 & -5 \end{array}\right] \xrightarrow{-\frac{1}{10}R_3}$

$\left[\begin{array}{rrr|rrr} 1 & 0 & -9 & 5 & -4 & -4 \\ 0 & 1 & 2 & -1 & 1 & 1 \\ 0 & 0 & 1 & -\frac{7}{10} & \frac{3}{5} & \frac{1}{2} \end{array}\right] \xrightarrow[R_2-2R_3]{R_1+9R_3} \left[\begin{array}{rrr|rrr} 1 & 0 & 0 & -\frac{13}{10} & \frac{7}{5} & \frac{1}{2} \\ 0 & 1 & 0 & \frac{2}{5} & -\frac{1}{5} & 0 \\ 0 & 0 & 1 & -\frac{7}{10} & \frac{3}{5} & \frac{1}{2} \end{array}\right]$

So $A^{-1} = \begin{bmatrix} -\frac{13}{10} & \frac{7}{5} & \frac{1}{2} \\ \frac{2}{5} & -\frac{1}{5} & 0 \\ -\frac{7}{10} & \frac{3}{5} & \frac{1}{2} \end{bmatrix}$.

15. $\left[\begin{array}{rrrr|rrrr} 1 & 1 & -1 & 1 & 1 & 0 & 0 & 0 \\ 2 & 1 & 1 & 0 & 0 & 1 & 0 & 0 \\ 2 & 1 & 0 & 1 & 0 & 0 & 1 & 0 \\ 2 & -1 & -1 & 3 & 0 & 0 & 0 & 1 \end{array}\right] \xrightarrow[\substack{R_3-2R_1 \\ R_4-2R_1}]{R_2-2R_1} \left[\begin{array}{rrrr|rrrr} 1 & 1 & -1 & 1 & 1 & 0 & 0 & 0 \\ 0 & -1 & 3 & -2 & -2 & 1 & 0 & 0 \\ 0 & -1 & 2 & -1 & -2 & 0 & 1 & 0 \\ 0 & -3 & 1 & 1 & -2 & 0 & 0 & 1 \end{array}\right] \xrightarrow{-R_2}$

$\left[\begin{array}{rrrr|rrrr} 1 & 1 & -1 & 1 & 1 & 0 & 0 & 0 \\ 0 & 1 & -3 & 2 & 2 & -1 & 0 & 0 \\ 0 & -1 & 2 & -1 & -2 & 0 & 1 & 0 \\ 0 & -3 & 1 & 1 & -2 & 0 & 0 & 1 \end{array}\right] \xrightarrow[\substack{R_3+R_2 \\ R_4+3R_2}]{R_1-R_2} \left[\begin{array}{rrrr|rrrr} 1 & 0 & 2 & -1 & -1 & 1 & 0 & 0 \\ 0 & 1 & -3 & 2 & 2 & -1 & 0 & 0 \\ 0 & 0 & -1 & 1 & 0 & -1 & 1 & 0 \\ 0 & 0 & -8 & 7 & 4 & -3 & 0 & 1 \end{array}\right] \xrightarrow{-R_3}$

$\left[\begin{array}{rrrr|rrrr} 1 & 0 & 2 & -1 & -1 & 1 & 0 & 0 \\ 0 & 1 & -3 & 2 & 2 & -1 & 0 & 0 \\ 0 & 0 & 1 & -1 & 0 & 1 & -1 & 0 \\ 0 & 0 & -8 & 7 & 4 & -3 & 0 & 1 \end{array}\right] \xrightarrow[\substack{R_2+3R_3 \\ R_4+8R_3}]{R_1-2R_3} \left[\begin{array}{rrrr|rrrr} 1 & 0 & 0 & 1 & -1 & -1 & 2 & 0 \\ 0 & 1 & 0 & -1 & 2 & 2 & -3 & 0 \\ 0 & 0 & 1 & -1 & 0 & 1 & -1 & 0 \\ 0 & 0 & 0 & -1 & 4 & 5 & -8 & 1 \end{array}\right]$

$$\xrightarrow[-R_4]{\substack{R_1+R_4\\R_2-R_4\\R_3-R_4}} \left[\begin{array}{cccc|cccc} 1 & 0 & 0 & 0 & 3 & 4 & -6 & 1 \\ 0 & 1 & 0 & 0 & -2 & -3 & 5 & -1 \\ 0 & 0 & 1 & 0 & -4 & -4 & 7 & -1 \\ 0 & 0 & 0 & 1 & -4 & -5 & 8 & -1 \end{array}\right].$$

So the required inverse is

$$A^{-1} = \begin{bmatrix} 3 & 4 & -6 & 1 \\ -2 & -3 & 5 & -1 \\ -4 & -4 & 7 & -1 \\ -4 & -5 & 8 & -1 \end{bmatrix}.$$

We can verify our result by showing that $A^{-1}A = A$. Thus,

$$\begin{bmatrix} 3 & 4 & -6 & 1 \\ -2 & -3 & 5 & -1 \\ -4 & -4 & 7 & -1 \\ -4 & -5 & 8 & -1 \end{bmatrix} \begin{bmatrix} 1 & 1 & -1 & 1 \\ 2 & 1 & 1 & 0 \\ 2 & 1 & 0 & 1 \\ 2 & -1 & -1 & 3 \end{bmatrix} = \begin{bmatrix} 1 & 0 & 0 & 0 \\ 0 & 1 & 0 & 0 \\ 0 & 0 & 1 & 0 \\ 0 & 0 & 0 & 1 \end{bmatrix}.$$

17. a. $A = \begin{bmatrix} 2 & 5 \\ 1 & 3 \end{bmatrix}, \; X = \begin{bmatrix} x \\ y \end{bmatrix}, \; B = \begin{bmatrix} 3 \\ 2 \end{bmatrix};$

b. $X = A^{-1}B = \begin{bmatrix} 3 & -5 \\ -1 & 2 \end{bmatrix} \begin{bmatrix} 3 \\ 2 \end{bmatrix} = \begin{bmatrix} -1 \\ 1 \end{bmatrix};$

19. a. $A = \begin{bmatrix} 2 & -3 & -4 \\ 0 & 0 & -1 \\ 1 & -2 & 1 \end{bmatrix}, \; X = \begin{bmatrix} x \\ y \\ z \end{bmatrix}, \; B = \begin{bmatrix} 4 \\ 3 \\ -8 \end{bmatrix}$

$$X = A^{-1}B = \begin{bmatrix} 2 & -11 & -3 \\ 1 & -6 & -2 \\ 0 & -1 & 0 \end{bmatrix} \begin{bmatrix} 4 \\ 3 \\ -8 \end{bmatrix} = \begin{bmatrix} -1 \\ 2 \\ -3 \end{bmatrix}$$

21. a. $A=\begin{bmatrix}1 & 4 & -1\\ 2 & 3 & -2\\ -1 & 2 & 3\end{bmatrix}$, $X=\begin{bmatrix}x\\ y\\ z\end{bmatrix}$, $B=\begin{bmatrix}3\\ 1\\ 7\end{bmatrix}$; b. $X=A^{-1}B=\begin{bmatrix}-\frac{13}{10} & \frac{7}{5} & \frac{1}{2}\\ \frac{2}{5} & -\frac{1}{5} & 0\\ -\frac{7}{10} & \frac{3}{5} & \frac{1}{2}\end{bmatrix}\begin{bmatrix}3\\ 1\\ 7\end{bmatrix}=\begin{bmatrix}1\\ 1\\ 2\end{bmatrix}$.

23. a. $A=\begin{bmatrix}1 & 1 & -1 & 1\\ 2 & 1 & 1 & 0\\ 2 & 1 & 0 & 1\\ 2 & -1 & -1 & 3\end{bmatrix}$, $X=\begin{bmatrix}x_1\\ x_2\\ x_3\\ x_4\end{bmatrix}$, $B=\begin{bmatrix}6\\ 4\\ 7\\ 9\end{bmatrix}$.

b. $X=A^{-1}B=\begin{bmatrix}3 & 4 & -6 & 1\\ -2 & -3 & 5 & -1\\ -4 & -4 & 7 & -1\\ -4 & -5 & 8 & -1\end{bmatrix}\begin{bmatrix}6\\ 4\\ 7\\ 9\end{bmatrix}=\begin{bmatrix}1\\ 2\\ 0\\ 3\end{bmatrix}$.

25. a. $A=\begin{bmatrix}1 & 2\\ 2 & -1\end{bmatrix}$, $X=\begin{bmatrix}x\\ y\end{bmatrix}$, $B=\begin{bmatrix}b_1\\ b_2\end{bmatrix}$;

b (i). $X=A^{-1}B=\begin{bmatrix}0.2 & 0.4\\ 0.4 & -0.2\end{bmatrix}\begin{bmatrix}14\\ 5\end{bmatrix}=\begin{bmatrix}4.8\\ 4.6\end{bmatrix}$ and we conclude that $x=4.8$ and $y=4.6$.

(ii). $X=A^{-1}B=\begin{bmatrix}0.2 & 0.4\\ 0.4 & -0.2\end{bmatrix}\begin{bmatrix}4\\ -1\end{bmatrix}=\begin{bmatrix}0.4\\ 1.8\end{bmatrix}$ and we conclude that $x=0.4$ and $y=1.8$.

27. a. First we find A^{-1}.

$$\left[\begin{array}{ccc|ccc}1 & 2 & 1 & 1 & 0 & 0\\ 1 & 1 & 1 & 0 & 1 & 0\\ 3 & 1 & 1 & 0 & 0 & 1\end{array}\right]\xrightarrow[R_3-3R_1]{R_2-R_1}\left[\begin{array}{ccc|ccc}1 & 2 & 1 & 1 & 0 & 0\\ 0 & -1 & 0 & -1 & 1 & 0\\ 0 & -5 & -2 & -3 & 0 & 1\end{array}\right]\xrightarrow{-R_2}$$

$$\left[\begin{array}{ccc|ccc}1 & 2 & 1 & 1 & 0 & 0\\ 0 & 1 & 0 & 1 & -1 & 0\\ 0 & -5 & -2 & -3 & 0 & 1\end{array}\right]\xrightarrow[R_3+5R_2]{R_1-2R_2}\left[\begin{array}{ccc|ccc}1 & 0 & 1 & -1 & 2 & 0\\ 0 & 1 & 0 & 1 & -1 & 0\\ 0 & 0 & -2 & 2 & -5 & 1\end{array}\right]\xrightarrow{-\frac{1}{2}R_3}$$

$$\left[\begin{array}{ccc|ccc} 1 & 0 & 1 & -1 & 2 & 0 \\ 0 & 1 & 0 & 1 & -1 & 0 \\ 0 & 0 & 1 & -1 & \frac{5}{2} & -\frac{1}{2} \end{array}\right] \xrightarrow{R_1-R_3} \left[\begin{array}{ccc|ccc} 1 & 0 & 0 & 0 & -\frac{1}{2} & \frac{1}{2} \\ 0 & 1 & 0 & 1 & -1 & 0 \\ 0 & 0 & 1 & -1 & \frac{5}{2} & -\frac{1}{2} \end{array}\right]$$

$$\begin{bmatrix} 1 & 2 & 1 \\ 1 & 1 & 1 \\ 3 & 1 & 1 \end{bmatrix} \begin{bmatrix} x \\ y \\ z \end{bmatrix} = \begin{bmatrix} b_1 \\ b_2 \\ b_3 \end{bmatrix}$$

b. (i). $\begin{bmatrix} x \\ y \\ z \end{bmatrix} = \begin{bmatrix} 0 & -\frac{1}{2} & \frac{1}{2} \\ 1 & -1 & 0 \\ -1 & \frac{5}{2} & -\frac{1}{2} \end{bmatrix} \begin{bmatrix} 7 \\ 4 \\ 2 \end{bmatrix} = \begin{bmatrix} -1 \\ 3 \\ 2 \end{bmatrix}$

and we conclude that $x = -1$, $y = 3$, and $z = 2$

(ii). $\begin{bmatrix} x \\ y \\ z \end{bmatrix} = \begin{bmatrix} 0 & -\frac{1}{2} & \frac{1}{2} \\ 1 & -1 & 0 \\ -1 & \frac{5}{2} & -\frac{1}{2} \end{bmatrix} \begin{bmatrix} 5 \\ -3 \\ -1 \end{bmatrix} = \begin{bmatrix} 1 \\ 8 \\ -12 \end{bmatrix}$

and we conclude that $x = 1$, $y = 8$, and $z = -12$.

29. a. $\left[\begin{array}{ccc|ccc} 3 & 2 & -1 & 1 & 0 & 0 \\ 2 & -3 & 1 & 0 & 1 & 0 \\ 1 & -1 & -1 & 0 & 0 & 1 \end{array}\right] \xrightarrow{R_1 \leftrightarrow R_3} \left[\begin{array}{ccc|ccc} 1 & -1 & -1 & 0 & 0 & 1 \\ 2 & -3 & 1 & 0 & 1 & 0 \\ 3 & 2 & -1 & 1 & 0 & 0 \end{array}\right] \xrightarrow[R_3-3R_1]{R_2-2R_1}$

$$\left[\begin{array}{ccc|ccc} 1 & -1 & -1 & 0 & 0 & 1 \\ 0 & -1 & 3 & 0 & 1 & -2 \\ 0 & 5 & 2 & 1 & 0 & -3 \end{array}\right] \xrightarrow{-R_2} \left[\begin{array}{ccc|ccc} 1 & -1 & -1 & 0 & 0 & 1 \\ 0 & 1 & -3 & 0 & -1 & 2 \\ 0 & 5 & 2 & 1 & 0 & -3 \end{array}\right] \xrightarrow[R_3-5R_2]{R_1+R_2}$$

$$\left[\begin{array}{ccc|ccc} 1 & 0 & -4 & 0 & -1 & 3 \\ 0 & 1 & -3 & 0 & -1 & 2 \\ 0 & 0 & 17 & 1 & 5 & -13 \end{array}\right] \xrightarrow{\frac{1}{17}R_3} \left[\begin{array}{ccc|ccc} 1 & 0 & -4 & 0 & -1 & 3 \\ 0 & 1 & -3 & 0 & -1 & 2 \\ 0 & 0 & 1 & \frac{1}{17} & \frac{5}{17} & -\frac{13}{17} \end{array}\right]$$

$$\xrightarrow[R_2+3R_3]{R_1+4R_3}\left[\begin{array}{ccc|ccc}1&0&0&\frac{4}{17}&\frac{3}{17}&-\frac{1}{17}\\0&1&0&\frac{3}{17}&-\frac{2}{17}&-\frac{5}{17}\\0&0&1&\frac{1}{17}&\frac{5}{17}&-\frac{13}{17}\end{array}\right].\text{ Therefore } A^{-1}=\begin{bmatrix}\frac{4}{17}&\frac{3}{17}&-\frac{1}{17}\\\frac{3}{17}&-\frac{2}{17}&-\frac{5}{17}\\\frac{1}{17}&\frac{5}{17}&-\frac{13}{17}\end{bmatrix}.$$

Next, $\begin{bmatrix}3&2&-1\\2&-3&1\\1&-1&-1\end{bmatrix}\begin{bmatrix}x\\y\\z\end{bmatrix}=\begin{bmatrix}b_1\\b_2\\b_3\end{bmatrix}$

b. (i) $\begin{bmatrix}x\\y\\z\end{bmatrix}=\begin{bmatrix}\frac{4}{17}&\frac{3}{17}&-\frac{1}{17}\\\frac{3}{17}&-\frac{2}{17}&-\frac{5}{17}\\\frac{1}{17}&\frac{5}{17}&-\frac{13}{17}\end{bmatrix}\begin{bmatrix}2\\-2\\4\end{bmatrix}=\begin{bmatrix}-\frac{2}{17}\\-\frac{10}{17}\\-\frac{60}{17}\end{bmatrix}$

We conclude that $x=-2/17$, $y=-10/17$, and $z=-60/17$.

(ii) $\begin{bmatrix}x\\y\\z\end{bmatrix}=\begin{bmatrix}\frac{4}{17}&\frac{3}{17}&-\frac{1}{17}\\\frac{3}{17}&-\frac{2}{17}&-\frac{5}{17}\\\frac{1}{17}&\frac{5}{17}&-\frac{13}{17}\end{bmatrix}\begin{bmatrix}8\\-3\\6\end{bmatrix}=\begin{bmatrix}1\\0\\-5\end{bmatrix}$. We conclude that $x=1, y=0$, and $z=-5$.

31. a. $AX=B_1$ and $AX=B_2$, where

$$A=\begin{bmatrix}1&1&1&1\\1&-1&-1&1\\0&1&2&2\\1&2&1&-2\end{bmatrix},\ X=\begin{bmatrix}x_1\\x_2\\x_3\\x_4\end{bmatrix},\ B_1=\begin{bmatrix}1\\-1\\4\\0\end{bmatrix}\text{ and } B_2=\begin{bmatrix}2\\8\\4\\-1\end{bmatrix}.$$

We first find A^{-1}.

$$\left[\begin{array}{cccc|cccc}1&1&1&1&1&0&0&0\\1&-1&-1&1&0&1&0&0\\0&1&2&2&0&0&1&0\\1&2&1&-2&0&0&0&1\end{array}\right]\xrightarrow[R_4-R_1]{R_2-R_1}\left[\begin{array}{cccc|cccc}1&1&1&1&1&0&0&0\\0&-2&-2&0&-1&1&0&0\\0&1&2&2&0&0&1&0\\0&1&0&-3&-1&0&0&1\end{array}\right]$$

$$\xrightarrow{R_2 \leftrightarrow R_3} \left[\begin{array}{cccc|cccc} 1 & 1 & 1 & 1 & 1 & 0 & 0 & 0 \\ 0 & 1 & 2 & 2 & 0 & 0 & 1 & 0 \\ 0 & -2 & -2 & 0 & -1 & 1 & 0 & 0 \\ 0 & 1 & 0 & -3 & -1 & 0 & 0 & 1 \end{array}\right] \xrightarrow[R_4 - R_2]{\substack{R_1 - R_2 \\ R_3 + 2R_2}}$$

$$\left[\begin{array}{cccc|cccc} 1 & 0 & -1 & -1 & 1 & 0 & -1 & 0 \\ 0 & 1 & 2 & 2 & 0 & 0 & 1 & 0 \\ 0 & 0 & 2 & 4 & -1 & 1 & 2 & 0 \\ 0 & 0 & -2 & -5 & -1 & 0 & -1 & 1 \end{array}\right] \xrightarrow{\frac{1}{2}R_3} \left[\begin{array}{cccc|cccc} 1 & 0 & -1 & -1 & 1 & 0 & -1 & 0 \\ 0 & 1 & 2 & 2 & 0 & 0 & 1 & 0 \\ 0 & 0 & 1 & 2 & -\frac{1}{2} & \frac{1}{2} & 1 & 0 \\ 0 & 0 & -2 & -5 & -1 & 0 & -1 & 1 \end{array}\right]$$

$$\xrightarrow[R_4 + 2R_3]{\substack{R_1 + R_3 \\ R_2 - 2R_3}} \left[\begin{array}{cccc|cccc} 1 & 0 & 0 & 1 & \frac{1}{2} & \frac{1}{2} & 0 & 0 \\ 0 & 1 & 0 & -2 & 1 & -1 & -1 & 0 \\ 0 & 0 & 1 & 2 & -\frac{1}{2} & \frac{1}{2} & 1 & 0 \\ 0 & 0 & 0 & -1 & -2 & 1 & 1 & 1 \end{array}\right] \xrightarrow[-R_4]{\substack{R_1 + R_4 \\ R_2 - 2R_4 \\ R_3 + 2R_4}}$$

$$\left[\begin{array}{cccc|cccc} 1 & 0 & 0 & 0 & -\frac{3}{2} & \frac{3}{2} & 1 & 1 \\ 0 & 1 & 0 & 0 & 5 & -3 & -3 & -2 \\ 0 & 0 & 1 & 0 & -\frac{9}{2} & \frac{5}{2} & 3 & 2 \\ 0 & 0 & 0 & 1 & 2 & -1 & -1 & -1 \end{array}\right]. \quad \text{So } A^{-1} = \begin{bmatrix} -\frac{3}{2} & \frac{3}{2} & 1 & 1 \\ 5 & -3 & -3 & -2 \\ -\frac{9}{2} & \frac{5}{2} & 3 & 2 \\ 2 & -1 & -1 & -1 \end{bmatrix}.$$

b. (i). $\begin{bmatrix} x_1 \\ x_2 \\ x_3 \\ x_4 \end{bmatrix} = \begin{bmatrix} -\frac{3}{2} & \frac{3}{2} & 1 & 1 \\ 5 & -3 & -3 & -2 \\ -\frac{9}{2} & \frac{5}{2} & 3 & 2 \\ 2 & -1 & -1 & -1 \end{bmatrix} \begin{bmatrix} 1 \\ -1 \\ 4 \\ 0 \end{bmatrix} = \begin{bmatrix} 1 \\ -4 \\ 5 \\ -1 \end{bmatrix}$

and we conclude that $x_1 = 1$, $x_2 = -4$, $x_3 = 5$, and $x_4 = -1$.

(ii). $\begin{bmatrix} x_1 \\ x_2 \\ x_3 \\ x_4 \end{bmatrix} = \begin{bmatrix} -\frac{3}{2} & \frac{3}{2} & 1 & 1 \\ 5 & -3 & -3 & -2 \\ -\frac{9}{2} & \frac{5}{2} & 3 & 2 \\ 2 & -1 & -1 & -1 \end{bmatrix} \begin{bmatrix} 2 \\ 8 \\ 4 \\ -1 \end{bmatrix} = \begin{bmatrix} 12 \\ -24 \\ 21 \\ -7 \end{bmatrix}$

and we conclude that $x_1 = 12$, $x_2 = -24$, $x_3 = 21$, and $x_4 = -7$.

33. a. Using Formula (13), we find $A^{-1} = \dfrac{1}{(2)(-5)-(-4)(3)}\begin{bmatrix} -5 & -3 \\ 4 & 2 \end{bmatrix} = \begin{bmatrix} -\frac{5}{2} & -\frac{3}{2} \\ 2 & 1 \end{bmatrix}$.

b. Using Formula (13) once again, we find

$$\left(A^{-1}\right)^{-1} = \frac{1}{(-\frac{5}{2})(1)-2(-\frac{3}{2})}\begin{bmatrix} 1 & \frac{3}{2} \\ -2 & -\frac{5}{2} \end{bmatrix} = \begin{bmatrix} 2 & 3 \\ -4 & -5 \end{bmatrix} = A.$$

35. a. $ABC = \begin{bmatrix} 2 & -5 \\ 1 & -3 \end{bmatrix}\begin{bmatrix} 4 & 3 \\ 1 & 1 \end{bmatrix}\begin{bmatrix} 2 & 3 \\ -2 & 1 \end{bmatrix} = \begin{bmatrix} 2 & -5 \\ 1 & -3 \end{bmatrix}\begin{bmatrix} 2 & 15 \\ 0 & 4 \end{bmatrix} = \begin{bmatrix} 4 & 10 \\ 2 & 3 \end{bmatrix}$.

Using the formula for finding the inverse of a 2 × 2 matrix, we find

$$A^{-1} = \begin{bmatrix} 3 & -5 \\ 1 & -2 \end{bmatrix},\ B^{-1} = \begin{bmatrix} 1 & -3 \\ -1 & 4 \end{bmatrix},\ C^{-1} = \begin{bmatrix} \frac{1}{8} & -\frac{3}{8} \\ \frac{1}{4} & \frac{1}{4} \end{bmatrix}.$$

b. Using the formula for finding the inverse of a 2 × 2 matrix, we find

$$(ABC)^{-1} = \begin{bmatrix} -\frac{3}{8} & \frac{5}{4} \\ \frac{1}{4} & -\frac{1}{2} \end{bmatrix}$$

$$C^{-1}B^{-1}A^{-1} = \begin{bmatrix} \frac{1}{8} & -\frac{3}{8} \\ \frac{1}{4} & \frac{1}{4} \end{bmatrix}\begin{bmatrix} 1 & -3 \\ -1 & 4 \end{bmatrix}\begin{bmatrix} 3 & -5 \\ 1 & -2 \end{bmatrix}$$

$$= \begin{bmatrix} \frac{1}{8} & -\frac{3}{8} \\ \frac{1}{4} & \frac{1}{4} \end{bmatrix}\begin{bmatrix} 0 & 1 \\ 1 & -3 \end{bmatrix} = \begin{bmatrix} -\frac{3}{8} & \frac{5}{4} \\ \frac{1}{4} & -\frac{1}{2} \end{bmatrix}.$$

Therefore, $(ABC)^{-1} = C^{-1}B^{-1}A^{-1}$.

37. Let x denote the number of copies of the deluxe edition and y the number of copies of the standard edition demanded per month when the unit prices are p and q dollars, respectively. Then the three systems of linear equations

$$\begin{aligned} 5x + y &= 20000 \\ x + 3y &= 15000 \end{aligned} \qquad \begin{aligned} 5x + y &= 25000 \\ x + 3y &= 15000 \end{aligned} \qquad \begin{aligned} 5x + y &= 25000 \\ x + 3y &= 20000 \end{aligned}$$

give the quantity demanded of each edition at the stated price. These systems may be written in the form $AX = B_1$, $AX = B_2$, and $AX = B_3$, where

$$A=\begin{bmatrix}5 & 1\\ 1 & 3\end{bmatrix},\quad B_1=\begin{bmatrix}20000\\ 15000\end{bmatrix},\quad B_2=\begin{bmatrix}25000\\ 15000\end{bmatrix},\quad \text{and}\quad B_3=\begin{bmatrix}25000\\ 20000\end{bmatrix}$$

Using the formula for finding the inverse of a 2 × 2 matrix, with $a = 5$, $b = 1$, $c = 1$, $d = 3$, and $D = ad - bc = (5)(3) - (1)(1) = 14$, we find that $A^{-1} = \begin{bmatrix}\frac{3}{14} & -\frac{1}{14}\\ -\frac{1}{14} & \frac{5}{14}\end{bmatrix}$.

a. $\begin{bmatrix}x\\ y\end{bmatrix}=\begin{bmatrix}\frac{3}{14} & -\frac{1}{14}\\ -\frac{1}{14} & \frac{5}{14}\end{bmatrix}\begin{bmatrix}20{,}000\\ 15{,}000\end{bmatrix}=\begin{bmatrix}3{,}214\\ 3{,}929\end{bmatrix}$ b. $\begin{bmatrix}x\\ y\end{bmatrix}=\begin{bmatrix}\frac{3}{14} & -\frac{1}{14}\\ -\frac{1}{14} & \frac{5}{14}\end{bmatrix}\begin{bmatrix}25{,}000\\ 15{,}000\end{bmatrix}=\begin{bmatrix}4{,}286\\ 3{,}571\end{bmatrix}$

c. $\begin{bmatrix}x\\ y\end{bmatrix}=\begin{bmatrix}\frac{3}{14} & -\frac{1}{14}\\ -\frac{1}{14} & \frac{5}{14}\end{bmatrix}\begin{bmatrix}25{,}000\\ 20{,}000\end{bmatrix}=\begin{bmatrix}3{,}929\\ 5{,}357\end{bmatrix}$.

39. Let x, y, and z denote the number of acres of soybeans, corn, and wheat to be cultivated, respectively. Furthermore, let a, b, and c denote the amount of land available; the amount of labor available, and the amount of money available for seeds, respectively. Then we have the system

$$\begin{aligned} x + y + z &= a \quad &\text{(land)}\\ 2x + 6y + 6z &= b \quad &\text{(labor)}\\ 12x + 20y + 8z &= c \quad &\text{(seeds)} \end{aligned}$$

The system can be written in the form $AX = B$, where

$$A=\begin{bmatrix}1 & 1 & 1\\ 2 & 6 & 6\\ 12 & 20 & 8\end{bmatrix},\quad X=\begin{bmatrix}x\\ y\\ z\end{bmatrix},\quad \text{and}\quad B=\begin{bmatrix}a\\ b\\ c\end{bmatrix}$$

Using the technique for finding A^{-1} developed in this section, we find

$$A^{-1}=\begin{bmatrix}\frac{3}{2} & -\frac{1}{4} & 0\\ -\frac{7}{6} & \frac{1}{12} & \frac{1}{12}\\ \frac{2}{3} & \frac{1}{6} & -\frac{1}{12}\end{bmatrix}.$$

a. Here $a = 1000$, $b = 4400$, and $c = 13{,}200$. Therefore

$$X = A^{-1}B = \begin{bmatrix} \frac{3}{2} & -\frac{1}{4} & 0 \\ -\frac{7}{6} & \frac{1}{12} & \frac{1}{12} \\ \frac{2}{3} & \frac{1}{6} & -\frac{1}{12} \end{bmatrix} \begin{bmatrix} 1000 \\ 4400 \\ 13200 \end{bmatrix} = \begin{bmatrix} 400 \\ 300 \\ 300 \end{bmatrix}$$

So, Jackson Farms should cultivate 400, 300, and 300 acres of soybeans, corn, and wheat, respectively.

b. Here $a = 1200$, $b = 5200$, and $c = 16{,}400$. Therefore,

$$X = A^{-1}B = \begin{bmatrix} \frac{3}{2} & -\frac{1}{4} & 0 \\ -\frac{7}{6} & \frac{1}{12} & \frac{1}{12} \\ \frac{2}{3} & \frac{1}{6} & -\frac{1}{12} \end{bmatrix} \begin{bmatrix} 1200 \\ 5200 \\ 16400 \end{bmatrix} = \begin{bmatrix} 500 \\ 400 \\ 300 \end{bmatrix}$$

So, Jackson Farms should cultivate 500, 400, and 300 acres of soybeans, corn, and wheat, respectively.

41. Let x, y, and z denote the amount to be invested in high-risk, medium-risk, and low-risk stocks, respectively. Next, let a denote the amount to be invested and let c denote the return on the investments. Then, we have the system

$$\begin{aligned} x + \quad y + \quad z &= a \\ x + \quad y - \quad z &= 0 \qquad \text{(since } z = x + y) \\ 0.15x + 0.1y + 0.06z &= c \end{aligned}$$

The system is equivalent to the matrix equation $AX = B$, where

$$A = \begin{bmatrix} 1 & 1 & 1 \\ 1 & 1 & -1 \\ .15 & .10 & .06 \end{bmatrix}, \quad X = \begin{bmatrix} x \\ y \\ z \end{bmatrix}, \quad \text{and} \quad B = \begin{bmatrix} a \\ 0 \\ c \end{bmatrix}.$$

We find $A^{-1} = \begin{bmatrix} -1.6 & -0.4 & 20 \\ 2.1 & 0.9 & -20 \\ 0.5 & -0.5 & 0 \end{bmatrix}$.

a. Here $a = 200{,}000$ and $c = 20{,}000$. Therefore,

$$X = A^{-1}B = \begin{bmatrix} -1.6 & -0.4 & 20 \\ 2.1 & 0.9 & -20 \\ 0.5 & -0.5 & 0 \end{bmatrix} \begin{bmatrix} 200{,}000 \\ 0 \\ 20{,}000 \end{bmatrix} = \begin{bmatrix} 80{,}000 \\ 20{,}000 \\ 100{,}000 \end{bmatrix}.$$

So, the club should invest \$80,000 in high-risk, \$20,000 in medium risk, and \$100,000 in low risk stocks.

b. Here $a = 220{,}000$ and $c = 22{,}000$. The solution is $x = 88{,}000$, $y = 22{,}000$, and $z = 110{,}000$; that is, the club should invest $88,000 in high-risk, $22,000 in medium-risk, and $110,000 in low-risk stocks.
c. Here $a = 240{,}000$ and $c = 22{,}000$. The result is $56,000 in high-risk stocks, $64,000 in medium-risk stocks, and $120,000 in low-risk stocks.

43. In order for the inverse of A to exist, $D = ad - bc \neq 0$.
Here $a = 1$, $b = 2$, $c = k$, and $d = 3$. So (1)(3)-(2)(k) $\neq 0$, or $k \neq \frac{3}{2}$
So A^{-1} has an inverse provided $k \neq \frac{3}{2}$. Using Formula 13, we have

$$A^{-1} = \frac{1}{3-2k}\begin{bmatrix} 3 & -2 \\ -k & 1 \end{bmatrix}.$$

45. True. Multiplying both sides of the equation by cA yields

$$I = (cA)(cA)^{-1} = (cA)\left[\frac{1}{c}(A^{-1})\right] = c\left(\frac{1}{c}\right)AA^{-1} = I.$$

47. True. $AX = B$ can have a unique solution only if A^{-1} exists, in which case the solution is found as follows:

$$A^{-1}(AX) = A^{-1}B$$
$$(A^{-1}A)X = A^{-1}B$$
$$IX = A^{-1}B$$
$$X = A^{-1}B$$

USING TECHNOLOGY EXERCISES 2.6, page 152

1. $\begin{bmatrix} 0.36 & 0.04 & -0.36 \\ 0.06 & 0.05 & 0.20 \\ -0.19 & 0.10 & 0.09 \end{bmatrix}$

3. $\begin{bmatrix} 0.01 & -0.09 & 0.31 & -0.11 \\ -0.25 & 0.58 & -0.15 & -0.02 \\ 0.86 & -0.42 & 0.07 & -0.37 \\ -0.27 & 0.01 & -0.05 & 0.31 \end{bmatrix}$

5. $\begin{bmatrix} 0.30 & 0.85 & -0.10 & -0.77 & -0.11 \\ -0.21 & 0.10 & 0.01 & -0.26 & 0.21 \\ 0.03 & -0.16 & 0.12 & -0.01 & 0.03 \\ -0.14 & -0.46 & 0.13 & 0.71 & -0.05 \\ 0.10 & -0.05 & -0.10 & -0.03 & 0.11 \end{bmatrix}$

7. $x = 1.2$, $y = 3.6$, and $z = 2.7$.

9. $x_1 = 2.50$, $x_2 = -0.88$, $x_3 = 0.70$, and $x_4 = 0.51$.

2.7 CONCEPT QUESTIONS, page 158

1. X represents total output, AX represent internal consumption, and D represents consumer demand.

EXERCISES 2.7, page 158

1. a. The amount of agricultural products consumed in the production of $100 million worth of manufactured goods is given by (100)(0.10), or $10 million.
b. The amount of manufactured goods required to produce $200 million of all goods in the economy is given by 200(0.1 + 0.4 + 0.3) = 160, or $160 million.
c. From the input-output matrix, we see that the agricultural sector consumes the greatest amount of agricultural products, namely, 0.4 units, in the production of each unit of goods in that sector. The manufacturing and transportation sectors consume the least, 0.1 units each.

3. Multiplying both sides of the given equation on the left by $(I - A)^{-1}$, we see that
$$X = (I - A)^{-1}D.$$
Now, $(I - A) = \begin{bmatrix} 1 & 0 \\ 0 & 1 \end{bmatrix} - \begin{bmatrix} 0.4 & 0.2 \\ 0.3 & 0.1 \end{bmatrix} = \begin{bmatrix} 0.6 & -0.2 \\ -0.3 & 0.9 \end{bmatrix}.$

Using the formula for finding the inverse of a 2 × 2 matrix, we find
$$(I - A)^{-1} = \begin{bmatrix} 1.875 & 0.417 \\ 0.625 & 1.25 \end{bmatrix}.$$

Then, $(I - A)^{-1}X = \begin{bmatrix} 1.875 & 0.417 \\ 0.625 & 1.25 \end{bmatrix}\begin{bmatrix} 10 \\ 12 \end{bmatrix} = \begin{bmatrix} 23.754 \\ 21.25 \end{bmatrix}.$

5. We first compute $(I - A) = \begin{bmatrix} 1 & 0 \\ 0 & 1 \end{bmatrix} - \begin{bmatrix} 0.5 & 0.2 \\ 0.2 & 0.5 \end{bmatrix} = \begin{bmatrix} 0.5 & -0.2 \\ -0.2 & 0.5 \end{bmatrix}$
Using the formula for finding the inverse of a 2 × 2 matrix, we find

$$(I-A)^{-1}=\begin{bmatrix}2.381 & 0.952\\ 0.952 & 2.381\end{bmatrix}.$$

Then $\begin{bmatrix}x\\ y\end{bmatrix}=\begin{bmatrix}2.381 & 0.952\\ 0.952 & 2.381\end{bmatrix}\begin{bmatrix}10\\ 20\end{bmatrix}=\begin{bmatrix}42.85\\ 57.14\end{bmatrix}$

7. We verify

$$(I-A)(I-A)^{-1}=\begin{bmatrix}0.92 & -0.60 & -0.30\\ -0.04 & 0.98 & -0.01\\ -0.02 & 0 & 0.94\end{bmatrix}\begin{bmatrix}1.13 & 0.69 & 0.37\\ 0.05 & 1.05 & 0.03\\ 0.02 & 0.02 & 1.07\end{bmatrix}=\begin{bmatrix}1 & 0 & 0\\ 0 & 1 & 0\\ 0 & 0 & 1\end{bmatrix}$$

9. a. $A=\begin{bmatrix}0.2 & 0.4\\ 0.3 & 0.3\end{bmatrix}$ and

$$(I-A)=\begin{bmatrix}1 & 0\\ 0 & 1\end{bmatrix}-\begin{bmatrix}0.2 & 0.4\\ 0.3 & 0.3\end{bmatrix}=\begin{bmatrix}0.8 & -0.4\\ -0.3 & 0.7\end{bmatrix}.$$

Using the formula for finding the inverse of a 2×2 matrix, we find

$$(I-A)^{-1}=\begin{bmatrix}1.591 & 0.909\\ 0.682 & 1.818\end{bmatrix}.$$

Then $\begin{bmatrix}x\\ y\end{bmatrix}=\begin{bmatrix}1.591 & 0.909\\ 0.682 & 1.818\end{bmatrix}\begin{bmatrix}120\\ 140\end{bmatrix}=\begin{bmatrix}318.18\\ 336.36\end{bmatrix}$

To fullfill consumer demand, \$318.2 million worth of agricultural goods and \$336.4 million worth of manufactured goods should be produced.

b. The net value of goods consumed in the internal process of production is

$$AX=X-D=\begin{bmatrix}318.18\\ 336.36\end{bmatrix}-\begin{bmatrix}120\\ 140\end{bmatrix}=\begin{bmatrix}198.18\\ 196.36\end{bmatrix}.$$

or \$198.2 million of agricultural goods and \$196.4 million worth of manufactured goods.

11. a.

$$(I-A)=\begin{bmatrix}1 & 0 & 0\\ 0 & 1 & 0\\ 0 & 0 & 1\end{bmatrix}-\begin{bmatrix}0.4 & 0.1 & 0.1\\ 0.1 & 0.4 & 0.3\\ 0.2 & 0.2 & 0.2\end{bmatrix}=\begin{bmatrix}0.6 & -0.1 & -0.1\\ -0.1 & 0.6 & -0.3\\ -0.2 & -0.2 & 0.8\end{bmatrix}$$

Using the methods of Section 2.6 we next compute the inverse of $(1-A)^{-1}$ and use this value to find

$$X=(1-A)^{-1}D=\begin{bmatrix}1.875 & 0.446 & 0.402\\ 0.625 & 2.054 & 0.848\\ 0.625 & 0.625 & 1.563\end{bmatrix}\begin{bmatrix}200\\ 100\\ 60\end{bmatrix}=\begin{bmatrix}443.7\\ 381.3\\ 281.3\end{bmatrix}.$$

Therefore, to fulfull demand, \$443.7 million worth of agricultural products, \$381.3 million worth of manufactured products, and \$281.3 million worth of transportation services should be produced.

b. To meet the gross output, the value of goods and transportation consumed in the internal process of production is

$$AX=X-D=\begin{bmatrix}443.7\\ 381.3\\ 281.3\end{bmatrix}-\begin{bmatrix}200\\ 100\\ 60\end{bmatrix}=\begin{bmatrix}243.7\\ 281.3\\ 221.3\end{bmatrix},$$

or \$243.7 million worth of agricultural products, \$281.3 million worth of manufactured services, and \$221.3 million worth of transportation services.

13. We want to solve the equation $(I-A)X=D$ for X, the total output matrix. First, we compute

$$(I-A)=\begin{bmatrix}1 & 0\\ 0 & 1\end{bmatrix}-\begin{bmatrix}0.4 & 0.2\\ 0.3 & 0.5\end{bmatrix}=\begin{bmatrix}0.6 & -0.2\\ -0.3 & 0.5\end{bmatrix}.$$

Using the formula for finding the inverse of a 2×2 matrix, we find

$$(I-A)^{-1}=\begin{bmatrix}2.08 & 0.833\\ 1.25 & 2.5\end{bmatrix}$$

Therefore,

$$X=(I-A)^{-1}D=\begin{bmatrix}2.08 & 0.833\\ 1.25 & 2.5\end{bmatrix}\begin{bmatrix}12\\ 24\end{bmatrix}=\begin{bmatrix}45\\ 75\end{bmatrix}.$$

We conclude that \$45 million worth of goods of one industry and \$75 million worth of goods of the other industry must be produced.

15. First, we compute

$$I-A=\begin{bmatrix}1 & 0 & 0\\ 0 & 1 & 0\\ 0 & 0 & 1\end{bmatrix}-\begin{bmatrix}0.2 & 0.4 & 0.2\\ 0.5 & 0 & 0.5\\ 0 & 0.2 & 0\end{bmatrix}=\begin{bmatrix}0.8 & -0.4 & -0.2\\ -0.5 & 1 & -0.5\\ 0 & -0.2 & 1\end{bmatrix}.$$

Next, using the Gauss-Jordan method, we find

$$(I-A)^{-1}=\begin{bmatrix}1.8 & 0.88 & 0.80\\ 1 & 1.6 & 1\\ 0.2 & 0.32 & 1.20\end{bmatrix}$$

Then $\begin{bmatrix}x\\ y\\ z\end{bmatrix}=\begin{bmatrix}1.8 & 0.88 & 0.80\\ 1 & 1.6 & 1\\ 0.2 & 0.32 & 1.20\end{bmatrix}\begin{bmatrix}10\\ 5\\ 15\end{bmatrix}=\begin{bmatrix}34.4\\ 33\\ 21.6\end{bmatrix}$.

We conclude that \$34.4 million worth of goods of one industry, \$33 million worth of a second industry, and \$21.6 million worth of a third industry should be produced.

USING TECHNOLOGY EXERCISES 2.7, page 162

1. The final outputs of the first, second, third, and fourth industries are 602.62, 502.30, 572.57, and 523.46 units, respectively.

3. The final outputs of the first, second, third, and fourth industries are 143.06, 132.98, 188.59, and 125.53 units, respectively.

CHAPTER 2, REVIEW QUESTIONS, page 163

1. a. one; many; no b. one; many; no 3. $R_i \leftrightarrow R_j$; cR_i; $R_i + aR_j$; solution

5. Size; entries 7. $m\times n$; $n\times m$; a_{ji} 9. a. Columns; rows b. $m\times p$

11. $A^{-1}A$; AA^{-1}; singular

CHAPTER 2, REVIEW EXERCISES, page 164

1. $\begin{bmatrix}1 & 2\\ -1 & 3\\ 2 & 1\end{bmatrix}+\begin{bmatrix}1 & 0\\ 0 & 1\\ 1 & 2\end{bmatrix}=\begin{bmatrix}2 & 2\\ -1 & 4\\ 3 & 3\end{bmatrix}$.

3. $\begin{bmatrix}-3 & 2 & 1\end{bmatrix}\begin{bmatrix}2 & 1\\ -1 & 0\\ 2 & 1\end{bmatrix}=\begin{bmatrix}-6 & -2\end{bmatrix}$.

5. By the equality of matrices, $x=2$, $z=1$, $y=3$ and $w=3$.

7. By the equality of matrices,
$a+3=6$, or $a=3$.

$-1 = e+2$, or $e = -3; b = 4$

$c+1 = -1,$ or $c = -2; d = 2$

$e+2 = -1,$ and $e = -3.$

9.

$$2A+3B = 2\begin{bmatrix} 1 & 3 & 1 \\ -2 & 1 & 3 \\ 4 & 0 & 2 \end{bmatrix} + 3\begin{bmatrix} 2 & 1 & 3 \\ -2 & -1 & -1 \\ 1 & 4 & 2 \end{bmatrix} = \begin{bmatrix} 2 & 6 & 2 \\ -4 & 2 & 6 \\ 8 & 0 & 4 \end{bmatrix} + \begin{bmatrix} 6 & 3 & 9 \\ -6 & -3 & -3 \\ 3 & 12 & 6 \end{bmatrix}$$

$$= \begin{bmatrix} 8 & 9 & 11 \\ -10 & -1 & 3 \\ 11 & 12 & 10 \end{bmatrix}.$$

11.

$$3A = 3\begin{bmatrix} 1 & 3 & 1 \\ -2 & 1 & 3 \\ 4 & 0 & 2 \end{bmatrix} = \begin{bmatrix} 3 & 9 & 3 \\ -6 & 3 & 9 \\ 12 & 0 & 6 \end{bmatrix}$$

and $$2(3A) = 2\begin{bmatrix} 3 & 9 & 3 \\ -6 & 3 & 9 \\ 12 & 0 & 6 \end{bmatrix} = \begin{bmatrix} 6 & 18 & 6 \\ -12 & 6 & 18 \\ 24 & 0 & 12 \end{bmatrix}.$$

13. $$B-C = \begin{bmatrix} 2 & 1 & 3 \\ -2 & -1 & -1 \\ 1 & 4 & 2 \end{bmatrix} - \begin{bmatrix} 3 & -1 & 2 \\ 1 & 6 & 4 \\ 2 & 1 & 3 \end{bmatrix} = \begin{bmatrix} -1 & 2 & 1 \\ -3 & -7 & -5 \\ -1 & 3 & -1 \end{bmatrix}$$

and so $$A(B-C) = \begin{bmatrix} 1 & 3 & 1 \\ -2 & 1 & 3 \\ 4 & 0 & 2 \end{bmatrix}\begin{bmatrix} -1 & 2 & 1 \\ -3 & -7 & -5 \\ -1 & 3 & -1 \end{bmatrix} = \begin{bmatrix} -11 & -16 & -15 \\ -4 & -2 & -10 \\ -6 & 14 & 2 \end{bmatrix}.$$

15. $$BC=\begin{bmatrix}2&1&3\\-2&-1&-1\\1&4&2\end{bmatrix}\begin{bmatrix}3&-1&2\\1&6&4\\2&1&3\end{bmatrix}=\begin{bmatrix}13&7&17\\-9&-5&-11\\11&25&24\end{bmatrix}$$

$$ABC=\begin{bmatrix}1&3&1\\-2&1&3\\4&0&2\end{bmatrix}\begin{bmatrix}13&7&17\\-9&-5&-11\\11&25&24\end{bmatrix}=\begin{bmatrix}-3&17&8\\-2&56&27\\74&78&116\end{bmatrix}.$$

17. Using the Gauss-Jordan elimination method, we find

$$\left[\begin{array}{cc|c}2&-3&5\\3&4&-1\end{array}\right]\xrightarrow{\frac{1}{2}R_1}\left[\begin{array}{cc|c}1&-\frac{3}{2}&\frac{5}{2}\\3&4&-1\end{array}\right]\xrightarrow{R_2-3R_1}\left[\begin{array}{cc|c}1&-\frac{3}{2}&\frac{5}{2}\\0&\frac{17}{2}&-\frac{17}{2}\end{array}\right]$$

$$\xrightarrow{\frac{2}{17}R_2}\left[\begin{array}{cc|c}1&-\frac{3}{2}&\frac{5}{2}\\0&1&-1\end{array}\right]\xrightarrow{R_1+\frac{3}{2}R_2}\left[\begin{array}{cc|c}1&0&1\\0&1&-1\end{array}\right].$$ We conclude that $x=1$ and $y=-1$.

19. $$\left[\begin{array}{ccc|c}1&-1&2&5\\3&2&1&10\\2&-3&-2&-10\end{array}\right]\xrightarrow[R_3-2R_1]{R_2-3R_1}\left[\begin{array}{ccc|c}1&-1&2&5\\0&5&-5&-5\\0&-1&-6&-20\end{array}\right]\xrightarrow{\frac{1}{5}R_2}\left[\begin{array}{ccc|c}1&-1&2&5\\0&1&-1&-1\\0&-1&-6&-20\end{array}\right]$$

$$\xrightarrow[R_3+R_2]{R_1+R_2}\left[\begin{array}{ccc|c}1&0&1&4\\0&1&-1&-1\\0&0&-7&-21\end{array}\right]\xrightarrow{-\frac{1}{7}R_3}\left[\begin{array}{ccc|c}1&0&1&4\\0&1&-1&-1\\0&0&1&3\end{array}\right]\xrightarrow[R_2+R_3]{R_1-R_3}$$

$$=\left[\begin{array}{ccc|c}1&0&0&1\\0&1&0&2\\0&0&1&3\end{array}\right].$$ Therefore, $x=1, y=2$, and $z=3$.

21. $$\left[\begin{array}{ccc|c}3&-2&4&11\\2&-4&5&4\\1&2&-1&10\end{array}\right]\xrightarrow{R_1-R_2}\left[\begin{array}{ccc|c}1&2&-1&7\\2&-4&5&4\\1&2&-1&10\end{array}\right]\xrightarrow[R_3-R_1]{R_2-2R_1}\left[\begin{array}{ccc|c}1&2&-1&7\\0&-8&7&-10\\0&0&0&3\end{array}\right].$$

Since this last row implies that 0 = 3!, we conclude that the system has no solution.

23. $$\left[\begin{array}{ccc|c} 3 & -2 & 1 & 4 \\ 1 & 3 & -4 & -3 \\ 2 & -3 & 5 & 7 \\ 1 & -8 & 9 & 10 \end{array}\right] \xrightarrow{R_1-R_3} \left[\begin{array}{ccc|c} 1 & 1 & -4 & -3 \\ 1 & 3 & -4 & -3 \\ 2 & -3 & 5 & 7 \\ 1 & -8 & 9 & 10 \end{array}\right] \xrightarrow[\substack{R_3-2R_1 \\ R_4-R_1}]{R_2-R_1} \left[\begin{array}{ccc|c} 1 & 1 & -4 & -3 \\ 0 & 2 & 0 & 0 \\ 0 & -5 & 13 & 13 \\ 0 & -9 & 13 & 13 \end{array}\right]$$

$$\xrightarrow{\frac{1}{2}R_2} \left[\begin{array}{ccc|c} 1 & 1 & -4 & -3 \\ 0 & 1 & 0 & 0 \\ 0 & -5 & 13 & 13 \\ 0 & -9 & 13 & 13 \end{array}\right] \xrightarrow[\substack{R_3+5R_2 \\ R_4+9R_2}]{R_1-R_2} \left[\begin{array}{ccc|c} 1 & 0 & -4 & -3 \\ 0 & 1 & 0 & 0 \\ 0 & 0 & 13 & 13 \\ 0 & 0 & 13 & 13 \end{array}\right] \xrightarrow{\frac{1}{13}R_3} \left[\begin{array}{ccc|c} 1 & 0 & -4 & -3 \\ 0 & 1 & 0 & 0 \\ 0 & 0 & 1 & 1 \\ 0 & 0 & 13 & 13 \end{array}\right]$$

$$\xrightarrow[R_4-13R_3]{R_1+4R_3} \left[\begin{array}{ccc|c} 1 & 0 & 0 & 1 \\ 0 & 1 & 0 & 0 \\ 0 & 0 & 1 & 1 \\ 0 & 0 & 0 & 0 \end{array}\right].$$ Therefore, $x = 1, y = 0$, and $z = 1$.

25. $$A^{-1} = \frac{1}{(3)(2)-(1)(1)} \begin{bmatrix} 2 & -1 \\ -1 & 3 \end{bmatrix} = \begin{bmatrix} \frac{2}{5} & -\frac{1}{5} \\ -\frac{1}{5} & \frac{3}{5} \end{bmatrix}.$$

27. $$A^{-1} = \frac{1}{(3)(2)-(2)(4)} \begin{bmatrix} 2 & -4 \\ -2 & 3 \end{bmatrix} = \begin{bmatrix} -1 & 2 \\ 1 & -\frac{3}{2} \end{bmatrix}.$$

29. $$\left[\begin{array}{ccc|ccc} 2 & 3 & 1 & 1 & 0 & 0 \\ 1 & -1 & 2 & 0 & 1 & 0 \\ 1 & 2 & 1 & 0 & 0 & 1 \end{array}\right] \xrightarrow{R_1-R_2} \left[\begin{array}{ccc|ccc} 1 & 4 & -1 & 1 & -1 & 0 \\ 1 & -1 & 2 & 0 & 1 & 0 \\ 1 & 2 & 1 & 0 & 0 & 1 \end{array}\right] \xrightarrow[R_3-R_1]{R_2-R_1}$$

$$\left[\begin{array}{ccc|ccc} 1 & 4 & -1 & 1 & -1 & 0 \\ 0 & -5 & 3 & -1 & 2 & 0 \\ 0 & -2 & 2 & -1 & 1 & 1 \end{array}\right] \xrightarrow{R_2-3R_3} \left[\begin{array}{ccc|ccc} 1 & 4 & -1 & 1 & -1 & 0 \\ 0 & 1 & -3 & 2 & -1 & -3 \\ 0 & -2 & 2 & -1 & 1 & 1 \end{array}\right] \xrightarrow[R_3+2R_2]{R_1-4R_2}$$

$$\left[\begin{array}{ccc|ccc} 1 & 0 & 11 & -7 & 3 & 12 \\ 0 & 1 & -3 & 2 & -1 & -3 \\ 0 & 0 & -4 & 3 & -1 & -5 \end{array}\right] \xrightarrow{-\frac{1}{4}R_3} \left[\begin{array}{ccc|ccc} 1 & 0 & 11 & -7 & 3 & 12 \\ 0 & 1 & -3 & 2 & -1 & -3 \\ 0 & 0 & 1 & -\frac{3}{4} & \frac{1}{4} & \frac{5}{4} \end{array}\right] \xrightarrow[R_2+3R_3]{R_1-11R_3}$$

$$\left[\begin{array}{ccc|ccc} 1 & 0 & 0 & \frac{5}{4} & \frac{1}{4} & -\frac{7}{4} \\ 0 & 1 & 0 & -\frac{1}{4} & -\frac{1}{4} & \frac{3}{4} \\ 0 & 0 & 1 & -\frac{3}{4} & \frac{1}{4} & \frac{5}{4} \end{array}\right]. \text{ So } A^{-1} = \begin{bmatrix} \frac{5}{4} & \frac{1}{4} & -\frac{7}{4} \\ -\frac{1}{4} & -\frac{1}{4} & \frac{3}{4} \\ -\frac{3}{4} & \frac{1}{4} & \frac{5}{4} \end{bmatrix}.$$

31. $$\left[\begin{array}{ccc|ccc} 1 & 2 & 4 & 1 & 0 & 0 \\ 3 & 1 & 2 & 0 & 1 & 0 \\ 1 & 0 & -6 & 0 & 0 & 1 \end{array}\right] \xrightarrow[R_3-R_1]{R_2-3R_1} \left[\begin{array}{ccc|ccc} 1 & 2 & 4 & 1 & 0 & 0 \\ 0 & -5 & -10 & -3 & 1 & 0 \\ 0 & -2 & -10 & -1 & 0 & 1 \end{array}\right] \xrightarrow{R_2-3R_3}$$

$$\left[\begin{array}{ccc|ccc} 1 & 2 & 4 & 1 & 0 & 0 \\ 0 & 1 & 20 & 0 & 1 & -3 \\ 0 & -2 & -10 & -1 & 0 & 1 \end{array}\right] \xrightarrow[R_3+2R_2]{R_1-2R_2} \left[\begin{array}{ccc|ccc} 1 & 0 & -36 & 1 & -2 & 6 \\ 0 & 1 & 20 & 0 & 1 & -3 \\ 0 & 0 & 30 & -1 & 2 & -5 \end{array}\right] \xrightarrow{\frac{1}{30}R_3}$$

$$\left[\begin{array}{ccc|ccc} 1 & 0 & -36 & 1 & -2 & 6 \\ 0 & 1 & 20 & 0 & 1 & -3 \\ 0 & 0 & 1 & -\frac{1}{30} & \frac{1}{15} & -\frac{1}{6} \end{array}\right] \xrightarrow[R_2-20R_3]{R_1+36R_3} \left[\begin{array}{ccc|ccc} 1 & 0 & 0 & -\frac{1}{5} & \frac{2}{5} & 0 \\ 0 & 1 & 0 & \frac{2}{3} & -\frac{1}{3} & \frac{1}{3} \\ 0 & 0 & 1 & -\frac{1}{30} & \frac{1}{15} & -\frac{1}{6} \end{array}\right]$$

So $$A^{-1} = \begin{bmatrix} -\frac{1}{5} & \frac{2}{5} & 0 \\ \frac{2}{3} & -\frac{1}{3} & \frac{1}{3} \\ -\frac{1}{30} & \frac{1}{15} & -\frac{1}{6} \end{bmatrix}.$$

33. $(A^{-1}B)^{-1} = B^{-1}(A^{-1})^{-1} = B^{-1}A$. Now

$$B^{-1} = \frac{1}{(3)(2)-4(1)}\begin{bmatrix} 2 & -1 \\ -4 & 3 \end{bmatrix} = \begin{bmatrix} 1 & -\frac{1}{2} \\ -2 & \frac{3}{2} \end{bmatrix}. \quad B^{-1}A = \begin{bmatrix} 1 & -\frac{1}{2} \\ -2 & \frac{3}{2} \end{bmatrix}\begin{bmatrix} 1 & 2 \\ -1 & 2 \end{bmatrix} = \begin{bmatrix} \frac{3}{2} & 1 \\ -\frac{7}{2} & -1 \end{bmatrix}.$$

35. $$2A - C = \begin{bmatrix} 2 & 4 \\ -2 & 4 \end{bmatrix} - \begin{bmatrix} 1 & 1 \\ -1 & 2 \end{bmatrix} = \begin{bmatrix} 1 & 3 \\ -1 & 2 \end{bmatrix}.$$

$$(2A-C)^{-1}=\frac{1}{(1)(2)-(-1)(3)}\begin{bmatrix}2 & -3\\ 1 & 1\end{bmatrix}=\begin{bmatrix}\frac{2}{5} & -\frac{3}{5}\\ \frac{1}{5} & \frac{1}{5}\end{bmatrix}.$$

37. $A=\begin{bmatrix}2 & 3\\ 1 & -2\end{bmatrix},\ X=\begin{bmatrix}x\\ y\end{bmatrix},\ C=\begin{bmatrix}-8\\ 3\end{bmatrix};\ A^{-1}=\frac{1}{(-2)(2)-(1)(3)}\begin{bmatrix}-2 & -3\\ -1 & 2\end{bmatrix}=\begin{bmatrix}\frac{2}{7} & \frac{3}{7}\\ \frac{1}{7} & -\frac{2}{7}\end{bmatrix}$

$$\begin{bmatrix}x\\ y\end{bmatrix}=A^{-1}B=\begin{bmatrix}\frac{2}{7} & \frac{3}{7}\\ \frac{1}{7} & -\frac{2}{7}\end{bmatrix}\begin{bmatrix}-8\\ 3\end{bmatrix}=\begin{bmatrix}-1\\ -2\end{bmatrix}.$$

39. Put

$$X=\begin{bmatrix}x\\ y\\ z\end{bmatrix},\ A=\begin{bmatrix}1 & -2 & 4\\ 2 & 3 & -2\\ 1 & 4 & -6\end{bmatrix},\ C=\begin{bmatrix}13\\ 0\\ -15\end{bmatrix}.$$

Then $AX=C$ and $X=A^{-1}C$. To find A^{-1},

$$\left[\begin{array}{ccc|ccc}1 & -2 & 4 & 1 & 0 & 0\\ 2 & 3 & -2 & 0 & 1 & 0\\ 1 & 4 & -6 & 0 & 0 & 1\end{array}\right]\xrightarrow[R_3-R_1]{R_2-2R_1}\left[\begin{array}{ccc|ccc}1 & -2 & 4 & 1 & 0 & 0\\ 0 & 7 & -10 & -2 & 1 & 0\\ 0 & 6 & -10 & -1 & 0 & 1\end{array}\right]\xrightarrow{R_2-R_3}$$

$$\left[\begin{array}{ccc|ccc}1 & -2 & 4 & 1 & 0 & 0\\ 0 & 1 & 0 & -1 & 1 & -1\\ 0 & 6 & -10 & -1 & 0 & 1\end{array}\right]\xrightarrow[R_3-6R_2]{R_1+2R_2}\left[\begin{array}{ccc|ccc}1 & 0 & 4 & -1 & 2 & -2\\ 0 & 1 & 0 & -1 & 1 & -1\\ 0 & 0 & -10 & 5 & -6 & 7\end{array}\right]\xrightarrow{-\frac{1}{10}R_3}$$

$$\left[\begin{array}{ccc|ccc}1 & 0 & 4 & -1 & 2 & -2\\ 0 & 1 & 0 & -1 & 1 & -1\\ 0 & 0 & 1 & -\frac{1}{2} & \frac{3}{5} & -\frac{7}{10}\end{array}\right]\xrightarrow{R_1-4R_3}\left[\begin{array}{ccc|ccc}1 & 0 & 0 & 1 & -\frac{2}{5} & \frac{4}{5}\\ 0 & 1 & 0 & -1 & 1 & -1\\ 0 & 0 & 1 & -\frac{1}{2} & \frac{3}{5} & -\frac{7}{10}\end{array}\right].$$

So $A^{-1}=\begin{bmatrix}1 & -\frac{2}{5} & \frac{4}{5}\\ -1 & 1 & -1\\ -\frac{1}{2} & \frac{3}{5} & -\frac{7}{10}\end{bmatrix}$. Therefore, $X=A^{-1}C=\begin{bmatrix}1 & -\frac{2}{5} & \frac{4}{5}\\ -1 & 1 & -1\\ -\frac{1}{2} & \frac{3}{5} & -\frac{7}{10}\end{bmatrix}\begin{bmatrix}13\\ 0\\ -15\end{bmatrix}=\begin{bmatrix}1\\ 2\\ 4\end{bmatrix}$

that is, $x=1$, $y=2$, and $z=4$.

41. Let

$$S = \begin{array}{c} \begin{array}{cccc} \text{Prem} & \text{Sup} & \text{Reg} & \text{Dies} \end{array} \\ \begin{bmatrix} 600 & 800 & 1000 & 700 \\ 700 & 600 & 1200 & 400 \\ 900 & 700 & 1400 & 800 \end{bmatrix} \end{array} \quad \text{and} \quad T = \begin{array}{c} \text{Prem} \\ \text{Sup} \\ \text{Reg} \\ \text{Dies} \end{array} \begin{bmatrix} 2.60 \\ 2.40 \\ 2.20 \\ 2.50 \end{bmatrix}$$

be the matrices representing the sales in the three gasoline stations and the unit prices for the various fuels, respectively. Then the total revenue at each station is found by computing

$$ST = \begin{bmatrix} 600 & 800 & 1000 & 700 \\ 700 & 600 & 1200 & 400 \\ 900 & 700 & 1400 & 800 \end{bmatrix} \begin{bmatrix} 2.60 \\ 2.40 \\ 2.20 \\ 2.50 \end{bmatrix} = \begin{bmatrix} 7430 \\ 6900 \\ 9100 \end{bmatrix}.$$

We conclude that the total revenue of station A is \$7430, that of station B is \$6900, and that of station C is \$9100.

43. We wish to solve the system of equations

$$\begin{aligned} 2x + 2y + 3z &= 210 \\ 2x + 3y + 4z &= 270 \\ 3x + 4y + 3z &= 300. \end{aligned}$$

Using the Gauss–Jordan method of elimination, we find

$$\left[\begin{array}{ccc|c} 2 & 2 & 3 & 210 \\ 2 & 3 & 4 & 270 \\ 3 & 4 & 3 & 300 \end{array}\right] \xrightarrow{\frac{1}{2}R_1} \left[\begin{array}{ccc|c} 1 & 1 & \frac{3}{2} & 105 \\ 2 & 3 & 4 & 270 \\ 3 & 4 & 3 & 300 \end{array}\right] \xrightarrow[R_3-3R_1]{R_2-2R_1} \left[\begin{array}{ccc|c} 1 & 1 & \frac{3}{2} & 105 \\ 0 & 1 & 1 & 60 \\ 0 & 1 & -\frac{3}{2} & -15 \end{array}\right]$$

$$\xrightarrow[R_3-R_2]{R_1-R_2} \left[\begin{array}{ccc|c} 1 & 0 & \frac{1}{2} & 45 \\ 0 & 1 & 1 & 60 \\ 0 & 0 & -\frac{5}{2} & -75 \end{array}\right] \xrightarrow{-\frac{2}{5}R_3} \left[\begin{array}{ccc|c} 1 & 0 & \frac{1}{2} & 45 \\ 0 & 1 & 1 & 60 \\ 0 & 0 & 1 & 30 \end{array}\right] \xrightarrow[R_2-R_3]{R_1-\frac{1}{2}R_3} \left[\begin{array}{ccc|c} 1 & 0 & 0 & 30 \\ 0 & 1 & 0 & 30 \\ 0 & 0 & 1 & 30 \end{array}\right]$$

So $x = y = z = 30$. Therefore, Desmond should produce 30 of each type of pendant.

CHAPTER 2 BEFORE MOVING ON, page 166

1. $\left[\begin{array}{ccc|c} 2 & 1 & -1 & -1 \\ 1 & 3 & 2 & 2 \\ 3 & 3 & -3 & -5 \end{array}\right] \xrightarrow{R_1 \leftrightarrow R_2} \left[\begin{array}{ccc|c} 1 & 3 & 2 & 2 \\ 2 & 1 & -1 & -1 \\ 3 & 3 & -3 & -5 \end{array}\right] \xrightarrow[R_3-3R_1]{R_2-2R_1} \left[\begin{array}{ccc|c} 1 & 3 & 2 & 2 \\ 0 & -5 & -5 & -5 \\ 0 & -6 & -9 & -11 \end{array}\right] \xrightarrow{-\frac{1}{5}R_2}$

$\left[\begin{array}{ccc|c} 1 & 3 & 2 & 2 \\ 0 & 1 & 1 & 1 \\ 0 & -6 & -9 & -11 \end{array}\right] \xrightarrow[R_3+6R_2]{R_1-3R_2} \left[\begin{array}{ccc|c} 1 & 0 & -1 & -1 \\ 0 & 1 & 1 & 1 \\ 0 & 0 & -3 & -5 \end{array}\right] \xrightarrow{-\frac{1}{3}R_3} \left[\begin{array}{ccc|c} 1 & 0 & -1 & -1 \\ 0 & 1 & 1 & 1 \\ 0 & 0 & 1 & \frac{5}{3} \end{array}\right] \xrightarrow[R_2-R_3]{R_1+R_3}$

$\left[\begin{array}{ccc|c} 1 & 0 & 0 & \frac{2}{3} \\ 0 & 1 & 0 & -\frac{2}{3} \\ 0 & 0 & 1 & \frac{5}{3} \end{array}\right]$. The solution is $x = \frac{2}{3}$, $y = -\frac{2}{3}$, $z = \frac{5}{3}$.

2. a. $x = 2$, $y = -3$, $z = 1$ b. No solution c. $x = 2$, $y = 1 - 3t$, $x = t$ (t, a parameter)
d. $x = y = z = w = 0$ e. $x = 2 + t$, $y = 3 - 2t$, $z = t$ (t, a parameter)

3. a.

$\left[\begin{array}{cc|c} 1 & 2 & 3 \\ 3 & -1 & -5 \\ 4 & 1 & -2 \end{array}\right] \xrightarrow[R_3-4R_1]{R_2-3R_1} \left[\begin{array}{cc|c} 1 & 2 & 3 \\ 0 & -7 & -14 \\ 0 & -7 & -14 \end{array}\right] \xrightarrow{-\frac{1}{7}R_2} \left[\begin{array}{cc|c} 1 & 2 & 3 \\ 0 & 1 & 2 \\ 0 & -7 & -14 \end{array}\right] \xrightarrow[R_3+7R_2]{R_1-2R_2}$

$\left[\begin{array}{cc|c} 1 & 0 & -1 \\ 0 & 1 & 2 \\ 0 & 0 & 0 \end{array}\right]$. The solution is $x = -1, y = 2$.

b.

$\left[\begin{array}{ccc|c} 1 & -2 & 4 & 2 \\ 3 & 1 & -2 & 1 \end{array}\right] \xrightarrow{R_2-3R_1} \left[\begin{array}{ccc|c} 1 & -2 & 4 & 2 \\ 0 & 7 & -14 & -5 \end{array}\right] \xrightarrow{\frac{1}{7}R_2}$

$\left[\begin{array}{ccc|c} 1 & -2 & 4 & 2 \\ 0 & 1 & -2 & -\frac{5}{7} \end{array}\right] \xrightarrow{R_1+2R_2} \left[\begin{array}{ccc|c} 1 & 0 & 0 & \frac{4}{7} \\ 0 & 1 & -2 & -\frac{5}{2} \end{array}\right]$.

The solution is $x = \frac{4}{7}$, $y = -\frac{5}{7} + 2t$, $z = t$ where t is a parameter.

4. a. $AB = \begin{bmatrix} 1 & -2 & 4 \\ 3 & 0 & 1 \end{bmatrix} \begin{bmatrix} 1 & -1 & 2 \\ 3 & 1 & -1 \\ 2 & 1 & 0 \end{bmatrix} = \begin{bmatrix} 3 & 1 & 4 \\ 5 & -2 & 6 \end{bmatrix}.$

b. $A + C^T = \begin{bmatrix} 1 & -2 & 4 \\ 3 & 0 & 1 \end{bmatrix} + \begin{bmatrix} 2 & 1 & 3 \\ -2 & 1 & 4 \end{bmatrix} = \begin{bmatrix} 3 & -1 & 7 \\ 1 & 1 & 5 \end{bmatrix}.$

$$(A + C^T)B = \begin{bmatrix} 3 & -1 & 7 \\ 1 & 1 & 5 \end{bmatrix} \begin{bmatrix} 1 & -1 & 2 \\ 3 & 1 & -1 \\ 2 & 1 & 0 \end{bmatrix} = \begin{bmatrix} 14 & 3 & 7 \\ 14 & 5 & 1 \end{bmatrix}$$

c. $C^T B - AB^T = \begin{bmatrix} 2 & 1 & 3 \\ -2 & 1 & 4 \end{bmatrix} \begin{bmatrix} 1 & -1 & 2 \\ 3 & 1 & -1 \\ 2 & 1 & 0 \end{bmatrix} - \begin{bmatrix} 1 & -2 & 4 \\ 3 & 0 & 1 \end{bmatrix} \begin{bmatrix} 1 & 3 & 2 \\ -1 & 1 & 1 \\ 2 & -1 & 0 \end{bmatrix}$

$$= \begin{bmatrix} 11 & 2 & 3 \\ 9 & 7 & -5 \end{bmatrix} - \begin{bmatrix} 11 & -3 & 0 \\ 5 & 8 & 6 \end{bmatrix} = \begin{bmatrix} 0 & 5 & 3 \\ 4 & -1 & -11 \end{bmatrix}.$$

5.

$$\left[\begin{array}{ccc|ccc} 2 & 1 & 2 & 1 & 0 & 0 \\ 0 & -1 & 3 & 0 & 1 & 0 \\ 1 & 1 & 0 & 0 & 0 & 1 \end{array}\right] \xrightarrow{R_1 \leftrightarrow R_3} \left[\begin{array}{ccc|ccc} 1 & 1 & 0 & 0 & 0 & 1 \\ 0 & -1 & 3 & 0 & 1 & 0 \\ 2 & 1 & 2 & 1 & 0 & 0 \end{array}\right] \xrightarrow[\substack{-R_2 \\ R_3 - 2R_1}]{R_1 + R_2}$$

$$\left[\begin{array}{ccc|ccc} 1 & 0 & 3 & 0 & 1 & 1 \\ 0 & 1 & -3 & 0 & -1 & 0 \\ 0 & -1 & 2 & 1 & 0 & -2 \end{array}\right] \xrightarrow[R_2 + 3R_3]{R_1 - 3R_3} \left[\begin{array}{ccc|ccc} 1 & 0 & 0 & 3 & -2 & -5 \\ 0 & 1 & 0 & -3 & 2 & 6 \\ 0 & 0 & 1 & -1 & 1 & 2 \end{array}\right].$$

$$A^{-1} = \begin{bmatrix} 3 & -2 & -5 \\ -3 & 2 & 6 \\ -1 & 1 & 2 \end{bmatrix}$$

6. $A = \begin{bmatrix} 2 & 0 & 1 \\ 2 & 1 & -1 \\ 3 & 1 & -1 \end{bmatrix}$, $B = \begin{bmatrix} 4 \\ -1 \\ 0 \end{bmatrix}$, and $X = \begin{bmatrix} x \\ y \\ z \end{bmatrix}$

To find A^{-1}:

$$\left[\begin{array}{ccc|ccc} 2 & 0 & 1 & 1 & 0 & 0 \\ 2 & 1 & -1 & 0 & 1 & 0 \\ 3 & 1 & -1 & 0 & 0 & 1 \end{array}\right] \xrightarrow{R_1 \leftrightarrow R_3} \left[\begin{array}{ccc|ccc} 3 & 1 & -1 & 0 & 0 & 1 \\ 2 & 1 & -1 & 0 & 1 & 0 \\ 2 & 0 & 1 & 1 & 0 & 0 \end{array}\right] \xrightarrow{R_1 - R_2}$$

$$\left[\begin{array}{ccc|ccc} 1 & 0 & 0 & 0 & -1 & 1 \\ 2 & 1 & -1 & 0 & 1 & 0 \\ 2 & 0 & 1 & 1 & 0 & 0 \end{array}\right] \xrightarrow[R_3 - 2R_1]{R_2 - 2R_1} \left[\begin{array}{ccc|ccc} 1 & 0 & 0 & 0 & -1 & 1 \\ 0 & 1 & -1 & 0 & 3 & -2 \\ 0 & 0 & 1 & 1 & 2 & -2 \end{array}\right] \xrightarrow{R_2 + R_3}$$

$$\left[\begin{array}{ccc|ccc} 1 & 0 & 0 & 0 & -1 & 1 \\ 0 & 1 & 0 & 1 & 5 & -4 \\ 0 & 0 & 1 & 1 & 2 & -2 \end{array}\right]$$

Therefore $A^{-1} = \begin{bmatrix} 0 & -1 & 1 \\ 1 & 5 & -4 \\ 1 & 2 & -2 \end{bmatrix}$. $X = \begin{bmatrix} x \\ y \\ z \end{bmatrix} = A^{-1}B = \begin{bmatrix} 0 & -1 & 1 \\ 1 & 5 & -4 \\ 1 & 2 & -2 \end{bmatrix} \begin{bmatrix} 4 \\ -1 \\ 0 \end{bmatrix} = \begin{bmatrix} 1 \\ -1 \\ 2 \end{bmatrix}$,

so $x = 1$, $y = -1$, and $z = 2$.

CHAPTER 3

3.1 Problem Solving Tips

Here are some hints for solving the problems in the exercises that follow.

1. When you graph an inequality, use a solid line to show that the line is included in the solution set ($\leq$, $=$, or $\geq$) and a dashed line to show that the line is not included in the solution set ($<$ or $>$).

2. Pick the point (0,0) as your test point if (0,0) does not lie on the line you are considering. Otherwise pick a point that makes it easy to evaluate the linear inequality you are working with.

3. If the solution set of a system of linear inequalities can be enclosed by a circle the solutions set is *bounded*. Otherwise it is *unbounded*.

3.1 CONCEPT QUESTIONS, p. 173

1. a. The solution set of $ax + by + c < 0$ is a half-plane that does not include the line with equation $ax + by + c = 0$. The solution of the set, on the other hand, includes the line.
   b. Its the line with equation $ax + by + c = 0$.

EXERCISES 3.1, page 173

1. $4x - 8 < 0$ implies $x < 2$. The graph of the inequality follows.

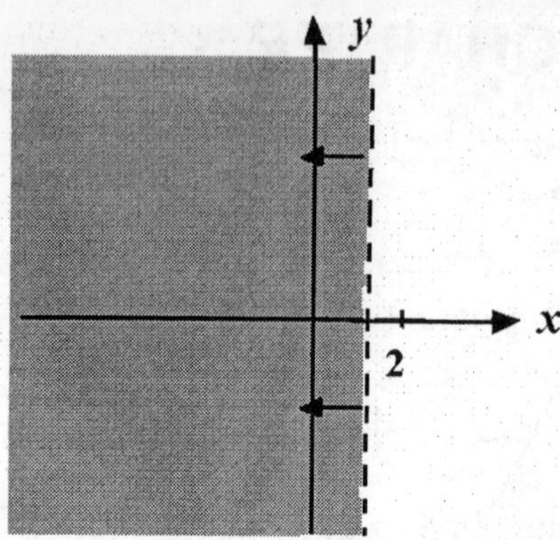

3. $x - y \leq 0$ implies $x \leq y$. The graph of the inequality follows.

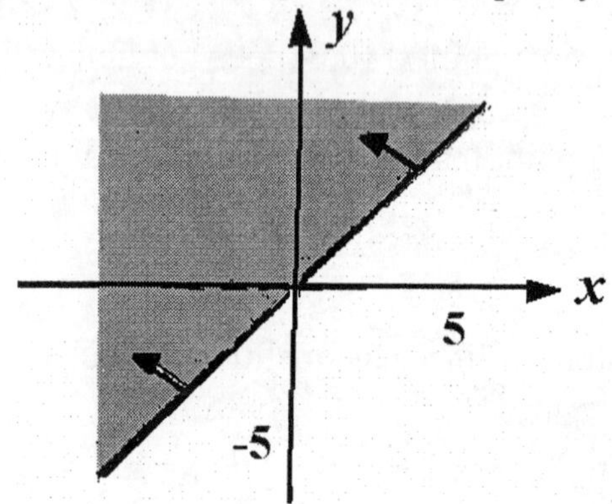

5. The graph of the inequality $x \leq -3$ follows.

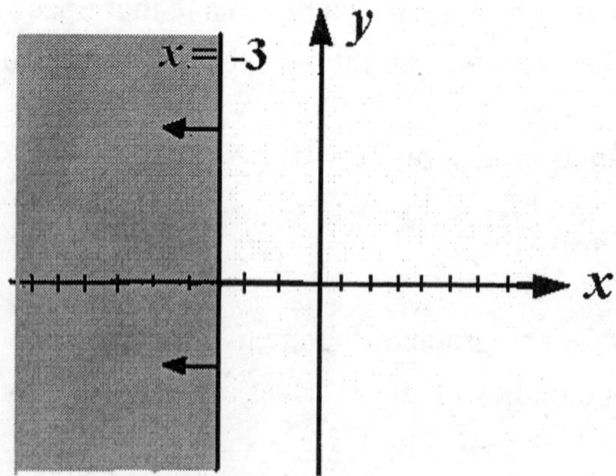

7. We first sketch the straight line with equation $2x + y = 4$. Next, picking the test point (0,0), we have $2(0) + (0) = 0 \le 4$. We conclude that the half-plane containing the origin is the required half-plane.

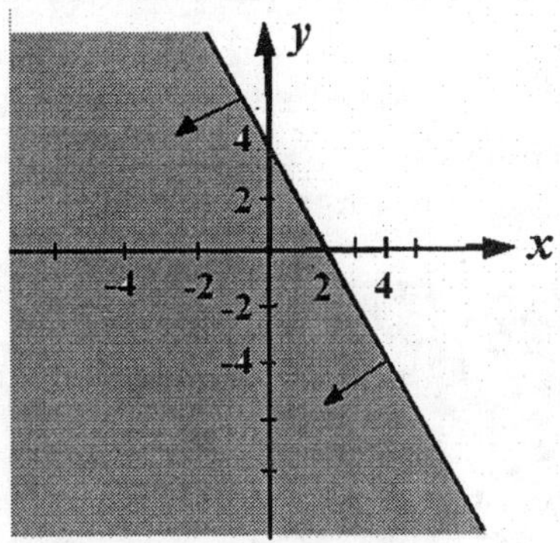

9. We first sketch the graph of the straight line $4x - 3y = -24$. Next, picking the test point (0,0), we see that $4(0) - 3(0) = 0 \not< -24$. We conclude that the half-plane not containing the origin is the required half-plane. The graph of this inequality follows.

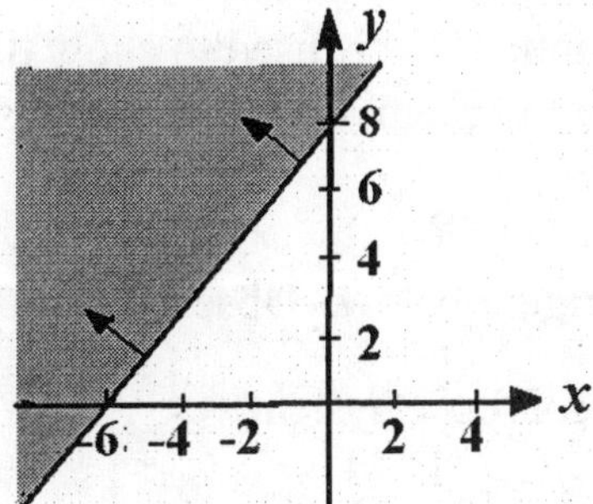

11. The system of linear inequalities that describes the shaded region is

$$x \ge 1, x \le 5, y \ge 2, \text{ and } y \le 4.$$

We may also combine the first and second inequalities and the third and fourth inequalities and write

$$1 \le x \le 5 \quad \text{and} \quad 2 \le y \le 4.$$

13. The system of linear inequalities that describes the shaded region is

$$2x - y \ge 2, \ 5x + 7y \ge 35, \text{ and } x \le 4.$$

15. The system of linear inequalities that describes the shaded region is $7x + 4y \leq 140$, $x + 3y \geq 30$, and $x - y \geq -10$.

17. The system of linear inequalities that describes the shaded region is $x + y \geq 7$, $x \geq 2$, $y \geq 3$, and $y \leq 7$.

19. The required solution set is shown below.

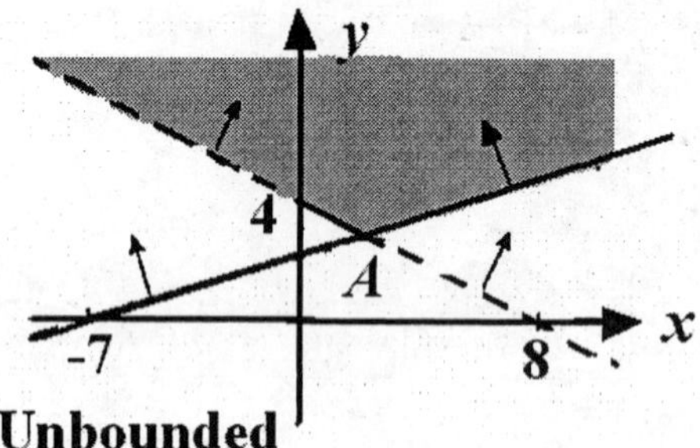

To find the coordinates of A, we solve the system

$$2x + 4y = 16$$
$$-x + 3y = 7,$$

giving $A = (2,3)$. Observe that a dotted line is used to show that no point on the line constitutes a solution to the given problem. Observe that this is an unbounded solution set.

21. The solution set is shown in the figure below. Observe that the set is unbounded.

To find the coordinates of A, we solve the system $\begin{cases} x - y = 0 \\ 2x + 3y = 10 \end{cases}$ giving $A = (2,2)$.

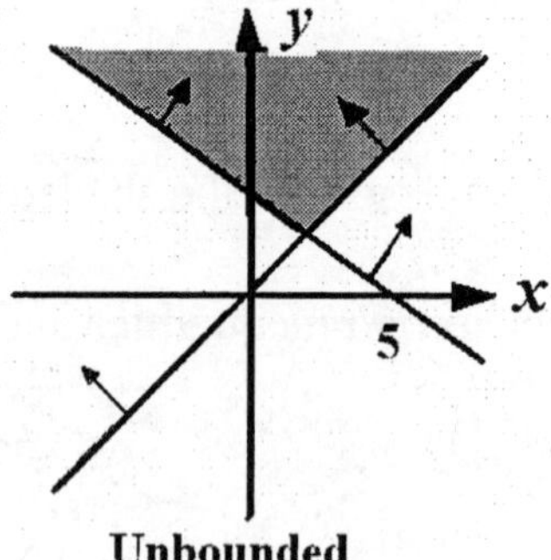

23. The half-planes defined by the two inequalities are shown in the figure that follows. Since the two half-planes have no points in common, we conclude that the given system of inequalities has no solution. (The empty set is a bounded set.)

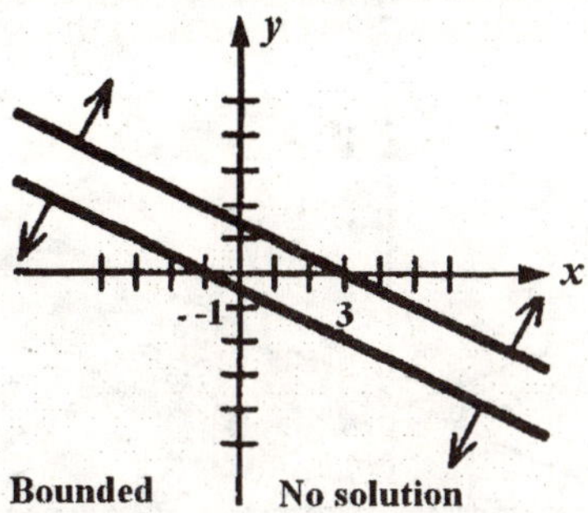

25. The half-planes defined by the three inequalities are shown below. The point A is found by solving the system $\begin{cases} x+y=6 \\ x \quad =3 \end{cases}$ giving $A = (3,3)$. Observe that this is a bounded solution set.

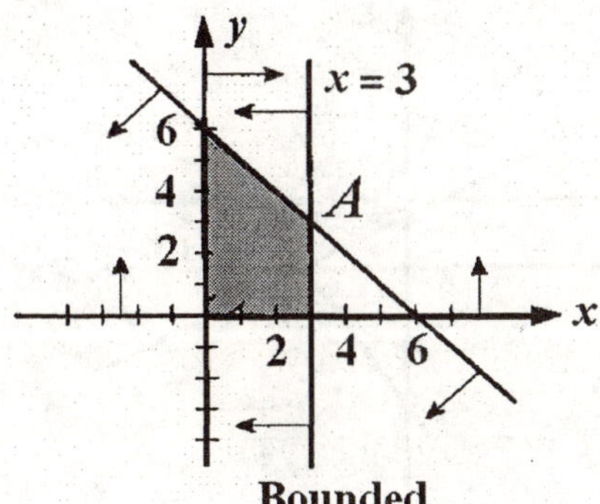

27. The half-planes defined by the given inequalities are shown at the right. Observe that the two lines described by the equations $3x - 6y = 12$ and $-x + 2y = 4$ do not intersect because they are parallel. The solution set is unbounded.

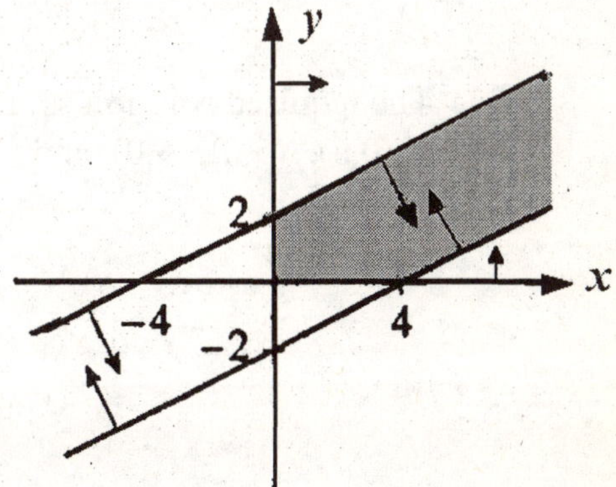

29. The required solution set is shown in the figure that follows. The coordinates of A are found by solving the system $\begin{cases} 3x-7y=-24 \\ x+3y=\ \ \ 8 \end{cases}$ giving $(-1,3)$. The solution set is unbounded.

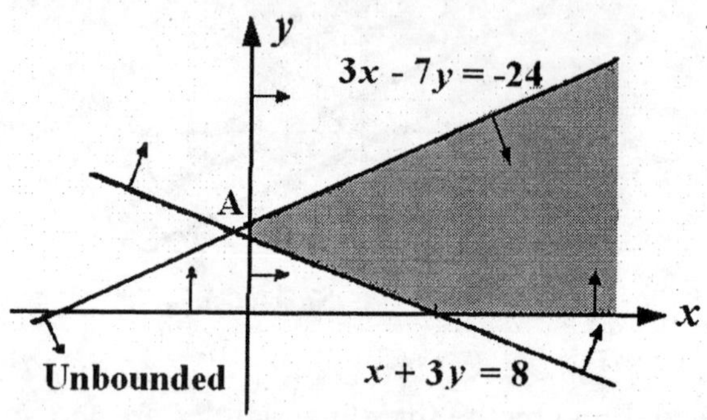

31. The required solution set is shown in the figure below. The solution set is bounded.

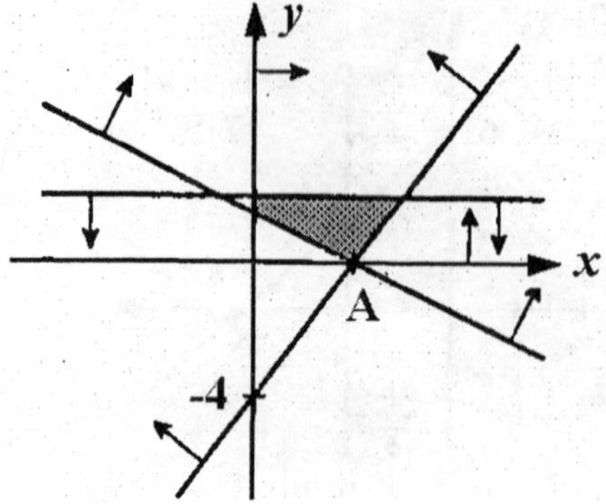

33. The required solution set is shown in the figure below. The solution set has vertices at (0,6), (5,0), (4,0), and (1,3). The solution set is bounded.

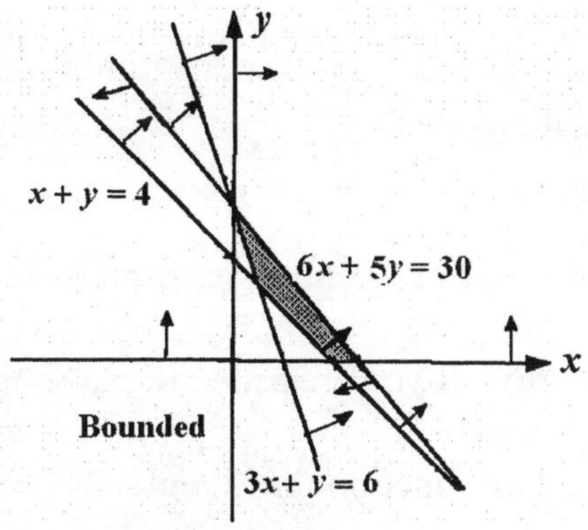

35. The required solution set is shown in the figure below. The unbounded solution set has vertices at (2,8), (0,6), (0,3),and (2,2).

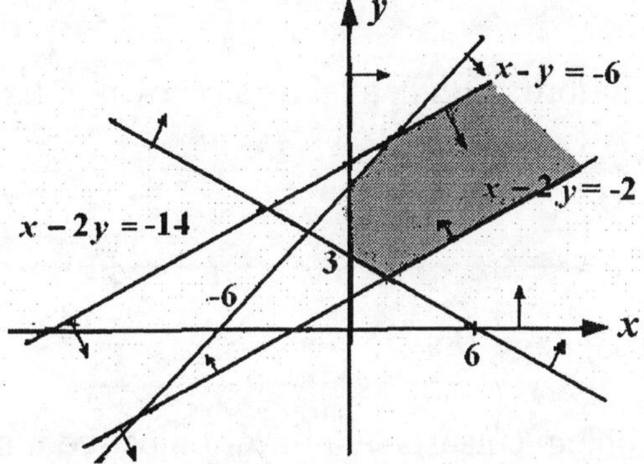

37. False. It is always a half-plane. A straight line is the graph of a linear equation and vice-versa.

39. True. Since a circle can always be enclosed by a rectangle, the solution set of such a system is bounded if it can be enclosed by a rectangle.

3.2 Problem Solving Tips

Here are some hints for solving the problems in the exercises that follow.

1. In a linear programming problem a linear objective function is maximized or

minimized subject to certain constraints. These constraints can be in the form of *linear equations* or *inequalities.*

2. When you solve an applied linear programming problem, it is important to understand the question in mathematical terms. If you are asked to maximize the profit, then the problem involves a linear objective function that should be maximized. If you are asked to minimize the costs, then the problem involves a linear objective function that should be minimized.

3. It's helpful to organize the information in a table, as shown in Examples 1-4, before you formulate the problem.

3.2 CONCEPT QUESTIONS, page 180

1. A linear programming problem consists of a linear objective function to be maximized or minimized subject to certain constraints in the form of linear equations or inequalities.

3. In a maximization linear programming problem, the objective function is to be maximized; whereas, in a minimization problem the objective function is to be minimized.

EXERCISES 3.2, page 181

1. We tabulate the given information:

	Product A	*Product B*	*Time Available*
Machine I	6	9	300
Machine II	5	4	180
Profit per unit ($)	3	4	

Let x and y denote the number of units of Product A and Product B to be produced. Then the required linear programming problem is:

Maximize $P = 3x + 4y$ subject to the constraints

$$6x + 9y \le 300$$
$$5x + 4y \le 180$$
$$x \ge 0, y \ge 0 \quad .$$

3. Let x denote the number of model A grates to be produced and y denote the number of model B grates to be produced. Since only 1000 pounds of cast iron are available, we must have

$$3x + 4y \le 1000.$$

The restriction that only 20 hours of labor are available per day implies that

$$6x + 3y \le 1200. \quad \text{(time in minutes)}$$

Then the profit on the production of these grates is given by

$$P = 2x + 1.5y.$$

Summarizing, we have the following linear programming problem:

Maximize $P = 2x + 1.5y$ subject to

$$3x + 4y \le 1000$$
$$6x + 3y \le 1200$$
$$x \ge 0, y \ge 0$$

5. Suppose the company extends x million dollars in homeowner loans and y million dollars in automobile loans. Then, the returns on these loans are given by $P = 0.1x + 0.12y$ million dollars. Since the company has a total of \$20 million for these loans, we have $x + y \le 20$. Furthermore, since the total amount of homeowner loans should be greater than or equal to four times the total amount of automobile loans, we have $x \ge 4y$. Therefore, the required linear programming problem is

Maximize $P = 0.1x + 0.12y$ subject to

$$x + \ y \le 20$$
$$x - 4y \ge 0$$
$$x \ge 0, y \ge 0$$

7. Let x denote the number of fully assembled units to be produced daily and let y denote the number of kits to be produced. Then the fraction of the day the fabrication department works on the fully assembled cabinets is $\frac{1}{200}x$. Similarly the fraction of the day the fabrication department works on kits is $\frac{1}{200}y$. Since

the fraction of the day during which the fabrication department is busy cannot exceed one, we must have

$$\frac{1}{200}x+\frac{1}{200}y\le 1.$$

Similarly, the restrictions place on the assembly department leads to the inequality

$$\frac{1}{100}x+\frac{1}{300}y\le 1.$$

The profit (objective) function is $P=50x+40y$. Summarizing, the required linear programming problem is

Maximize $P=50x+40y$ subject to

$$\frac{1}{200}x+\frac{1}{200}y\le 1$$

$$\frac{1}{100}x+\frac{1}{300}y\le 1$$

$$x\ge 0,\ y\ge 0.$$

9. Let x and y denote the number of days the Saddle Mine and the Horseshoe Mine are operated, respectively. Then the operating cost is $C = 14{,}000x + 16{,}000y$. The amount of gold produced in the two mines is $(50x+75y)$ oz, and this amount must be at least 650 oz. So we have $50x + 75y \ge 650$. Similarly, the requirement for silver production leads to the inequality $3000x + 1000y \ge 18{,}000$. So the problem is

Minimize $C = 14{,}000x + 16{,}000y$ subject to

$$50x+\ \ 75y\ge 650$$

$$3000x+1000y\ge 18{,}000$$

$$x\ge 0,\ y\ge 0$$

11. Let x and y denote the amount of food A and food B, respectively, used to prepare a meal. Then the requirement that the meal contain a minimum of 400 mg of calcium implies $30x + 25y \ge 400$. Similarly, the requirements that the meal contain at least 10 mg of iron and 40 mg of vitamin C imply that $\begin{cases} x+0.5y\ge 10 \\ 2x+5y\ge 40 \end{cases}$. The cholesterol content is given by $C = 2x + 5y$. Therefore, the linear programming problem is

Minimize $C = 2x + 5y$ subject to

$$30x + 25y \geq 400$$
$$x + 0.5y \geq 10$$
$$2x + \;\; 5y \geq 40$$
$$x \geq 0, y \geq 0$$

13. Let x and y denote the number of advertisements to be placed in newspaper I and newspaper II, respectively. Then the problem is

Minimize $C = 1000x + 800y$ subject to

$$70,000x + 10,000y \geq 2,000,000$$
$$40,000x + 20,000y \geq 1,400,000$$
$$20,000x + 40,000y \geq 1,000,000$$
$$x \geq 0, y \geq 0$$

15. Let x, y, and z denote the amount of money she invests in project A, project B, and project C, respectively. Since she plans to invest up to $2 million, we must have $x + y + z \leq 2,000,000$. Because she decides to put not more than 20% of her total investment in project C, we have

$$z \leq 0.2(x + y + z) \quad \text{or} \quad -0.2x - 0.2y + 0.8z \leq 0$$

Since her investments in project B and C should not exceed 60% of her total investment, we have

$$y + z \leq 0.6(x + y + z) \quad \text{or} \quad -0.6x + 0.4y + 0.4z \leq 0$$

Also, since her investment in project A should be at least 60% of her investments in projects B and C, we have

$$x \geq 0.6(y + z) \quad \text{or} \quad -x + 0.6y + 0.6z \leq 0$$

Finally, the returns on her investments are given by $P = 0.1x + 0.15y + 0.2z$.
To summarize, the problem is

Maximize $P = 0.1x + 0.15y + 0.2z$ subject to

$$x + \;\; y + \;\; z \leq 2,000,000$$
$$-2x - 2y + 8z \leq 0$$
$$-6x + 4y + 4z \leq 0$$
$$-10x + 6y + 6z \leq 0$$
$$x \geq 0, y \geq 0, \; z \geq 0$$

17. Let x, y, and z denote the number of units produced of products A, B, and C,

respectively. From the given information, we formulate the following linear programming problem:

Maximize $P = 18x + 12y + 15z$ subject to

$$2x + y + 2z \le 900$$
$$3x + y + 2z \le 1080$$
$$2x + 2y + z \le 840$$
$$x \ge 0, y \ge 0, z \ge 0$$

19. We first tabulate the given information:

Department	*Model A*	*Model B*	*Model C*	*Time Available*
Fabrication	$\frac{5}{4}$	$\frac{3}{2}$	$\frac{3}{2}$	310
Assembly	1	1	$\frac{3}{4}$	205
Finishing	1	1	$\frac{1}{2}$	190

Let x, y, and z denote the number of units of model A, model B, and model C to be produced, respectively. Then the required linear programming problem is

Maximize $P = 26x + 28y + 24z$ subject to

$$\tfrac{5}{4}x + \tfrac{3}{2}y + \tfrac{3}{2}z \le 310$$
$$x + y + \tfrac{3}{4}z \le 205$$
$$x + y + \tfrac{1}{2}z \le 190$$
$$x \ge 0, y \ge 0, z \ge 0$$

21. The shipping costs are tabulated in the following table.

	Warehouse A	*Warehouse B*	*Warehouse C*
Plant I	60	60	80
Plant II	80	70	50

Letting x_1 denote the number of pianos shipped from plant I to warehouse A, x_2 the number of pianos shipped from plant I to warehouse B, and so we have

	Warehouse A	*Warehouse B*	*Warehouse C*	*Maximum Production*
Plant I	x_1	x_2	x_3	300
Plant II	x_4	x_5	x_6	250
Minimum Requirement	200	150	200	

From the two tables we see that the total monthly shipping cost is given by

$$C = 60x_1 + 60x_2 + 80x_3 + 80x_4 + 70x_5 + 50x_6.$$

Next, the production constraints on plants *I* and *II* lead to the inequalities

$$x_1 + x_4 \geq 200$$
$$x_2 + x_5 \geq 150$$
$$x_3 + x_6 \geq 200$$

Summarizing we have the following linear programming problem:

Minimize $C = 60x_1 + 60x_2 + 80x_3 + 80x_4 + 70x_5 + 50x_6$ subject to

$$x_1 + x_2 + x_3 \leq 300$$
$$x_4 + x_5 + x_6 \leq 250$$
$$x_1 + x_4 \geq 200$$
$$x_2 + x_5 \geq 150$$
$$x_3 + x_6 \geq 200$$
$$x_1 \geq 0,\ x_2 \geq 0,\ \ldots,\ x_6 \geq 0$$

23. The given data can be summarized as follows:

	Concentrates			
	Pineapple	*Orange*	*Banana*	*Profit ($)*
Pineapple-orange	8	8	0	1
Orange-banana	0	12	4	0.80
Pineapple-orange-banana	4	8	4	0.90
Maximum available (oz)	16,000	24,000	5,000	

Suppose x, y, and z cartons of pineapple-orange, orange-banana, and pineapple-orange-banana juice are to be produced, respectively. The linear programming problem is

$$\begin{aligned} \text{Maximize } P = x + 0.8y + 0.9z \text{ subject to}& \\ 8x + \qquad 4z &\le 16{,}000 \\ 8x + 12y + 8z &\le 24{,}000 \\ 4y + 4z &\le 5{,}000 \\ z &\le 800 \\ x \ge 0, y \ge 0,\ z &\le 0 \end{aligned}$$

25. False. The objective function $P = xy$ is not a linear function in x and y.

3.3 Problem Solving Tips

Here are some hints for solving the problems in the exercises that follow.

1. To solve a linear programming problem using the Method of Corners, (a) graph the feasible set, (b) find the coordinates of all corner points (vertices) of the feasible set, and (c) evaluate the objective function at each corner point. Then check to see which vertex yields the maximum (or minimum). If only one vertex yields the maximum (or minimum) then the problem has a unique solution. If there are two adjacent vertices that yield the same maximum (or minimum), then any point lying on the line segment joining these two vertices is a solution.

2. It's helpful to set up a table, as shown in Examples 1-3, to evaluate the objective function at each vertex.

3.3 CONCEPT QUESTIONS, page 192

1. a. The feasible set is the set of points satisfying the constraints associated with the linear programming problem.

linear programming problem.

b. A feasible solution of a linear programming problem is a point in the feasible set.

c. An optimal solution of a linear programming problem is a feasible solution that also optimizes (maximizes or minimizes) the objective function.

EXERCISES 3.3, page 192

1. Evaluating the objective function at each of the corner points we obtain the following table.

Vertex	$Z = 2x + 3y$
(1,1)	$\boxed{5}$
(8,5)	31
(4,9)	$\boxed{35}$
(2,8)	28

From the table, we conclude that the maximum value of Z is 35 and it occurs at the vertex (4,9). The minimum value of Z is 5 and it occurs at the vertex (1,1).

3. Evaluating the objective function at each of the corner points we obtain the following table.

Vertex	$Z = 3x + 4y$
(0,20)	80
(3,10)	49
(4,6)	36
(9,0)	$\boxed{27}$

From the graph, we conclude that there is no maximum value since Z is unbounded. The minimum value of Z is 27 and it occurs at the vertex (9,0).

5. Evaluating the objective function at each of the corner points we obtain the following table.

From the table, we conclude that the maximum value of Z is 44 and it occurs at every point on the line segment joining the points (4,10) and (12,8). The minimum value of Z is 15 and it occurs at the vertex (15,0).

7. The problem is to maximize $P = 2x + 3y$ subject to

$$x + y \leq 6$$
$$x \leq 3$$
$$x \geq 0, y \geq 0$$

The feasible set S for the problem is shown in the following figure, and the values of the function P at the vertices of S are summarized in the accompanying table.

Vertex	$P = 2x + 3y$
A(0,0)	0
B(3,0)	6
C(3,3)	15
D(0,6)	18

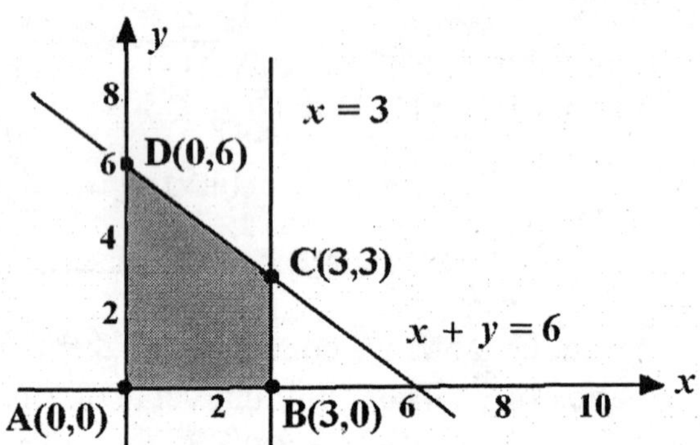

We conclude that P attains a maximum value of 18 when $x = 0$ and $y = 6$.

9. The problem is to maximize $P = 2x + y$ subject to

$$x + y \leq 4$$
$$2x + y \leq 5$$
$$x \geq 0, y \geq 0$$

Referring to the following figure and table

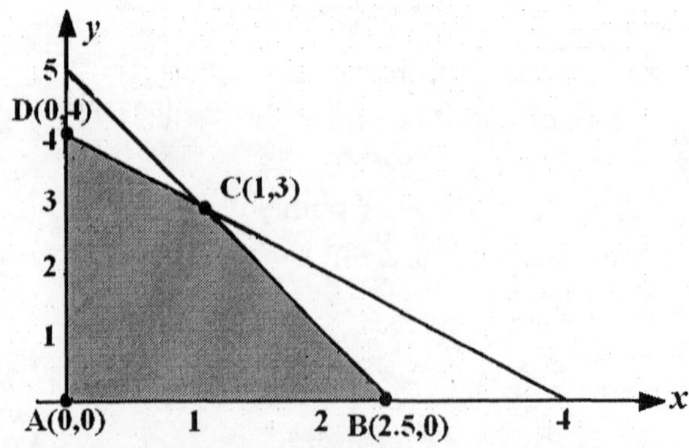

V	$P = 2x + y$
(0,0)	0
(2.5,0)	5
(1,3)	5
(0,4)	4

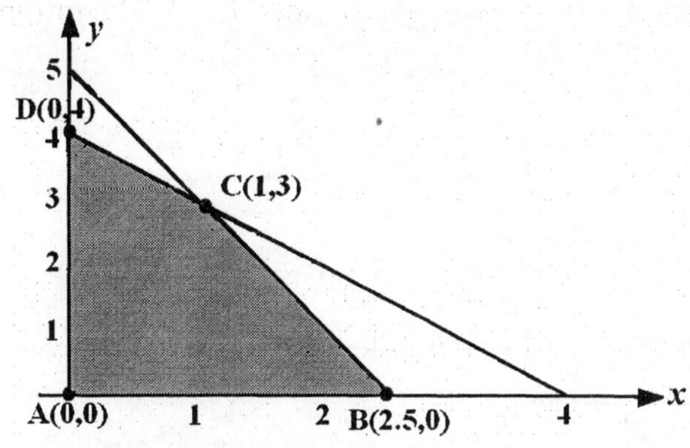

V	$P = 2x + y$
(0,0)	0
(2.5,0)	5
(1,3)	5
(0,4)	4

we conclude that P attains a maximum value of 5 at any point (x, y) lying on the line segment joining (1, 3) to (2.5,0).

11. The problem is

$$\text{Maximize } P = x + 8y \text{ subject to}$$
$$x + y \leq 8$$
$$2x + y \leq 10$$
$$x \geq 0, y \geq 0$$

From the following figure and table

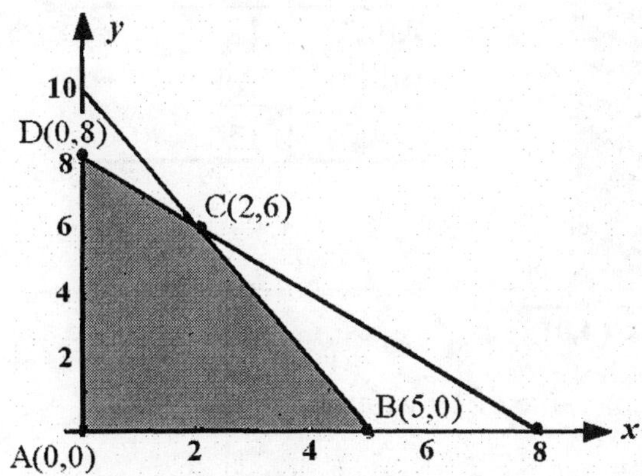

V	$P = x + 8y$
A(0,0)	0
B(5,0)	5
C(2,6)	50
D(0,8)	64

we conclude that P attains a maximum value of 64 when $x = 0$ and $y = 8$.

13. The linear programming problem is

$$\text{Maximize } P = x + 3y \text{ subject to}$$

V	$P = x + 3y$
A(0,0)	0
B(1,0)	1
C(1,3)	10
D(0,4)	12

we conclude that P attains a maximum value of 12 when $x = 0$ and $y = 4$.

15. The linear programming problem is

$$\text{Minimize } C = 3x + 4y \text{ subject to}$$
$$x + \ y \geq 3$$
$$x + 2y \geq 4$$
$$x \geq 0, y \geq 0$$

From the following figure and table,

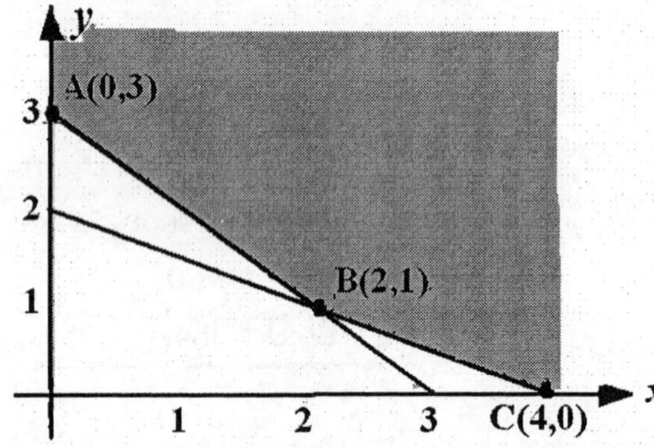

V	$C = 3x + 4y$
A(0,3)	12
B(2,1)	10
C(4,0)	12

we conclude that C attains a minimum value of 10 when $x = 2$ and $y = 1$.

17. The linear programming problem is

$$\text{Minimize } C = 3x + 6y \text{ subject to}$$

$$x + 2y \geq 40$$
$$x + y \geq 30$$
$$x \geq 0, y \geq 0$$

From the following figure and table,

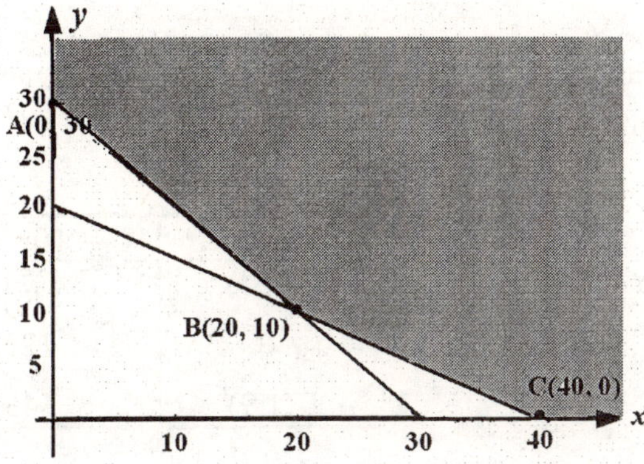

V	$C = 3x + 6y$
A(0,30)	180
B(20,10)	120
C(40,0)	120

we conclude that C attains a minimum value of 120 at any point on the line segment joining (20,10) to (40,0).

19. The problem is

Minimize $C = 2x + 10y$ subject to

$$5x + 2y \geq 40$$
$$x + 2y \geq 20$$
$$y \geq 3,\ x \geq 0$$

The feasible set S for the problem is shown in the following figure and the values of the function C at the vertices of S are summarized in the accompanying table.

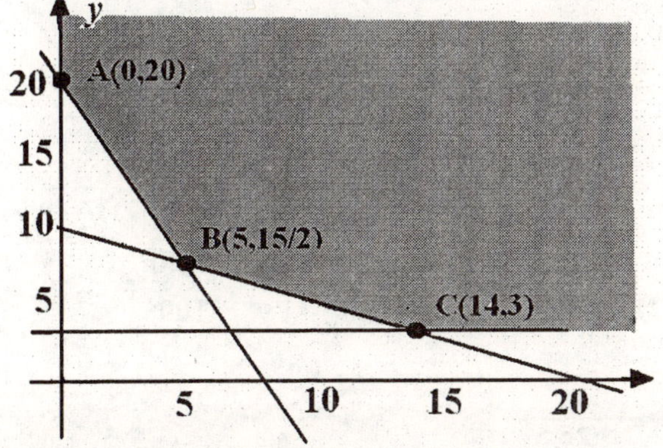

Vertex	$C = 2x + 10y$
A(0,20)	200
B(5, $\frac{15}{2}$)	85
C(14,3)	$\boxed{58}$

We conclude that C attains a minimum value of 58 when $x = 14$ and $y = 3$.

21. The problem is to minimize $C = 10x + 15y$ subject to

$$\begin{aligned} x + y &\le 10 \\ 3x + \ \ y &\ge 12 \\ -2x + 3y &\ge 3 \\ x \ge 0,\ y &\ge 0 \end{aligned}$$

The feasible set is shown in the following figure, and the values of C at each of the vertices of S are shown in the accompanying table.

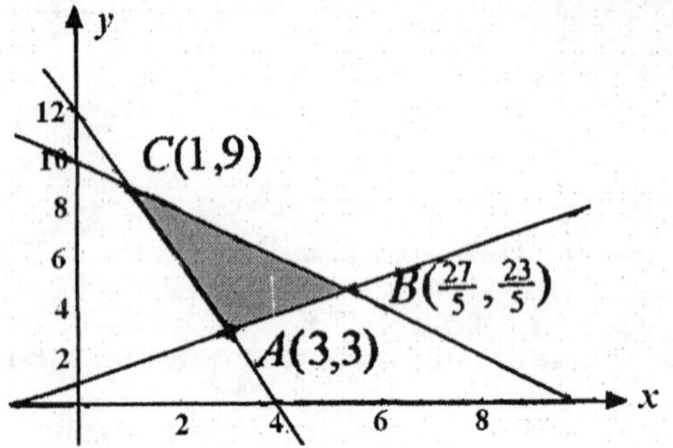

Vertex	$C = 10x + 15y$
A(3,3)	$\boxed{75}$
$B(\frac{27}{5}, \frac{23}{5})$	123
C(1,9)	145

We conclude that C attains a minimum value of 75 when $x = 3$ and $y = 3$.

23. The problem is to maximize $P = 3x + 4y$ subject to

$$\begin{aligned} x + 2y &\le 50 \\ 5x + 4y &\le 145 \\ 2x + \ \ y &\ge 25 \\ y \ge 5, x &\ge 0 \end{aligned}$$

The feasible set S is shown in the figure that follows, and the values of P at each of the vertices of S are shown in the accompanying table.

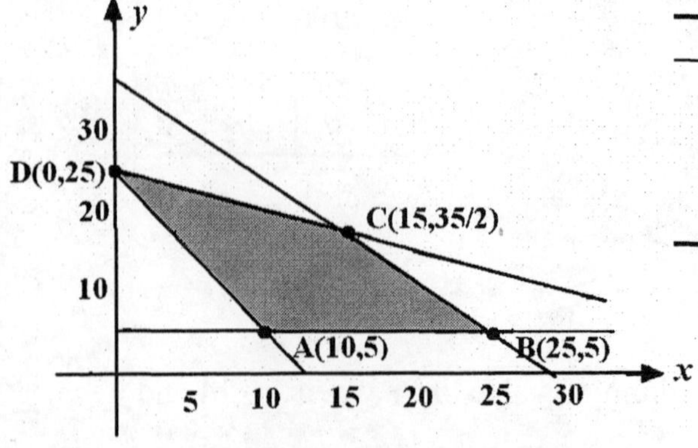

Vertex	$P = 3x + 4y$
A(10,5)	50
B(25,5)	95
$C(15, \frac{35}{2})$	$\boxed{115}$
D(0,25)	100

We conclude that P attains a maximum value of 115 when $x = 15$ and $y = 35/2$.

25. The problem is to maximize $P = 2x + 3y$
subject to

$$\begin{aligned} x + y &\le 48 \\ x + 3y &\ge 60 \\ 9x + 5y &\le 320 \\ x \ge 10, y &\ge 0 \end{aligned}$$

The feasible set S is shown in the figure that follows, and the values of P at each of the vertices of S are shown in the accompanying table.

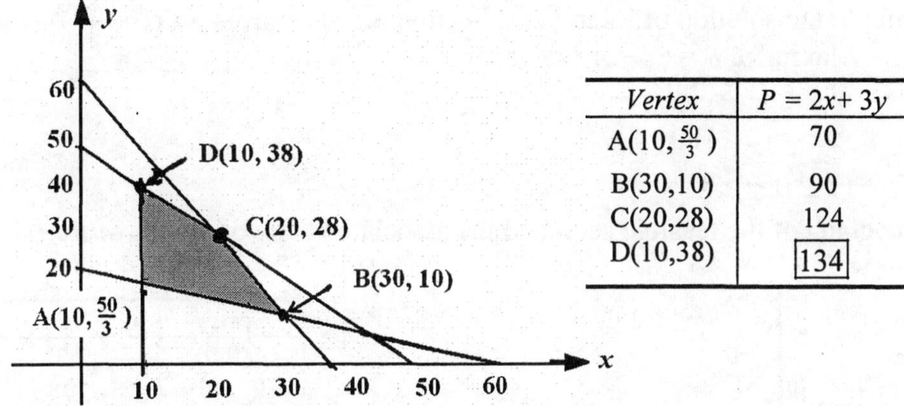

Vertex	$P = 2x + 3y$
A(10, $\frac{50}{3}$)	70
B(30,10)	90
C(20,28)	124
D(10,38)	134

We conclude that P attains a maximum value of 134 when $x = 10$ and $y = 38$.

27. The problem is to find the maximum and minimum value of $P = 10x + 12y$ subject to

$$\begin{aligned} 5x + 2y &\ge 63 \\ x + y &\ge 18 \\ 3x + 2y &\le 51 \\ x \ge 0, y &\ge 0 \end{aligned}$$

The feasible set follows and the values of P at each of the vertices of S are shown in the accompanying table.

Vertex	$P = 10x + 12y$
A(9,9)	198
B(15,3)	186
C(6, $\frac{33}{2}$)	258

P attains a maximum value of 258 when $x = 6$ and $y = 33/2$. The minimum value of P is 186. It is attained when $x = 15$ and $y = 3$.

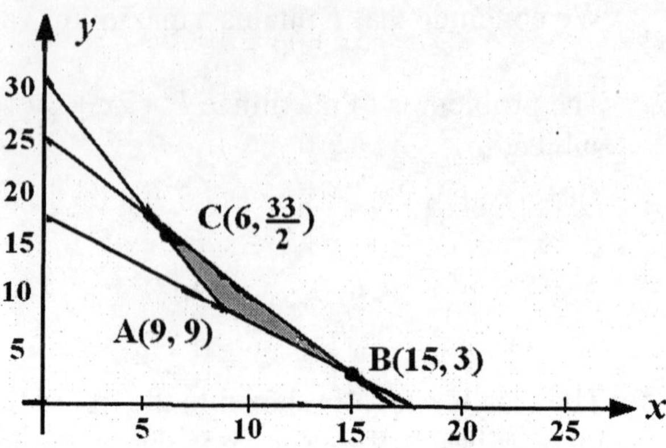

29. Refer to the solution of Exercise 1, Section 3.2, The problem is

Maximize $P = 3x + 4y$ subject to

$$6x + 9y \le 300$$

$$5x + 4y \le 180$$

$$x \ge 0, y \ge 0$$

The graph of the feasible set S and the associated table of values of P follow.

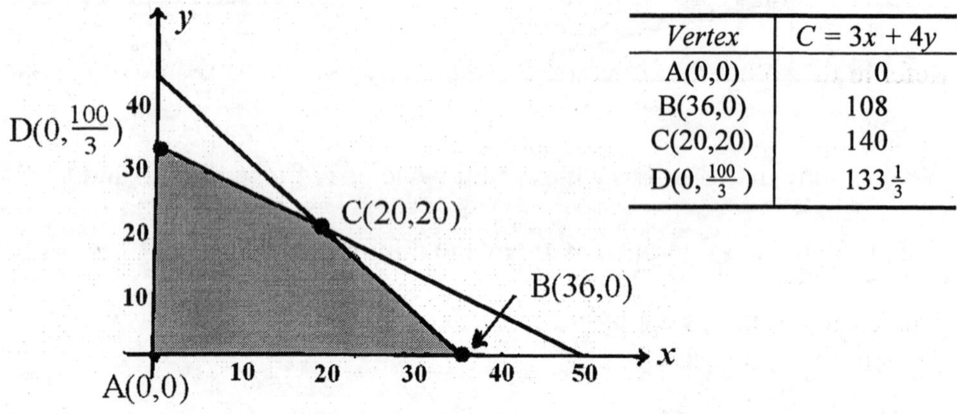

Vertex	$C = 3x + 4y$
A(0,0)	0
B(36,0)	108
C(20,20)	140
D(0, $\frac{100}{3}$)	$133\frac{1}{3}$

P attains a maximum value of 140 when $x = y = 20$. Thus, by producing 20 units of each product in each shift, the company will realize an optimal profit of \$140.

31. Refer to the solution of Exercise 3, Section 3.2, The problem is

Maximize $P = 2x + 1.5y$ subject to

$$3x + 4y \le 1000$$
$$6x + 3y \le 1200$$
$$x \ge 0,\ y \ge 0$$

The graph of the feasible set S and the associated table of values of P follow.

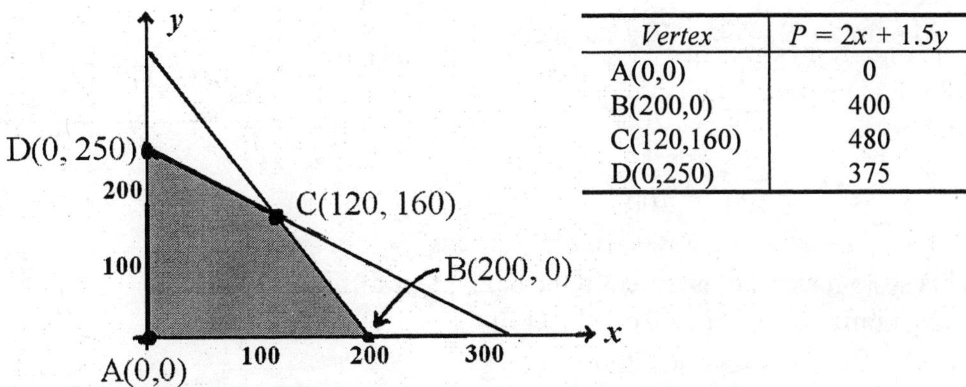

Vertex	$P = 2x + 1.5y$
A(0,0)	0
B(200,0)	400
C(120,160)	480
D(0,250)	375

P attains a maximum value of 480 when $x = 120$ and $y = 160$. Thus, by producing 120 model A grates and 160 model B grates in each shift, the company will realize an optimal profit of \$480.

33. Refer to the solution of Exercise 5, Section 3.2. The linear programming problem is

Maximize $P = 0.1x + 0.12y$ subject to
$$x + \ y \le 20$$
$$x - 4y \ge 0$$
$$x \ge 0,\ y \ge 0$$

The feasible set S for the problem is shown in the figure at the right, and the value of P at each of the vertices of S is shown in the accompanying table.

Vertex	$C = 0.1x + 0.12y$
A(0,0)	0
B(16,4)	2.08
C(20,0)	2.00

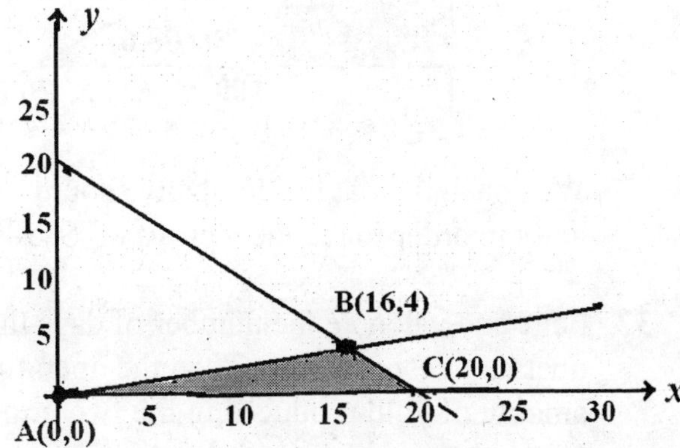

The maximum value of P is attained when $x = 16$ and $y = 4$. Thus, by extending \$16 million in housing loans and \$4 million in automobile loans, the company will realize a return of \$2.08 million on its loans.

35. Refer to Exercise 7, Section 3.2. The problem is

Maximize $P = 50x + 40y$ subject to

$$\frac{1}{200}x + \frac{1}{200}y \le 1$$
$$\frac{1}{100}x + \frac{1}{300}y \le 1$$
$$x \ge 0,\ y \ge 0.$$

This system may be rewritten in the equivalent form

Maximize $P = 50x + 40y$ subject to

$$x + y \le 200$$
$$3x + y \le 300$$
$$x \le 0,\ y \le 0$$

The graph of the feasible set S and the associated table of values of P follow.

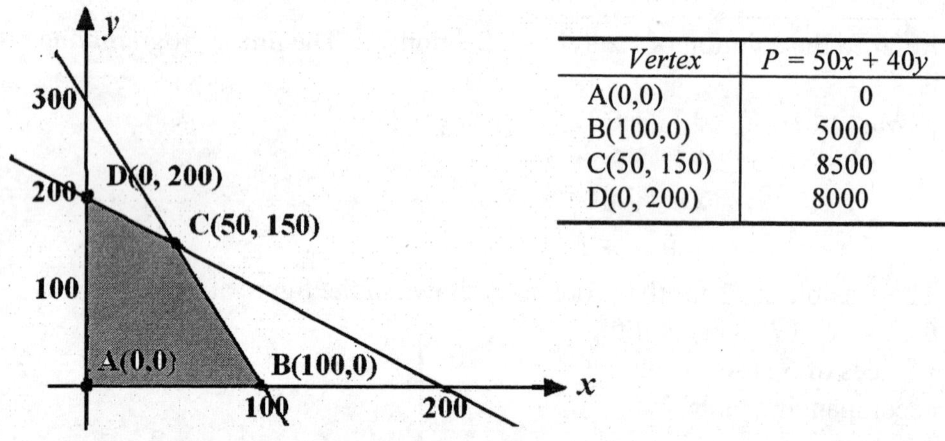

Vertex	$P = 50x + 40y$
A(0,0)	0
B(100,0)	5000
C(50, 150)	8500
D(0, 200)	8000

We conclude that the company should produce 50 fully assembled units and 150 kits daily in order to realize a profit of \$8500.

37. Let x and y denote the number of days the Saddle Mine and the Horseshoe Mine are operated, respectively. Then the operating cost is $C = 14{,}000x + 16{,}000y$. The amount of gold produced in the two mines is $(50x + 75y)$ oz, and this amount must

be at least 650 oz. So we have $50x + 75y \geq 650$. Similarly, the requirement for silver production leads to the inequality $3000x + 1000y \geq 18{,}000$. So the problem is

Minimize $C = 14{,}000x + 16{,}000y$ subject to

$$50x + 75y \geq 650$$
$$3000x + 1000y \geq 18{,}000$$
$$x \geq 0,\ y \geq 0$$

The feasible set is shown in the accompanying figure. From the table, we see that the minimum value of $C = 152{,}000$ is attained at $x = 4$ and $y = 6$. So, the Saddle Mine should be operated for 4 days and the Horseshoe Mine should be operated for 6 days at a minimum cost of \$152,000/day.

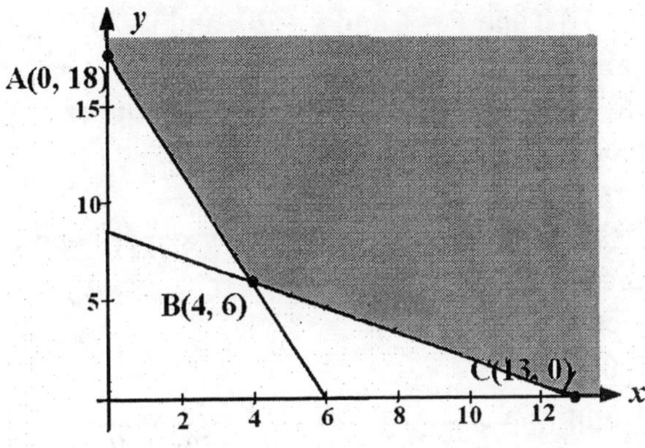

Vertex	$C = 14{,}000x + 16{,}000y$
A(0, 18)	288,000
B(4, 6)	152,000
C(13, 0)	182,000

39. Refer to Exercise 11, Section 3.2. The problem is

Minimize $C = 2x + 5y$ subject to

$$30x + 25y \geq 400$$
$$x + 0.5y \geq 10$$
$$2x + 5y \geq 40$$
$$x \geq 0,\ y \geq 0$$

The graph of the feasible set S and the associated table of values of C follow.

Vertex	$C = 2x + 5y$
A(0,20)	100
B(5,10)	60
C(10,4)	40
D(20,0)	40

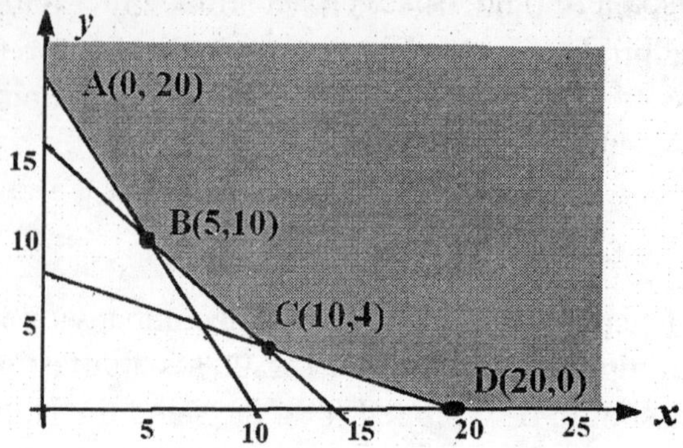

C attains a minimum value of 40 when $x = 10$ and $y = 4$ and $x = 20$, and $y = 0$. This means that any point lying on the line joining the points (10,4) and (20,0) will satisfy these constraints. For example, we could use 10 ounces of food A and 4 ounces of food B, or we could use 20 ounces of food A and zero ounces of food B.

41. Let x and y denote the number of advertisements to be placed in newspaper I and newspaper II, respectively. Then the problem is

Minimize $C = 1000x + 800y$ subject to

$$70,000x + 10,000y \geq 2,000,000$$
$$40,000x + 20,000y \geq 1,400,000$$
$$20,000x + 40,000y \geq 1,000,000$$
$$x \geq 0, y \geq 0$$

The feasible set is shown in the accompanying figure.

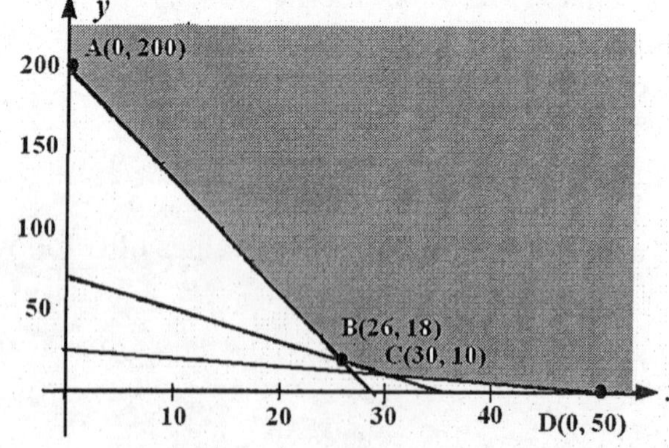

V	$C = 1000x + 800y$
A(0,200)	200,000
B(26,18)	40,400
C(30,10)	38,000
D(0,50)	40,000

From the table, we see that the minimum value of C of 38,000 is attained at $x = 30$ and $y = 10$. Thus, Everest Deluxe World Travel should place 30 advertisements in newspaper I, and 10 advertisements in newspaper II at a total (minimum) cost of $38,000.

43. The problem is

Minimize $C = 14{,}500 - 20x - 10y$ subject to

$$x + y \ge 40$$
$$x + y \le 100$$
$$0 \le x \le 80$$
$$0 \le y \le 70$$

The feasible set S for the problem is shown in the figure at the right, and the value of C at each of the vertices of S is given in the accompanying table.

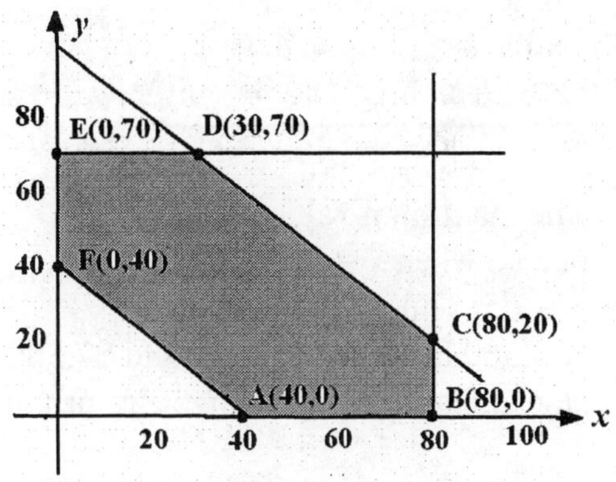

Vertex	$C = 14{,}500 - 20x - 10y$
A(40,0)	13,700
B(80,0)	12,900
C(80,20)	12,700
D(30,70)	13,200
E(0,70)	13,800
F(0,40)	14,100

We conclude that the minimum value of C occurs when $x = 80$ and $y = 20$. Thus, 80 engines should be shipped from plant I to assembly plant A, and 20 engines should be shipped from plant I to assembly plant B; whereas

$$(80 - x) = 80 - 80 = 0, \quad \text{and} \quad (70 - y) = 70 - 20 = 50$$

engines should be shipped from plant II to assembly plants A and B, respectively, at a total cost of $12,700.

45. Let x denote Patricia's investment in growth stocks and y denote the value of her investment in speculative stocks, where both x and y are measured in thousands of dollars. Then the return on her investments is given by $P = 0.15x + 0.25y$. Since her investment may not exceed $30,000, we have the constraint $x + y \le 30$. The condition that her investment in growth stocks be at least 3 times as much as her investment in speculative stocks translates into the inequality $x \ge 3y$. Thus, we have

the following linear programming problem:

$$\text{Maximize } P = 0.15x + 0.25y \text{ subject to}$$
$$x + y \le 30$$
$$x - 3y \ge 0$$
$$x \ge 0, \; y \ge 0$$

The graph of the feasible set S is shown in the figure at the right and the value of P at each of the vertices of S is shown in the accompanying table.

Vertex	$C = 0.15x + 0.25y$
A(0,0)	0
B(30,0)	4.5
$C(\frac{45}{2}, \frac{15}{2})$	5.25

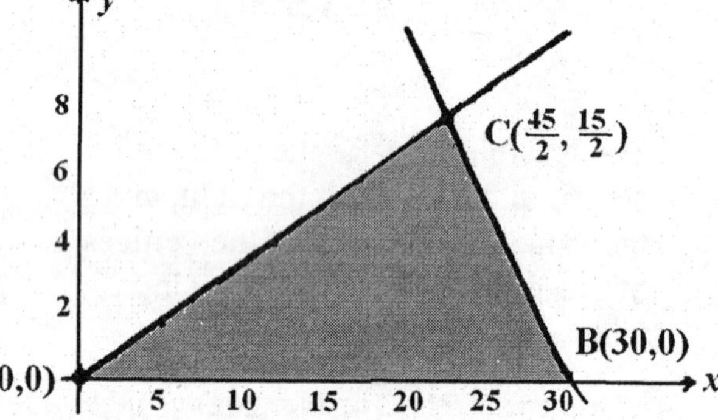

The maximum value of P occurs when $x = 22.5$ and $y = 7.5$. Thus, by investing \$22,500 in growth stocks and \$7,500 in speculative stocks. Patricia will realize a return of \$5250 on her investments.

47. Let x denote the number of urban families and let y denote the number of suburban families interviewed by the company. Then, the amount of money paid to Trendex will be

$$P = 6000 + 8(x + y) - 4.4x - 5y = 6000 + 3.6x + 3y.$$

Since a maximum of 1500 families are to be interviewed, we have

$$x + y \le 1500.$$

Next, the condition that at least 500 urban families are to be interviewed translates into the condition $x \ge 500$. Finally the condition that at least half of the families interviewed must be from the suburban area gives

$$y \ge \tfrac{1}{2}(x + y) \quad \text{or} \quad y - x \ge 0$$

Thus, we are led to the following programming problem:

$$\text{Maximize } P = 6000 + 3.6x + 3y \text{ subject to}$$
$$x + y \le 1500$$
$$y - x \ge 0$$
$$x \ge 500, y \ge 0$$

The graph of the feasible set S for this problem follows and the value of P at each of the vertices of S is given in the accompanying table.

Vertex	$P = 6000 + 3.6x + 3y$
A(500,500)	9,300
B(750,750)	10,950
C(500,1000)	10,800

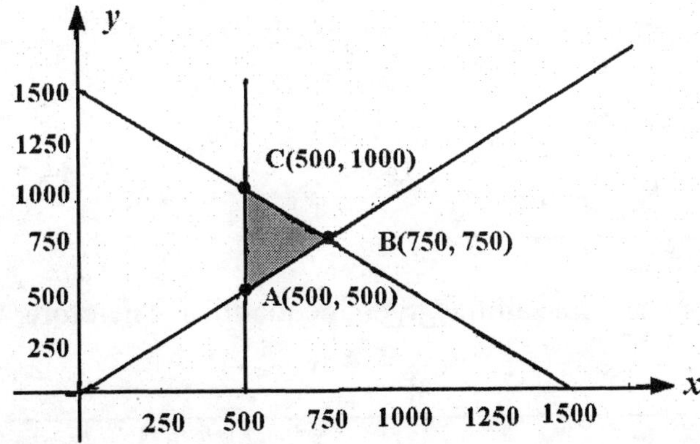

Using the method of corners, we conclude that the profit will be maximized when $x = 750$ and $y = 750$. Thus, a maximum profit of \$10,950 will be realized when 750 urban and 750 suburban families are interviewed.

49. False. It can have either one or infinitely many solutions.

51. a. True. Since $a > 0$, the term ax can be made as large as we please by taking x sufficiently large (because S is unbounded) and therefore P is unbounded as well.
b. True. Maximizing $P = ax + by$ on S is the same as minimizing

$$Q = -P = -(ax + by) = -ax - by = Ax + By,$$

where $A \geq 0$ and $B \geq 0$. Since $x \geq 0$ and $y \geq 0$, the linear function Q, and therefore P, has at least one optimal solution.53. Let $A(x_1, y_1)$ and $B(x_2, y_2)$. Then you can verify that $Q(\bar{x}, \bar{y})$, where

$$\bar{x} = x_1 + t(x_2 - x_1) \quad \text{and} \quad \bar{y} = y_1 + t(y_2 - y_1)$$

and t is a number satisfying $0 < t < 1$. Therefore, the value of P at Q is

$$P = a\bar{x} + b\bar{y} = a[x_1 + t(x_2 - x_1)] + b[y_1 + t(y_2 - y_1)]$$
$$= ax_1 + by_1 + [a(x_2 - x_1) + b(y_2 - y_1)]t.$$

Now, if $c = a(x_2 - x_1) + b(y_2 - y_1) = 0$, then P has the (maximum) value $ax_1 + by_1$ on the line segment joining A and B; that is, the infinitely many solutions lie on this line segment. If $c > 0$, then a point a little to the right of Q will give a larger value of P. Thus, P is not maximal at Q. (Such a point can be found because Q lies in the interior of the line segment). A similar statement holds for the case $c < 0$. Thus, the maximum of P cannot occur at Q unless it occurs in every point on the line segment joining A and B.

55. a.

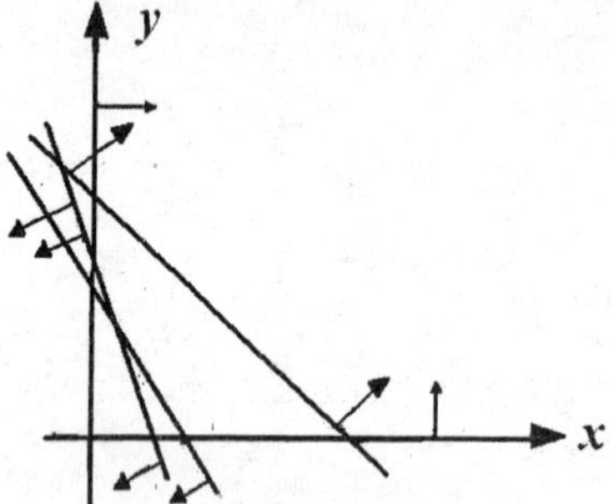

b. There is no point that satisfies all the given inequalities. Therefore, there is no solution.

3.4 Problem Solving Tips

In applied problems, it's important to interpret the mathematical solution of a problem in terms of the real-life problem that you are solving.

1. One consideration in sensitivity analysis is *how changes in the coefficients of the objective function* affect the optimal solution. In Example 1, we optimized the profit function $P = 2x + 1.5y$, and found that the optimal solution would not be changed if the contribution to the profit of a model A grate assumed values between \$1.125 and \$3.00 (changes in the coefficient of x) ; similarly, we found that the contribution to the profit of a model B grate could assume values between \$1.00 and \$2.67 (changes in the coefficient of y) without changing the optimal solution.

2. Another consideration is *how changes to the constants on the right-hand side of the*

constraint inequalities affect the optimal solution. In the example discussed (production problem for Ace Novelty), we found that the profit could be improved from $148.80 to $152.40 by increasing the time available on Machine I by 10 minutes (the change in constraint I).

3.4 CONCEPT QUESTIONS, page 209

1. The coefficients of x and y represent the cost for producing a unit of product A and a unit of product B, respectively.

3. a. The shadow price of the ith resource (associated with the ith constraint of a linear programming problem) is the amount by which the value of the objective function is improved if the right-hand side of the ith constraint is increased by 1 unit.
   b. Binding constraints are constraints which hold with equality at the optimal solution. They cannot be increased without increasing the resources.

EXERCISES 3.4, page 209

1. Refer to the discussion on pages 198-200 in the text.
   a. Suppose the contribution to the profit of each type-B souvenir is $c so that

$$P = x + cy\,,\ \text{or}\ \ y = -\frac{x}{c} + \frac{p}{c}$$

   and the slope of the isoprofit line is $-1/c$. In order for the current optimal solution to be unaffected, this slope must be less than or equal to the slope of the line associated with constraint 2. Thus,

$$-\frac{1}{c} \le -\frac{1}{3},\ \ \frac{1}{c} \ge \frac{1}{3}\ \ \text{, or}\ \ \ c \le 3.$$

   Similarly, we see that the slope of the isoprofit line must be greater than or equal to the slope of the line associated with constraint 1. Thus

$$-\frac{1}{c} \ge -2,\ \ \frac{1}{c} \le 2\ \ \text{or}\ \ c \ge \frac{1}{2}.$$

   Thus, the profit must lie between $0.50 and $3 as was to be shown.
   b. If the profit of a type-A souvenir is $1.50, then the results on page 200 tell us that

the current optimal solution holds. That is, we produce 48 type-A ($x = 48$) and 84 type B ($y = 84$) souvenirs giving a profit of

$$P = 1.5x + 1.2y = 1.5(48) + 1.2(84) = 172.8 \quad \text{or} \quad \$172.80.$$

c. Here the results of part (a) show that the current optimal solution holds. So, with $x = 48$ and $y = 84$, we find

$$P = x + 2y = 48 + 2(84) = 216 \quad \text{or} \quad \$216.$$

3. Refer to the discussion on pages 205-208 in the text.

a. Suppose resource 2 is changed from 1200 to 1200 + k. Then the constraint is changed to

$$6x + 3y = 1200 + k$$

The current optimal solution is shifted to the new optimal solution at the point D' (see the accompanying figure). To find the coordinates of D', we solve the system

$$3x + 4y = 1000$$
$$6x + 3y = 1200 + k$$

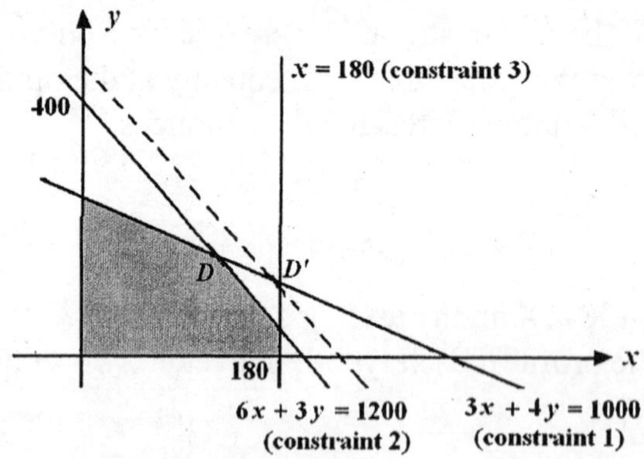

We obtain $x = \frac{4}{15}(450 + k)$ and $y = \frac{1}{5}(800 - k)$. The nonnegative requirements on x and y require that

$$450 + k \geq 0 \quad \text{or} \quad k \geq -450 \quad \text{and} \quad 800 - k \geq 0 \quad \text{or} \quad k \leq 800.$$

Next, the constraint $x \leq 180$ (constraint 3) requires that

$$\frac{4}{15}(450 + k) \leq 180, \quad 450 + k \leq 675, \quad \text{or} \quad k \leq 225.$$

These three inequalities imply that $-450 \leq k \leq 225$. Therefore, resource 2 must lie between 1200 – 450 and 1200 + 225; that is, between 750 and 1425.

b. Suppose resource 3 is changed to 180 + ℓ. Refer to the accompanying figure.

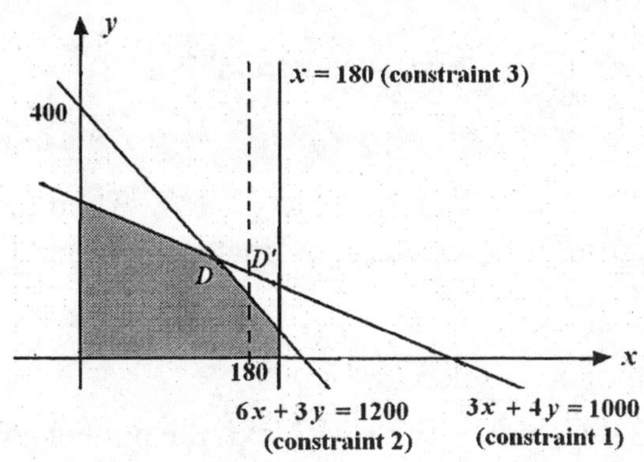

To find the coordinates of D', we solve

$$\begin{aligned} 6x + 3y &= 1200 \\ x &= 180 + \ell \end{aligned}$$

obtaining $x = 180 + \ell$ and $y = 40 - 2\ell$. Now for the current optimal solution to hold, $x = 180 + \ell \geq 120$ or $\ell \geq -60$. Therefore, resource 3 must be greater than or equal to $180 - 60 = 120$; that is, it cannot be decreased by more than 60.

5. a. The feasible set is shown in the accompanying figure. From the table

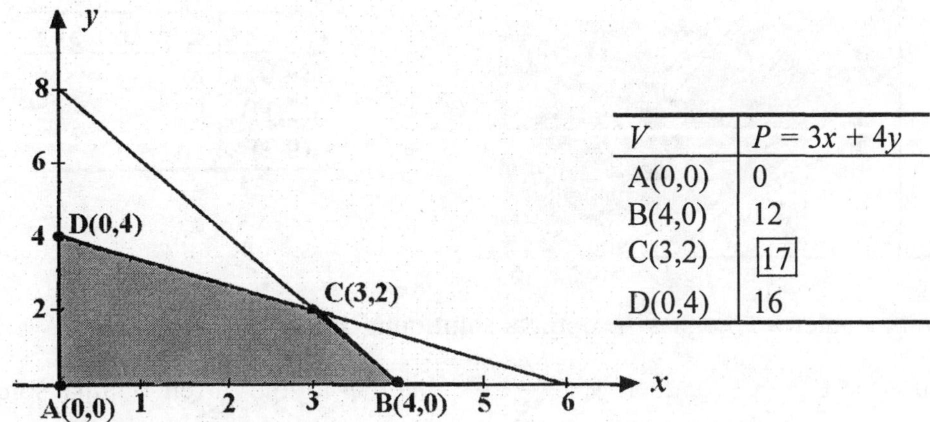

V	$P = 3x + 4y$
A(0,0)	0
B(4,0)	12
C(3,2)	17
D(0,4)	16

we see that the optimal solution is $x = 3, y = 2$, and $P = 17$.

b. Suppose $P = cx + 4y$. then $y = -\frac{c}{4}x + \frac{P}{4}$ and the slope of the isoprofit line is $-\frac{c}{4}$. In order for the current optimal solution to hold, we must have

$$-\frac{c}{4} \leq -\frac{2}{3} \quad \text{or} \quad c \geq \frac{8}{3}. \quad \text{(Slope of constraint 1 is } -\frac{2}{3}\text{)}$$

and $\quad -\frac{c}{4} \geq -2 \quad$ or $\quad c \leq 8.$ (Slope of constraint 2 is -2)

Therefore $\frac{8}{3} \leq c \leq 8$

c. Suppose resource 1 is changed from 12 to 12 + h. Then the optimal solution is moved to a new optimal solution whose coordinates are found by solving the system

$$2x + 3y = 12 + h$$
$$2x + \ y = 8$$

The solutions are $x = 3 - \frac{h}{4}$ and $y = 2 + \frac{h}{2}$. Next, the nonnegativity of x and y imply that $h \geq -4$ and $h \leq 12$. So, $-4 \leq h \leq 12$ and the resource can assume values between 12 – 4 and 12 + 12 or between 8 and 24.

d. If $h = 1$, then $x = 3 - \frac{1}{4}$ and $y = 2 + \frac{1}{2}$. The shadow price is

$$3[3 - \tfrac{1}{4}] + 4[2 + \tfrac{1}{2}] - [3(3) + 4(2)] = -\tfrac{3}{4} + 4(\tfrac{1}{2}) = \tfrac{5}{4}.$$

e. Since constraints 1 and 2 hold with equality at the optimal solution (3,2), they are both binding.

7. a. The feasible set is shown in the accompanying figure.

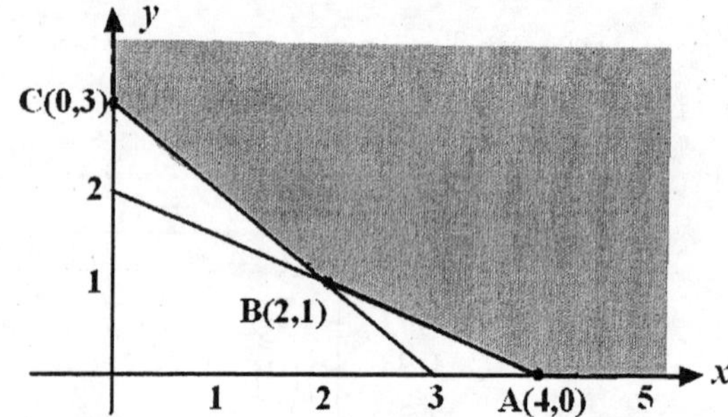

V	$C = 2x + 5y$
A(4,0)	8
B(2,1)	9
C(0,3)	15

From the table, we see that the optimal solution is $x = 4, y = 0$, and $C = 8$.

b. Suppose $C = cx + 5y$. then $y = -\frac{c}{5} + \frac{C}{5}$. In order for the current optimal solution to hold, we must have

$-\frac{c}{5} \geq -\frac{1}{2}, \frac{c}{5} \leq \frac{1}{2}$, or $c \leq \frac{5}{2}$ and $-\frac{c}{5} \geq -1$, $\frac{c}{5} \leq 1$, or $c \leq 5$. Therefore, $0 \leq c \leq \frac{5}{2}$.

NOTE: For a minimization problem the optimal solution is the intersection of the corner point in the feasible set with the isoprofit line nearest the origin.

c. Suppose resource 1 is changed from 4 to 4 + h. Then solving the system

$$x+2y=4+h$$
$$x+\ y=\ 3$$

we obtain $x=2-h$ and $y=1+h.$ We must have $y=1+h\geq 0$ or $h\geq -1$ (see the figure). Therefore, resource 1 can assume values greater than or equal to 4 – 1 or 3.

d. Put $h = 1$. Then we see that the shadow price for resource 1 is

$$[2(5) + 5(0)] - [2(4) + 5(0)] = 2 \qquad (\text{Since } C = 2x + 5y)$$

e. Constraint 1 is binding since equality holds at the optimal solution, but constraint 2 is nonbinding.

9. a. The feasible set is shown in the accompanying figure.

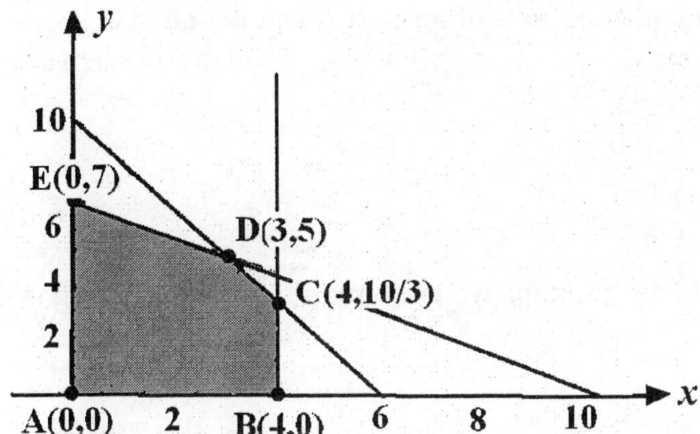

V	$P = 4x + 3y$
A(0,0)	0
B(4,0)	16
C(4, $\frac{10}{3}$)	26
D(3,5)	27
E(0,7)	21

From the table, we see that the optimal solution is $x = 3, y = 5$, and $P = 27$.

b. Suppose $P=cx+3y$ so that $y=-\frac{c}{3}x+\frac{P}{3}$. For the current optimal solution to be unaffected, we must have

$$-\frac{c}{3}\geq -\frac{5}{3},\ \frac{c}{3}\leq \frac{5}{3},\ \text{or}\ \ c\leq 5$$

and
$$-\frac{c}{3}\leq -\frac{2}{3},\ \frac{c}{3}\geq \frac{2}{3},\ \text{or}\ \ c\geq 2.\ \text{Therefore,}\ 2\leq c\leq 5.$$

c. Suppose the right-hand side of the constraint is replaced by $30 + h$. Then we have the system

$$5x+3y=30+h$$
$$2x+3y=21$$

The solutions are $x=3+\frac{h}{3}$ and $y=5-\frac{2h}{9}$. Since $x\geq 0$, we have $\frac{h}{3}+3\geq 0$ or $h\geq -9$. Also, $y\geq 0$ implies $5-\frac{2h}{9}\geq 0$, $\frac{2h}{9}\leq 5$ or $h\leq \frac{45}{2}$. Finally,

$x \le 4$ implies $\frac{h}{3}+3 \le 4$ or $h \le 3$. So $-9 \le h \le 3$ and we see that resource 1 can assume values between 30 – 9 and 30 + 3; that is, between 21 and 33.

d. Putting $h = 1$ in part (*c*) gives $x = 3+\frac{1}{3}=\frac{10}{3}$ and $y-5-\frac{2}{9}=\frac{43}{9}$. Therefore, the shadow price associated with constraint 1 is

$$[4(\tfrac{10}{3})+3(\tfrac{43}{9})]-[4(3)+3(5)]=\tfrac{2}{3}. \qquad \text{(Since } P=4x+3y\text{)}$$

e. The first two constraints are binding since equality holds at the optimal solution . Constraint 3 is nonbinding.

11. a. Let x and y denote the number of units of product A and the number of units of product B to be manufactured, respectively, Then the problem at hand is

$$\begin{aligned} \text{Maximize} \quad & P = 3x+4y \\ \text{subject to} \quad & 6x+9y \le 300 \\ & 5x+4y \le 180 \\ & x \ge 0, y \ge 0 \end{aligned}$$

The feasible set is shown in the accompanying figure. From the following table

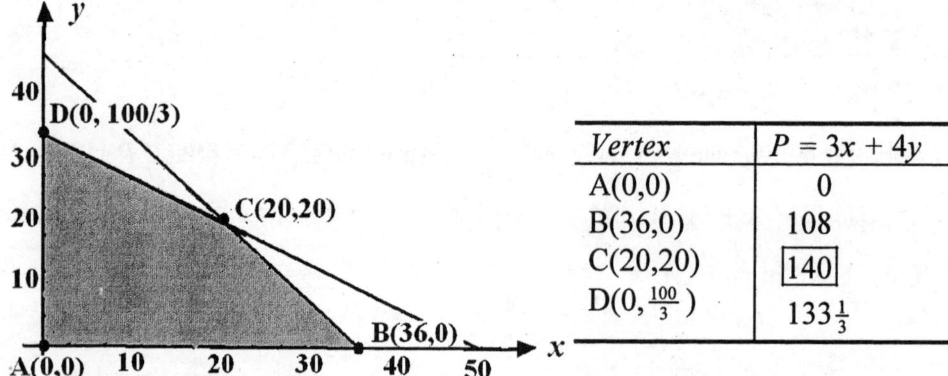

Vertex	$P = 3x + 4y$
A(0,0)	0
B(36,0)	108
C(20,20)	140
D(0, $\frac{100}{3}$)	$133\frac{1}{3}$

we see that the optimal solution is $x = y = 20$, and $P = 140$. So, the company should produce 20 units of each product giving a maximum profit of \$140.

b. Suppose $P = cx+4y$ so that $y=-\frac{c}{4}x+\frac{P}{4}$. In order for the current optimal solution to remain optimal, we must have

$$-\frac{c}{4} \le -\frac{2}{3}, \ \frac{c}{4} \ge \frac{2}{3}, \text{ or } c \ge \frac{8}{3} \text{ and } -\frac{c}{4} \ge -\frac{5}{4}, \ \frac{c}{4} \le \frac{5}{4}, \text{ or } c \le 5.$$

Therefore, $\frac{8}{3} \leq c \leq 5$.

c. Suppose the right-hand side of the constraint is replaced by $300 + h$. Then we have the system

$$6x + 9y = 300 + h$$
$$5x + 4y = 180$$

The solutions are $x = 20 - \dfrac{4}{21}h$ and $y = 20 + \dfrac{5}{21}h$. Now, $x \geq 0$ implies $20 - \frac{4}{21}h \geq 0$, or $h \leq 105$. Next, $y \geq 0$ implies $50 + \frac{5}{21}h \geq 0$, or $h \geq -84$. Therefore, $-84 \leq h \leq 105$. So the resource can assume values between 300 – 84 and 300 + 105; that is, between 216 and 405.

d. Put $h = 1$ in part (*c*) and we find $x = 20 - \frac{4}{21}$, and $y = 20 + \frac{5}{21}h$. Using this result and the result from part (a), we see that the required shadow price is

$$3(20 - \tfrac{4}{21}) + 4(20 + \tfrac{5}{21}) - [3(20) + 4(20)] = \tfrac{8}{21}.$$

13. a. Let x denote the number of days the Saddle Mine is operated, and let y denote the number of days the Horseshoe Mine is operated. Then the problem is

$$\begin{aligned} &\text{Minimize } C = 14{,}000x + 16{,}000y \\ &\text{subject to } \quad 50x + 75y \geq 650 \\ &\qquad 3000x + 1000y \geq 18{,}000 \\ &\qquad x \geq 0, y \geq 0 \end{aligned}$$

The feasible set is shown in the accompanying figure. From the following table

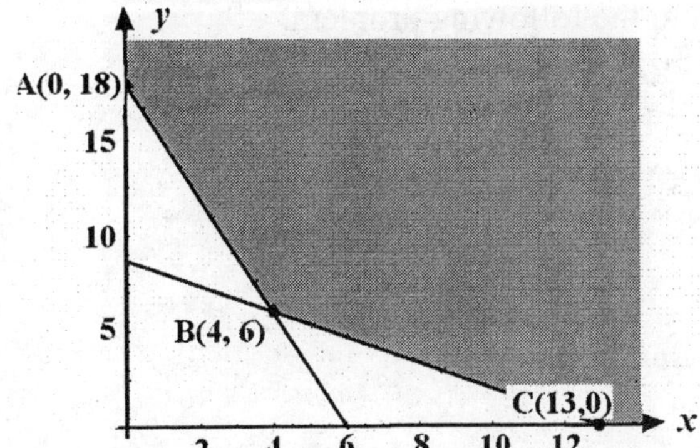

Vertex	$C = 14{,}000x + 16{,}000y$
A(0,18)	288,000
B(4,6)	152,000
C(13,0)	182,000

we see that the optimal solution is $x = 4$, $y = 6$, and $C = 152{,}000$. Therefore, the company should operate the Saddle Mine for 4 days and the Horseshoe Mine for 6 days at a cost of \$152,000.

b. Suppose $C = cx + 16{,}000y$. Then $y = -\dfrac{c}{16000}x + \dfrac{C}{16{,}000}$. In order for the current optimal solution to hold, we must have

$$-\frac{c}{16{,}000} \le -\frac{50}{75}, \frac{c}{16{,}000} \ge \frac{50}{75}, \text{ or } c \ge 10{,}666\tfrac{2}{3}$$

and $-\dfrac{c}{16{,}000} \ge -3, \dfrac{c}{16{,}000} \le 3$, or $c \le 48{,}000$. Therefore, $10{,}666\frac{2}{3} \le c \le 48{,}000$.

c. Suppose the right-hand side of the first constraint (that pertains to the requirements for gold) is changed to 650 + h. Then we have the system

$$\begin{aligned} 50x + 75y &= 650 + h \\ 2000x + 1000y &= 18{,}000 \end{aligned}$$

The solutions are $x = 4 - \frac{h}{175}$ and $y = 6 + \frac{3h}{175}$. Now, $x \ge 0$ implies $4 - \frac{h}{175} \ge 0$, or $h \le 700$. Next, $y \ge 0$ implies $6 + \frac{3h}{175} \ge 0$, or $h \ge -350$. Therefore, $-350 \le h \le 700$ and so the resource can assume values between $650 - 350$ and 650 + 700; that is, between 300 and 1350.

d. Put $h = 1$ in part (a), and we have $x = 4 - \frac{1}{175}$ and $y = 6 + \frac{3}{175}$. Using the fact that the optimal solution found in part (a) is $x = 4$ and $y = 6$, we see that the shadow price is

$$\begin{aligned} &14{,}000(4 - \tfrac{1}{175}) + 16{,}000(6 + \tfrac{3}{175}) - [14{,}000(4) + 16{,}000(6)] \\ &= 194.29, \text{ or } \$194.29. \end{aligned}$$

15. a. Let x denote the number of Model A satellite radios and y the number of Model B satellite radios to be produced. We are led to the following problem.

$$\begin{aligned} \text{Maximize } & P = 12x + 10y \\ \text{subject to } & 15x + 10y \le 1500 \\ & 10x + 12y \le 1320 \\ & x \le 80 \\ & x \ge 0, y \ge 0 \end{aligned}$$

The feasible set is shown in the accompanying figure.

V	$P = 12x + 10y$
A(0,0)	0
B(80,0)	960
C(80,30)	1260
D(60,60)	1320
E(0,110)	1100

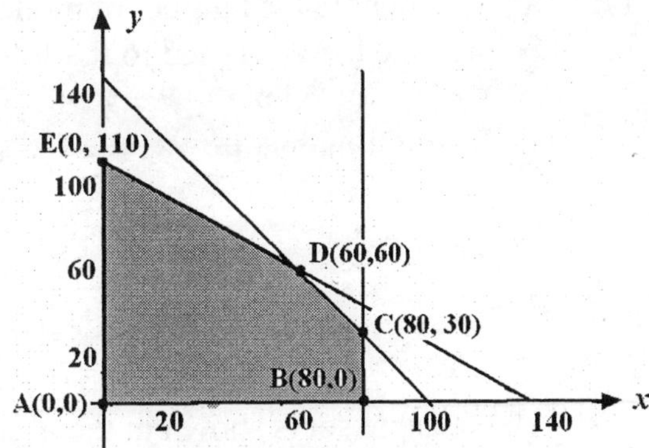

From the table we see that the optimal solution is $x = y = 60$, and $P = 1320$. So Soundex should produce 60 of each model so that they realize a profit of \$1320.

b. Suppose $P = cx + 10y$ so that $y = -\frac{c}{10}x + \frac{P}{10}$. In order for the current optimal solution to hold, we must have

$$-\frac{c}{10} \geq -\frac{3}{2}, \frac{c}{10} \leq \frac{3}{2}, \text{ or } c \leq 15 \text{ and } -\frac{c}{10} \leq -\frac{5}{6}, \frac{c}{10} \geq \frac{5}{6}, \text{ or } c \geq \frac{25}{3}.$$

Therefore, $8\frac{1}{3} \leq c \leq 15$.

c. Suppose the right-hand side of the first constraint (that pertains to the constraint on the use of Machine I) is changed to $1500 + h$. Then, we have the system

$$15x + 10y = 1500 + h$$
$$10x + 12y = 1320.$$

The solutions are $x = 60 + \frac{3}{20}h$ and $y = 60 - \frac{h}{8}$. Now, $x \geq 0$ implies $60 + \frac{3}{20}h \geq 0$, or $h \geq -400$. Next, $y \geq 0$ implies $60 - \frac{h}{8} \geq 0$ or $h \leq 480$. Therefore, $-400 \leq h \leq \frac{400}{3}$ and we see that the values assumed by this resource must lie between $1500 - 400$ and $1500 + \frac{400}{3}$; that is, between 1100 and $1633\frac{1}{3}$.

d. Using the result of part (c) with $h = 1$, we find $x = 60 + \frac{3}{20}$ and $y = 60 - \frac{1}{8}$. Next, using the result of part (c), we see that the shadow price is

$$12(60 + \tfrac{3}{20}) + 10(60 - \tfrac{1}{8}) - [12(60) + 10(60)]$$
$$= 12(\tfrac{3}{20}) + 10(-\tfrac{1}{8}) = 0.55 \quad \text{or} \quad \$0.55.$$

e. Constraints 1 and 2 are binding since equality holds at the optimal solution. Constraint 3 is nonbinding.

17. a. Let x denote the number of model A grates and y the number of Model B grates to be produced. We are led to the following problem:

$$\begin{aligned} \text{Maximize} \quad & P = 2x + 1.5y \\ \text{subject to} \quad & 3x + 4y \le 1000 \quad \text{(Constraint 1)} \\ & 6x + 3y \le 1200 \quad \text{(Constraint 2)} \\ & y \le 200 \quad \text{(Constraint 3)} \\ & x \ge 0, y \ge 0 \end{aligned}$$

The feasible set is shown in the accompanying figure. From the table,

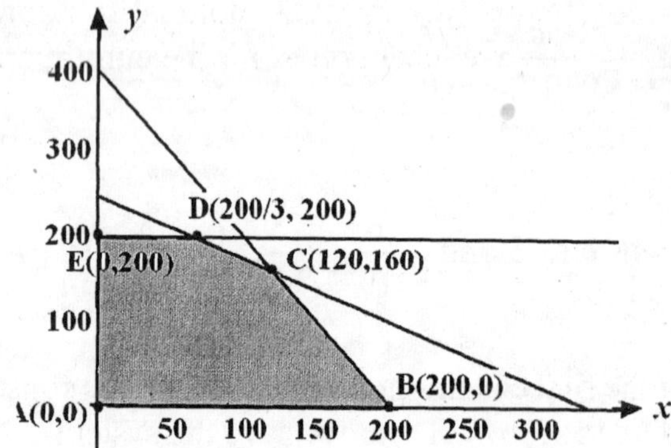

V	$P = 2x + 1.5y$
A(0,0)	0
B(200,0)	400
C(120,160)	**480**
D($\frac{200}{3}$,200)	433.33
E(0,200)	300

we see that the optimal solution is $x = 120$, $y = 160$, and $P = 480$. Therefore, Kane Manufacturing should produce 120 Model A grates and 160 Model B grates for a daily profit of $480.

b. Suppose $P = cx + \frac{3}{2}y$ so that $y = -\frac{2}{3}cx + \frac{2P}{3}$. In order for the current optimal solution to hold, we must have

$$-\frac{2c}{3} \le -\frac{3}{4}, \quad \frac{2c}{3} \ge \frac{3}{4} \quad \text{or} \quad c \ge 1.125,$$

$$-\frac{2c}{3} \ge -2, \quad \frac{2c}{3} \le 2 \quad \text{or} \quad c \le 3,$$

and $-\frac{3}{2}c \le 0$ or $c \ge 0$. Therefore, $1.125 \le c \le 3$.

c. The constraint of interest here is constraint 1. Suppose we replace the right-hand side of this constraint by $1000 + h$. Then we have the following system.

$$3x + 4y = 1000 + h$$

$$6x + 3y = 1200$$

The solutions are $x = 120 - \frac{1}{5}h$ and $y = 160 + \frac{2}{5}h$. since $x \geq 0$, we have $120 - \frac{1}{5}h \geq 0$ or $h \leq 600$. Next, $y \geq 0$ implies $160 + \frac{2}{5}h \geq 0$ or $h \geq -400$. Finally, $y \leq 200$ implies $160 + \frac{2}{5}h \leq 200$, or $h \leq 100$. Therefore, $-400 \leq h \leq 100$. So, the values assumed by this resource must lie between 1000 – 400 and 1000 + 100; that is, between 600 and 1100.

d. Let $h = 1$. Then the results of part (*c*) give $x = 120 - \frac{1}{5}$ and $y = 160 + \frac{2}{5}$. Recall that the current optimal solution is $x = 120$ and $y = 160$. We see that the shadow price associated with the resource for cast iron (constraint 1) is

$$2(120 - \tfrac{1}{5}) + \tfrac{3}{2}(160 + \tfrac{2}{5}) - [2(120) + \tfrac{3}{2}(160)]$$
$$= 2(-\tfrac{1}{5}) + \tfrac{3}{2}(\tfrac{2}{5}) = \tfrac{1}{5} \text{ or } \$0.20.$$

e. Constraints 1 and 2 are binding since equality holds at the optimal solution. Constraint 3 is nonbinding since constraint 3 reduces to $160 \leq 200$ when $x = 120$ and $y = 160$ and so equality does not hold.

CHAPTER 3 CONCEPT QUESTIONS, page 212

1. a. Half plane; line b. $ax + by \leq c$; $ax + by = c$

3. Objective function; maximized; minimized; linear; inequalities

5. Parameters; optimal

CHAPTER 3 REVIEW EXERCISES, page 213

1. Evaluating Z at each of the corner points of the feasible set S, we obtain the following table.

Vertex	$Z = 2x + 3y$
(0,0)	0
(5,0)	10
(3,4)	18
(0,6)	18

We conclude that Z attains a minimum value of 0 when $x = 0$ and $y = 0$, and a maximum value of 18 when x and y lie on the line segment joining (3,4) and (0,6).

3. The graph of the feasible set S follows.

Vertex	$P = 3x + 5y$
A(0,0)	0
B(5,0)	15
C(3,2)	19
D(0,4)	20

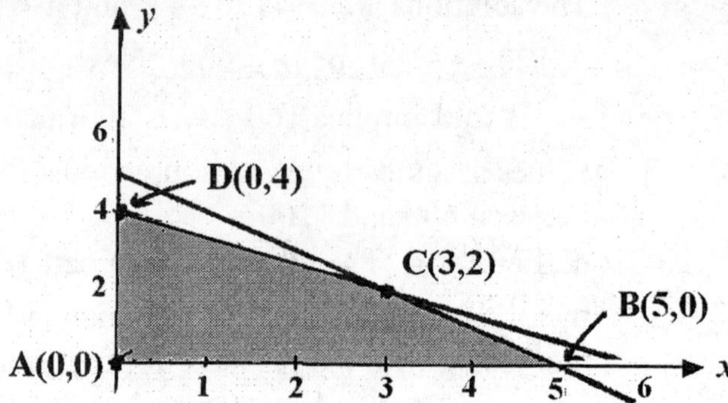

We conclude that the maximum value of P is 20 when $x = 0$ and $y = 4$.

5. The values of the objective function $C = 2x + 5y$ at the corner points of the feasible set are given in the following table. The graph of the feasible set S follows.

Vertex	$C = 2x + 5y$
A(0,16)	80
B(3,4))	26
C(15,0)	30

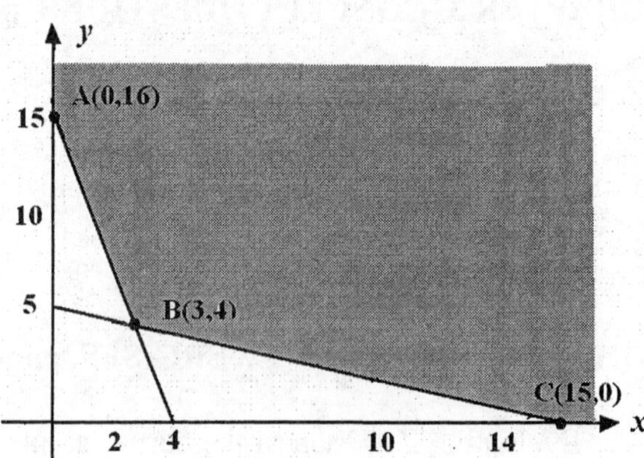

We conclude that the minimum value of C is 26 when $x = 3$ and $y = 4$.

7. The values of the objective function $P = 3x + 2y$ at the vertices of the feasible set are given in the following table. The graph of the feasible set follows.

Vertex	$P = 3x + 2y$
$A(0, \frac{28}{5})$	$\frac{56}{5}$
B(7,0)	21
C(8,0)	24
D(3,10)	$\boxed{29}$
E(0,12)	24

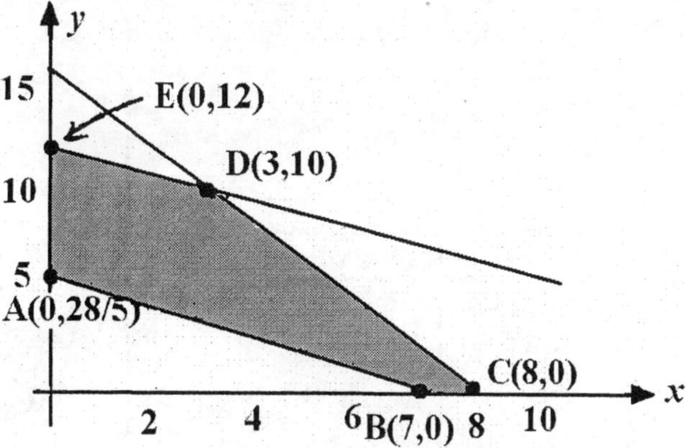

We conclude that P attains a maximum value of 29 when $x = 3$ and $y = 10$.

9. The graph of the feasible set S is shown at the right. The values of the objective function $C = 2x + 7y$ at each of the corner points of the feasible set S are shown in the table that follows.

Vertex	$C = 2x + 7y$
A(20,0)	$\boxed{40}$
B(10,3)	41
C(0,9)	63

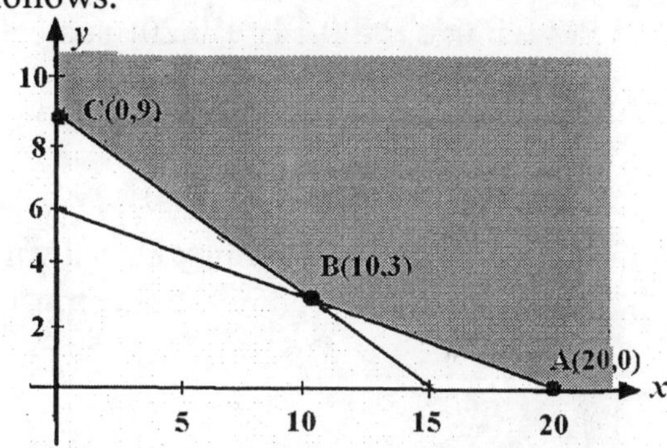

We conclude that C attains a minimum value of 40 when $x = 20$ and $y = 0$.

11. The graph of the feasible set S is shown in the following figure. We conclude that Q attains a maximum value of 22 when $x = 22$ and $y = 0$, and a minimum value of $5\frac{1}{2}$ when $x = 3$ and $y = \frac{5}{2}$.

Vertex	$Q = x + y$
A(2,5)	7
B($3, \frac{5}{2}$)	$5\frac{1}{2}$
C(8,0)	8
D(22,0)	22

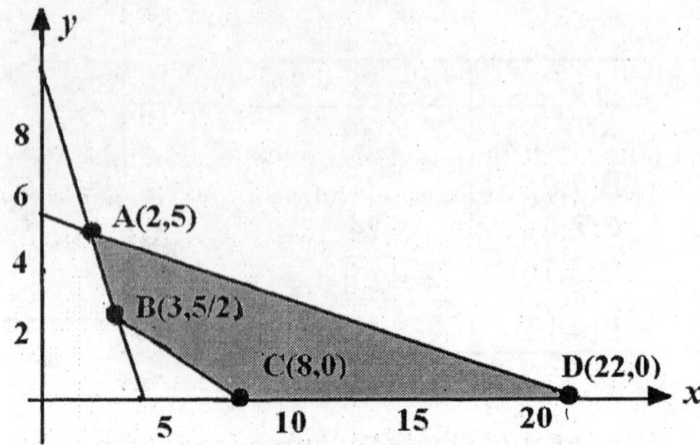

13. Suppose the investor puts x and y thousand dollars into the stocks of company *A* and company *B*, respectively. Then the mathematical formulation leads to the linear programming problem:

Maximize $P = 0.14x + 0.20y$ subject to

$$\begin{aligned} x + \quad y &\le 80 \\ 0.01x + 0.04y &\le 2 \\ x \ge 0,\ y &\ge 0 \end{aligned}$$

The feasible set *S* for this problem is shown in figure that follows and the values at each corner point are given in the accompanying table.

Vertex	$P = 0.14x + 0.20y$
A(0,0)	0
B(80,0)	11.2
C(40,40)	13.6
D(0,50)	10

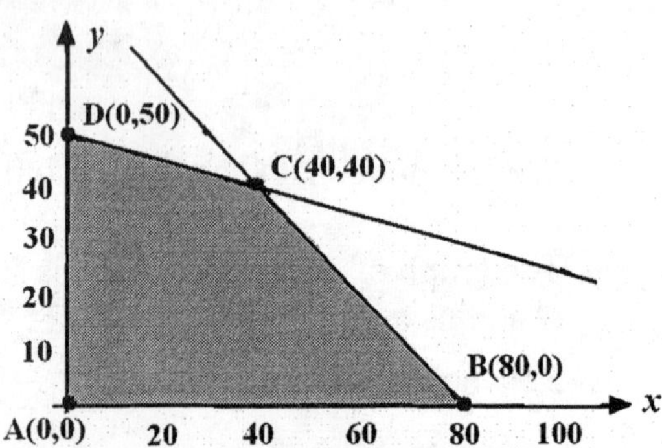

P attains a maximum value of 13.6 when $x = 40$ and $y = 40$. Thus, by investing $40,000 in the stocks of each company, the investor will achieve a maximum return of $13,600.

15. Let x denote the number of model A grates and y the number of model B grates to be produced. Then the constraint on the amount of cast iron available leads to

$$3x + 4y \le 1000$$

and the number of minutes of labor used each day leads to

$$6x + 3y \le 1200.$$

One additional constraint specifies that $y \ge 180$. The daily profit is $P = 2x + 1.5y$. Therefore, we have the following linear programming problem:

Maximize $P = 2x + 1.5y$ subject to

$$3x + 4y \le 1000$$
$$6x + 3y \le 1200$$
$$x \ge 0,\ y \ge 180$$

The graph of the feasible set S is shown in the figure that follows.

Vertex	$P = 2x + 1.5y$
A(0,180)	270
B(0,250)	375
C($93\frac{1}{3}$,180)	$\boxed{456\frac{2}{3}}$

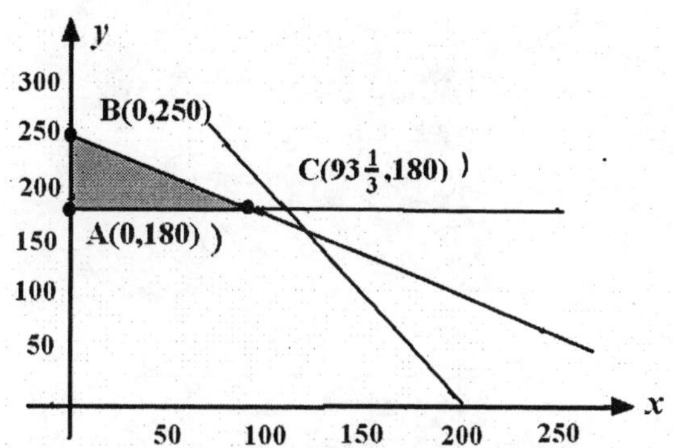

Thus, the optimal profit of $456 is realized when 93 units of model A grates and 180 units of model B grates are produced.

CHAPTER 3 BEFORE MOVING ON, page 214

1. a.

$$2x + \ y \le 10$$
$$x + 3y \le 15$$
$$x \le \ 4$$
$$x \le 0,\ y \ge 0$$

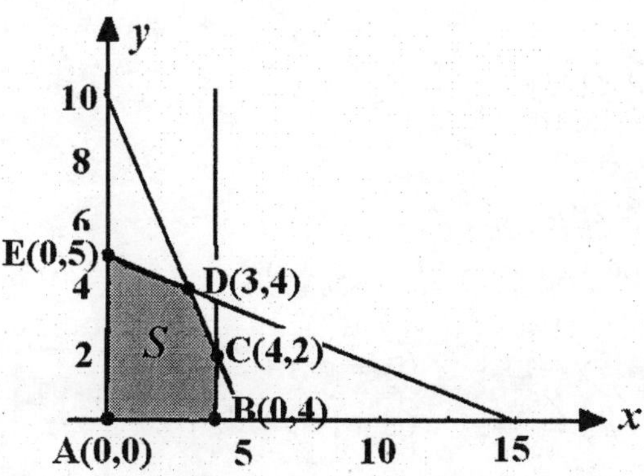

To find the vertex D, we solve $\begin{cases} 2x + \ \ y = 10 \\ x + 3y = 15 \end{cases} \Rightarrow \begin{matrix} 2x + \ \ y = 10 \\ 2x + 6y = 30 \end{matrix}$

This gives $5y = 20$ or $y = 4$ and $x = 15 - 3y = 15 - 3(4) = 3$.

b.

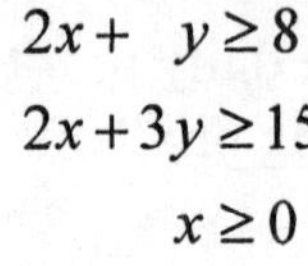

$$\begin{aligned} 2x + \ \ y &\geq 8 \\ 2x + 3y &\geq 15 \\ x &\geq 0 \\ y &\geq 2 \end{aligned}$$

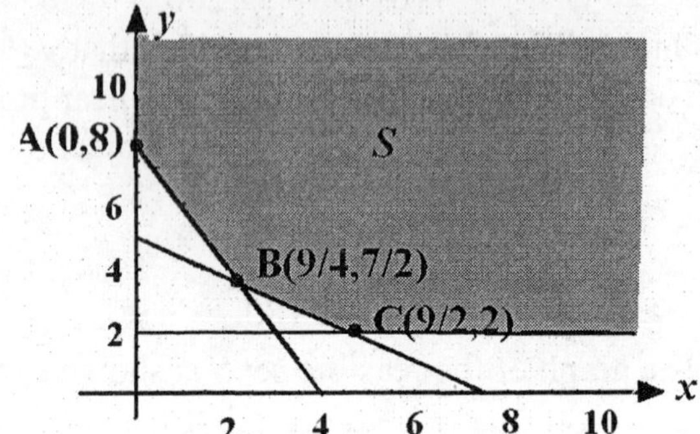

To find the vertex B, we solve

$$\left.\begin{matrix} 2x + \ \ y = 8 \\ 2x + 3y = 15 \end{matrix}\right\} \Rightarrow 2y = 7 \text{ or } y = \tfrac{7}{2}. \text{ So, } 2x + \tfrac{7}{2} = 8,\ 2x = \tfrac{9}{2} \text{ or } x = \tfrac{9}{4}.$$

2.

	$Z = 3x - y$
(8,2)	22
(28,8)	76
(16,24)	24
(3,16)	-7

The maximum value is $z = 76$ and the minimum value is $z = -7$.

3. Maximize $P = x + 3y$

subject to

$$\begin{aligned} 2x + 3y &\leq 11 \\ 3x + 7y &\leq 24 \\ x \geq 0,\ y &\geq 0 \end{aligned}$$

To find the coordinates of C, we solve

$$\left.\begin{matrix} 2x + 3y = 11 \\ 3x + 7y = 24 \end{matrix}\right\} \Rightarrow \left.\begin{matrix} 6x + \ \ 9y = 33 \\ 6x + 14y = 48 \end{matrix}\right\} \Rightarrow 5y = 15 \text{ or } y = 3. \text{ So } x = 1.$$

Vertex	$P = x + 3y$
A(0,0)	0
B($\frac{11}{2}$,0)	$\frac{11}{2}$
C(1,3)	10
D(0, $\frac{24}{7}$)	$\frac{72}{7} = 10\frac{2}{7}$

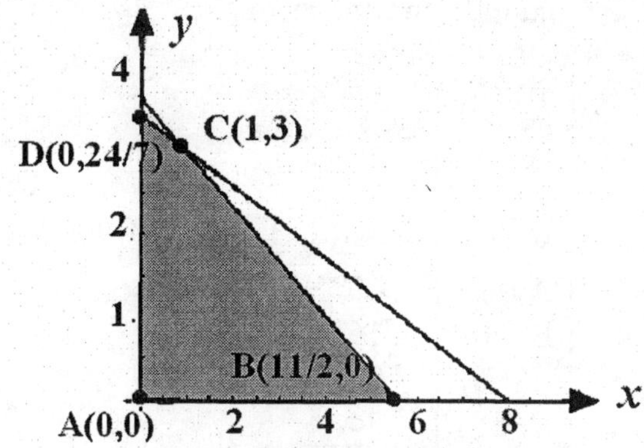

So the maximum value of P is $10\frac{2}{7}$ attained at $x = 0$ and $y = \frac{24}{7}$.

4. Minimize $C = 4x + y$ subject to

$$2x + y \geq 10$$
$$2x + 3y \geq 24$$
$$x + 3y \geq 15$$
$$x \geq 0,\ y \geq 0$$

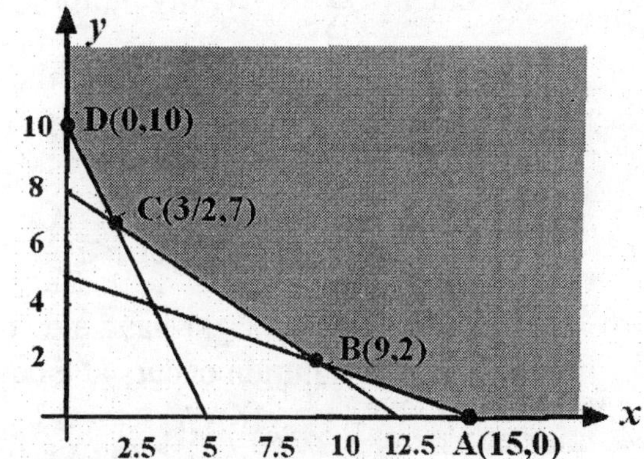

To find the coordinates of C, we solve

$$\left.\begin{aligned} 2x + y &= 10 \\ 2x + 3y &= 24 \end{aligned}\right\} \Rightarrow 2y = 14 \text{ or } y = 7, \text{ so } x = \tfrac{3}{2}$$

To find the coordinates of B, we solve

$$\left.\begin{aligned} 2x + 3y &= 24 \\ x + 3y &= 15 \end{aligned}\right\} \Rightarrow x = 9, \text{ so } y = 2$$

	$C = 4x + y$
A(15,0)	70
B(9,2)	38
C($\frac{3}{2}$,7)	13
D(0,10)	10

So the minimum value of C is 10 attained at $x = 0$, $y = 10$.

5. Maximize $P = 2x + 3y$ subject to

$$x + 2y \leq 16$$
$$3x + 2y \leq 24$$
$$x \geq 0,\ y \geq 0$$

a. To find the coordinates of the vertex C, we solve

$$\left.\begin{array}{r} x+2y=16 \\ 3x+2y=24 \end{array}\right\} \Rightarrow x=4 \text{ and } y=6.$$

Vertex	$P = 2x + 3y$
A(0,0)	0
B(8,0)	16
C(4,6)	26
D(0,8)	24

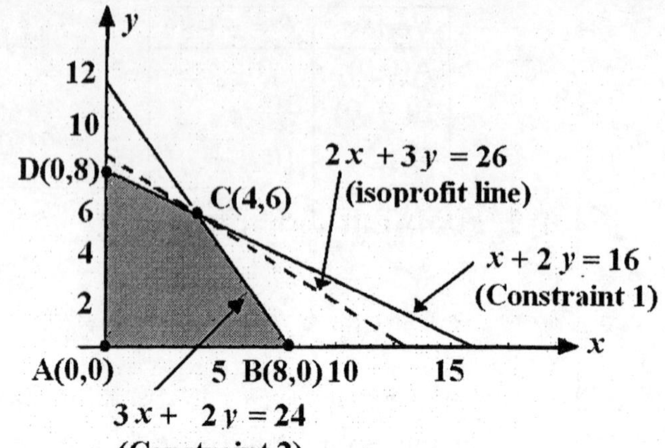

The solution is $x = 4$, $y = 6$, and $P = 26$.

b. Let $P = cx + 3y$. Solving for y, we have $y = \left(-\frac{c}{3}\right)x + \frac{P}{3}$. If the slope of the isoprofit line is greater than the slope of the line associated with constraint 1, then the optimal solution will shift from point C to point D. But the slope of the line associated with constraint 1 is $-\frac{1}{2}$. So we have

$-\frac{c}{3} \le -\frac{1}{2}$ or $\frac{c}{3} \ge \frac{1}{2}$ or $c \ge \frac{3}{2}$. Therefore, $\frac{3}{2} \le c \le \frac{9}{2}$.

c. Suppose the right-hand side of constraint 1 is replaced by $16 + h$. then the new optimal solution occurs at a new point c'. To find the coordinate of c', we solve

$$\left.\begin{array}{r} x+2y=16+h \\ 3x+2y=24 \end{array}\right\} \Rightarrow 2x = 8-h \text{ or } x = \tfrac{1}{2}(8-h)$$

Then $\quad 2y = 24 - 3x = 24 - \frac{3}{2}(8-h) = 12 + \frac{3}{2}h$, or $y = 6 + \frac{3}{4}h$.

The nonnegativity of x implies $\frac{1}{2}(8-h) \ge 0$ or $h \le 8$, and the nonnegativity of y implies $6 + \frac{3}{4}h \ge 0$ or $h \ge -8$. Therefore $-8 \le h \le 8$. So resource 1 must lie between 16 - 8, or 8 and 16 + 8, or 24, that is between 8 and 24.

d. If we set $h = 1$ in part (c), we obtain

$$x = \tfrac{1}{2}(8-1) = \tfrac{7}{2}$$
$$y = 6 + \tfrac{3}{4}(1) = 6\tfrac{3}{4}$$

Therefore, the profit realized at this level of production is

$$P = 2x + 3y = 2\left(\tfrac{7}{2}\right) + 3\left(\tfrac{27}{4}\right) = \tfrac{109}{4} = 27\tfrac{1}{4}.$$

So the shadow price is $27\frac{1}{4} - 26$ or \$1.25.

e. Both constraints are binding because they hold with equality at the optimal solution $C(4,6)$.

CHAPTER 4

4.1 Problem Solving Tips

Here are some hints for solving the problems in the exercises that follow.

1. Make sure that you set up the initial simplex tableau correctly. (a) First rewrite the linear inequalities as equalities by introducing slack variables. (b) Next rewrite the objective function so that all variables are on the left and the coefficient of P is 1. Then place this equation below the other equations.

2. In the simplex method, the optimal solution has been reached if *all the entries* in the last row to the left of the vertical line are *nonnegative.*

3. In the simplex method, you will know that the optimal solution has not been reached if there are *negative entries in the last row to the left of the vertical line*. If so, locate the most negative entry in that row. This will give you the pivot column. Proceed to find the pivot row by dividing each positive entry in the pivot column by its corresponding entry in the column of constants. Look for the smallest ratio. The corresponding entry will be in the pivot row. Proceed to make the pivot column a unit column by pivoting about the pivot element.

4.1 CONCEPT QUESTIONS, page 232

1. a. The objective function is to be maximized.
   b. All the variables involved in the problem are nonnegative.
   c. Each linear constraint may be written so that the expression involving the variables is less than or equal to a nonnegative constant.

3. To find the *pivot column* locate the most negative entry to the left of the vertical line in the last row. The column containing this entry is the pivot column. To find the *pivot row,* divide each positive entry in the pivot column into its corresponding entry in the column of constants. The pivot row is the row corresponding to the smallest ratio thus obtained. The *pivot element* is the element common to both the pivot column and the pivot row.

EXERCISES 4.1, page 233

1. All entries in the last row of the simplex tableau are nonnegative and an optimal solution has been reached. We find
$$x = 30/7,\ y = 20/7,\ u = 0,\ v = 0, \text{ and } P = 220/7.$$

3. The simplex tableau is not in final form because there is an entry in the last row that is negative. The entry in the first row, second column, is the next pivot element and has a value of 1/2.

5. The simplex tableau is in final form. We find
$$x = 1/3, y = 0, z = 13/3, u = 0, v = 6, w = 0 \text{ and } P = 17.$$

7. The simplex tableau is not in final form because there are two entries in the last row that are negative. The entry in the third row, second column, is the pivot element and has a value of 1.

9. The simplex tableau is in final form. The solutions are
$$x = 30,\ y = 10, z = 0, u = 0, v = 0, \text{ and } P = 60$$
and
$$x = 30,\ y = 0, z = 0, u = 10, v = 0, \text{ and } P = 60.$$
(There are infinitely many solutions.)

11. We obtain the following sequence of tableaus:

	x	y	u	v	P	Const
p.r. →	1	[1]	1	0	0	4
	2	1	0	1	0	5
	−3	−4	0	0	1	0
		↑ p.c.				

Ratio
4
5

$\xrightarrow[R_3+4R_1]{R_2-R_1}$

x	y	u	v	P	Const.
1	1	1	0	0	4
1	0	−1	1	0	1
1	0	4	0	1	16

The last tableau is in final form and we conclude that $x = 0$, $y = 4$, $u = 0$, $v = 1$, and $P = 16$.

13.

	x	y	u	v	P	Const
p.r. →	1	[2]	1	0	0	12
	3	2	0	1	0	24
	−10	−12	0	0	1	0
		↑ p.c.				

Ratio
$12/2 = 6$
$24/2 = 12$

$\xrightarrow{\frac{1}{2}R_1}$

x	y	u	v	P	Const
$\frac{1}{2}$	1	$\frac{1}{2}$	0	0	6
3	2	0	1	0	24
−10	−12	0	0	1	0

$\xrightarrow[R_3+12R_1]{R_2-2R_1}$

	x	y	u	v	P	Const
	$\frac{1}{2}$	1	$\frac{1}{2}$	0	0	6
p.r. →	[2]	0	−1	1	0	12
	−4	0	6	0	1	72
	↑ pc.					

Ratio
$6/(1/2) = 12$
$12/2 = 6$

$\xrightarrow{\frac{1}{2}R_2}$

x	y	u	v	P	Const
$\frac{1}{2}$	1	$\frac{1}{2}$	0	0	6
1	0	$-\frac{1}{2}$	$\frac{1}{2}$	0	6
−4	0	6	0	1	72

$\xrightarrow[R_3+4R_2]{R_1-\frac{1}{2}R_2}$

x	y	u	v	P	Const
0	1	$\frac{3}{4}$	$-\frac{1}{4}$	0	3
1	0	$-\frac{1}{2}$	$\frac{1}{2}$	0	6
0	0	4	2	1	96

The last tableau is in final form. We find that $x = 6$, $y = 3$, $u = 0$, $v = 0$, and $P = 96$.

15. We obtain the following sequence of tableaus:

	x	y	u	v	w	P	Const.	Ratio
	3	1	1	0	0	0	24	24
	2	1	0	1	0	0	18	18
p.r. →	1	[3]	0	0	1	0	24	8
	–4	–6	0	0	0	1	0	
		↑ p.c.						

$\xrightarrow{\frac{1}{3}R_3}$

x	y	u	v	w	P	Const.
3	1	1	0	0	0	24
2	1	0	1	0	0	18
$\frac{1}{3}$	1	0	0	$\frac{1}{3}$	0	8
–4	–6	0	0	0	1	0

$\xrightarrow[R_4+6R_3]{R_1-R_3,\ R_2-R_3}$

	x	y	u	v	w	P	Const.	Ratio
p.r. →	[$\frac{8}{3}$]	0	1	0	$-\frac{1}{3}$	0	16	6
	$\frac{5}{3}$	0	0	1	$-\frac{1}{3}$	0	10	6
	$\frac{1}{3}$	1	0	0	$\frac{1}{3}$	0	8	24
	–2	0	0	0	2	1	48	
	↑ p.c.							

(Observe that we have a choice here.)

$\xrightarrow{\frac{3}{8}R_1}$

x	y	u	v	w	P	Const.
1	0	$\frac{3}{8}$	0	$-\frac{1}{8}$	0	6
$\frac{5}{3}$	0	0	1	$-\frac{1}{3}$	0	10
$\frac{1}{3}$	1	0	0	$\frac{1}{3}$	0	8
–2	0	0	0	2	1	48

$\xrightarrow[R_4+2R_1]{R_2-\frac{5}{3}R_1,\ R_3-\frac{1}{3}R_1}$

x	y	u	v	w	P	Const.
1	0	$\frac{3}{8}$	0	$-\frac{1}{8}$	0	6
0	0	$-\frac{5}{8}$	1	$-\frac{1}{8}$	0	0
0	1	$-\frac{1}{8}$	0	$\frac{3}{8}$	0	6
0	0	$\frac{3}{4}$	0	$\frac{7}{4}$	1	60

We deduce that $x = 6$, $y = 6$, $u = 0$, $v = 0$, $w = 0$, and $P = 60$.

17. We obtain the following sequence of tableaus:

	x	y	z	u	v	P	Const.	Ratio
	1	1	1	1	0	0	8	$8/1 = 8$
p.r. →	3	2	[4]	0	1	0	24	$24/4 = 6$
	−3	−4	−5	0	0	1	0	
			↑ p.c.					

$\xrightarrow{\frac{1}{4}R_2}$

x	y	z	u	v	P	Const.
1	1	1	1	0	0	8
$\frac{3}{4}$	$\frac{1}{2}$	1	0	$\frac{1}{4}$	0	6
−3	−4	−5	0	0	1	0

$\xrightarrow[R_3+5R_2]{R_1-R_2}$

	x	y	z	u	v	P	Const.	Ratio
p.r. →	$\frac{1}{4}$	[$\frac{1}{2}$]	0	1	$-\frac{1}{4}$	0	2	$2/(1/2) = 4$
	$\frac{3}{4}$	$\frac{1}{2}$	1	0	$\frac{1}{4}$	0	6	$6/(1/2) = 12$
	$\frac{3}{4}$	$-\frac{3}{2}$	0	0	$\frac{5}{4}$	1	30	
		↑ p.c.						

$\xrightarrow{2R_1}$

x	y	z	u	v	P	Const.
$\frac{1}{2}$	1	0	2	$-\frac{1}{2}$	0	4
$\frac{3}{4}$	$\frac{1}{2}$	1	0	$\frac{1}{4}$	0	6
$\frac{3}{4}$	$-\frac{3}{2}$	0	0	$\frac{5}{4}$	1	30

$\xrightarrow[R_3+\frac{3}{2}R_1]{R_2-\frac{1}{2}R_1}$

x	y	z	u	v	P	Const.
$\frac{1}{2}$	1	0	2	$-\frac{1}{2}$	0	4
$\frac{1}{2}$	0	1	-1	$\frac{1}{2}$	0	4
$\frac{3}{2}$	0	0	3	$\frac{1}{2}$	1	36

This last tableau is in final form. We find that $x = 0, y = 4, z = 4, u = 0, v = 0$, and $P = 36$.

19.

	x	y	z	u	v	w	P	Const.	Ratio
	3	10	5	1	0	0	0	120	$120/10 = 12$
p.r. →	5	[2]	8	0	1	0	0	6	$6/2 = 3$
	8	10	3	0	0	1	0	105	$105/10 = 21/2$
	-3	-4	-1	0	0	0	1	0	
		↑ p.c.							

$\xrightarrow{\frac{1}{2}R_2}$

x	y	z	u	v	w	P	Const.
3	10	5	1	0	0	0	120
$\frac{5}{2}$	1	4	0	$\frac{1}{2}$	0	0	3
8	10	3	0	0	1	0	105
-3	-4	-1	0	0	0	1	0

$\xrightarrow[R_4+4R_2]{R_1-10R_2,\ R_3-10R_2}$

x	y	z	u	v	w	P	Const.
-22	0	-35	1	-5	0	0	90
$\frac{5}{2}$	1	4	0	$\frac{1}{2}$	0	0	3
-17	0	-37	0	-5	1	0	75
7	0	15	0	2	0	1	12

The last tableau is in final form. We find that $x = 0$, $y = 3, z = 0, u = 90, v = 0$, $w = 75$, and $P = 12$.

21. We obtain the following sequence of tableaus:

	x	y	z	u	v	w	P	Const.	Ratio
	1	1	1	1	0	0	0	20	20
p.r. →	2	[4]	3	0	1	0	0	42	$10\frac{1}{2}$
	2	0	3	0	0	1	0	30	---
	–4	–6	–5	0	0	0	1	0	
		↑ p.c.							

$\xrightarrow{\frac{1}{4}R_2}$

x	y	z	u	v	w	P	Const.
1	1	1	1	0	0	0	20
$\frac{1}{2}$	1	$\frac{3}{4}$	0	$\frac{1}{4}$	0	0	$\frac{21}{2}$
2	0	3	0	0	1	0	30
–4	–6	–5	0	0	0	1	0

$\xrightarrow[R_4+6R_2]{R_1-R_2}$

	x	y	z	u	v	w	P	Const.	Ratio
	$\frac{1}{2}$	0	$\frac{1}{4}$	1	$-\frac{1}{4}$	0	0	$\frac{19}{2}$	19
	$\frac{1}{2}$	1	$\frac{3}{4}$	0	$\frac{1}{4}$	0	0	$\frac{21}{2}$	21
p.r. →	[2]	0	3	0	0	1	0	30	15
	–1	0	$-\frac{1}{2}$	0	$\frac{3}{2}$	0	1	63	
	↑ p.c.								

$\xrightarrow{\frac{1}{2}R_3}$

x	y	z	u	v	w	P	Const.
$\frac{1}{2}$	0	$\frac{1}{4}$	1	$-\frac{1}{4}$	0	0	$\frac{19}{2}$
$\frac{1}{2}$	1	$\frac{3}{4}$	0	$\frac{1}{4}$	0	0	$\frac{21}{2}$
1	0	$\frac{3}{2}$	0	0	$\frac{1}{2}$	0	15
–1	0	$-\frac{1}{2}$	0	$\frac{3}{2}$	0	1	63

$R_1-\frac{1}{2}R_3$, $R_2-\frac{1}{2}R_3$, R_4+R_3 $\longrightarrow$

x	y	z	u	v	w	P	*Const.*
0	0	$-\frac{1}{2}$	1	$-\frac{1}{4}$	$-\frac{1}{4}$	0	2
0	1	0	0	$\frac{1}{4}$	$-\frac{1}{4}$	0	3
1	0	$\frac{3}{2}$	0	0	$\frac{1}{2}$	0	15
0	0	1	0	$\frac{3}{2}$	$\frac{1}{2}$	1	78

So the solution is $x = 15$, $y = 3$, $z = 0$, $u = 2$, $v = 0$, $w = 0$, and $P = 78$.

23. We obtain the following sequence of tableaus:

	x	y	z	u	v	w	P	*Const.*
$p.r. \rightarrow$	[2]	1	1	1	0	0	0	10
	3	5	1	0	1	0	0	45
	2	5	1	0	0	1	0	40
	−12	−10	−5	0	0	0	1	0
	↑ $p.c.$							

Ratio

$10/2 = 5$

$45/3 = 15$

$40/2 = 20$

$\xrightarrow{\frac{1}{2}R_1}$

x	y	z	u	v	w	P	*Const.*
1	$\frac{1}{2}$	$\frac{1}{2}$	$\frac{1}{2}$	0	0	0	5
3	5	1	0	1	0	0	45
2	5	1	0	0	1	0	40
−12	−10	−5	0	0	0	1	0

$R_2 - 3R_1$
$R_3 - 2R_1$
$\xrightarrow{R_4 + 12R_1}$

	x	y	z	u	v	w	P	*Const.*
	1	$\frac{1}{2}$	$\frac{1}{2}$	$\frac{1}{2}$	0	0	0	5
	0	$\frac{7}{2}$	$-\frac{1}{2}$	$-\frac{3}{2}$	1	0	0	30
$p.r. \rightarrow$	0	[4]	0	−1	0	1	0	30
	0	−4	1	6	0	0	1	60
		↑ $p.c$						

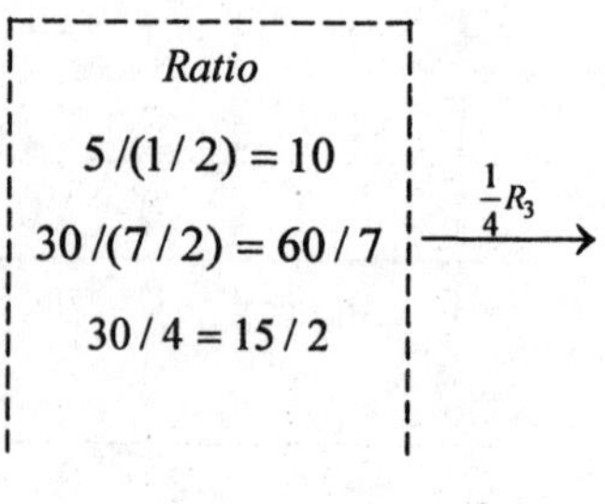

$\xrightarrow{\frac{1}{4}R_3}$

x	y	z	u	v	w	P	Const.
1	$\frac{1}{2}$	$\frac{1}{2}$	$\frac{1}{2}$	0	0	0	5
0	$\frac{7}{2}$	$-\frac{1}{2}$	$-\frac{3}{2}$	1	0	0	30
0	1	0	$-\frac{1}{4}$	0	$\frac{1}{4}$	0	$\frac{15}{2}$
0	–4	1	6	0	0	1	60

$\xrightarrow[R_4+4R_3]{R_1-\frac{1}{2}R_3,\ R_2-\frac{7}{2}R_3}$

x	y	z	u	v	w	P	Const.
1	0	$\frac{1}{2}$	$\frac{5}{8}$	0	$-\frac{1}{8}$	0	$\frac{5}{4}$
0	0	$-\frac{1}{2}$	$-\frac{5}{8}$	1	$-\frac{7}{8}$	0	$\frac{15}{4}$
0	1	0	$-\frac{1}{4}$	0	$\frac{1}{4}$	0	$\frac{15}{2}$
0	0	1	5	0	1	1	90

.

This last tableau is in final form, and we conclude that $x = 5/4, y = 15/2, z = 0,$ $u = 0, v = 15/4, w = 0,$ and $P = 90$.

25. We obtain the following sequence of tableaus, where u, v and w are slack variables.

	x	y	z	u	v	w	P	Const.	Ratio
p.r. →	[2]	1	2	1	0	0	0	7	$\frac{7}{2}$
	2	3	1	0	1	0	0	8	4
	1	2	3	0	0	1	0	7	7
	–24	–16	–23	0	0	0	1	0	
	↑ p.c.								

$\xrightarrow{\frac{1}{2}R_1}$

x	y	z	u	v	w	P	Const.
[1]	$\frac{1}{2}$	1	$\frac{1}{2}$	0	0	0	$\frac{7}{2}$
2	3	1	0	1	0	0	8
1	2	3	0	0	1	0	7
–24	–16	–23	0	0	0	1	0

$\xrightarrow[R_4+24R_1]{R_2-2R_1,\ R_3-R_1}$

	x	y	z	u	v	w	P	Const.	Ratio
	1	$\frac{1}{2}$	1	$\frac{1}{2}$	0	0	0	$\frac{7}{2}$	7
p.r. →	0	$\boxed{2}$	−1	−1	1	0	0	1	$\frac{1}{2}$
	0	$\frac{3}{2}$	2	$-\frac{1}{2}$	0	1	0	$\frac{7}{2}$	$\frac{7}{3}$
	0	−4	1	12	0	0	1	84	
		↑ p.c.							

$\xrightarrow{\frac{1}{2}R_2}$

x	y	z	u	v	w	P	Const.
1	$\frac{1}{2}$	1	$\frac{1}{2}$	0	0	0	$\frac{7}{2}$
0	1	$-\frac{1}{2}$	$-\frac{1}{2}$	$\frac{1}{2}$	0	0	$\frac{1}{2}$
0	$\frac{3}{2}$	2	$-\frac{1}{2}$	0	1	0	$\frac{7}{2}$
0	−4	1	12	0	0	1	84

$\xrightarrow[R_4+4R_2]{R_1-\frac{1}{2}R_2,\; R_3-\frac{3}{2}R_2}$

	x	y	z	u	v	w	P	Const.	Ratio
	1	0	$\frac{5}{4}$	$\frac{3}{4}$	$\frac{1}{4}$	0	0	$\frac{13}{4}$	$\frac{13}{5}$
	0	1	$-\frac{1}{2}$	$-\frac{1}{2}$	$\frac{1}{2}$	0	0	$\frac{1}{2}$	--
p.r. →	0	0	$\boxed{\frac{11}{4}}$	$\frac{1}{4}$	$-\frac{3}{4}$	1	0	$\frac{11}{4}$	1
	0	0	−1	10	2	0	1	86	
			↑ p.c.						

$\xrightarrow{\frac{4}{11}R_3}$

x	y	z	u	v	w	P	Const.
1	0	$\frac{5}{4}$	$\frac{3}{4}$	$\frac{1}{4}$	0	0	$\frac{13}{4}$
0	1	$-\frac{1}{2}$	$-\frac{1}{2}$	$\frac{1}{2}$	0	0	$\frac{1}{2}$
0	0	1	$\frac{1}{11}$	$-\frac{3}{11}$	$\frac{4}{11}$	0	1
0	0	−1	10	2	0	1	86

$\xrightarrow[R_4+R_3]{R_1-\frac{5}{4}R_3,\; R_2+\frac{1}{2}R_3}$

x	y	z	u	v	w	P	Const.
1	0	0	$\frac{7}{11}$	$\frac{13}{22}$	$-\frac{5}{11}$	0	2
0	1	0	$-\frac{5}{11}$	$\frac{4}{11}$	$\frac{2}{11}$	0	1
0	0	1	$\frac{1}{11}$	$-\frac{3}{11}$	$\frac{4}{11}$	0	1
0	0	0	$\frac{111}{11}$	$\frac{19}{11}$	$\frac{4}{11}$	1	87

This last tableau is in final form and we conclude that P attains a maximum value of 87 when $x = 2, y = 1$, and $z = 1$.

27. Pivoting about the x-column in the initial simplex tableau, we have

	x	y	z	u	v	P	Const.	Ratio
	3	3	−2	1	0	0	100	100/3
p.r. →	[5]	5	3	0	1	0	150	150/5
	−2	−2	4	0	0	1	0	
	↑ p.c.							

$\xrightarrow{\frac{1}{5}R_2}$

x	y	z	u	v	P	Const.
3	3	−2	1	0	0	100
1	1	$\frac{3}{5}$	0	$\frac{1}{5}$	0	30
−2	−2	4	0	0	1	0

$\xrightarrow{\substack{R_1-3R_2\\R_3+2R_2}}$

x	y	z	u	v	P	Const.
0	0	$-\frac{19}{5}$	1	$-\frac{3}{5}$	0	10
1	1	$\frac{3}{5}$	0	$\frac{1}{5}$	0	30
0	0	$\frac{26}{5}$	0	$\frac{2}{5}$	1	60

and we see that one optimal solution occurs when $x = 30, y = 0, z = 0$, and $P = 60$. Similarly, pivoting about the y-column, we obtain another optimal solution: $x = 0, y = 30, z = 0$, and $P = 60$.

29. Let the number of model A and model B fax machines made each month be x and y, respectively. Then we have the following linear programming problem:

Maximize $P = 30x + 40y$ subject to

$$\begin{aligned} 100x + 150y &\le 600{,}000 \\ x + \quad y &\le \;\; 2{,}500 \\ x \ge 0, \; y &\ge 0 \end{aligned}$$

Using the simplex method, we obtain the following sequenceof tableaus:

x	y	u	v	P	Const	Ratio
100	150	1	0	0	600,000	4000
1	[1]	0	1	0	2,500	2500
–30	–40	0	0	1	0	

$\xrightarrow[R_3+40R_2]{R_1-150R_2}$

x	y	u	v	P	Const
–50	0	1	–150	0	225000
1	1	0	1	0	2500
10	0	0	40	1	100000

We conclude that the maximum monthly profit is \$100,000, and this occurs when 0 model *A* and 2500 model *B* fax machines are produced.

31. Suppose the farmer plants x acres of Crop *A* and y acres of Crop *B*. Then the problem is

Maximize $P = 150x + 200y$ subject to

$$x + y \le 150$$
$$40x + 60y \le 7400$$
$$20x + 25y \le 3300$$
$$x \ge 0,\ y \ge 0$$

Using the simplex method, we obtain the following sequence of tableaus:

x	y	u	v	w	P	Const
1	1	1	0	0	0	150
40	60	0	1	0	0	7400
20	25	0	0	1	0	3300
–150	–200	0	0	0	1	0

$\xrightarrow[R_4+150R_1]{R_2-40R_1,\ R_3-20R_1}$

	x	y	u	v	w	P	Const	Ratio
	1	1	1	0	0	0	150	150
	0	20	–40	1	0	0	1400	700
p.r. →	0	[5]	–20	0	1	0	300	60
	0	–50	150	0	0	1	22500	
		↑ p.c.						

$\xrightarrow{\frac{1}{5}R_3}$

x	y	u	v	w	P	Const
1	1	1	0	0	0	150
0	20	–40	1	0	0	1400
0	1	–4	0	$\frac{1}{5}$	0	60
0	–50	150	0	0	1	22500

$\xrightarrow[R_4+50R_3]{R_1-R_3,\ R_2-20R_3}$

	x	y	u	v	w	P	Const	Ratio
	1	0	5	0	$-\frac{1}{5}$	0	90	18
p.r. →	0	0	[40]	1	–4	0	200	5
	0	1	–4	0	$\frac{1}{5}$	0	60	--
	0	0	–50	0	10	1	25500	
			↑ p.c.					

$\xrightarrow{\frac{1}{40}R_2}$

x	y	u	v	w	P	Const
1	0	5	0	$-\frac{1}{5}$	0	90
0	0	1	$\frac{1}{40}$	$-\frac{1}{10}$	0	5
0	1	–4	0	$\frac{1}{5}$	0	60
0	0	–50	0	10	1	25500

$\xrightarrow[R_4+50R_2]{R_1-5R_2,\ R_3+4R_2}$

x	y	u	v	w	P	Const
1	0	0	$-\frac{1}{8}$	$\frac{3}{10}$	0	65
0	0	1	$\frac{1}{40}$	$-\frac{1}{10}$	0	5
0	1	0	$\frac{1}{10}$	$-\frac{1}{5}$	0	80
0	0	0	$\frac{5}{4}$	5	1	25750

The last tableau is in final form. We find $x = 65$, $y = 80$, and $P = 25{,}750$. So the maximum profit of \$25,750 is realized by planting 65 acres of Crop *A* and 80 acres of Crop *B*. Since $u = 5$, we see that there are 5 acres of land left unused.

33. Suppose Ashley invests x, y, and z dollars in the money market fund, the international equity fund, and the growth-and-income fund, respectively. Then the objective function is $P = 0.06x + 0.1y + 0.15z$. The constraints are

$$x + y + z \leq 250{,}000; \quad z \leq 0.25(x + y + z); \text{and} \quad y \leq 0.5(x + y + z).$$

The last two inequalities simplify to

$$-\tfrac{1}{4}x - \tfrac{1}{4}y + \tfrac{3}{4}z \leq 0 \quad \text{or} \quad -x - y + 3z \leq 0$$

and $$-\tfrac{1}{2}x + \tfrac{1}{2}y - \tfrac{1}{2}z \leq 0, \quad \text{or} \quad -x + y - z \leq 0.$$

So the required linear programming problem is

Maximize $P = 0.06x + 0.1y + 0.15z = \frac{3}{50}x + \frac{1}{10}y + \frac{3}{20}z$ subject to

$$\begin{aligned} x + y + \ z &\leq 250000 \\ -x - y + 3z &\leq 0 \\ -x + y - \ z &\leq 0 \\ x \geq 0, y \geq 0&, z \geq 0 \end{aligned}$$

Let u, v, and w be slack variables. We obtain the following tableaus:

	x	y	z	u	v	w	P	Const.	Ratio	
p.r. →	1	1	1	1	0	0	0	250000	250000	
	−1	−1	[3]	0	1	0	0	0	0	$\xrightarrow{\frac{1}{3}R_2}$
	−1	1	−1	0	0	1	0	0	--	
	$-\frac{3}{50}$	$-\frac{1}{10}$	$-\frac{3}{20}$	0	0	0	1	0		
			↑ p.c							

x	y	z	u	v	w	P	*Const.*	
1	1	1	1	0	0	0	250000	
$-\frac{1}{3}$	$-\frac{1}{3}$	$\boxed{1}$	0	$\frac{1}{3}$	0	0	0	$\xrightarrow[R_3+R_2]{R_1-R_2}$
-1	1	-1	0	0	1	0	0	$R_4+\frac{3}{20}R_2$
$-\frac{3}{50}$	$-\frac{1}{10}$	$-\frac{3}{20}$	0	0	0	1	0	

x	y	z	u	v	w	P	*Const.*	*Ratio*	
$\frac{4}{3}$	$\frac{4}{3}$	0	1	$-\frac{1}{3}$	0	0	250000	$\frac{250000}{4/3}=187500$	
$-\frac{1}{3}$	$-\frac{1}{3}$	1	0	$\frac{1}{3}$	0	0	0	--	$\xrightarrow{\frac{3}{2}R_3}$
$-\frac{4}{3}$	$\boxed{\frac{2}{3}}$	0	0	$\frac{1}{3}$	1	0	0	0	
$-\frac{11}{100}$	$-\frac{3}{20}$	0	0	$\frac{1}{20}$	0	1	0		
	↑ *p.c.*								

x	y	z	u	v	w	P	*Const.*	
$\frac{4}{3}$	$\frac{4}{3}$	0	1	$-\frac{1}{3}$	0	0	250000	
$-\frac{1}{3}$	$-\frac{1}{3}$	1	0	$\frac{1}{3}$	0	0	0	$\xrightarrow[R_2+\frac{1}{3}R_3]{R_1-\frac{4}{3}R_3}$
-2	$\boxed{1}$	0	0	$\frac{1}{2}$	$\frac{3}{2}$	0	0	$R_4+\frac{3}{20}R_2$
$-\frac{11}{100}$	$-\frac{3}{20}$	0	0	$\frac{1}{20}$	0	1	0	

x	y	z	u	v	w	P	*Const.*	*Ratio*	
$\boxed{4}$	0	0	1	-1	-2	0	250000	62500	
-1	0	1	0	$\frac{1}{2}$	$\frac{1}{2}$	0	0	--	$\xrightarrow{\frac{1}{4}R_1}$
-2	1	0	0	$\frac{1}{2}$	$\frac{3}{2}$	0	0	0	
$-\frac{41}{100}$	0	0	0	$\frac{3}{20}$	$\frac{9}{40}$	1	0		
↑ *p.c.*									

x	y	z	u	v	w	P	Const.
$\boxed{1}$	0	0	$\frac{1}{4}$	$-\frac{1}{4}$	$-\frac{1}{2}$	0	62500
−1	0	1	0	$\frac{1}{2}$	$\frac{1}{2}$	0	0
−2	1	0	0	$\frac{1}{2}$	$\frac{3}{2}$	0	0
$-\frac{41}{100}$	0	0	0	$\frac{3}{20}$	$\frac{9}{40}$	1	0

$\xrightarrow[R_4+\frac{41}{100}R_1]{R_2+R_1,\ R_3+2R_2}$

x	y	z	u	v	w	P	Const.
1	0	0	$\frac{1}{4}$	$-\frac{1}{4}$	$-\frac{1}{2}$	0	62500
0	0	1	$\frac{1}{4}$	$\frac{1}{4}$	0	0	62500
0	1	0	$\frac{1}{2}$	0	$\frac{1}{2}$	0	125000
0	0	0	$\frac{41}{400}$	$\frac{19}{400}$	$\frac{1}{50}$	1	25625

The last tableau is in final form, and we see that $x = 62{,}500$, $y = 125{,}000$, $z = 62{,}500$, and $P = 25{,}625$. So, Ashley should invest \$62,500 in the money market fund, \$125,000 in the international equity fund, and \$62,500 in the growth-and-income fund. Her maximum return will be \$25,625.

35. Suppose the Excelsior Company buys x, y, and z minutes of morning, afternoon, and evening commercials, respectively. Then we wish to maximize

$$P = 200{,}000x + 100{,}000y + 600{,}000z \text{ subject to}$$

$$\begin{aligned} 3000x + 1000y + 12{,}000z &\le 102{,}000 \\ z &\le 6 \\ x + y + z &\le 25 \\ x \ge 0, y \ge 0, z &\ge 0 \end{aligned}$$

Using the simplex method, we obtain the following sequence of tableaus.

x	y	z	u	v	w	P	Const.	Ratio
$\boxed{3000}$	1000	12,000	1	0	0	0	102,000	17/2
0	0	1	0	1	0	0	6	6
1	1	1	0	0	1	0	25	25
−200,000	−100,000	−600,000	0	0	0	1	0	

$\xrightarrow[R_4+600{,}000R_2]{R_1-12{,}000R_2,\ R_3-R_2}$

x	y	z	u	v	w	P	Const.
3000	1000	0	1	−12,000	0	0	30,000
0	0	1	0	1	0	0	6
1	1	0	0	−1	1	0	19
−200,000	−100,000	0	0	600,000	0	1	3,600,000

Ratio
10
--
19

$\xrightarrow{\frac{1}{3000}R_1}$

x	y	z	u	v	w	P	Const.
1	$\frac{1}{3}$	0	$\frac{1}{3000}$	−4	0	0	10
0	0	1	0	1	0	0	6
1	1	0	0	−1	1	0	19
−200,000	−100,000	0	0	600,000	0	1	3,600,000

$\xrightarrow[R_4+200{,}000R_1]{R_3-R_1}$

x	y	z	u	v	w	P	Const.
1	$\frac{1}{3}$	0	$\frac{1}{3000}$	−4	0	0	10
0	0	1	0	1	0	0	6
0	$\frac{2}{3}$	0	$-\frac{1}{3000}$	$\boxed{3}$	1	0	9
0	$-\frac{100{,}000}{3}$	0	$\frac{200}{3}$	−200,000	0	1	5,600,000

$\xrightarrow{\frac{1}{3}R_3}$

x	y	z	u	v	w	P	Const.
1	$\frac{1}{3}$	0	$\frac{1}{3000}$	−4	0	0	10
0	0	1	0	1	0	0	6
0	$\frac{2}{9}$	0	$-\frac{1}{9000}$	1	$\frac{1}{3}$	0	3
0	$-\frac{100{,}000}{3}$	0	$\frac{200}{3}$	−200,000	0	1	5,600,000

$\xrightarrow{\substack{R_1+4R_3\\R_2-R_3\\R_4+200{,}000R_3}}$

x	y	z	u	v	w	P	Const.
1	$\frac{11}{9}$	0	$-\frac{1}{9000}$	0	$\frac{4}{3}$	0	22
0	$-\frac{2}{9}$	1	$\frac{1}{9000}$	0	$-\frac{1}{3}$	0	3
0	$\frac{2}{9}$	0	$-\frac{1}{9000}$	1	$\frac{1}{3}$	0	3
0	$\frac{100{,}000}{9}$	0	$\frac{400}{9}$	0	$\frac{200{,}000}{3}$	1	6,200,000

We conclude that $x = 22$, $y = 0$, $z = 3$, $u = 0$, $v = 3$, and $P = 6{,}200{,}000$. Therefore, the company should buy 22 minutes of morning and 3 minutes of evening advertising time, thereby maximizing their exposure to 6,200,000 viewers.

37. We first tabulate the given information:

Department	*Model A*	*Model B*	*Model C*	*Time available*
Fabrication	$\frac{5}{4}$	$\frac{3}{2}$	$\frac{3}{2}$	310
Assembly	1	1	$\frac{3}{4}$	205
Finishing	1	1	$\frac{1}{2}$	190
Profit	26	28	24	

Let x, y, and z denote the number of units of model A, model B, and model C to be produced, respectively. Then the required linear programming problem is

$$\begin{aligned}
&\text{Maximize } P = 26x + 28y + 24z \text{ subject to}\\
&\tfrac{5}{4}x + \tfrac{3}{2}y + \tfrac{3}{2}z \le 310\\
&x + y + \tfrac{3}{4}z \le 205\\
&x + y + \tfrac{1}{2}z \le 190\\
&x \ge 0, y \ge 0, z \ge 0
\end{aligned}$$

Using the simplex method, we obtain the following tableaus:

	x	y	z	u	v	w	P	*Const.*	*Ratio*
	$\frac{5}{4}$	$\frac{3}{2}$	$\frac{3}{2}$	1	0	0	0	310	$206\frac{2}{3}$
	1	1	$\frac{3}{4}$	0	1	0	0	205	205
p.r. →	1	[1]	$\frac{1}{2}$	0	0	1	0	190	190
	−26	−28	−24	0	0	0	1	0	
		↑ p.c.							

$R_1 - \frac{3}{2}R_3$
$R_2 - R_3$
$R_4 + 28R_3$ →

	x	y	z	u	v	w	P	Const.	Ratio
p.r. →	$-\frac{1}{4}$	0	$\boxed{\frac{3}{4}}$	1	0	$-\frac{3}{2}$	0	25	$33\frac{1}{3}$
	0	0	$\frac{1}{4}$	0	1	-1	0	15	60
	1	1	$\frac{1}{2}$	0	0	1	0	190	380
	2	0	-10	0	0	28	1	5320	
			↑ p.c.						

$\xrightarrow{\frac{4}{3}R_1}$

x	y	z	u	v	w	P	Const.
$-\frac{1}{3}$	0	1	$\frac{4}{3}$	0	-2	0	$\frac{100}{3}$
0	0	$\frac{1}{4}$	0	1	-1	0	15
1	1	$\frac{1}{2}$	0	0	1	0	190
2	0	-10	0	0	28	1	5320

$\xrightarrow[R_4+10R_1]{R_2-\frac{1}{4}R_1,\; R_3-\frac{1}{2}R_1}$

	x	y	z	u	v	w	P	Const.	Ratio
	$-\frac{1}{3}$	0	1	$\frac{4}{3}$	0	-2	0	$\frac{100}{3}$	--
p.r. →	$\boxed{\frac{1}{12}}$	0	0	$-\frac{1}{3}$	1	$-\frac{1}{2}$	0	$\frac{20}{3}$	80
	$\frac{7}{6}$	1	0	$-\frac{2}{3}$	0	2	0	$\frac{520}{3}$	$148\frac{4}{7}$
	$-\frac{4}{3}$	0	0	$\frac{40}{3}$	0	8	1	$\frac{16960}{3}$	
	↑ p.c.								

$\xrightarrow{12R_2}$

x	y	z	u	v	w	P	Const.
$-\frac{1}{3}$	0	1	$\frac{4}{3}$	0	-2	0	$\frac{100}{3}$
1	0	0	-4	12	-6	0	80
$\frac{7}{6}$	1	0	$-\frac{2}{3}$	0	2	0	$\frac{520}{3}$
$-\frac{4}{3}$	0	0	$\frac{40}{3}$	0	8	1	$\frac{16960}{3}$

$\xrightarrow[R_4+\frac{4}{3}R_2]{R_1+\frac{1}{3}R_2,\; R_3-\frac{7}{6}R_2}$

x	y	z	u	v	w	P	Const.
0	0	1	0	4	−4	0	60
1	0	0	−4	12	−6	0	80
0	1	0	4	−14	9	0	80
0	0	0	8	16	0	1	5760

The last tableau is in final form. We see that $x = 80$, $y = 80$, $z = 60$, $u = 0$, $v = 0$, $w = 0$, and $P = 5760$. So, by producing 80 units each of Models A and B, and 60 units of Model C, the company stands to make a profit of \$5760. Since $u = v = w = 0$, there are no resources left over.

39. Let x, y, and z denote the number (in thousands) of bottles of formula *I*, formula *II*, and formula *III*, respectively, produced. The resulting linear programming problem is

$$\text{Maximize } P = 180x + 200y + 300z \text{ subject to}$$
$$\tfrac{5}{2}x + 3y + 4z \le 70$$
$$x \le 9$$
$$y \le 12$$
$$z \le 6$$
$$x \ge 0,\ y \ge 0,\ z \ge 0$$

Using the simplex method, we have

	x	y	z	s	t	u	v	P	Const.	Ratio
	$\frac{5}{2}$	3	4	1	0	0	0	0	70	$17\frac{1}{2}$
	1	0	0	0	1	0	0	0	9	--
	0	1	0	0	0	1	0	0	12	--
p.r. →	0	0	[1]	0	0	0	1	0	6	6
	−180	−200	−300	0	0	0	0	1	0	
			↑ p.c.							

$\xrightarrow[R_5 + 300R_4]{R_1 - 4R_4}$

	x	y	z	s	t	u	v	P	Const.	Ratio
	$\frac{5}{2}$	3	0	1	0	0	-4	0	46	$15\frac{1}{3}$
	1	0	0	0	1	0	0	0	9	--
p.r. →	0	[1]	0	0	0	1	0	0	12	12
	0	0	1	0	0	0	1	0	6	--
	-180	-200	0	0	0	0	300	1	1800	
		↑ p.c.								

$\xrightarrow[R_5+200R_3]{R_1-3R_3}$

x	y	z	s	t	u	v	P	Const.	Ratio
$\frac{5}{2}$	0	0	1	0	-3	-4	0	10	4
1	0	0	0	1	0	0	0	9	9
0	1	0	0	0	1	0	0	12	--
0	0	1	0	0	0	1	0	6	--
-180	0	0	0	0	200	300	1	4200	

$\xrightarrow{\frac{2}{5}R_1}$

x	y	z	s	t	u	v	P	Const.
1	0	0	$\frac{2}{5}$	0	$-\frac{6}{5}$	$-\frac{8}{5}$	0	4
1	0	0	0	1	0	0	0	9
0	1	0	0	0	1	0	0	12
0	0	1	0	0	0	1	0	6
-180	0	0	0	0	200	300	1	4200

$\xrightarrow[R_5+180R_1]{R_2-R_1}$

x	y	z	s	t	u	v	P	Const.	Ratio
1	0	0	$\frac{2}{5}$	0	$-\frac{6}{5}$	$-\frac{8}{5}$	0	4	--
1	0	0	$-\frac{2}{5}$	1	[$\frac{6}{5}$]	$\frac{8}{5}$	0	5	$\frac{25}{6}$
0	1	0	0	0	1	0	0	12	12
0	0	1	0	0	0	1	0	6	--
0	0	0	72	0	-16	12	1	4920	

$\xrightarrow{\frac{5}{6}R_2}$

x	y	z	s	t	u	v	P	Const.
1	0	0	$\frac{2}{5}$	0	$-\frac{6}{5}$	$-\frac{8}{5}$	0	4
0	0	0	$-\frac{1}{3}$	$\frac{5}{6}$	1	$\frac{4}{3}$	0	$\frac{25}{6}$
0	1	0	0	0	1	0	0	12
0	0	1	0	0	0	1	0	6
0	0	0	72	0	−16	12	1	4920

$\xrightarrow{R_1+\frac{6}{5}R_2,\ R_3-R_2,\ R_5+16R_2}$

x	y	z	s	t	u	v	P	Const.
1	0	0	0	1	0	0	0	9
0	0	0	$-\frac{1}{3}$	$\frac{5}{6}$	1	$\frac{4}{3}$	0	$\frac{25}{6}$
0	1	0	$\frac{1}{3}$	$-\frac{5}{6}$	0	$-\frac{4}{3}$	0	$\frac{47}{6}$
0	0	1	0	0	0	1	0	6
0	0	0	$\frac{200}{3}$	$\frac{40}{3}$	0	$\frac{100}{3}$	1	$4986\frac{2}{3}$

Therefore, $x = 9, y = 47/6, z = 6, s = 0, t = 0, u = \frac{25}{6}$ and $P \approx 4986.67$; that is, the company should manufacture 9000 bottles of formula *I*, 7833 bottles of formula *II*, and 6000 bottles of formula *III* for a maximum profit of \$4986.67. Yes, ingredients for 4167 bottles of formula *II*.

41. Refer to Section 3.2, Exercise 15, of this manual. Let u, v, w, and s be slack variables. We obtain the following tableaus.

	x	y	z	u	v	w	s	P	Const.	Ratio
	1	1	1	1	0	0	0	0	2000000	2000000
p.r. →	−2	−2	[8]	0	1	0	0	0	0	0
	−6	4	4	0	0	1	0	0	0	0
	−10	6	6	0	0	0	1	0	0	0
	$-\frac{1}{10}$	$-\frac{3}{20}$	$-\frac{1}{5}$	0	0	0	0	1	0	
			↑ p.c.							

$\xrightarrow{\frac{1}{8}R_2}$

x	y	z	u	v	w	s	P	Const.
1	1	1	1	0	0	0	0	2000000
$-\frac{1}{4}$	$-\frac{1}{4}$	$\boxed{1}$	0	$\frac{1}{8}$	0	0	0	0
-6	4	4	0	0	1	0	0	0
-10	6	6	0	0	0	1	0	0
$-\frac{1}{10}$	$-\frac{3}{20}$	$-\frac{1}{5}$	0	0	0	0	1	0

$\xrightarrow[R_4-6R_2 \quad R_5+\frac{1}{5}R_2]{R_1-R_2 \quad R_3-4R_2}$

x	y	z	u	v	w	s	P	Const.	Ratio
$\frac{5}{4}$	$\frac{5}{4}$	0	1	$-\frac{1}{8}$	0	0	0	2000000	1600000
$-\frac{1}{4}$	$-\frac{1}{4}$	1	0	$\frac{1}{8}$	0	0	0	0	--
-5	$\boxed{5}$	0	0	$-\frac{1}{2}$	1	0	0	0	0
$-\frac{17}{2}$	$\frac{15}{2}$	0	0	$-\frac{3}{4}$	0	1	0	0	0
$-\frac{3}{20}$	$-\frac{1}{5}$	0	0	$\frac{1}{40}$	0	0	1	0	

$\xrightarrow{\frac{1}{5}R_3}$

x	y	z	u	v	w	s	P	Const.
$\frac{5}{4}$	$\frac{5}{4}$	0	1	$-\frac{1}{8}$	0	0	0	2000000
$-\frac{1}{4}$	$-\frac{1}{4}$	1	0	$\frac{1}{8}$	0	0	0	0
-1	$\boxed{1}$	0	0	$-\frac{1}{10}$	$\frac{1}{5}$	0	0	0
$-\frac{17}{2}$	$\frac{15}{2}$	0	0	$-\frac{3}{4}$	0	1	0	0
$-\frac{3}{20}$	$-\frac{1}{5}$	0	0	$\frac{1}{40}$	0	0	1	0

$\xrightarrow[R_4-\frac{15}{2}R_3 \quad R_5+\frac{1}{5}R_3]{R_1-\frac{5}{4}R_3 \quad R_2+\frac{1}{4}R_3}$

x	y	z	u	v	w	s	P	Const.	Ratio
$\boxed{\frac{5}{2}}$	0	0	1	$-\frac{1}{4}$	0	0	0	2000000	800000
$-\frac{1}{2}$	0	1	0	$\frac{1}{10}$	$\frac{1}{20}$	0	0	0	--
-1	1	0	0	$-\frac{1}{10}$	$\frac{1}{5}$	0	0	0	0
-1	0	0	0	0	$-\frac{3}{2}$	1	0	0	0
$-\frac{7}{20}$	0	0	0	$\frac{1}{200}$	$\frac{1}{25}$	0	1	0	

$\xrightarrow{\frac{2}{5}R_1}$

x	y	z	u	v	w	s	P	Const.
$\boxed{1}$	0	0	$\frac{2}{5}$	0	$-\frac{1}{10}$	0	0	800000
$-\frac{1}{2}$	0	1	0	$\frac{1}{10}$	$\frac{1}{20}$	0	0	0
-1	1	0	0	$-\frac{1}{10}$	$\frac{1}{5}$	0	0	0
-1	0	0	0	0	$-\frac{3}{2}$	1	0	0
$-\frac{7}{20}$	0	0	0	$\frac{1}{200}$	$\frac{1}{25}$	0	1	0

$\xrightarrow[\substack{R_4+R_1 \\ R_5+\frac{7}{20}R_1}]{\substack{R_2+\frac{1}{2}R_1 \\ R_3+R_1}}$

x	y	z	u	v	w	s	P	Const.
1	0	0	$\frac{2}{5}$	0	$-\frac{1}{10}$	0	0	800000
0	0	1	$\frac{1}{5}$	$\frac{1}{10}$	0	0	0	400000
0	1	0	$\frac{2}{5}$	$-\frac{1}{10}$	$\frac{1}{10}$	0	0	800000
0	0	0	$\frac{2}{5}$	0	$-\frac{8}{5}$	1	0	800000
0	0	0	$\frac{7}{50}$	$\frac{1}{200}$	$\frac{1}{200}$	0	1	280000

The last tableau is in final form, and we see that $x = 800{,}000$, $y = 800{,}000$, $z = 400{,}000$, and $P = 280{,}000$. Thus, the financier should invest $800,000 each in Projects A and B, and $400,000 in Project C. The maximum returns are $280,000.

43. False. Consider the linear programming problem

Maximize $P = 2x + 3y$ subject to

$$-x + y \le 0$$
$$x \ge 0,\ y \ge 0$$

45. True. Consider the objective function $P = c_1x + c_2x_2 + \cdots c_nx_n$, which may be written in the form

$$-c_1x_1 - c_2x_2 - \cdots - c_nx_n + P = 0$$

Observe that the most negative of the numbers $-c_1, -c_2, \cdots, -c_n$ (which are the numbers comprising the last row of the simplex tableau) is just the largest coefficient of x_i in the expression for *P*. Thus, moving in the direction of the variable with this coefficient ensures that *P* increases most.

USING TECHNOLOGY EXERCISES 4.1, page 242

1. $x = 1.2, y = 0, z = 1.6, w = 0$, and $P = 8.8$

3. $x = 1.6, y = 0, z = 0, w = 3.6$, and $P = 12.4$

4.2 Problem Solving Tips

Here are some hints for solving the problems in the exercises that follow.

1. Dual problems are standard minimization problems. The given problem is called the primal problem and the problem related to it is called the dual problem.

2. To solve a dual problem, first *write down the tableau* for the primal problem. (Note that this is not a simplex tableau as there are no slack variables.) Then *interchange the columns and rows* of this tableau. Use this tableau to *write the dual problem* and then use the simplex method to complete the solution to the problem. The minimum value of C will appear in the lower right corner of the final simplex tableau.

4.2 CONCEPT QUESTIONS, page 252

1. Maximize $P = -C = 3x + 5y$
   subject to
   $$5x + 2y \leq 30$$
   $$x + 3y \leq 21$$
   $$x \geq 0,\ y \geq 0$$

3. The primal problem is the linear programming (maximization) problem associated with a minimization linear programming problem. The dual problem is the linear programming (minimization) problem associated with the maximization linear programming problem.

EXERCISES 4.2, page 252

1. We solve the associated regular problem:

Maximize $P = -C = 2x - y$ subject to

$$x + 2y \le 6$$
$$3x + 2y \le 12$$
$$x \ge 0,\ y \ge 0$$

Using the simplex method where u and v are slack variables, we have

	x	y	u	v	P	Const.	Ratio
	1	2	1	0	0	6	6
p.r. →	[3]	2	0	1	0	12	4
	−2	1	0	0	1	0	
	↑ p.c.						

$\xrightarrow{\frac{1}{3}R_2}$

x	y	u	v	P	Const.
1	2	1	0	0	6
1	$\frac{2}{3}$	0	$\frac{1}{3}$	0	4
−2	1	0	0	1	0

$\xrightarrow{\substack{R_1 - R_2 \\ R_3 + 2R_2}}$

x	y	u	v	P	Const.
0	$\frac{4}{3}$	1	$-\frac{1}{3}$	0	2
1	$\frac{2}{3}$	0	$\frac{1}{3}$	0	4
0	$\frac{7}{3}$	0	$\frac{2}{3}$	1	8

Therefore, $x = 4$, $y = 0$, and $C = -P = -8$.

3. We maximize $P = -C = 3x + 2y$. Using the simplex method, we obtain

	x	y	u	v	P	Const.	Ratio
	3	4	1	0	0	24	8
p.r. →	[7]	−4	0	1	0	16	$\frac{16}{7}$
	−3	−2	0	0	1	0	
	↑ p.c.						

$\xrightarrow{\frac{1}{7}R_2}$

x	y	u	v	P	Const.
3	4	1	0	0	24
1	$-\frac{4}{7}$	0	$\frac{1}{7}$	0	$\frac{16}{7}$
−3	−2	0	0	1	0

$\xrightarrow{\substack{R_1 - 3R_2 \\ R_3 + 3R_2}}$

	x	y	u	v	P	Const.	Ratio	
p.r. →	0	$\boxed{\frac{40}{7}}$	1	$-\frac{3}{7}$	0	$\frac{120}{7}$	3	$\xrightarrow{\frac{7}{40}R_1}$
	1	$-\frac{4}{7}$	0	$\frac{1}{7}$	0	$\frac{16}{7}$	--	
	0	$-\frac{26}{7}$	0	$\frac{3}{7}$	1	$\frac{48}{7}$		
		↑ p.c.						

x	y	u	v	P	Const.
0	1	$\frac{7}{40}$	$-\frac{3}{40}$	0	3
1	$-\frac{4}{7}$	0	$\frac{1}{7}$	0	$\frac{16}{7}$
0	$-\frac{26}{7}$	0	$\frac{3}{7}$	1	$\frac{48}{7}$

$\xrightarrow[R_3+\frac{26}{7}R_1]{R_2+\frac{4}{7}R_1}$

x	y	u	v	P	Const.
0	1	$\frac{7}{40}$	$-\frac{3}{40}$	0	3
1	0	$\frac{1}{10}$	$\frac{1}{10}$	0	4
0	0	$\frac{13}{20}$	$\frac{3}{20}$	1	18

The last tableau is in final form. We find $x = 4$, $y = 3$, and $C = -P = -18$.

5. We maximize $P = -C = -2x + 3y + 4z$ subject to the given constraints. Using the simplex method we obtain

	x	y	z	u	v	w	P	Const.	Ratio	
	−1	2	−1	1	0	0	0	8	--	
p.r. →	1	−2	$\boxed{2}$	0	1	0	0	10	5	$\xrightarrow{\frac{1}{2}R_2}$
	2	4	−3	0	0	1	0	12	--	
	2	−3	−4	0	0	0	1	0		
			↑ p.c.							

x	y	z	u	v	w	P	Const.
−1	2	−1	1	0	0	0	8
$\frac{1}{2}$	−1	1	0	$\frac{1}{2}$	0	0	5
2	4	−3	0	0	1	0	12
2	−3	−4	0	0	0	1	0

$\xrightarrow{R_1+R_2,\ R_3+3R_2,\ R_4+4R_2}$

	x	y	z	u	v	w	P	Const.	Ratio
p.r. →	$-\frac{1}{2}$	$\boxed{1}$	0	1	$\frac{1}{2}$	0	0	13	13
	$\frac{1}{2}$	-1	1	0	$\frac{1}{2}$	0	0	5	--
	$\frac{7}{2}$	1	0	0	$\frac{3}{2}$	1	0	27	27
	4	-7	0	0	2	0	1	20	
		↑ p.c.							

$\xrightarrow{\substack{R_2+R_1 \\ R_3-R_1 \\ R_4+7R_1}}$

x	y	z	u	v	w	P	Const.
$-\frac{1}{2}$	1	0	1	$\frac{1}{2}$	0	0	13
0	0	1	1	1	0	0	18
4	0	0	-1	1	1	0	14
$\frac{1}{2}$	0	0	7	$\frac{11}{2}$	0	1	111

The last tableau is in final form. We see that $x = 0$, $y = 13$, $z = 18$, $w = 14$, and $C = -P = -111$.

7. $x = 5/4$, $y = 1/4$, $u = 2$, $v = 3$, and $C = P = 13$.

9. $x = 5$, $y = 10$, $z = 0$, $u = 1$, $v = 2$, and $C = P = 80$.

11. We first write the tableau

x	y	Const.
1	2	4
3	2	6
2	5	

Then obtain the following by interchanging rows and columns:

u	v	Const.
1	3	2
2	2	5
4	6	

From this table we construct the dual problem:

Maximize the objective function

$$P = 4u + 6v \text{ subject to}$$

$$u + 3v \le 2$$

$$2u + 2v \le 5$$

$$u \ge 0,\ v \ge 0$$

Solving the dual problem using the simplex method with x and y as the slack variables, we obtain

	u	v	x	y	P	Const.	Ratio
p.r. →	1	[3]	1	0	0	2	$\frac{2}{3}$
	2	2	0	1	0	5	$\frac{5}{2}$
	−4	−6	0	0	1	0	
		↑ p.c.					

$\xrightarrow{\frac{1}{3}R_1}$

u	v	x	y	P	Const.
$\frac{1}{3}$	1	$\frac{1}{3}$	0	0	$\frac{2}{3}$
2	2	0	1	0	5
−4	−6	0	0	1	0

$\xrightarrow[R_3+6R_1]{R_2-2R_1}$

	u	v	x	y	P	Const.	Ratio
p.r. →	[$\frac{1}{3}$]	1	$\frac{1}{3}$	0	0	$\frac{2}{3}$	2
	$\frac{4}{3}$	0	$-\frac{2}{3}$	1	0	$\frac{11}{3}$	$2\frac{3}{4}$
	−2	0	2	0	1	4	
	↑ p.c.						

$\xrightarrow{3R_1}$

u	v	x	y	P	Const.
1	3	1	0	0	2
$\frac{4}{3}$	0	$-\frac{2}{3}$	1	0	$\frac{11}{3}$
−2	0	2	0	1	4

$\xrightarrow[R_3+2R_1]{R_2-\frac{4}{3}R_1}$

u	v	x	y	P	Const.
1	3	1	0	0	2
0	–4	–2	1	0	1
0	6	4	0	1	8

Interpreting the final tableau, we see that $x = 4$, $y = 0$, and $P = C = 8$.

13. We first write the tableau

x	y	Const.
6	1	60
2	1	40
1	1	30
6	4	

Then we obtain the following tableau by interchanging rows and columns:

u	v	w	Const.
6	2	1	6
1	1	1	4
60	40	30	

From this table we construct the dual problem:

Maximize $P = 60u + 40v + 30w$ subject to

$$6u + 2v + w \le 6$$

$$u + v + w \le 4$$

$$u \ge 0,\ v \ge 0,\ w \ge 0$$

We solve the problem as follows.

	u	v	w	x	y	P	Const.	Ratio	
p.r. →	[6]	2	1	1	0	0	6	1	$\xrightarrow{\frac{1}{6}R_1}$
	1	1	1	0	1	0	4	4	
	–60	–40	–30	0	0	1	0	--	
	↑ p.c.								

u	v	w	x	y	P	Const.
1	$\frac{1}{3}$	$\frac{1}{6}$	$\frac{1}{6}$	0	0	1
1	1	1	0	1	0	4
−60	−40	−30	0	0	1	0

$\xrightarrow[R_3+60R_1]{R_2-R_1}$

	u	v	w	x	y	P	Const.	Ratio
	1	$\frac{1}{3}$	$\frac{1}{6}$	$\frac{1}{6}$	0	0	1	6
p.r. →	0	$\frac{2}{3}$	$\boxed{\frac{5}{6}}$	$-\frac{1}{6}$	1	0	3	18 / 5
	0	−20	−20	10	0	1	60	--
			↑ p.c.					

$\xrightarrow{\frac{6}{5}R_2}$

u	v	w	x	y	P	Const.
1	$\frac{1}{3}$	$\frac{1}{6}$	$\frac{1}{6}$	0	0	1
0	$\frac{4}{5}$	1	$-\frac{1}{5}$	$\frac{6}{5}$	0	$\frac{18}{5}$
0	−20	−20	10	0	1	60

$\xrightarrow[R_3+20R_2]{R_1-\frac{1}{6}R_2}$

	u	v	w	x	y	P	Const.	Ratio
p.r. →	1	$\boxed{\frac{1}{5}}$	0	$\frac{1}{5}$	$-\frac{1}{5}$	0	$\frac{2}{5}$	2
	0	$\frac{4}{5}$	1	$-\frac{1}{5}$	$\frac{6}{5}$	0	$\frac{18}{5}$	9 / 2
	0	−4	0	6	24	1	132	--
		↑ p.c.						

$\xrightarrow{5R_1}$

u	v	w	x	y	P	Const.
5	1	0	1	−1	0	2
0	$\frac{4}{5}$	1	$-\frac{1}{5}$	$\frac{6}{5}$	0	$\frac{18}{5}$
0	−4	0	6	24	1	132

$\xrightarrow[R_3+4R_1]{R_2-\frac{4}{5}R_1}$

u	v	w	x	y	P	Const.
5	1	0	1	–1	0	2
–4	0	1	–1	2	0	2
20	0	0	10	20	1	140

The last tableau is in final form. We find that $x = 10, y = 20$, and $C = 140$.

15. We first write the tableau

x	y	z	Const.
20	10	1	10
1	1	2	20
200	150	120	

Then obtain the following by interchanging rows and columns:

u	v	Const.
20	1	200
10	1	150
1	2	120
10	20	

From this table we construct the dual problem:

Maximize $P = 10u + 20v$ subject to

$$20u + v \leq 200$$
$$10u + v \leq 150$$
$$u + 2v \leq 120$$
$$u \geq 0, v \geq 0$$

Solving this problem, we obtain the following tableaus:

	u	v	x	y	z	P	Const.	Ratio	
	20	1	1	0	0	0	200	200	
	10	1	0	1	0	0	150	150	$\xrightarrow{\frac{1}{2}R_3}$
p.r. →	1	[2]	0	0	1	0	120	60	
	–10	–20	0	0	0	1	0		
		↑ p.c.							

u	v	x	y	z	P	Const.
20	1	1	0	0	0	200
10	1	0	1	0	0	150
$\frac{1}{2}$	1	0	0	$\frac{1}{2}$	0	60
−10	−20	0	0	0	1	0

$\xrightarrow{\substack{R_1-R_3 \\ R_2-R_3 \\ R_4+20R_3}}$

u	v	x	y	z	P	Const.
$\frac{39}{2}$	0	1	0	$-\frac{1}{2}$	0	140
$\frac{19}{2}$	0	0	1	$-\frac{1}{2}$	0	90
$\frac{1}{2}$	1	0	0	$\frac{1}{2}$	0	60
0	0	0	0	10	1	1200

This last tableau is in final form. We find that $x = 0$, $y = 0$, $z = 10$, and $C = 1200$.

17. We first write the tableau

x	y	z	Const.
1	2	2	10
2	1	1	24
1	1	1	16
6	8	4	

Then we obtain the following tableau by interchanging rows and columns:

u	v	w	Const.
1	2	1	6
2	1	1	8
2	1	1	4
10	24	16	

From this table we construct the dual problem:

Maximize the objective function

$P = 10u + 24v + 16w$ subject to

$$u + 2v + w \le 6$$
$$2u + v + w \le 8$$
$$2u + v + w \le 4$$
$$u \ge 0, v \ge 0, w \ge 0$$

Solving the dual problem using the simplex method with x, y, and z as slack variables, we obtain

	u	v	w	x	y	z	P	Const.	Ratio
p.r. →	1	[2]	1	1	0	0	0	6	3
	2	1	1	0	1	0	0	8	8
	2	1	1	0	0	1	0	4	4
	−10	−24	−16	0	0	0	1	0	
		↑ p.c.							

$\xrightarrow{\frac{1}{2}R_1}$

u	v	w	x	y	z	P	Const.
$\frac{1}{2}$	1	$\frac{1}{2}$	$\frac{1}{2}$	0	0	0	3
2	1	1	0	1	0	0	8
2	1	1	0	0	1	0	4
−10	−24	−16	0	0	0	1	0

$\xrightarrow[R_4 + 24R_1]{R_2 - R_1,\ R_3 - R_1}$

	u	v	w	x	y	z	P	Const.	Ratio
	$\frac{1}{2}$	1	$\frac{1}{2}$	$\frac{1}{2}$	0	0	0	3	6
	$\frac{3}{2}$	0	$\frac{1}{2}$	$-\frac{1}{2}$	1	0	0	5	10
p.r. →	$\frac{3}{2}$	0	[$\frac{1}{2}$]	$-\frac{1}{2}$	0	1	0	1	2
	2	0	−4	12	0	0	1	72	
			↑ p.c.						

$\xrightarrow{2R_3}$

u	v	w	x	y	z	P	Const.
$\frac{1}{2}$	1	$\frac{1}{2}$	$\frac{1}{2}$	0	0	0	3
$\frac{3}{2}$	0	$\frac{1}{2}$	$-\frac{1}{2}$	1	0	0	5
3	0	1	–1	0	2	0	2
2	0	–4	12	0	0	1	72

$\xrightarrow[R_4+4R_3]{R_1-\frac{1}{2}R_3,\; R_2-\frac{1}{2}R_3}$

u	v	w	x	y	z	P	Const.
–1	1	0	1	0	–1	0	2
0	0	0	0	1	–1	0	4
3	0	1	–1	0	2	0	2
14	0	0	8	0	8	1	80

The solution to the primal problem is $x = 8, y = 0, z = 8$, and $C = 80$.

19. We first write

x	y	z	Const.
2	4	3	6
6	0	1	2
0	6	2	4
30	12	20	

Then obtain the following by interchanging rows and columns:

u	v	w	Const.
2	6	0	30
4	0	6	12
3	1	2	20
6	2	4	

From this table we construct the dual problem:

Maximize $P = 6u + 2v + 4w$ subject to

$$2u + 6v \leq 30$$

$$4u + 6w \leq 12$$

$$3u + v + 2w \leq 20$$

$$u \geq 0,\ v \geq 0,\ w \geq 0$$

Using the simplex method, we obtain

	u	v	w	x	y	z	P	Const.	Ratio
	2	6	0	1	0	0	0	30	15
p.r. →	$\boxed{4}$	0	6	0	1	0	0	12	3
	3	1	2	0	0	1	0	20	$\frac{20}{3}$
	−6	−2	−4	0	0	0	1	0	
	↑ p.c.								

$\xrightarrow{\frac{1}{4}R_2}$

u	v	w	x	y	z	P	Const.
2	6	0	1	0	0	0	30
1	0	$\frac{3}{2}$	0	$\frac{1}{4}$	0	0	3
3	1	2	0	0	1	0	20
−6	−2	−4	0	0	0	1	0

$\xrightarrow[R_4+6R_2]{R_1-2R_2,\ R_3-3R_2}$

	u	v	w	x	y	z	P	Const.	Ratio
p.r. →	0	$\boxed{6}$	−3	1	$-\frac{1}{2}$	0	0	24	4
	1	0	$\frac{3}{2}$	0	$\frac{1}{4}$	0	0	3	---
	0	1	$-\frac{5}{2}$	0	$-\frac{3}{4}$	1	0	11	11
	0	−2	5	0	$\frac{3}{2}$	0	1	18	
		↑ p.c.							

$\xrightarrow{\frac{1}{6}R_1}$

u	v	w	x	y	z	P	Const.
0	1	$-\frac{1}{2}$	$\frac{1}{6}$	$-\frac{1}{12}$	0	0	4
1	0	$\frac{3}{2}$	0	$\frac{1}{4}$	0	0	3
0	1	$-\frac{5}{2}$	0	$-\frac{3}{4}$	1	0	11
0	−2	5	0	$\frac{3}{2}$	0	1	18

$\xrightarrow[R_4+2R_1]{R_3-R_1}$

u	v	w	x	y	z	P	$Const.$
0	1	$-\frac{1}{2}$	$\frac{1}{6}$	$-\frac{1}{12}$	0	0	4
1	0	$\frac{3}{2}$	0	$\frac{1}{4}$	0	0	3
0	0	−2	$-\frac{1}{6}$	$-\frac{2}{3}$	1	0	7
0	0	4	$\frac{1}{3}$	$\frac{4}{3}$	0	1	26

The last tableau is in final form. We find $x = 1/3$, $y = 4/3$, $z = 0$, and $C = 26$.

21. Let x denote the number of type-A vessels and y the number of type-B vessels to be operated. Then the problem is

Maximize $C = 44{,}000x + 54{,}000y$ subject to

$$60x + 80y \geq 360$$

$$160x + 120y \geq 680$$

$$x \geq 0, y \geq 0$$

We first write down the following tableau for the primal problem:

x	y	Constant
60	80	360
160	120	680
44,000	54,000	

Next, we interchange the columns and rows of the frequency tableaus obtaining

u	v	Constant
60	160	44,000
80	120	54,000
360	680	

Proceeding, we are led to the dual problem

Maximize $P = 360u + 680v$ subject to

$$60u + 160v \leq 44000$$

$$80u + 120v \leq 54000$$

$$u \geq 0, v \geq 0$$

Let x and y be slack variables. We obtain the following tableaus:

	u	v	x	y	P	Const.	Ratio
p.r. →	60	[160]	1	0	0	44,000	275
	80	120	0	1	0	54,000	450
	−360	−680	0	0	1	0	
		↑ p.c.					

$\xrightarrow{\frac{1}{160}R_1}$

u	v	x	y	P	Const.
$\frac{3}{8}$	[1]	$\frac{1}{160}$	0	0	275
80	120	0	1	0	54,000
−360	−680	0	0	1	0

$\xrightarrow[R_3+680R_1]{R_2-120R_1}$

	u	v	x	y	P	Const.	Ratio
	$\frac{3}{8}$	1	$\frac{1}{160}$	0	0	275	$733\frac{1}{3}$
p.r. →	[35]	0	$-\frac{3}{4}$	1	0	21000	600
	−105	0	$\frac{17}{4}$	0	1	187000	
	↑ p.c.						

$\xrightarrow{\frac{1}{35}R_1}$

u	v	x	y	P	Const.
$\frac{3}{8}$	1	$\frac{1}{160}$	0	0	275
[1]	0	$-\frac{3}{140}$	$\frac{1}{35}$	0	600
−105	0	$\frac{17}{4}$	0	1	187000

$\xrightarrow[R_3+105R_2]{R_1-\frac{3}{8}R_2}$

u	v	x	y	P	Const.
0	1	$\frac{1}{70}$	$-\frac{3}{280}$	0	50
1	0	$-\frac{3}{140}$	$\frac{1}{35}$	0	600
0	0	2	3	1	250000

The last tableau is in final form. The fundamental theorem of duality tells us that the solution to the primal problem is $x = 2, y = 3$ with a minimum value for C of 250,000. Thus, Deluxe River Cruises should use 2 type-A vessels and 3 type-B vessels. The minimum operating cost is \$250,000.

23. Let x and y denote, respectively, the number of advertisements to be placed in newspaper I and newspaper II. Then the problem is

$$\text{Minimize } C = 1000x + 800y \text{ subject to}$$
$$70000x + 10000y \geq 2000000$$
$$40000x + 20000y \geq 1400000$$
$$20000x + 40000y \geq 1000000$$
$$x \geq 0, y \geq 0$$

or upon simplification:

$$\text{Minimize } C = 1000x + 800y \text{ subject to}$$
$$7x + y \geq 200$$
$$2x + y \geq 70$$
$$x + 2y \geq 50$$
$$x \geq 0, y \geq 0$$

We first write down the following tableau for the primal problem

x	y	*Const.*
7	1	200
2	1	70
1	2	50
1000	800	

Next, we interchange the columns and rows of the foregoing tableau:

u	v	w	*Const.*
7	2	1	1000
1	1	2	800
200	70	50	

Proceeding, we obtain the following dual problem
Maximize $P = 200u + 70v + 50w$ subject to

$$7u + 2v + w \leq 1000$$
$$u + v + 2w \leq 800$$
$$u \geq 0, v \geq 0, w \geq 0$$

Let x and y denote the slack variables. We obtain the following tableaus:

	u	v	w	x	y	P	Const.	Ratio
p.r. →	$\boxed{7}$	2	1	1	0	0	1000	$142\frac{6}{7}$
	1	1	2	0	1	0	800	800
	–200	–70	–50	0	0	1	0	
	↑ p.c.							

$\xrightarrow{\frac{1}{7}R_1}$

u	v	w	x	y	P	Const.
$\boxed{1}$	$\frac{2}{7}$	$\frac{1}{7}$	$\frac{1}{7}$	0	0	$\frac{1000}{7}$
1	1	2	0	1	0	800
–200	–70	–50	0	0	1	0

$\xrightarrow[R_3+200R_1]{R_2-R_1}$

	u	v	w	x	y	P	Const.	Ratio
	1	$\frac{2}{7}$	$\frac{1}{7}$	$\frac{1}{7}$	0	0	$\frac{1000}{7}$	1000
p.r. →	0	$\frac{5}{7}$	$\boxed{\frac{13}{7}}$	$-\frac{1}{7}$	1	0	$\frac{4600}{7}$	$657\frac{1}{7}$
	0	$-\frac{90}{7}$	$-\frac{150}{7}$	$\frac{200}{7}$	0	1	$\frac{200000}{7}$	
			↑ p.c.					

$\xrightarrow{\frac{7}{13}R_2}$

u	v	w	x	y	P	Const.
1	$\frac{2}{7}$	$\frac{1}{7}$	$\frac{1}{7}$	0	0	$\frac{1000}{7}$
0	$\frac{5}{13}$	$\boxed{1}$	$-\frac{1}{13}$	$\frac{7}{13}$	0	$\frac{4600}{7}$
0	$-\frac{90}{7}$	$-\frac{150}{7}$	$\frac{200}{7}$	0	1	$\frac{200000}{7}$

$\xrightarrow[R_1+\frac{150}{7}R_2]{R_1-\frac{1}{7}R_2}$

	u	v	w	x	y	P	Const.	Ratio
p.r. →	1	$\boxed{\frac{3}{13}}$	0	$\frac{2}{13}$	$-\frac{1}{13}$	0	$\frac{1200}{13}$	400
	0	$\frac{5}{13}$	1	$-\frac{1}{13}$	$\frac{7}{13}$	0	$\frac{4600}{13}$	920
	0	$-\frac{60}{13}$	0	$\frac{300}{13}$	$\frac{150}{13}$	1	$\frac{470000}{13}$	
		↑ p.c.						

$\xrightarrow{\frac{13}{3}R_1}$

u	v	w	x	y	P	Const.
$\frac{13}{3}$	$\boxed{1}$	0	$\frac{2}{3}$	$-\frac{1}{3}$	0	400
0	$\frac{5}{13}$	1	$-\frac{1}{13}$	$\frac{7}{13}$	0	$\frac{4600}{13}$
0	$-\frac{60}{13}$	0	$\frac{300}{13}$	$\frac{150}{13}$	1	$\frac{470000}{13}$

$$\xrightarrow[R_3+\frac{60}{13}R_1]{R_2-\frac{5}{13}R_1}$$

u	v	w	x	y	P	Const.
$\frac{13}{3}$	1	0	$\frac{2}{3}$	$-\frac{1}{3}$	0	400
$-\frac{5}{2}$	0	1	$-\frac{1}{3}$	$\frac{2}{3}$	0	200
20	0	0	30	10	1	38000

The last tableau is in final form. The fundamental theorem of duality tells us that the solution to the primal problem is $x = 30$, $y = 10$, and $C = 38000$. Therefore, Everest Deluxe World Travel should place 30 advertisements in newspaper I and 10 advertisements in newspaper II at a minimum cost of $38,000.

25. The given data may be summarized as follows:

	Orange Juice	*Grapefruit Juice*
Vitamin A	60 I.U.	120 I.U.
Vitamin C	16 I.U.	12 I.U.
Calories	14	11

Suppose x ounces of orange juice and y ounces of pink-grapefruit juice are required for each glass of the blend. Then the problem is

Minimize $C = 14x + 11y$ subject to

$$60x + 120y \geq 1200$$

$$16x + 12y \geq 200$$

$$x \geq 0, y \geq 0$$

To construct the dual problem, we first write down the tableau

x	y	Const.
60	120	1200
16	12	200
14	11	

Then obtain the following by interchanging rows and columns:

u	v	Const.
60	16	14
120	12	11
1200	200	

From this table we construct the dual problem:

Maximize $P = 1200u + 200v$ subject to

$$60u + 16v \le 14$$

$$120u + 12v \le 11$$

$$u \ge 0, v \ge 0$$

The initial tableau is

u	v	x	y	P	Const.
60	16	1	0	0	14
120	12	0	1	0	11
−1200	−200	0	0	1	0

Using the following sequence of row operations,

1. $\frac{1}{120}R_2$ 2. $R_1 - 60R_2$, $R_3 + 1200R_2$ 3. $\frac{1}{10}R_1$ 4. $R_2 - \frac{1}{10}R_1$, $R_3 + 80R_1$

we obtain the final tableau

u	v	x	y	P	Const.
0	1	$\frac{1}{10}$	$-\frac{1}{20}$	0	$\frac{17}{20}$
0	0	$-\frac{1}{100}$	$\frac{1}{75}$	0	$\frac{1}{150}$
0	0	8	6	1	178

We conclude that the owner should use 8 ounces of orange juice and 6 ounces of pink grapefruit juice per glass of the blend for a minimal calorie count of 178.

27. True. To maximize P, one maximizes $-C$. Since the minimization problem has a unique solution, the negative of that solution is the solution of the maximization problem.

USING TECHNOLOGY EXERCISES 4.2, page 259

1. $x = \frac{4}{3}$, $y = \frac{10}{3}$, $z = 0$, and $C = \frac{14}{3}$

3. $x = 0.9524$, $y = 4.2857$, $z = 0$, and $C = 6.0952$.

5. a. $x = 3, y = 2,$ and $P = 17$ b. $\frac{8}{3} \le c_1 \le 8; \frac{3}{2} \le c_2 \le \frac{9}{2}$
c. $8 \le b_1 \le 24; 4 \le b_2 \le 12$ d. $\frac{5}{4}; \frac{1}{4}$ *e.* Both constraints are binding

7. a. $x = 4, y = 0,$ and $C = 8$ b. $0 \le c_1 \le \frac{5}{2}; 4 \le c_2 < \infty$
c. $3 \le b_1 < \infty; -\infty < b_2 \le 4$ d. 2;0 e. Both constraints are binding.

4.3 Problem Solving Tips

Here are some hints for solving the problems in the exercises that follow.

1. If you are solving a problem involving mixed constraints, make sure that the problem is written as a maximization problem. All constraints in a mixed constraint problem except $x \ge 0,\ y \ge 0,\ z \ge 0, \ldots$ should be written as $\ge$ constraints. If a constraint is in the form of an equality, rewrite it in the form of two equivalaent inequalities.

2. To find the *pivot element in a mixed constraint problem*, first check to see if there are any negative entries in the column of constants. If so, pick any negative entry in the row in which a negative entry in the column of constants occurs. (If there are no negative entries, use the simplex method to solve the problem.) The column containing this entry is the pivot column. Now locate the pivot row by computing the *positive* ratios of the numbers in the column of constants to the corresponding numbers in the pivot column (excluding the last row). The smallest ratio corresponds to the pivot row. The pivot element occurs at the intersection of the pivot row and the pivot column.

4.3 CONCEPT QUESTIONS, page 269

1. It is not a standard maximization problem because the second inequality in the system of constraints cannot be written in a form in which the expression involving the variables is less than or equal to a nonnegative constant.

3. It is not a standard maximization problem because the second constraint in the system of constraints is an equation. It cannot be rewritten as a restricted minimization problem because if the problem is written as a minimization problem, the objective function $C = -P = -x - 3y$ has coefficients that are not all nonnegative.

EXERCISES 4.3, page 270

1. Maximize $P = -C = -2x + 3y$ subject to

$$\begin{aligned} -3x - 5y &\le -20 \\ 3x + y &\le 16 \\ -2x + y &\le 1 \\ x \ge 0, y &\ge 0 \end{aligned}$$

3. Maximize $P = -C = -5x - 10y - z$ subject to

$$\begin{aligned} -2x - y - z &\le -4 \\ -x - 2y - 2z &\le -2 \\ 2x + 4y + 3z &\le 12 \\ x \ge 0,\ y \ge 0, \text{ and } z &\ge 0 \end{aligned}$$

5. We set up the tableau and solve the problem using the simplex method:

	x	y	u	v	P	Const	Ratio
	2	[5]	1	0	0	20	4
p.r. →	1	–5	0	1	0	–5	1
	–1	–2	0	0	1	0	
		↑ p.c.					

$\xrightarrow{-\frac{1}{5}R_2}$

x	y	u	v	P	Const
2	5	1	0	0	20
$-\frac{1}{5}$	1	0	$-\frac{1}{5}$	0	1
–1	–2	0	0	1	0

$\xrightarrow[R_3+2R_2]{R_1-5R_2}$

x	y	u	v	P	Const	Ratio
$\boxed{3}$	0	1	1	0	15	5
$-\frac{1}{5}$	1	0	$-\frac{1}{5}$	0	1	--
$-\frac{7}{5}$	0	0	$-\frac{2}{5}$	1	2	

$\xrightarrow{\frac{1}{3}R_1}$

x	y	u	v	P	Const
1	0	$\frac{1}{3}$	$\frac{1}{3}$	0	5
$-\frac{1}{5}$	1	0	$-\frac{1}{5}$	0	1
$-\frac{7}{5}$	0	0	$-\frac{2}{5}$	1	2

$\xrightarrow[R_3+\frac{7}{5}R_1]{R_2+\frac{1}{5}R_1}$

x	y	u	v	P	Const
1	0	$\frac{1}{3}$	$\frac{1}{3}$	0	5
0	1	$\frac{1}{15}$	$-\frac{2}{15}$	0	2
0	0	$\frac{7}{15}$	$\frac{1}{15}$	1	9

The maximum value of P is 9 when $x = 5$ and $y = 2$.

7. We first rewrite the problem as a maximization problem with inequality constraints using $\le$, obtaining the following equivalent problem:

Maximize $P = -C = 2x - y$ subject to

$$x + 2y \le 6$$
$$3x + 2y \le 12$$
$$x \ge 0, y \ge 0$$

Following the procedure outlined for nonstandard problems, we have

	x	y	u	v	P	Const	Ratio
	1	2	1	0	0	6	6
p.r. →	$\boxed{3}$	2	0	1	0	12	4
	−2	1	0	0	1	0	
	↑ p.c.						

$\xrightarrow{\frac{1}{3}R_2}$

x	y	u	v	P	Const
1	2	1	0	0	6
1	$\frac{2}{3}$	0	$\frac{1}{3}$	0	4
−2	1	0	0	1	0

$\xrightarrow[R_3+2R_2]{R_1-R_2}$

x	y	u	v	P	Const
0	$\frac{4}{3}$	1	$-\frac{1}{3}$	0	2
1	$\frac{2}{3}$	0	$\frac{1}{3}$	0	4
0	$\frac{7}{3}$	0	$\frac{2}{3}$	1	8

We conclude that C attains a minimum value of -8 when $x = 4$ and $y = 0$.

9. Using the simplex method we have

	x	y	u	v	P	Const	Ratio
	1	3	1	0	0	6	6
p.r. →	[−2]	3	0	1	0	−6	3
	−1	−4	0	0	1	0	
	↑ p.c.						

$\xrightarrow{-\frac{1}{2}R_2}$

x	y	u	v	P	Const
1	3	1	0	0	6
1	$-\frac{3}{2}$	0	$-\frac{1}{2}$	0	3
−1	−4	0	0	1	0

$\xrightarrow[R_3+R_2]{R_1-R_2}$

x	y	u	v	P	Const
0	$\frac{9}{2}$	1	$\frac{1}{2}$	0	3
1	$-\frac{3}{2}$	0	$-\frac{1}{2}$	0	3
0	$-\frac{11}{2}$	0	$-\frac{1}{2}$	1	3

$\xrightarrow{\frac{2}{9}R_1}$

x	y	u	v	P	Const
0	1	$\frac{2}{9}$	$\frac{1}{9}$	0	$\frac{2}{3}$
1	$-\frac{3}{2}$	0	$-\frac{1}{2}$	0	3
0	$-\frac{11}{2}$	0	$-\frac{1}{2}$	1	3

$\xrightarrow[R_3+\frac{11}{2}R_1]{R_2+\frac{3}{2}R_1}$

x	y	u	v	P	Const
0	1	$\frac{2}{9}$	$\frac{1}{9}$	0	$\frac{2}{3}$
1	0	$\frac{1}{3}$	$-\frac{1}{3}$	0	4
0	0	$\frac{11}{9}$	$\frac{1}{9}$	1	$\frac{20}{3}$

We conclude that P attains a maximum value of 20/3, when $x = 4$ and $y = 2/3$.

11. We rewrite the problem as

Maximize $P = x + 2y$ subject to

$$2x + 3y \le 12$$
$$-x + 3y \le 3$$
$$-x + 3y \ge 3$$
$$x \ge 0, y \ge 0$$

The initial tableau is

x	y	u	v	w	P	Const.
2	3	1	0	0	0	12
–1	3	0	1	0	0	3
1	–3	0	0	1	0	–3
–1	–2	0	0	0	1	0

Using the following sequence of row operations
1. $-\frac{1}{3}R_3$ 2. $R_1 - 3R_3$, $R_2 - 3R_3$, $R_4 + 2R_3$ 3. $\frac{1}{3}R_1$
4. $R_3 + \frac{1}{3}R_1$, $R_4 + \frac{5}{3}R_1$ 5. $R_1 - \frac{1}{3}R_2$, $R_3 + \frac{2}{9}R_2$, $R_4 + \frac{1}{9}R_2$
we obtain the final tableau

x	y	u	v	w	P	Const.
1	0	$\frac{1}{3}$	$-\frac{1}{3}$	0	0	3
0	0	0	1	1	0	0
0	1	$\frac{1}{9}$	$\frac{2}{9}$	0	0	2
0	0	$\frac{5}{9}$	$\frac{1}{9}$	0	1	7

We conclude that P attains a maximum value of 7 when $x = 3$ and $y = 2$.

13. We rewrite the problem as

Maximize $P = 5x + 4y + 2z$ subject to
$$x + 2y + 3z \le 24$$
$$-x + y - z \le -6$$
$$x \ge 0, y \ge 0, z \ge 0$$

The initial tableau is

x	y	z	u	v	P	Const.
1	2	3	1	0	0	24
–1	1	–1	0	1	0	–6
–5	–4	–2	0	0	1	0

Using the following sequence of row operations
1. $-R_2$ 2. $R_1 - 2R_2$, $R_3 + 5R_2$ 3. $\frac{1}{3}R_1$ 4. $R_2 + R_1$, $R_3 + 9R_1$
5. $3R_1$ 6, $R_2 + \frac{2}{3}R_1$; $R_3 + 2R_1$
we obtain the final tableau

x	y	z	u	v	P	Const.
0	3	2	1	1	0	18
1	2	3	1	0	0	24
0	6	13	5	0	1	120

We deduce that P attains a maximum value of 120 when $x = 24$, $y = 0$, and $z = 0$.

15. The problem is to maximize $P = -C = -x + 2y - z$ subject to the given constraints. The initial tableau is

x	y	z	u	v	w	P	Const.
1	−2	3	1	0	0	0	10
2	1	−2	0	1	0	0	15
2	1	3	0	0	1	0	20
1	−2	1	0	0	0	1	0

Using the following sequence of row operations,
1. $R_1 + 2R_2$, $R_3 - R_2$, $R_4 + 2R_2$ 2. $\frac{1}{5}R_3$ 3. $R_1 + R_3$, $R_2 + 2R_3$, $R_4 + 3R_3$
we obtain the final tableau

x	y	z	u	v	w	P	Const.
5	0	0	1	$\frac{9}{5}$	$\frac{1}{5}$	0	41
2	1	0	0	$\frac{3}{5}$	$\frac{2}{5}$	0	17
0	0	1	0	$-\frac{1}{5}$	$\frac{1}{5}$	0	1
5	0	0	0	$\frac{7}{5}$	$\frac{3}{5}$	1	33

We conclude that C attains a minimum value of -33 when $x = 0$, $y = 17$, $z = 1$, and $C = -P = -33$.

17. Rewriting the third constraint as $-x + 2y - z \leq -4$, we obtain the following initial tableau

x	y	z	u	v	w	P	*Const.*
1	2	3	1	0	0	0	28
2	3	−1	0	1	0	0	6
−1	2	−1	0	0	1	0	−4
−2	−1	−1	0	0	0	1	0

Using the following sequence of row operations,

1. $\frac{1}{2}R_2$ 2. $R_1 - R_2$, $R_3 + R_2$, $R_4 + 2R_2$ 3. $-\frac{2}{3}R_3$
4. $R_1 - \frac{7}{2}R_3$, $R_2 + \frac{1}{2}R_3$, $R_4 + 2R_3$ 5. $\frac{3}{26}R_1$ 6. $R_2 - \frac{1}{3}R_1$; $R_3 + \frac{7}{3}R_1$, $R_4 + \frac{8}{3}R_1$
7. $\frac{26}{7}R_1$ 8. $R_2 + \frac{11}{26}R_1$, $R_3 + \frac{1}{26}R_1$, $R_4 + \frac{8}{13}R_1$

we obtain the final tableau

x	y	z	u	v	w	P	*Const.*
0	$\frac{26}{7}$	0	$\frac{3}{7}$	$\frac{2}{7}$	1	0	$\frac{68}{7}$
1	$\frac{11}{7}$	0	$\frac{1}{7}$	$\frac{3}{7}$	1	0	$\frac{46}{7}$
0	$\frac{1}{7}$	1	$\frac{2}{7}$	$-\frac{1}{7}$	0	0	$\frac{50}{7}$
0	$\frac{16}{7}$	0	$\frac{4}{7}$	$\frac{5}{7}$	0	1	$\frac{142}{7}$

from which we deduce that P attains a maximum value of 142/7 when $x = 46/7$, $y = 0$, and $z = 50/7$.

19. Rewriting the third constraint $(2x + y + z = 10)$ in the form

$$2x + y + z \geq 10 \quad \text{and} \quad -2x - y - z \leq -10,$$

we obtain the following initial tableau.

x	y	z	t	u	v	w	P	*Const.*
1	2	1	1	0	0	0	0	20
3	1	0	0	1	0	0	0	30
2	1	1	0	0	1	0	0	10
−2	−1	−1	0	0	0	1	0	−10
−1	−2	−3	0	0	0	0	1	0

Using the following sequence of row operations

1. $-R_4$ 2. $R_1 - R_4$, $R_3 - R_4$, $R_5 + 3R_4$ 3. $R_1 - R_3$, $R_4 + R_3$, $R_5 + 3R_3$

we obtain the final tableau

x	y	z	t	u	v	w	P	Const.
−1	1	0	1	0	−1	0	0	10
3	1	0	0	1	0	0	0	30
0	0	0	0	0	1	1	0	0
2	1	1	0	0	1	0	0	10
5	1	0	0	0	3	0	1	30

We conclude that P attains a maximum value of 30 when $x = 0$, $y = 0$, and $z = 10$.

21. Let x and y denote the number of acres of crops A and B, respectively to be planted. Then the problem is

Maximize $P = 150x + 200y$ subject to the constraints

$$x + \quad y \le 150$$
$$40x + 60y \le 7400$$
$$20x + 25y \le 3300$$
$$x \ge 80$$
$$x \ge 0, y \ge 0$$

Using the simplex method, we obtain

x	y	u	v	w	z	P	Const.	Ratio
1	1	1	0	0	0	0	150	150
40	60	0	1	0	0	0	7400	185
20	25	0	0	1	0	0	3300	165
[−1]	0	0	0	0	1	0	−80	--
−150	−200	0	0	0	0	1	0	

$\xrightarrow{-R_4}$

x	y	u	v	w	z	P	Const.
1	1	1	0	0	0	0	150
40	60	0	1	0	0	0	7400
20	25	0	0	1	0	0	3300
1	0	0	0	0	−1	0	80
−150	−200	0	0	0	0	1	0

$\xrightarrow[R_5+150R_4]{R_1-R_4,\ R_2-40R_4,\ R_3-20R_4}$

x	y	u	v	w	z	P	Const.
0	1	1	0	0	1	0	70
0	60	0	1	0	40	0	4200
0	25	0	0	1	20	0	1700
1	0	0	0	0	−1	0	80
0	−200	0	0	0	−150	1	12,000

$\xrightarrow{\frac{1}{25}R_3}$

x	y	u	v	w	z	P	Const.
0	1	1	0	0	1	0	70
0	60	0	1	0	40	0	4200
0	1	0	0	$\frac{1}{25}$	$\frac{4}{5}$	0	68
1	0	0	0	0	−1	0	80
0	−200	0	0	0	−150	1	12,000

$\xrightarrow[R_5+200R_3]{R_1-R_3,\ R_2-60R_3}$

x	y	u	v	w	z	P	Const.
0	0	1	0	$-\frac{1}{25}$	$\frac{1}{5}$	0	2
0	0	0	1	$-\frac{12}{5}$	−8	0	120
0	1	0	0	$\frac{1}{25}$	$\frac{4}{5}$	0	68
1	0	0	0	0	−1	0	80
0	0	0	0	8	10	1	25,600

We conclude that the farmer should plant 80 acres of crop *A* and 68 acres of crop *B* to realize a maximum profit of $25,600.

23. Let x and y denote the amount (in dollars) invested in home loans and commercial-development loans, respectively. Then the problem is

Maximize $P = 0.08x + 0.06y$ subject to

$$-x+3y \le 0$$
$$y \ge 10{,}000{,}000$$
$$x+y = 60{,}000{,}000$$
$$x \ge 0,\ y \ge 0$$

Substituting $x = 60{,}000{,}000 - y$ into the first equation and the first and second inequalities, we have

$$\text{Maximize } P = 0.08(60{,}000{,}000 - y) + 0.06y$$
$$= 4{,}800{,}000 - 0.02y \text{ subject to}$$
$$y \le 15{,}000{,}000$$
$$y \ge 10{,}000{,}000$$
$$x \ge 0, y \ge 0.$$

Using the simplex method, we have

$$\begin{array}{r|r} y \quad u \quad v \quad P & Const. \\ \hline 1 \quad 1 \quad 0 \quad 0 & 15{,}000{,}000 \\ p.r.\rightarrow \; -1 \quad 0 \quad 1 \quad 0 & -10{,}000{,}000 \\ \hline 0.02 \quad 0 \quad 0 \quad 1 & 4{,}800{,}000 \\ \uparrow & \\ p.c. & \end{array} \xrightarrow{-R_2}$$

$$\begin{array}{cccc|c} y & u & v & P & Const. \\ \hline 1 & 1 & 0 & 0 & 15{,}000{,}000 \\ 1 & 0 & -1 & 0 & 10{,}000{,}000 \\ \hline 0.02 & 0 & 0 & 1 & 4{,}800{,}000 \end{array} \xrightarrow[R_3 - 0.02R_2]{R_1 - R_2} \begin{array}{ccccc|c} x & y & u & v & P & Const. \\ \hline 1 & 1 & 1 & 0 & 0 & 50{,}000 \\ -1 & 1 & 0 & 1 & 0 & -20{,}000 \\ \hline -0.10 & -0.20 & 0 & 0 & 1 & 0 \end{array}$$

We conclude that the bank should extend \$50 million in home loans, \$10 million of commercial-development loans to attain a maximum return of \$4.6 million.

25. Let x, y, and z denote the number of units of products A, B, and C manufactured by the company. Then the linear programming problem is

$$\text{Maximize } P = 18z + 12y + 15z \text{ subject to}$$
$$2x + y + 2z \le 900$$
$$3x + y + 2z \le 1080$$
$$2x + 2y + z \le 840$$
$$x - y + z \le 0$$
$$x \ge 0, \; y \ge 0, \; z \ge 0$$

The initial tableau is

x	y	z	t	u	v	w	P	Const.
2	1	2	1	0	0	0	0	900
3	1	2	0	1	0	0	0	1080
2	2	1	0	0	1	0	0	840
1	–1	1	0	0	0	1	0	0
–18	–12	–15	0	0	0	0	1	0

Using the following sequence of row operations,

1. $R_1 - 2R_4,\ R_2 - 3R_4,\ R_3 - 2R_4,\ R_5 + 18R_4$ 2. $\frac{1}{4}R_3$
3. $R_1 - 3R_3,\ R_2 - 4R_3,\ R_4 + R_3,\ R_5 + 30R_3$ 4. $\frac{4}{3}R_4$
5. $R_1 - \frac{3}{4}R_4,\ R_3 + \frac{1}{4}R_4,\ R_5 + \frac{9}{2}R_4$

we obtain the final tableau

x	y	z	t	u	v	w	P	Const.
–1	0	0	1	0	–1	–1	0	60
0	0	0	0	1	–1	–1	0	240
$\frac{1}{3}$	1	0	0	0	$\frac{1}{3}$	$-\frac{1}{3}$	0	280
$\frac{4}{3}$	0	1	0	0	$\frac{1}{3}$	$\frac{2}{3}$	0	280
–6	0	0	0	0	9	6	1	7560

and conclude that the company should produce 0 units of product *A*, 280 units of product *B*, and 280 units of product *C* to realize a maximum profit of $7,560.

27. Let x denote the number of ounces of food *A* and y denote the number of ounces of food *B* used in the meal. Then the problem is to minimize the amount of cholesterol in the meal. Thus, the linear programming problem is

Maximize $P = -C = -2x - 5y$ subject to

$$30x + 25y \geq 400$$
$$x + \tfrac{1}{2}y \geq 10$$
$$2x + 5y \geq 40$$
$$x \geq 0, y \geq 0$$

The initial tableau is

x	y	u	v	w	C	Const.
–30	–25	1	0	0	0	–400
–1	$-\frac{1}{2}$	0	1	0	0	–10
–2	–5	0	0	1	0	–40
2	5	0	0	0	1	0

Using the following sequence of row operations

1. $-R_2$ 2. R_1+30R_2; R_3+2R_2; R_4-2R_2 3. $-\frac{1}{4}R_3$
4. R_1+10R_2; $R_2-\frac{1}{2}R_3$; R_4-4R_3 5. $-\frac{1}{25}R_1$ 6. $R_2+\frac{5}{4}R_1$; $R_3-\frac{1}{2}R_1$

we obtain the final tableau

x	y	u	v	w	C	Const.
0	0	$-\frac{1}{25}$	1	$\frac{1}{10}$	0	2
1	0	$-\frac{1}{20}$	1	$\frac{1}{4}$	0	10
0	1	$\frac{1}{50}$	0	$-\frac{3}{10}$	0	4
0	0	0	0	1	1	–40

Thus, the minimum content of cholesterol is 40mg when 10 ounces of food A and 4 ounces of food B are used. (Since the u-column is not in unit form, we see that the problem has multiple solutions.)

CHAPTER 4, CONCEPT REVIEW QUESTIONS, page 274

1. Maximized; nonnegative; less than; equal to.

3. Minimized; nonnegative; greater than; equal to

CHAPTER 4 REVIEW EXERCISES, page 274

1. This is a regular linear programming problem. Using the simplex method with u and v as slack variables, we obtain the following sequence of tableaus:

	x	y	u	v	P	Const	Ratio
p.r. →	1	[3]	1	0	0	15	5
	4	1	0	1	0	16	16
	–3	–4	0	0	1	0	
		↑ p.c.					

$\xrightarrow{\frac{1}{3}R_1}$

x	y	u	v	P	Const.
$\frac{1}{3}$	1	$\frac{1}{3}$	0	0	5
4	1	0	1	0	16
–3	–4	0	0	1	0

$\xrightarrow{\substack{R_2-R_1\\R_3+4R_1}}$

	x	y	u	v	P	Const.	Ratio
	$\frac{1}{3}$	1	$\frac{1}{3}$	0	0	5	15
p.c. →	$\boxed{\frac{11}{3}}$	0	$-\frac{1}{3}$	1	0	11	3
	$-\frac{5}{3}$	0	$\frac{4}{3}$	0	1	20	
	↑ p.c.						

$\xrightarrow{\frac{3}{11}R_2}$

x	y	u	v	P	Const.
$\frac{1}{3}$	1	$\frac{1}{3}$	0	0	5
1	0	$-\frac{1}{11}$	$\frac{3}{11}$	0	3
$-\frac{5}{3}$	0	$\frac{4}{3}$	0	1	20

$\xrightarrow[R_3+\frac{5}{3}R_2]{R_1-\frac{1}{3}R_2}$

x	y	u	v	P	Const.
0	1	$\frac{4}{11}$	$-\frac{1}{11}$	0	4
1	0	$-\frac{1}{11}$	$\frac{3}{11}$	0	3
0	0	$\frac{13}{11}$	$\frac{5}{11}$	1	25

and conclude that $x = 3$, $y = 4$, $u = 0$, $v = 0$, and $P = 25$.

3. Using the simplex method to solve this regular linear programming problem we have

	x	y	z	u	v	P	Const.	Ratio
p.r. →	1	2	$\boxed{3}$	1	0	0	12	4
	1	-3	2	0	1	0	10	5
	-2	-3	-5	0	0	1	0	
			↑ p.c.					

$\xrightarrow{\frac{1}{3}R_1}$

x	y	z	u	v	P	Const.
$\frac{1}{3}$	$\frac{2}{3}$	1	$\frac{1}{3}$	0	0	4
1	-3	2	0	1	0	10
-2	-3	-5	0	0	1	0

$\xrightarrow[R_3+5R_1]{R_2-2R_1}$

	x	y	z	u	v	P	Const.	Ratio
	$\frac{1}{3}$	$\frac{2}{3}$	1	$\frac{1}{3}$	0	0	4	12
p.r. →	$\boxed{\frac{1}{3}}$	$-\frac{13}{3}$	0	$-\frac{2}{3}$	1	0	2	6
	$-\frac{1}{3}$	$\frac{1}{3}$	0	$\frac{5}{3}$	0	1	20	
	↑ p.c.							

$\xrightarrow{3R_2}$

x	y	z	u	v	P	Const.
$\frac{1}{3}$	$\frac{2}{3}$	1	$\frac{1}{3}$	0	0	4
1	−13	0	−2	3	0	6
$-\frac{1}{3}$	$\frac{1}{3}$	0	$\frac{5}{3}$	0	1	20

$\xrightarrow[R_3+\frac{1}{3}R_2]{R_1-\frac{1}{3}R_2}$

x	y	z	u	v	P	Const.
0	1	$\frac{1}{5}$	$\frac{1}{5}$	$-\frac{1}{5}$	0	$\frac{2}{5}$
0	−13	0	−2	3	0	6
0	−4	0	1	1	1	22

$\xrightarrow[R_3+4R_1]{R_2+13R_1}$

x	y	z	u	v	P	Const.
0	1	$\frac{1}{5}$	$\frac{1}{5}$	$-\frac{1}{5}$	0	$\frac{2}{5}$
1	0	$\frac{13}{5}$	$\frac{3}{5}$	$\frac{2}{5}$	0	$\frac{56}{5}$
0	0	$\frac{4}{5}$	$\frac{9}{5}$	$\frac{1}{5}$	1	$\frac{118}{5}$

We conclude that the P attains a maximum value of 23.6 when $x = 11.2$, $y = 0.4$, $z = 0$, $u = 0$, and $v = 0$.

5. We first write the tableau

x	y	Const.
2	3	6
2	1	4
3	2	

Then obtain the following by interchanging rows and columns:

u	v	Const.
2	2	3
3	1	2
6	4	

From this table we construct the dual problem:

Maximize the objective function $P = 6u + 4v$ subject to the constraints

$$2u + 2v \leq 3$$

$$3u + v \leq 2$$

$$u \geq 0,\ v \geq 0$$

Using the simplex method, we have

	u	v	x	y	P	Const	Ratio
	2	2	1	0	0	3	3/2
p.r. →	[3]	1	0	1	0	2	2/3
	−6	−4	0	0	1	0	
	↑ p.c.						

$\xrightarrow{\frac{1}{3}R_2}$

u	v	x	y	P	Const
2	2	1	0	0	3
1	$\frac{1}{3}$	0	$\frac{1}{3}$	0	$\frac{2}{3}$
−6	−4	0	0	1	0

$\xrightarrow[R_3+6R_2]{R_1-2R_2}$

u	v	x	y	P	Const	Ratio
0	[$\frac{4}{3}$]	1	$-\frac{2}{3}$	0	$\frac{5}{3}$	5/4
1	$\frac{1}{3}$	0	$\frac{1}{3}$	0	$\frac{2}{3}$	2
0	−2	0	2	1	4	

$\xrightarrow{\frac{3}{4}R_1}$

u	v	x	y	P	Const
0	1	$\frac{3}{4}$	$-\frac{1}{2}$	0	$\frac{5}{4}$
1	$\frac{1}{3}$	0	$\frac{1}{3}$	0	$\frac{2}{3}$
0	−2	0	2	1	4

$\xrightarrow[R_3+2R_1]{R_2-\frac{1}{3}R_1}$

u	v	x	y	P	Const
0	1	$\frac{3}{4}$	$-\frac{1}{2}$	0	$\frac{5}{4}$
1	0	$-\frac{1}{4}$	$\frac{1}{2}$	0	$\frac{1}{4}$
0	0	$\frac{3}{2}$	1	1	$\frac{13}{2}$

Therefore, C attains a minimum value of 13/2 when $x = 3/2$, $y = 1$, $u = \frac{1}{4}$ and $v = \frac{5}{4}$.

7. We first write the tableau

x	y	z	Const.
3	2	1	4
1	1	3	6
24	18	24	

Then obtain the following by interchanging rows and columns:

u	v	Const.
3	1	24
2	1	18
1	3	24
4	6	

From this table we construct the dual problem:

Maximize the objective function $P = 4u + 6v$ subject to

$$\begin{aligned} 3u + v &\le 24 \\ 2u + v &\le 18 \\ u + 3v &\le 24 \\ u \ge 0,\ v &\ge 0 \end{aligned}$$

The initial tableau is

u	v	x	y	z	P	Const.
3	1	1	0	0	0	24
2	1	0	1	0	0	18
1	3	0	0	1	0	24
–4	–6	0	0	0	1	0

Using the following sequence of row operations

1. $\frac{1}{3}R_3$ 2. $R_1 - R_3$, $R_2 - R_3$, $R_4 + 6R_3$ 3. $\frac{3}{8}R_1$ 4. $R_2 - \frac{5}{3}R_1$, $R_3 - \frac{1}{3}R_1$, $R_4 + 2R_1$

we obtain the final tableau

u	v	x	y	z	P	Const.
1	0	$\frac{3}{8}$	0	$-\frac{1}{8}$	0	6
0	0	$-\frac{5}{8}$	1	$-\frac{1}{8}$	0	0
0	1	$-\frac{1}{8}$	0	$\frac{3}{8}$	0	6
0	0	$\frac{3}{4}$	0	$\frac{7}{4}$	0	60

We conclude that C attains a minimum value of 60 when $x = 3/4$, $y = 0$, $z = 7/4$, $u = 6$, and $v = 6$.

9. Rewriting the problem, we have

Maximize $P = 3x - 4y$ subject to

$$\begin{aligned} x + y &\le 45 \\ -x + 2y &\le -10 \\ x \ge 0,\ y &\ge 0 \end{aligned}$$

Using the simplex method, we have

x	y	u	v	P	Const
1	1	1	0	0	45
–1	2	0	1	0	–10
–3	4	0	0	1	0

Ratio
45
10

$\xrightarrow{-R_2}$

x	y	u	v	P	Const
1	1	1	0	0	45
1	–2	0	–1	0	10
–3	4	0	0	1	0

$\xrightarrow[R_3+3R_2]{R_1-R_2}$

x	y	u	v	P	Const
0	3	1	1	0	35
1	–2	0	–1	0	10
0	–2	0	–3	1	30

$\xrightarrow[R_3+3R_1]{R_2+R_1}$

x	y	u	v	P	Const
0	3	1	1	0	35
1	1	1	0	0	45
0	7	3	0	1	135

We conclude that P attains a maximum value of 135 when $x = 45$ and $y = 0$.

11. We first write the problem in the form

Maximize $P = 2x + 3y$ subject to

$$2x+5y \le 20$$
$$x-5y \le -5$$
$$x \ge 0,\ y \ge 0$$

The initial tableau is

x	y	u	v	P	Const
2	5	1	0	0	20
1	–5	0	1	0	–5
–2	–3	0	0	1	0

Using the sequence of row operations

1. $\frac{1}{5}R_1$ 2. R_2+5R_1, R_3+3R_1 3. $\frac{1}{3}R_2$ 4. $R_1-\frac{2}{5}R_2$, $R_3+\frac{4}{5}R_2$

we obtain the final tableau

x	y	u	v	P	Const
0	1	$\frac{1}{15}$	$-\frac{2}{15}$	0	2
1	0	$\frac{1}{3}$	$\frac{1}{3}$	0	5
0	0	$\frac{13}{15}$	$\frac{4}{15}$	1	16

We conclude that P attains a maximum value of 16 when $x = 5$ and $y = 2$.

13. Refer to Section 3.2, Exercise 9, of this manual. The problem simplifies to

Minimize $C = 14000x+16000y$ subject to

$$2x + 3y \geq 26$$
$$3x + y \geq 18$$
$$x \geq 0, y \geq 0$$

We first write down the following tableaus for the primal problem:

x	y	*Const.*
2	3	26
3	1	18
14000	16000	

Next, we interchange the columns and rows of the foregoing tableau:

u	v	*Const.*
2	3	14000
3	1	16000
26	18	

This leads to the dual problem:

Maximize $P = 26u + 18v$ subject to

$$2u + 3v \leq 14000$$
$$3u + v \leq 16000$$
$$u \geq 0, v \geq 0$$

Let x and y be slack variables. We obtain the following tableaus:

	u	v	x	y	P	*Const*	*Ratio*
	2	3	1	0	0	14000	7000
p.r. →	$\boxed{3}$	1	0	1	0	16000	$5333\frac{1}{3}$
	−26	−18	0	0	1	0	
	↑ *p.c.*						

$\xrightarrow{\frac{1}{3}R_2}$

u	v	x	y	P	*Const*
2	3	1	0	0	14000
1	$\frac{1}{3}$	0	$\frac{1}{3}$	0	$\frac{16000}{3}$
−26	−18	0	0	1	0

$\xrightarrow[R_3+26R_2]{R_1-R_2}$

	u	v	x	y	P	*Const*	*Ratio*
p.r. →	0	$\boxed{\frac{7}{3}}$	1	$-\frac{2}{3}$	0	$\frac{10000}{3}$	$\frac{10000}{7}$
	1	$\frac{1}{3}$	0	$\frac{1}{3}$	0	$\frac{16000}{3}$	16000
	0	$-\frac{28}{3}$	0	$\frac{26}{3}$	1	$\frac{41600}{3}$	
		↑ *p.c.*					

$\xrightarrow{\frac{3}{7}R_1}$

u	v	x	y	P	*Const*
0	$\boxed{1}$	$\frac{3}{7}$	$-\frac{2}{7}$	0	$\frac{10000}{7}$
1	$\frac{1}{3}$	0	$\frac{1}{3}$	0	$\frac{16000}{3}$
0	$-\frac{28}{3}$	0	$\frac{26}{3}$	1	$\frac{416000}{3}$

$\xrightarrow[R_3-\frac{28}{3}R_1]{R_3-\frac{1}{3}R_1}$

u	v	x	y	P	*Const*
0	1	$\frac{3}{7}$	$-\frac{2}{7}$	0	$\frac{10000}{7}$
1	0	$-\frac{1}{7}$	$\frac{3}{7}$	0	$\frac{34000}{7}$
0	0	4	6	1	152000

The last tableau is in final form. The fundamental theorem of duality tells us that the solution to the primal problem is $x = 4, y = 6$, and $C = 152{,}000$. So, the Saddle Mine should be operated for 4 days and the Horseshoe Mine should be operated for 6 days at a minimum cost of \$152,000/day.

15. Let x, y, and z denote the number of units of products A, B, and C made, respectively. Then the problem is to maximize the profit

$$P = 4x + 6y + 8z \text{ subject to}$$
$$9x + 12y + 18z \le 360$$
$$6x + 6y + 10z \le 240$$
$$x \ge 0,\ y \ge 0,\ z \ge 0$$

The initial tableau is

x	y	z	u	v	P	*Const.*
9	12	18	1	0	0	360
6	6	10	0	1	0	240
–4	–6	–8	0	0	1	0

Using the sequence of row operations
1. $\frac{1}{18}R_1$ 2. $R_2 - 10R_1$ 3. $R_3 + 8R_1$ 4. $\frac{3}{2}R_1$ 5. $R_2 + \frac{2}{3}R_1$, $R_3 + \frac{2}{3}R_1$
we obtain the final tableau

x	y	z	u	v	P	*Const.*
$\frac{3}{4}$	1	$\frac{3}{2}$	$\frac{1}{12}$	0	0	30
$\frac{3}{2}$	0	1	$-\frac{1}{2}$	1	0	60
$\frac{1}{2}$	0	1	$\frac{1}{2}$	0	1	180

and conclude that the company should produce 0 units of product A, 30 units of product B, and 0 units of product C to realize a maximum profit of \$180.

CHAPTER 4 BEFORE MOVING ON, page 275

1. We introduce slack variables u, v, and w.

x	y	z	u	v	w	P	Constant	Ratio
2	[1]	-1	1	0	0	0	3	3
1	-2	3	0	1	0	0	1	–
3	2	4	0	0	1	0	17	$\frac{17}{2}$
-1	-2	3	0	0	0	1	0	
	↑ p.c.							

The pivot element is 1 as shown.

2. $x = 2, y = 0, z = 11, u = 2, v = w = 0$, and $P = 28$.

3. Introduce the slack variables u and v.

x	y	u	v	P	Constant	Ratio
4	3	1	0	0	30	$\frac{15}{2}$
2	-3	0	1	0	6	3
-5	-2	0	0	1	0	

$\xrightarrow{\frac{1}{2}R_2}$

x	y	u	v	P	Constant
4	3	1	0	0	30
1	$-\frac{3}{2}$	0	$\frac{1}{2}$	0	3
−5	−2	0	0	1	0

$\xrightarrow[R_3+5R_2]{R_1-4R_2}$

x	y	u	v	P	Constant	Ratio
0	9	1	−2	0	18	2
1	$-\frac{3}{2}$	0	$\frac{1}{2}$	0	3	--
0	$-\frac{19}{2}$	0	$\frac{5}{2}$	1	15	

$\xrightarrow{\frac{1}{9}R_1}$

x	y	u	v	P	Constant
0	1	$\frac{1}{9}$	$-\frac{2}{9}$	0	2
1	$-\frac{3}{2}$	0	$\frac{1}{2}$	0	3
0	$-\frac{19}{2}$	0	$\frac{5}{2}$	1	15

$\xrightarrow[R_3+\frac{19}{2}R_1]{R_2+\frac{3}{2}R_1}$

x	y	u	v	P	Constant
0	1	$\frac{1}{9}$	$-\frac{2}{9}$	0	2
1	0	$\frac{1}{6}$	$\frac{1}{6}$	0	6
0	0	$\frac{19}{8}$	$\frac{7}{18}$	1	34

The optimal solution is $x = 6, y = 2, u = v = 0$, and $P = 34$.

4. We first write the following tableau for the primal problem

x	y	Constant
1	1	3
2	3	6
1	2	

and then construct the following tableau for the dual problem

u	v	Constant
1	2	1
1	3	2
3	6	

The dual problem is

Maximize $P = 3u + 6v$

subject to

$$u + 2v \leq 1$$
$$u + 3v \leq 2$$
$$u \geq 0,\ v \geq 0$$

Introduce the slack variables x and y and then use the simplex method:

u	v	x	y	P	Constant	Ratio
1	2	1	0	0	1	$\frac{1}{2}$
1	3	0	1	0	2	$\frac{2}{3}$
-3	–6	0	0	1	0	
	↑ p.c.					

$\xrightarrow{\frac{1}{2}R_1}$

u	v	x	y	P	Constant
$\frac{1}{2}$	1	$\frac{1}{2}$	0	0	$\frac{1}{2}$
1	3	0	1	0	2
–3	–6	0	0	1	0

$\xrightarrow[R_3+6R_1]{R_2-3R_1}$

u	v	x	y	P	Constant
$\frac{1}{2}$	1	$\frac{1}{2}$	0	0	$\frac{1}{2}$
$-\frac{1}{2}$	0	$-\frac{3}{2}$	1	0	$\frac{1}{2}$
0	0	3	0	1	3

The solution (to the primal problem) is $x = 3$, $y = 0$, and $C = 3$.

5. Maximize $P = 2x + y$
subject to

$$2x + 5y \le 20$$
$$4x + 3y \ge 16$$
$$x \ge 0,\ y \ge 0$$

We rewrite the problem as
Maximize $p = 2x + y$
subject to

$$2x + 5y \le 20$$
$$-4x - 3y \le -16$$
$$x \ge 0,\ y \ge 0$$

Introduce slack variables u and $v$and then use the simplex method obtaining

x	y	u	v	P	Constant	Ratio
2	5	1	0	0	20	10
−4	−3	0	1	0	−11	$\frac{11}{4}$
−2	−1	0	0	1	0	
↑ p.c.						

$\xrightarrow{-\frac{1}{4}R_2}$

x	y	u	v	P	Constant
2	5	1	0	0	20
1	$\frac{3}{4}$	0	$-\frac{1}{4}$	0	$\frac{11}{4}$
−2	−1	0	0	1	0

$\xrightarrow[R_3+2R_2]{R_1-2R_2}$

x	y	u	v	P	Constant	Ratio
0	$\frac{7}{2}$	1	$\boxed{\frac{1}{2}}$	0	$\frac{29}{2}$	29
1	$\frac{3}{4}$	0	$-\frac{1}{4}$	0	$\frac{11}{4}$	--
0	$\frac{1}{2}$	0	$-\frac{1}{2}$	1	$\frac{11}{2}$	
			↑ p.c.			

$\xrightarrow{2R_1}$

x	y	u	v	P	Constant
0	7	2	1	0	29
1	$\frac{3}{4}$	0	$-\frac{1}{4}$	0	$\frac{11}{4}$
0	$\frac{1}{2}$	0	$-\frac{1}{2}$	1	$\frac{11}{2}$

$\xrightarrow[R_3+\frac{1}{2}R_1]{R_2+\frac{1}{4}R_1}$

x	y	u	v	P	Constant
0	7	2	1	0	29
1	$\frac{5}{2}$	$\frac{1}{2}$	0	0	10
0	4	1	0	1	20

The optimal solution is $x = 10, y = 0, u = 0, v = 29$, and $P = 20$.

CHAPTER 5

5.1 Problem Solving Tips

In this section, you were given several formulas for computing interest. As you work through the exercises that follow, first decide which formula you need to solve the problem. Then write out your solution. After doing this a few times, you should have the formulas memorized. The key here is to try not to look at the formula in the text, and to work the problem just as if you were taking a test. If you train yourself to work in this manner, test-taking will be a lot easier.

Here are some hints for solving the problems in the exercises that follow.

1. First decide if the problem involves *simple interest* or *compound interest.* This will be stated in the problem.

2. Determine whether the problem is asking for the *present value* or *future value* of an amount. For example, if you are asked to determine the value of an investment 5 years from now with interest compounded each year, then use a compound interest formula giving the accumulated amount. If you are asked to determine the current value of an investment that will have a value of $50,000 five years from now with interest compounded each year, then use a present value formula for compound interest. (If interest is compounded continuously it will be stated in the problem.)

3. The effective rate of interest is the same as the APR rate that you see in

advertisements involving loans. Since the interest for different loans may be compounded over different periods (every day, every month, bi-annually, annually, ...) it provides the consumer with a method of comparing rates. It is the simple interest rate that would produce the same accumulated amount in 1 year as the nominal rate compounded *m* times per year.

5.1 CONCEPT QUESTIONS, page 288

1. In simple interest, the interest is based on the original principal. In compound interest, interest earned is periodically added to the principal and thereafter earns interest at the same rate.

3. The effective rate of interest is the simple interest that would produce the same amount in 1 year as the nominal rate compounded *m* times a year.

EXERCISES 5.1, page 289

1. The interest is given by $I = (500)(2)(0.08) = 80$, or \$80.
   The accumulated amount is 500 + 80, or \$580.

3. The interest is given by $I = (800)(0.06)(0.75) = 36$, or \$36.
   The accumulated amount is 800 + 36, or \$836.

5. We are given that $A = 1160$, $t = 2$, and $r = 0.08$, and we are asked to find P. Since

$$A = P(1 + rt)$$

we see that $$P = \frac{A}{1+rt} = \frac{1160}{1+(0.08)(2)} = 1000, \text{ or } \$1000.$$

7. We use the formula $I = Prt$ and solve for t when $I = 20$, $P = 1000$, and $r = 0.05$. Thus,

$$20 = 1000(0.05)\left(\frac{t}{365}\right), \quad \text{and} \quad t = \frac{365(20)}{50} = 146, \quad \text{or 146 days.}$$

9. We use the formula $A = P(1 + rt)$ with $A = 1075$, $P = 1000$, $t = 0.75$, and solve for r. Thus,

$$1075 = 1000(1 + 0.75r)$$
$$75 = 750r$$

or $$r = 0.10.$$

Therefore, the interest rate is 10 percent per year.

11. $A = 1000(1 + 0.07)^8 \approx 1718.19$, or \$1718.19.

13. $A = 2500\left(1+\dfrac{0.07}{2}\right)^{20} \approx 4974.47$, or \$4974.47.

15. $A = 12{,}000\left(1+\dfrac{0.08}{4}\right)^{42} \approx 27{,}566.93$, or \$27,566.93.

17. $A = 150{,}000\left(1+\dfrac{0.14}{12}\right)^{48} \approx 261{,}751.04$, or \$261,751.04.

19. $A = 150{,}000\left(1+\dfrac{0.12}{365}\right)^{1095} = 214{,}986.69$, or \$214,986.69.

21. Using the formula

$$r_{eff} = \left(1+\frac{r}{m}\right)^m - 1$$

with $r = 0.10$ and $m = 2$, we have

$$r_{eff} = \left(1+\frac{0.10}{2}\right)^2 - 1 = 0.1025, \quad \text{or 10.25 percent..}$$

23. Using the formula $r_{eff} = \left(1+\dfrac{r}{m}\right)^m - 1$

with $r = 0.08$ and $m = 12$, we have

$$r_{eff} = \left(1+\frac{0.08}{12}\right)^{12} - 1 \approx 0.08300, \quad \text{or 8.3 percent per year.}$$

25. The present value is given by $P = 40{,}000\left(1+\dfrac{0.08}{2}\right)^{-8} \approx 29{,}227.61$, or \$29,227.61.

27. The present value is given by

$$P = 40,000\left(1+\frac{0.07}{12}\right)^{-48} \approx 30,255.95, \quad \text{or } \$30,255.95.$$

29. $A = 5000e^{0.08(4)} \approx 6885.64$, or \$6,885.64.

31. We use formula (5) with $A = 6000$, $P = 5000$, and $t = 3$. Thus,

$$6000 = 5000e^{3r}$$

$$e^{3r} = \frac{6000}{5000} = 1.2;$$

Next, taking the logarithm of each side of the equation, we have

$$3r = \ln 1.2 \qquad \text{[The natural logarithm of } e^{3r} \text{ is } 3r.]$$

$$r = \frac{\ln 1.2}{3} \approx 0.6077$$

So the interest rate is 6.08% per year.

33. We use formula (5) with $A = 7000$, $P = 6000$, and $r = 0.075$. Thus

$$7000 = 6000e^{0.075t}; \quad e^{0.075t} = \tfrac{7000}{6000} = \tfrac{7}{6}$$

Next, taking the logarithm of each side, we have

$$0.075t \ln e = \ln\tfrac{7}{6} \qquad \text{[The natural logarithm of } e^{0.075t} \text{ is } 0.075t.]$$

and $t = \dfrac{\ln\frac{7}{6}}{0.075} \approx 2.055$. So, it will take 2.06 years.

35. Think of \$300 as the principal and \$306 as the accumulated amount at the end of 30 days. If r denotes the simple interest rate per annum, then we have $P = 300$, $A = 306$, $t = 1/12$, and we are required to find r. Using (1b) we have

$$306 = 300\left(1+\frac{r}{12}\right) = 300 + r\left(\frac{300}{12}\right)$$

and $r = \left(\dfrac{12}{300}\right)6 = 0.24$, or 24 percent per year.

37. The Abdullahs will owe $A = P(1+rt) = 120,000[1+(0.12)(\frac{3}{12})] = 123,600$, or \$123,600.

39. Here $P = 10{,}000$, $I = 3500$, and $t = 7$, and so from Formula (1a), we have

$$3500 = 10000(r)7 \quad \text{and so } r = \frac{3500}{70000} = 0.05$$

So the bond pays interest at the rate of 5% per year.

41. The rate that you would expect to pay is

$$A = 480(1 + 0.08)^5 \approx 705.28, \text{ or } \$705.28 \text{ per day.}$$

43. The amount that they can expect to pay is given by

$$A = 210{,}000(1 + 0.05)^4 \approx 255{,}256, \quad \text{or approximately } \$255{,}256.$$

45. The investment will be worth $A = 1.5\left(1+\frac{0.055}{2}\right)^{20} = 2.58064$, or approximately

$2.58 million dollars.

47. We use Formula (3) with $P = 15{,}000$, $r = 0.098$, $m = 12$, and $t = 4$ giving the worth of Jodie's account as

$$A = 15{,}000\left(1+\frac{0.098}{12}\right)^{(12)(4)} \approx 22{,}163.753, \text{ or approximately } \$22{,}163.75.$$

49. Using the formula $P = A\left(1+\frac{r}{m}\right)^{-mt}$, we have

$$P = 40{,}000\left(1+\frac{0.085}{4}\right)^{-20} \approx 26{,}267.49, \text{ or } \$26{,}267.49.$$

51. a. They should set aside

$$P = 100{,}000(1 + 0.085)^{-13} \approx 34{,}626.88, \text{ or } \$34{,}626.88.$$

b. They should set aside

$$P = 100{,}000\left(1+\frac{0.085}{2}\right)^{-26} \approx 33{,}886.16, \quad \text{or } \$33{,}886.16.$$

c. They should set aside

$$P = 100{,}000\left(1+\frac{0.085}{4}\right)^{-52} \approx 33{,}506.76, \text{ or } \$33{,}506.76.$$

53. The effective rate of interest for the Bendix Mutual Fund is

$$r_{eff} = \left(1+\frac{0.104}{4}\right)^4 - 1 \approx 0.1081 \quad \text{or} \quad 10.81\%/\text{yr}$$

whereas the effective rate of interest for the Acme Mutual fund is

$$r_{eff} = \left(1+\frac{0.106}{2}\right)^4 - 1 \approx 0.1088 \quad \text{or} \quad 10.88\%/\text{yr}.$$

We conclude that the Acme Mutual Fund has a better rate of return.

55. The present value of the \$8000 loan due in 3 years is given by

$$P = 8000\left(1+\frac{0.10}{2}\right)^{-6} = 5969.72, \text{ or } \$5969.72.$$

The present value of the \$15,000 loan due in 6 years is given by

$$P = 15,000\left(1+\frac{0.10}{2}\right)^{-12} = 8352.56, \text{ or } \$8352.56.$$

Therefore, the amount the proprietors of the inn will be required to pay at the end of 5 years is given by

$$A = 14,322.28\left(1+\frac{0.10}{2}\right)^{10} = 23,329.48, \quad \text{or } \$23,329.48.$$

57. Using the compound interest formula with $A = 256{,}000$, $P = 200{,}000$ and $t = 6$, we have

$$256,000 = 200,000(1+R)^6$$
$$(1+R)^{1/6} = (1.28)^{1/6}$$
$$1+R = 1.042,$$
$$R = 0.042, \quad \text{or 4.2 percent.}$$

59. Let the effective rate of interest be R. Then R satisfies

$$A = P(1+R)^t$$

or $$10,000 = 6724.53\left(1+\frac{R}{2}\right)^{14}$$

$$1+\frac{R}{2} = (1.48709278)^{1/14} = 1.028750032$$

and $$R = 2(2.875) = 0.0575, \text{ or 5.75 percent.}$$

61. Let the effective rate of interest be R. Then R satisfies

$$A = P(1+R)^t$$

or $$5170.42 = 5000(1+R)^{245/365}$$

$$1 + R = (1.034084)^{365/245} = 1.051199629$$

and $$R = 0.5119962, \text{ or approximately } 5.12\ \%/\text{year},$$

63. The projected online retail sales for 2008 are

$$(1.332)(1.278)(1.305)(1.199)(1.243)(1.14)(1.176)(1.105)(23.5)$$
$$= 115.26, \text{ or approximately } \$115.3 \text{ billion.}$$

65. Suppose $1 is invested in each investment.

Investment A: Accumulated amount is $\left(1+\frac{0.1}{2}\right)^8 \approx 1.47746$.

Investment B: Accumulated amount is $e^{0.0975(4)} \approx 1.47698$.

So Investment A has a higher rate of return.

67. a. If they invest the money at 10.5 percent compounded quarterly, they should set aside $P = 70{,}000\left(1+\frac{0.105}{4}\right)^{-28} \approx 33{,}885.14$, or $33,885.14.

b. If they invest the money at 10.5 percent compounded continuously, they should set aside $P = 70{,}000e^{-0.735} = 33{,}565.38$, or $33,565.38.

69. $P(t) = V(t)e^{-rt} = 80{,}000e^{\sqrt{t}/2}e^{-rt} = 80{,}000e^{(\sqrt{t}/2-0.09t)}$

$P(4) = 80{,}000e^{1-0.09(4)} \approx 151{,}718.47$, or approximately $151,718.

71. By definition, $A = P(1+r_{eff})^t$. So

$$(1+r_{eff})^t = \frac{A}{P}, \quad 1+r_{eff} = \left(\frac{A}{P}\right)^{1/t} \text{ and } \quad r_{eff} = \left(\frac{A}{P}\right)^{1/t} - 1.$$

73. True. $A = P(1+rt) = Prt$ is a linear function of t.

75. True. With $m = 1$, the effective rate is $r_{eff} = \left(1+\frac{r}{1}\right)^1 - 1 = r$.

USING TECHNOLOGY EXERCISES 5.1, page 296

1. $5872.78 3. $475.49 5. 8.95%/yr 7. 10.20%/yr 9. $29,743.30
11. $53,303.25

5.2 Problem Solving Tips

1. Note the difference between annuities and the compound interest problems solved in Section 5.1. An annuity is a *sequence of payments* made at regular intervals. If we are asked to find the value of a sequence of payments at some future time, then we use the formula for the future value of an annuity $S = R\left[\frac{(1+i)^n - 1}{i}\right]$. If we are asked to find the current value of a sequence of payments that will be made over a certain period of time, then we use the formula for the present value of an annuity $P = R\left[\frac{1-(1+i)^{-n}}{i}\right]$.

2. Note that the problems in this section deal with *ordinary annuities*—annuities in which the payments are made at the end of each payment period.

5.2 CONCEPT QUESTIONS, page 303

1. The term is fixed. The periodic payments are of the same size. The payments are made at the end of the payment period. The payments coincide with the interest conversion periods.

3. The future value S of an annuity of n payments of R dollars each, paid at the end of each investment period into an account that earns interest at the rate of i per period, is

$$S = R\left[\frac{(1+i)^n - 1}{i}\right].$$

An example is given by a retirement fund in which an employee makes a monthly deposit of a fixed amount for a certain period of time.

EXERCISES 5.2, page 304

1. $S = 1000\left[\frac{(1+0.1)^{10}-1}{0.1}\right] = 15{,}937.42$, or \$15,937.42.

3. $S = 1800\left[\frac{\left(1+\frac{0.08}{4}\right)^{24}-1}{\frac{0.08}{4}}\right] \approx 54{,}759.35$, or \$54,759.35.

5. $S = 600\left[\frac{\left(1+\frac{0.12}{4}\right)^{36}-1}{\frac{0.12}{4}}\right] \approx 37{,}965.57$, or \$37,965.57.

7. $S = 200\left[\frac{\left(1+\frac{0.09}{12}\right)^{243}-1}{\frac{0.09}{12}}\right] \approx 137{,}209.97$, or \$137,209.97.

9. $P = 5000\left[\frac{1-(1+0.08)^{-8}}{0.08}\right] \approx 28{,}733.19$, or \$28,733.19.

11. $P = 4000\left[\frac{1-(1+0.09)^{-5}}{0.09}\right] \approx 15{,}558.61$, or \$15,558.61.

13. $P = 800\left[\frac{1-\left(1+\frac{0.12}{4}\right)^{-28}}{\frac{0.12}{4}}\right] \approx 15{,}011.29$, or \$15,011.29.

15. She will have $S = 1500\left[\frac{(1+0.08)^{25}-1}{0.08}\right] = 109{,}658.91$, or $109,658.91.

17. On October 31, Linda's account will be worth

$$S = 40\left[\frac{\left(1+\frac{0.07}{12}\right)^{11}-1}{\frac{0.07}{12}}\right] \approx 453.06\text{, or \$453.06.}$$

One month later, this account will be worth $A = (453.06)\left(1+\frac{0.07}{12}\right) = 455.70$, or $455.70.

19. The amount in Colin's employee retirement account is given by

$$S = 100\left[\frac{\left(1+\frac{0.07}{12}\right)^{144}-1}{\frac{0.07}{12}}\right] \approx 22{,}469.50\text{, or \$22,469.50.}$$

The amount in Colin's IRA is given by

$$S = 2000\left[\frac{(1+0.09)^{8}-1}{0.09}\right] \approx 22{,}056.95\text{, or \$22,056.95.}$$

Therefore, the total amount in his retirement fund is given by
$22{,}469.50 + 22{,}056.95 = 44{,}526.45$, or $44,526.45.

21. To find how much Karen has at age 65, we use formula (6) with $R = 150$, $i = \frac{r}{m} = \frac{0.05}{12}$, and $n = mt = (12)(40) = 480$, giving

$$S = 150\left[\frac{\left(1+\frac{0.05}{12}\right)^{480}-1}{\frac{0.05}{12}}\right] \approx 228{,}903.0235$$

or $228,903.02. To find how much Matt will have upon attaining the age of 65, use

Formula (6) with $R = 250$, $i = \frac{r}{m} = \frac{0.05}{12}$, and $n = mt = (12)(30) = 360$ giving

$$S = 250\left[\frac{\left(1+\frac{0.05}{12}\right)^{360} - 1}{\frac{0.05}{12}}\right] \approx 208{,}064.6588, \text{ or } \$208{,}064.66.$$

So Karen will have the bigger nest egg.

23. The equivalent cash payment is given by

$$P = 450\left[\frac{1-\left(1+\frac{0.09}{12}\right)^{-24}}{\frac{0.09}{12}}\right] \approx 9850.12, \text{ or } \$9850.12.$$

25. We use the formula for the present value of an annuity obtaining

$$P = 22\left[\frac{1-\left(1+\frac{0.18}{12}\right)^{-36}}{\frac{0.18}{12}}\right] \approx 608.54, \text{ or } \$608.54.$$

27. With an $2400 monthly payment, the present value of their loan would be

$$P = 2400\left[\frac{1-\left(1+\frac{0.075}{12}\right)^{-360}}{\frac{0.075}{12}}\right] \approx 343{,}242.31, \text{ or } \$343{,}242.31.$$

With a $3000 monthly payment, the present value of their loan would be

$$P = 3000\left[\frac{1-\left(1+\frac{0.075}{12}\right)^{-360}}{\frac{0.075}{12}}\right] \approx 429{,}052.88 \text{ or } \$429{,}052.88.$$

Since they intend to make a $40,000 down payment, the range of homes they should consider is $383,242 to $469,053.

29. The lower limit of their investment is

$$A = 2400\left[\frac{1-\left(1+\frac{0.07}{12}\right)^{-180}}{\frac{0.07}{12}}\right] + 40,000 \approx 307,014.30$$

or approximately $307,104. The upper limit of their investment is

$$A = 3000\left[\frac{1-\left(1+\frac{0.07}{12}\right)^{-180}}{\frac{0.07}{12}}\right] + 45,000 \approx 373,767.87$$

or approximately $373,768. Therefore, the price range of houses they should consider is $347,014 to $413,768.

31. The deposits of $200/month into the bank account for a period of 2 years will grow into a sum of

$$A_1 = 200\frac{\left(1+\frac{0.06}{12}\right)^{24} - 1}{\frac{0.06}{12}} \approx 5086.391, \text{ or } \$5086.39.$$

For the next 3 years, this amount will grow into a sum of

$$A_2 = A_1\left(1+\frac{0.06}{12}\right)^{36} \approx 6086.785, \text{ or } \$6086.79.$$

The deposits of $300/month will grow into a sum of

$$A_3 = 300\left[\frac{\left(1+\frac{0.06}{12}\right)^{36} - 1}{\frac{0.06}{12}}\right] \approx 11800.831, \text{ or } \$11,800.83.$$

Therefore, at the end of 5 years. He will have

$$A_2 + A_3 = 6086.785 + 11800.831 \approx 17887.616, \text{ or approximately } \$17,887.62.$$

33. False. This statement would be true only if the interest rate is equal to zero.

USING TECHNOLOGY EXERCISES 5.2, page 308

1. \$59,622.15 3. \$8453.59 5. \$35,607.23 7. \$13,828.60

5.3 Problem Solving Tips

1. If a problem asks for the periodic payment that will amortize a loan over n periods, then use the amortization formula $R = \dfrac{Pi}{1-(1+i)^{-n}}$. For example, if you want to calculate the payment for a home mortgage use this formula to find the payment.

2. If a problem asks for the periodic payment required to accumulate a certain sum of money over n periods, then use the formula $R = \dfrac{iS}{(1+i)^n - 1}$. For example, if a business man wants to set aside a certain sum of money through periodic payments for the purchase of new equipment, then use this formula to find the payment.

5.3 CONCEPT QUESTIONS, page 315

1. $R = \dfrac{Pi}{1-(1+i)^{-n}}$.

   a. We rewrite $R = \dfrac{Pi}{1-\dfrac{1}{(1+i)^n}}$. If n increases, then $(1+i)^n$ increases and $1/(1+i)^n$ decreases. Therefore $1-1/(1+i)^n$ increases and so R decreases.
   b. If the principal and interest rate are fixed, and the number of payments are allowed to increase, then the size of the monthly payments gets smaller.

EXERCISES 5.3, page 315

1. The size of each installment is given by

$$R = \frac{100{,}000(0.08)}{1-(1+0.08)^{-10}} \approx 14{,}902.95, \text{ or } \$14{,}902.95.$$

3. The size of each installment is given by

$$R = \frac{5000(0.01)}{1-(1+0.01)^{-12}} \approx 444.24, \text{ or } \$444.24.$$

5. The size of each installment is given by

$$R = \frac{25{,}000(0.0075)}{1-(1+0.0075)^{-48}} \approx 622.13, \text{ or } \$622.13.$$

7. The size of each installment is

$$R = \frac{80{,}000(0.00875)}{1-(1+0.00875)^{-360}} \approx 731.79, \text{ or } \$731.79.$$

9. The periodic payment that is required is

$$R = \frac{20{,}000(0.02)}{(1+0.02)^{12}-1} \approx 1491.19, \text{ or } \$1491.19.$$

11. The periodic payment that is required is

$$R = \frac{100{,}000(0.0075)}{(1+0.0075)^{120}-1} \approx 6.76, \text{ or } \$516.76.$$

13. The periodic payment that is required is

$$R = \frac{250{,}000(0.00875)}{(1+0.00875)^{300}-1} \approx 172.95, \text{ or } \$172.95.$$

15. The periodic payment that is required is

$$R = \frac{50000\left(\frac{.10}{4}\right)}{\left(1+\frac{.10}{4}\right)^{20}-1} = 1957.36, \text{ or } \$1957.36.$$

17. The periodic payment that is required is

$$R = \frac{35000\left(\frac{0.075}{2}\right)}{1-\left(1+\frac{0.075}{2}\right)^{-13}} = 3450.87, \text{ or } \$3450.87.$$

19. The size of each installment is given by

$$R = \frac{100{,}000(0.10)}{1-(1+0.10)^{-10}} \approx 16{,}274.54, \text{ or } \$16{,}274.54.$$

21. The monthly payment in each case is given by $R = \dfrac{100{,}000\left(\frac{r}{12}\right)}{1-\left(1+\frac{r}{12}\right)^{-360}}$.

Thus, if $r = 0.08$, then $R = \dfrac{100{,}000\left(\frac{0.08}{12}\right)}{1-\left(1+\frac{0.08}{12}\right)^{-360}} \approx 733.76$, or \$733.76

If $r = 0.09$, then $R = \dfrac{100{,}000\left(\frac{0.09}{12}\right)}{1-\left(1+\frac{0.09}{12}\right)^{-360}} \approx 804.62$, or \$804.62

If $r = 0.10$, then $R = \dfrac{100{,}000\left(\frac{0.10}{12}\right)}{1-\left(1+\frac{0.10}{12}\right)^{-360}} \approx 877.57$, or \$877.57.

If $r = 0.11$, then $R = \dfrac{100{,}000\left(\frac{0.11}{12}\right)}{1-\left(1+\frac{0.11}{12}\right)^{-360}} \approx 952.32$, or \$952.32.

a. The difference in monthly payments in the two loans is
\$877.57 - \$665.30 = \$212.27.

b. The monthly mortgage payment on a $150,000 mortgage would be
$1.5(\$877.57) = \$1316.36$.

The monthly mortgage payment on a $50,000 mortgage would be
$0.5(\$877.57) = \$438.79$.

23. a. The amount of the loan required is 16000 - (0.25)(16000) or 12,000 dollars. If the car is financed over 36 months, the payment will be

$$R = \frac{12{,}000\left(\frac{0.10}{12}\right)}{1-\left(1+\frac{0.10}{12}\right)^{-36}} \approx 387.21, \text{ or } \$387.21 \text{ per month.}$$

If the car is financed over 48 months, the payment will be

$$R = \frac{12{,}000\left(\frac{0.10}{12}\right)}{1-\left(1+\frac{0.10}{12}\right)^{-48}} \approx 304.35, \text{ or } \$304.35 \text{ per month.}$$

b. The interest charges for the 36-month plan are

$$36(387.21) - 12000 = 1939.56,$$

or $1939.56. The interest charges for the 48-month plan are

$$48(304.35) - 12000 = 2608.80, \text{ or } \$2608.80.$$

25. The amount borrowed is 270,000 - 30,000 = 240,000 dollars. The size of the monthly installment is

$$R = \frac{240{,}000\left(\frac{0.08}{12}\right)}{1-\left(1+\frac{0.08}{12}\right)^{-360}} \approx 1761.03, \text{ or } \$1761.03.$$

To find their equity after five years, we compute

$$P = 1761.03\left[\frac{1-\left(1+\frac{0.08}{12}\right)^{-300}}{\frac{0.08}{12}}\right] \approx 228{,}167$$

or $228,167, and so their equity is $270{,}000 - 228{,}167 = 41{,}833$, or $41,833.
To find their equity after ten years, we compute

$$P = 1761.03\left[\frac{1-\left(1+\frac{0.08}{12}\right)^{-240}}{\frac{0.08}{12}}\right] \approx 210{,}539, \text{ or } \$210{,}539.$$

and their equity is $270{,}000 - 210{,}539 = 59{,}461$, or \$59,461.
To find their equity after twenty years, we compute

$$P = 1761.03\left[\frac{1-\left(1+\frac{0.08}{12}\right)^{-120}}{\frac{0.08}{12}}\right] \approx 145{,}147, \text{ or } \$145{,}147,$$

and their equity is 270,000 – 145,147, or \$124,853.

27. The amount that must be deposited annually into this fund is given by
$$R = \frac{(0.07)(2.5)}{(1+0.07)^{20}-1} = 0.06098231 \text{ million, or approximately } \$60{,}982.31 \text{ annually.}$$

29. The amount that must be deposited quarterly into this fund is
$$R = \frac{\left(\frac{0.09}{4}\right)200{,}000}{\left(1+\frac{0.09}{4}\right)^{40}-1} \approx 3{,}135.48, \text{ or } \$3{,}135.48.$$

31. The size of each monthly installment is given by
$$R = \frac{\left(\frac{0.085}{12}\right)250{,}000}{\left(1+\frac{0.085}{12}\right)^{300}-1} \approx 242.23, \quad \text{or } \$242.23.$$

33. Here $S = 450{,}000$, $i = \frac{0.10}{12}$, and $n = mt = (12)(30) = 360$. Thus, Formula (9) gives

$$450{,}000 = R\left[\frac{\left(1+\frac{0.10}{12}\right)^{360}-1}{\left(\frac{0.10}{12}\right)}\right].$$

Therefore, $R = \dfrac{450,000\left(\dfrac{0.10}{12}\right)}{\left(1+\dfrac{0.10}{12}\right)^{360}-1} \approx 199.07$, and her monthly payment is \$199.07.

35. The value of the IRA account after 20 years is

$$S = 375\left[\frac{\left(1+\dfrac{0.08}{4}\right)^{80}-1}{\dfrac{0.08}{4}}\right] \approx 72{,}664.48, \text{ or } \$72{,}664.48.$$

The payment he would receive at the end of each quarter for the next 15 years is given by

$$R = \frac{\left(\dfrac{0.08}{4}\right)72{,}664.48}{1-\left(1+\dfrac{0.08}{4}\right)^{-60}} \approx 2090.41, \text{ or } \$2090.41.$$

If he continues working and makes quarterly payments until age 65, the value of the IRA account would be

$$S = 375\left[\frac{\left(1+\dfrac{0.08}{4}\right)^{100}-1}{\dfrac{0.08}{4}}\right] \approx 117{,}087.11, \text{ or } \$117{,}087.11.$$

The payment he would receive at the end of each quarter for the next 10 years is given by

$$R = \frac{\left(\dfrac{0.08}{4}\right)117{,}087.11}{1-\left(1+\dfrac{0.08}{4}\right)^{-40}} \approx 4280.21, \text{ or } \$4280.21.$$

37. The monthly payment the Sandersons are required to make under the terms of their original loan is given by

$$R = \frac{100,000\left(\frac{0.10}{12}\right)}{1-\left(1+\frac{0.10}{12}\right)^{-240}} \approx 965.02, \text{ or } \$965.02.$$

The monthly payment the Sandersons are required to make under the terms of their new loan is given by

$$R = \frac{100,000\left(\frac{0.078}{12}\right)}{1-\left(1+\frac{0.078}{12}\right)^{-240}} \approx 824.04, \text{ or } \$824.04.$$

The amount of money that the Sandersons can expect to save over the life of the loan by refinancing is given by 240(965.02 - 824.04) = 33,835.20, or \$33,835.20.

39. Kim's monthly payment is found using Formula (13) with $P = 180{,}000$, $i = \frac{r}{m} = \frac{0.095}{12}$, and $n = (12)(30) = 360$. Thus,

$$R = 180,000\left[\frac{\frac{0.095}{12}}{1-\left(1+\frac{0.095}{12}\right)^{-360}}\right] \approx 1513.5376.$$

After 8 years, he has paid (8)(12) or 96 payments. His outstanding principal is given by the sum of the remaining installments, (360 – 96), or 264. Using Formula 11, we find

$$P = 1513.5376\left[\frac{1-\left(1+\frac{0.095}{12}\right)^{-264}}{\frac{0.095}{12}}\right] \approx 167,341.3271.$$

So his outstanding principal is \$167,341.33.

41. Sarah's monthly payment is found using Formula (13) with $P = 200{,}000$, $i = \frac{r}{m} = \frac{0.06}{12}$, and $n = (12)(15) = 180$. Thus,

$$R = 200,000\left[\frac{\frac{0.06}{12}}{1-\left(1+\frac{0.06}{12}\right)^{-180}}\right] \approx 1687.7137.$$

After 85 years, she has paid (5)(12) or 60 payments. Her outstanding principal is given by the sum of the remaining installments, (180 – 60), or 120. Using Formula 11, we find

$$P = 1687.7137\left[\frac{1-\left(1+\frac{0.06}{12}\right)^{-120}}{\frac{0.06}{12}}\right] \approx 152,018.2012$$

So her outstanding principal is $152,018.20.

43. a. Here $P = 200{,}000$, $i = \frac{r}{m} = \frac{0.095}{12}$, and $n = mt = (12)(30) = 360$. Therefore,

$$R = \frac{200,000\left(\frac{0.095}{12}\right)}{1-\left(1+\frac{0.095}{12}\right)^{-360}} \approx 1681.7084, \text{ and so her monthly payment is \$1681.71.}$$

b. After $(4)(12) = 48$ monthly payments have been made, her outstanding principal is given by the sum of the present values of the remaining installments (which is $360 - 48 = 312$). Using Formula (11), we find it to be

$$P = 1681.7084\left[\frac{1-\left(1+\frac{0.095}{12}\right)^{-312}}{\frac{0.095}{12}}\right] \approx 194,282.6675, \text{ or approximately \$194,282.67.}$$

c. Here $P = 194{,}282.6675$, $i = \frac{r}{m} = \frac{0.0675}{12}$, and $n = (12)(30) = 360$. So

$$R = \frac{194282.67\left(\frac{0.0675}{12}\right)}{1-\left(1+\frac{0.0675}{12}\right)^{-360}} \approx 1260.1137$$

and so her new monthly payment is $1260.11.

d. Emily will save 1681.71 – 1260.11, or $421.60 per month.

45. The amount of the loan the Meyers need to secure is \$280,000. Using the bank's financing, the monthly payment would be

$$R = \frac{280,000\left(\frac{0.11}{12}\right)}{1-\left(1+\frac{0.11}{12}\right)^{-300}} \approx 2744.32, \text{ or } \$2744.32.$$

Using the seller's financing, the monthly payment would be

$$R = \frac{280,000\left(\frac{0.098}{12}\right)}{1-\left(1+\frac{0.098}{12}\right)^{-300}} \approx 2504.99, \text{ or } \$2504.99.$$

By choosing the seller's financing rather than the bank's, the Meyers would save

$(2744.32 - 2504.99)(300) = 71,799$, or \$71,799 in interest.

USING TECHNOLOGY EXERCISES 5.3, page 321

1. \$628.02 3. \$1685.47 5. \$2234.40 7. \$1436.41

9. \$18,288.92. The amortization schedule follows.

End of Period	*Interest charged*	*Repayment made*	*Payment toward Principal*	*Outstanding Principal*
0				120,000.00
1	10,200.00	18,288.92	8,088.92	111,911.08
2	9,512.44	18,288.92	8,776.48	103,134.60
3	8,766.44	18,288.92	9,522.48	93,612.12
4	7,957.03	18,288.92	10,331.89	83,280.23
5	7,078.82	18,288.92	11,210.10	72,070.13
6	6,125.96	18,288.92	12,162.96	59,907.17
7	5,092.11	18,288.92	13,196.81	46,710.36
8	3,970.38	18,288.92	14,318.54	32,391.82
9	2,753.30	18,288.92	15,535.62	16,856.20
10	1,432.78	18,288.98	16,856.14	.00

5.4 Problem Solving Tips

1. Note the difference between an arithmetic progression and a geometric progression.

An *arithmetic progression* is a sequence of numbers in which each term after the first is obtained by *adding a constant d to the preceding term.* A *geometric progression* is a sequence of numbers in which each term after the first is obtained by *multiplying the preceding term by a constant r.*

5.4 CONCEPT QUESTIONS, page 328

1. a. $a_n = a + (n-1)d$ b. $S_n = \frac{n}{2}[2a + (n-1)d]$

EXERCISES 5.4, page 328

1. $a_9 = 6 + (9 - 1)3 = 30$

3. $a_8 = -15 + (8 - 1)\left(\frac{3}{2}\right) = -\frac{9}{2} = -4.5.$

5. $a_{11} - a_4 = (a_1 + 10d) - (a_1 + 3d) = 7d$. Also, $a_{11} - a_4 = 107 - 30 = 77$.
Therefore, $7d = 77$, and $d = 11$. Next,

$$a_4 = a + 3d = a + 3(11) = a + 33 = 30.$$

and $a = -3$. Therefore, the first five terms are -3, 8, 19, 30, 41.

7. Here $a = x$, $n = 7$, and $d = y$. Therefore, the required term is

$$a_7 = x + (7 - 1)y = x + 6y.$$

9. Using the formula for the sum of the terms of an arithmetic progression with $a = 4$, $d = 7$ and $n = 15$, we have

$$S_n = \frac{n}{2}[2a + (n-1)d]]$$

$$S_{15} = \frac{15}{2}[2(4) + (15-1)7] = \frac{15}{2}(106) = 795.$$

11. The common difference is $d = 2$ and the first term is $a = 15$. Using the formula for the nth term

$$a_n = a + (n - 1)d,$$

we have $57 = 15 + (n - 1)(2) = 13 + 2n$

$2n = 44$, and $n = 22$.

Using the formula for the sum of the terms of an arithmetic progression with $a = 15$, $d = 2$ and $n = 22$, we have

$$S_n = \frac{n}{2}[2a + (n-1)d]$$

$$S_{22} = \frac{22}{2}[2(15) + (22-1)2] = 11(72) = 792.$$

13. $$f(1) + f(2) + f(3) + \cdots + f(20)$$
$$= [3(1) - 4] + [3(2) - 4] + [3(3) - 4] + \cdots + [3(20) - 4]$$
$$= 3(1 + 2 + 3 + \cdots + 20) + 20(-4)$$
$$= 3\left(\frac{20}{2}\right)[2(1) + (20-1)1] - 80$$
$$= 550.$$

15. $$S_n = \frac{n}{2}[2a_1 + (n-1)d] = \frac{n}{2}(a_1 + a_1 + (n-1)d]$$
$$= \frac{n}{2}(a_1 + a_n)$$

a. $S_{11} = \frac{11}{2}(3 + 47) = 275.$ b. $S_{20} = \frac{20}{2}[5 + (-33)] = -280.$

17. Let n be the number of weeks till she reaches 10 miles. Then

$$a_n = 1 + (n-1)\frac{1}{4} = 1 + \frac{1}{4}n - \frac{1}{4} = \frac{1}{4}n + \frac{3}{4} = 10$$

Therefore, $n + 3 = 40$, and $n = 37$; that is, it will take Karen 37 weeks to meet her goal.

19. To compute the Kunwoo's fare by taxi, take $a = 1$, $d = 0.60$, and $n = 25$. Then the required fare is given by

$$a_{25} = 1 + (25 - 1)0.60 = 15.4,$$

or \$15.40. Therefore, by taking the airport limousine, the Kunwoo will save

$$15.40 - 7.50 = 7.90, \text{ or } \$7.90.$$

21. a. Using the formula for the sum of an arithmetic progression, we have

$$S_n = \frac{n}{2}[2a+(n-1)d]]$$
$$= \frac{N}{2}[2(1)+(N-1)(1)] = \frac{N}{2}(N+1).$$

b. $S_{10} = \frac{10}{2}(10+1) = 5(11) = 55$

$$D_3 = (C-S)\frac{N-(n-1)}{S_N} = (6000-500)\frac{10-(3-1)}{55} = 5500\left(\frac{8}{55}\right)$$
$$= 800, \text{ or } \$800.$$

23. This is a geometric progression with $a = 4$ and $r = 2$. Next, $a_7 = 4(2)^6 = 256$,

and $S_7 = \frac{4(1-2^7)}{1-2} = 508.$

25. If we compute the ratios

$$\frac{a_2}{a_1} = \frac{-\frac{3}{8}}{\frac{1}{2}} = -\frac{3}{4} \quad \text{and} \quad \frac{a_3}{a_2} = \frac{\frac{1}{4}}{-\frac{3}{8}} = -\frac{2}{3},$$

we see that the given sequence is not geometric since the ratios are not equal.

27. This is a geometric progression with $a = 243$, and $r = 1/3$.

$$a_7 = 243(\tfrac{1}{3})^6 = \tfrac{1}{3}$$
$$S_7 = \frac{243(1-(\frac{1}{3})^7)}{1-\frac{1}{3}} = 364\tfrac{1}{3}.$$

29. First, we compute

$$r = \frac{a_2}{a_1} = \frac{3}{-3} = -1.$$

Next, $a_{20} = -3(-1)^{19} = 3$ and so $S_{20} = \frac{-3[1-(-1)^{20}]}{[1-(-1)]} = 0$.

31. The population in five years is expected to be

$$200{,}000(1.08)^{6-1} = 200{,}000\,(1.08)^5 \approx 293{,}866.$$

33. The salary of a union member whose salary was \$42,000 six years ago is given by the 7th term of a geometric progression whose first term is 42,000 and whose common ratio is 1.03. Thus

$a_7 = (42{,}000)(1.03)^6 = 50{,}150.20$, or $50,150.20.

35. With 8 percent raises per year, the employee would make

$$S_4 = 48{,}000\left[\frac{1-(1.08)^4}{1-1.08}\right] \approx 216{,}293.38,$$

or $216,293.38 over the next four years.

With $4000 raises per year, the employee would make

$$S_4 = \frac{4}{2}[2(48{,}000)+(4-1)4000] = 216{,}000$$

or $216,000 over the next four years. We conclude that the employee should choose annual raises of 8 percent per year.

37. *a*. During the sixth year, she will receive

$a_6 = 10{,}000(1.15)^5 \approx 20{,}113.57$, or $20,113.57.

b. The total amount of the six payments will be given by

$$S_6 = \frac{10{,}000[1-(1.15)^6]}{1-1.15} \approx 87{,}537.38, \text{ or } \$87{,}537.38.$$

39. The book value of the office equipment at the end of the eighth year is given by

$$V(8) = 150{,}000\left(1-\frac{2}{10}\right)^8 \approx 25{,}165.82, \text{ or } \$25{,}165.82.$$

41. The book value of the restaurant equipment at the end of six years is given by

$V(6) = 150{,}000(0.8)^6 \approx 39{,}321.60$,

or $39,321.60. By the end of the sixth year, the equipment will have depreciated by

$D(n) = 150{,}000 - 39{,}321.60 = 110{,}678.40$, or $110,678.40.

43. True. Suppose d is the common difference of $a_1, a_2, ..., a_n$ and e is the common difference of $b_1, b_2, ..., b_n$. Then $d + e$ is the common difference of $a_1 + b_1, a_2 + b_2, ..., a_n + b_n$ is obtained by adding the constant $c + d$ to it, we see that it is indeed an arithmetic progression.

CHAPTER 5 CONCEPT REVIEW, page 331

1. a. Original; $P(1+rt)$ b. Interest; $P(1+i)^n$; $A(1+i)^{-n}$

3. Annuity; ordinary annuity; simple annuity

5. $\dfrac{Pi}{1-(1+i)^{-n}}$

7. Constant d; $a+(n-1)d$; $\dfrac{n}{2}[2a+(n-1)d]$

CHAPTER 5, REVIEW EXERCISES, page 332

1. a. Here $P = 5000$, $r = 0.1$, and $m = 1$. Thus, $i = r = 0.1$ and $n = 4$. So
$$A = 5000(1.1)^4 = 7320.5, \text{ or } \$7320.50.$$
b. Here $m = 2$ so that $i = 0.1/2 = 0.05$ and $n = (4)(2) = 8$. So
$$A = 5000(1.05)^8 \approx 7387.28 \text{ or } \$7387.28.$$
c. Here $m = 4$, so that $i = 0.1/4 = 0.025$ and $n = (4)(4) = 16$. So
$$A = 5000(1.025)^{16} \approx 7{,}422.53, \text{ or } \$7422.53.$$
d. Here $m = 12$, so that $i = 0.1/12$ and $n = (4)(12) = 48$. So
$$A = 5000\left(1+\frac{0.10}{12}\right)^{48} \approx 7446.77, \text{ or } \$7446.77.$$

3. a. The effective rate of interest is given by
$$r_{eff} = \left(1+\frac{r}{m}\right)^m - 1 = (1 + 0.12) - 1 = 0.12, \text{ or 12 percent.}$$
b. The effective rate of interest is given by
$$r_{eff} = \left(1+\frac{r}{m}\right)^m - 1 = \left(1+\frac{0.12}{2}\right)^2 - 1 = 0.1236, \text{ or 12.36 percent.}$$
c. The effective rate of interest is given by
$$r_{eff} = \left(1+\frac{r}{m}\right)^m - 1 = \left(1+\frac{0.12}{4}\right)^4 - 1 \approx 0.125509, \text{ or 12.5509 percent.}$$
d. The effective rate of interest is given by
$$r_{eff} = \left(1+\frac{r}{m}\right)^m - 1 = \left(1+\frac{0.12}{12}\right)^{12} - 1 \approx 0.126825, \text{ or 12.6825 percent.}$$

5. The present value is given by

$$P = 41{,}413\left(1+\frac{0.065}{4}\right)^{-20} \approx 30{,}000.29, \text{ or approximately } \$30{,}000.$$

7.

$$S = 150\left[\frac{\left(1+\frac{0.08}{4}\right)^{28} - 1}{\frac{0.08}{4}}\right] \approx 5557.68, \text{ or } \$5557.68.$$

9. Using the formula for the present value of an annuity with $R = 250$, $n = 36$, $i = 0.09/12 = 0.0075$, we have

$$P = 250\left[\frac{1-(1.0075)^{-36}}{0.0075}\right] \approx 7861.70, \text{ or } \$7861.70.$$

11. Using the amortization formula with $P = 22{,}000$, $n = 36$, and $i = 0.085/12$, we find

$$R = \frac{22{,}000\left(\frac{0.085}{12}\right)}{1-\left(1+\frac{0.085}{12}\right)^{-36}} \approx 694.49, \text{ or } \$694.49.$$

13. Using the sinking fund formula with $S = 18{,}000$, $n = 48$, and $i = 0.06/12$, we have

$$R = \frac{\left(\frac{0.06}{12}\right)18{,}000}{\left(1+\frac{0.06}{12}\right)^{48} - 1} \approx 332.73, \text{ or } \$332.73.$$

15. We are asked to find r such that

$$\left(1+\frac{r}{365}\right)^{365} = \left(1+\frac{0.072}{12}\right)^{12} = 1.074424168.$$

Then

$$1+\frac{r}{365} = (1.074424168)^{1/365}$$

$$\frac{r}{365} = (1.074424168)^{1/365} - 1$$

and

$$r = 365[1.074424168)^{1/365} - 1]$$

$$\approx 0.071791919, \text{ or approximately } 7.179 \text{ percent.}$$

17. We want to find r where r satisfies the equation $8.2 = 4.5\,e^{r(5)}$. We have

$$e^{5r} = \frac{8.2}{4.5} \quad \text{or} \quad 5r = \ln\frac{8.2}{4.5}$$

and $r = \frac{1}{5}\ln\left(\frac{8.2}{4.5}\right) \approx 0.12$. So the annual rate of return is 12 percent per year.

19. Let a_n denote the sales during the nth year of operation. Then

$$\frac{a_{n+1}}{a_n} = 1.14 = r.$$

Therefore, the sales during the fourth year of operation were

$$a_4 = a_1 r^{n-1} = 1{,}750{,}000(1.14)^{4-1} = 2{,}592{,}702, \text{ or } \$2{,}592{,}702.$$

The total sales over the first four years of operation are given by

$$S_4 = \frac{a(1-r^4)}{1-r} = \frac{1{,}750{,}000[1-(1.14)^4]}{1-(1.14)} = 8{,}612{,}002, \text{ or } \$8{,}612{,}002.$$

21. We use the compound interest formula with $P = 2000$, $t = 5$, $r = 0.08$, and $m = 52$, obtaining $A = P(1+i)^n = 2000\left(1+\frac{0.08}{52}\right)^{5(52)} \approx 2982.73$, or \$2982.73.

23. We use formula (5), $A = Pe^{rt}$, with $A = 5986.09$, $r = 0.06$, and $t = 3$, obtaining

$$Pe^{3(0.06)} = 5986.09, \quad \text{or} \quad P = 5986.09e^{-0.18} \approx 5000$$

So Andrew originally deposited \$5000 into the account .

25. Using the sinking fund formula with $S = 40{,}000$, $n = 120$, and $i = 0.08/12$, we find

$$R = \frac{\left(\frac{0.08}{12}\right)40{,}000}{\left(1+\frac{0.08}{12}\right)^{120} - 1} = 218.64, \text{ or } \$218.64.$$

27. Using the formula for the present value of an annuity, we see that the equivalent cash payment of Maria's auto lease is

$$P = 300\left[\frac{1-\left(1+\frac{0.05}{12}\right)^{-48}}{\frac{0.05}{12}}\right] = 13{,}026.89, \text{ or } \$13{,}026.89.$$

29. a. The monthly payment is given by

$$P = \frac{(120{,}000)(0.0075)}{1-(1+0.0075)^{-360}} \approx 965.55, \text{ or } \$965.55.$$

b. We can find the total interest payment by computing

$360(965.55) - 120{,}000 \approx 227{,}598$, or \$227,598.

c. We first compute the present value of their remaining payments. Thus,

$$P = 965.55\left[\frac{1-(1+0.0075)^{-240}}{0.0075}\right] \approx 107{,}316.01.$$

or \$107,316.01. Then their equity is

150,000 - 107,316.01, or approximately \$42,684.

31. Using the sinking fund formula with $S = 500{,}000$, $n = 20$, and $I = 0.10/4$, we find that the amount of each installment should be

$$R = \frac{\left(\frac{0.10}{4}\right)500{,}000}{\left(1+\frac{0.10}{4}\right)^{20} - 1} \approx 19{,}573.56, \text{ or } \$19{,}573.56.$$

33. a. Using the amortization formula, we find that Bill's monthly payment will be

$$R = \frac{3200\left(\frac{0.186}{12}\right)}{1-\left(1+\frac{0.186}{12}\right)^{-18}} \approx 205.09, \text{ or } \$205.09.$$

b. Using the formula for computing the effective rate of interest with $r = 0.186$ and $m = 12$, we have

$$r_{eff} = \left(1+\frac{0.186}{12}\right)^{12} - 1 = 0.2027, \text{ or } 20.27 \text{ percent per year.}$$

CHAPTER 5, BEFORE MOVING ON, page 334

1. Here $P = 2000$, $r = 0.08$, $t = 3$, and $m = 12$. So $i = \frac{0.08}{12}$. Therefore

$$A = 2000\left(1+\frac{0.08}{12}\right)^{(12)(3)} \approx 2540.47, \text{ or } \$2{,}540.47.$$

2. Here $r = 0.06$ and $m = 365$. So,

$$r_{eff} = \left(1+\frac{0.06}{365}\right)^{365} - 1 \approx 0.0618,$$ or approximately 6.2% per year.

3. Here $R = 800$, $n = (1)(52)$, $r = 0.06$, and $m = 52$. So

$$S = \frac{800\left[\left(1+\frac{0.06}{52}\right)^{520} - 1\right]}{\frac{0.06}{52}} \approx 569{,}565.47,$$ or \$569,565.47.

4. Here $P = 100{,}000$, $t = 10$, $r = 0.08$, $m = 12$. So

$$R = \frac{100{,}000\left(\frac{0.08}{12}\right)}{1-\left(1+\frac{0.08}{12}\right)^{-120}} \approx 1213.276,$$ or approximately \$1213.28.

5. Here $S = 15{,}000$, $t = 6$, $r = 0.1$, and $m = 52$. So

$$R = \frac{\left(\frac{0.1}{52}\right)15{,}000}{\left(1+\frac{0.1}{52}\right)^{(6)(52)} - 1} \approx 35.132,$$ or \$35.13.

6. a. Here $a_1 = a = 3$ and $d = 4$. So, with $n = 10$,

$$S_{10} = \frac{10}{2}[2(2)+9(4)] = 210$$

b. Here $a = 2$ and $r = 2$. So with $n = 8$

$$S_8 = \frac{\frac{1}{2}\left[1-(2)^8\right]}{1-2} = 127.5$$

CHAPTER 6

6.1 Problem Solving Tips

1. It's often easier to remember a formula if you learn to express the formula in words.

Thus DeMorgan's Laws $(A \cup B)^c = A^c \cap B^c$ says that "the complement of the union of two sets is equal to the intersection of their complements," and $(A \cap B)^c = A^c \cup B^c$ says that "the complement of the intersection of two sets is equal to the union of their complements."

6.1 CONCEPTS QUESTIONS, page 342

1. a. A set is a well-defined collection of objects. As an example, consider the set of all freshmen students in a college.
   b. Two sets A and B are equal if they have exactly the same elements.
   c. The empty set is the set that contains no elements.

3. a. If $A \subset B$, then $B^c \subset A^c$.
   b. If $A^c = \varnothing$, then $A = U$ where U is the universal set.

EXERCISES 6.1, page 342

1. $\{x \mid x$ is a gold medalist in the 2006 Winter Olympic Games$\}$

3. $\{x \mid x$ is an integer greater than 2 and less than 8$\}$

5. $\{2,3,4,5,6\}$

7. $\{-2\}$

9. a. True--the order in which the elements are listed is not important.
   b. False-- A is a set, not an element.

11. a. False. The empty set has no elements. b. False. 0 is an element and $\varnothing$ is a set.

13. True.

15. a. True. 2 belongs to A. b. False. For example, 5 belongs to A but $5 \notin \{2,4,6\}$.

17. a. and b.

19. a. $\varnothing$, {1}, {2}, {1,2}
    b. $\varnothing$, {1}, {2}, {3}, {1,2}, {1,3}, {2,3}, {1,2,3}
    c. $\varnothing$, {1}, {2}, {3}, {4}, {1,2}, {1,3}, {1,4}, {2,3}, {2,4}, {3,4}, {1,2,3}, {1,2,4}, {2,3,4}, {1,3,4}, {1,2,3,4}

21. {1, 2, 3, 4, 6, 8, 10} 22. {1, 2, 4, a, b} 23. {Jill, John, Jack, Susan, Sharon}

25. a.

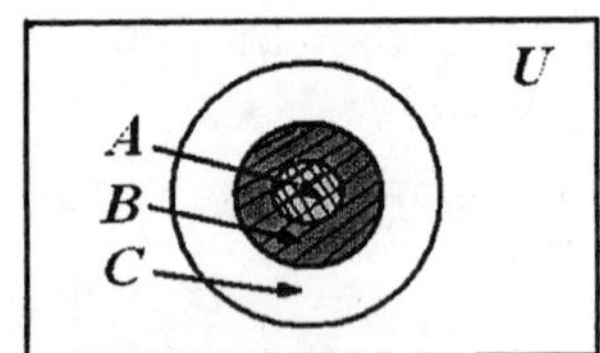

b.

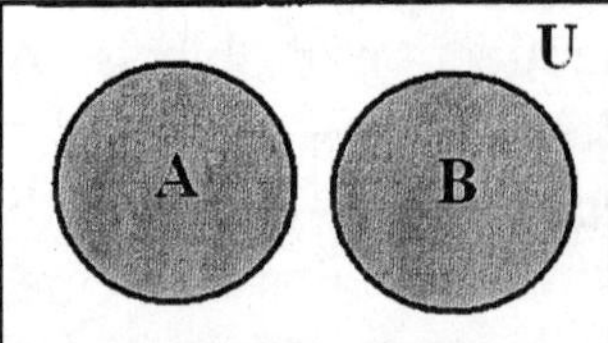

c.

27. a.

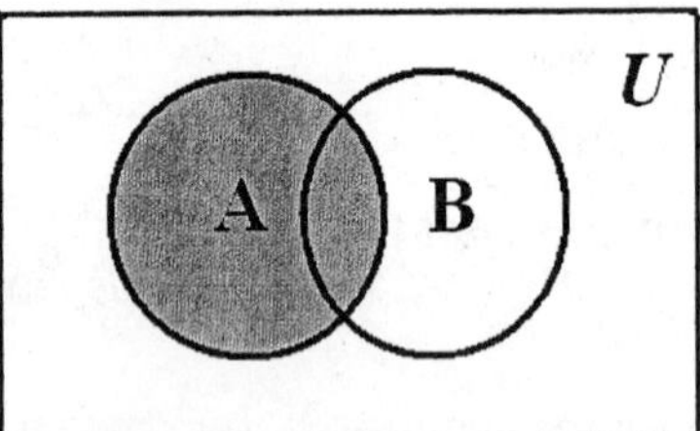

b.

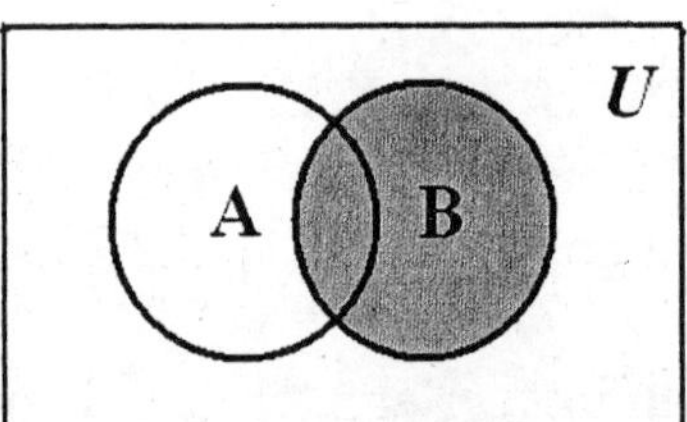

29. a.

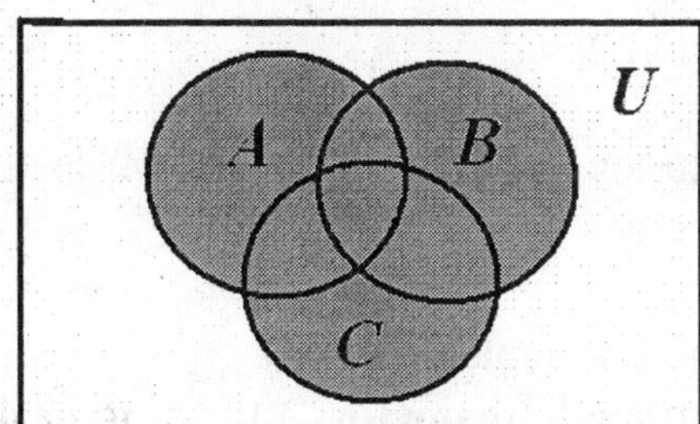

b.

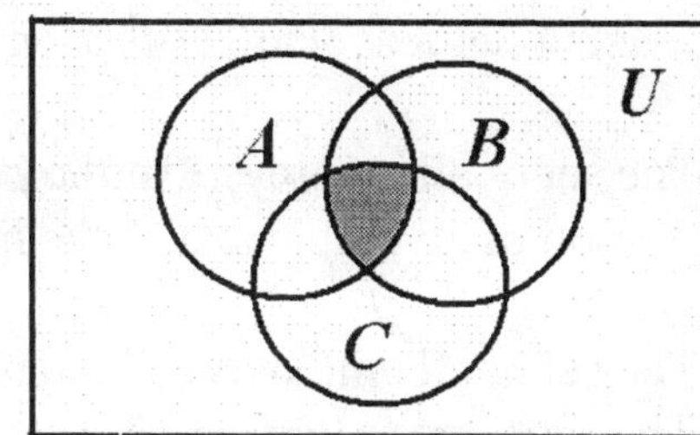

31. a.

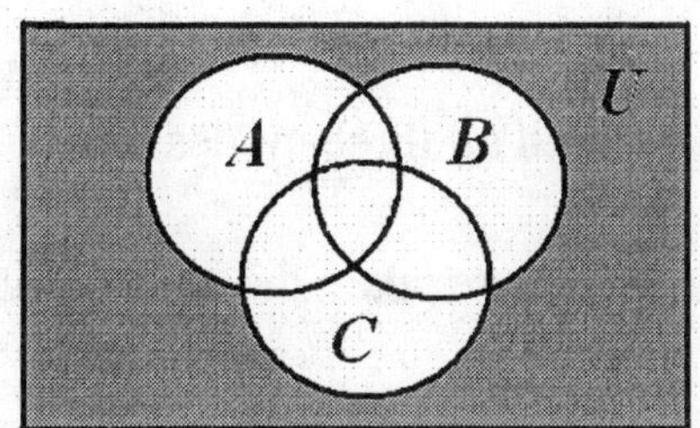

b.

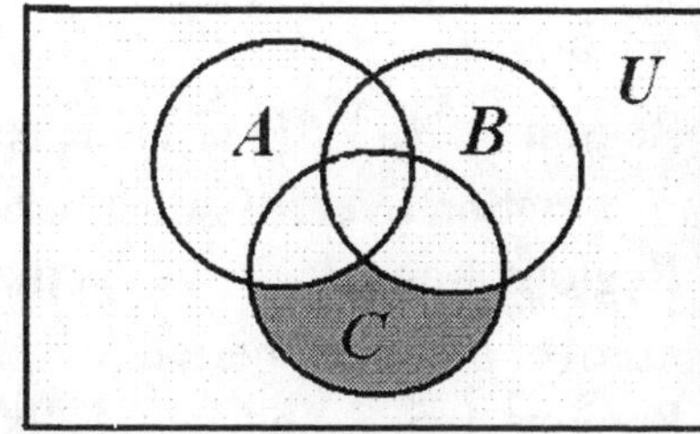

33. a. $A^C = \{2, 4, 6, 8, 10\}$
b. $B \cup C = \{2, 4, 6, 8, 10\} \cup \{1, 2, 4, 5, 8, 9\} = \{1, 2, 4, 5, 6, 8, 9, 10\}$
c. $C \cup C^C = U = \{1, 2, 3, 4, 5, 6, 7, 8, 9, 10\}$

35. a. $(A \cap B) \cup C = C = \{1, 2, 4, 5, 8, 9\}$
b. $(A \cup B \cup C)^C = \varnothing$
c. $(A \cap B \cap C)^C = U = \{1, 2, 3, 4, 5, 6, 7, 8, 9, 10\}$

37. a. The sets are not disjoint. 4 is an element of both sets.
b. The sets are disjoint as they have no common elements.

39. a. The set of all employees at the Universal Life Insurance Company who do not drink tea.
b. The set of all employees at the Universal Life Insurance Company who do not drink coffee.

41. a. The set of all employees at the Universal Life Insurance Company who drink tea but not coffee.

b. The set of all employees at the Universal Life Insurance Company who drink coffee but not tea.

43. a. The set of all employees at the hospital who are not doctors.
b. The set of all employees at the hospital who are not nurses.

45. a. The set of all employees at the hospital who are female doctors.
b. The set of all employees at the hospital who are both doctors and administrators.

47. a. $D \cap F$ b. $R \cap F^C \cap L^C$

49. a. B^C b. $A \cap B$ c. $A \cap B \cap C^C$

51. a. Region 1: $A \cap B \cap C$ is the set of tourists who used all three modes of transportation over a 1-week period in London.
b. Regions 1 and 4: $A \cap C$ is the set of tourists who have taken the underground and a bus over a 1-week period in London.
c. Regions 4, 5, 7, and 8: B^c is the set of tourists who have not taken a cab over a 1-week period in London.

53. $A \subset A \cup B$ $B \subset A \cup B$

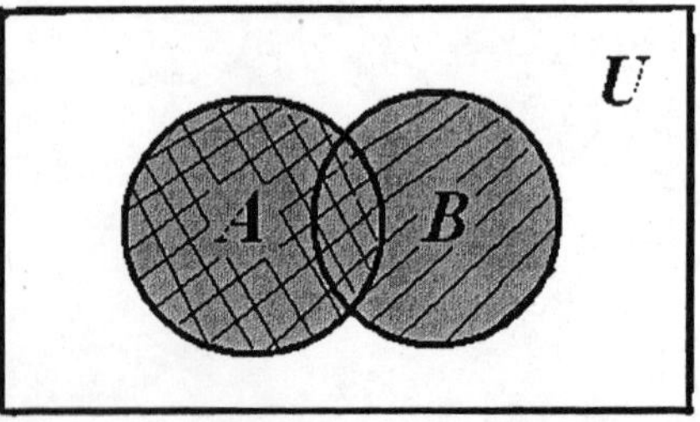

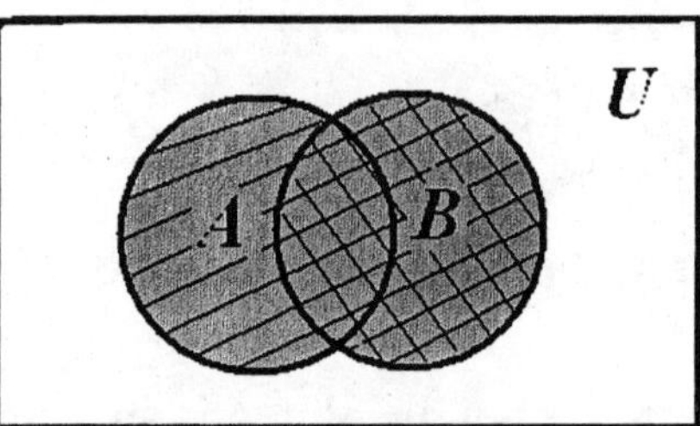

55. $A \cup (B \cup C) = (A \cup B) \cup C$

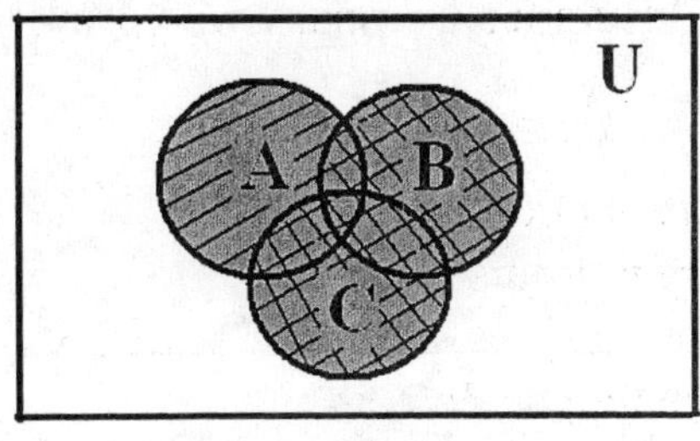

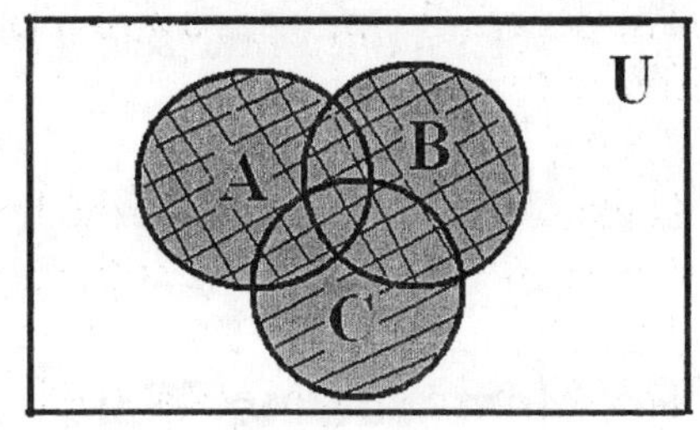

57. $A \cap (B \cup C) = (A \cap B) \cup (A \cap C)$

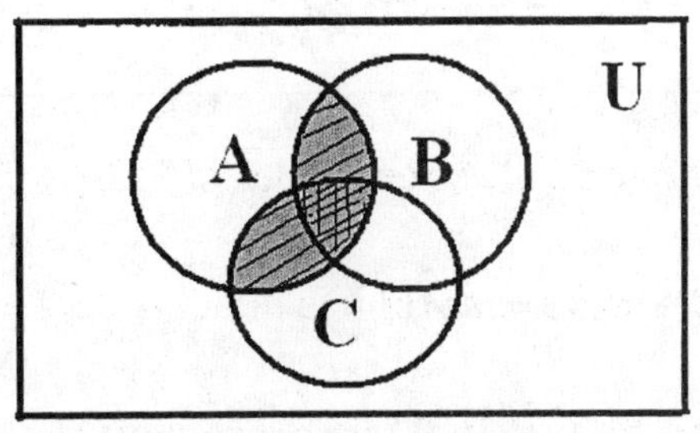

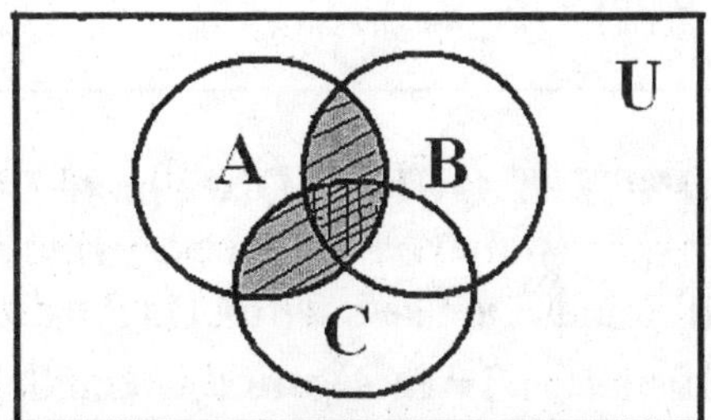

59. a. $A \cup (B \cup C) = \{1, 3, 5, 7, 9\} \cup (\{1, 2, 4, 7, 8\} \cup \{2, 4, 6, 8\})$
$= \{1, 3, 5, 7, 9\} \cup \{1, 2, 4, 6, 7, 8\}$
$= \{1, 2, 3, 4, 5, 6, 7, 8, 9\}$
$(A \cup B) \cup C = (\{1, 3, 5, 7, 9\} \cup (\{1, 2, 4, 7, 8\}) \cup \{2, 4, 6, 8\})$
$= \{1, 2, 3, 4, 5, 7, 8, 9\} \cup \{2, 4, 6, 8\}$
$= \{1, 2, 3, 4, 5, 6, 7, 8, 9\}$

b. $A \cap (B \cap C) = \{1, 3, 5, 7, 9\} \cap (\{1, 2, 4, 7, 8\} \cap \{2, 4, 6, 8\})$
$= \{1, 3, 5, 7, 9\} \cap (\{2, 4, 8\}$
$= \varnothing$
$(A \cap B) \cap C = (\{1, 3, 5, 7, 9\} \cap \{1, 2, 4, 7, 8\}) \cap \{2, 4, 6, 8\}$
$= \{1, 7\} \cap \{2, 4, 6, 8\}$
$= \varnothing.$

61. a. *r, u, v, w, x y* b. *v, r*

63. a. *t, y, s* b. *t ,s, w, x, z* 65. $A \subset C$

67. False. Since every element in a set *A* belongs to *A*, *A* is a subset of itself.

69. True. If at least one of the sets A or B is nonempty, then $A \cup B \neq \varnothing$.

71. True. $(A \cup A^c)^c = U^c = \varnothing$.

6.2 Problem Solving Tips

1. In the problems that follow it will often be helpful to draw a Venn Diagram to solve the problem.

6.2 CONCEPT QUESTIONS, page 349

1. a. If A and B are sets with $A \cap B = \varnothing$, then $n(A) + n(B) = n(A \cup B)$.
   b. If $n(A \cup B) \neq n(A) + n(B)$, then $A \cap B \neq \varnothing$.

EXERCISES 6.2, page 349

1. $A \cup B = \{a, e, g, h, i, k, l, m, o, u\}$, and so $n(A \cup B) = 10$. Next, $n(A) + n(B) = 5 + 5 = 10$.

3. a. $A = \{2, 4, 6, 8\}$ and $n(A) = 4$. b. $B = \{6, 7, 8, 9, 10\}$ and $n(B) = 5$
   c. $A \cup B = \{2, 4, 6, 7, 8, 9, 10\}$ and $n(A \cup B) = 7$.
   d. $A \cap B = \{6, 8\}$ and $n(A \cap B) = 2$.

5. Using the results of Exercise 3, we see that $n(A \cup B) = 7$ and $n(A) + n(B) - n(A \cap B) = 4 + 5 - 2 = 7$.

7. Since $n(A \cup B) = n(A) + n(B) - n(A \cap B)$
   $$n(B) = n(A \cup B) + n(A \cap B) - n(A) = 30 + 5 - 15 = 20.$$

9. Refer to the Venn diagram at the right.
   a. $n(A \cup B) = 60 + 40 + 40 = 140$.
   b. $n(A^C) = 40 + 60 = 100$.
   c. $n(A \cap B^C) = 60$.

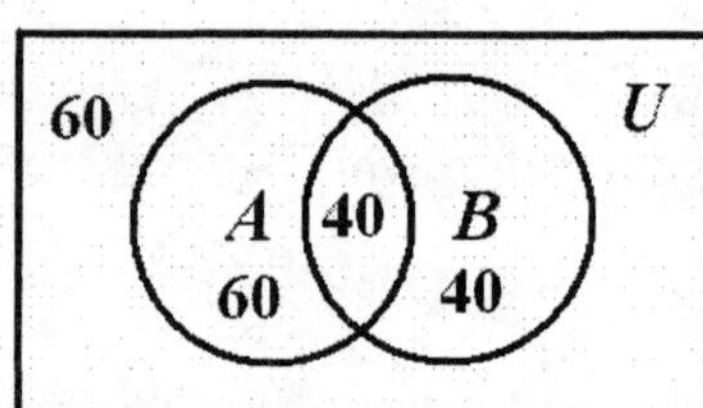

11. $n(A \cup B) = n(A) + n(B) - n(A \cap B) = 6 + 10 - 3 = 13.$

13. $n(A \cup B) = n(A) + n(B) - n(A \cap B)$ so
$n(A \cap B) = n(A) + n(B) - n(A \cup B) = 4 + 5 - 9 = 0.$

15. $n(A \cap B \cap C)$
$= n(A \cup B \cup C) - n(A) - n(B) - n(C) + n(A \cap B) + n(A \cap C) + n(B \cap C)$
so $n(C) = n(A \cup B \cup C) - n(A \cap B \cap C) - n(A) - n(B)$
$+ n(A \cap B) + n(A \cap C) + n(B \cap C$
$= 25 - 2 - 12 - 12 + 5 + 5 + 4 = 13.$

17. Let A denote the set of prisoners in the Wilton County Jail who were accused of a felony and B the set of prisoners in that jail who were accused of a misdemeanor. Then we are given that

$$n(A \cup B) = 190$$

Refer to the diagram at the right.
Then the number of prisoners who were accused of both a felony and a misdemeanor is given by $(A \cap B) = n(A) + n(B) - n(A \cup B)$
$= 130 + 121 - 190 = 61.$

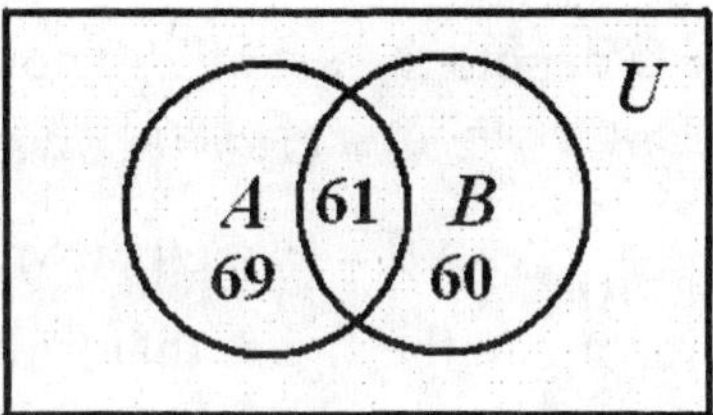

19. Let U denote the set of all customers surveyed, and let
$A = \{x \in U \mid x \text{ buys brand } A\}$
$B = \{x \in U \mid x \text{ buys brand } B\}.$
Then $n(U) = 120$, $n(A) = 80$,
$n(B) = 68$, and $n(A \cap B) = 42$.
Refer to the diagram at the right.

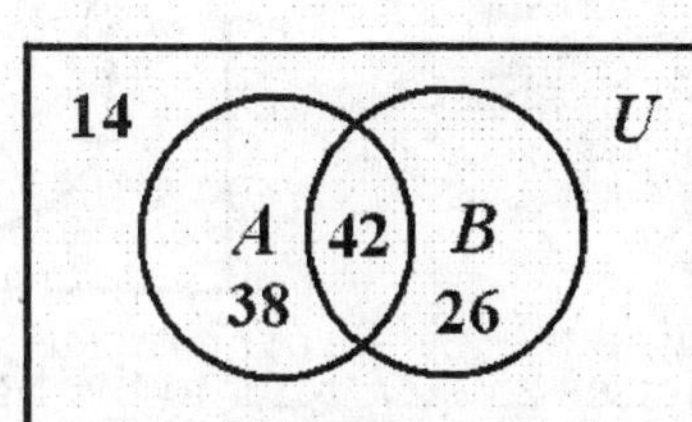

a. The number of customers who buy at least one of these brands is
$n(A \cup B) = 80 + 68 - 42 = 106.$

b. The number who buy exactly one of these brands is
$n(A \cap B^C) + n(A^C \cap B) = 38 + 26 = 64$

c. The number who buy only brand A is $n(A \cap B^C) = 38.$

d. The number who buy none of these brands is $n[(A \cup B)^C] = 120 - 106 = 14.$

21. Let U denote the set of 200 investors and let

$A = \{ x \in U | x \text{ uses a discount broker}\}$

$B = \{ x \in U | x \text{ uses a full-service broker}\}.$

Refer to the diagram at the right.

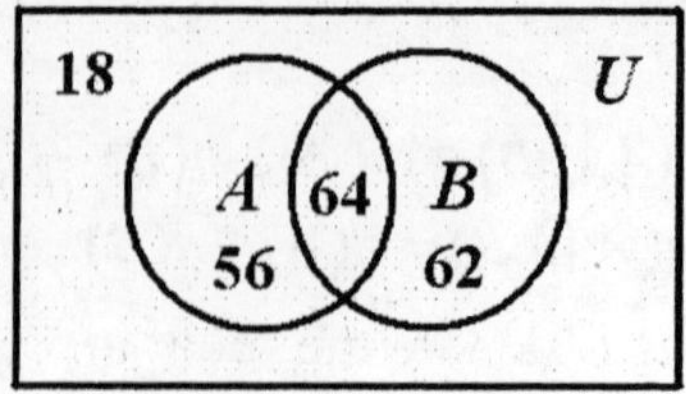

a. The number of investors who use at least one kind of broker is
$n(A \cup B) = n(A) + n(B) - n(A \cap B) = 120 + 126 - 64 = 182.$

b. The number of investors who use exactly one kind of broker is
$n(A \cap B^C) + n(A^C \cap B) = 56 + 62 = 118.$

c. The number of investors who use only discount brokers is $n(A \cap B^C) = 56$.

d. The number of investors who don't use a broker is
$n(A \cup B)^C = n(U) - n(A \cup B) = 200 - 182 = 18.$

23. Let U denote the set of 200 households in the survey and let

$A = \{x \in U | x \text{ owns a desktop computer}\}$

$B = \{x \in U | x \text{ owns a laptop computer}\}$

Referring to the figure that follows, we see that the number of households that own both desktop and laptop computers is $n(A \cap B) = 200 - 120 - 10 - 40 = 30.$

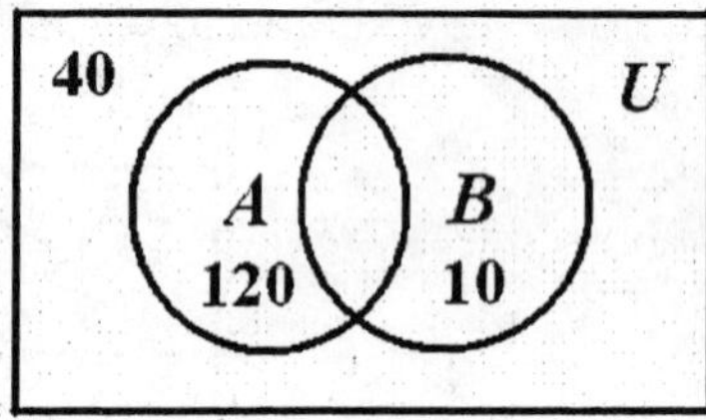

In Exercises 25 - 27, refer to the figure that follows.

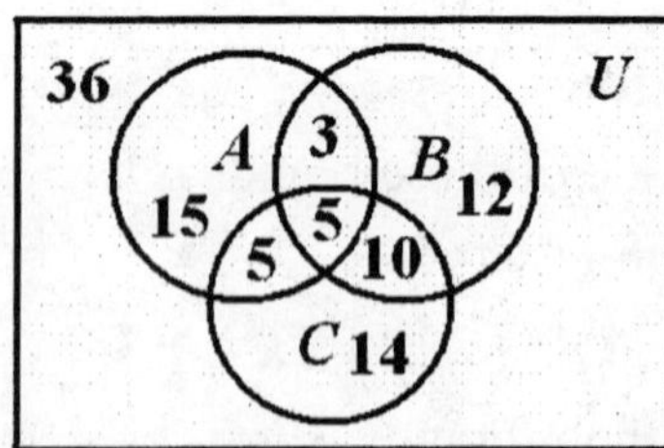

25. a. $n(A \cup B \cup C) = 64$ b. $n(A^C \cap B \cap C) = 10$

27. a. $n(A^C \cap B^C \cap C^C) = n[(A \cup B \cup C)^C] = 36$ b. $n[A^C \cap (B \cup C)] = 36$

29. Let U denote the set of all economists surveyed, and let

$A = \{ x \in U |\ x \text{ had lowered his estimate of the consumer inflation rate}\}$

$B = \{ x \in U |\ x \text{ had raised his estimate of the } GDP \text{ growth rate}\}$.

Refer to the diagram that follows.

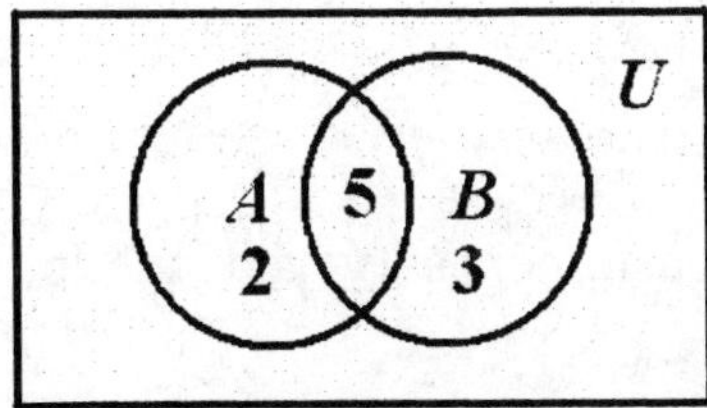

Then $n(U) = 10$, $n(A) = 7$, $n(B) = 8$, and $n(A \cap B^C) = 2$. Then the number of economists who had both lowered their estimate of the consumer inflation rate and raised their estimate of the GDP rate is given by $n(A \cap B) = 5$.

31. Let U denote the set of 100 college students who were surveyed and let

$A = \{x \in U |\ x$ is a student who reads *Time* magazine$\}$

$B = \{x \in U |\ x$ is a student who reads *Newsweek* magazine$\}$

and $C = \{x \in U |\ x$ is a student who reads *U.S. News and World Report* magazine$\}$

Refer to the diagram that follows.

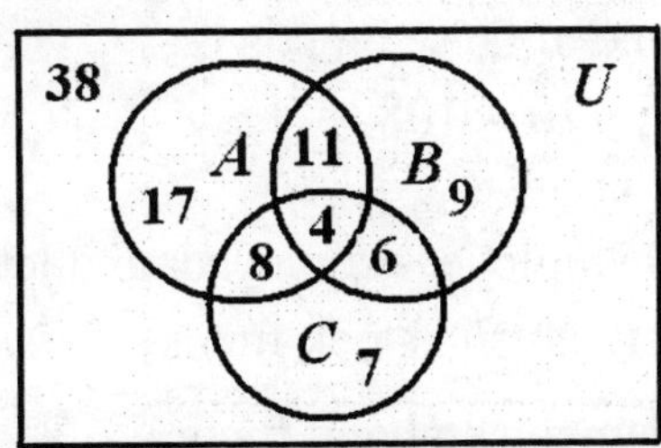

Then $n(A) = 40$, $n(B) = 30$, $n(C) = 25$, $n(A \cap B) = 15$, $n(A \cap C) = 12$, $n(B \cap C) = 10$, and $n(A \cap B \cap C) = 4$.

a. The number of students surveyed who read at least one magazine is $n(A \cup B \cup C) = 17 + 11 + 4 + 8 + 6 + 7 + 9 = 62$

b. The number of students surveyed who read exactly one magazine is

$$n(A \cap B^C \cap C^C) + n(A^C \cap B \cap C^C) + n(A^C \cap B^C \cap C)$$
$$= 17 + 9 + 7 = 33.$$

c. The number of students surveyed who read exactly two magazines is

$$n(A \cap B \cap C^C) + n(A^C \cap B \cap C) + n(A \cap B^C \cap C)$$
$$= 11 + 6 + 8 = 25.$$

d. The number of students surveyed who did not read any of these magazines is

$$n(A \cup B \cup C)^C = 100 - 62 = 38.$$

33. Let U denote the set of all customers surveyed, and let

$A = \{x \in U |\ x \text{ buys brand } A\}$

$B = \{x \in U |\ x \text{ buys brand } B\}$.

$C = \{x \in U |\ x \text{ buys brand } C\}$.

Refer to the figure at the right. Then
$n(U) = 120$, $n(A \cap B \cap C^C) = 15$,
$n(A^C \cap B \cap C^C) = 25$,
$n(A^C \cap B^C \cap C) = 26$,
$n(A \cap B \cap C^C) = 15$, $n(A \cap B^C \cap C) = 10$,
$n(A^C \cap B \cap C) = 12$, and $n(A \cap B \cap C) = 8$.

a. The number of customers who buy at least one of these brands is

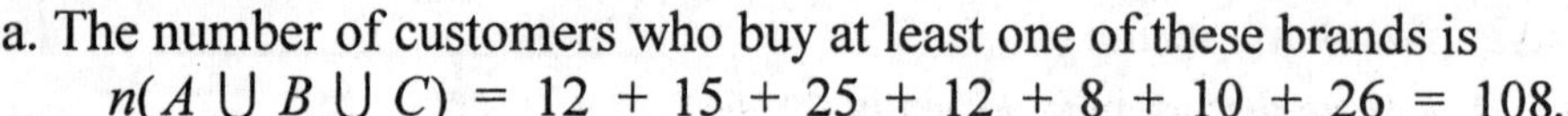

$$n(A \cup B \cup C) = 12 + 15 + 25 + 12 + 8 + 10 + 26 = 108.$$

b. The number who buy labels A and B but not C is $n(A \cap B \cap C^C) = 15$

c. The number who buy brand A is $n(A) = 12 + 10 + 15 + 8 = 45$.

d. The number who buy none of these brands is

$$n[(A \cup B \cup C)^C] = 120 - 108 = 12.$$

35. Let U denote the set of 200 employees surveyed, and let

$A = \{x \in U | x \text{ had investments in stock funds}\}$

$B = \{x \in U | x \text{ had investments in bond funds}\}$

$C = \{x \in U | x \text{ had investments in money market funds}\}$

Then $n(U)=200,\ n(A)=141,\ n(B)=91,\ n(C)=60,\ n(A\cap B)=47,$

$n(A\cap C)=36,\ n(B\cap C)=36,$ and $n(A^c\cap B^c\cap C^c)=n[(A\cup B\cup C)^c]=5$

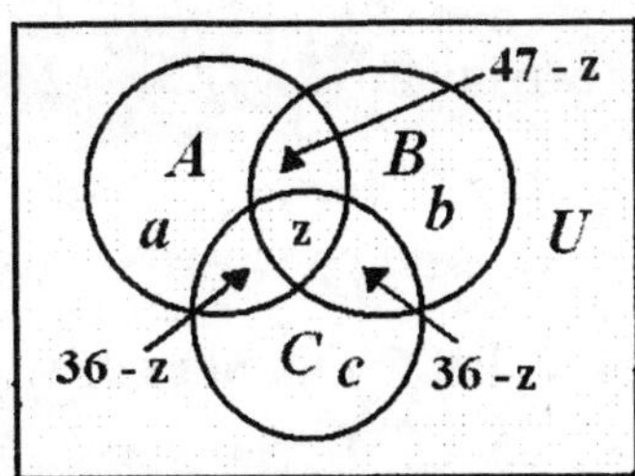

Letting $n(A\cap B\cap C)=z$ and using the fact that $n(A\cap B)=47,\ n(A\cap C)=36,$ and $n(B\cap C)=36,$ leads to the Venn diagram shown. Next, using the fact that $n(A)=141,\ n(B)=91,$ and $n(C)=60$ leads to

$$a+(36-z)+(47-z)+z=141$$
$$b+(47-z)+(36-z)+z=91$$
$$c+(36-z)+(36-z)+z=60$$
$$a+b+c+(36-z)+(47-z)+(36-z)+z+5=200$$

which simplifies to $a-z=58,\ b-z=8,\ c-z=-12,\ a+b+c-2z=76.$
Solving, we find $a=80,\ b=30,\ c=10,$ and $z=22.$ Therefore,

a. The number of employees surveyed who had invested in all three investments is $n(A\cap B\cap C)=z=22.$

b. The number who had invested in stock funds only is given by $n(A\cap B^c\cap C^c)=a=80.$

37. True. $n(A\cup B)=n(A)+n(B)-n(A\cap B)$.

39. True. If $A\cap B\neq\varnothing$, then $n(A\cup B)=n(A)+n(B)-n(A\cap B)$.

6.3 CONCEPT QUESTIONS, page 357

1. If a task T_1 can be performed in N_1 ways, a task T_2 can be performed in N_2 ways, ..., and finally a task T_n can be performed in N_n ways, then the number of ways of performing the tasks $T_1,T_2,...,T_n$ in succession is given by $N_1N_2\cdots N_n$.

EXERCISES 6.3, page 357

1. By the multiplication principle, the number of rates is given by (4)(3) = 12.

3. By the multiplication principle, the number of ways that a blackjack hand can be dealt is (4)(16) = 64.

5. By the multiplication principle, she can create (2)(4)(3) = 24 different ensembles.

7. The number of paths is $2 \times 4 \times 3$, or 24.

9. By the multiplication principle, we see that the number of ways a health-care plan can be selected is (10)(3)(2) = 60.

11. $10^9 = 1{,}000{,}000{,}000$.

13. The number of different responses is $\underbrace{(5)\,(5)\,...\,(5)}_{50 \text{ terms}} = 5^{50}$.

15. The number of selections is given by (2)(5)(3), or 30 selections.

17. The number of different selections is (10)(10)(10)(10) - 10 = 10000 - 10 = 9990.

19. a. The number of license plate numbers that may be formed is (26)(26)(26)(10)(10)(10), or 17,576,000.
    b. The number of license plate numbers that may be formed is (10)(10)(10)(26)(26)(26) =17,576,000.

21. If every question is answered, there are 2^{10}, or 1024, ways. In the second case, there are 3 ways to answer each question, and so we have 3^{10}, or 59,049, ways.

23. The number of ways the first, second and third prizes can be awarded is (15)(14)(13) = 2730.

25. The number of ways in which the nine symbols on the wheels can appear in the window slot is (9)(9)(9), or 729. The number of ways in which the eight symbols other than the "lucky dollar" can appear in the window slot is (8)(8)(8) or 512. Therefore, the number of ways in which the "lucky dollars" can appear in the window slot is 729 - 512, or 217.

27. True. There are 4 choices for the digit in the hundreds position, 4 choices in the tens position and 2 choices in the units position, or $4 \bullet 4 \bullet 2$, ore 32 such numbers.

6.4 Problem Solving Tips

1. Note the difference between a permutation and a combination. A permutation is an arrangement of a set of distinct objects in a *definite order* whereas a combination is an arrangement of a set of distinct objects without regard to order. In a permutation of two distinct objects A and B, we distinguish between the selections AB and BA, whereas in a combination these selections would be considered the same.

2. Sometimes the solution of an applied problem involves the multiplication principle and a permutation and/or a combination. (See Example 12 on page 367 in the text.)

6.4 CONCEPT QUESTIONS, page 369

1. a. Given a set of distinct objects, a permutation of the set is an arrangement of these objects in a *definite order*.

   b. $P(n,r)=\dfrac{n!}{(n-r)!}$ so $P(5,3)=\dfrac{5!}{(5-3)!}=\dfrac{5\cdot 4}{2\cdot 1}=10$

3. a. $C(n,r)=\dfrac{n!}{r!(n-r)!}$ b. $C(6,3)=\dfrac{6!}{3!(6-3)!}=\dfrac{6\cdot 5\cdot 4}{3\cdot 2}=20$

EXERCISES 6.4, page 369

1. $3(5!)=3(5)(4)(3)(2)(1)=360.$

3. $\dfrac{5!}{2!3!}=5(2)=10.$

5. $P(5,5)=\dfrac{5!}{(5\text{-}5)!}=\dfrac{5!}{0!}=120$

7. $P(5,2)=\dfrac{5!}{(5-2)!}=\dfrac{5!}{3!}=(5)(4)=20$

9. $P(n,1)=\dfrac{n!}{(n-1)!}=n$

11. $C(6,6)=\dfrac{6!}{6!0!}=1$

13. $C(7,4)=\dfrac{7!}{4!3!}=\dfrac{7\cdot 6\cdot 5}{3\cdot 2}=35$

15. $C(5,0)=\dfrac{5!}{5!0!}=1$

17. $C(9,6)=\dfrac{9!}{3!6!}=\dfrac{9\cdot 8\cdot 7}{3\cdot 2}=84$

19. $C(n,2)=\dfrac{n!}{(n-2)!2!}=\dfrac{n(n-1)}{2}$

21. $P(n,n-2)=\dfrac{n!}{(n-(n-2))!}=\dfrac{n!}{(n-n+2)!}=\dfrac{n!}{2}$

23. Order is important here since the word "*glacier*" is different from "*reicalg*", so this is a permutation.

25. Order is not important here. Therefore, we are dealing with a combination. If we consider a sample of three cellphones of which one is defective, it does not matter whether the defective cellphone is the first member of our sample, the second member of our sample, or the third member of our sample. The net result is a sample of three cellphones of which one is defective.

27. The order is important here. Therefore, we are dealing with a permutation. Consider, for example, 9 books on a library shelf. Each of the 9 books would have a call number, and the books would be placed in order of their call numbers; that is, a call number of 902 would come before a call number of 910.

29. The order is not important here, and consequently we are dealing with a combination. It would not matter if the hand *Q Q Q 5 5* were dealt or the hand *5 5 Q Q Q*. In each case the hand would consist of three queens and a pair.

31. The number of 4-letter permutations is $P(4,4)=\dfrac{4!}{0!}=4\cdot 3\cdot 2\cdot 1=24$.

33. The number of seating arrangements is $P(4,4)=\dfrac{4!}{0!}=24$.

35. The number of different batting orders is $P(9,9)=\dfrac{9!}{0!}=362{,}880$.

37. The number of different ways the 3 candidates can be selected is

$$C(12,3)=\frac{12!}{9!3!}=\frac{12\cdot 11\cdot 10}{3\cdot 2\cdot 1}=220.$$

39. There are 10 letters in the word *ANTARCTICA*, 3*A*s, 1*N*, 2*T*s, 1*R*, 2*C*s, and 1*I*. Therefore, we use the formula for the permutation of n objects, not all distinct:

$$\frac{n!}{n_1!n_2!\cdots n_r!} = \frac{10!}{3!2!2!} = 151{,}200\,.$$

41. The number of ways the 3 sites can be selected is

$$C(12,3) = \frac{12!}{9!3!} = \frac{12\cdot 11\cdot 10}{3\cdot 2\cdot 1} = 220\,.$$

43. The number of ways in which the sample of 3 microprocessors can be selected is

$$C(100,3) = \frac{100!}{97!3!} = \frac{100\cdot 99\cdot 98}{3\cdot 2\cdot 1} = 161{,}700.$$

45. In this case order is important, as it makes a difference whether a commercial is shown first, last, or in between. The number of ways that the director can schedule the commercials is given by $P(6,6) = 6! = 720$.

47. The inquiries can be directed in

$$P(12,6) = \frac{12!}{6!} = 12\cdot 11\cdot 10\cdot 9\cdot 8\cdot 7 = 665{,}280, \quad \text{or } 665{,}280 \text{ ways.}$$

49. a. The ten books can be arranged in $P(10,10) = 10! = 3{,}628{,}800$ ways.
b. If books on the same subject are placed together, then they can be arranged on the shelf

$$P(3,3) \times P(4,4) \times P(3,3) \times P(3,3) = 5184 \text{ ways.}$$

Here we have computed the number of ways the mathematics books can be arranged times the number of ways the social science books can be arranged times the number of ways the biology books can be arranged times the number of ways the 3 sets of books can be arranged.

51. Notice that order is certainly important here.
a. The number of ways that the 20 featured items can be arranged is given by

$$P(20,20) = 20! = 2.43 \times 10^{18}.$$

b. If items from the same department must appear in the same row, then the number of ways they can be arranged on the page is

Number of ways of arranging the rows	x	Number of ways of arranging the items in each of the 5 rows
$P(5,5)$	•	$P(4,4)\times P(4,4)\times P(4,4)\times P(4,4)\times P(4,4)$

$= 5! \times (4!)^5 = 955{,}514{,}880.$

53. The number of ways is given by

$$2\{C(2,2) + [C(3,2) - C(2,2)]\} = 2[1 + (3 - 1)] = 2 \times 3 = 6$$

(number of players)[number of ways to win in exactly 2 sets + number of ways to win in exactly 3 sets]

55. The number of ways the measure can be passed is

$$C(3,3) \times [C(8,6) + C(8,7) + C(8,8)] = 37.$$

Here three of the three permanent members must vote for passage of the bill and this can be done in $C(3,3) = 1$ way. Of the 8 nonpermanent members who are voting 6 can vote for passage of the bill, or 7 can vote for passage, or 8 can vote for passage. Therefore, there are

$$C(8,6) + C(8,7) + C(8,8) = 37 \text{ ways}$$

that the nonpermanent members can vote to ensure passage of the measure. This gives $1 \times 37 = 37$ ways that the members can vote so that the bill is passed.

57. a. If no preference is given to any student, then the number of ways of awarding the 3 teaching assistantships is $C(12,3) = \dfrac{12!}{3!9!} = 220.$

b. If it is stipulated that one particular student receive one of the assistantships, then the remaining two assistantships must be awarded to two of the remaining 11 students. Thus, the number of ways is $C(11,2) = \dfrac{11!}{2!9!} = 55.$

c. If at least one woman is to be awarded one of the assistantships, and the group of students consists of seven men and five women, then the number of ways the assistantships can be awarded is given by

$$C(5,1) \times C(7,2) + C(5,2) \times C(7,1) + C(5,3)$$

$$= \frac{5!}{4!1!} \cdot \frac{7!}{5!2!} + \frac{5!}{3!2!} \cdot \frac{7!}{6!1!} + \frac{5!}{3!2!} = 105 + 70 + 10 = 185.$$

59. The number of ways of awarding 3 contracts to 7 different firms is given by

$$P(7,3) = \frac{7!}{4!} = 210.$$

The number of ways of awarding the 3 contracts to 2 different firms (one firm gets 2 contracts) from a choice of 7 different firms is

$$C(7,2) \times P(3,2) = 126.$$ (First pick the two firms, and then award the 3 contracts.)

Therefore, the number of ways the contracts can be awarded if no firm is to receive more than 2 contracts is given by $210 + 126 = 336$.

61. The number of different curricula that are available for the student's consideration is given by

$C(5,1) \times C(3,1) \times C(6,2) \times C(4,1) + C(5,1) \times C(3,1) \times C(6,2) \times C(3,1)$

$$= \frac{5!}{4!1!} \cdot \frac{3!}{2!1!} \cdot \frac{6!}{4!2!} \cdot \frac{4!}{3!1!} + \frac{5!}{4!1!} \cdot \frac{3!}{2!1!} \cdot \frac{6!}{4!2!} \cdot \frac{3!}{2!1!}$$

$$= (5)(3)(15)(4) + (5)(3)(15)(3) = 900 + 675 = 1575.$$

63. The number of ways is given by $C(10,2) + C(10,1) + C(10,0) = 45 + 10 + 1 = 56$ or $C(10,8) + C(10,9) + C(10,10) = 56.$

65. The number of ways of dealing a straight flush (5 cards in sequence in the same suit is given by

the number of ways of selecting 5 cards in sequence in the same suit	×	the number of ways of selecting a suit	
10	•	$C(4,1) =$	40.

67. The number of ways of dealing a flush (5 cards in one suit that are not all in sequence) is given by

the number of ways of selecting 5 cards in one suit	-	the number of ways of selecting 5 cards in one suit in sequence
$4C(13,5)$	-	$4(10)$

$$= 5148 - 40 = 5108.$$

69. The number of ways of dealing a full house (3 of a kind and a pair) is given by

the number of ways of picking 3 of a kind from a given rank	×	the number of ways of picking a pair from the 12 remaining ranks
$13C(4,3)$	•	$12C(4,2)$

$$= 13(4) \bullet (12)(6) = 3744.$$

71. The bus will travel a total of 6 blocks. Each route must include 2 blocks running north and south and 4 blocks running east and west. To compute the total number of possible routes, it suffices to compute the number of ways the 2 blocks running north and south can be selected from the six blocks. Thus,

$$C(6,2) = \frac{6!}{2!4!} = 15.$$

73. The number of ways that the quorum can be formed is given by

$$C(12,6)+C(12,7)+C(12,8)+C(12,9)+C(12,10)+C(12,11)+C(12,12)$$
$$=\frac{12!}{6!6!}+\frac{12!}{7!5!}+\frac{12!}{8!4!}+\frac{12!}{9!3!}+\frac{12!}{10!2!}+\frac{12!}{11!1!}+\frac{12!}{12!0!}$$
$$=924+792+495+220+66+12+1=2510.$$

75. Using the formula given in Exercise 74, we see that the number of ways of seating the 5 commentators at a round table is
$$(5-1)! = 4! = 24.$$

77. The number of possible corner points is $C(8,3)=\dfrac{8!}{5!3!}=56.$

79. True.

81. True. $C(n,r)=\dfrac{n!}{(n-r)!r!}$ and $C(n,n-r)=\dfrac{n!}{[n-(n-r)]!(n-r)!}=\dfrac{n!}{r!(n-r)!}$.
So, $C(n,r)=C(n,n-r)$.

USING TECHNOLOGY EXERCISES 6.4, page 374

1. $1.307674368 \times 10^{12}$

3. $2.56094948229 \times 10^{16}$

5. 674,274,182,400

7. 133,784,560

9. 4,656,960

11. Using the multiplication principle, the number of 10-question exams she can set is given by $C(25,3) \times C(40,5) \times C(30,2) = 658{,}337{,}004{,}000$.

CHAPTER 6 CONCEPT REVIEW, page 375

1. Set; elements; set b. Equal 3. Subset 5. Union; intersection 7. $A^C \cap B^C \cap C^C$

CHAPTER 6, REVIEW EXERCISES, page 375

1. {3}. The set consists of all solutions to the equation $3x - 2 = 7$.

3. {4, 6, 8, 10}

5. Yes.

7. Yes.

9.

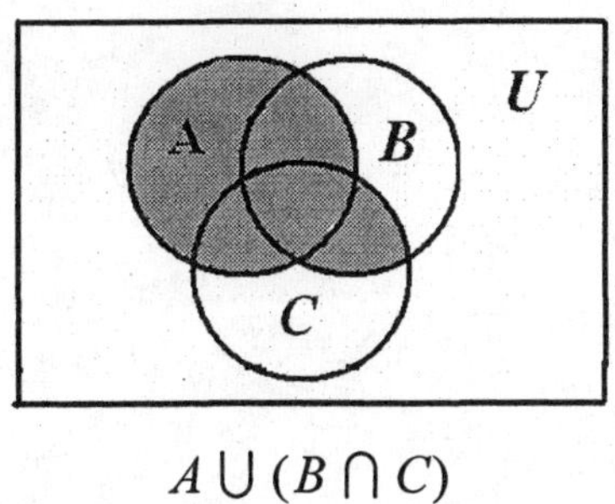

$A \cup (B \cap C)$

11.

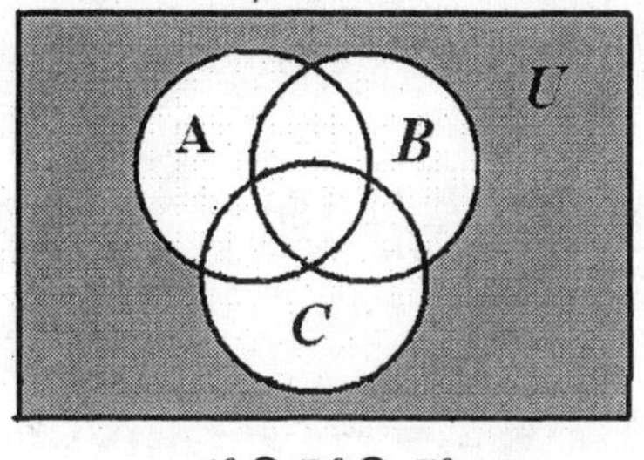

$A^c \cap B^c \cap C^c$

13. $A \cup (B \cup C) = \{a,b\} \cup [\{b,c,d\} \cup \{a,d,e\}]$
$= \{a,b\} \cup \{a,b,c,d,e\} = \{a,b,c,d,e\}.$
$(A \cup B) \cup C = [\{a,b\} \cup \{b,c,d\}] \cup \{a,d,e\}$
$= \{a,b,c,d\} \cup \{a,d,e\} = \{a,b,c,d,e\}.$

15. $A \cap (B \cup C) = \{a,b\} \cap [\{b,c,d\} \cup \{a,d,e\}] = \{a,b\} \cap \{a,b,c,d,e\} = \{a,b\}.$
$(A \cap B) \cup (A \cap C) = [\{a,b\} \cap \{b,c,d\}] \cup [\{a,b\} \cap \{a,d,e\} = \{b\} \cup \{a\} = \{a,b\}.$

17. The set of all participants in a consumer behavior survey who both avoided buying a product because it is not recyclable and boycotted a company's products because of its record on the environment.

19. The set of all participants in a consumer behavior survey who both did not use cloth diapers rather than disposable diapers and voluntarily recycled their garbage.

In Exercises 21-25, refer to the following Venn diagram.

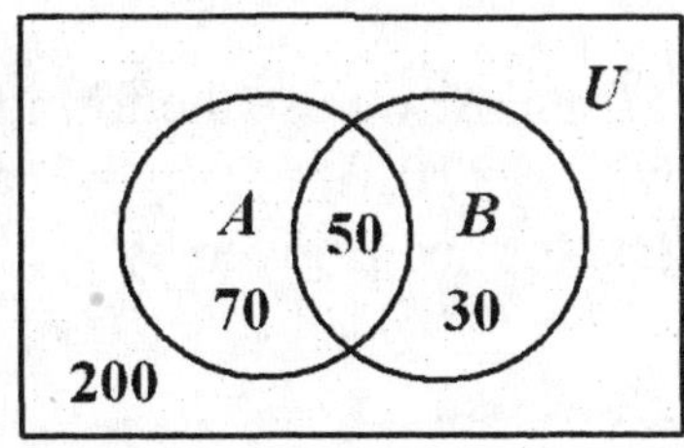

21. $n(A \cup B) = n(A) + n(B) - n(A \cap B) = 120 + 80 - 50 = 150.$

23. $n(B^c) = n(U) - n(B) = 350 - 80 = 270.$

25. $n(A \cap B^c) = n(A) - n(A \cap B) = 120 - 50 = 70.$

27. $C(20, 18) = \dfrac{20!}{18!2!} = 190.$

29. $C(5, 3) \cdot P(4, 2) = \dfrac{5!}{3!2!} \cdot \dfrac{4!}{2!} = 10 \cdot 12 = 120.$

31. Let U denote the set of 5 major cards, and let

$A = \{x \in U \mid x \text{ offered cash advances}\}$

$B = \{x \in U \mid x \text{ offered extended payments for all goods and services purchased}\}$

$C = \{x \in U \mid x \text{ required an annual fee that was less than \$35}\}$

Thus, $n(A) = 3$, $n(B) = 3$, $n(C) = 2$

$n(A \cap B) = 2$, $n(B \cap C) = 1$

and $n(A \cap B \cap C) = 0$

Using the Venn diagram on the right, we have

$$x + y + 2 = 3$$
$$y + 2 = 2$$

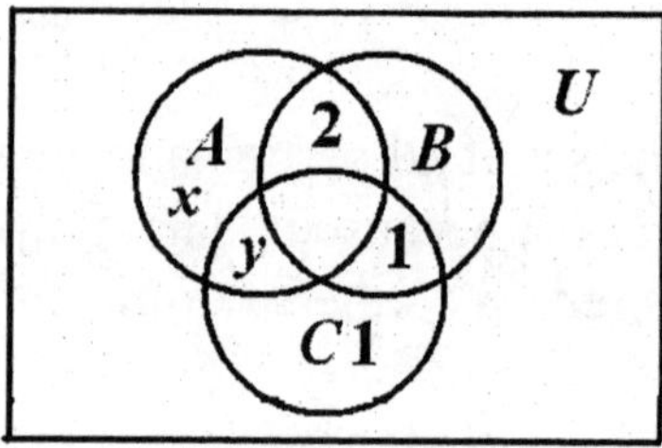

Solving the system, we find $x = 1$, and $y = 0$. Therefore, the number of cards that offer cash advances and have an annual fee that is less than \$35 is given by $n(A \cap C) = y = 0$.

33. The number of ways the compact discs can be arranged on a shelf is

$P(6,6) = 6! = 720$ ways.

35. a. Since there is repetition of the letters C, I, and N, we use the formula for the permutation of n objects, not all distinct, with $n = 10$, $n_1 = 2$, $n_2 = 3$, and $n_3 = 3$. Then the number of permutations that can be formed is given by

$$\frac{10!}{2!3!3!} = 50{,}400.$$

b. Here, again, we use the formula for the permutation of n objects, not all distinct, this time with $n = 8$, $n_1 = 2$, $n_2 = 2$, and $n_3 = 2$. Then the number of permutations is given by $\frac{8!}{2!2!2!} = 5040$.

37. Let U denote the set comprising Halina's clients, and let

$$A = \{x \in U \mid x \text{ owns stocks}\}$$

$$B = \{x \in U \mid x \text{ owns bonds}\}$$

$$C = \{x \in U \mid x \text{ owns mutual funds}\}$$

Then $n(A) = 300$, $n(B) = 180$, $n(C) = 160$, $n(A \cap B) = 110$, $n(A \cap C) = 120$, $n(B \cap C) = 90$. Let $n(A \cap B \cap C) = z$. Then we are lead to the following Venn diagram.

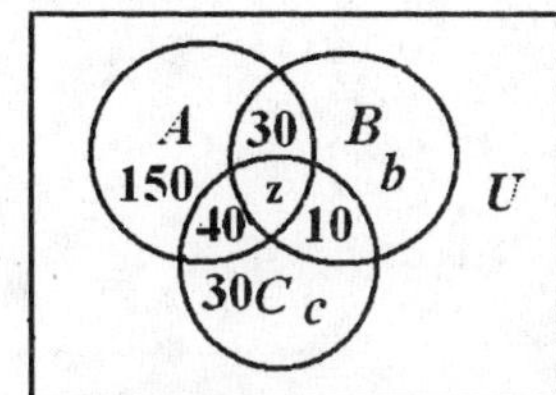

Next, using the fact that $n(A) = 300$, $n(B) = 180$, and $n(C) = 160$ leads to

$$a + (110 - z) + (120 - z) + z = 300$$

$$b + (110 - z) + (90 - z) + z = 180$$

$$c + (120 - z) + (90 - z) + z = 160$$

$$a + b + c + (110 - z) + (120 - z) + (90 - z) + z = 400$$

which simplifies to $a - z = 70$, $b - z = -20$, $c - z = -50$, and $a + b + c - 2z = 80$. Solving, we find $a = 150$, $b = 60$, $c = 30$ and $z = 80$. Therefore, the number who own stocks, bonds and mutual funds is $n(A \cap B \cap C) = z = 80$.

39. The number of possible outcomes is (6)(4)(5)(6) = 720.

41. a. The number of ways the 7 students can be assigned to seats is $P(7,7) = 7! = 5040$.

b. The number of ways 2 specified students can be seated next to each other is $2(6) = 12$.

(Think of seven numbered seats. Then the students can be seated in seats 1-2, or 2-3, or 3-4, or 4-5, or 5-6, or 6-7. Since there are 6 such possibilities and the pair of students can be seated in 2 ways, we conclude that there are 2(6) possible arrangements.) Then the remaining 5 students can be seated in $P(5,5) = 5!$ ways. Therefore, the number of ways the 7 students can be seated if two specified students sit next to each other is $2(6)5! = 1440$. Finally, the number of ways the students can be seated if the two students do not sit next to each other is

$$P(7,7) - 2(6)5! = 5040 - 1440 = 3600.$$

43. a. The number of samples that can be selected is $C(15,4) = \dfrac{15!}{4!11!} = 1365$.

b. There are $C(10,4) = \dfrac{10!}{4!6!} = 210$ ways of selecting 4 balls none of which are white. Therefore, there are 1365 - 210, or 1155 ways of selecting 4 balls of which at least one is white.

CHAPTER 6, BEFORE MOVING ON, page 377

1. a. $B \cup C = \{b,c,d,e,f,g\}$. So, $A \cap (B \cup C) = \{d,f,g\}$.
   b. $A \cap C = \{f\}$ and so $(A \cap C) \cup (B \cup C) = \{b,c,d,e,f,g\}$
   c. $A^c = \{b,c,e\}$.

2. Refer to the Venn diagram at the right.

$$A \cap (B \cup C)^c$$

is the shaded area and

$$n[A \cap (B \cup C)^c] = 20 - (7 + 4 + 6) = 3$$

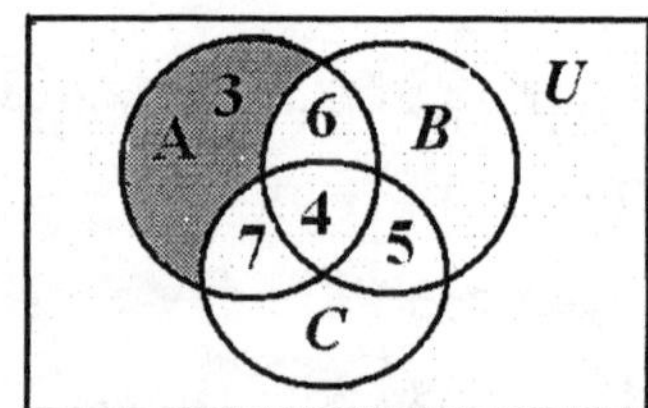

3. The number of ways is $P(6,4) = \dfrac{6!}{4!} = 360$.

4. The number of ways of obtaining the three deuces is $C(4,3) = \dfrac{4!}{3!1!} = 4$.

The number of ways of obtaining the 2 face cards is $C(12,2) = \dfrac{12!}{2!10!} = 66$.

Therefore, the number of hands with 3 deuces and 2 face cards is $4(66) = 264$.

5. There are $C(6,3) = \dfrac{6!}{3!3!} = 20$ ways of picking the 3 seniors and $C(5,2) = \dfrac{5!}{2!3!} = 10$ ways of picking the 2 juniors. Therefore, there are (20)(10) or 200 possible teams.

CHAPTER 7

7.1 Problem Solving Tips

1. The *union of two events A* and *B*, written $A \cup B$, is the set of outcomes of A and/or B. The *intersection of two events A* and *B*, written $A \cap B$, is the set of outcomes of both A and B. The *complement of an event A*, written A^c, is the set of outcomes in the sample space S that are not in A.

2. Two events A and B are mutually exclusive if $A \cap B = \varnothing$. In other words, the two events cannot occur at the same time.

7.1 CONCEPT QUESTIONS, page 385

1. An experiment is an activity with observable results. Examples vary.

EXERCISES 7.1, page 385

1. $E \cup F = \{a, b, d, f\}$; $E \cap F = \{a\}$.

3. $F^C = \{b, c, e\}$; $E \cap G^C = \{a, b\} \cap \{a, d, f\} = \{a\}$.

5. Since $E \cap F = \{a\}$ is not a null set, we conclude that E and F are not mutually exclusive.

7. $E \cup F \cup G = \{2, 4, 6\} \cup \{1, 3, 5\} \cup \{5, 6\} = \{1, 2, 3, 4, 5, 6\}$.

9. $(E \cup F \cup G)^C = \{1, 2, 3, 4, 5, 6\}^C = \varnothing$.

11. Yes, $E \cap F = \varnothing$; that is, E and F do not contain any common elements.

13. $E^c = \{2, 4, 6\}^c = \{1, 3, 5\} = F$ and so E and F are complementary.

15. $E \cup F$ 17. G^C 19. $(E \cup F \cup G)^C$

21. a. Refer to Example 4, page 382.

$$E = \{(2, 1), (3, 1), (4, 1), (5, 1), (6, 1), (3, 2), (4, 2), (5, 2), (6, 2), (4, 3), (5, 3), (6, 3), (5, 4), (6, 4), (6, 5)\}$$

b. $E = \{(1, 2), (2, 4), (3, 6)\}$

23. $\varnothing$, $\{a\}$, $\{b\}$, $\{c\}$, $\{a, b\}$, $\{a, c\}$, $\{b, c\}$, $\{a, b, c\}$.

25. a. $S = \{R, B\}$ b. $\varnothing$, $\{B\}$, $\{R\}$, $\{B, R\}$

27. a. $S = \{(H, 1), (H, 2), (H, 3), (H, 4), (H, 5), (H, 6), (T, 1), (T, 2), (T, 3), (T, 4), (T, 5), (T, 6)\}$

b. $E = \{(H, 2), (H, 4), (H, 6)\}$

29. a. Here $S = \{1, 2, 3, 4, 5, 6\}$, $E = \{2\}$, and $F = \{2, 4, 6\}$. Since $E \cap F = \{2\} \neq \varnothing$, we conclude that E and F are not mutually exclusive.
b. $E^c = \{1, 3, 4, 5, 6\} \neq F$ and so E and F are not complementary.

31. $S = \{(d, d, d), (d, d, n), (d, n, d), (n, d, d), (d, n, n), (n, d, n), (n, n, d), (n, n, n)\}$

33. a. $\{ABC, ABD, ABE, ACD, ACE, ADE, BCD, BCE, BDE, CDE\}$;
b. 6 c. 3 d. 6

35. a. E^C b. $E^C \cap F^C$ c. $E \cup F$ d. $(E \cap F^C) \cup (E^C \cap F)$

37. a. $S = \{t \mid t > 0\}$ b. $E = \{t \mid 0 < t \leq 2\}$ c. $F = \{t \mid t > 2\}$

39. a. $S = \{0, 1, 2, 3, \ldots, 10\}$ b. $E = \{0, 1, 2, 3\}$ c. $F = \{5, 6, 7, 8, 9, 10\}$

41. a. $S = \{0, 1, 2, \ldots, 20\}$ b. $E = \{0, 1, 2, \ldots, 9\}$ c. $F = \{20\}$

43. Let S denote the sample space of the experiment that is the set of 52 cards. Then $E = \{x \in S \mid x \text{ is an ace}\}$ and $F = \{x \in S \mid x \text{ is a spade}\}$ and

$E \cap F = \{x \in S \mid x \text{ is the ace of spades}\}$. Now $n(E) = 4$, $n(F) = 13$, and $n(E \cap F) = 1$. Also, $E \cup F = \{x \in S \mid x \text{ is an ace or a spade}\}$ and $n(E \cup F) = 16$, and

$$n(E) + n(F) - n(E \cap F) = 4 + 13 - 1 = 16 = n(E \cup F).$$

45. $E^C \cap F^C = (E \cup F)^C$ by DeMorgan's Law. Since $(E \cup F) \cap (E \cup F)^C = \varnothing$, they are mutually exclusive.

47. False. Let $E = \{1, 2, 3\}$, $F = \{4, 5, 6\}$, and $G = \{4, 5\}$. Then $E \cap F = \varnothing$ and $E \cap G = \varnothing$, but $F \cap G = \{4,5\} \neq \varnothing$.

7.2 Problem Solving Tips

1. *Uniform sample spaces* are sample spaces in which the outcomes are equally likely.

2. Events consisting of a single outcome are called *simple events.*

3. To find the probability of an event E, (a) determine an appropriate space S associated with the experiment and (b) assign probabilities to the simple events of the experiment. Then $P(E) = P(s_1) + P(s_2) + P(s_3) + \cdots + P(s_n)$, where $E = \{s_1, s_2, s_3, \ldots, s_n\}$ and $\{s_1\}, \{s_2\}, \{s_3\}, \ldots, \{s_n\}$ are the simple events of S.

7.2 CONCEPT QUESTIONS, page 393

1. a. By assigning probabilities to each simple event of an experiment, we obtain a *probability distribution* that gives the probability of each simple event.

   b. The function P that assigns a probability to each of the simple events is called a *probability function.*

3. $P(E) = P(s_1) + P(s_2) + \cdots + P(s_n)$

   $P(\varnothing) = 0.$

EXERCISES 7.2, page 393

1. $\{(H,H)\}$, $\{(H,T)\}$, $\{(T,H)\}$, $\{(T,T)\}$.

3. $\{(D,m)\}$, $\{(D,f)\}$, $\{(R,m)\}$, $\{(R,f)\}$, $\{(I,m)\}$, $\{(I,f)\}$

5. $\{(1,i)\}$, $\{(1,d)\}$, $\{(1,s)\}$, $\{(2,i)\}$, $\{(2,d)\}$, $\{(2,s)\}$, ..., $\{(5,i)\}$, $\{(5,d)\}$,$\{(5,s)\}$

7. $\{(\text{A},\text{Rh}^+)\}$, $\{(\text{A}, \text{Rh}^-)\}$, $\{(\text{B}, \text{Rh}^+)\}$, $\{\text{B}, \text{Rh}^-)\}$,$\{(\text{AB}, \text{Rh}^+)\}$,$\{(\text{AB},\text{Rh}^-)\}$,
$\{(\text{O},\text{Rh}^+)\}$,$\{(\text{O},\text{Rh}^-)\}$

9. The probability distribution associated with this data is

Grade	A	B	C	D	F
Probability	0.10	0.25	0.45	0.15	0.05

11. a. $S = \{(0 < x \leq 200), (200 < x \leq 400), (400 < x \leq 600), (600 < x \leq 800), (800 < x \leq 1000), (x > 1000)\}$

b.

Number of cars (x)	**Probability**
$0 < x \leq 200$	0.075
$200 < x \leq 400$	0.1
$400 < x \leq 600$	0.175
$600 < x \leq 800$	0.35
$800 < x \leq 1000$	0.225
$x > 1000$	0.075

13.

Outcome	Favor	Oppose	Don't Know
Probability	$\frac{910}{1936}$	$\frac{891}{1936}$	$\frac{135}{1936}$

or

Opinion	Favor	Oppose	Don't Know
Probability	0.47	0.46	0.07

15. The probability distribution associated with this data is

Rating	A	B	C	D	E
Probability	0.026	0.199	0.570	0.193	0.012

17. The probability distribution is

Number of figures produced (in dozens)	30	31	32	33	34	35	36
Probability	0.125	0	0.1875	0.25	0.1875	0.125	0.125

19. The probability is $\dfrac{84,000,000}{179,000,000} \approx 0.469$.

21.

Income ($)	0-24,999	25,000-49,999	50,000-74,999	75,000-99,999
Probability	.287	.293	.195	.102

Income ($)	100,000-124,999	125,000-149,999	150,000-199,999	200,000 or more
Probability	.052	.025	.022	0.024

23. a. The probability that a person killed by lightning is a male is $\dfrac{376}{439} \approx 0.856$.

b. The probability that a person killed by lightning is a female is

$$\frac{439-376}{439} = \frac{63}{439} \approx 0.144.$$

25. The probability that the retailer uses electronic tags as antitheft devices is

$$\frac{81}{179} \approx 0.46.$$

27. a. $P(D) = \frac{13}{52} = \frac{1}{4}$ b. $P(B) = \frac{26}{52} = \frac{1}{2}$ c. $P(A) = \frac{4}{52} = \frac{1}{13}$

29. The probability of arriving at the traffic light when it is red is

$$\frac{30}{30+5+45} = \frac{30}{80} = 0.375.$$

31. The probability is

$$P(D) + P(C) + P(B) + P(A) = 0.15 + 0.45 + 0.25 + 0.10 = 0.95.$$

33. a. $P(E) = \frac{62}{9+62+27} = \frac{62}{98} \approx 0.633$ b. $P(E) = \frac{27}{98} \approx 0.276$

35. There are two ways of getting a 7, one die showing a 3 and the other die showing a 4, and vice versa.

37. No. Since the die is loaded the outcomes are not equally likely.

39. No. Since the coin is weighted, the outcomes are not equally likely.

41. Let G denote a female birth and let B denote a male birth. Then the eight equally likely outcomes of this experiment are

$$GGG \ GGB \ GBG \ BGG \ BGB \ BBG \ GBB \ BBB.$$

a. The event that there are two girls and a boy in the family is

$$E = \{GGB, GBG, BGG\}.$$

Since there are three favorable outcomes, $P(E) = 3/8$.

b. The event that the oldest child is a girl is

$$F = \{GGG, GGB, GBG, GBB\}.$$

Since there are 4 favorable outcomes, $P(F) = 1/2$.

c. The event that the oldest child is a girl and the youngest child is a boy is

$$G = \{GGB, GBB\}.$$

Since there are two favorable outcomes, $P(F) = 1/4$.

43. The probability that the primary cause of the crash was due to pilot error or bad weather is given by $\dfrac{327+22}{327+49+14+22+19+15} = \dfrac{349}{446} \approx 0.7825.$

45. False. $P(s_1)+P(s_2)+\cdots+P(s_n)=1$

7.3 Problem Solving Tips

If S is a sample space of an experiment and E and F are events of the experiment then

(1) $P(E)\geq 0$ for any E.

(2) $P(S)=1$

(3) If E and F are mutually exclusive, then $P(E \cup F)= P(E)+P(F)$. More generally, if E and F are any two events of an experiment, then
$$P(E\cup F) = P(E)+P(F)-P(E\cap F).$$

(4) $P(E^c)=1-P(E)$

7.3 CONCEPT QUESTIONS, page 403

1. a. The event E cannot occur.
   b. There is a 50 percent chance that the event F will occur.
   c. The probability that an event of S will occur is a certainty.
   d. The probability of the event $E\cup F$ occurring is given by the sum of the probabilities of E and F minus the probability of $E\cap F$.

EXERCISES 7.3, page 403

1. Refer to Example 4, page 382. Let E denote the event of interest. Then
$$P(E)=\frac{18}{36}=\frac{1}{2}$$

3. Refer to Example 4, page 382. The event of interest is $E = \{1,1\}$, and $P(E) = 1/36$.

5. Let E denote the event of interest. Then $E = \{(6,2),(6,1),(1,6),(2,6)\}$ and $$P(E)=\frac{4}{36}=\frac{1}{9}.$$

7. Let E denote the event that the card drawn is a king, and let F denote the event that the card drawn is a diamond. Then the required probability is $P(E \cap F) = \frac{1}{52}$.

9. Let E denote the event that a face card is drawn. Then $P(E) = \frac{12}{52} = \frac{3}{13}$.

11. Let E denote the event that an ace is drawn. Then $P(E) = 1/13$. Then E^c is the event that an ace is not drawn and $P(E^C) = 1 - P(E) = \frac{12}{13}$.

13. Let E denote the event that a ticket holder will win first prize, then

$$P(E) = \frac{1}{500} = 0.002,$$

and the probability of the event that a ticket holder will not win first prize is

$$P(E^c) = 1 - 0.002 = 0.998.$$

15. Property 2 of the laws of probability is violated. The sum of the probabilities must add up to 1. In this case $P(S) = 1.1$, which is not possible.

17. The five events are not mutually exclusive; the probability of winning at least one purse is

$$1 - \text{probability of losing all 5 times} = 1 - \frac{9^5}{10^5} = 1 - 0.5905 = 0.4095.$$

19. The two events are not mutually exclusive; hence, the probability of the given event is $\frac{1}{6} + \frac{1}{6} - \frac{1}{36} = \frac{11}{36}$.

21. $E^C \cap F^C = \{c, d, e\} \cap \{a, b, e\} = \{e\} \neq \varnothing$.

23. Let G denote the event that a customer purchases a pair of glasses and let C denote the event that the customer purchases a pair of contact lenses. Then

$$P(G \cup C)^C \neq 1 - P(G) - P(C).$$

Mr. Owens has not considered the case in which the customer buy both glasses and contact lenses.

25. a. $P(E \cap F) = 0$ since E and F are mutually exclusive.
b. $P(E \cup F) = P(E) + P(F) - P(E \cap F) = 0.2 + 0.5 = 0.7.$
c. $P(E^c) = 1 - P(E) = 1 - 0.2 = 0.8.$
d. $P(E^C \cap F^C) = P[(E \cup F)^C] = 1 - P(E \cup F) = 1 - 0.7 = 0.3.$

27. a. $P(A) = P(s_1) + P(s_2) = \frac{1}{8} + \frac{3}{8} = \frac{1}{2}$; $P(B) = P(s_1) + P(s_3) = \frac{1}{8} + \frac{1}{4} = \frac{3}{8}$
b. $P(A^C) = 1 - P(A) = 1 - \frac{1}{2} = \frac{1}{2}$; $P(B^C) = 1 - P(B) = 1 - \frac{3}{8} = \frac{5}{8}$
c. $P(A \cap B) = P(s_1) = \frac{1}{8}$
d. $P(A \cup B) = P(A) + P(B) - P(A \cap B) = \frac{1}{2} + \frac{3}{8} - \frac{1}{8} = \frac{3}{4}$

29. Referring to the following diagram we see that

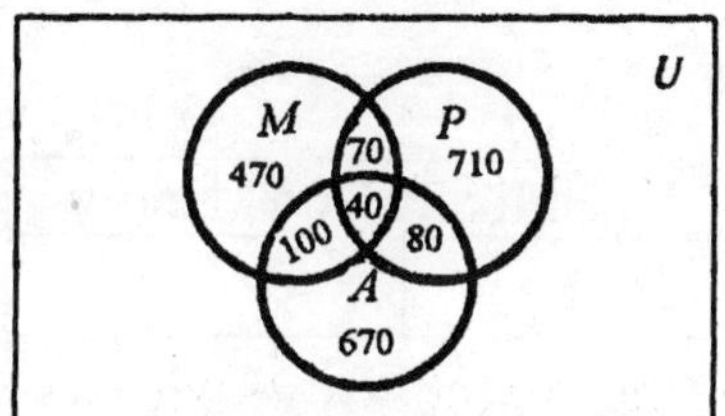

P: lack of parental support
M: malnutrition
A: abused or neglected

the probability that a teacher selected at random from this group said that lack of parental support is the only problem hampering a student's schooling is

$$\frac{710}{2140} = 0.33.$$

31. Let E and F denote the events that the person surveyed learned of the products from *Good Housekeeping* and *The Ladies Home Journal*, respectively.

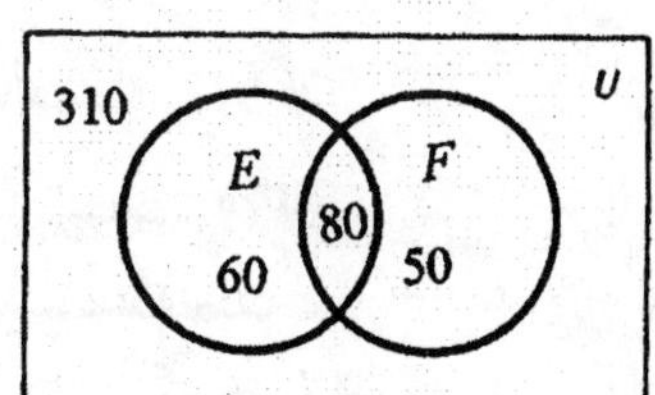

Then $P(E) = \frac{140}{500} = \frac{7}{25}$, $P(F) = \frac{130}{500} = \frac{13}{50}$

and $P(E \cap F) = \frac{80}{500} = 0.16$

a. $P(E \cap F) = \frac{80}{500} = 0.16$

b. $P(E \cup F) = \frac{14}{50} + \frac{13}{50} - \frac{8}{50} = \frac{19}{50} = 0.38$

c. $P(E \cap F^C) + P(E^C \cap F) = \frac{60}{500} + \frac{50}{500} = \frac{110}{500} = 0.22$.

33. The probability distribution is

Less	The Same	More	No Answer
$\frac{27}{160}$	$\frac{66}{160}$	$\frac{61}{160}$	$\frac{6}{160}$

a. The required probability is $\frac{61}{160}$ or approxinately .38.
b. The required probability is $\frac{66}{160} + \frac{27}{160} = \frac{93}{160}$ or approximately .58.

35. The probability distribution is

V. Likely	S. Likely	S. Unlikely	V. Unlikely	Don't Know
$\frac{40}{200}$	$\frac{28}{200}$	$\frac{26}{200}$	$\frac{104}{200}$	$\frac{2}{200}$

a. The required probability is $\frac{40}{200} = 0.2$.

b. The required probability is $\frac{28}{200} + \frac{40}{200} = \frac{68}{200} = 0.34$.

37. a. The required percent is (46 + 22), or .68.
b. The required percent is 100 - (8 + 5), or .87.

39. Let $A = \{t \mid t < 3\}$, $B = \{t \mid t \leq 4\}$, $C = \{t \mid t > 5\}$

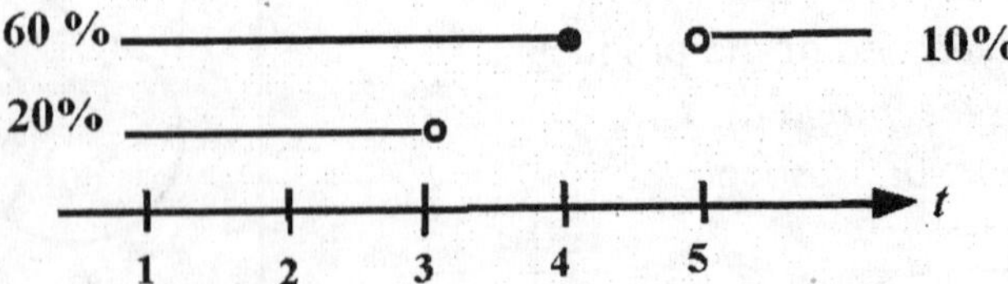

a. $D = \{t \mid t \leq 5\}$ and $P(D) = 1 - P(C) = 1 - 0.1 = 0.9$.

b. $E = \{t \mid t > 4\}$ and $P(E) = 1 - P(B) = 1 - 0.6 = 0.4$.

c. $F = \{t \mid 3 \le t \le 4\}$ and $P(F) = P(A^C \cap B) = 0.4$.

41. a. The probability that the participant favors tougher gun-control laws is $\frac{150}{250} = 0.6$.

b. The probability that the participant owns a handgun is $\frac{58+25}{250} = 0.332$.

c. The probability that the participant owns a handgun but not a rifle is

$$\frac{58}{250} = 0.232.$$

d. The probability that the participant favors tougher gun-control laws and does not own a handgun is $\frac{12+138}{250} = 0.6$.

43. The probability that Bill will fail to solve the problem is $1 - p_1$, and the probability that Mike will fail to solve the problem is $1 - p_2$. Therefore, the probability that both Bill and Mike will fail to solve the problem is $(1 - p_1)(1 - p_2)$. So, the probability that at least one of them will solve the problem is

$$1 - (1 - p_1)(1 - p_2) = 1 - (1 - p_2 - p_1 + p_1 p_2) = p_1 + p_2 - p_1 p_2$$

45. True. Write $B = A \cup (B - A)$. Since A and $B - A$ are mutually exclusive, we have

$$P(B) = P(A) + P(B - A).$$

Since $P(B) = 0$, we have $P(A) + P(B - A) = 0$. If $P(A) > 0$, then $P(B - A) < 0$ and this is not possible. Therefore, $P(A) = 0$.

47. False. Take $E_1 = \{1, 2\}$ and $E_2 = \{2, 3\}$ where $S = \{1, 2, 3\}$. Then $P(E_1) = \frac{2}{3}$, $P(E_2) = \frac{2}{3}$, but $P(E_1 \cup E_2) = P(S) = 1$.

7.4 Problem Solving Tips

1. If S is a uniform sample space and E is any event in S, then

$$P(E) = \frac{\text{Number of favorable outcomes in } E}{\text{Number of possible outcomes in } S} = \frac{n(E)}{n(S)}$$

7.4 CONCEPT QUESTIONS, page 412

1. If S is a uniform sample space and E is any event, then the probability of an event occurring in a uniform sample space is

$$P(E) = \frac{\text{Number of favorable outcomes in } E}{\text{Number of possible outcomes in } S} = \frac{n(E)}{n(S)}.$$

EXERCISES 7.4, page 412

1. Let E denote the event that the coin lands heads all five times. Then

$$P(E) = \frac{1}{2^5} = \frac{1}{32}.$$

3. Let E denote the event that the coin lands tails all 5 times, then

$$P(E^c) = 1 - P(E) = 1 - \frac{1}{32} = \frac{31}{32},$$

where E^c is the event that the coin lands heads at least once.

5. $P(E) = \dfrac{13 \cdot C(4,2)}{C(52,2)} = \dfrac{78}{1326} \approx 0.0588.$

7. $P(E) = \dfrac{C(26,2)}{C(52,2)} = \dfrac{325}{1326} \approx 0.2451.$

9. The probability of the event that two of the balls will be white and two will be blue is $P(E) = \dfrac{n(E)}{n(S)} = \dfrac{C(3,2) \cdot C(5,2)}{C(8,4)} = \dfrac{(3)(10)}{70} = \dfrac{3}{7}.$

11. The probability of the event that exactly three of the balls are blue is

$$P(E) = \frac{n(E)}{n(S)} = \frac{C(5,3)C(3,1)}{C(8,4)} = \frac{30}{70} = \frac{3}{7}.$$

13. $$P(E) = \frac{C(3,2) \cdot C(1,1)}{8} = \frac{3}{8}.$$

15. $$P(E) = \frac{C(3,3)}{8} = \frac{1}{8}.$$

17. The number of elements in the sample space is 2^{10}. There are $C(10,6) = \frac{10!}{6!4!}$, or 210 ways of answering exactly six questions correctly. Therefore, the required probability is $\frac{210}{2^{10}} = \frac{210}{1024} \approx 0.205$.

19. a. Let E denote the event that both of the bulbs are defective.Then

$$P(E) = \frac{C(4,2)}{C(24,2)} = \frac{\frac{4!}{2!2!}}{\frac{24!}{22!2!}} = \frac{4 \cdot 3}{24 \cdot 23} = \frac{1}{46} \approx 0.022.$$

b. Let F denote the event that none of the bulbs are defective. Then

$$P(F) = \frac{C(20,2)}{C(24,2)} = \frac{20!}{18!2!} \cdot \frac{22!2!}{24!} = \frac{20}{24} \cdot \frac{19}{23} = 0.6884.$$

Therefore, the probability that at least one of the light bulbs is defective is given by $1 - P(F) = 1 - 0.6884 = 0.3116$.

21. a. The probability that both of the cartridges are defective is

$$P(E) = \frac{C(6,2)}{C(80,2)} = \frac{15}{3160} = 0.0047.$$

b. Let F denote the event that none of the cartridges are defective. Then

$$P(F) = \frac{C(74,2)}{C(80,2)} = \frac{2701}{3160} = 0.855,$$

and $P(F^c) = 1 - P(F) = 1 - 0.855 = 0.145$ is the probability that at least 1 of the cartridges is defective.

23. a. The probability that Mary's name will be selected is $P(E) = \frac{12}{100} = 0.12$;
The probability that both Mary's and John's names will be selected is

$$P(F) = \frac{C(98,10)}{C(100,12)} = \frac{\frac{98!}{88!10!}}{\frac{100!}{88!12!}} = \frac{12 \cdot 11}{100 \cdot 99} \approx 0.013.$$

b. The probability that Mary's name will be selected is $P(M) = \frac{6}{40} = 0.15$.
The probability that both Mary's and John's names will be selected is

$$P(M)\cdot P(J)=\frac{6}{60}\cdot\frac{6}{40}=\frac{36}{2400}=0.015.$$

25. The probability is given by

$$\frac{C(12,8)\cdot C(8,2)}{C(20,10)}+\frac{C(12,9)C(8,1)}{C(20,10)}+\frac{C(12,10)}{C(20,10)}$$
$$=\frac{(28)(495)+(220)(8)+66}{184,756}\approx 0.085$$

27. a. The probability that he will select brand B is

$$\frac{C(4,2)}{C(5,3)}=\frac{6}{10}=\frac{3}{5}.\qquad \left(\frac{\text{the number of selections that include brand } B}{\text{the number of possible selections}}\right)$$

b. The probability that he will select brands B and C is

$$\frac{C(3,1)}{C(5,3)}=0.3$$

c. The probability that he will select at least one of the two brands, B and C is

$$1-\frac{C(3,3)}{C(5,3)}=0.9.$$ (1 - probability that he does not select brands B and C.)

29. The probability that the three "Lucky Dollar" symbols will appear in the window of the slot machine is

$$P(E)=\frac{n(E)}{n(S)}=\frac{(1)(1)(1)}{C(9,1)C(9,1)C(9,1)}=\frac{1}{729}.$$

31. The probability of a ticket holder having all four digits in exact order is

$$\frac{1}{C(10,1)\cdot C(10,1)\cdot C(10,1)\cdot C(10,1)}=\frac{1}{10,000}=0.0001.$$

33. The probability of a ticket holder having a specified digit in exact order is

$$\frac{C(1,1)C(10,1)C(10,1)C(10,1)}{10^4}=0.10.$$

35. The number of ways of selecting a 5-card hand from 52 cards is given by

$$C(52,5)=2,598,960.$$

The number of straight flushes that can be dealt in each suit is 10, so there are 4(10) possible straight flushes. Therefore, the probability of being dealt a straight flush is

$$\frac{4(10)}{C(52,5)} = \frac{40}{2{,}598{,}960} = 0.0000154.$$

37. The number of ways of being dealt a flush in one suit is $C(13,5)$, and, since there are four suits, the number of ways of being dealt a flush is $4 \cdot C(13,5)$. Since we wish to exclude the hands that are straight flushes we subtract the number of possible straight flushes from $4 \cdot C(13,5)$. Therefore, the probability of being drawn a flush, but not a straight flush, is

$$\frac{4 \cdot C(13,5) - 40}{C(52,5)} = \frac{5108}{2{,}598{,}960} = 0.0019654.$$

39. The total number of ways to select three cards of one rank is $13 \cdot C(4,3)$. The remaining two cards must form a pair of another rank and there are

$$12 \cdot C(4,2)$$

ways of selecting these pairs. Next, the total number of ways to be dealt a full house is $13 \cdot C(4,3) \cdot 12 \cdot C(4,2) = 3744$.

Hence, the probability of being dealt a full house is $\frac{3{,}744}{2{,}598{,}960} \approx 0.0014406$.

41. Let E denote the event that in a group of 5, no two will have the same sign. Then

$$P(E) = \frac{12 \cdot 11 \cdot 10 \cdot 9 \cdot 8}{12^5} \approx 0.381944.$$

Therefore, the probability that at least two will have the same sign is given by

$$1 - P(E) = 1 - 0.381944 \approx 0.618.$$

b. $P(\text{no Aries}) = \frac{11 \cdot 11 \cdot 11 \cdot 11 \cdot 11}{12^5} \approx 0.647228.$

$$P(\text{1 Aries}) = \frac{C(5,1) \cdot (1)(11)(11)(11)(11)}{12^5} \approx 0.2941945.$$

Therefore, the probability that at least two will have the sign Aries is given by

$$1 - [P(\text{no Aries}) + P(\text{1 Aries})] = 1 - 0.9414225 \approx 0.059.$$

43. Referring to the table on page 411, we see that in a group of 50 people, the probability that none of the people will have the same birthday is $1 - 0.970 = 0.03$.

7.5 Problem Solving Tips

1. The probability that the event B will occur given that the event A has already occurred is called the *conditional probability* of B given A, written $P(B|A)$.

2. If $P(A) \neq 0$, then $P(B|A) = \dfrac{P(A \cap B)}{P(A)}$.

3. The *Product Rule* states that $P(A \cap B) = P(A) \cdot P(B|A)$.

4. Two events are *independent* if the outcome of one does not affect the outcome of the other; that is $P(A|B) = P(A)$ and $P(B|A) = P(B)$. Do not confuse independent events with mutually exclusive events. (The latter cannot occur at the same time.)

5. Two events are independent if and only if $P(A \cap B) = P(A) \cdot P(B)$.

7.5 CONCEPT QUESTIONS, page 425

1. The conditional probability of an event is the probability of an event occurring given that another event has already occurred. Examples will vary.

3. $P(A \cap B) = P(A)P(B|A)$.

EXERCISES 7.5, page 425

1. a. $P(A|B) = \dfrac{P(A \cap B)}{P(B)} = \dfrac{0.2}{0.5} = \dfrac{2}{5}$. b. $P(B|A) = \dfrac{P(A \cap B)}{P(A)} = \dfrac{0.2}{0.6} = \dfrac{1}{3}$.

3. $P(A \cap B) = P(A)P(B|A) = (0.6)(0.5) = 0.3.$

5. $P(A) \cdot P(B) = (0.3)(0.6) = 0.18 = P(A \cap B)$. Therefore the events are independent.
7. $P(A \cap B) = P(A) + P(B) - P(A \cup B) = 0.5 + 0.7 - 0.85 = 0.35 = P(A) \cdot P(B)$

so they are independent events.

9. a. $P(A \cap B) = P(A)P(B) = (0.4)(0.6) = 0.24$.
   b. $P(A \cup B) = P(A) + P(B) - P(A \cap B) = 0.4 + 0.6 - 0.24 = 0.76$.

11. a. $P(A) = 0.5$ b. $P(E|A) = 0.4$
    c. $P(A \cap E) = P(A)P(E|A) = (0.5)(0.4) = 0.2$
    d. $P(E) = (0.5)(0.4) + (0.5)(0.3) = 0.35$.
    e. No. $P(A \cap E) \neq P(A) \cdot P(E) = (0.5)(0.35)$
    f. A and E are not independent events.

13.

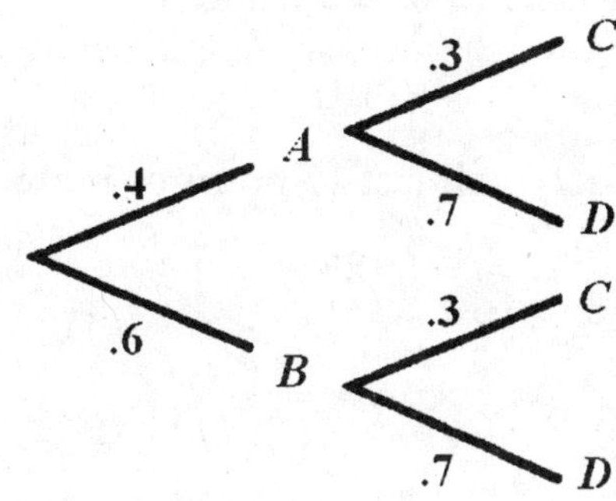

a. $P(A) = 0.4$ b. $P(C|A) = 0.3$ c. $P(A \cap C) = P(A)P(C|A) = (0.4)(0.3) = 0.12$
d. $P(C) = (0.4)(0.3) + (0.6)(0.3) = 0.3$
e. Yes. $P(A \cap C) = 0.12 = P(A)P(C)$ f. Yes.

15. a. Refer to Figure 15, page 416. Here $E = \{(5,1),(5,2),(5,3),(5,4),(5,5),(5,6)\}$ and $F = \{(6,4),(5,5),(4,6)\}$. So $P(F) = \frac{3}{36} = \frac{1}{12}$.
    b. $P(E \cap F) = \frac{1}{36}$ since $E \cap F = \{(5,5)\}$. c. $P(F|E) = \frac{1}{6}$. d. $P(E) = \frac{6}{36} = \frac{1}{6}$.
    e. $P(F|E) = \dfrac{P(E \cap F)}{P(E)} = \dfrac{\frac{1}{36}}{\frac{1}{6}} = \dfrac{1}{6} \neq P(F) = \dfrac{1}{12}$ and so the events are not independent.

17. Let A denote the event that the sum of the numbers is less than 9 and let B denote the event that one of the numbers is a 6. Then, $P(A|B) = \dfrac{P(A \cap B)}{P(B)} = \dfrac{\frac{4}{36}}{\frac{11}{36}} = \dfrac{4}{11}$.

19. Refer to Figure 15, page 416 in the text. Here
    $$E = \{(3,1),(3,2),(3,3),(3,4),(3,5),(3,6)\}$$

and $F = \{(1,6),(6,1),(2,5),(5,2),(3,4),(4,3)\}$. Then $E \cap F = \{(3,4)\}$.

Now, $P(E \cap F) = \frac{1}{36}$ and this is equal to $P(E) \cdot P(F) = \left(\frac{6}{36}\right)\left(\frac{6}{36}\right) = \frac{1}{36}$.

So E and F are independent events.

21. $P(E \cap F) = \frac{13}{24} = \frac{1}{4}$, $P(E) = \frac{26}{52} = \frac{1}{2}$, and $P(F) = \frac{13}{52} = \frac{1}{4}$. Now,

$$P(E) \cdot P(F) = (\tfrac{1}{2})(\tfrac{1}{4}) = \tfrac{1}{8} \neq P(E \cap F) = \tfrac{1}{4}.$$

So E and F are not independent events. The knowledge that the card drawn is black increases the probability that it is a spade.

23. Let A denote the event that the battery lasts 10 or more hours and let B denote the event that the battery lasts 15 or more hours.

Then $P(A) = 0.8$, $P(B) = 0.15$

and $P(A \cap B) = 0.15$.

Therefore, the probability that the battery will last 15 hours or more is

$$P(B|A) = \frac{P(A \cap B)}{P(A)} = \frac{0.15}{0.8} = \frac{3}{16} = 0.1875.$$

25. Refer to the following tree diagram:

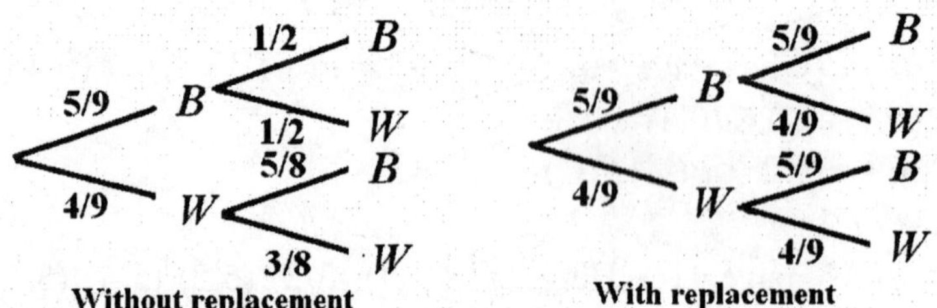

a. The probability that the second ball drawn is a white ball if the second ball is drawn without replacing the first is

$$P(B)P(W|B) + P(W)P(W|W) = (\frac{5}{9})(\frac{1}{2}) + (\frac{4}{9})(\frac{3}{8}) = \frac{4}{9}.$$

b. The probability that the second ball drawn is a white ball if the first ball is replaced before the second is drawn is $(\frac{5}{9})(\frac{4}{9}) + (\frac{4}{9})(\frac{4}{9}) = \frac{4}{9}$.

27. Refer to the following tree diagram:

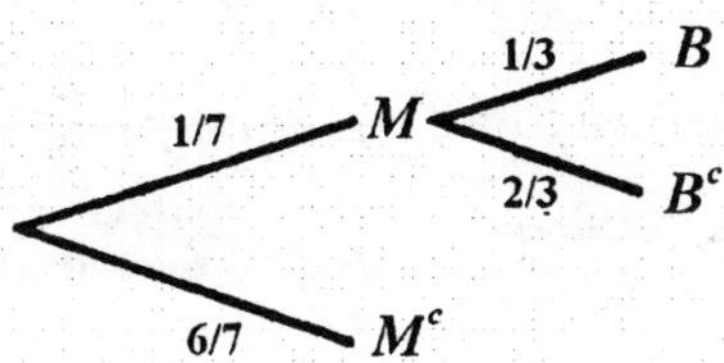

a. The probability that a student selected at random from this medical school is black is $(\frac{1}{7})(\frac{1}{3}) = \frac{1}{21}$

b. The probability that a student selected at random from this medical school is black if it is known that the student is a member of a minority group is $P(B|M) = 1/3$.

29. Let D denote the event that the card drawn is a diamond. Consider the tree diagram that follows.

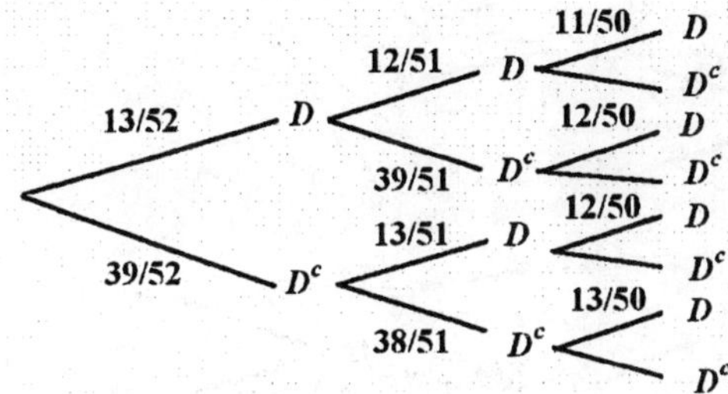

Then the required probability is

$$(\frac{13}{52})(\frac{12}{51})(\frac{11}{50}) + (\frac{13}{52})(\frac{39}{51})(\frac{12}{50}) + (\frac{39}{52})(\frac{13}{51})(\frac{12}{50}) + (\frac{39}{52})(\frac{38}{51})(\frac{13}{50}) = 0.25.$$

31. The sample space for a three-child family is

S = {*GGG*, *GGB*, *GBG*, *GBB*, *BGG*, *BGB*, *BBG*, *BBB*}.

Since we know that there is at least one girl in the three-child family we are dealing with a reduced sample space

S_1 = {*GGG*, *GGB*, *GBG*, *GBB*, *BGG*, *BGB*, *BBG*}

in which there are 7 outcomes. Then the probability that all three children are girls is

$$P(E) = \frac{n(E)}{n(S)} = \frac{1}{7}.$$

33. a.

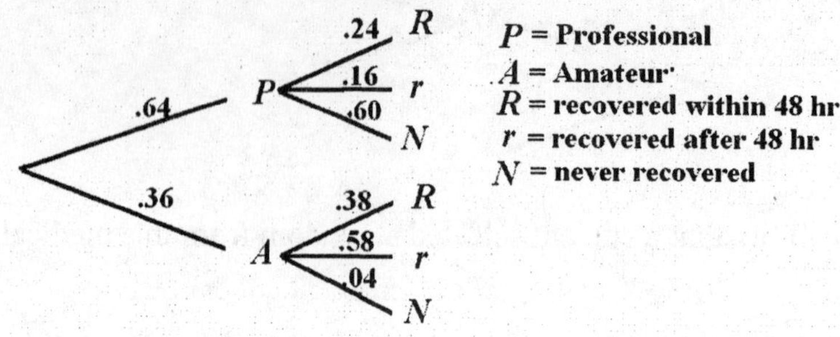

b. The required probability is 0.24.
c. The required probability is $(0.64)(0.60) + (0.36)(0.04) \approx 0.40$.

35. Refer to the following tree diagram.

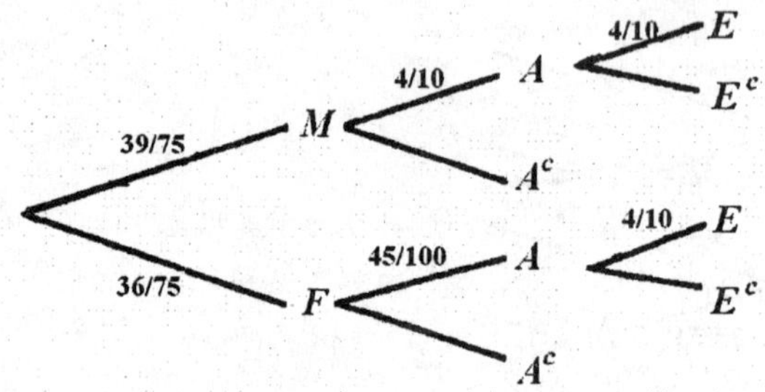

a. $P(A \cap E|M) = (0.4)(0.4) = 0.16$.

b. $P(A) = P(M \cap A) + P(F \cap A) = \frac{39}{75} \cdot \frac{4}{10} + \frac{36}{75} \cdot \frac{45}{100} = 0.424$

c. $P(M \cap A \cap E) + P(F \cap A \cap E) = \frac{39}{75} \cdot \frac{4}{10} \cdot \frac{4}{10} + \frac{36}{75} \cdot \frac{45}{100} \cdot \frac{4}{10}$
$= 0.0832 + 0.0864 = 0.1696$.

37. a. The probability that none of the dozen eggs is broken is $(0.992)^{12} = 0.908$. Therefore, the probability that at least one egg is broken is $1 - 0.908 = 0.092$.
b. Using the results of (a), we see that the required probability is
$(0.092)(0.092)(0.908) \approx 0.008$.

39. a. $P(A) = \frac{1120}{4000} = 0.280$; $P(B) = \frac{1560}{4000} = 0.390$;

$$P(A \cap B) = \frac{720}{4000} = 0.180\ ;$$

$$P(B|A) = \frac{P(A \cap B)}{P(A)} = \frac{n(A \cap B)}{n(A)} = \frac{720}{1120} \approx 0.643$$

$$P(B|A^C) = \frac{P(A^C \cap B)}{P(A^C)} = \frac{n(A^C \cap B)}{n(A^C)} = \frac{840}{2880} \approx 0.292\,.$$

b. $P(B|A) \neq P(B)$ so A and B are not independent events.

41. Let C denote the event that a person in the survey was a heavy coffee drinker and Pa denote the event that a person in the survey had cancer of the pancreas. Then

$$P(C) = \frac{3200}{10000} = 0.32 \text{ and } P(Pa) = \frac{160}{10000} = 0.016$$

$$P(C \cap Pa) = \frac{132}{160} = 0.825$$

and $P(C) \cdot P(Pa) = 0.00512 \neq P(C \cap Pa)$. Therefore the events are not independent.

43. The probability that the first test will fail is 0.03, that the second test will fail is 0.015, and that the third test will fail is 0.015. Since these are independent events the probability that all three tests will fail is

$$(0.03)(0.015)(0.015) = 0.0000068.$$

45. a. Let $P(A)$, $P(B)$, $P(C)$ denote the probability that the first, second, and third patient suffer a rejection, respectively. Then $P(A) = \frac{1}{2}$, $P(B) = \frac{1}{3}$, and $P(C) = \frac{1}{10}$. Therefore, the probabilities that each patient does not suffer a rejection are given by $P(A^c) = \frac{1}{2}$, $P(B^c) = \frac{2}{3}$, and $P(C^c) = \frac{9}{10}$. Then, the probability that none of the 3 patients suffers a rejection is given by

$$P(A^c) \cdot P(B^c) \cdot P(C^c) = \frac{1}{2} \cdot \frac{2}{3} \cdot \frac{9}{10} = \frac{18}{60} = \frac{3}{10}.$$

Therefore, the probability that at least one patient will suffer rejection is

$$1 - P(A^c) \cdot P(B^c) \cdot P(C^c) = 1 - \frac{3}{10} = \frac{7}{10}.$$

b. The probability that exactly two patients will suffer rejection is

$$P(A)P(B)P(C^c) + P(A)P(B^c)P(C) + P(A^c)P(B)P(C)$$

$$= \frac{1}{2}\cdot\frac{1}{3}\cdot\frac{9}{10}+\frac{1}{2}\cdot\frac{2}{3}\cdot\frac{1}{10}+\frac{1}{2}\cdot\frac{1}{3}\cdot\frac{1}{10}=\frac{9+2+1}{60}=\frac{12}{60}=\frac{1}{5}.$$

47. Let A denote the event that at least one of the floodlights remain functional over the one-year period. Then

$$P(A) = 0.99999 \text{ and } P(A^c) = 1 - P(A) = 0.00001.$$

Letting n represent the minimum number of floodlights needed, we have

$$(0.01)^n = 0.00001$$
$$n\log(0.01) = -5$$
$$n(-2) = -5$$
$$n = \frac{5}{2} = 2.5.$$

Therefore, the minimum number of floodlights needed is 3.

49. The probability that the event will not occur in one trial is $1 - p$. Therefore, the probability that it will not occur in n independent trials is $(1 - p)^n$. Therefore, the probability that it will occur at least once in n independent trials is $1 - (1 - p)^n$.

51. $P(E|F) = \frac{P(E\cap F)}{P(F)} = \frac{P(F)}{P(F)} = 1 \quad (E\cap F = F \text{ since } F \subset E).$

Interpretation: Since $F \subset E$, an occurrence of F implies an occurrence of E. In other words, given that F has occurred, it is a certainty that E will occur, that is, $P(E|F) = 1$.

53. True. Since A and B are mutually exclusive, $A\cap B = \varnothing$, and

$$P(A|B) = \frac{P(A\cap B)}{P(B)} = \frac{P(\varnothing)}{P(B)} = 0.$$

55. True. This follows from Formula 3, $P(A\cap B) = P(A)\cdot P(B|A)$.

7.6 Problem Solving Tips

1. If $A_1, A_2, \ldots, A_n$ is a partition of a sample space S and E is an event of the experiment such that $P(E) \neq 0$, then

$$P(A_i|E) = \frac{P(A_i)\cdot P(E|A_i)}{P(A_1)\cdot P(E|A_1) + P(A_2)\cdot P(E|A_2) + \cdots + P(A_n)\cdot P(E|A_n)}$$

If you draw a tree diagram to represent the experiment, then this formula can also be remembered by noting that

$$P(A_i|E) = \frac{\text{The product of the probabilities along the limb through } A_i}{\text{The sum of the products of the probabilities along each limb terminating at } E}$$

7.6 CONCEPT QUESTIONS, page 433

1. An *a priori probability* gives the likelihood that an event *will* occur and an *a posteriori probability* givse the probability of an event occurring *after* the outcomes of an experiment have been observed.
   Examples will vary.

3. It represents the a posteriori probability that the component having the property described by E was produced in factory A.

EXERCISES 7.6, page 433

1.

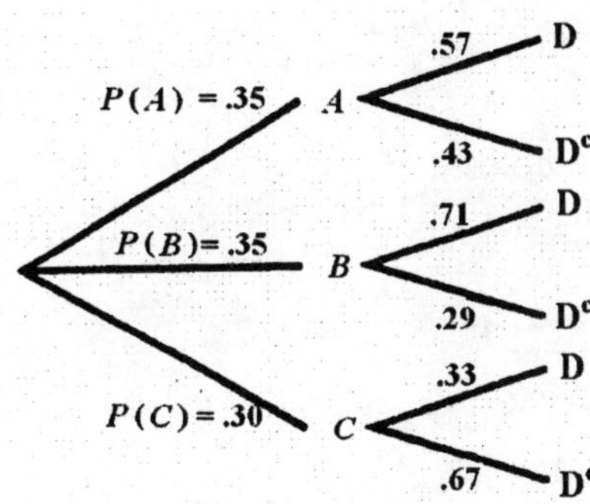

3. a. $P(D^C) = \dfrac{15+10+20}{35+35+30} = 0.45$; b. $P(B|D^C) = \dfrac{10}{15+10+20} = 0.22$.

5. a. $P(D)=\dfrac{25+20+15}{50+40+35}=0.48$ b. $P(B|D)=\dfrac{20}{25+20+15}=0.33$

7. a. $P(A)\cdot P(D|A)=(0.4)(0.2)=0.08$ b. $P(B)\cdot P(D|B)=(0.6)(0.25)=0.15$

c. $P(A|D)=\dfrac{P(A)\cdot P(D|A)}{P(A)\cdot P(D|A)+P(B)\cdot P(D|B)}=\dfrac{(0.4)(0.2)}{0.08+0.15}\approx 0.35$

9. a. $P(A)\cdot P(D|A)=\dfrac{1}{3}\cdot\dfrac{1}{4}=\dfrac{1}{12}$ b. $P(B)\cdot P(D|B)=\dfrac{1}{2}\cdot\dfrac{1}{2}=\dfrac{1}{4}$

c. $P(C)\cdot P(D|C)=\dfrac{1}{6}\cdot\dfrac{1}{3}=\dfrac{1}{18}$

d. $P(A|D)=\dfrac{P(A)\cdot P(D|A)}{P(A)\cdot P(D|A)+P(B)\cdot P(D|B)+P(C)\cdot P(C|B)}$

$$=\frac{\frac{1}{12}}{\frac{1}{12}+\frac{1}{4}+\frac{1}{18}}=\frac{1}{12}\cdot\frac{36}{14}=\frac{3}{14}$$

11. Let A denote the event that the first card drawn is a heart and B the event that the second card drawn is a heart. Then

$$P(A|B)=\frac{P(A)\cdot P(B|A)}{P(A)\cdot P(B|A)+P(A^C)\cdot P(B|A^C)}$$

$$=\frac{\frac{1}{4}\cdot\frac{12}{51}}{\frac{1}{4}\cdot\frac{12}{51}+\frac{3}{4}\cdot\frac{13}{51}}=\frac{4}{17}.$$

13. Using the following tree diagram, we see that

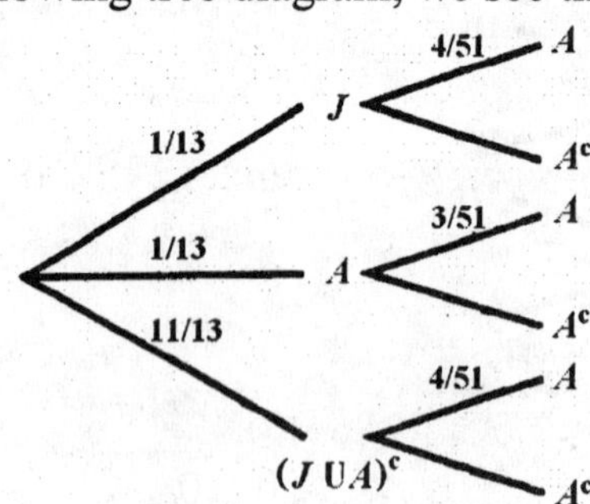

$$P(J|A)=\frac{\frac{1}{13}\cdot\frac{4}{51}}{\frac{1}{13}\cdot\frac{4}{51}+\frac{1}{13}\cdot\frac{3}{51}+\frac{11}{13}\cdot\frac{4}{51}}=\frac{\frac{4}{13\cdot 51}}{\frac{51}{13\cdot 51}}=0.0784\,.$$

15. The probabilities associated with this experiment are represented in the following tree diagram.

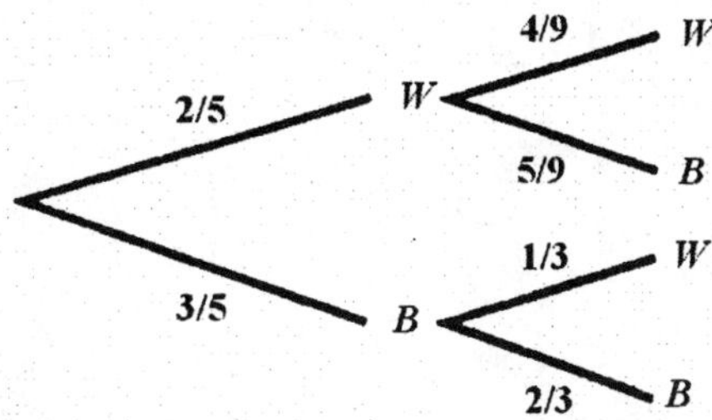

17. Referring to the tree diagram in Exercise 15, we see that the probability that the transferred ball was black given that the second ball was white is

$$P(B|W)=\frac{\frac{3}{5}\cdot\frac{1}{3}}{\frac{2}{5}\cdot\frac{4}{9}+\frac{3}{5}\cdot\frac{1}{3}}=\frac{9}{17}.$$

19. Let D denote the event that a senator selected at random is a Democrat, R denote the event that a senator selected at random is a Republican, and M the event that a senator has served in the military. From the following tree diagram

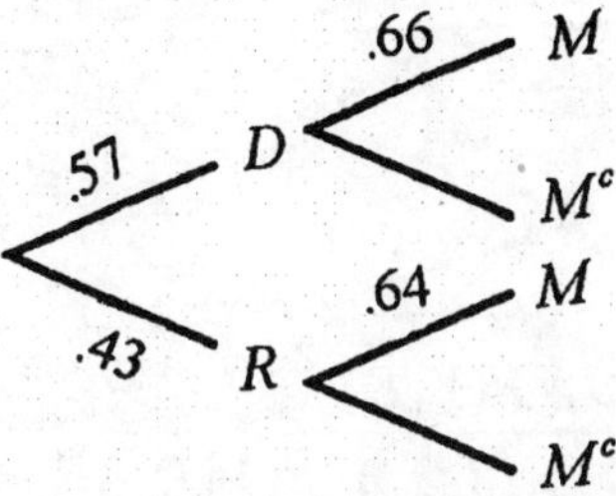

we see that the probability that a senator selected at random who has served in the military is a Republican is

$$P(R|M)=\frac{P(R)P(M|R)}{P(M)}=\frac{(0.64)(0.43)}{(0.66)(0.57)+(0.64)(0.43)}$$
$$\approx 0.422.$$

21. Let H_2 denote the event that the coin tossed is the two-headed coin, H_B denote the event that the coin tossed is the biased coin, and H_F denote the event that the coin tossed is the fair coin. Referring to the following tree diagram, we see that

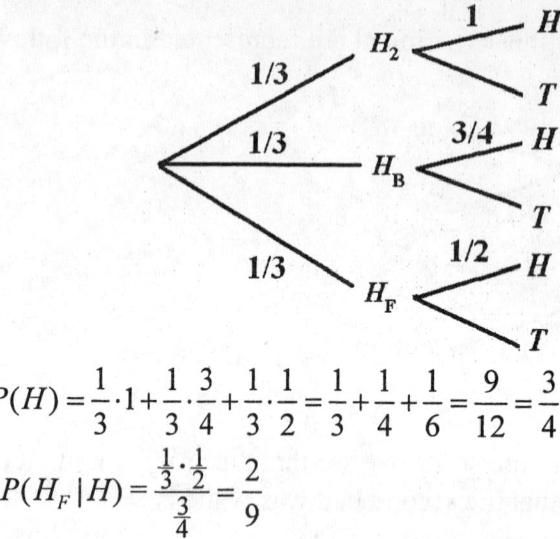

a. $P(H) = \frac{1}{3} \cdot 1 + \frac{1}{3} \cdot \frac{3}{4} + \frac{1}{3} \cdot \frac{1}{2} = \frac{1}{3} + \frac{1}{4} + \frac{1}{6} = \frac{9}{12} = \frac{3}{4}$

b. $P(H_F \mid H) = \dfrac{\frac{1}{3} \cdot \frac{1}{2}}{\frac{3}{4}} = \frac{2}{9}$

23. Let D denote the event that the person has the disease, and let Y denote the event that the test is positive. Referring to the following tree diagram, we see that the required probability is

.95 Y
D
.20
.04 Y
.80
D^c

$$P(D|Y) = \frac{P(D) \cdot P(Y|D)}{P(D) \cdot P(Y|D) + P(D^C) \cdot P(Y|D^C)}$$
$$= \frac{(0.2)(0.95)}{(0.2)(0.95) + (0.8)(0.04)} \approx 0.856.$$

25. Let x denote the age of an insured driver, and let A denote the event that an insured driver is in an accident. Using the tree diagram we find,

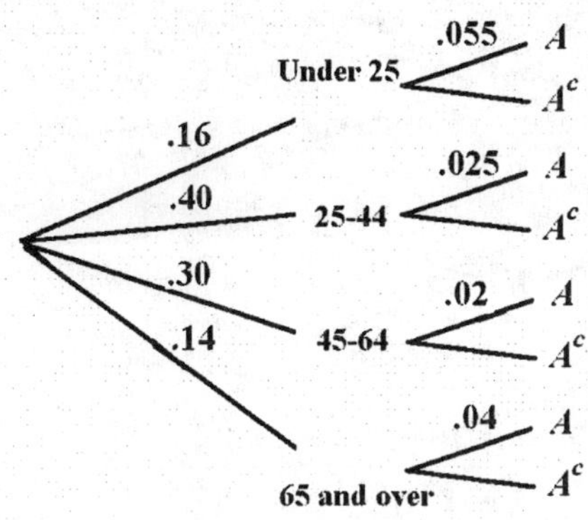

a. $P(A) = (0.16)(0.055) + (0.4)(0.025) + (0.3)(0.02) + (0.14)(0.04)$
$\approx 0.03.$

b. $P(x < 25 | A) = \dfrac{(0.16)(0.055)}{(0.03)} \approx 0.29$

27. Let E, F, and G denote the events that the child selected at random is 12 years old, 13 years old, or 14 years old, respectively; and let C^c denote the event that the child does not have a cavity. Using the tree diagram that follows, we see that

$$P(G|C^C) = \frac{\frac{5}{12}(0.28)}{\frac{1}{4}(0.42) + \frac{1}{3}(0.34) + \frac{5}{12}(0.28)} = 0.348$$

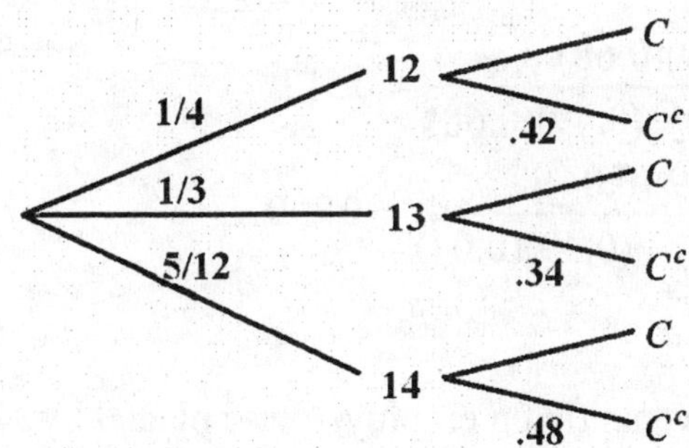

29. Let M and F denote the events that a person arrested for crime in 1988 was male or female, respectively; and let U denote the event that the person was under the age of 18. Using the following tree diagram, we have

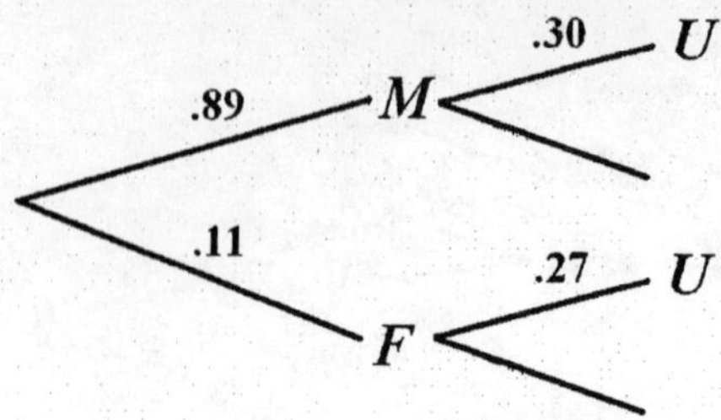

a. $P(U) = (0.89)(0.30) + (0.11)(0.27) = 0.2967.$

b. $P(F|U) = \dfrac{(0.11)(0.27)}{(0.89)(0.30) + (0.11)(0.27)} = 0.1001.$

31. Using the tree diagram that follows, we see that

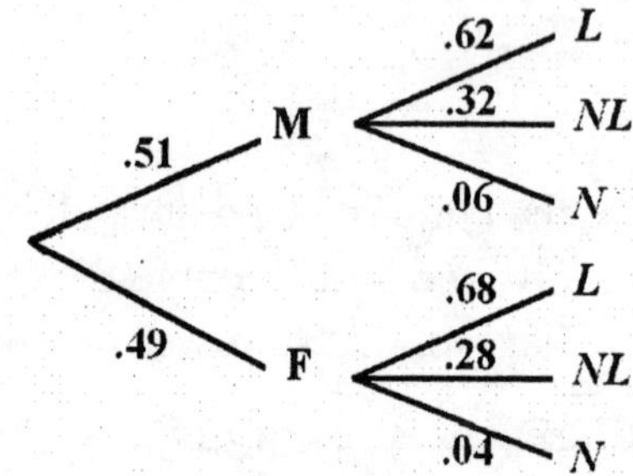

a. $P(F|L) = \dfrac{(0.49)(0.68)}{(0.51)(0.62) + (0.49)(0.68)} = 0.513.$

b. $P(F|N) = \dfrac{(0.49)(0.04)}{(0.51)(0.06) + (0.49)(0.04)} = 0.390.$

33. Let N and D denote the events that a employee was placed by Nancy or Darla, respectively; and let S denote the event that the employee placed by one of these women was satisfactory. Using the tree diagram that follows, we see that

$$P(D|S^C) = \frac{(0.55)(0.3)}{(0.45)(0.2)+(0.55)(0.3)} = 0.647$$

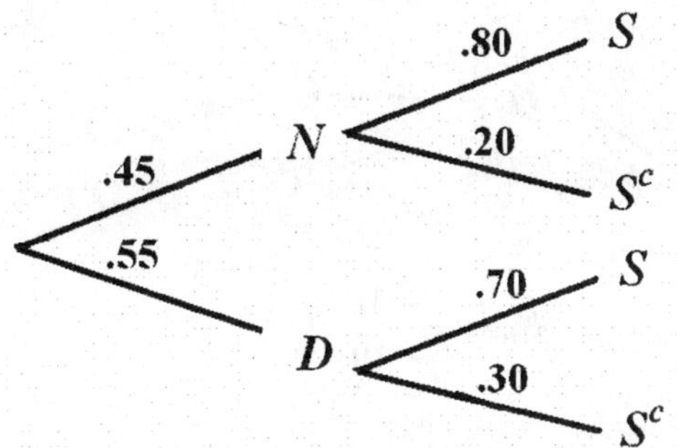

35. Using the tree diagram shown at the right, we see that

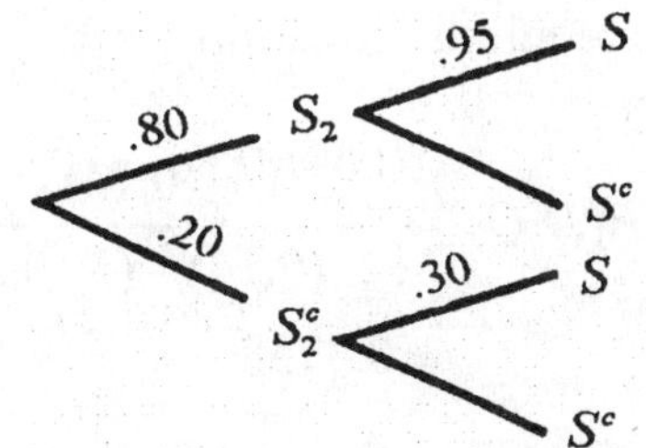

$$P(S_2 \mid S) = \frac{(0.95)(0.8)}{(0.95)(0.8) + (0.2)(0.3)} \approx 0.93.$$

37. Let A, B, C, and D denote the event that the age of a guest is between 21 and 34, between 35 and 44, …, 55 and over, respectively, and let O denote the event that a man keeps his paper money in order of denomination. Refer to the tree diagram that follows.

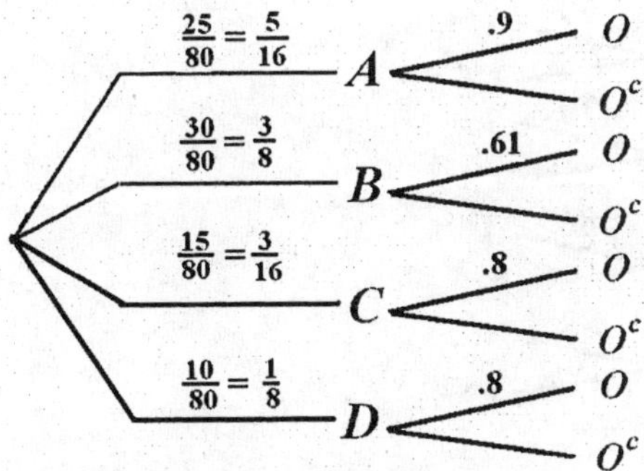

The required probability is

$$P(B \mid O) = \frac{\left(\frac{3}{8}\right)\left(\frac{61}{100}\right)}{\left(\frac{5}{16}\right)\left(\frac{9}{10}\right) + \left(\frac{3}{8}\right)\left(\frac{61}{100}\right) + \left(\frac{3}{16}\right)\left(\frac{8}{10}\right) + \left(\frac{1}{8}\right)\left(\frac{8}{10}\right)} \approx 0.3010.$$

39. Let D and R denote the event that the respondent is a Democrat or a Republican voter, respectively. Next, let S, O, and R denote the event that the respondent supports, opposes, or either doesn't know or refuses, respectively. Refer to the following diagram

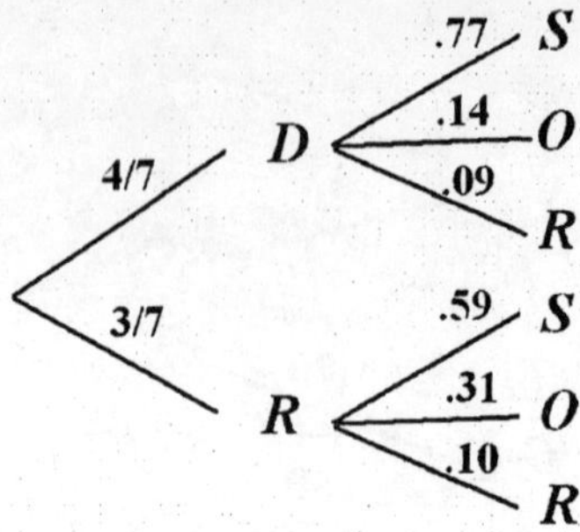

The required probability is

$$P(D|O) = \frac{(\frac{4}{7})(0.14)}{(\frac{4}{7})(0.14) + (\frac{3}{7})(0.31)} = \frac{(0.571428)(0.14)}{(0.571428)(0.14) + (0.428571)(0.31)}$$
$$\approx 0.3758.$$

41. Let A, B, C, D, and E denote the event that the income level is 0-16, 17-33, ..., 96-100 percentile, respectively, and let V denote the event that a person voted in the election . Refer to the following diagram.

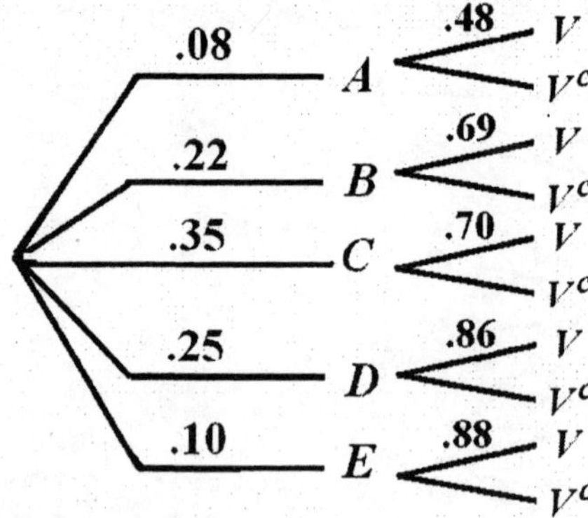

We see that the probability is

$$P(B|V) = \frac{(0.22)(0.69)}{(0.08)(0.48) + (0.22)(0.60) + (0.35)(0.70) + (0.25)(0.86) + (0.10)(0.88)}$$
$$\approx 0.2056.$$

CHAPTER 7, CONCEPT REVIEW, page 439

1. Experiment,; sample; space; event 3. Uniform; $1/n$ 5. Independent

CHAPTER 7, REVIEW EXERCISES, page 439

1. a. $P(E \cap F) = 0$ since E and F are mutually exclusive.
   b. $P(E \cup F) = P(E) + P(F) - P(E \cap F) = 0.4 + 0.2 = 0.6$
   c. $P(E^c) = 1 - P(E) = 1 - 0.4 = 0.6$.
   d. $P(E^C \cap F^C) = P(E \cup F)^C = 1 - P(E \cup F) = 1 - 0.6 = 0.4$.
   e. $P(E^C \cup F^C) = P(E \cap F)^C = 1 - P(E \cap F) = 1 - 0 = 1$.

3. a. The probability of the number being even is
   $P(2) + P(4) + P(6) = 0.12 + 0.18 + 0.19 = 0.49$.
   b. The probability that the number is either a 1 or a 6 is
   $P(1) + P(6) = 0.20 + 0.19 = 0.39$.
   c. The probability that the number is less than 4 is
   $P(1) + P(2) + P(3) = 0.20 + 0.12 + 0.16 = 0.48$.

5. Let A denote the event that a video-game cartridge has an audio defect and let V

   denote the event that a video-game cartridge has a video defect. Then using the following Venn diagram, we have

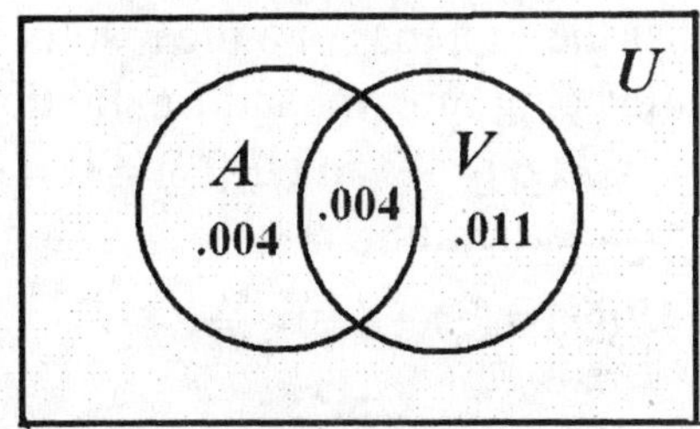

   a. the probability that a cartridge purchased by a customer will have a video or audio defect is

$$P(A \cup V) = P(A \cap V^C) + P(V \cap A^C) + P(A \cap V)$$
$$= 0.004 + 0.004 + 0.011 = 0.019$$

   b. the probability that a cartridge purchased by a customer will have not have a video or audio defect is $P(A \cup V)^C = 1 - P(A \cup V) = 1 - 0.019 = 0.981$.

7. $P(A \cap E) = (0.3)(0.6) = 0.18$.

9. $P(C \cap E) = (0.2)(0.3) = 0.06$.

11. $P(E) = 0.18 + 0.25 + 0.06 = 0.49$

13. a. The probability that none of the pens in the sample are defective is

$$\frac{C(18,3)}{C(20,3)} = \frac{\frac{18!}{15!3!}}{\frac{20!}{17!3!}} = \frac{18!}{15!} \cdot \frac{17!}{20!} = \frac{68}{95} \approx 0.71579.$$

Therefore, the probability that at least one is defective is given by

$1 - 0.71579 \approx 0.284.$

b. The probability that two are defective is given by

$$\frac{C(2,2) \cdot C(18,1)}{C(20,3)} = \frac{18}{\frac{20!}{17!3!}} = \frac{6}{380} \approx 0.0158 .$$

Therefore, the probability that no more than 1 is defective is given by

$$1 - \frac{6}{380} = \frac{374}{380} \approx 0.984 .$$

15. Let E denote the event that the sum of the numbers is 8 and let D denote the event that the numbers appearing on the face of the two dice are different. Then

$$P(E|D) = \frac{P(E \cap D)}{P(D)} = \frac{4}{30} = \frac{2}{15}.$$

17. Let E, F, and G denote the events that the first toss of a die results in a 6 being thrown, the second toss results in tails being thrown, and finally that a face card is selected, respectively. Then $P(E) = \frac{1}{6}$, $P(F) = \frac{1}{2}$, and $P(G) = \frac{12}{52} = \frac{3}{13}$. Since the outcomes are independent, the required probability is

$$P(E \cap F \cap G) = P(E)P(F)P(G) = (\tfrac{1}{6})(\tfrac{1}{2})(\tfrac{3}{13}) = \tfrac{1}{52}.$$

19. The probability that all three cards are face cards is $\frac{C(12,3)}{C(52,3)} = 0.00995.$

21. Referring to the tree diagram at the right, we see that the probability that the second card is black, given that the first card was red is

$$\frac{26}{51} = 0.510 .$$

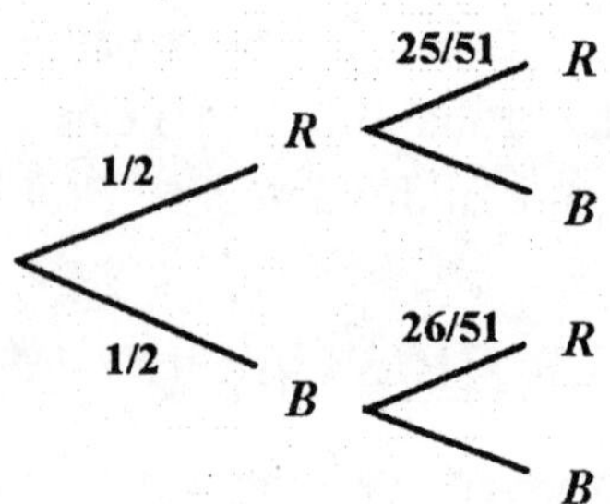

23. a. The required probability is $\frac{50}{295} \approx 0.17$.

b. The required probability is $\frac{50+40+31+29+25}{295+325+167+50+248} \approx 0.16$.

25. Referring to the tree diagram that follows, we see that

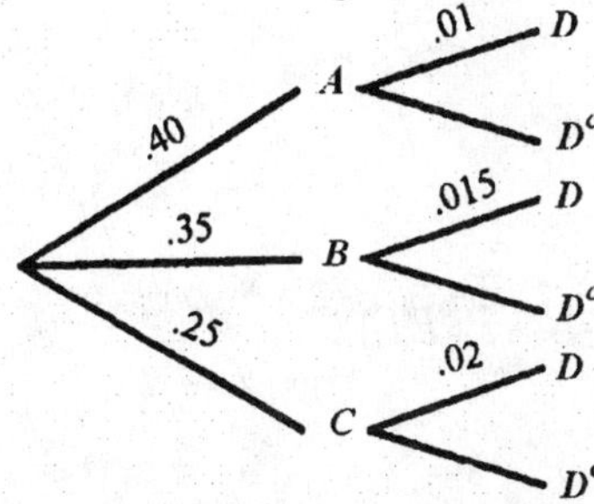

$$P(B|D) = \frac{(0.35)(0.015)}{(0.40)(0.01)+(0.35)(0.015)+(0.25)(0.02)}$$

$$= \frac{(0.35)(0.015)}{0.01425} \approx 0.368.$$

27. Refer to the following diagram.

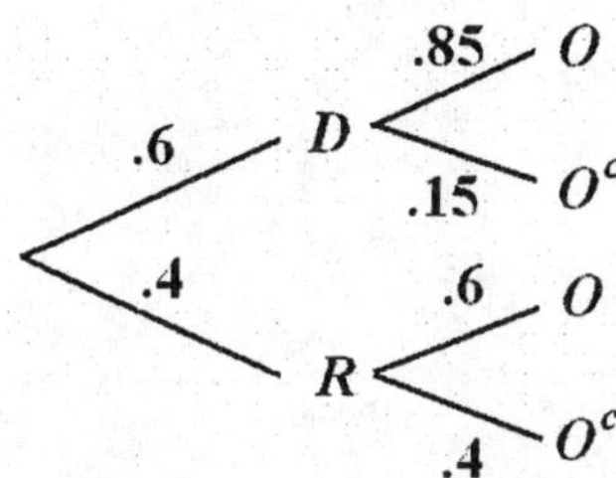

The required probability is $P(D|O) = \frac{(0.4)(0.6)}{(0.6)(0.85)+(0.4)(0.6)} = 0.32.$

CHAPTER 7, BEFORE MOVING ON, page 441

1. $P(s_1, s_3, s_6\} = \frac{1}{12} + \frac{3}{12} + \frac{1}{12} = \frac{5}{12}.$

2. The number of ways of drawing a deuce or face card is 16. Therefore, the required probability is $\frac{16}{52} = \frac{4}{13}$.

3. Refer to the following Venn diagram.

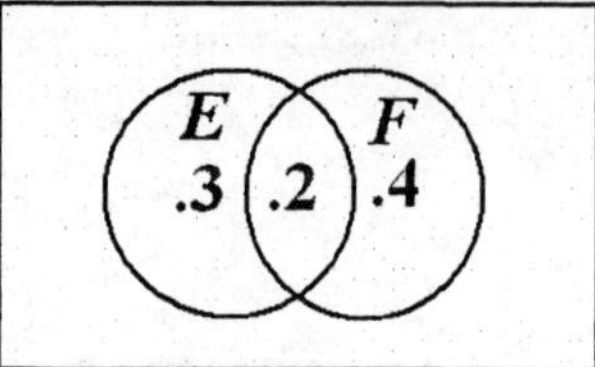

a. $P(E \cup F) = 0.3 + 0.2 + 0.4 = 0.9$

b. $P(E \cap F^c) = 0.3$

4. Since A and B are independent, $P(A \cap B) = P(A) \cdot P(B) = .3 \times .6 = .18$.
$P(A \cup B) = P(A) + P(B) - P(A \cap B) = .3 - .6 - .18 = .72$

5. $P(A|D) = \dfrac{(0.4)(0.2)}{(0.4)(0.2) + (0.6)(0.3)} \approx 0.3077.$

CHAPTER 8

8.1 Problem Solving Tips

1. A *random variable* is a rule that assigns a number to each outcome of a chance experiment.

2. A *probability distribution of a random variable* gives the distinct values of the random variable X and the probabilities associated with these values.

3. A *histogram* is the graph of the probability distribution of a random variable.

8.1 CONCEPT QUESTIONS, page 449

1. A random variable is a rule that assigns a number to each outcome of a chance experiment. Examples vary.
3. To construct a histogram for a probability distribution follow these steps.
   a. Locate the values of the random variable on a number line.
   b. Above each such number on the number line, erect a rectangle with width 1 and height equal to the probability associated with that value of the random variable.

EXERCISES 8.1, page 449

1. a. See part (b).
   b.
   c. {GGG}

Outcome	GGG	GGR	GRG	RGG	GRR	RGR	RRG	RRR
Value	3	2	2	2	1	1	1	0

3. X may assume the values in the set $S = \{1, 2, 3, ...\}$.

5. The event that the sum of the dice is 7 is $E = \{(1,6),(2,5),(3,4),(4,3),(5,2),(6,1)\}$

and $P(E) = \frac{6}{36} = \frac{1}{6}$.

7. X may assume the value of any positive integer. The random variable is infinite discrete.

9. $\{d \mid d \geq 0\}$. The random variable is continuous.

11. X may assume the value of any positive integer. The random variable is infinite discrete.

13. a. $P(X = -10) = 0.20$

b. $P(X \geq 5) = 0.1 + 0.25 + 0.1 + 0.15 = 0.60$

c. $P(-5 \leq X \leq 5) = 0.15 + 0.05 + 0.1 = 0.30$

d. $P(X \leq 20) = 0.20 + 0.15 + 0.05 + 0.1 + 0.25 + 0.1 + 0.15 = 1$

15.

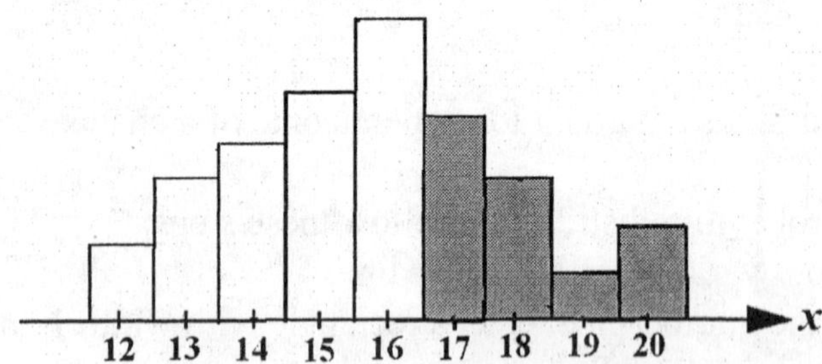

17. a.

x	1	2	3	4	5	6
$P(X = x)$	$\frac{1}{6}$	$\frac{1}{6}$	$\frac{1}{6}$	$\frac{1}{6}$	$\frac{1}{6}$	$\frac{1}{6}$

y	1	2	3	4	5	6
$P(Y = y)$	$\frac{1}{6}$	$\frac{1}{6}$	$\frac{1}{6}$	$\frac{1}{6}$	$\frac{1}{6}$	$\frac{1}{6}$

b.

$x+y$	2	3	4	5	6	7	8	9	10	11	12
$P(X+Y=x+y)$	$\frac{1}{36}$	$\frac{2}{36}$	$\frac{3}{36}$	$\frac{4}{36}$	$\frac{5}{36}$	$\frac{6}{36}$	$\frac{5}{36}$	$\frac{4}{36}$	$\frac{3}{36}$	$\frac{2}{36}$	$\frac{1}{36}$

19. a.

x	0	1	2	3	4	5
$P(X=x)$	0.017	0.067	0.033	0.117	0.233	0.133

x	6	7	8	9	10
$P(X=x)$	0.167	0.1	0.05	0.067	0.017

b.

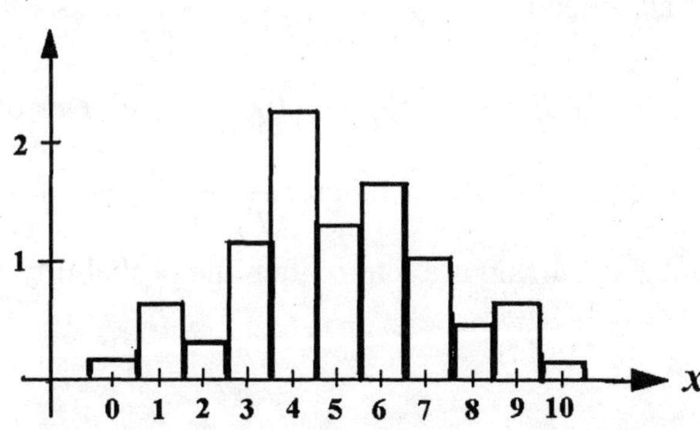

21.

x	1	2	3	4	5
$P(X=x)$	0.007	0.029	0.021	0.079	0.164

x	6	7	8	9	10
$P(X=x)$	0.15	0.20	0.207	0.114	0.029

23. True. This follows from the definition.

USING TECHNOLOGY EXERCISES 8.1, page 453

1.

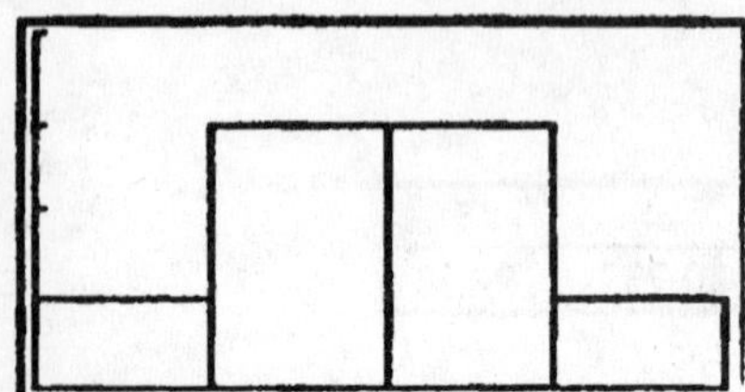

3.

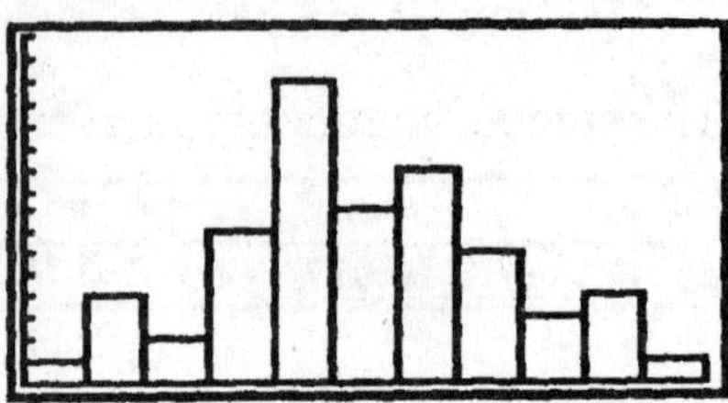

8.2 Problem Solving Tips

1. The *expected value* of a random variable X is given by

$$E(X) = x_1 p_1 + x_2 p_2 + \cdots + x_n p_n$$

where $x_1, x_2, \ldots, x_n$ are the values assumed by X and $p_1, p_2, \ldots, p_n$ are the associated probabilities.

2. It $P(E)$ is the probability of an event E occurring, then the *odds in favor* of E occurring are $\dfrac{P(E)}{P(E^c)}$ and the *odds against* E occurring are $\dfrac{P(E^c)}{P(E)}$.

3. If the odds in favor of an event E occurring are a to b, then the probability of E occurring is

$$P(E) = \frac{a}{a+b}.$$

8.2 CONCEPT QUESTIONS, page 463

1. The expected value of a random variable X is given by

$$E(X) = x_1 p_1 + x_2 p_2 + \cdots + x_n p_n$$

Example varies.

3. a. The odds in favor of E occurring are $\dfrac{P(E)}{P(E^c)}$.

b. The odds in favor of E occurring are $\frac{a}{b}$.

EXERCISES 8.2, page 463

1. a. The student's grade-point average is given by

$$\frac{(2)(4)(3)+(3)(3)(3)+(4)(2)(3)+(1)(1)(3)}{(10)(3)} \quad \text{or } 2.6.$$

b.

X	0	1	2	3	4
$P(X=x)$	0	0.1	0.4	0.3	0.2

$$E(X) = 1(0.1) + 2(0.4) + 3(0.3) + 4(0.2) = 2.6.$$

3. $E(X) = -5(0.12) + -1(0.16) + 0(0.28) + 1(0.22) + 5(0.12) + 8(0.1)$
$= 0.86.$

5. $E(X) = 0(0.07) + 25(0.12) + 50(0.17) + 75(0.14) + 100(0.28)$
$+ 125(0.18) + 150(0.04) = 78.5,$ or \$78.50.

7. A customer entering the store is expected to buy
$E(X) = (0)(0.42) + (1)(0.36) + (2)(0.14) + (3)(0.05) + (4)(0.03)$
$= 0.91$, or 0.91 videocassettes.

9. The expected number of accidents is given by
$E(X) = (0)(0.935) + (1)(0.03) + (2)(0.02) + (3)(0.01) + (4)(0.005) = 0.12$

11. The expected number of machines that will break down on a given day is given by
$E(X) = (0)(0.43) + (1)(0.19) + (2)(0.12) + (3)(0.09) + (4)(0.04)$
$+ (5)(0.03) + (6)(0.03) + (7)(0.02) + (8)(0.05)$
$= 1.73.$

13. The associated probabilities are $\frac{3}{50}, \frac{8}{50}, \ldots, \frac{5}{50}$, respectively. Therefore, the expected interest rate is
$(4.9)(\frac{3}{50}) + 5(\frac{8}{50}) + 5.1(\frac{12}{50}) + 5.2(\frac{14}{50}) + 5.3(\frac{8}{50}) + 5.4(\frac{5}{50})$
≈ 5.162, or 5.16%

15. The expected net earnings of a person who buys one ticket are

$$-1(0.997) + 24(0.002) + 99(0.0006) + 499(0.0002) + 1999(0.0002)$$
$$= -0.39,$$ or a loss of \$0.39 per ticket.

17. The expected gain of the insurance company is given by

$$E(X) = 0.992(130) - (9870)(0.008) = 50,$$ or \$50.

19. The company's expected gross profit is given by

$$E(X) = (0.3)(80{,}000) + (0.60)(75{,}000) + (0.10)(70{,}000) - 64{,}000$$
$$= 12{,}000,$$ or \$12,000.

21. City A: $E(X) = (10{,}000{,}000)(0.2) - 250{,}000 = 1{,}750{,}000$, or \$1.75 million.
City B: $E(X) = (7{,}000{,}000)(0.3) - 200{,}000 = 1{,}900{,}000$, or \$1.9 million.
We see that the company should bid for the rights in city B.

23. The expected number of houses sold per year at company A is given by

$$E(X) = (12)(.02) + (13)(.03) + (14)(.05) + (15)(.07) + (16)(.07)$$
$$+ (17)(.16) + (18)(.17) + (19)(.13) + (20)(.11)$$
$$+ (21)(.09) + (22)(.06) + (23)(.03) + (24)(.01)$$
$$= 18.09.$$

The expected number of houses sold per year at company B is given by

$$E(X) = (6)(.01) + (7)(.04) + (8)(.07) + (9)(.06) + (10)(.11)$$
$$+ (11)(.12) + (12)(.19) + (13)(.17) + (14)(.13)$$
$$+ (15)(.04) + (16)(.03) + (17)(.02) + (18)(.01)$$
$$= 11.77.$$

Then, Sally's expected commission at company A is given by

$$(0.03)(308{,}000)(18.09) = 167{,}151.60,$$ or \$167,151.60

Her expected commission at company B is given by

$$(0.03)(474{,}000)(11.77) = 167{,}369.40,$$

or \$167,369.40. Based on these expectations, she should accept the job offer with company B.

25. Maria might expect her business to grow at the rate of

$$(5)(0.12) + (4.5)(0.24) + (3)(0.4) + (0)(0.2) + (-0.5)(0.04) = 2.86$$

or 2.86%/year for the upcoming year.

27. The expected value of the winnings on a $1 bet placed on a split is

$$E(X) = 17 \cdot \frac{2}{38} + (-1) \cdot \frac{36}{38} \approx -0.0526, \text{ or a loss of 5.3 cents.}$$

29. The expected value of a player's winnings are

$$(1)(\frac{18}{37}) + (-1)(\frac{19}{37}) = -\frac{1}{37} \approx -0.027, \text{ or a loss of 2.7 cents per bet.}$$

31. The odds in favor of E occurring are $\frac{P(E)}{P(E^C)} = \frac{0.4}{0.6}$, or 2 to 3. The odds against E occurring are 3 to 2.

33. The probability of E not occurring is given by $P(E) = \frac{2}{3+2} = \frac{2}{5} = 0.4$.

35. The probability that she will win her match is $P(E) = \frac{7}{7+5} = \frac{7}{12} = 0.5833$.

37. The probability that the business deal will not go through is

$$P(E) = \frac{5}{5+9} = \frac{5}{14} \approx 0.3571.$$

39. a. The mean is given by

$$\frac{40 + 45 + 2(50) + 55 + 2(60) + 2(75) + 2(80) + 4(85) + 2(90) + 2(95) + 100}{20}$$

$= 74.$

The mode is 85 (the value that appears most frequently).

The median is 80 (the middle value).

b. The mode is the least representative of this set of test scores.

41. We first arrange the numbers in increasing order:

$$0,0,\underbrace{1,1,\cdots,1}_{9\text{ times}},\underbrace{2,2,\cdots,2}_{15\text{ times}},\underbrace{3,3,\cdots,3}_{12\text{ times}},\underbrace{4,4,\cdots,4}_{8\text{ times}},\underbrace{5,5,\cdots,5}_{6\text{ times}},\underbrace{6,6,\cdots,6}_{4\text{ times}},7,7,8$$

There are 69 numbers. So the median is 3. This is close to the mean of 3.1 obtained in Example 1, Section 8.2.

43. The average is $\frac{1}{10}(16.1+16+\cdots+16.2)=16$, or 16 oz. Next, we arrange the numbers in increasing order:

15.8, 15.9, 15.9, 16, 16, 16, 16, 16.1, 16.1, 16.2

The median is $\frac{16+16}{2}=16$, or 16 oz. The mode is 16.

45. True This follows from the definition.

8.3 Problem Solving Tips

1. The *variance* of a random variable X is a measure of the spread of a probability distribution about its mean. The variance of a random variable X is given by

$$\text{Var}(X)=p_1(x_1-\mu)^2+p_2(x_2-\mu)^2+\cdots+p_n(x_n-\mu)^2$$

where $x_1,x_2,\ldots,x_n$ denote the values assumed by X and

$p_1=P(X=x_1), p_2=P(X=x_2),\ldots, p_n=P(X=x_n)$. The *standard deviation* of a random variable X is $\sigma=\sqrt{\text{Var}(X)}$.

2. *Chebychev's Inequality* gives the proportion of values of a random variable X lying within k standard deviations of the expected value of X. The probability that a randomly chosen outcome of the experiment lies between $\mu-k\sigma$ and $\mu+k\sigma$ is

$$P(\mu-k\sigma\le X\le\mu+k\sigma)\ge 1-\frac{1}{k^2}.$$

8.3 CONCEPT QUESTIONS, page 473

1. If a random variable has the probability distribution

x	x_1	x_2	x_3	$\cdots$	x_n
$P(X=x)$	p_1	p_2	p_3	$\cdots$	p_n

and expected value $E(X)=\mu$, then the variance of the random variable X is

$$\text{Var}(X)=p_1(x_1-\mu)^2+p_2(x_2-\mu)^2+\cdots+p_n(x_n-\mu)^2$$

and the standard variation of the random variable X is given by $\sigma=\sqrt{\text{Var}(X)}$.

EXERCISES 8.3, page 473

1. $\mu=(1)(.4)+(2)(.3)+3(.2)+(4)(.1)=2.$
$\text{Var}(X)=(.4)(1-2)^2+(.3)(2-2)^2+(.2)(3-2)^2+(.1)(4-2)^2$
$=.4+0+.2+.4=1$
$\sigma=\sqrt{1}=1.$

3. $\mu=-2(\frac{1}{16})+-1(\frac{4}{16})+0(\frac{6}{16})+1(\frac{4}{16})+2(\frac{1}{16})=\frac{0}{16}=0.$

$\text{Var}(X)=\frac{1}{16}(-2-0)^2+\frac{4}{16}(-1-0)^2+\frac{6}{16}(0-0)^2+\frac{4}{16}(1-0)^2+\frac{1}{16}(2-0)^2$
$=1$

$\sigma=\sqrt{1}=1.$

5. $\mu=0.1(430)+(0.2)(480)+(0.4)(520)+(0.2)(565)+(0.1)(580)$
$=518.$
$\text{Var}(X)=.1(430-518)^2+(.2)(480-518)^2+(.4)(520-518)^2$
$+(.2)(565-518)^2+(.1)(580-518)^2$
$=1891.$
$\sigma=\sqrt{1891}\approx 43.49.$

7. The mean of the histogram in Figure (b) is more concentrated about its mean than the histogram in Figure (a). Therefore, the histogram in Figure (a) has the larger variance.

9. $E(X)=1(.1)+2(.2)+3(.3)+4(.2)+5(.2)=3.2.$
$\text{Var}(X)=(.1)(1-3.2)^2+(.2)(2-3.2)^2+(.3)(3-3.2)^2$

$$+ (.2)(4 - 3.2)^2 + (.2)(5 - 3.2)^2$$
$$= 1.56$$

11. $\mu = \dfrac{1+2+3+\cdots+8}{8} = 4.5$

$$V(X) = \tfrac{1}{8}(1-4.5)^2 + \tfrac{1}{8}(2-4.5)^2 + \cdots + \tfrac{1}{8}(8-4.5)^2 = 5.25$$

13. a. Let X be the annual birth rate during the years 1991 - 2000.
b.

x	15	16	17	18	19	21
$P(X=x)$	.02	.30	.08	.56	.02	.02

x	14.5	14.6	14.7	14.8	15.2	15.5	15.9	16.3
$P(X=x)$	.2	.1	.2	.1	.1	.1	.1	.1

c. $E(X) = (.2)(14.5) + (.1)(14.6) + (.2)(14.7) + (.1)(14.8) + (.1)(15.2) + (.1)(15.5) + (.1)(15.9) + (.1)(16.3)$
$= 15.07.$

$V(X) = (.2)(14.5 - 15.07)^2 + (.1)(14.6 - 15.07)^2 + (.2)(14.7 - 15.07)^2 + (.1)(14.8 - 15.07)^2 + (.1)(15.2 - 15.07)^2 + (.1)(15.5 - 15.07)^2 + (.1)(15.9 - 15.07)^2 + (.1)(16.3 - 15.07)^2$
$= .3621$

$\sigma = \sqrt{.3621} \approx .602$.

15. a. Mutual Fund A

$\mu = (.2)(-4) + (.5)(8) + (.3)(10) = 6.2$, or \$620.

$V(X) = (.2)(-4 - 6.2)^2 + (.5)(8 - 6.2)^2 + (.3)(10 - 6.2)^2$
$= 26.76$, or \$267,600.

Mutual Fund B

$\mu = (.2)(-2) + (.4)(6) + (.4)(8)$
$= 5.2$, or \$520.

$V(X) = (.2)(-2 - 5.2)^2 + (.4)(6 - 5.2)^2 + (.4)(8 - 5.2)^2$
$= 13.76$, or \$137,600.

b. Mutual Fund A

c. Mutual Fund B

17. $\text{Var}(X) = (.4)(1)^2 + (.3)(2)^2 + (.2)(3)^2 + (.1)(4)^2 - (2)^2 = 1.$

19. $\mu = [\frac{10}{500}(280) + \frac{20}{500}(290) + \cdots + \frac{5}{500}(450)] = 339.6$, or \$339,600.

$V(X) = [\frac{10}{500}(280-339.6)^2 + \frac{20}{500}(290-339.6)^2 + \cdots + \frac{5}{500}(450-339.6)^2][(1000)^2]$

$= 1443.84 \times 10^6$ dollars.

$\sigma = \sqrt{1443.84 \times 10^6} = 37.998 \times 10^3$, or \$37,998.

21. The mean is given by

$$\frac{(94.5)(4)+(84.5)(8)+(74.5)(12)+(64.5)(4)+(54.5)(2)}{30} \approx 77.167,$$

or approximately 77.2.

$$\text{Var}(x) = \left(\tfrac{4}{30}\right)(94.5-77.167)^2 + \left(\tfrac{8}{30}\right)(84.5-77.167)^2 + \left(\tfrac{12}{30}\right)(74.5-77.167)^2 + \left(\tfrac{4}{30}\right)(64.5-77.167)^2 + \left(\tfrac{2}{30}\right)(54.5-77.167)^2 \approx 112.88889$$

Therefore, $\sigma = \sqrt{112.8889} \approx 10.62$.

23.

x	1342	1428	1545	1707	1807	1815
Rel. Freq.	1	1	1	1	1	1
$P(X=x)$	$\frac{1}{6}$	$\frac{1}{6}$	$\frac{1}{6}$	$\frac{1}{6}$	$\frac{1}{6}$	$\frac{1}{6}$

$\mu_x = \frac{1}{6}(1342+1428+1545+1707+1807+1815) = 1607.33$.

So the average of the average hours worked, per worker, is 1607.33 hr.

$$\text{Var}(X) = \tfrac{1}{6}[1342-1607.33)^2 + (1428-1607.33)^2 + (1545-1607.33)^2 + (1707-1607.33)^2 + (1807-1607.33)^2 + (1815-1607.33)^2$$

$$\approx 33228.8889$$

$$\sigma_X = \sqrt{\text{Var}(X)} \approx 182.2879$$

The standard deviation is 182.2879 hr.

25.

x	5.22	5.23	5.24	5.31	5.55	5.56	5.57	5.59	5.7
Rel Freq.	1	1	1	1	2	1	1	1	1
$P(X=x)$	$\frac{1}{10}$	$\frac{2}{10}$	$\frac{1}{10}$	$\frac{1}{10}$	$\frac{1}{10}$	$\frac{1}{10}$	$\frac{1}{10}$	$\frac{1}{10}$	$\frac{1}{10}$

$$\mu_X = \frac{1}{10}(5.22+5.23+5.24+5.31+2(5.55)+5.56+5.57+5.59+5.7)$$

$$= \frac{54.52}{10} = 5.452.$$

$$\begin{aligned}\text{Var}(X) &= \tfrac{1}{10}[(5.22-5.452)^2+(5.23-5.452)^2+(5.24-5.452)^2+(5.31-5.452)^2\\ &\quad +2(5.55-5.452)^2+(5.56-5.452)^2+(5.57-5.452)^2\\ &\quad +(5.59-5.452)^2+(5.7-5.452)^2]\\ &\approx 0.29356\end{aligned}$$

$\sigma_X \approx 0.171337$

So the mean is 5.452 and the standard deviation is 0.1713.

27. The probability distribution is

x	16.0	16.3	16.5	16.8	17.0	18.0	18.5
Rel. Freq of Occurrence	1	1	4	1	3	1	1
$P(X=x)$	$\frac{1}{12}$	$\frac{1}{12}$	$\frac{1}{3}$	$\frac{1}{12}$	$\frac{1}{4}$	$\frac{1}{12}$	$\frac{1}{12}$

The mean of X is

$$\mu = \tfrac{1}{12}(16)+\tfrac{1}{12}(16.3)+\tfrac{1}{3}(16.5)+\tfrac{1}{12}(16.8)+\tfrac{1}{4}(17.0)+\tfrac{1}{12}(18.0)+\tfrac{1}{12}(18.5)\approx 16.8833.$$

So the average monthly seasonally adjusted annualized sales rate is approximately 6.9 million. The variance of X is

$$\begin{aligned}\text{Var}(X) &= \tfrac{1}{12}(16-16.8833)^2+\tfrac{1}{12}(16.3-16.8833)^2+\tfrac{1}{3}(16.5-16.8833)^2\\ &\quad +\tfrac{1}{12}(16.8-16.8833)^2+\tfrac{1}{4}(17.0-16.8833)^2+\tfrac{1}{12}(18.0-16.8833)^2\\ &\quad +\tfrac{1}{12}(18.5-16.8833)^2\\ &\approx 0.4681.\end{aligned}$$

So $\quad \sigma = \sqrt{0.4681} \approx 0.68$.

This says that the monthly sales do not differ much from the average of 16.9 million units.

29. a. Using Chebychev's inequality we have

$P(\mu - k\sigma \le X \le \mu + k\sigma) \ge 1 - 1/k^2$.

$\mu - k\sigma = 42 - k(2) = 38$, and $k = 2$,

and $P(\mu - k\sigma \le X \le \mu + k\sigma) \ge 1 - 1/(2)^2$

$\ge 1 - 1/4$

$\ge 3/4$, or at least .75.

b. Using Chebychev's inequality we have

$P(\mu - k\sigma \le X \le \mu + k\sigma) \ge 1 - 1/k^2$.

$\mu - k\sigma = 42 - k(2) = 32$, and $k = 5$,

and $P(\mu - k\sigma \le X \le \mu + k\sigma) \ge 1 - 1/(5)^2$
$\ge 1 - 1/25 \ge 24/25$, or at least .96.

31. Here $\mu = 50$ and $\sigma = 1.4$. Now, we require that $c = k\sigma$, or $k = \dfrac{c}{1.4}$.

Next, we solve $0.96 = 1 - \left(\dfrac{1.4}{c}\right)^2$; $\dfrac{1.96}{c^2} = 0.04$; $c^2 = \dfrac{1.96}{0.04} = 49$, or $c = 7$.

33. Using Chebychev's inequality we have $P(\mu - k\sigma \le X \le \mu + k\sigma) \ge 1 - 1/k^2$.
Here, $\mu - k\sigma = 24 - k(3) = 20$, and $k = 4/3$.
So $P(\mu - k\sigma \le X \le \mu + k\sigma) \ge 1 - 1/(4/3)^2 \ge 1 - 9/16 = 7/16$, or at least .4375.

35. Using Chebychev's inequality we have
$P(\mu - k\sigma \le X \le \mu + k\sigma) \ge 1 - 1/k^2$.
Here, $\mu - k\sigma = 52{,}000 - k(500) = 50{,}000$, and $k = 4$.
So $P(\mu - k\sigma \le X \le \mu + k\sigma) \ge 1 - 1/(4)^2 \ge 1 - 1/16 \ge 15/16$, or at least .9375.

37. True. This follows from the definition.

USING TECHNOLOGY EXERCISES 8.3, page 479

1. a.

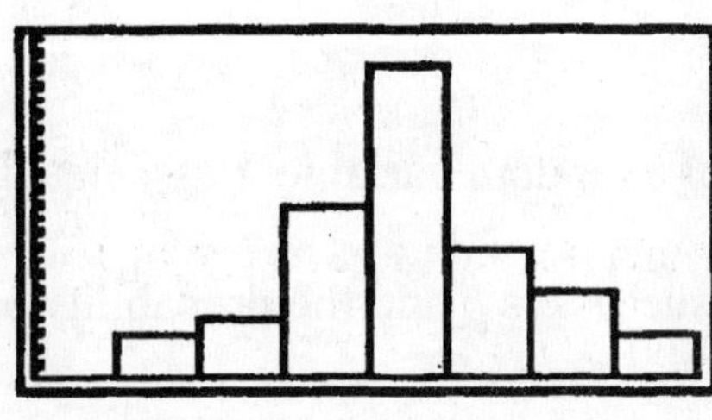

b. $\mu = 4$ and $\sigma \approx 1.40$

3. a.

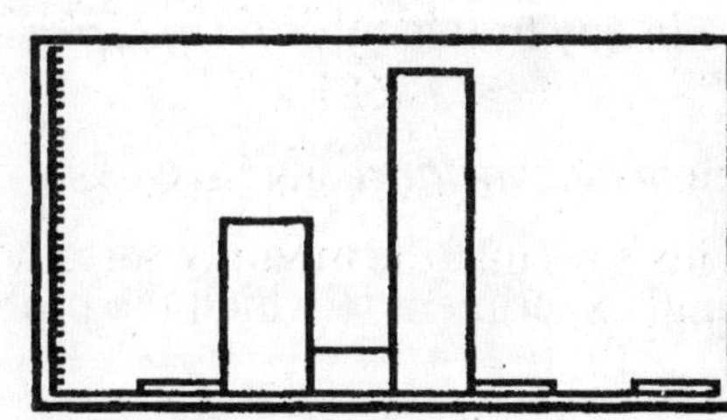

b. $\mu = 17.34$ and $\sigma \approx 1.11$

5. a. Let X denote the random variable that gives the weight of a carton of sugar.
b. The probability distribution for the random variable X is

x	4.96	4.97	4.98	4.99	5.00	5.01	5.02	5.03	5.04	5.05	5.06
$P(X=x)$	$\frac{3}{30}$	$\frac{4}{30}$	$\frac{4}{30}$	$\frac{1}{30}$	$\frac{1}{30}$	$\frac{5}{30}$	$\frac{3}{30}$	$\frac{3}{30}$	$\frac{4}{30}$	$\frac{1}{30}$	$\frac{1}{30}$

$\mu = 5.00467 \approx 5.00$; $V(X) = 0.0009$; $\sigma = \sqrt{0.0009} = 0.03$

7. a.

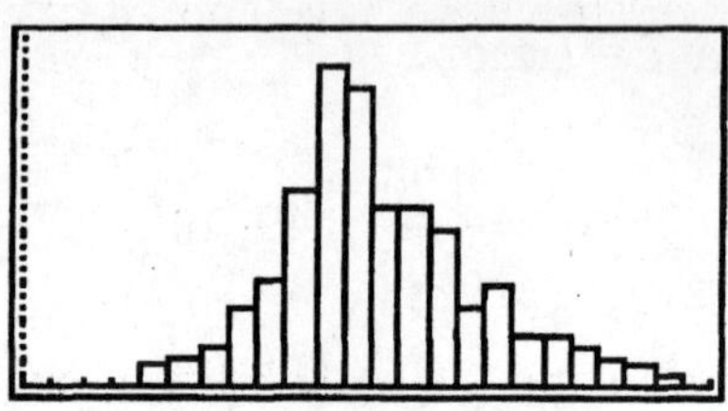

b. $\mu = 65.875$ and $\sigma = 1.73$.

8.4 Problem Solving Tips

1. In a *binomial experiment*, (a) the number of trials are fixed, (b) there are two outcomes, (c) the probability of success in each trial is the same, and (d) the trials are independent of each other.

2. The *probability of exactly x successes in n independent trials* in a binomial experiment, in which the probability of success in any trial is p and the probability of failure in any trial is q, is $C(n,x)p^x q^{n-x}$.

3. The *mean*, *variance*, and *standard deviation* of a random variable X associated with a binomial experiment in which the probability of success is p and the probability of failure is q are $\mu = E(X) = np$, $\text{Var}(X) = npq$, and $\sigma_X = \sqrt{npq}$.

8.4 CONCEPT QUESTIONS, page 487

1. a. 2 b. It is fixed. c. They are independent d. $C(n,x)p^x q^{n-x}$

EXERCISES 8.4, page 487

1. Yes. The number of trials is fixed, there are two outcomes of the experiment, the probability in each trial is fixed $(p = \frac{1}{6})$, and the trials are independent of each other.

3. No. There are more than 2 outcomes in each trial.

5. No. There are more than 2 outcomes in each trial and the probability of success (an accident) in each trial is not the same.

7. $C(4,2)(\frac{1}{3})^2(\frac{2}{3})^2 = \dfrac{4!}{2!2!}(\frac{4}{81}) \approx .296.$

9. $C(5,3)(.2)^3(.8)^2 = (\frac{5!}{2!3!})(.2)^3(.8)^2 \approx .0512.$

11. The required probability is given by $P(X = 0) = C(5,0)(\frac{1}{3})^0(\frac{2}{3})^5 \approx .132.$

13. The required probability is given by

$$P(X \geq 3) = C(6,3)(\tfrac{1}{2})^3(\tfrac{1}{2})^{6-3} + C(6,4)(\tfrac{1}{2})^4(\tfrac{1}{2})^{6-4} + C(6,5)(\tfrac{1}{2})^5(\tfrac{1}{2})^{6-5}$$
$$+ C(6,6)(\tfrac{1}{2})^6(\tfrac{1}{2})^{6-6}$$
$$= \tfrac{6!}{3!3!}(\tfrac{1}{2})^6 + \tfrac{6!}{4!2!}(\tfrac{1}{2})^6 + \tfrac{6!}{5!1!}(\tfrac{1}{2})^6 + \tfrac{6!}{6!0!}(\tfrac{1}{2})^6 = \tfrac{1}{64}(20 + 15 + 6 + 1) = \tfrac{21}{32}.$$

15. The probability of no failures, or, equivalently, the probability of five successes is

$$P(X = 5) = C(5,5)(\tfrac{1}{3})^5(\tfrac{2}{3})^{5-5} = \tfrac{1}{243} \approx .00412.$$

17. Here $n = 4$, and $p = 1/6$. Then $P(X = 2) = C(4,2)(\frac{1}{6})^2(\frac{5}{6})^2 = (\frac{25}{216}) \approx .116.$

19. Here $n = 5$, $p = .4$, and therefore, $q = 1 - .4 = .6$.

a. $P(X = 0) = C(5,0)(.4)^0(.6)^5 \approx .078$; $P(X = 1) = C(5,1)(.4)(.6)^4 \approx .259$

$P(X = 2) = C(5,2)(.4)^2(.6)^3 \approx .346$; $P(X = 3) = C(5,3)(.4)^3(.6)^3 \approx .230$

$P(X = 4) = C(5,4)(.4)^4(.6)^2 \approx .077$; $P(X = 5) = C(5,5)(.4)^5(.6) \approx .010$

b.

X	0	1	2	3	4	5
$P(X=x)$	.078	.259	.346	.230	.077	.010

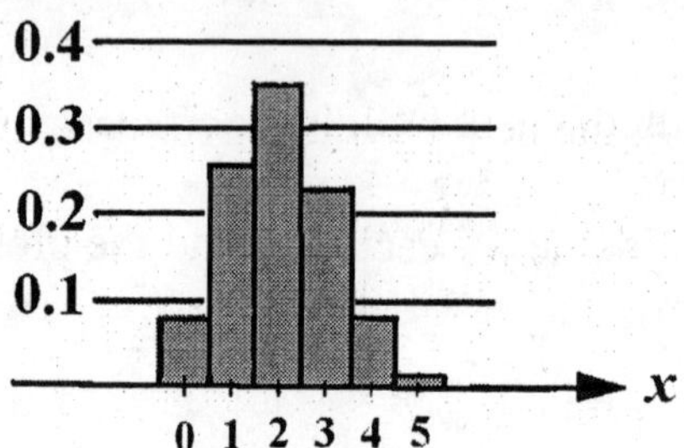

c. $\mu = np = 5(.4) = 2$; $\sigma = \sqrt{npq} = \sqrt{5(.4)(.6)} = \sqrt{1.2} \approx 1.095.$

21. Here $1 - p = 1/50$ or $p = 49/50$. So the probability of obtaining 49 or 50 nondefective fuses is

$$P(X=49) + P(X=50) = C(50,49)(\tfrac{49}{50})^{49}(\tfrac{1}{50}) + C(50,50)(\tfrac{49}{50})^{50}(\tfrac{1}{50})^0 \approx .74.$$

This is also the probability of at most one defective fuse. So the inference is incorrect.

23. The probability that she will serve 0 or 1 ace is given by

$$P(X=0) + P(X=1) = C(5,0)(.15)^0(.85)^5 + C(5,1)(.15)^1(.85)^4$$

$$= \frac{5!}{0!5!}(.85)^5 + \frac{5!}{1!4!}(.15)(.85)^4 \approx .83521.$$

Therefore, the probability that she will serve at least two aces is

$$1 - P(X=0) - P(X=1) = 1 - .83521 = .16479, \text{ or approximately } .165.$$

25. The required probability is given by $P(X=6) = C(6,6)(\tfrac{1}{4})^6(\tfrac{3}{4})^0 \approx .0002.$

27. a. The probability that six or more people stated a preference for brand A is

$$\begin{aligned} P(X \geq 6) &= C(10,6)(.6)^6(.4)^4 + C(10,7)(.6)^7(.4)^3 \\ &\quad + C(10,8)(.6)^8(.4)^2 + C(10,9)(.6)^9(.4)^1 \\ &\quad + C(10,10)(.6)^{10}(.4)^0 \\ &\approx .251 + .215 + .121 + .040 + .006 = .633. \end{aligned}$$

b. The required probability is 1 - .633 = .367.

29. a. Let X denote the number of new buildings that are in violation of the building code. Then $p = \frac{1}{3}$ and $p = \frac{2}{3}$. Therefore, the probability that the first 3 new buildings will pass inspection and the remaining 2 will fail the inspection is

$$\left(\frac{2}{3}\right)\left(\frac{2}{3}\right)\left(\frac{2}{3}\right)\left(\frac{1}{3}\right)\left(\frac{1}{3}\right) = \frac{2^3}{3^5},$$

or approximately .0329.

b. The probability that just 3 of the new buildings will pass inspection (and so 2 will fail the inspection) is $C(5,2)\left(\frac{2}{3}\right)^3\left(\frac{1}{3}\right)^2 = \frac{5!}{3!2!} \cdot \frac{2^3}{3^5}$, or approximately .3292.

31. This is a binomial experiment with $n = 9$, $p = 1/3$, and $q = 2/3$.

a. The probability is given by $P(X = 3) = C(9,3)(\frac{1}{3})^3(\frac{2}{3})^6 = (\frac{9!}{6!3!})(\frac{1}{3})^3(\frac{2}{3})^6 \approx 0.273.$

b. The probability is given by

$$\begin{aligned}&P(X = 0) + P(X = 1) + P(X = 2) + P(X = 3)\\ &= C(9,0)(\tfrac{1}{3})^0(\tfrac{2}{3})^9 + C(9,1)(\tfrac{1}{3})(\tfrac{2}{3})^8 + C(9,2)(\tfrac{1}{3})^2(\tfrac{2}{3})^7 + C(9,3)(\tfrac{1}{3})^3(\tfrac{2}{3})^6\\ &= .026 + .117 + .234 + .273 \approx .650.\end{aligned}$$

33. This is a binomial experiment with $n = 10$, $p = .02$, and $q = .98$.

a. The probability that the sample contains no defectives is given by

$$P(X = 0) = C(10,0)(.02)^0(.98)^{10} \approx .817.$$

b. The probability that the sample contains at most 2 defective sets is given by

$$\begin{aligned}P(X \le 2) &= C(10,0)(.02)^0(.98)^{10} + C(10,1)((.02)(.98)^9\\ &\quad + C(10,2)(.02)^2(.98)^8\\ &\approx .817 + .167 + .015 = .999.\end{aligned}$$

35. a. The required probability is $P(X = 2) = C(10,2)(.05)^2(.95)^8 \approx .075.$

b. The required probability is

$$\begin{aligned}P(X \ge 2) &= 1 - P(X \le 2)\\ &= 1 - [C(10,2)(.05)^2(.95)^8 + C(10,1)(.05)(.95)^9\\ &\quad + C(10,0)(.05)^0(.95)^{10}]\\ &\approx .012.\end{aligned}$$

37. a. The required probability is

$$P(X = 0) = C(20,0)(.1)^0(.9)^{20} \approx .1216.$$

b. The required probability is

$$P(X = 0) = C(20,0)(.05)^0(.95)^{20} \approx .3585.$$

39. The required probability is $P(X=0)=C(10,0)(.1)^0(.9)^{10}\approx .3487$.

41. The mean number of people for whom the drug is effective is

$$\mu = np = (500)(.75) = 375.$$

The standard deviation of the number of people for whom the drug can be expected to be effective is $\sigma=\sqrt{npq}=\sqrt{(500)(0.75)(0.25)}\approx 9.68$.

43. False. There are exactly two outcomes.

45. False. Here $p=\frac{1}{4}$ and $q=1-\frac{1}{4}=\frac{3}{4}$.

The probability that the batter will get a hit if he bats four times is

$$1-P(X=0)=1-C(4,0)(\tfrac{1}{4})^0(\tfrac{3}{4})^4=1-\frac{4!}{4!0!}\cdot 1\cdot\frac{81}{256}=1-\frac{81}{256}=\frac{175}{256}\approx .68,$$

which is far from guaranteeing that he gets a hit.

8.5 Problem Solving Tips

1. *Normal distributions* are a special class of continuous probability distributions. A *normal curve* is the bell-shaped graph of a normal distribution. The *standard normal curve* has mean $\mu=0$ and standard deviation $\sigma=1$. The random variable associated with a standard normal distribution is called a *standard normal variable* and is denoted by Z. The areas under the standard normal curve to the left of the number z corresponding to the probabilities $P(Z<z)$ or $P(Z\le z)$ are given in Table 2, Appendix C.

2. The area of the region under the normal curve between $x=a$ and $x=b$ is equal to the area of the region under the standard normal curve between

$z=\dfrac{a-\mu}{\sigma}$ and $z=\dfrac{b-\mu}{\sigma}$.

The probability of the random variable X associated with this area is

$$P(a < X < b) = \left(\frac{a-\mu}{\sigma} < Z < \frac{b-\mu}{\sigma}\right).$$

8.5 CONCEPT QUESTIONS, page 497

1. a. μ b. It is symmetric about the line $x = \mu$.
   c. Yes. It approaches the x-axis. d. 1 e. Between $\mu - 5$ and $\mu + 5$.

EXERCISES 8.5, page 497

1. $P(Z < 1.45) = .9265.$

3. $P(Z < -1.75) = .0401.$

5. $P(-1.32 < Z < 1.74) = P(Z < 1.74) - P(Z < -1.32) = .9591 - .0934 = .8657.$

7. $P(Z < 1.37) = .9147.$

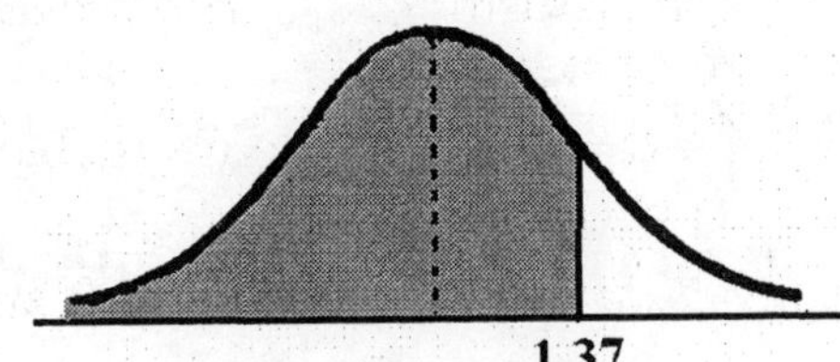

9. $P(Z < -0.65) = .2578.$

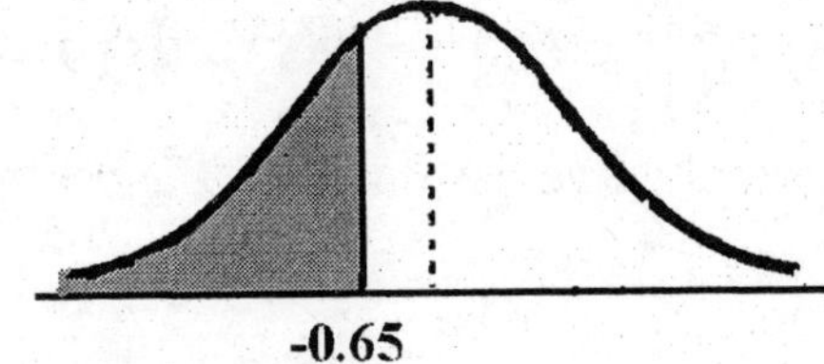

11. $P(Z > -1.25) = 1 - P(Z < -1.25) = 1 - .1056 = .8944$

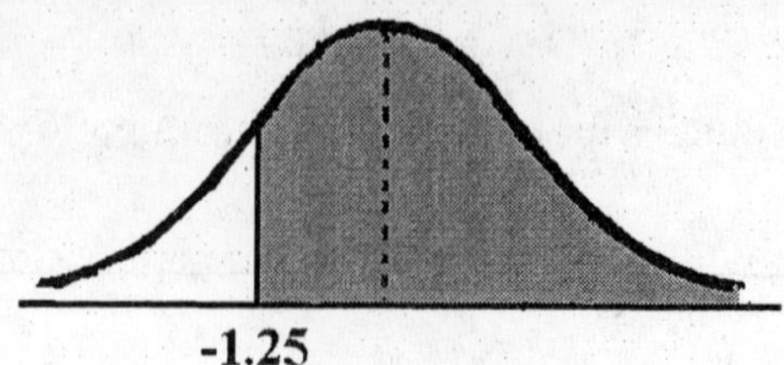

13. $P(0.68 < Z < 2.02) = P(Z < 2.02) - P(Z < 0.68)$
$= .9783 - .7517 = .2266.$

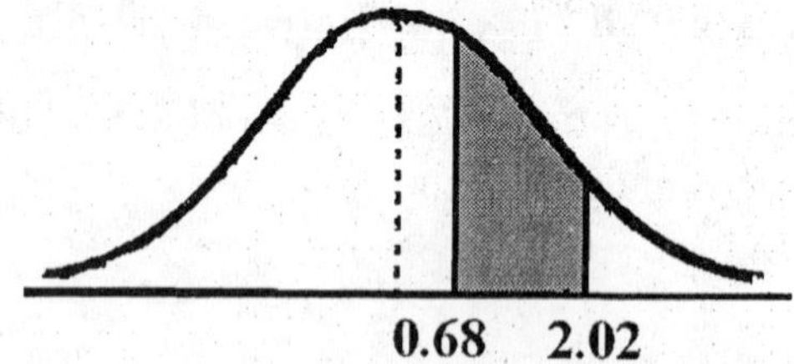

15. a. Referring to Table 2, we see that $P(Z < z) = .8907$ implies that $z = 1.23$.
b. Referring to Table 2, we see that $P(Z < z) = .2090$ implies that $z = -0.81$.

17. a. $P(Z > -z) = 1 - P(Z < -z) = 1 - .9713 = .0287$ implies $z = 1.9$.
b. $P(Z < -z) = .9713$ implies that $z = -1.9$.

19. a. $P(X < 60) = P(Z < \frac{60-50}{5}) = P(Z < 2) = .9772.$

b. $P(X > 43) = P(Z > \frac{43-50}{5}) = P(Z > -1.4) = P(Z < 1.4) = .9192.$

c. $P(46 < X < 58) = P(\frac{46-50}{5} < Z < \frac{58-50}{5}) = P(-0.8 < Z < 1.6)$
$= P(Z < 1.6) - P(Z < -0.8) = .9452 - .2119 = .7333.$

8.6 Problem Solving Tips

1. A binomial distribution associated with a binomial experiment involving n trials with probability of success p and probability of failure q may be approximated by a normal distribution (if n is large and p is not close to 0 or 1) with $\mu = np$ and $\sigma = \sqrt{npq}$.

8.6 CONCEPT QUESTIONS, page 504

1. The central limit theorem allows us to approximate a binomial distribution by a normal distribution under certain conditions.

EXERCISES 8.6, page 505

1. $\mu = 20$ and $\sigma = 2.6$.

a. $P(X > 22) = P(Z > \frac{22-20}{2.6}) = P(Z > 0.77) = P(Z < -0.77) = .2206.$

b. $P(X < 18) = P(Z < \frac{18-20}{2.6}) = P(Z < -0.77) = .2206.$

c. $P(19 < X < 21) = P(\frac{19-20}{2.6} < Z < \frac{21-20}{2.6}) = P(-0.38 < Z < 0.38)$

$= P(Z < 0.38) - P(Z < -0.38) = .6480 - .3520 = .2960.$

3. $\mu = 750$ and $\sigma = 75$.

a. $P(X > 900) = P(Z > \frac{900-750}{75}) = P(Z > 2) = P(Z < -2) = .0228.$

b. $P(X < 600) = P(Z < \frac{600-750}{75}) = P(Z < -2) = .0228.$

c. $P(750 < X < 900) = P(Z < \frac{750-750}{75} < Z < \frac{900-750}{75})$

$= P(0 < Z < 2) = P(Z < 2) - P(Z < 0)$

$= .9772 - .5000 = .4772.$

d. $P(600 < X < 800) = P(\frac{600-750}{75} < Z < \frac{800-750}{75})$
$= P(-2 < Z < .67) = P(Z < .67) - P(Z < -2)$
$= .7486 - .0228 = .7258.$

5. $\mu = 100$ and $\sigma = 15$.

a. $P(X > 140) = P(Z > \frac{140-100}{15}) = P(Z > 2.67) = P(Z < -2.67) = .0038.$

b. $P(X > 120) = P(Z > \frac{120-100}{15}) = P(Z > 1.33) = P(Z < -1.33) = .0918.$

c. $P(100 < X < 120) = P(\frac{100-100}{15} < Z < \frac{120-100}{15}) = P(0 < Z < 1.33)$
$= P(Z < 0) - P(Z < 1.33) = .9082 - .5000 = .4082.$

d. $P(X < 90) = P(Z < \frac{90-100}{15}) = P(Z < -0.67) = .2514.$

7. Here $\mu = 575$ and $\sigma = 50$.
$P(550 < X < 650) = P(\frac{550-575}{50} < Z < \frac{650-575}{50}) = P(-0.5 < Z < 1.5)$
$= P(Z < 1.5) - P(Z < -0.5) = .9332 - .3085 = .6247.$

9. Here $\mu = 22$ and $\sigma = 4$.
$P(X < 12) = P(Z < \frac{12-22}{4}) = P(Z < -2.5) = .0062$, or 0.62 percent.

11. $\mu = 70$ and $\sigma = 10$.
To find the cut-off point for an A, we solve $P(Y < y) = .85$ for y. Now
$P(Y < y) = P\left(Z < \frac{y-70}{10}\right) = .85$ implies $\frac{y-70}{10} = 1.04$, or $y = 80.4 \approx 80$.

For a B: $P(Y < y) = P\left(Z < \frac{y-70}{10}\right) = .75$ implies $\frac{y-70}{10} = .67$, or $y \approx 77$.

For a C: $P(Y < y) = P\left(Z < \frac{y-70}{10}\right) = .60$ implies $\frac{y-70}{10} = .25$, or $y \approx 73$.

For a D: $P\left(Z \leq \frac{y-70}{10}\right) = .2$ implies $\frac{y-70}{10} = -.84$ or $y \approx 62$.

For a F: $P\left(Z < \dfrac{y-70}{10}\right) = .05$ implies $\dfrac{y-70}{10} = -1.65$, or $y \approx 54$.

13. Let X denote the number of heads in 25 tosses of the coin. Then X is a binomial random variable. Also, $n = 25, p = .4$, and $q = .6$. So

$$\mu = (25)(.4) = 10; \quad \sigma = \sqrt{(25)(.4)(.6)} \approx 2.45.$$

Approximating the binomial distribution by a normal distribution with a mean of 10 and a standard deviation of 2.45, we find upon letting Y denote the associated normal random variable,

a. $P(X < 10) \approx P(Y < 9.5) = P\left(Z < \dfrac{9.5-10}{2.45}\right) = P(Z < \text{-}0.20) = .4207.$

b. $P(10 \le X \le 12) \approx P(9.5 < Y < 12.5)$

$$= P\left(\frac{9.5-10}{2.45} < Z < \frac{12.5-10}{2.45}\right) = P(Z < 1.02) - P(Z < -0.20)$$

$$= P(Z < 1.02) - P(Z < \text{-}0.20) = .8461 - .4207 = .4254.$$

c. $P(X > 15) \approx P(Y \ge 15)$

$$= P(Z > \frac{15.5-10}{2.45}) = P(Z > 2.24) = P(Z < -2.24) = .0125.$$

15. Let X denote the number of times the marksman hits his target. Then X has a binomial distribution with $n = 30, p = .6$ and $q = .4$. Therefore,

$$\mu = (30)(.6) = 18, \ \sigma = \sqrt{(30)(.6)(.4)} = 2.68.$$

a. $P(X \ge 20) \approx P(Y \ge 19.5)$

$$= P\left(Z > \frac{19.5-18}{2.68}\right) = P(Z > 0.56) = P(Z < \text{-}0.56) = .2877.$$

b. $P(X < 10) \approx P(Y < 9.5) = P\left(Z < \dfrac{9.5-18}{2.68}\right) = P(Z < -3.17) = .0008.$

c. $P(15 \le X \le 20) \approx P(14.5 < Y < 20.5) = P\left(\dfrac{14.5-18}{2.68} < Z < \dfrac{20.5-18}{2.68}\right)$

$$= P(Z < 0.93) - P(Z < -1.31) = .8238 - .0951 = .7287.$$

17. Let X denote the number of "seconds." Then X has a binomial distribution with $n = 200, p = .03$, and $q = .97$. Then

$$\mu = (200)(.03) = 6; \ \sigma = \sqrt{(200)(.03)(.97)} \approx 2.41,$$

and $P(X < 10) \approx P(Y < 9.5) = P(Z < \frac{9.5-6}{2.41}) = P(Z < 1.45) = .9265.$

19. Let X denote the number of workers who meet with an accident during a 1-year period. Then $\mu = (800)(.1) = 80$; $\sigma = \sqrt{(800)(.1)(.9)} \approx 8.49$,

and $$P(X > 70) \approx P(Y > 70.5)$$
$$= P\left(Z > \frac{70.5-80}{8.49}\right) = P(Z > -1.12) = P(Z < 1.12) = .8686.$$

21. a. Let X denote the number of mice that recovered from the disease. Then X has a binomial distribution with $n = 50$, $p = .5$, and $q = .5$, so
$$\mu = (50)(.5) = 25;\ \sigma = \sqrt{(50)(.5)(.5)} \approx 3.54,$$
Approximating the binomial distribution by a normal distribution with a mean of 25 and a standard deviation of 3.54, we find that the probability that 35 or more of the mice would recover from the disease without benefit of the drug is
$$P(X \geq 35) \approx P(Y > 34.5)$$
$$= P(Z > \frac{34.5-25}{3.54}) = P(Z > 2.68) = P(Z < -2.68) = .0037.$$
b. The drug is very effective.

23. Let n denote the number of reservations the company should accept. Then we need to find
$$P(X \geq 2000) \approx P(Y > 1999.5) = .01$$
or equivalently, $P(Z \geq \frac{1999.5 - np}{\sqrt{npq}}) = .01$ [Here p = .92 and q = .08.]

or $P(Z \leq \frac{np - 1999.5}{\sqrt{npq}}) = .01$.

Next, $\frac{.92n - 1999.5}{\sqrt{0.0736n}} = -2.33$
$$(0.92n - 1999.5)^2 = (-2.33)^2(0.0736n)$$
$$0.8464n^2 - 3679.08n + 3{,}998{,}000.25 = 0.39956704n,$$
or $$0.8464n^2 - 3679.479567n + 3{,}998{,}000.25 = 0.$$
Using the quadratic formula, we obtain $n = \frac{3679.479567 \pm \sqrt{2940.2376}}{1.6928} \approx 2142,$
or 2142. [You can verify that 2206 is not a root of the original equation (before squaring).] Therefore, the company should accept no more than 2142 reservations.

CHAPTER 8, CONCEPT REVIEW, page 508

1. Random

3. Sum; $(\frac{1}{2})(-2)+(\frac{1}{4})(3)+(\frac{1}{4})(4)=\frac{3}{4}$

5 $p_1(x_1-\mu)^2+p_2(x_2-\mu)^2+\cdots+p_n(x_n-\mu)^2; \sqrt{\text{Var}(X)}$

7. Continuous; probability density function; set

CHAPTER 8, REVIEW EXERCISES, page 508

1. a. $S = \{$WWW, BWW, WBW, WWB, BBW, BWB, WBB, BBB$\}$

b.

Outcome	WWW	BWW	WBW	WWB	BBW	BWB	WBB	BBB
Value	0	1	1	1	2	2	2	3

c.

x	0	1	2	3
$P(X=x)$	$\frac{1}{35}$	$\frac{12}{35}$	$\frac{18}{35}$	$\frac{4}{35}$

d.

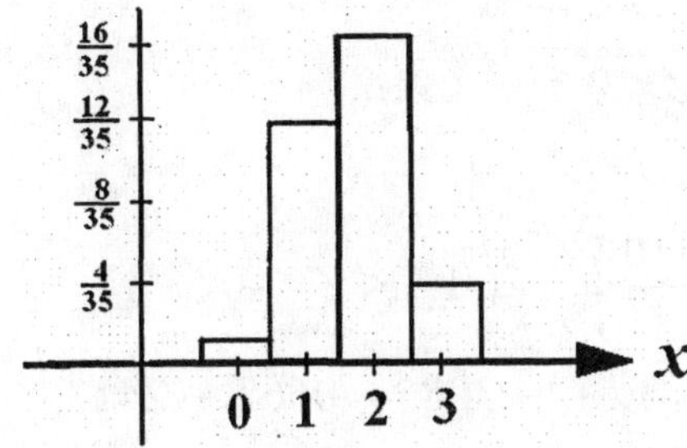

3. a. $P(1 \le X \le 4) = .1 + .2 + .3 + .2 = .8.$

b. $\mu = 0(.1) + 1(.1) + 2(.2) + 3(.3) + 4(.2) + 5(.1) = 2.7.$

$$V(X) = .1(0 - 2.7)^2 + .1(1 - 2.7)^2 + .2(2 - 2.7)^2 + .3(3 - 2.7)^2 + .2(4 - 2.7)^2 + .1(5 - 2.7)^2$$
$$= 2.01$$

$\sigma = \sqrt{2.01} \approx 1.418.$

5. $P(Z < 0.5) = .6915.$

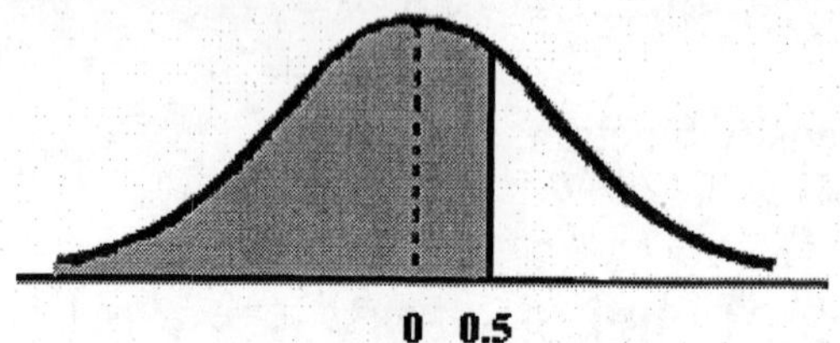

7. $P(-0.75 < Z < 0.5) = P(Z < 0.5) - P(Z < -0.75) = .6915 - .2266 = .4649.$

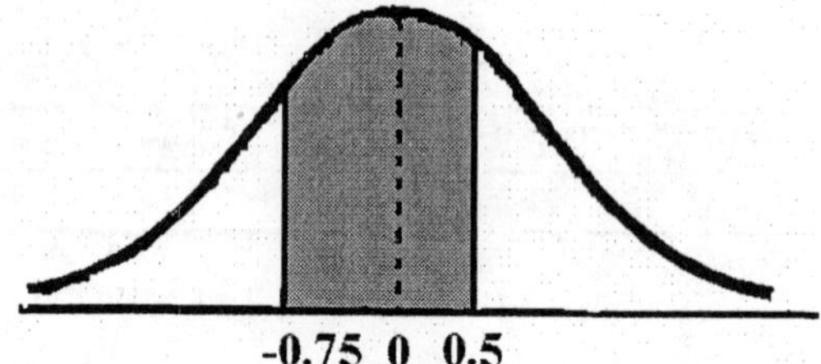

9. If $P(Z < z) = .9922$, then $z = 2.42$.

11. If $P(Z > z) = .9788$, then $P(Z < -z) = .9788$, and $-z = 2.03$, or $z = -2.03$.

13. $P(X < 11) = P\left(Z < \frac{11-10}{2}\right) = P(Z < 0.5) = .6915.$

15. $P(7 < X < 9) = P\left(\frac{7-10}{2} < Z < \frac{9-10}{2}\right) = P(-1.5 < Z < -0.5)$

$= P(Z < -0.5) - P(Z < -1.5) = .3085 - .0668 = .2417.$

17. This is a binomial experiment with $p = .7$, and so $q = .3$. The probability that he will get exactly two strikes in four attempts is given by

$$P(X = 2) = C(4,2)(.7)^2(.3)^2 \approx .2646.$$

The probability that he will get at least two strikes in four attempts is given by

$$P(X = 2) + P(X = 3) + P(X = 4)$$
$$= C(4,2)(.7)^2(.3)^2 + C(4,3)(.7)^3(.3) + C(4,4)(.7)^4(.3)^0$$

$= .2646 + .4116 + .2401 \approx .9163.$

19. Here $\mu = 64.5$ and $\sigma = 2.5$. Next, $64.5 - 2.5k = 59.5$ and $64.5 + 2.5k = 69.5$ and $k = 2$. Therefore, the required probability is given by

$$P(59.5 \leq X \leq 69.5) \geq 1 - \frac{1}{2^2} = .75.$$

21. This is a binomial experiment with $n = 20$, $p = .1$, and $q = .9$. We compute

$$\begin{aligned} P(X \leq 2) &= P(X = 0) + P(X = 1) + P(X = 2) \\ &= C(20,0)(.1)^0(.9)^{20} + C(20,1)(.1)(.9)^{19} + C(20,2)(.1)^2(.9)^{18} \\ &= .12158 + .27017 + .28518 \approx .677. \end{aligned}$$

23. Let X denote the diameter of the steel rods. Then the probability that a rod will be accepted by the buyer is

$$\begin{aligned} P(0.995 < X < 1.005) &= P\left(\frac{0.9945 - 1}{0.002} < Z < \frac{1.0055 - 1}{0.002}\right) \\ &= P(Z < 2.75) - P(Z < -2.75) = .9970 - .0030 = .9940. \end{aligned}$$

Then the probability that a rod will be rejected by the buyer is

$$1 - .9940 = .006,$$

and the percentage of rods that will be rejected by the buyer is .6 percent.

25. This is a binomial experiment with $n = 200$, $p = .05$, and $q = .95$. Next,

$$\mu = (.05)(200) = 10; \quad \sigma = \sqrt{(.05)(200)(.95)} = 3.08$$

Then $P(X \leq 20) = P\left(Z < \frac{20.5 - 10}{3.08}\right) = P(Z < 3.41) = .9997.$

CHAPTER 8 BEFORE MOVING ON, page 509

1.

X	-3	-2	0	1	2	3
$P(X = x)$	.05	.10	.25	.3	.2	.1

2. a.

$$\begin{aligned} P(X \leq 0) &= P(X = 0) + P(X = -1) + P(X = -3) + P(X = -4) \\ &= .28 + .32 + .14 + .06 = .8 \end{aligned}$$

b. $P(-4 \leq X \leq 1) = 1 - P(X = 3) = 1 - .08 = .92$

3. $\mu = (-3)(.08) + (-1)(.24) + 0(.32) + 1(.16) + 3(.12) + 5(.08) = 0.44$

$$\begin{aligned}\text{Var}(X) &= .08(-3-.44)^2 + .24(-1-.44)^2 + .32(0-.44)^2 \\ &\quad + .16(1-.44)^2 + .12(3-.44)^2 + .08(5-.44)^2 \\ &\approx 4.0064.\end{aligned}$$

$\sigma_X \approx 2$

4. a. $P(X=0) = C(4,0)(.3)^0(.7)^4 = \dfrac{4!}{0!4!} \cdot 1 \cdot (0.7)^4 \approx 0.2401$

$P(X=1) = C(4,1)(.3)^1(.7)^3 = \dfrac{4!}{1!3!}(.3)(.7)^3 \approx .4116$

$P(X=2) = C(4,2)(.3)^2(.7)^2 = \dfrac{4!}{2!2!}(.3)^2(.7)^2 \approx .2646$

$P(X=3) = C(4,3)(.3)^3(.7)^1 = \dfrac{4!}{3!1!}(.3)^3(.7) \approx .0756$

$P(X=4) = C(4,4)(.3)^4(.7)^0 = \dfrac{4!}{4!0!}(.3)^4 \cdot 1 \approx .0081$

b. $\mu = E(X) = np = 4(.3) = .12$

$\sigma_X = \sqrt{npq} = \sqrt{4(.3)(.7)} \approx .917$

5. Here $\mu = 60$ and $\sigma = 5$. Therefore,

a. $P(X < 70) = P(Z < \frac{70-60}{5}) = P(Z < 2) \approx .9772$

b. $P(X > 50) = P(Z > \frac{50-60}{5}) = P(Z > -2) = P(Z < 2) \approx .9772$

c. $P(50 < X < 70) = P(\frac{50-60}{5} < Z < \frac{70-60}{5})$

$= P(-2 < Z < 2) = P(Z < 2) - P(Z < -2)$

$= .9772 - .0228 = .9544$

6. Here $n = 30$, $p = .5$ and so $q = .5$. Therefore

$\mu = np = 30(.5) = 15$ and $\sigma = \sqrt{npq} = \sqrt{(30)(.5)(.5)} \approx 2.7386$

a. $P(X < 10) \approx P(Y < 9.5) = P(Z < \frac{9.5-15}{2.7386}) = P(Z < -2) = .0228$

b. $P(12 \le X \le 16) \approx P(12 < Y < 16)$

$= P(\frac{12-15}{2.7386} < Z < \frac{16-15}{2.7386}) = P(-1.10 < Z < .37)$

$= P(Z < .37) - P(Z < -1.10) = .6443 - .1357 = .5086$

c. $P(X > 20) \approx P(Y > 20.5) = P(Z > \frac{20.5-15}{2.7386})$

$= P(Z > 2) = P(Z < -2) \approx .0228$

CHAPTER 9

9.1 Problem Solving Tips

1. A *transition matrix* associated with a Markov chain with *n* states is an $n \times n$ matrix *T* with entries a_{ij} $(1 \le i \le n; 1 \le j \le n)$. The transition probability associated with the transition from state 1 to state 2 is represented by the entry a_{21}.

2. Each entry in the transition matrix is nonnegative. The sum of the entries in each column of *T* is 1.

3. If *T* represents the $n \times n$ transition matrix associated with a Markov process, then the probability distribution of the system after *m* observations is given by $X_m = T^m X_0$.

9.1 CONCEPT QUESTIONS, page 519

1. A *finite stochastic process* is an experiment consisting of a finite number of stages in which the outcomes and associated probabilities at each stage depend on the outcomes and associated probabilities of the *preceding stages*. In a *Markov chain*, the probabilities associated with the outcomes at any stage of the experiment depend only on the outcomes of the *preceding stage*.

3. a. $n \times n$ b. $a_{ij} = P(\text{state } i | \text{state } j)$; no c. 1

EXERCISES 9.1, page 519

1. Yes. All entries are nonnegative and the sum of the entries in each column is equal to 1.

3. Yes.

5. No. The sum of the entries of the third column is not 1.

7. Yes.

9. No. It is not a square ($n \times n$) matrix.

11. a. The conditional probability that the outcome state 1 will occur given that the outcome state 1 has occurred is .3.
    b. .7
    c. We compute $X_1 = TX_0 = \begin{bmatrix} .3 & .6 \\ .7 & .4 \end{bmatrix}\begin{bmatrix} .4 \\ .6 \end{bmatrix} = \begin{bmatrix} .48 \\ .52 \end{bmatrix}$.

13. We compute $TX_0 = \begin{bmatrix} .6 & .2 \\ .4 & .8 \end{bmatrix}\begin{bmatrix} .5 \\ .5 \end{bmatrix} = \begin{bmatrix} .4 \\ .6 \end{bmatrix}$

    Thus, after 1 stage of the experiment, the probability of state 1 occurring is 0.4 and the probability of state 2 occurring is 0.6. The tree diagram describing this process follows.

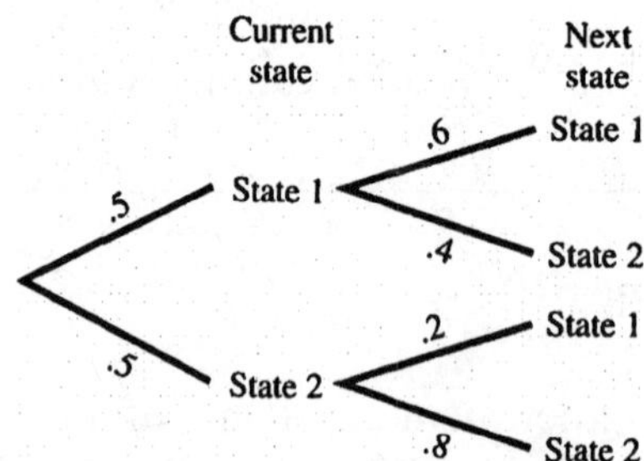

Using this diagram, we see that the probabilities of state 1 and state 2 occurring in the next stage of the experiment are given by

$$P(S_1) = (.5)(.6) + (.5)(.2) = .4$$
$$P(S_2) = (.5)(.4) + (.5)(.8) = .6$$

Observe that these probabilities are precisely those represented in the probability distribution vector X_0T.

15. $X_1 = TX_0 = \begin{bmatrix} .4 & .8 \\ .6 & .2 \end{bmatrix}\begin{bmatrix} .6 \\ .4 \end{bmatrix} = \begin{bmatrix} .56 \\ .44 \end{bmatrix}$.

    $X_2 = TX_1 = \begin{bmatrix} .4 & .8 \\ .8 & .2 \end{bmatrix}\begin{bmatrix} .56 \\ .44 \end{bmatrix} = \begin{bmatrix} .576 \\ .424 \end{bmatrix}$.

17. $X_1 = TX_0 = \begin{bmatrix} \frac{1}{4} & \frac{1}{4} & \frac{1}{2} \\ \frac{1}{4} & \frac{1}{2} & \frac{1}{2} \\ \frac{1}{2} & \frac{1}{4} & 0 \end{bmatrix}\begin{bmatrix} \frac{1}{4} \\ \frac{1}{2} \\ \frac{1}{4} \end{bmatrix} = \begin{bmatrix} \frac{5}{16} \\ \frac{7}{16} \\ \frac{1}{4} \end{bmatrix}$; $X_2 = TX_0 = \begin{bmatrix} \frac{1}{4} & \frac{1}{4} & \frac{1}{2} \\ \frac{1}{4} & \frac{1}{2} & \frac{1}{2} \\ \frac{1}{2} & \frac{1}{4} & 0 \end{bmatrix}\begin{bmatrix} \frac{5}{16} \\ \frac{7}{16} \\ \frac{1}{4} \end{bmatrix} = \begin{bmatrix} \frac{5}{16} \\ \frac{27}{64} \\ \frac{17}{64} \end{bmatrix}$.

19. a.

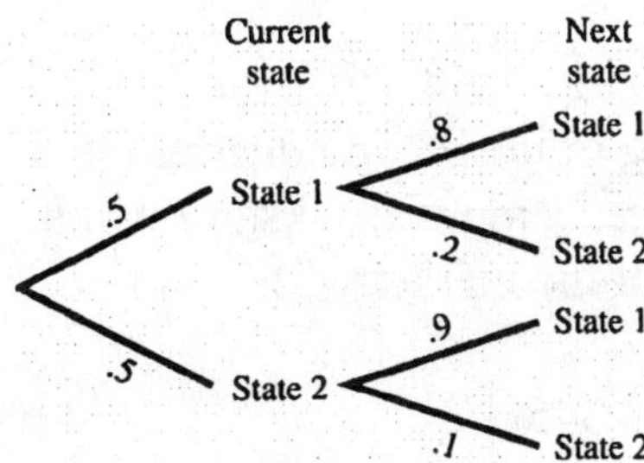

b. $T = \begin{array}{c c} & \begin{array}{cc} L & R \end{array} \\ \begin{array}{c} L \\ R \end{array} & \begin{bmatrix} .8 & .9 \\ .2 & .1 \end{bmatrix} \end{array}$

c. $X_0 = \begin{array}{c} L \\ R \end{array}\begin{bmatrix} .5 \\ .5 \end{bmatrix}$

d. $X_1 = \begin{array}{c c} & \begin{array}{cc} L & R \end{array} \\ \begin{array}{c} L \\ R \end{array} & \begin{bmatrix} .8 & .9 \\ .2 & .1 \end{bmatrix} \end{array}\begin{bmatrix} .5 \\ .5 \end{bmatrix} = \begin{array}{c} L \\ R \end{array}\begin{bmatrix} .85 \\ .15 \end{bmatrix}$

21. a. $X_1 = TX_0 = \begin{array}{c c} & \begin{array}{cc} R & D \end{array} \\ \begin{array}{c} R \\ D \end{array} & \begin{bmatrix} .7 & .2 \\ .3 & .8 \end{bmatrix} \end{array}\begin{bmatrix} .6 \\ .4 \end{bmatrix} = \begin{array}{c} R \\ D \end{array}\begin{bmatrix} .5 \\ .5 \end{bmatrix}$

so if the election were held now, it would be a tie.

b. $X_1 = TX_0 = \begin{array}{c c} & \begin{array}{cc} R & D \end{array} \\ \begin{array}{c} R \\ D \end{array} & \begin{bmatrix} .7 & .2 \\ .3 & .8 \end{bmatrix} \end{array}\begin{bmatrix} .5 \\ .5 \end{bmatrix} = \begin{array}{c} R \\ D \end{array}\begin{bmatrix} .45 \\ .55 \end{bmatrix}$ so the Democratic candidate would win.

23. After one pickup and discharge the distribution will be

$$X_1 = TX_0 = \begin{bmatrix} .6 & .4 & .3 \\ .3 & .3 & .3 \\ .1 & .3 & .4 \end{bmatrix} \begin{bmatrix} .6 \\ .2 \\ .2 \end{bmatrix} = \begin{bmatrix} .5 \\ .3 \\ .2 \end{bmatrix}$$

or 50% will be in Zone I, 30% will be in Zone II, and 20 percent will be in Zone III.

25. The expected distribution is given by

$$X_1 = TX_0 = \begin{bmatrix} .80 & .10 & .05 \\ .10 & .75 & .05 \\ .10 & .15 & .90 \end{bmatrix} \begin{bmatrix} .4 \\ .4 \\ .2 \end{bmatrix} = \begin{bmatrix} .37 \\ .35 \\ .28 \end{bmatrix}$$

and we conclude that at the beginning of the second quarter the university Bookstore will have 37 percent of the market, the Campus Bookstore will have 35 percent, and the Book Mart will have 28 percent of the market.
Similarly,

$$X_2 = TX_2 = \begin{bmatrix} .80 & .10 & .05 \\ .10 & .75 & .05 \\ .10 & .15 & .90 \end{bmatrix} \begin{bmatrix} .37 \\ .35 \\ .28 \end{bmatrix} = \begin{bmatrix} .345 \\ .3135 \\ .3415 \end{bmatrix}$$

implies that the University Bookstore will have 34.5% of the market, the Campus Bookstore will have 31.35% of the market, and the Book Mart will have 34.15% of the market at the beginning of the third quarter.

27. $$X_1 = TX_0 = \begin{bmatrix} .80 & .10 & .20 & .10 \\ .10 & .70 & .10 & .05 \\ .05 & .10 & .60 & .05 \\ .05 & .10 & .10 & .80 \end{bmatrix} \begin{bmatrix} .3 \\ .3 \\ .2 \\ .2 \end{bmatrix} = \begin{bmatrix} .33 \\ .27 \\ .175 \\ .225 \end{bmatrix}$$

Similarly

$$X_2 = TX_1 = \begin{bmatrix} .3485 \\ .25075 \\ .15975 \\ .241 \end{bmatrix} \quad \text{and } X_3 = TX_2 = \begin{bmatrix} .3599 \\ .2384 \\ .1504 \\ .2513 \end{bmatrix}$$

Assuming that the present trend continues, 36% of the students in their senior year will major in business, 23.8% will major in the humanities, 15% will major in education, and 25.1% will major in the natural sciences.

29. False. In a Markov Chain, an outcome depends only on the preceding stage.

USING TECHNOLOGY EXERCISES 9.1, page 523

1. $X_5 = \begin{bmatrix} .204489 \\ .131869 \\ .261028 \\ .186814 \\ .2158 \end{bmatrix}$

3. Manufacturer A will have 23.95% of the market, Manufacturer B will have 49.71% of the market share, and manufacturer C will have 26.34 percent of the market share.

9.2 Problem Solving Tips

1. A stochastic matrix T is *regular* if and only if some power of T has entries that are all positive.

2. To find the *steady-state distribution vector* X for a transition matrix T, solve the vector equation $TX = X$ together with the condition that the sum of the elements of the vector X is 1.

9.2 CONCEPT QUESTIONS, page 530

1. a. Let T be an $n \times n$ transition matrix and let X_0 be an $(n \times 1)$ initial distribution vector. If the sequence of vectors $X_1, X_2, \ldots, X_n, \ldots,$ defined by $X_i = TX_{i-1}$ $(i = 1, 2, 3, \ldots)$ converges to a vector X as n gets larger and larger, then T is called the *steady-state distribution vector.*
   b. If $T, T^2, T^3, \ldots, T^m, \ldots,$ converges to a matrix L as m increases, then L is called the *steady-state matrix*.
   c. A stochastic matrix T is a regular Markov chain if the sequence $T, T^2, T^3, \ldots$ approaches a steady-state matrix in which the rows of the limiting matrix are all equal and all the entries are positive.

EXERCISES 9.2, page 530

1. Since all entries in the matrix are positive, it is regular.

3. $T^2 = \begin{bmatrix} 1 & .8 \\ 0 & .2 \end{bmatrix}\begin{bmatrix} 1 & .8 \\ 0 & .2 \end{bmatrix} = \begin{bmatrix} 1 & .96 \\ 0 & .04 \end{bmatrix}$; $T^3 = \begin{bmatrix} 1 & .96 \\ 0 & .04 \end{bmatrix}\begin{bmatrix} 1 & .8 \\ 0 & .2 \end{bmatrix} = \begin{bmatrix} 1 & .992 \\ 0 & .008 \end{bmatrix}$

   and we see that the a_{21} entry will always be zero, so T is not regular.

5. $T^2 = \begin{bmatrix} \frac{1}{2} & \frac{3}{4} & 0 \\ \frac{1}{2} & 0 & \frac{1}{2} \\ 0 & \frac{1}{4} & \frac{1}{2} \end{bmatrix}\begin{bmatrix} \frac{1}{2} & \frac{3}{4} & 0 \\ \frac{1}{2} & 0 & \frac{1}{2} \\ 0 & \frac{1}{4} & \frac{1}{2} \end{bmatrix} = \begin{bmatrix} \frac{5}{8} & \frac{3}{8} & \frac{3}{8} \\ \frac{1}{4} & \frac{1}{2} & \frac{1}{4} \\ \frac{1}{8} & \frac{1}{8} & \frac{3}{8} \end{bmatrix}$

   and so this matrix is regular.

7. $T^2 = \begin{bmatrix} .7 & .2 & .3 \\ .3 & .8 & .3 \\ 0 & 0 & .4 \end{bmatrix}\begin{bmatrix} .7 & .2 & .3 \\ .3 & .8 & .3 \\ 0 & 0 & .4 \end{bmatrix} = \begin{bmatrix} .55 & .3 & .39 \\ .45 & .7 & .45 \\ 0 & 0 & .16 \end{bmatrix}$

   and so forth.

   Continuing, we see that $T^3, T^4, \ldots,$ will have the a_{31} and a_{32} entries equal to zero and T is not regular.

9. We solve the matrix equation

$$\begin{bmatrix} \frac{1}{3} & \frac{1}{4} \\ \frac{2}{3} & \frac{3}{4} \end{bmatrix}\begin{bmatrix} x \\ y \end{bmatrix} = \begin{bmatrix} x \\ y \end{bmatrix}$$

   or equivalently, the system of equations

$$\begin{aligned} \tfrac{1}{3}x + \tfrac{1}{4}y &= x \\ \tfrac{2}{3}x + \tfrac{3}{4}y &= y \\ x + \quad y &= 1. \end{aligned}$$

   Solving this system of equations, we find the required vector to be $\begin{bmatrix} \frac{3}{11} \\ \frac{8}{11} \end{bmatrix}$.

11. We have $TX = X$, that is, $\begin{bmatrix} .5 & .2 \\ .5 & .8 \end{bmatrix}\begin{bmatrix} x \\ y \end{bmatrix} = \begin{bmatrix} x \\ y \end{bmatrix}$

   or equivalently, the system of equations

$$.5x + .2y = x$$
$$.5x + .8y = y.$$

These two equations are equivalent to the single equation $0.5x - 0.2y = 0$. We must also have $x + y = 1$. So we have the system

$$.5x - .2y = x$$
$$x + \quad y = 1$$

The second equation gives $y = 1 - x$, which when substituted into the first equation yields,

$$0.5x - 0.2(1 - x) = 0, \ 0.7x - 0.2 = 0, \text{ or } x = 2/7.$$

Therefore, $y = 5/7$ and the steady-state distribution vector is $\begin{bmatrix} \frac{2}{7} \\ \frac{5}{7} \end{bmatrix}$.

13. We solve the system

$$\begin{bmatrix} 0 & \frac{1}{8} & 1 \\ 1 & \frac{5}{8} & 0 \\ 0 & \frac{1}{4} & 0 \end{bmatrix} \begin{bmatrix} x \\ y \\ z \end{bmatrix} = \begin{bmatrix} x \\ y \\ z \end{bmatrix}$$

together with the equation $x + y + z = 1$; that is, the system

$$-x + \tfrac{1}{8}y + z = 0$$
$$x - \tfrac{3}{8}y \quad = 0$$
$$\tfrac{1}{4}y - z = 0$$
$$x + y + z = 1.$$

Using the Gauss-Jordan method, we find that the required steady-state vector is

$$\begin{bmatrix} \frac{3}{13} \\ \frac{8}{13} \\ \frac{2}{13} \end{bmatrix}.$$

15. We solve the system

$$\begin{bmatrix} .2 & 0 & .3 \\ 0 & .6 & .4 \\ .8 & .4 & .3 \end{bmatrix} \begin{bmatrix} x \\ y \\ z \end{bmatrix} = \begin{bmatrix} x \\ y \\ z \end{bmatrix}$$

together with the equation $x+y+z=1$, or equivalently, the system

$$\begin{aligned} -0.8x \qquad\quad +0.3z &= 0 \\ -0.4y+0.4z &= 0 \\ 0.8x+0.4y-0.7z &= 0 \\ x \quad +y \quad\; +z &= 1. \end{aligned}$$

Using the Gauss-Jordan method, we find that the required steady-state vector is

$$\begin{bmatrix} \frac{3}{19} \\ \frac{8}{19} \\ \frac{8}{19} \end{bmatrix}.$$

17. a. We want to solve

$$\begin{bmatrix} .8 & .9 \\ .2 & .1 \end{bmatrix}\begin{bmatrix} x \\ y \end{bmatrix} = \begin{bmatrix} x \\ y \end{bmatrix},$$

or equivalently

$$\begin{aligned} .8x+.9y &= x \\ .2x+.1y &= y \\ x+\;\; y &= 1 \end{aligned}$$

Solving this system, we find that the required steady-state vector is $\begin{bmatrix} \frac{2}{11} \\ \frac{9}{11} \end{bmatrix}$ and conclude that in the long run, the mouse will turn left 81.8% of the time.

19. We compute

$$X_1 = \begin{bmatrix} .72 & .12 \\ .28 & .88 \end{bmatrix}\begin{bmatrix} .48 \\ .52 \end{bmatrix} = \begin{bmatrix} .408 \\ .592 \end{bmatrix},$$

and conclude that, ten years from now, there will be 40.8 percent 1-wage-earners and 59.2% 2-wage earners.

b. We solve the system

$$\begin{bmatrix} .72 & .12 \\ .28 & .88 \end{bmatrix}\begin{bmatrix} x \\ y \end{bmatrix} = \begin{bmatrix} x \\ y \end{bmatrix}$$

together with the equation $x + y = 1$.

$$\begin{aligned} -0.28x + 0.12y &= 0 \\ 0.28x - 0.12y &= 0 \\ x + \quad y &= 1 \end{aligned}$$

Solving, we find $x = 0.3$ and $y = 0.7$, and conclude that in the long run, there will be 30% 1-wage earners and 70% 2-wage earners.

21. a. If this trend continues, the percentage of homeowners in this city who will own single-family homes or condominiums two years from now will be given by $X_2 = TX_1$. Thus,

$$X_1 = TX_0 = \begin{bmatrix} .85 & .35 \\ .15 & .65 \end{bmatrix} \begin{bmatrix} .8 \\ .2 \end{bmatrix} = \begin{bmatrix} .75 \\ .25 \end{bmatrix}$$

$$X_2 = TX_1 = \begin{bmatrix} .85 & .35 \\ .15 & .65 \end{bmatrix} \begin{bmatrix} .75 \\ .25 \end{bmatrix} = \begin{bmatrix} .725 \\ .275 \end{bmatrix}$$

and we conclude that 72.5% will own single-family homes and 27.5% will own condominiums at that time.

b. We solve the system

$$\begin{bmatrix} .85 & .35 \\ .15 & .65 \end{bmatrix} \begin{bmatrix} x \\ y \end{bmatrix} = \begin{bmatrix} x \\ y \end{bmatrix}$$

together with the equation $x + y = 1$. Thus,

$$\begin{aligned} -0.15x + 0.35y &= 0 \\ 0.15x - 0.35y &= 0 \\ x + \quad y &= 1 \end{aligned}$$

Solving, we find $x = 0.7$ and $y = 0.3$, and conclude that in the long run 70% will own single family homes and 30% will own condominiums.

23. a.

$$X_1 = TX_0 = \begin{bmatrix} .8 & .1 & .1 \\ .1 & .85 & .05 \\ .1 & .05 & .85 \end{bmatrix} \begin{bmatrix} .3 \\ .4 \\ .3 \end{bmatrix} = \begin{bmatrix} .31 \\ .385 \\ .305 \end{bmatrix}$$

$$X_2 = TX_1 = \begin{bmatrix} .8 & .1 & .1 \\ .1 & .85 & .05 \\ .1 & .05 & .85 \end{bmatrix} \begin{bmatrix} .31 \\ .385 \\ .305 \end{bmatrix} = \begin{bmatrix} .317 \\ .3735 \\ .3095 \end{bmatrix}$$

From our computations, we conclude that after two weeks 31.7% of the viewers will watch the *ABC* news, 37.35% will watch the *CBS* news, and 30.95% will watch the *NBC* news.

b. We solve the system

$$\begin{bmatrix} .8 & .1 & .1 \\ .1 & .85 & .05 \\ .1 & .05 & .85 \end{bmatrix} \begin{bmatrix} x \\ y \\ z \end{bmatrix} = \begin{bmatrix} x \\ y \\ z \end{bmatrix}$$

together with the equation $x + y + z = 1$, or equivalently, the system

$$\begin{aligned} -0.2x + \ 0.1y + \ 0.1z &= 0 \\ 0.1x - 0.15y + 0.05z &= 0 \\ 0.1x + 0.05y - 0.15z &= 0 \\ x + \quad y + \quad z &= 1 \end{aligned}$$

Using the Gauss-Jordan elimination method, we find that the required steady-state vector is $\begin{bmatrix} \frac{1}{3} \\ \frac{1}{3} \\ \frac{1}{3} \end{bmatrix}$, and conclude that each network will comand 33 1/3 % of the audience in the long run.

25. We wish to solve

$$\begin{bmatrix} \frac{1}{2} & \frac{1}{4} & 0 \\ \frac{1}{2} & \frac{1}{2} & \frac{1}{2} \\ 0 & \frac{1}{4} & \frac{1}{2} \end{bmatrix} \begin{bmatrix} x \\ y \\ z \end{bmatrix} = \begin{bmatrix} x \\ y \\ z \end{bmatrix}$$

together with the equation $x + y + z = 1$, or, equivalently, the system of equations

$$\begin{aligned} \tfrac{1}{2}x + \tfrac{1}{4}y \qquad &= x \\ \tfrac{1}{2}x + \tfrac{1}{2}y + \tfrac{1}{2}z &= y \\ \tfrac{1}{4}y + \tfrac{1}{2}z &= z \\ x + \ y + \ z &= 1 \end{aligned}$$

Solving this system, we find that

$$x = \tfrac{1}{4},\ y = \tfrac{1}{2}, \text{ and } z = \tfrac{1}{4}.$$

Thus, in the long run, 25% of the plants will have red flowers, 50% will have pink flowers and 25% will have white flowers.

27. False. All the entries of the limiting matrix must be positive as well.

29. Let T be a regular stochastic matrix and X the steady-state distribution vector that satisfies the equation $TX = T$ and assume that the sum of the elements of X are equal to 1. Then $TX = X$ implies that $TX = T^2X$, or $X = T^2X$,.... So we have $X = T^nX$. Next, let L be the steady-state distribution vector, then

$$L = \lim_{m\to\infty} X_m = \lim_{m\to\infty} T^m X_0 = T^m X_0$$

when m is large. Multiplying both sides by T, we obtain

$$TL = T^{m+1}X_0 \approx L.$$

Thus, L also satisfies $TL = L$ together with the condition that the sum of the elements in L be equal to 1. Since the matrix equation $TX = X$ has a unique solution, we conclude that $X = L$.

USING TECHNOLOGY EXERCISES 9.2, page 534

1. $X_5 = \begin{bmatrix} 0.2045 \\ 0.1319 \\ 0.2610 \\ 0.1868 \\ 0.2158 \end{bmatrix}$

9.3 Problem Solving Tips

1. An *absorbing stochastic matrix* has at least one absorbing state and it is possible to go from any nonabsorbing state to an absorbing state in one or more stages.

2. To find the steady-state matrix of an absorbing stochastic matrix A partition the matrix A into submatrices $\left[\begin{array}{c|c} I & S \\ \hline O & R \end{array}\right]$. Then the steady-state matrix of A is given by

$\left[\begin{array}{c|c} I & S(I-R)^{-1} \\ \hline O & O \end{array}\right]$, where the order of I in the expression $(I-R)^{-1}$ is the same as the order of R.

9.3 CONCEPT QUESTIONS, page 540

1. An absorbing stochastic matrix has the following properties:
   a. There is at least one absorbing state. b. It is possible to go from any nonabsorbing state to an absorbing state in one or more stages.

EXERCISES 9.3, page 540

1. The given matrix is an absorbing stochastic matrix

$$T = \begin{array}{c} \\ 1 \\ 2 \end{array}\begin{array}{c} \begin{array}{cc} 1 & 2 \end{array} \\ \begin{bmatrix} \frac{2}{5} & 0 \\ \frac{3}{5} & 1 \end{bmatrix} \end{array}$$

   State 2 is an absorbing state. State 1 is nonabsorbing, but an object in this state has a probability of 3/5 of going to the absorbing state 2.

3. The given matrix is

$$\begin{array}{c} \\ 1 \\ 2 \\ 3 \end{array}\begin{array}{c} \begin{array}{ccc} 1 & 2 & 3 \end{array} \\ \begin{bmatrix} 1 & .5 & 0 \\ 0 & 0 & 1 \\ 0 & .5 & 0 \end{bmatrix} \end{array}$$

   States 1 and 3 are absorbing states. State 2 is not absorbing, but an object in this state has a probability of .5 of going to the absorbing state 1 and .5 of going to the absorbing state 3. Thus, the matrix is an absorbing matrix.

5. Yes. It is an absorbing stochastic matrix since it is possible to go from state 1 to the absorbing states 2 and 3.

7. The given matrix is

$$\begin{array}{c} \\ 1 \\ 2 \\ 3 \\ 4 \end{array}\begin{array}{c} \begin{array}{cccc} 1 & 2 & 3 & 4 \end{array} \\ \begin{bmatrix} 1 & 0 & .3 & 0 \\ 0 & 1 & .2 & 0 \\ 0 & 0 & .1 & .5 \\ 0 & 0 & .4 & .5 \end{bmatrix} \end{array}$$

States 1 and 2 are absorbing states. States 3 and 4 are not. However, it is possible for an object to go from state 3 to state 1 with probability 0.3. Furthermore, it is also possible for an object to go from the non-absorbing state 4 to an absorbing state. For example, via state 3 with a probability of 0.5. Therefore, the given matrix is an absorbing matrix.

9. The required matrix is $\begin{array}{c} \\ 2 \\ 1 \end{array}\begin{array}{c} \begin{array}{cc} 2 & 1 \end{array} \\ \left[\begin{array}{c|c} 1 & .4 \\ \hline 0 & .6 \end{array}\right] \end{array}$ where $S = [.4]$ and $R = [.6]$

11. $\begin{array}{c} \\ 3 \\ 2 \\ 1 \end{array}\begin{array}{c} \begin{array}{ccc} 3 & 2 & 1 \end{array} \\ \left[\begin{array}{c|cc} 1 & .4 & .5 \\ \hline 0 & .4 & .5 \\ 0 & .2 & 0 \end{array}\right] \end{array}$ where $S = [.4 \quad .5]$ and $R = \begin{bmatrix} .4 & .5 \\ .2 & 0 \end{bmatrix}$

or $\begin{array}{c} \\ 3 \\ 1 \\ 2 \end{array}\begin{array}{c} \begin{array}{ccc} 3 & 1 & 2 \end{array} \\ \left[\begin{array}{c|cc} 1 & .5 & .4 \\ \hline 0 & 0 & .2 \\ 0 & .5 & .4 \end{array}\right] \end{array}$ where $S = [.5 \quad .4]$ and $R = \begin{bmatrix} 0 & .2 \\ .5 & .4 \end{bmatrix}$.

13. $\left[\begin{array}{cc|cc} 1 & 0 & .2 & .4 \\ 0 & 1 & .3 & 0 \\ \hline 0 & 0 & .3 & .2 \\ 0 & 0 & .2 & .4 \end{array}\right]$, $S = \begin{bmatrix} .2 & .4 \\ .3 & 0 \end{bmatrix}$, $R = \begin{bmatrix} .3 & .2 \\ .2 & .4 \end{bmatrix}$

or $\left[\begin{array}{cc|cc} 1 & 0 & .4 & .2 \\ 0 & 1 & 0 & .3 \\ \hline 0 & 0 & .4 & .2 \\ 0 & 0 & .2 & .3 \end{array}\right]$, $S = \begin{bmatrix} .4 & .2 \\ 0 & .3 \end{bmatrix}$, $R = \begin{bmatrix} .4 & .2 \\ .2 & .3 \end{bmatrix}$ and so forth.

15. Rewriting the matrix so that the absorbing states appear first, we have

$$\begin{array}{c} \\ 2 \\ 1 \end{array}\begin{array}{c} \begin{array}{cc} 2 & 1 \end{array} \\ \left[\begin{array}{c|c} 1 & .45 \\ \hline 0 & .55 \end{array}\right] \end{array}$$ where $S = [.45]$ and $R = [.55]$. Then

$(I - R) = [.45]$ and $(I - R)^{-1} = [.45]\left[\dfrac{1}{.45}\right] = 1$

Therefore the steady-state matrix is $\begin{array}{c} \\ 2 \\ 1 \end{array}\begin{array}{c} \begin{array}{cc} 2 & 1 \end{array} \\ \left[\begin{array}{c|c} 1 & 1 \\ \hline 0 & 0 \end{array}\right] \end{array}$.

17. Here we have $\left[\begin{array}{c|cc} 1 & .2 & .3 \\ \hline 0 & .4 & .2 \\ 0 & .4 & .5 \end{array}\right]$ where $S = [.2 \quad .3]$ and $R = \begin{bmatrix} .4 & .4 \\ .2 & .5 \end{bmatrix}$

Next, $I - R = \begin{bmatrix} 1 & 0 \\ 0 & 1 \end{bmatrix} - \begin{bmatrix} .4 & .2 \\ .4 & .5 \end{bmatrix} = \begin{bmatrix} .6 & -.2 \\ -.4 & .5 \end{bmatrix}$

Using the formula for finding the inverse of a 2×2 matrix, we have

$$(I - R)^{-1} = \begin{bmatrix} 2.27 & .91 \\ 1.8 & 2.73 \end{bmatrix}$$

Then $S(I - R)^{-1} = [.2 \quad .3]\begin{bmatrix} 2.27 & .91 \\ 1.8 & 2.73 \end{bmatrix} = [.994 \quad 1] \approx [1 \quad 1]$

We conclude that the steady-state matrix is $\left[\begin{array}{c|cc} 1 & 1 & 1 \\ \hline 0 & 0 & 0 \\ 0 & 0 & 0 \end{array}\right]$.

19. Upon rewriting the given matrix so that the absorbing states appear first, we have

$$\begin{array}{c} \\ 2 \\ 4 \\ 1 \\ 3 \end{array}\begin{array}{c} \begin{array}{cccc} 2 & 4 & 1 & 3 \end{array} \\ \left[\begin{array}{cc|cc} 1 & 0 & \frac{1}{2} & 0 \\ 0 & 1 & 0 & 0 \\ \hline 0 & 0 & \frac{1}{2} & \frac{1}{3} \\ 0 & 0 & 0 & \frac{2}{3} \end{array}\right] \end{array}$$

where $S = \begin{bmatrix} \frac{1}{2} & 0 \\ 0 & 0 \end{bmatrix}$ and $R = \begin{bmatrix} \frac{1}{2} & \frac{1}{3} \\ 0 & \frac{2}{3} \end{bmatrix}$. Next, we compute

$$I - R = \begin{bmatrix} 1 & 0 \\ 0 & 1 \end{bmatrix} - \begin{bmatrix} \frac{1}{2} & \frac{1}{3} \\ 0 & \frac{2}{3} \end{bmatrix} = \begin{bmatrix} \frac{1}{2} & -\frac{1}{3} \\ 0 & \frac{1}{3} \end{bmatrix}.$$

Using the formula for finding the inverse of a 2×2 matrix, we have

$$(I - R)^{-1} = \begin{bmatrix} 2 & 2 \\ 0 & 3 \end{bmatrix} \text{ and so } S(I - R)^{-1} = \begin{bmatrix} 2 & 2 \\ 0 & 3 \end{bmatrix}\begin{bmatrix} \frac{1}{2} & 0 \\ 0 & 0 \end{bmatrix} = \begin{bmatrix} 1 & 1 \\ 0 & 0 \end{bmatrix}.$$

Therefore, the steady-state matrix is $\left[\begin{array}{cc|cc} 1 & 0 & 1 & 1 \\ 0 & 1 & 0 & 0 \\ \hline 0 & 0 & 0 & 0 \\ 0 & 0 & 0 & 0 \end{array}\right]$.

21. Here $\left[\begin{array}{cc|cc} 1 & 0 & \frac{1}{4} & \frac{1}{3} \\ 0 & 1 & \frac{1}{4} & \frac{1}{3} \\ \hline 0 & 0 & \frac{1}{2} & 0 \\ 0 & 0 & 0 & \frac{1}{3} \end{array}\right]$, $S = \begin{bmatrix} \frac{1}{4} & \frac{1}{3} \\ \frac{1}{4} & \frac{1}{3} \end{bmatrix}$, $R = \begin{bmatrix} \frac{1}{2} & 0 \\ 0 & \frac{1}{3} \end{bmatrix}$

and $I - R = \begin{bmatrix} 1 & 0 \\ 0 & 1 \end{bmatrix} - \begin{bmatrix} \frac{1}{2} & 0 \\ 0 & \frac{1}{3} \end{bmatrix} = \begin{bmatrix} \frac{1}{2} & 0 \\ 0 & \frac{2}{3} \end{bmatrix}$.

Using the formula for finding the inverse of a 2×2 matrix, we find

$$(I - R)^{-1} = \begin{bmatrix} 2 & 0 \\ 0 & \frac{3}{2} \end{bmatrix} \quad \text{and} \quad S(I - R)^{-1} = \begin{bmatrix} \frac{1}{4} & \frac{1}{3} \\ \frac{1}{4} & \frac{1}{3} \end{bmatrix}\begin{bmatrix} 2 & 0 \\ 0 & \frac{3}{2} \end{bmatrix} = \begin{bmatrix} \frac{1}{2} & \frac{1}{2} \\ \frac{1}{2} & \frac{1}{2} \end{bmatrix}.$$

The steady-state matrix is given by $\left[\begin{array}{cc|cc} 1 & 0 & \frac{1}{2} & \frac{1}{2} \\ 0 & 1 & \frac{1}{2} & \frac{1}{2} \\ \hline 0 & 0 & 0 & 0 \\ 0 & 0 & 0 & 0 \end{array}\right]$.

23. The absorbing states already appear first in the matrix, so it need not be rewritten. Next, $(I-R)=\begin{bmatrix} .8 & -.2 \\ -.2 & .6 \end{bmatrix}$ and $(I-R)^{-1}=\begin{bmatrix} \frac{15}{11} & \frac{5}{11} \\ \frac{5}{11} & \frac{20}{11} \end{bmatrix}$ so that

$$S(I-R)^{-1}=\begin{bmatrix} \frac{2}{10} & \frac{1}{10} \\ \frac{1}{10} & \frac{2}{10} \\ \frac{3}{10} & \frac{1}{10} \end{bmatrix}\begin{bmatrix} \frac{15}{11} & \frac{5}{11} \\ \frac{5}{11} & \frac{20}{11} \end{bmatrix}=\begin{bmatrix} \frac{7}{22} & \frac{3}{11} \\ \frac{5}{22} & \frac{9}{22} \\ \frac{5}{11} & \frac{7}{22} \end{bmatrix}$$

Therefore, the steady-state matrix is given by

$$\left[\begin{array}{ccc|cc} 1 & 0 & 0 & \frac{7}{22} & \frac{3}{11} \\ 0 & 1 & 0 & \frac{5}{22} & \frac{9}{22} \\ 0 & 0 & 1 & \frac{5}{11} & \frac{7}{22} \\ \hline 0 & 0 & 0 & 0 & 0 \\ 0 & 0 & 0 & 0 & 0 \end{array}\right].$$

25. a. State 2 is absorbing. State 1 is not absorbing, but it is possible for an object to go from state 1 to state 2 with probability .8. Therefore, the matrix is absorbing. Rewriting, we obtain

$$\begin{array}{c} \\ 1 \\ 2 \end{array}\begin{array}{c} \begin{array}{cc} 2 & 1 \end{array} \\ \left[\begin{array}{c|c} 1 & .2 \\ \hline 0 & .8 \end{array}\right] \end{array}$$ where $S=[.2]$ and $R=[.8]$.

b. We compute $I-R=[1]-[.2]=[.8]$. So $(I-R)^{-1}=[1.25]$. Therefore,

$S(I-R)^{-1}=[.8][1.25]=[1]$ and the steady state matrix is $\left[\begin{array}{c|c} 1 & 1 \\ \hline 0 & 0 \end{array}\right]$.

This result tells us that in the long run only unleaded gas will be used.

27. Here

$$\begin{array}{c} \\ \$0 \\ \$4 \\ \$1 \\ \$2 \\ \$3 \end{array}\begin{array}{c} \begin{array}{ccccc} \$0 & \$4 & \$1 & \$2 & \$3 \end{array} \\ \left[\begin{array}{cc|ccc} 1 & 0 & \frac{1}{2} & 0 & 0 \\ 0 & 1 & 0 & 0 & \frac{1}{2} \\ \hline 0 & 0 & 0 & \frac{1}{2} & 0 \\ 0 & 0 & \frac{1}{2} & 0 & \frac{1}{2} \\ 0 & 0 & 0 & \frac{1}{2} & 0 \end{array}\right] \end{array} \text{ where } S = \begin{bmatrix} \frac{1}{2} & 0 & 0 \\ 0 & 0 & \frac{1}{2} \end{bmatrix} \text{ and } R = \begin{bmatrix} 0 & \frac{1}{2} & 0 \\ \frac{1}{2} & 0 & \frac{1}{2} \\ 0 & \frac{1}{2} & 0 \end{bmatrix}.$$

Next, $I - R = \begin{bmatrix} 1 & -\frac{1}{2} & 0 \\ -\frac{1}{2} & 1 & -\frac{1}{2} \\ 0 & -\frac{1}{2} & 1 \end{bmatrix}$ and $(I - R)^{-1} = \begin{bmatrix} \frac{3}{2} & 1 & \frac{3}{2} \\ 1 & 2 & 1 \\ \frac{3}{2} & 1 & \frac{3}{2} \end{bmatrix}$ and

$$S(I - R)^{-1} = \begin{bmatrix} \frac{3}{4} & \frac{1}{2} & \frac{1}{4} \\ \frac{1}{4} & \frac{1}{2} & \frac{3}{4} \end{bmatrix}.$$

Therefore, the steady-state matrix is given by $\left[\begin{array}{cc|ccc} 1 & 0 & \frac{3}{4} & \frac{1}{2} & \frac{1}{4} \\ 0 & 1 & \frac{1}{4} & \frac{1}{2} & \frac{3}{4} \\ \hline 0 & 0 & 0 & 0 & 0 \\ 0 & 0 & 0 & 0 & 0 \\ 0 & 0 & 0 & 0 & 0 \end{array}\right]$

We conclude that if Diane started out with \$1, the probability that she would leave the game a winner is 1/4. Similarly, if she started out with \$2, the probability that she would leave the game a winner is 1/2, and if she started out with \$3, the probability that she would leave as a winner is 3/4.

29. a.

$$\left[\begin{array}{cc|cc} 1 & 0 & .25 & .1 \\ 0 & 1 & 0 & .9 \\ \hline 0 & 0 & 0 & 0 \\ 0 & 0 & .75 & 0 \end{array}\right]$$

b. $I - R = \begin{bmatrix} 1 & 0 \\ -.75 & 1 \end{bmatrix}$ and $(I - R)^{-1} = \begin{bmatrix} 1 & 0 \\ .75 & 1 \end{bmatrix}$ and

$$S(I-R)^{-1}=\begin{bmatrix} .25 & .1 \\ 0 & .9 \end{bmatrix}\begin{bmatrix} 1 & 0 \\ .75 & 1 \end{bmatrix}=\begin{bmatrix} .325 & .1 \\ .675 & .9 \end{bmatrix}$$. Therefore, the steady-state matrix is

$$\left[\begin{array}{cc|cc} 1 & 0 & .325 & .1 \\ 0 & 1 & .675 & .9 \\ \hline 0 & 0 & 0 & 0 \\ 0 & 0 & 0 & 0 \end{array}\right]$$

c. From the steady-state matrix, we see that the probability that a beginning student enrolled in the program will compete the course successfullly is 0.675.

31. False. It must be possible to go from any nonabsorbing state to an absorbing state in one or more stages.

33. The transition matrix is $\begin{array}{c} \\ aa \\ Aa \\ AA \end{array}\begin{array}{c} \begin{array}{ccc} aa & Aa & AA \end{array} \\ \left[\begin{array}{c|cc} 1 & \frac{1}{2} & 0 \\ \hline 0 & \frac{1}{2} & 1 \\ 0 & 0 & 0 \end{array}\right] \end{array}$. Since the entries in T are exactly the same as those in Example 4, the steady-state matrix is $\begin{array}{c} \\ aa \\ Aa \\ AA \end{array}\begin{array}{c} \begin{array}{ccc} aa & Aa & AA \end{array} \\ \left[\begin{array}{c|cc} 1 & 1 & 1 \\ \hline 0 & 0 & 0 \\ 0 & 0 & 0 \end{array}\right] \end{array}$

Interpreting the steady-state matrix, we see that in the long run all the flowers produced by the plants will be white.

CHAPTER 9, CONCEPT REVIEW, page 543

1. Probabilities; preceding
3. Transition
5. Distribution; steady-state
7. Absorbing; leave; stages

CHAPTER 9, REVIEW EXERCISES, page 543

1. Since the entries $a_{12}=-2$ and $a_{22}=-8$ are negative, the given matrix is not stochastic and is hence not a regular stochastic matrix.

3. $T^2 = \begin{bmatrix} \frac{1}{2} & 0 & \frac{1}{3} \\ 0 & 0 & \frac{1}{3} \\ \frac{1}{2} & 1 & \frac{1}{3} \end{bmatrix} \begin{bmatrix} \frac{1}{2} & 0 & \frac{1}{3} \\ 0 & 0 & \frac{1}{3} \\ \frac{1}{2} & 1 & \frac{1}{3} \end{bmatrix} = \begin{bmatrix} \frac{5}{12} & \frac{1}{3} & \frac{5}{18} \\ \frac{1}{6} & \frac{1}{3} & \frac{1}{9} \\ \frac{5}{12} & \frac{1}{3} & \frac{11}{18} \end{bmatrix}$ and so the matrix is regular.

5. $X_1 = \begin{bmatrix} 0 & \frac{1}{4} & \frac{3}{5} \\ \frac{2}{5} & \frac{1}{2} & \frac{1}{5} \\ \frac{3}{5} & \frac{1}{4} & \frac{1}{5} \end{bmatrix} \begin{bmatrix} \frac{1}{2} \\ \frac{1}{2} \\ 0 \end{bmatrix} = \begin{bmatrix} \frac{1}{8} \\ \frac{9}{20} \\ \frac{17}{40} \end{bmatrix}$. $X_2 = \begin{bmatrix} 0 & \frac{1}{4} & \frac{3}{5} \\ \frac{2}{5} & \frac{1}{2} & \frac{1}{5} \\ \frac{3}{5} & \frac{1}{4} & \frac{1}{5} \end{bmatrix} \begin{bmatrix} \frac{1}{8} \\ \frac{9}{20} \\ \frac{17}{40} \end{bmatrix} = \begin{bmatrix} \frac{147}{400} \\ \frac{9}{25} \\ \frac{109}{400} \end{bmatrix} = \begin{bmatrix} .3675 \\ .36 \\ .2725 \end{bmatrix}$.

7. This is an absorbing matrix since state 1 is an absorbing state and it is possible to go from any nonabsorbing state to state 1.

9. This is not an absorbing stochastic matrix since there is no absorbing matrix.

11. We solve the matrix equation

$$\begin{bmatrix} .6 & .3 \\ .4 & .7 \end{bmatrix} \begin{bmatrix} x \\ y \end{bmatrix} = \begin{bmatrix} x \\ y \end{bmatrix}$$

or equivalently, the system of equations

$$\begin{aligned} -.4x + .3y &= 0 \\ .4x - .3y &= 0 \\ x + \quad y &= 1 \end{aligned}$$

Solving this system of equations, we find the steady-state distribution vector to be

$\begin{bmatrix} \frac{3}{7} \\ \frac{4}{7} \end{bmatrix}$ and the steady-state matrix to be $\begin{bmatrix} \frac{3}{7} & \frac{3}{7} \\ \frac{4}{7} & \frac{4}{7} \end{bmatrix}$.

13. We solve the system

$$\begin{bmatrix} .6 & .4 & .3 \\ .2 & .2 & .2 \\ .2 & .4 & .5 \end{bmatrix} \begin{bmatrix} x \\ y \\ z \end{bmatrix} = \begin{bmatrix} x \\ y \\ z \end{bmatrix}$$

together with the equation $x + y + z = 1$, or equivalently, the system

$$\begin{aligned} .6x + .4y + .3z &= x \\ .2x + .2y + .2z &= y \\ .2x + .4y + .5z &= z \\ x + \quad y \quad + z &= 1 \end{aligned}$$

upon solving the system, we find the $x = .457,\ y = .20,$ and $z = .343,$

and the steady-state distribution vector is given by $\begin{bmatrix} .457 \\ .20 \\ .343 \end{bmatrix}$ and the steady-state matrix

is $\begin{bmatrix} .457 & .457 & .457 \\ .20 & .20 & .20 \\ .343 & .343 & .343 \end{bmatrix}$.

15. a. The transition matrix for the Markov Chain is given by

$$T = \begin{matrix} & \begin{matrix} A & U & N \end{matrix} \\ \begin{matrix} A \\ U \\ N \end{matrix} & \begin{bmatrix} .85 & 0 & .10 \\ .10 & .95 & .05 \\ .05 & .05 & .85 \end{bmatrix} \end{matrix}$$

b. The probability vector describing the distribution of land 10 years ago is given by

$$\begin{matrix} A \\ U \\ N \end{matrix} \begin{bmatrix} .50 \\ .15 \\ .35 \end{bmatrix}.$$

To find the required probability vector, we compute

$$TX_0 = \begin{bmatrix} .85 & 0 & .10 \\ .10 & .95 & .05 \\ .05 & .05 & .85 \end{bmatrix} \begin{bmatrix} .50 \\ .15 \\ .35 \end{bmatrix} = \begin{bmatrix} .46 \\ .21 \\ .33 \end{bmatrix}$$

$$TX_1 = \begin{bmatrix} .85 & 0 & .10 \\ .10 & .95 & .05 \\ .05 & .05 & .85 \end{bmatrix} \begin{bmatrix} .46 \\ .21 \\ .33 \end{bmatrix} = \begin{bmatrix} .424 \\ .262 \\ .314 \end{bmatrix}.$$

Thus, the probability vector describing the distribution of land 10 years from now is

$$\begin{bmatrix} .424 \\ .262 \\ .314 \end{bmatrix}.$$

CHAPTER 9 BEFORE MOVING ON, page 544

1.

$$X_1 = TX_0 = \begin{bmatrix} .3 & .4 \\ .7 & .6 \end{bmatrix}\begin{bmatrix} .6 \\ .4 \end{bmatrix} = \begin{bmatrix} .34 \\ .66 \end{bmatrix}$$

$$X_2 = TX_1 = \begin{bmatrix} .3 & .4 \\ ..7 & .6 \end{bmatrix}\begin{bmatrix} .34 \\ .66 \end{bmatrix} = \begin{bmatrix} .366 \\ .634 \end{bmatrix}$$

2. We solve the equation $TX = X$ or

$$\begin{bmatrix} \frac{1}{3} & \frac{1}{4} \\ \frac{2}{3} & \frac{3}{4} \end{bmatrix}\begin{bmatrix} x \\ y \end{bmatrix} = \begin{bmatrix} x \\ y \end{bmatrix}$$

which is equivalent to the system

$$\begin{aligned} \tfrac{1}{3}x + \tfrac{1}{4}y &= x \\ \tfrac{2}{3}x + \tfrac{3}{4}y &= y \end{aligned}$$

This system is equivalent to $\frac{2}{3}x - \frac{1}{4}y = 0$. Solving

$$\begin{aligned} \tfrac{2}{3}x - \tfrac{1}{4}y &= 0 \\ x + \quad y &= 1 \end{aligned}$$

we find $x = \frac{3}{11}$ and $y = \frac{8}{11}$. Therefore, the steady-state distribution vector is $\begin{bmatrix} \frac{3}{11} \\ \frac{8}{11} \end{bmatrix}$.

3. Rewriting the matrix so that the absorbing state appears first, we have

$$\begin{array}{c} \\ 2 \\ 3 \\ 1 \end{array}\begin{array}{c} \begin{array}{ccc} 2 & 3 & 1 \end{array} \\ \left[\begin{array}{c|cc} 1 & \frac{1}{4} & 0 \\ \hline 0 & \frac{3}{4} & \frac{2}{3} \\ 0 & 0 & \frac{1}{3} \end{array}\right] \end{array}$$

We see that $S = \begin{bmatrix} \frac{1}{4} & 0 \end{bmatrix}$ and $R = \begin{bmatrix} \frac{3}{4} & \frac{2}{3} \\ 0 & \frac{1}{3} \end{bmatrix}$.

$$I-R=\begin{bmatrix}1 & 0\\0 & 1\end{bmatrix}-\begin{bmatrix}\frac{3}{4} & \frac{2}{3}\\0 & \frac{1}{3}\end{bmatrix}=\begin{bmatrix}\frac{1}{4} & -\frac{2}{3}\\0 & \frac{2}{3}\end{bmatrix};\ (I-R)^{-1}=\begin{bmatrix}4 & 4\\0 & \frac{3}{2}\end{bmatrix}$$ and so

$$S(I-R)^{-1}=\begin{bmatrix}\frac{1}{4} & 0\end{bmatrix}\begin{bmatrix}4 & 4\\0 & \frac{3}{2}\end{bmatrix}=\begin{bmatrix}1 & 1\end{bmatrix}.$$

Therefore, the steady-state matrix of T is

$$\left[\begin{array}{c|cc}1 & 1 & 1\\\hline 0 & 0 & 0\\0 & 0 & 0\end{array}\right]$$

CHAPTER 10

EXERCISES 10.1, page 551

1. $27^{2/3} = (3^3)^{2/3} = 3^2 = 9.$

3. $\left(\frac{1}{\sqrt{3}}\right)^0 = 1$. Recall that any number raised to the zero power is 1.

5. $\left[\left(\frac{1}{8}\right)^{1/3}\right]^{-2} = \left(\frac{1}{2}\right)^{-2} = (2^2) = 4.$

7. $\left(\frac{7^{-5}\cdot 7^2}{7^{-2}}\right)^{-1} = (7^{-5+2+2})^{-1} = (7^{-1})^{-1} = 7^1 = 7.$

9. $(125^{2/3})^{-1/2} = 125^{(2/3)(-1/2)} = 125^{-1/3} = \frac{1}{125^{1/3}} = \frac{1}{5}.$

11. $\frac{\sqrt{32}}{\sqrt{8}} = \sqrt{\frac{32}{8}} = \sqrt{4} = 2.$

13. $\frac{16^{5/8}16^{1/2}}{16^{7/8}} = 16^{(5/8+1/2-7/8)} = 16^{1/4} = 2.$

15. $16^{1/4}\cdot 8^{-1/3} = 2\cdot\left(\frac{1}{8}\right)^{1/3} = 2\cdot\frac{1}{2} = 1.$

17. True.

19. False. $x^3 \times 2x^2 = 2x^{3+2} = 2x^5 \neq 2x^6$.

21. False. $\frac{2^{4x}}{1^{3x}} = \frac{2^{4x}}{1} = 2^{4x}$.

23. False. $\frac{1}{4^{-3}} = 4^3 = 64.$

25. False. $(1.2^{1/2})^{-1/2} = (1.2)^{-1/4} \neq 1.$

27. $(xy)^{-2} = \frac{1}{(xy)^2}$.

29. $\dfrac{x^{-1/3}}{x^{1/2}}=x^{(-1/3)-(1/2)}=x^{-5/6}=\dfrac{1}{x^{5/6}}.$

31. $12^0(s+t)^{-3}=1\cdot\dfrac{1}{(s+t)^3}=\dfrac{1}{(s+t)^3}.$

33. $\dfrac{x^{7/3}}{x^{-2}}=x^{(7/3)+2}=x^{(7/3)+(6/3)}=x^{13/3}.$

35. $(x^2y^{-3})(x^{-5}y^3)=(x^{2-5}y^{-3+3})=x^{-3}y^0=x^{-3}=\dfrac{1}{x^3}.$

37. $\dfrac{x^{3/4}}{x^{-1/4}}=x^{(3/4)-(-1/4)}=x^{4/4}=x.$

39. $\left(\dfrac{x^3}{-27y^{-6}}\right)^{-2/3}=x^{3(-2/3)}\left(-\dfrac{1}{27}\right)^{-2/3}y^{6(-2/3)}=x^{-2}\left(-\dfrac{1}{3}\right)^{-2}y^{-4}=\dfrac{9}{x^2y^4}.$

41. $\left(\dfrac{x^{-3}}{y^{-2}}\right)^2\left(\dfrac{y}{x}\right)^4=\dfrac{x^{-3(2)}y^4}{y^{-2(2)}x^4}=\left(\dfrac{y^{4+4}}{x^{4+6}}\right)=\dfrac{y^8}{x^{10}}.$

43. $\sqrt[3]{x^{-2}}\cdot\sqrt{4x^5}=x^{-2/3}\cdot 4^{1/2}\cdot x^{5/2}=x^{-(2/3)+(5/2)}\cdot 2=2x^{11/6}.$

45. $-\sqrt[4]{16x^4y^8}=-(16^{1/4}\cdot x^{4/4}\cdot y^{8/4})=-2xy^2.$

47. $\sqrt[6]{64x^8y^3}=(64)^{1/6}\cdot x^{8/6}y^{3/6}=2x^{4/3}y^{1/2}.$

49. $2^{3/2}=(2)(2^{1/2})=2(1.414)=2.828.$

51. $9^{3/4}=(3^2)^{3/4}=3^{6/4}=3^{3/2}=3\cdot 3^{1/2}=3(1.732)=5.196.$

53. $10^{3/2}=10^{1/2}\cdot 10=(3.162)(10)=31.62.$

55. $10^{2.5}=10^2\cdot 10^{1/2}=100(3.162)=316.2.$

57. $\dfrac{3}{2\sqrt{x}}\cdot\dfrac{\sqrt{x}}{\sqrt{x}}=\dfrac{3\sqrt{x}}{2x}.$

59. $\frac{2y}{\sqrt{3y}} \cdot \frac{\sqrt{3y}}{\sqrt{3y}} = \frac{2y\sqrt{3y}}{3y} = \frac{2}{3}\sqrt{3y}.$

61. $\frac{1}{\sqrt[3]{x}} \cdot \frac{\sqrt[3]{x^2}}{\sqrt[3]{x^2}} = \frac{\sqrt[3]{x^2}}{\sqrt[3]{x^3}} = \frac{\sqrt[3]{x^2}}{x}.$

63. $\frac{2\sqrt{x}}{3} \cdot \frac{\sqrt{x}}{\sqrt{x}} = \frac{2x}{3\sqrt{x}}.$

65. $\sqrt{\frac{2y}{x}} = \frac{\sqrt{2y}}{\sqrt{x}} \cdot \frac{\sqrt{2y}}{\sqrt{2y}} = \frac{2y}{\sqrt{2xy}}.$

67. $\frac{\sqrt[3]{x^2z}}{y} \cdot \frac{\sqrt[3]{xz^2}}{\sqrt[3]{xz^2}} = \frac{\sqrt[3]{x^3z^3}}{y\sqrt[3]{xz^2}} = \frac{xz}{y\sqrt[3]{xz^2}}.$

EXERCISES 10.2, page 559

1. $(7x^2 - 2x + 5) + (2x^2 + 5x - 4) = 7x^2 - 2x + 5 + 2x^2 + 5x - 4 = 9x^2 + 3x + 1.$

3. $(5y^2 - 2y + 1) - (y^2 - 3y - 7) = 5y^2 - 2y + 1 - y^2 + 3y + 7 = 4y^2 + y + 8.$

5. $x - \{2x - [-x - (1 - x)]\} = x - \{2x - [-x - 1 + x]\} = x - \{2x + 1\} = x - 2x - 1 = -x - 1.$

7. $(\frac{1}{3} - 1 + e) - (-\frac{1}{3} - 1 + e^{-1}) = \frac{1}{3} - 1 + e + \frac{1}{3} + 1 - \frac{1}{e} = \frac{2}{3} + e - \frac{1}{e} = \frac{3e^2 + 2e - 3}{3e}.$

9. $$3\sqrt{8} + 8 - 2\sqrt{y} + \frac{1}{2}\sqrt{x} - \frac{3}{4}\sqrt{y} = 3\sqrt{4 \cdot 2} + 8 + \frac{1}{2}\sqrt{x} - \frac{11}{4}\sqrt{y}$$
$$= 6\sqrt{2} + 8 + \frac{1}{2}\sqrt{x} - \frac{11}{4}\sqrt{y}.$$

11. $(x + 8)(x - 2) = x(x - 2) + 8(x - 2) = x^2 - 2x + 8x - 16 = x^2 + 6x - 16.$

13. $(a + 5)^2 = (a + 5)(a + 5) = a(a + 5) + 5(a + 5) = a^2 + 5a + 5a + 25$
$= a^2 + 10a + 25.$

15. $(x + 2y)^2 = (x + 2y)(x + 2y) = x(x + 2y) + 2y(x + 2y)$
$= x^2 + 2xy + 2yx + 4y^2 = x^2 + 4xy + 4y^2.$

17. $(2x + y)(2x - y) = 2x(2x - y) + y(2x - y) = 4x^2 - 2xy + 2xy - y^2 = 4x^2 - y^2$.

19. $(x^2 - 1)(2x) - x^2(2x) = 2x^3 - 2x - 2x^3 = -2x$.

21. $$2\left(t+\sqrt{t}\right)^2 - 2t^2 = 2(t+\sqrt{t})(t+\sqrt{t}) - 2t^2 = 2(t^2 + 2t\sqrt{t} + t) - 2t^2$$
$$= 2t^2 + 4t\sqrt{t} + 2t - 2t^2 = 4t\sqrt{t} + 2t = 2t(2\sqrt{t} + 1).$$

23. $4x^5 - 12x^4 - 6x^3 = 2x^3(2x^2 - 6x - 3)$.

25. $7a^4 - 42a^2b^2 + 49a^3b = 7a^2(a^2 - 6b^2 + 7ab)$.

27. $e^{-x} - xe^{-x} = e^{-x}(1 - x)$.

29. $2x^{-5/2} - \frac{3}{2}x^{-3/2} = \frac{1}{2}x^{-5/2}(4 - 3x)$.

31. $6ac + 3bc - 4ad - 2bd = 3c(2a + b) - 2d(2a + b) = (2a + b)(3c - 2d)$.

33. $4a^2 - b^2 = (2a + b)(2a - b)$. (Difference of two squares)

35. $10 - 14x - 12x^2 = -2(6x^2 + 7x - 5) = -2(3x + 5)(2x - 1)$.

37. $3x^2 - 6x - 24 = 3(x^2 - 2x - 8) = 3(x - 4)(x + 2)$.

39. $12x^2 - 2x - 30 = 2(6x^2 - x - 15) = 2(3x - 5)(2x + 3)$.

41. $9x^2 - 16y^2 = (3x)^2 - (4y)^2 = (3x - 4y)(3x + 4y)$.

43. $x^6 + 125 = (x^2)^3 + (5)^3 = (x^2 + 5)(x^4 - 5x^2 + 25)$.

45. $(x^2 + y^2)x - xy(2y) = x^3 + xy^2 - 2xy^2 = x^3 - xy^2$.

47. $$\begin{aligned} &2(x - 1)(2x + 2)^3[4(x - 1) + (2x + 2)] \\ &= 2(x - 1)(2x + 2)^3[4x - 4 + 2x + 2] \\ &= 2(x - 1)(2x + 2)^3[6x - 2] \\ &= 4(x - 1)(3x - 1)(2x + 2)^3. \end{aligned}$$

49. $4(x-1)^2(2x+2)^3(2)+(2x+2)^4(2)(x-1)$
$= 2(x-1)(2x+2)^3[4(x-1)+(2x+2)] = 2(x-1)(2x+2)^3(6x-2)$
$= 4(x-1)(3x-1)(2x+2)^3$.

51. $(x^2+2)^2[5(x^2+2)^2-3](2x) = (x^2+2)^2[5(x^4+4x^2+4)-3](2x)$
$= (2x)(x^2+2)^2(5x^4+20x^2+17)$.

53. $x^2+x-12=0$, or $(x+4)(x-3)=0$, so that $x=-4$ or $x=3$. We conclude that the roots are $x=-4$ and $x=3$.

55. $4t^2+2t-2=(2t-1)(2t+2)=0$. Thus, $t=1/2$ and $t=-1$ are the roots.

57. $\frac{1}{4}x^2-x+1=(\frac{1}{2}x-1)(\frac{1}{2}x-1)=0$. Thus $\frac{1}{2}x=1$, and $x=2$ is a double root of the equation.

59. Here we use the quadratic formula to solve the equation $4x^2+5x-6=0$. Then, $a=4$, $b=5$, and $c=-6$. Therefore,

$$x=\frac{-b\pm\sqrt{b^2-4ac}}{2a}=\frac{-(5)\pm\sqrt{(5)^2-4(4)(-6)}}{2(4)}=\frac{-5\pm\sqrt{121}}{8}=\frac{-5\pm 11}{8}.$$

Thus, $x=-\frac{16}{8}=-2$ and $x=\frac{6}{8}=\frac{3}{4}$ are the roots of the equation.

61. We use the quadratic formula to solve the equation $8x^2-8x-3=0$. Here $a=8$, $b=-8$, and $c=-3$. Therefore,

$$x=\frac{-b\pm\sqrt{b^2-4ac}}{2a}=\frac{-(-8)\pm\sqrt{(-8)^2-4(8)(-3)}}{2(8)}=\frac{8\pm\sqrt{160}}{16}$$

$$=\frac{8\pm 4\sqrt{10}}{16}=\frac{2\pm\sqrt{10}}{4}.$$

Thus, $x=\frac{1}{2}+\frac{1}{4}\sqrt{10}$ and $x=\frac{1}{2}-\frac{1}{4}\sqrt{10}$ are the roots of the equation.

63. We use the quadratic formula to solve $2x^2+4x-3=0$. Here, $a=2$, $b=4$, and $c=-3$. Therefore

$$x=\frac{-b\pm\sqrt{b^2-4ac}}{2a}=\frac{-(4)\pm\sqrt{(4)^2-4(2)(-3)}}{2(2)}=\frac{-4\pm\sqrt{40}}{4}$$

$$=\frac{-4\pm 2\sqrt{10}}{4}=\frac{-2\pm\sqrt{10}}{2}.$$

Thus, $x=-1+\frac{1}{2}\sqrt{10}$ and $x=-1-\frac{1}{2}\sqrt{10}$ are the roots of the equation.

EXERCISES 10.3, page 566

1. $$\frac{x^2+x-2}{x^2-4}=\frac{(x+2)(x-1)}{(x+2)(x-2)}=\frac{x-1}{x-2}.$$

3. $$\frac{12t^2+12t+3}{4t^2-1}=\frac{3(4t^2+4t+1)}{4t^2-1}=\frac{3(2t+1)(2t+1)}{(2t+1)(2t-1)}=\frac{3(2t+1)}{2t-1}.$$

5. $$\frac{(4x-1)(3)-(3x+1)(4)}{(4x-1)^2}=\frac{12x-3-12x-4}{(4x-1)^2}=-\frac{7}{(4x-1)^2}.$$

7. $$\frac{(2x+3)(1)-(x+1)(2)}{(2x+3)^2}=\frac{2x+3-2x-2}{(2x+3)^2}=\frac{1}{(2x+3)^2}.$$

9. $$\frac{(e^x+1)e^x-e^x(2)(e^x+1)(e^x)}{(e^x+1)^4}=\frac{e^x(e^x+1)(1-2e^x)}{(e^x+1)^4}=\frac{e^x(1-2e^x)}{(e^x+1)^3}.$$

11. $$\frac{2a^2-2b^2}{b-a}\cdot\frac{4a+4b}{a^2+2ab+b^2}=\frac{2(a+b)(a-b)4(a+b)}{-(a-b)(a+b)(a+b)}=-8.$$

13. $$\frac{3x^2+2x-1}{2x+6}\div\frac{x^2-1}{x^2+2x-3}=\frac{(3x-1)(x+1)}{2(x+3)}\cdot\frac{(x+3)(x-1)}{(x+1)(x-1)}=\frac{3x-1}{2}.$$

15. $$\frac{58}{3(3t+2)}+\frac{1}{3}=\frac{58+3t+2}{3(3t+2)}=\frac{3t+60}{3(3t+2)}=\frac{t+20}{3t+2}.$$

17. $$\frac{2x}{2x-1}-\frac{3x}{2x+5}=\frac{2x(2x+5)-3x(2x-1)}{(2x-1)(2x+5)}=\frac{4x^2+10x-6x^2+3x}{(2x-1)(2x+5)}$$

$$= \frac{-2x^2 + 13x}{(2x-1)(2x+5)} = -\frac{x(2x-13)}{(2x-1)(2x+5)}.$$

19. $(x + \frac{1}{x})(x^2 - 1) = x^3 + x - x - \frac{1}{x} = \frac{x^4 - 1}{x}.$

21. $$\frac{4}{x^2 - 9} - \frac{5}{x^2 - 6x + 9} = \frac{4}{(x+3)(x-3)} - \frac{5}{(x-3)^2}$$

$$= \frac{4(x-3) - 5(x+3)}{(x-3)^2(x+3)} = -\frac{x+27}{(x-3)^2(x+3)}.$$

23. $$2 + \frac{1}{a+2} - \frac{2a}{a-2} = \frac{2(a+2)(a-2) + a - 2 - 2a(a+2)}{(a+2)(a-2)}$$

$$= \frac{2a^2 - 8 + a - 2 - 2a^2 - 4a}{(a+2)(a-2)} = -\frac{3a+10}{(a+2)(a-2)}.$$

25. $$\frac{1 + \frac{1}{x}}{1 - \frac{1}{x}} = \frac{\frac{x+1}{x}}{\frac{x-1}{x}} = \frac{x+1}{x} \cdot \frac{x}{x-1} = \frac{x+1}{x-1}.$$

27. $$\frac{x^{-2} - y^{-2}}{x+y} = \frac{\frac{1}{x^2} - \frac{1}{y^2}}{x+y} = \frac{\frac{y^2 - x^2}{x^2 y^2}}{x+y} = \frac{(y+x)(y-x)}{x^2 y^2} \cdot \frac{1}{x+y} = \frac{y-x}{x^2 y^2}.$$

29. $$\frac{4x^2}{2\sqrt{2x^2+7}} + \sqrt{2x^2+7} = \frac{4x^2 + 2\sqrt{2x^2+7}\sqrt{2x^2+7}}{2\sqrt{2x^2+7}} = \frac{4x^2 + 4x^2 + 14}{2\sqrt{2x^2+7}}$$

$$= \frac{4x^2 + 7}{\sqrt{2x^2+7}}.$$

31. $$\frac{2x(x+1)^{-1/2}-(x+1)^{1/2}}{x^2}=\frac{(x+1)^{-1/2}(2x-x-1)}{x^2}=\frac{(x+1)^{-1/2}(x-1)}{x^2}$$
$$=\frac{x-1}{x^2\sqrt{x+1}}.$$

33. $$\frac{(2x+1)^{1/2}-(x+2)(2x+1)^{-1/2}}{2x+1}=\frac{(2x+1)^{-1/2}(2x+1-x-2)}{2x+1}$$
$$=\frac{(2x+1)^{-1/2}(x-1)}{2x+1}=\frac{x-1}{(2x+1)^{3/2}}.$$

35. (b) is equivalent to $\dfrac{a}{1+\dfrac{b}{x}}=\dfrac{a}{\dfrac{x+b}{x}}=\dfrac{ax}{x+b}$ which is (a).

37. (b) is equivalent to $\dfrac{x+\dfrac{1}{x^3}}{1-\dfrac{1}{x^3}}=\dfrac{x^4+1}{x^3}\cdot\dfrac{x^3}{x^3-1}=\dfrac{x^4+1}{x^3-1}$ which is (a).

39. $$\frac{1}{\sqrt{3}-1}\cdot\frac{\sqrt{3}+1}{\sqrt{3}+1}=\frac{\sqrt{3}+1}{3-1}=\frac{\sqrt{3}+1}{2}.$$

41. $$\frac{1}{\sqrt{x}-\sqrt{y}}\cdot\frac{\sqrt{x}+\sqrt{y}}{\sqrt{x}+\sqrt{y}}=\frac{\sqrt{x}+\sqrt{y}}{x-y}.$$

43. $$\frac{\sqrt{a}+\sqrt{b}}{\sqrt{a}-\sqrt{b}}\cdot\frac{\sqrt{a}+\sqrt{b}}{\sqrt{a}+\sqrt{b}}=\frac{(\sqrt{a}+\sqrt{b})^2}{a-b}.$$

45. $$\frac{\sqrt{x}}{3}\cdot\frac{\sqrt{x}}{\sqrt{x}}=\frac{x}{3\sqrt{x}}.$$

47. $$\frac{1-\sqrt{3}}{3}\cdot\frac{1+\sqrt{3}}{1+\sqrt{3}}=\frac{1^2-(\sqrt{3})^2}{3(1+\sqrt{3})}=-\frac{2}{3(1+\sqrt{3})}.$$

49. $$\frac{1+\sqrt{x+2}}{\sqrt{x+2}}\cdot\frac{1-\sqrt{x+2}}{1-\sqrt{x+2}}=\frac{1-(x+2)}{\sqrt{x+2}(1-\sqrt{x+2})}=-\frac{x+1}{\sqrt{x+2}(1-\sqrt{x+2})}.$$

EXERCISES 10.4, page 572

1. The statement is false because -3 is greater than -20. (See the number line that follows).

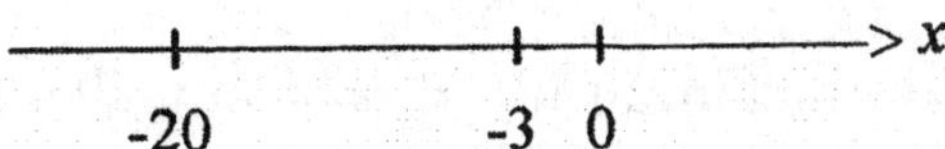

3. The statement is false because 2/3 [which is equal to (4/6)] is less than 5/6.

5. The interval (3,6) is shown on the number line that follows. Note that this is an open interval indicated by (and)

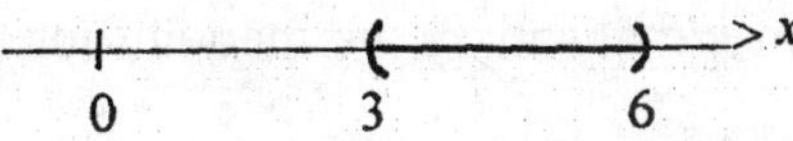

7. The interval [-1,4) is shown on the number line that follows. Note that this is a half-open interval indicated by [(closed) and) (open).

9. The infinite interval $(0,\infty)$ is shown on the number line that follows.

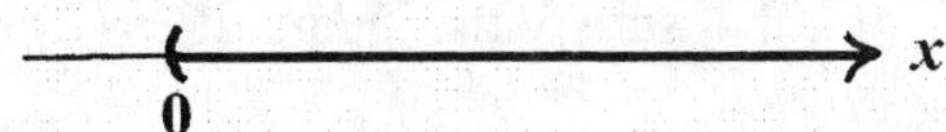

11. First, $2x + 4 < 8$ (Add -4 to each side of the inequality.)
Next, $2x < 4$ (Multiply each side of the inequality by 1/2)
and $x < 2$.

13. We are given the inequality $-4x \geq 20$.
Then $x \leq -5$. (Multiply both sides of the inequality by -1/4 and reverse the sign of the inequality.)

We write this in interval notation as $(-\infty,-5]$.

15. We are given the inequality $-6 < x - 2 < 4$.
First $\quad -6 + 2 < x < 4 + 2$ (Add +2 to each member of the inequality.)
and $\quad -4 < x < 6$, so the solution set is the open interval $(-4,6)$.

17. We want to find the values of x that satisfy the inequalities $x + 1 > 4$ or $x + 2 < -1$. Adding -1 to both sides of the first inequality, we obtain $x + 1 - 1 > 4 - 1$, or $x > 3$. Similarly, adding -2 to both sides of the second inequality, we obtain $x + 2 - 2 < -1 - 2$, or $x < -3$. Therefore, the solution set is $(-\infty,-3) \cup (3,\infty)$.

19. We want to find the values of x that satisfy the inequalities

$$x + 3 > 1 \text{ and } x - 2 < 1.$$

Adding -3 to both sides of the first inequality, we obtain

$$x + 3 - 3 > 1 - 3, \text{ or } \quad x > -2.$$

Similarly, adding 2 to each side of the second inequality, we obtain $x < 3$, and the solution set is $(-2,3)$.

21. We want to find the values of x that satisfy the inequalities $(x + 3)(x - 5) \leq 0$. From the sign diagram

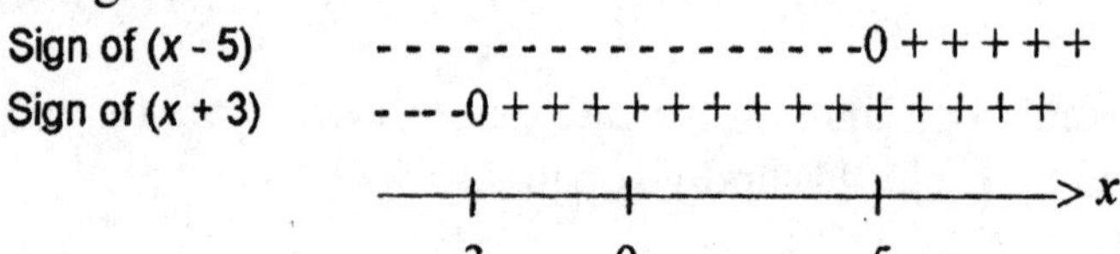

we see that the given inequality is satisfied when $-3 \leq x \leq 5$, that is, when the signs of the two factors are different or when one of the factors is equal to zero.

23. We want to find the values of x that satisfy the inequalities $(2x - 3)(x - 1) \geq 0$. From the sign diagram

Sign of $(2x - 3)$ ------------------0++
Sign of $(x - 1)$ --------------0+++++

>x

0 1 $\frac{3}{2}$

we see that the given inequality is satisfied when $x \leq 1$ or $x \geq \frac{3}{2}$; that is,when the signs of both factors are the same, or one of the factors is equal to zero.

25. We want to find the values of x that satisfy the inequalities $\dfrac{x+3}{x-2} \geq 0$. From the sign diagram we see that the given inequality is satisfied when $x \leq -3$ or $x > 2$, that is, when the signs of the two factors are the same. Notice that $x = 2$ is not included

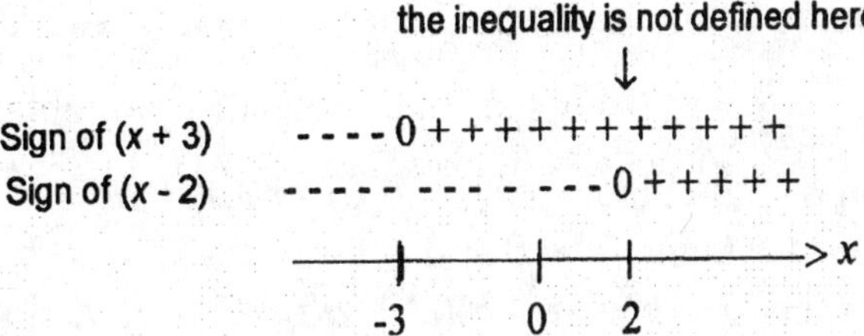

because the inequality is not defined at that value of x.

27. We want to find the values of x that satisfy the inequalities $\dfrac{x-2}{x-1} \leq 2$. Subtracting from each side of the given inequality gives

$$\frac{x-2}{x-1} - 2 \leq 0, \quad \frac{x-2-2(x-1)}{x-1} \leq 0, \text{ or } \quad -\frac{x}{x-1} \leq 0.$$

From the sign diagram

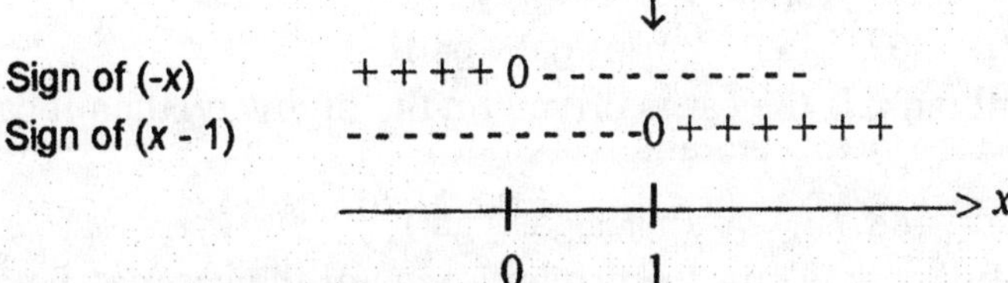

we see that the given inequality is satisfied when $x \leq 0$ or $x > 1$; that is, when the signs of the two factors differ. Notice that $x = 1$ is not included because the inequality is undefined at that value of x.

29. $|-6+2| = 4.$

31. $\dfrac{|-12+4|}{|16-12|} = \dfrac{|-8|}{4} = 2.$

33. $\sqrt{3}|-2| + 3|-\sqrt{3}| = \sqrt{3}(2) + 3\sqrt{3} = 5\sqrt{3}.$

35. $|\pi - 1| + 2 = \pi - 1 + 2 = \pi + 1.$

37. $\left|\sqrt{2}-1\right|+\left|3-\sqrt{2}\right|=\sqrt{2}-1+3-\sqrt{2}=2.$

39. False. If $a > b$, then $-a < -b$, $-a + b < -b + b$, and $b - a < 0$.

41. False. Let $a = -2$ and $b = -3$. Then $a^2 = 4$ and $b^2 = 9$, and $4 < 9$. Note that we only need to provide a counterexample to show that the statement is not always true.

43. True. There are three possible cases.
   Case 1 If $a > 0, b > 0$, then $a^3 > b^3$, since $a^3 - b^3 = (a - b)(a^2 + ab + b^2) > 0$.
   Case 2 If $a > 0, b < 0$, then $a^3 > 0$ and $b^3 < 0$ and it follows that $a^3 > b^3$.
   Case 3 If $a < 0$ and $b < 0$, then $a^3 - b^3 = (a - b)(a^2 + ab + b^2) > 0$, and we see that $a^3 > b^3$. (Note that $(a - b) > 0$ and $ab > 0$.)

45. False. Take $a = -2$, then $\left|-a\right| = \left|-(-2)\right| = \left|2\right| = 2 \neq a$.

47. True. If $a - 4 < 0$, then $\left|a-4\right| = 4 - a = \left|4-a\right|$. If $a - 4 > 0$, then
$$\left|4-a\right| = a - 4 = \left|a-4\right|.$$

49. False. Take $a = 3, b = -1$. Then $\left|a+b\right| = \left|3-1\right| = 2 \neq \left|a\right| + \left|b\right| = 3 + 1 = 4.$

51. If the car is driven in the city, then it can be expected to cover
$$(18.1)(20) = 362 \qquad \text{(miles/gal} \cdot \text{gal)}$$
or 362 miles on a full tank. If the car is driven on the highway, then it can be expected to cover
$$(18.1)(27) = 488.7 \qquad \text{(miles/gal} \cdot \text{gal)}$$
or 488.7 miles on a full tank. Thus, the driving range of the car may be described by the interval [362, 488.7].

53.
$$6(P - 2500) \leq 4(P + 2400)$$
$$6P - 15000 \leq 4P + 9600$$
$$2P \leq 24600, \text{ or } P \leq 12300.$$
Therefore, the maximum profit is \$12,300.

55. Let x represent the salesman's monthly sales in dollars. Then
$$0.15(x - 24000) \geq 6000$$
$$15(x - 24000) \geq 600000$$
$$15x - 360000 \geq 600000$$
$$15x \geq 960000$$

$$x \geq 64{,}000.$$

We conclude that the salesman must attain sales of at least \$64,000 to reach his goal.

57. The rod is acceptable if $0.49 < 0.51$ or $-0.01 < x - 0.5 < 0.01$. This gives the required inequality $|x - 0.5| < 0.01$.

59. We want to solve the inequality

$$-6x^2 + 30x - 10 \geq 14. \qquad \text{(Remember } x \text{ is expressed in thousands.)}$$

Adding -14 to both sides of this inequality, we have

$$-6x^2 + 30x - 10 - 14 \geq 14 - 14,$$

or

$$-6x^2 + 30x - 24 \geq 0.$$

Dividing both sides of the inequality by -6 (which reverses the sign of the inequality), we have $x^2 - 5x + 4 \leq 0$. Factoring this last expression, we have

$$(x - 4)(x - 1) \leq 0.$$

From the following sign diagram,

Sign of $(x - 4)$ - - - - - - - - - - - - - - - - 0 + + + + + + + +

Sign of $(x - 1)$ - - - - - - 0 + + + + + + + + + + + + + + + +

0 1 4 → x

we see that x must lie between 1 and 4. (The inequality is only satisfied when the two factors have opposite signs.) Since x is expressed in thousands of units, we see that the manufacturer must produce between 1000 and 4000 units of the commodity.

CHAPTER 10 REVIEW EXERCISES, page 574

1. $\left(\dfrac{9}{4}\right)^{3/2} = \dfrac{9^{3/2}}{4^{3/2}} = \dfrac{27}{8}.$

3. $(3 \cdot 4)^{-2} = 12^{-2} = \dfrac{1}{12^2} = \dfrac{1}{144}.$

5. $\dfrac{(3 \cdot 2^{-3})(4 \cdot 3^5)}{2 \cdot 9^3} = \dfrac{3 \cdot 2^{-3} \cdot 2^2 \cdot 3^5}{2 \cdot (3^2)^3} = \dfrac{2^{-1} \cdot 3^6}{2 \cdot 3^6} = \dfrac{1}{4}.$

7. $\dfrac{4(x^2 + y)^3}{x^2 + y} = 4(x^2 + y)^2.$

9. $\dfrac{\sqrt[4]{16x^5yz}}{\sqrt[4]{81xyz^5}} = \dfrac{(2^4x^5yz)^{1/4}}{(3^4xyz^5)^{1/4}} = \dfrac{2x^{5/4}y^{1/4}z^{1/4}}{3x^{1/4}y^{1/4}z^{5/4}} = \dfrac{2x}{3z}.$

11. $\left(\dfrac{3xy^2}{4x^3y}\right)^{-2}\left(\dfrac{3xy^3}{2x^2}\right)^3 = \left(\dfrac{3y}{4x^2}\right)^{-2}\left(\dfrac{3y^3}{2x}\right)^3 = \left(\dfrac{4x^2}{3y}\right)^2\left(\dfrac{3y^3}{2x}\right)^3 = \dfrac{(16x^4)(27y^9)}{(9y^2)(8x^3)} = 6xy^7.$

13. $\sqrt[3]{81x^5y^{10}} \cdot \sqrt[3]{9xy^2} = 3^{4/3}x^{5/3}y^{10/3} \cdot 3^{2/3}x^{1/3}y^{2/3} = 3^2x^2y^4 = 9x^2y^4.$

15. $-2\pi^2r^3 + 100\pi r^2 = -2\pi r^2(\pi r - 50).$

17. $16 - x^2 = 4^2 - x^2 = (4 - x)(4 + x).$

19. $-2x^2 - 4x + 6 = (-2x - 6)(x - 1) = -2(x + 3)(x - 1).$

21. $\dfrac{(t+6)(60)-(60t+180)}{(t+6)^2} = \dfrac{60t+360-60t-180}{(t+6)^2} = \dfrac{180}{(t+6)^2}.$

23. $\dfrac{2}{3}\left(\dfrac{4x}{2x^2-1}\right)+3\left(\dfrac{3}{3x-1}\right) = \dfrac{8x}{3(2x^2-1)}+\dfrac{9}{3x-1} = \dfrac{8x(3x-1)+27(2x^2-1)}{3(2x^2-1)(3x-1)}$

$$= \frac{78x^2-8x-27}{3(2x^2-1)(3x-1)}.$$

25. $8x^2 + 2x - 3 = (4x + 3)(2x - 1) = 0$ and $x = -3/4$ and $x = 1/2$ are the roots of the equation.

27. $-x^3 - 2x^2 + 3x = -x(x^2 + 2x - 3) = -x(x + 3)(x - 1) = 0$ and the roots of the equation are \ $x = 0$, $x = -3$, and $x = 1$.

29. Adding x to both sides yields $3 \le 3x + 9$ or $3x \ge -6$, $\Rightarrow x \ge -2$. We conclude that the solution set is $[-2,\infty)$.

31. The inequalities imply $x > 5$ or $x < -4$. So the solution is $(-\infty,-4) \cup (5,\infty)$.

33. $|-5+7|+|-2| = |2|+|-2| = 2+2 = 4.$

35. $|2\pi-6|-\pi = 2\pi-6-\pi = \pi-6.$

37. Factoring the given expression, we have $(2x - 1)(x + 2) \le 0$. From the sign diagram we conclude that the given inequality is satisfied when $-2 \le x \le \frac{1}{2}$.

Sign of (2x - 1) - - - - - - - - - - - - - - - - 0 + + + +
Sign of (x + 2) - - - - - -0 + ++ + + + + + + + + +

$$\xrightarrow{\quad -2 \qquad 0 \qquad \frac{1}{2} \quad} x$$

39. The given inequality is equivalent to $|2x - 3| < 5$ or $-5 < 2x - 3 < 5$. Thus, $-2 < 2x < 8$, or $-1 < x < 4$.

41. $$\frac{\sqrt{x}-1}{x-1} = \frac{\sqrt{x}-1}{x-1}\cdot\frac{\sqrt{x}+1}{\sqrt{x}+1} = \frac{(\sqrt{x})^2-1}{(x-1)(\sqrt{x}+1)} = \frac{x-1}{(x-1)(\sqrt{x}+1)} = \frac{1}{\sqrt{x}+1}.$$

43. Here we use the quadratic formula to solve the equation $x^2 - 2x - 5$. Then $a = 1$, $b = -2$, and $c = -5$. Thus,
$$x = \frac{-b \pm \sqrt{b^2-4ac}}{2a} = \frac{-(-2) \pm \sqrt{(-2)^2-4(1)(-5)}}{2(1)} = \frac{2 \pm \sqrt{24}}{2} = 1 \pm \sqrt{6}.$$

45. $2(1.5C + 80) \le 2(2.5C - 20) \Rightarrow 1.5C + 80 \le 2.5C - 20$, so $C \ge 100$ and the minimum cost is \$100.

CHAPTER 10, Before Moving On, page 575

1. a. $\left|\pi - 2\sqrt{3}\right| - \left|\sqrt{3} - \sqrt{2}\right| = -(\pi - 2\sqrt{3}) - (\sqrt{3} - \sqrt{2}) = \sqrt{3} + \sqrt{2} - \pi$.

   b. $\left[\left(-\frac{1}{3}\right)^{-3}\right]^{1/3} = \left(-\frac{1}{3}\right)^{(-3)(\frac{1}{3})} = \left(-\frac{1}{3}\right)^{-1} = -3$.

2. a. $\sqrt[3]{64x^6} \cdot \sqrt{9y^2x^6} = (4x^2)(3yx^3) = 12x^5y$

   b. $$\left(\frac{a^{-3}}{b^{-4}}\right)^2\left(\frac{b}{a}\right)^{-3} = \frac{a^{-6}}{b^{-8}}\cdot\frac{b^{-3}}{a^{-3}} = \frac{b^8}{a^6}\cdot\frac{a^3}{b^3} = \frac{b^5}{a^3}$$

3. a. $$\frac{2x}{3\sqrt{y}}\cdot\frac{\sqrt{y}}{\sqrt{y}} = \frac{2x\sqrt{y}}{3y}$$

   b. $$\frac{x}{\sqrt{x}-4}\cdot\frac{\sqrt{x}+4}{\sqrt{x}+4} = \frac{x(\sqrt{x}+4)}{x-16}$$

4. a. $\dfrac{(x^2+1)(\frac{1}{2}x^{-1/2})-x^{1/2}(2x)}{(x^2+1)^2}=\dfrac{\frac{1}{2}x^{-1/2}[(x^2+1)-4x^2]}{(x^2+1)^2}=\dfrac{1-3x^2}{2x^{1/2}(x^2+1)^2}$

b. $-\dfrac{3x}{\sqrt{x+2}}+3\sqrt{x+2}=-\dfrac{3x+3(x+2)}{\sqrt{x+2}}=\dfrac{6}{\sqrt{x+2}}=\dfrac{6\sqrt{x+2}}{x+2}$

5. $\dfrac{\sqrt{x}+\sqrt{y}}{\sqrt{x}-\sqrt{y}}=\dfrac{\sqrt{x}+\sqrt{y}}{\sqrt{x}-\sqrt{y}}\cdot\dfrac{\sqrt{x}-\sqrt{y}}{\sqrt{x}-\sqrt{y}}=\dfrac{x-y}{(\sqrt{x}-\sqrt{y})^2}$

6. a. $12x^3-10x^2-12x=2x(6x^2-5x-6)=2x(2x-3)(3x+2)$

b. $2bx-2by+3cx-3cy=2b(x-y)+3c(x-y)=(2b+3c)(x-y)$

7. a. $12x^2-9x-3=0$; $3(4x^2-3x-1)=0$; $3(4x+1)(x-1)=0$ so $x=-\frac{1}{4}$ or $x=1$.

b. $3x^2-5x+1=0$. Using the quadratic equations, with $a=3$, $b=-5$, and $c=1$, we have $x=\dfrac{-(-5)\pm\sqrt{25-12}}{2(3)}=\dfrac{5\pm\sqrt{13}}{6}$.

8. Factoring the given expression, we have $(3x+2)(2x-3)\le 0$. From the sign diagram we conclude that the given inequality is satisfied when $-\frac{2}{3}\le x\le\frac{3}{2}$.

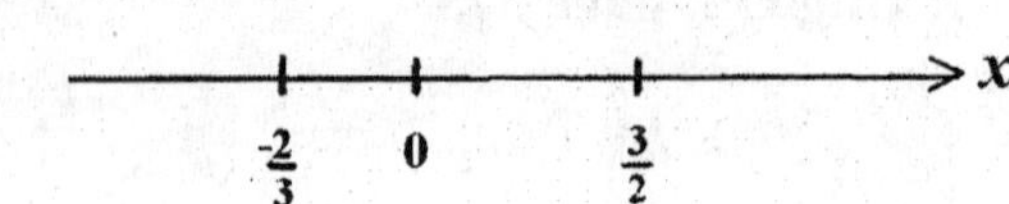

CHAPTER 11

11.1 Problem Solving Tips

New mathematical terms in each section appear in blue bold-faced type along with their definition or they are boxed (the green boxes). Each time you encounter a new term, read through the definition and then try to express the definition in your own words without looking at the book. Once you understand these definitions, it will be easier for you to work the exercise sets that follow each section.

Here are some hints for solving the problems in the exercises that follow:

1. **To find the domain of a function** $f(x)$, find all values of x for which $f(x)$ is a real number.

   a. *If the function involves a quotient*, check to see if there are any values of x at which the denominator is equal to zero. (Remember, division by zero is not allowed.) Then exclude those points from the domain.

   b. *If the function involves the root of a real number*, check to see if the root is an even or an odd root. If n is even, the nth root of a negative number is not defined and, consequently, those values of x yielding the nth root of a negative number must be excluded from the domain of f. For example, $\sqrt{x-1}$ is only defined for $x \geq 1$, so the domain of $f(x) = \sqrt{x-1}$ is $[1, \infty)$.

2. **To evaluate a piecewise-defined function** $f(x)$ at a specific value of x, check to see which subdomain x lies in. Then evaluate the function using the rule for that subdomain.

3. **To determine whether a curve is the graph of a function**, use the vertical-line test. If you can draw a vertical line through the curve that intersects the curve in more than one point, then the curve is *not* a function.

CONCEPT QUESTIONS, page 586

1. a. A function is a rule that associates with each element in a set A exactly one element in a set B.
   b. The domain of a function f is the set of all elements x in the set such that $f(x)$ is an element in B. The range of f is the set of all elements $f(x)$ whenever x is an element in its domain.
   c. An independent variable is a variable in the domain of a function f. The dependent variable is $y = f(x)$.

EXERCISES 11.1, page 586

1. $f(x) = 5x + 6$. Therefore $f(3) = 5(3) + 6 = 21$; $\quad f(-3) = 5(-3) + 6 = -9$;

   $f(a) = 5(a) + 6 = 5a + 6$; $f(-a) = 5(-a) + 6 = -5a + 6$; and

   $f(a + 3) = 5(a + 3) + 6 = 5a + 15 + 6 = 5a + 21$.

3. $g(x) = 3x^2 - 6x - 3$; $g(0) = 3(0) - 6(0) - 3 = -3$;

   $g(-1) = 3(-1)^2 - 6(-1) - 3 = 3 + 6 - 3 = 6$; $g(a) = 3(a)^2 - 6(a) - 3 = 3a^2 - 6a - 3$

   $g(-a) = 3(-a)^2 - 6(-a) - 3 = 3a^2 + 6a - 3$

   $g(x + 1) = 3(x + 1)^2 - 6(x + 1) - 3 = 3(x^2 + 2x + 1) - 6x - 6 - 3$

   $= 3x^2 + 6x + 3 - 6x - 9 = 3x^2 - 6$.

5. $f(x) = 2x + 5$; $f(a + h) = 2(a + h) + 5 = 2a + 2h + 5$. $f(-a) = 2(-a) + 5 = -2a + 5$

$f(a^2) = 2(a^2) + 5 = 2a^2 + 5$; $f(a - 2h) = 2(a - 2h) + 5 = 2a - 4h + 5$
$f(2a - h) = 2(2a - h) + 5 = 4a - 2h + 5$

7. $s(t) = \dfrac{2t}{t^2 - 1}$. Therefore, $s(4) = \dfrac{2(4)}{(4)^2 - 1} = \dfrac{8}{15}$. $s(0) = \dfrac{2(0)}{0^2 - 1} = 0$

$s(a) = \dfrac{2(a)}{a^2 - 1} = \dfrac{2a}{a^2 - 1}$; $s(2 + a) = \dfrac{2(2 + a)}{(2 + a)^2 - 1} = \dfrac{2(2 + a)}{a^2 + 4a + 4 - 1} = \dfrac{2(2 + a)}{a^2 + 4a + 3}$

$s(t + 1) = \dfrac{2(t + 1)}{(t + 1)^2 - 1} = \dfrac{2(t + 1)}{t^2 + 2t + 1 - 1} = \dfrac{2(t + 1)}{t(t + 2)}$.

9. $f(t) = \dfrac{2t^2}{\sqrt{t - 1}}$. Therefore, $f(2) = \dfrac{2(2^2)}{\sqrt{2 - 1}} = 8$; $f(a) = \dfrac{2a^2}{\sqrt{a - 1}}$;

$f(x + 1) = \dfrac{2(x + 1)^2}{\sqrt{(x + 1) - 1}} = \dfrac{2(x + 1)^2}{\sqrt{x}}$; $f(x - 1) = \dfrac{2(x - 1)^2}{\sqrt{(x - 1) - 1}} = \dfrac{2(x - 1)^2}{\sqrt{x - 2}}$.

11. Since $x = -2 \le 0$, we see that $f(-2) = (-2)^2 + 1 = 4 + 1 = 5$. Since $x = 0 \le 0$, we see that $f(0) = (0)^2 + 1 = 1$. Since $x = 1 > 0$, we see that $f(1) = \sqrt{1} = 1$.

13. Since $x = -1 < 1$, $f(-1) = -\frac{1}{2}(-1)^2 + 3 = \frac{5}{2}$. Since $x = 0 < 1$,

$f(0) = -\frac{1}{2}(0)^2 + 3 = 3$. Since $x = 1 \ge 1$, $f(1) = 2(1^2) + 1 = 3$.

Since $x = 2 \ge 1$, $f(2) = 2(2^2) + 1 = 9$.

15. a. $f(0) = -2$ b. (i) $f(x) = 3$ when $x \approx 2$ (ii) $f(x) = 0$ when $x = 1$
c. $[0,6]$ d. $[-2, 6]$

17. $g(2) = \sqrt{2^2 - 1} = \sqrt{3}$ and the point $(2, \sqrt{3})$ lies on the graph of g.

19. $f(-2) = \dfrac{|-2 - 1|}{-2 + 1} = \dfrac{|-3|}{-1} = -3$ and the point (-2,-3) does lie on the graph of f.

21. Since $f(x)$ is a real number for any value of x, the domain of f is $(-\infty, \infty)$.

23. $f(x)$ is not defined at $x = 0$ and so the domain of f is $(-\infty,0) \cup (0,\infty)$.

25. $f(x)$ is a real number for all values of x. Note that $x^2 + 1 \geq 1$ for all x. Therefore, the domain of f is $(-\infty, \infty)$.

27. Since the square root of a number is defined for all real numbers greater than or equal to zero, we have $5 - x \geq 0$, or $-x \geq -5$ and so $x \leq 5$. (Recall that multiplying by -1 reverses the sign of an inequality.)
Therefore, the domain of g is $(-\infty,5]$.

29. The denominator of f is zero when $x^2 - 1 = 0$ or $x = \pm 1$. Therefore, the domain of f is $(-\infty,-1) \cup (-1,1) \cup (1,\infty)$.

31. f is defined when $x + 3 \geq 0$, that is, when $x \geq -3$. Therefore, the domain of f is $[-3,\infty)$.

33. The numerator is defined when $1 - x \geq 0$, $-x \geq -1$ or $x \leq 1$. Furthermore, the denominator is zero when $x = \pm 2$. Therefore, the domain is the set of all real numbers in $(-\infty,-2) \cup (-2,1]$.

35. a. The domain of f is the set of all real numbers.
b. $f(x) = x^2 - x - 6$. Therefore,

$f(-3) = (-3)^2 - (-3) - 6 = 9 + 3 - 6 = 6$; $f(-2) = (-2)^2 - (-2) - 6 = 4 + 2 - 6 = 0$.

$f(-1) = (-1)^2 - (-1) - 6 = 1 + 1 - 6 = -4$; $f(0) = (0)^2 - (0) - 6 = -6$.

$f(\frac{1}{2}) = (\frac{1}{2})^2 - (\frac{1}{2}) - 6 = \frac{1}{4} - \frac{2}{4} - \frac{24}{4} = -\frac{25}{4}$; $f(1) = (1)^2 - 1 - 6 = -6$.

$f(2) = (2)^2 - 2 - 6 = 4 - 2 - 6 = -4$; $f(3) = (3)^2 - 3 - 6 = 9 - 3 - 6 = 0$.

c.

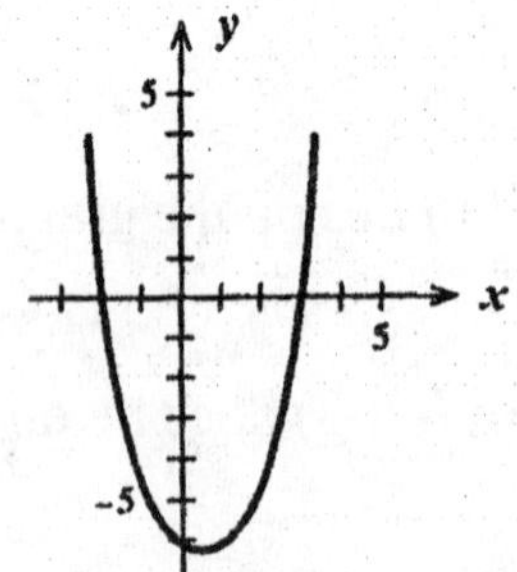

37. $f(x) = 2x^2 + 1$;

x	-3	-2	-1	0	1	2	3
$f(x)$	19	9	3	1	3	9	19

$(-\infty, \infty)$; $[1, \infty)$

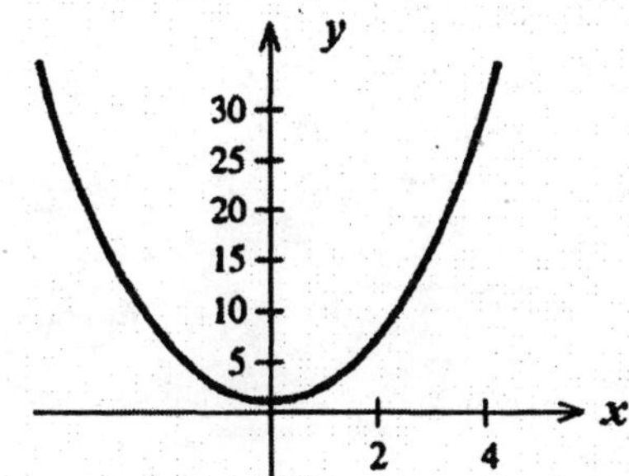

39. $f(x) = 2 + \sqrt{x}$ $[0, \infty)$; $[2, \infty)$

x	0	1	2	4	9	16
$f(x)$	2	3	3.41	4	5	6

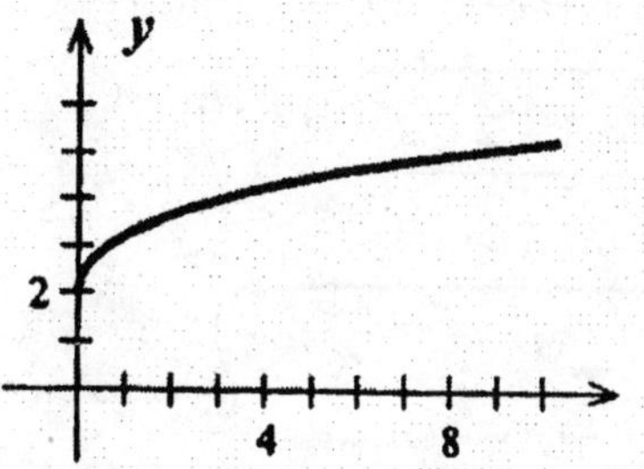

41. $f(x) = \sqrt{1 - x}$

x	0	-1	-3	-8	-15
$f(x)$	1	1.4	2	3	4

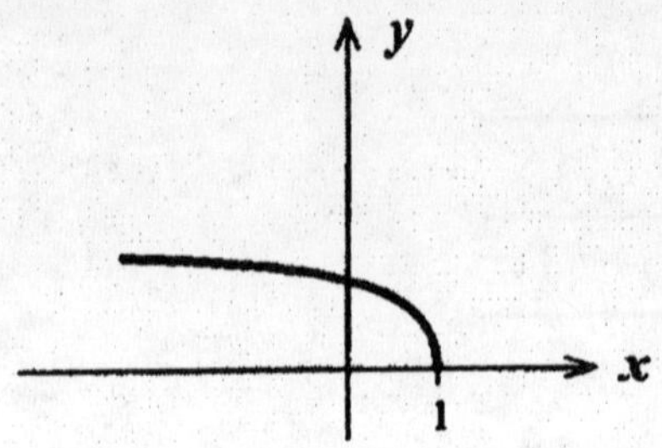

$(-\infty, 1]; [0, \infty)$

43. $f(x) = |x| - 1$

x	-3	-2	-1	0	1	2	3
$f(x)$	2	1	0	-1	0	1	2

$(-\infty, \infty); [-1, \infty)$

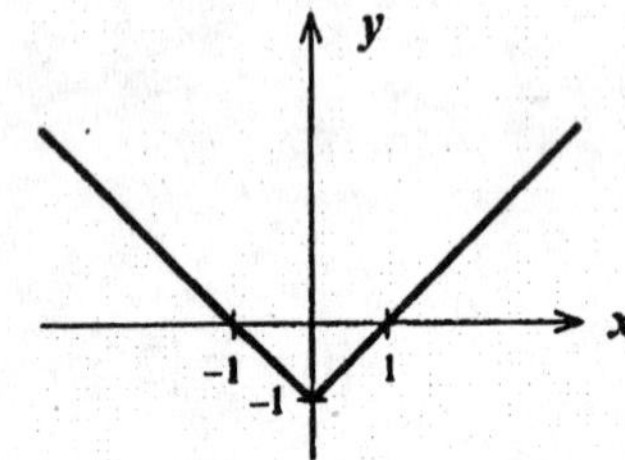

45. $f(x) = \begin{cases} x & \text{if } x < 0 \\ 2x+1 & \text{if } x \geq 0 \end{cases}$

x	-3	-2	-1	0	1	2	3
$f(x)$	-3	-2	-1	1	3	5	7

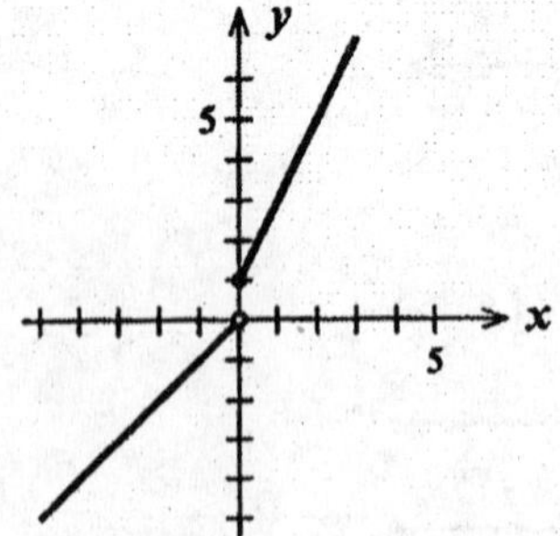

$(-\infty, \infty); (-\infty, 0) \cup [1, \infty)$

47. If $x \leq 1$, the graph of f is the half-line $y = -x + 1$. For $x > 1$, use the table

x	2	3	4
$f(x)$	3	8	15

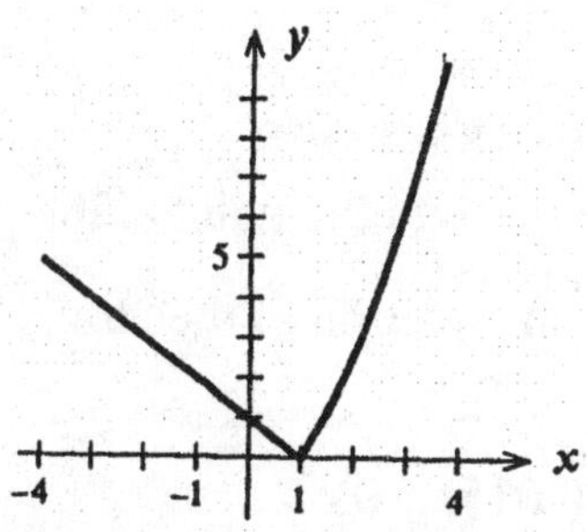

$(-\infty, \infty)$; $[0, \infty)$

49. Each vertical line cuts the given graph at exactly one point, and so the graph represents y as a function of x.

51. Since there is a vertical line that intersects the graph at three points, the graph does not represent y as a function of x.

53. Each vertical line intersects the graph of f at exactly one point, and so the graph represents y as a function of x.

55. Each vertical line intersects the graph of f at exactly one point, and so the graph represents y as a function of x.

57. The circumference of a circle with a 5-inch radius is given by

$$C(5) = 2\pi(5) = 10\pi, \text{ or } 10\pi \text{ inches.}$$

59. $\frac{4}{3}(\pi)(2r)^3 = \frac{4}{3}\pi 8r^3 = 8(\frac{4}{3}\pi r^3)$. Therefore, the volume of the tumor is increased by a factor of 8.

61. a. From $t = 0$ to $t = 5$, the graph for cassettes lies above that for CDs so from 1985 to 1990, sales of prerecorded cassettes were greater than that of CDs.
b. Sales of prerecorded CDs were greater than that of prerecorded cassettes from 1990 on.

c. The graphs intersect at the point with coordinates $x = 5$ and $y \approx 3.5$, and this tells us that the sales of the two formats were the same in 1990 with the level of sales at approximately \$3.5 billion.

63. a. The slope of the straight line passing through the points (0, 0.58) and (20, 0.95) is $m = \dfrac{0.95 - 0.58}{20 - 0} = 0.0185$, and so an equation of the straight line passing through these two points is

$$y - 0.58 = 0.0185(t - 0) \text{ or } y = 0.0185t + 0.58$$

Next, the slope of the straight line passing through the points (20, 0.95) and (30, 1.1) is $m = \dfrac{1.1 - 0.95}{30 - 20} = 0.015$, and so an equation of the straight line passing through the two points is

$$y - 0.95 = 0.015(t - 20) \text{ or } y = 0.015t + 0.65.$$

Therefore, the rule for f is

$$f(t) = \begin{cases} 0.0185t + 0.58 & 0 \le t \le 20 \\ 0.015t + 0.65 & 20 < t \le 30 \end{cases}$$

b. The ratios were changing at the rates of 0.0185/yr and 0.015/yr from 1960 through 1980, and from 1980 through 1990, respectively.

c. The ratio was 1 when $t \approx 20.3$. This shows that the number of bachelor's degrees earned by women equaled the number earned by men for the first time around 1983.

65. a. $T(x) = 0.06x$

b. $T(200) = 0.06(200) = 12$, or \$12.00; $T(5.65) = 0.06(5.65) = 0.34$, or \$0.34.

67. The child should receive $D(4) = \frac{2}{25}(500)(4) = 160$, or 160 mg.

69. a. Take $m = 7.5$ and $b = 20$, then $f(t) = 7.5t + 20 \quad (0 \le t \le 6)$.

b. $f(6) = 7.5(6) + 20 = 65$, or 65 million households.

71. a. The graph of the function is a straight line passing through (0, 120,000) and (10,0). Its slope is $m = \dfrac{0 - 120{,}000}{10 - 0} = -12{,}000$. The required equation is

$$V = -12{,}000n + 120{,}000.$$

b.

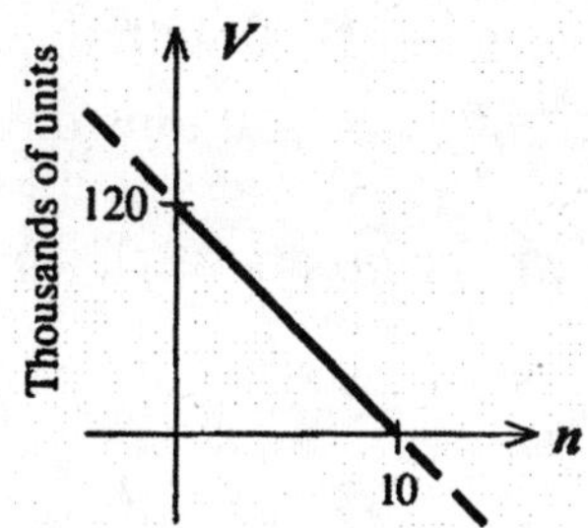

c. $V = -12{,}000(6) + 120{,}000 = 48{,}000$, or \$48,000.
d. This is given by the slope, that is, \$12,000 per year.

73. The domain of the function f is the set of all real positive numbers where $V \neq 0$; that is, $(0,\infty)$. The graph of f follows.

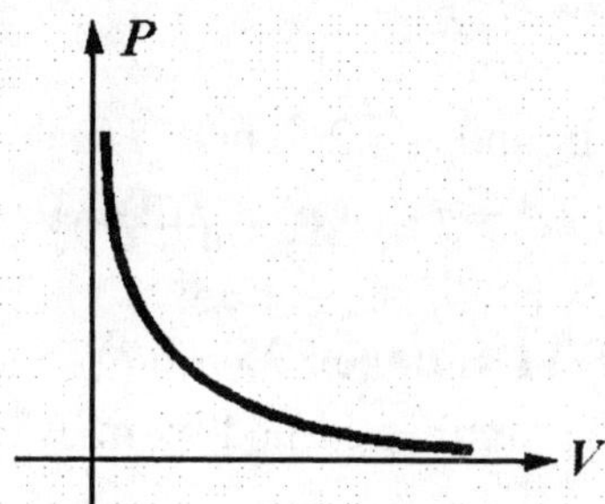

75. a. $N(0) = 3.6$ or 3.6 million people;
$N(25) = 0.0031(25)^2 + 0.16(25) + 3.6 = 9.5375$, or approximately 9.5 million people.
b. $N(30) = 0.0031(30)^2 + 0.16(30) + 3.6 = 11.19$, or approximately 11.2 million people.

77. When the proportion of popular votes won by the Democratic presidential candidate is 0.60, the proportion of seats in the House of Representatives won by Democratic candidates is given by

$$s(0.6) = \frac{(0.6)^3}{(0.6)^3 + (1-0.6)^3} = \frac{0.216}{0.216 + 0.064} = \frac{0.216}{0.280} \approx 0.77.$$

79. $N(t) = -0.0014t^3 + 0.027t^2 - 0.008t + 4.1$
a. $N(0) = 4.1$, or 4.1 million.

b. $N(12) = -0.0014(12)^3 + 0.027(12)^2 - 0.008(12) + 4.1 = 5.4728$, or 5.47 million.

81. a. The amount of solids discharged in 1989 ($t = 0$) was 130 tons/day; in 1992 ($t = 3$), it was 100 tons/day; and in 1996 ($t = 7$), it was $f(7) = 1.25(7)^2 - 26.25(7) + 162.5 = 40$, or 40 tons/day.

b.

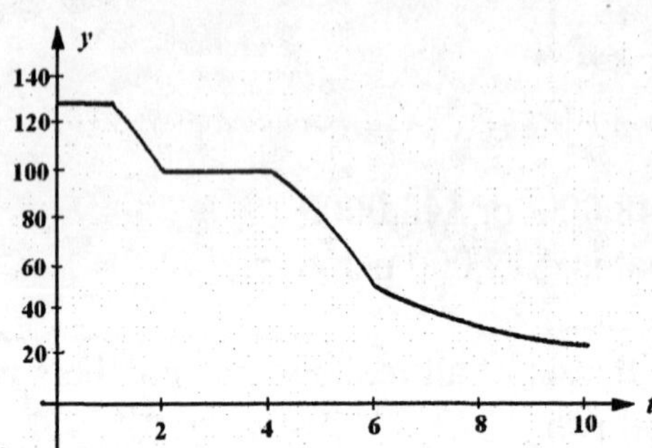

83. True, by definition of a function (page 578).

85. False. Let $f(x) = x^2$, then take $a = 1$, and $b = 2$. Then $f(a) = f(1) = 1$ and $f(b) = f(2) = 4$ and $f(a) + f(b) = 1 + 4 \neq f(a+b) = f(3) = 9$.

USING TECHNOLOGY EXERCISES 11.1, page 595

1.

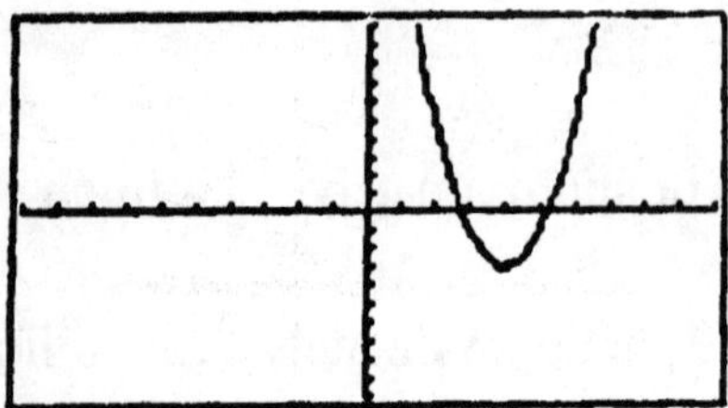

3.

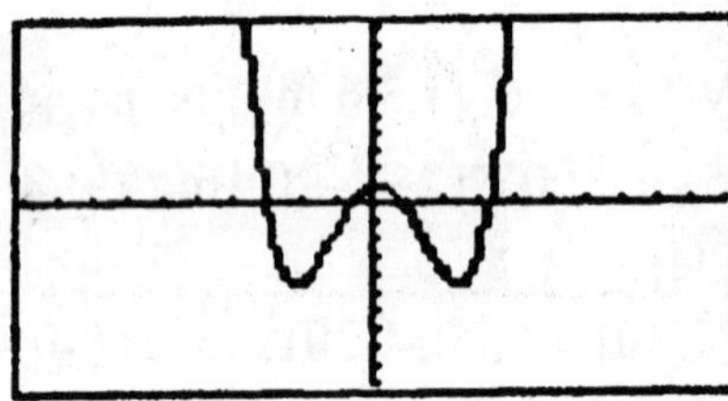

5. a.

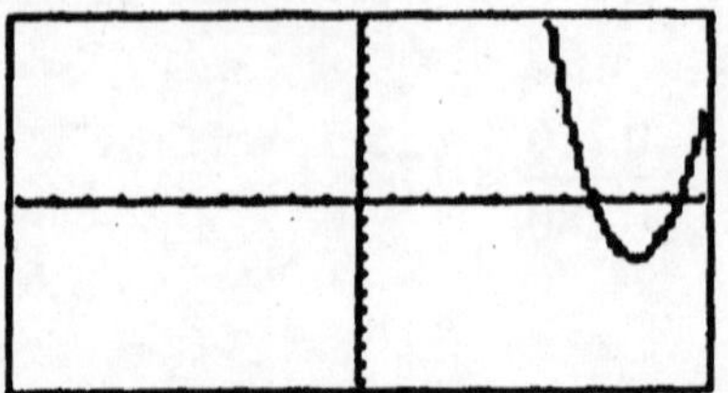

b.

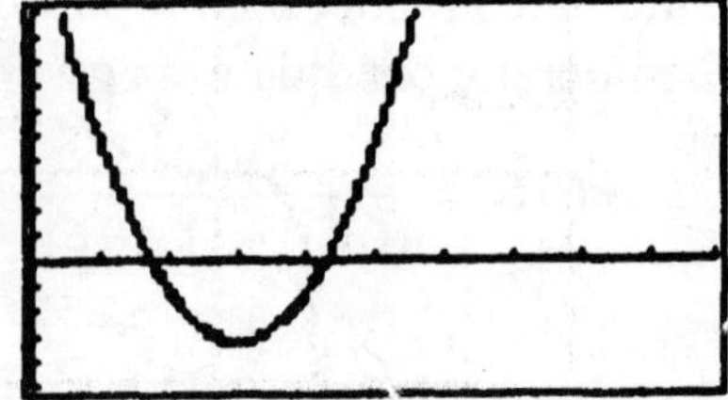

7. a.

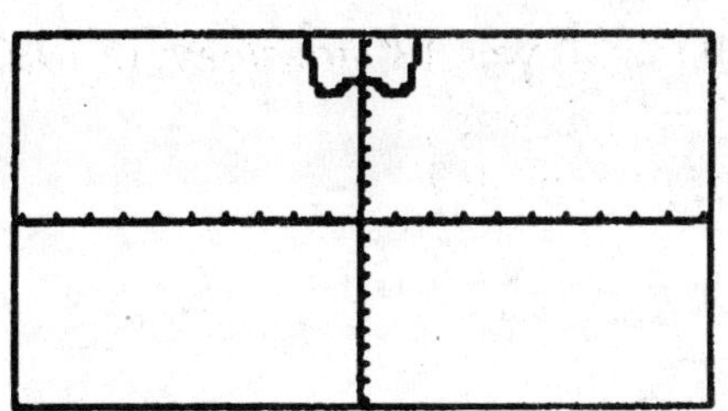

b.

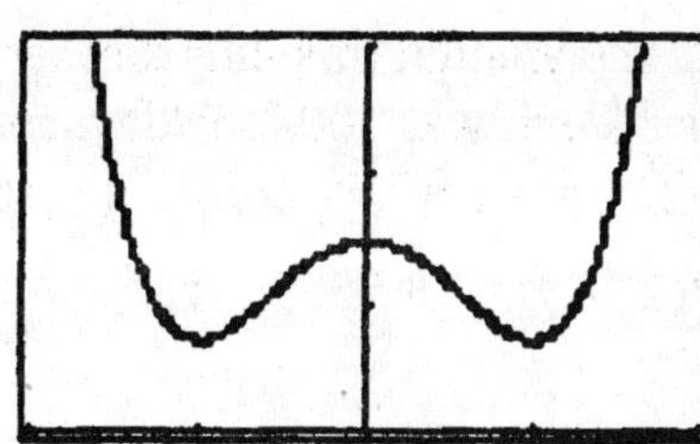

9. a.

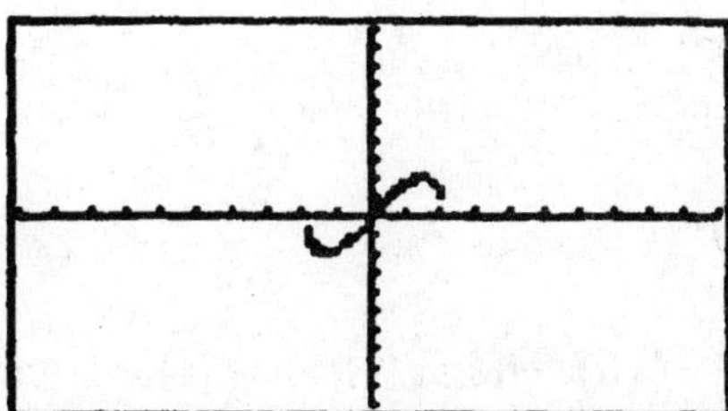

b.

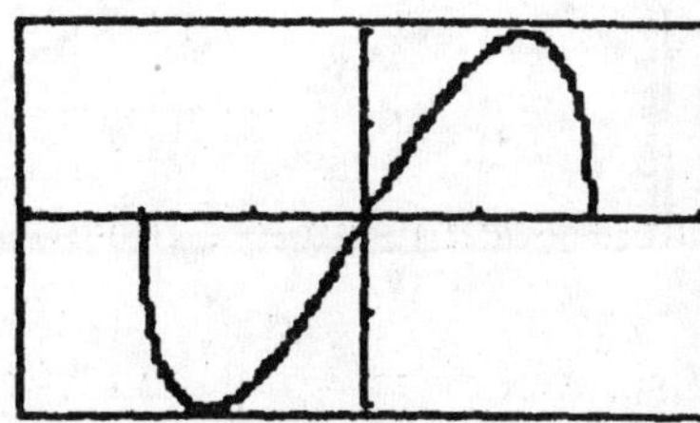

11.

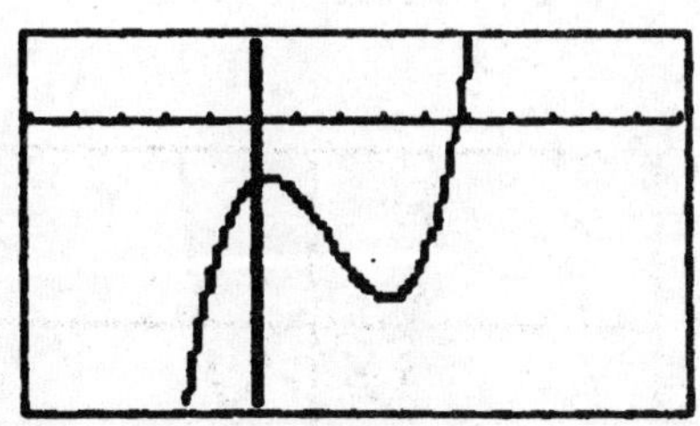

13.

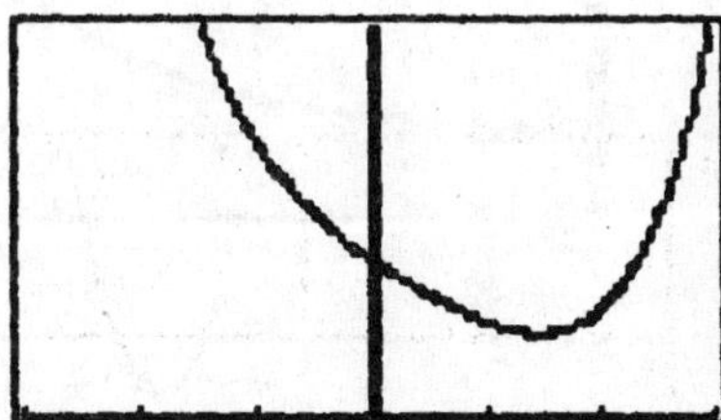

15.

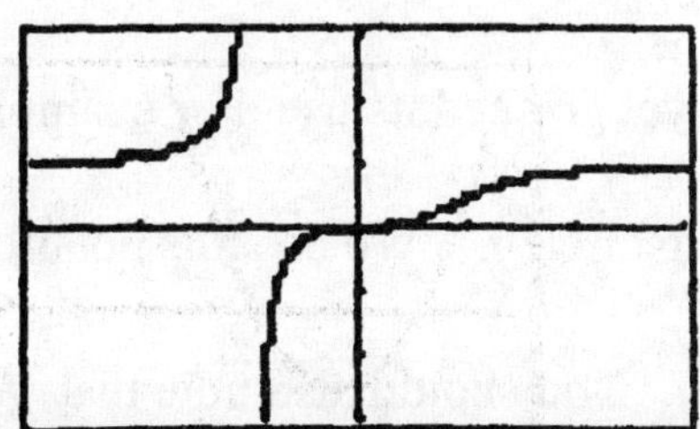

17. 18; $f(-1) = -3(-1)^3 + 5(-1)^2 - 2(-1) + 8 = 3 + 5 + 2 + 8 = 18.$

19. 2; $f(1) = \dfrac{(1)^4 - 3(1)^2}{1-2} = \dfrac{1-3}{-1} = 2.$

21. $f(2.145) \approx 18.5505$

23. $f(2.41) \approx 4.1616$

25. a.

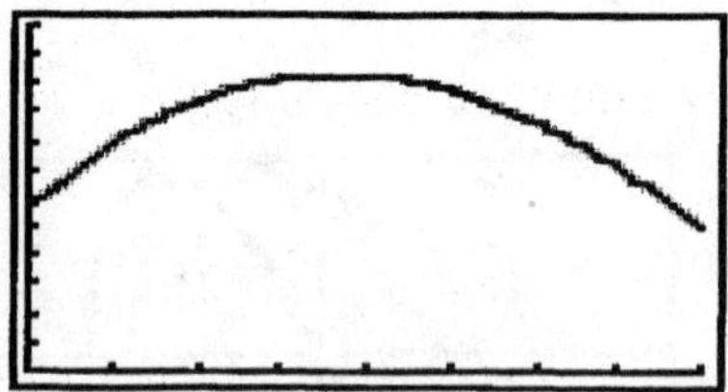

b. $f(2) \approx 9.4066$, or approximately 9.41%/yr
$f(6) \approx 8.7062$, or approximately 8.71%/yr.

27. a.

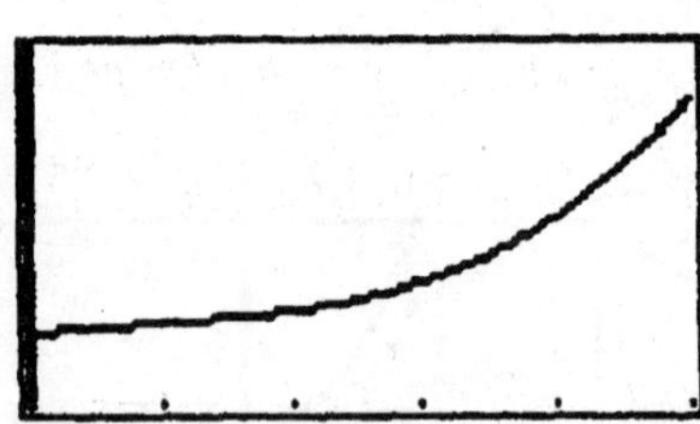

b. $f(6) = 44.7;$

$f(8) = 52.7;$r

$f(11) = 129.2.$

11.2 Problem Solving Tips

When you come across new notation, make sure that you understand that notation. If you can't express the notation verbally, you haven't yet grasped its use. For example, in this section we introduced the notation $g \circ f$, read "*g* circle *f*." We use this notation to describe the composition of the functions *g* and *f*. You should also note that $g \circ f$ is different from $f \circ g$.

Here are some hints for solving the problems in the exercises that follow:

1. If f and g are functions with domains A and B, respectively, then **the domain of** $f+g, f-g,$ **and** fg is $A \cap B$. The **domain of the quotient** f/g is $A \cap B$ excluding all numbers x such that $g(x)=0$.

2. **To find the rule for the composite function** $g \circ f$, evaluate the function g at $f(x)$. Similarly, to find $f \circ g$ evaluate the function f at $g(x)$.

11.2 CONCEPT QUESTIONS, page 601

1. a. $(f+g)(x)=f(x)+g(x)$; $(f-g)(x)=f(x)-g(x)$; $(fg)(x)=f(x)g(x)$; all with domain $A \cap B$. $(f/g)(x)=\dfrac{f(x)}{g(x)}$, domain $A \cap B$ excluding $x \in A \cap B$ such that $g(x)=0$.

   b. $(f+g)(2)=f(2)+g(2)=3+(-2)=1; (f-g)(2)=f(2)-g(2)=3-(-2)=5;$;
   $(fg)(2)=f(2)g(2)=3(-2)=-6; (f/g)(2)=\dfrac{f(2)}{g(2)}=\dfrac{3}{-2}=-\dfrac{3}{2}$

EXERCISES 11.2, page 601

1. $(f+g)(x)=f(x)+g(x)=(x^3+5)+(x^2-2)=x^3+x^2+3.$

3. $fg(x)=f(x)g(x)=(x^3+5)(x^2-2)=x^5-2x^3+5x^2-10.$

5. $\dfrac{f}{g}(x)=\dfrac{f(x)}{g(x)}=\dfrac{x^3+5}{x^2-2}.$

7. $\frac{fg}{h}(x) = \frac{f(x)g(x)}{h(x)} = \frac{(x^3+5)(x^2-2)}{2x+4} = \frac{x^5-2x^3+5x^2-10}{2x+4}$

9. $(f+g)(x) = f(x)+g(x) = x-1+\sqrt{x+1}.$

11. $(fg)(x) = f(x)g(x) = (x-1)\sqrt{x+1}$

13. $\frac{g}{h}(x) = \frac{g(x)}{h(x)} = \frac{\sqrt{x+1}}{2x^3-1}.$

15. $\frac{fg}{h}(x) = \frac{(x-1)(\sqrt{x+1})}{2x^3-1}$

17. $\frac{f-h}{g}(x) = \frac{x-1-(2x^3-1)}{\sqrt{x+1}} = \frac{x-2x^3}{\sqrt{x+1}}.$

19. $(f+g)(x) = x^2+5+\sqrt{x}-2 = x^2+\sqrt{x}+3.$

$(f-g)(x) = x^2+5-(\sqrt{x}-2) = x^2-\sqrt{x}+7.$

$(fg)(x) = (x^2+5)(\sqrt{x}-2); (\frac{f}{g})(x) = \frac{x^2+5}{\sqrt{x}-2}.$

21. $(f+g)(x) = \sqrt{x+3}+\frac{1}{x-1} = \frac{(x-1)\sqrt{x+3}+1}{x-1}.$

$(f-g)(x) = \sqrt{x+3}-\frac{1}{x-1} = \frac{(x-1)\sqrt{x+3}-1}{x-1}.$

$(fg)(x) = \sqrt{x+3}\left(\frac{1}{x-1}\right) = \frac{\sqrt{x+3}}{x-1}. \quad (\frac{f}{g}) = \sqrt{x+3}(x-1).$

23. $(f+g)(x) = \frac{x+1}{x-1}+\frac{x+2}{x-2} = \frac{(x+1)(x-2)+(x+2)(x-1)}{(x-1)(x-2)}$

$= \frac{x^2-x-2+x^2+x-2}{(x-1)(x-2)} = \frac{2x^2-4}{(x-1)(x-2)} = \frac{2(x^2-2)}{(x-1)(x-2)}.$

$(f-g)(x) = \frac{x+1}{x-1}-\frac{x+2}{x-2} = \frac{(x+1)(x-2)-(x+2)(x-1)}{(x-1)(x-2)}$

$= \frac{x^2-x-2-x^2-x+2}{(x-1)(x-2)} = \frac{-2x}{(x-1)(x-2)}.$

$$(fg)(x)=\frac{(x+1)(x+2)}{(x-1)(x-2)};\ (\frac{f}{g})(x)=\frac{(x+1)(x-2)}{(x-1)(x+2)}.$$

25. $(f\circ g)(x)=f(g(x))=f(x^2)=(x^2)^2+x^2+1=x^4+x^2+1.$
$(g\circ f)(x)=g(f(x))=g(x^2+x+1)=(x^2+x+1)^2.$

27. $(f\circ g)(x)=f(g(x))=f(x^2-1)=\sqrt{x^2-1}+1.$
$(g\circ f)(x)=g(f(x))=g(\sqrt{x}+1)=(\sqrt{x}+1)^2-1=x+2\sqrt{x}+1-1=x+2\sqrt{x}.$

29. $(f\circ g)(x)=f(g(x))=f\left(\frac{1}{x}\right)=\frac{1}{x}\div\left(\frac{1}{x^2}+1\right)=\frac{1}{x}\cdot\frac{x^2}{x^2+1}=\frac{x}{x^2+1}.$
$(g\circ f)(x)=g(f(x))=g\left(\frac{x}{x^2+1}\right)=\frac{x^2+1}{x}.$

31. $h(2)=g[f(2)]$. But $f(2)=2^2+2+1=7$, so $h(2)=g(7)=49.$

33. $h(2)=g[f(2)]$. But $f(2)=\frac{1}{2(2)+1}=\frac{1}{5}$, so $h(2)=g(\frac{1}{5})=\frac{1}{\sqrt{5}}=\frac{\sqrt{5}}{5}.$

35. $f(x)=2x^3+x^2+1,\ g(x)=x^5.$

37. $f(x)=x^2-1,\ g(x)=\sqrt{x}.$

39. $f(x)=x^2-1,\ g(x)=\frac{1}{x}.$

41. $f(x)=3x^2+2,\ g(x)=\frac{1}{x^{3/2}}.$

43. $f(a+h)-f(a)=[3(a+h)+4]-(3a+4)=3a+3h+4-3a-4=3h.$

45. $f(a+h)-f(a)=4-(a+h)^2-(4-a^2)$
$=4-a^2-2ah-h^2-4+a^2=-2ah-h^2=-h(2a+h).$

47. $\frac{f(a+h)-f(a)}{h}=\frac{[(a+h)^2+1]-(a^2+1)}{h}=\frac{a^2+2ah+h^2+1-a^2-1}{h}=\frac{2ah+h^2}{h}$

$$= \frac{h(2a+h)}{h} = 2a+h.$$

49. $\frac{f(a+h)-f(a)}{h} = \frac{[(a+h)^3-(a+h)]-(a^3-a)}{h}$

$$= \frac{a^3+3a^2h+3ah^2+h^3-a-h-a^3+a}{h}$$

$$= \frac{3a^2h+3ah^2+h^3-h}{h} = 3a^2+3ah+h^2-1.$$

51. $\frac{f(a+h)-f(a)}{h} = \frac{\frac{1}{a+h}-\frac{1}{a}}{h} = \frac{\frac{a-(a+h)}{a(a+h)}}{h} = -\frac{1}{a(a+h)}.$

53. $F(t)$ represents the total revenue for the two restaurants at time t.

55. $f(t)g(t)$ represents the (dollar) value of Nancy's holdings at time t.

57. $g \circ f$ is the function giving the amount of carbon monoxide pollution at time t.

59. $C(x) = 0.6x + 12{,}100$.

61. a. $f(t) = 267;\ g(t) = 2t^2 + 46t + 733$
   b. $h(t) = (f+g)(t) = f(t) + g(t) = 267 + (2t^2 + 46t + 733) = 2t^2 + 46t + 1000$
   c. $h(13) = 2(13)^2 + 46(13) + 1000 = 1936$, or 1936 tons.

63. a. $P(x) = R(x) - C(x)$
$= -0.1x^2 + 500x - (0.000003x^3 - 0.03x^2 + 200x + 100{,}000)$

$= -0.000003x^3 - 0.07x^2 + 300x - 100{,}000.$
   b. $P(1500) = -0.000003(1500)^3 - 0.07(1500)^2 + 300(1500) - 100{,}000$
$= 182{,}375$ or \$182,375.

65. a. The gap is
$$G(t) - C(t) = (3.5t^2 + 26.7t + 436.2) - (24.3t + 365)$$
$$= 3.5t^2 + 2.4t + 71.2.$$

b. At the beginning of 1983, the gap was

$$G(0) = 3.5(0)^2 + 2.4(0) + 71.2 = 71.2, \text{ or } 71{,}200.$$

At the beginning of 1986, the gap was

$$G(3) = 3.5(3)^2 + 2.4(3) + 71.2 = 109.9, \text{ or } 109{,}900.$$

67. a. The occupancy rate at the beginning of January is

$$r(0) = \frac{10}{81}(0)^3 - \frac{10}{3}(0)^2 + \frac{200}{9}(0) + 55 = 55, \text{ or 55 percent.}$$

$$r(5) = \frac{10}{81}(5)^3 - \frac{10}{3}(5)^2 + \frac{200}{9}(5) + 55 = 98.2, \text{ or 98.2 percent.}$$

b. The monthly revenue at the beginning of January is

$$R(55) = -\frac{3}{5000}(55)^3 + \frac{9}{50}(55)^2 = 444.68, \text{ or } \$444{,}700.$$

The monthly revenue at the beginning of June is

$$R(98.2) = -\frac{3}{5000}(98.2)^3 + \frac{9}{50}(98.2)^2 = 1167.6, \text{ or } \$1{,}167{,}600.$$

69. True. $(f+g)(x) = f(x) + g(x) = g(x) + f(x) = (g+f)(x)$.

71. False. Take $f(x) = \sqrt{x}$ and $g(x) = x+1$. Then $(g \circ f)(x) = \sqrt{x}+1$, but $(f \circ g)(x) = \sqrt{x+1}$.

11.3 Problem Solving Tips

When you solve a problem involving a function, it is helpful to identify the type of function you are working with. For example, if you wish to find the domain of a *polynomial function* you know that there are no restrictions on the domain, since a polynomial is defined for all real numbers. If you want to find the domain of a *rational function*, you know that you have to check to see if there are any values for which the denominator is equal to 0.

Here are some hints for solving the problems in the exercises that follow:

1. To find the market equilibrium of a commodity, find the point of intersection of the supply and demand equations for the commodity. (Market equilibrium prevails when the quantity produced is equal to the quantity demanded.).

2. To construct a mathematical model, follow the guidelines given in the text on page 84. First try solving Examples 5 and 6 in the text without looking at the solution. Then go on to try a few similar problems (#72-80, on page 90-91 of the text).

11.3 CONCEPT QUESTIONS, page 611

1. (Answers will vary).

3. a. A demand function $p = D(x)$ gives the relationship between the unit price of a commodity, *p*, and the quantity, *x*, demanded. A supply function $p = S(x)$ gives the relationship between the unit price of a commodity, *p*, and the quantity, *x*, the supplier will make available in the market place.

   b. Market equilibrium occurs when the quantity produced is equal to the quantity demanded. To find the market equilibrium, we solve the equations $p = D(x)$ and $p = S(x)$ simultaneously.

EXERCISES 11.3, page 611

1. *f* is a polynomial function in *x* of degree 6.

3. Expanding $G(x) = 2(x^2 - 3)^3$, we have $G(x) = 2x^6 - 18x^4 + 54x^2 - 54$, and we conclude that *G* is a polynomial function in *x* of degree 6.

5. *f* is neither a polynomial nor a rational function.

7. The individual's disposable income is $D = (1 - 0.28)60{,}000 = 43{,}200$, or \$43,200.

9. $P(28) = -\frac{1}{8}(28)^2 + 7(28) + 30 = 128$, or \$128,000.

11. $S(6) = 0.73(6)^2 + 15.8(6) + 2.7 = 123.78$ (million)
    $S(8) = 0.73(8)^2 + 15.8(8) + 2.7 = 175.82$ (million).

13. $N(0) = 0.7$ (per 100 million vehicle miles driven)
    $N(7) = 0.0336(7)^3 - 0.118(7)^2 + 0.215(7) + 0.7 = 7.9478$ per 100 million vehicle miles driven.

15. a. $N(0) = 0.32$ or 320,000
    b. $N(4) = -0.0675(4)^4 + 0.5083(4)^3 - 0.893(4)^2 + 0.66(4) + 0.32 = 3.9232$ or 3,923,200.

17. $N(5) = 0.0018425(10)^{2.5} \approx 0.58265$, or approximately 0.583 million.
    $N(10) = 0.0018425(15)^{2.5} \approx 1.6056$, or approximately 1.606 million.

19. a.

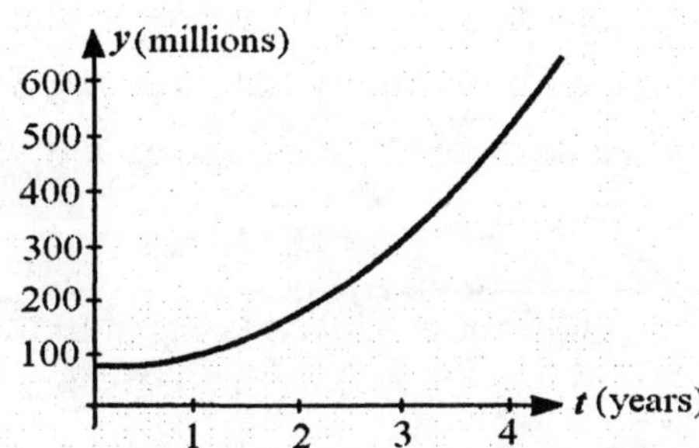

b. $f(4) = 38.57(4^2) - 24.29(4) + 79.14 = 599.1$, or 599,100,000.

21. a. The given data implies that $R(40) = 50$, that is,

$$\frac{100(40)}{b+40} = 50$$

$$50(b+40) = 4000, \text{ or } b = 40.$$

Therefore, the required response function is $R(x) = \dfrac{100x}{40+x}$.

b. The response will be $R(60) = \dfrac{100(60)}{40+60} = 60$, or approximately 60 percent.

23. The average U.S. credit card debt at the beginning of 1994 was

$$D(0) = 4.77(1+0)^{0.2676} = 4.77 \text{ or } \$4770.$$

At the beginning of 1996, it was $D(2) = 4.77(1+2)^{0.2676} = 6.400$ or \$6400. At the beginning of 1999, it was

$$D(5) = 5.6423(5^{0.1818}) \approx 7.560 \text{ or } \$7560.$$

25. a. $A(0) = 16.4$, or \$16.4 billion.

$A(1) = 16.4(1+1)^{0.1} \approx 17.58$, or \$17.58 billion.

$A(2) = 16.4(2+1)^{0.1} \approx 18.30$, or \$18.3 billion.

$A(3) = 16.4(3+1)^{0.1} \approx 18.84$, ot \$18.84 billion.

$A(4) = 16.4(4+1)^{0.1} \approx 19.26$, or \$19.26 billion.

The nutritional market has been growing over the years from 1999 through 2003.

b.

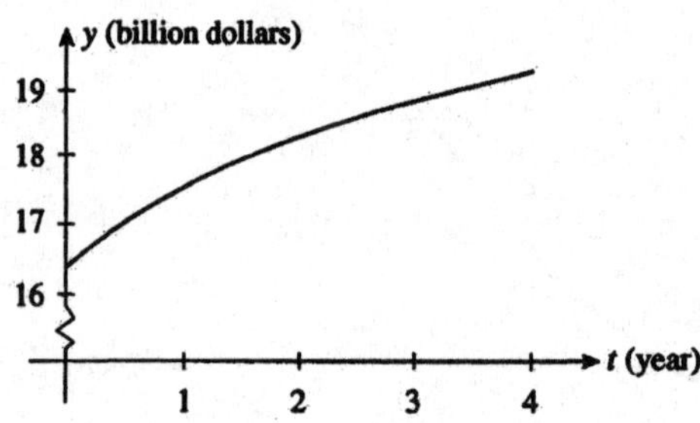

27. a.

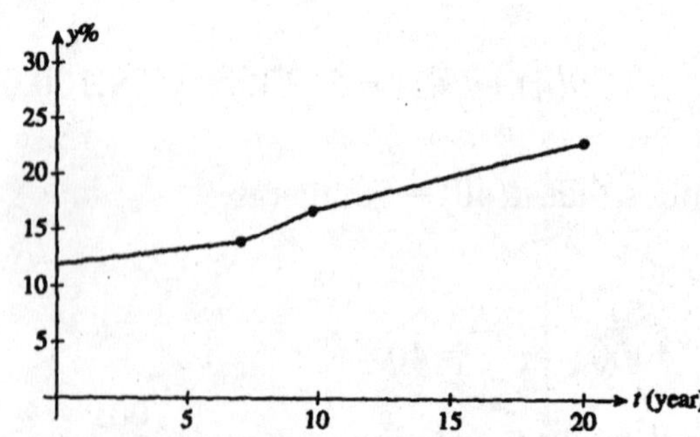

b. $f(5) = \frac{2}{7}(5) + 12 = \frac{10}{7} + 12 \approx 13.43$, or 13.43 percent

$f(15) = \frac{3}{5}(15) + 11 = 20$, or 20 percent.

29. $h(t) = f(t) - g(t) = \dfrac{110}{\frac{1}{2}t + 1} - 26(\frac{1}{4}t^2 - 1)^2 - 52.$

$h(0) = f(0) - g(0) = \dfrac{110}{\frac{1}{2}(0) + 1} - 26\left[\frac{1}{4}(0)^2 - 1\right]^2 - 52 = 110 - 26 - 52 = 32$, or \$32.

$h(1) = f(1) - g(1) = \dfrac{110}{\frac{1}{2}(1) + 1} - 26\left[\frac{1}{4}(1)^2 - 1\right]^2 - 52 = 6.71$, or \$6.71.

$h(2) = f(2) - g(2) = \dfrac{110}{\frac{1}{2}(2) + 1} - 26\left[\frac{1}{4}(2)^2 - 1\right]^2 - 52 = 3$, or \$3.

We conclude that the price gap was narrowing.

31. a. $P(0) = 59.8$; $P(1) = 0.3(1) + 58.6 = 58.9$; $P(2) = 56.79(2)^{0.06} = 59.2$;
$P(3) = 56.79(3)^{0.06} = 60.7$; $P(4) = 56.79(4)^{0.06} = 61.7$

b.

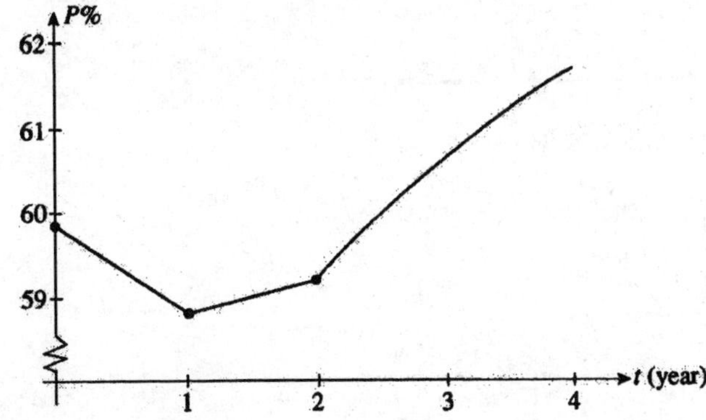

c. $P(3) = 60.7$, or 60.7%.

33. a.

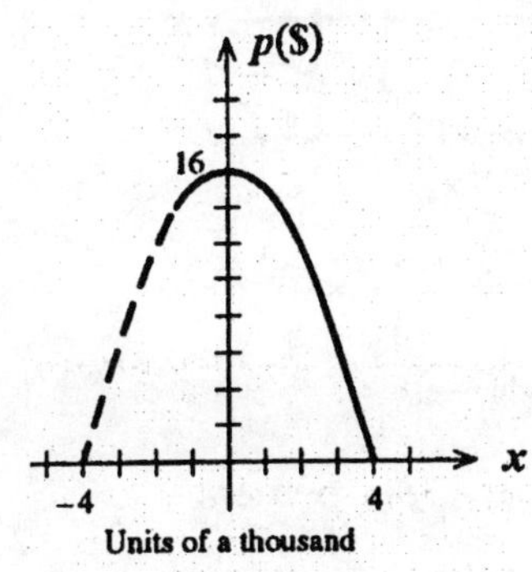

b. If $p = 7$, we have $7 = -x^2 + 16$, or $x^2 = 9$, so that $x = \pm 3$. Therefore, the quantity demanded when the unit price is \$7 is 3000 units.

35. a.

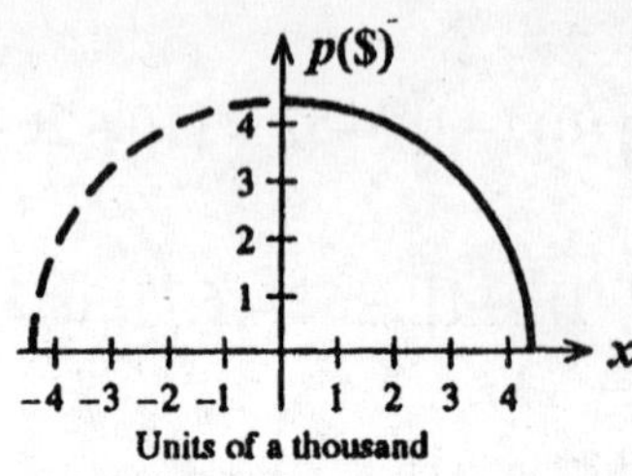

b. If $p = 3$, then $3 = \sqrt{18 - x^2}$, and $9 = 18 - x^2$, so that $x^2 = 9$ and $x = \pm 3$. Therefore, the quantity demanded when the unit price is \$3 is 3000 units.

37. a.

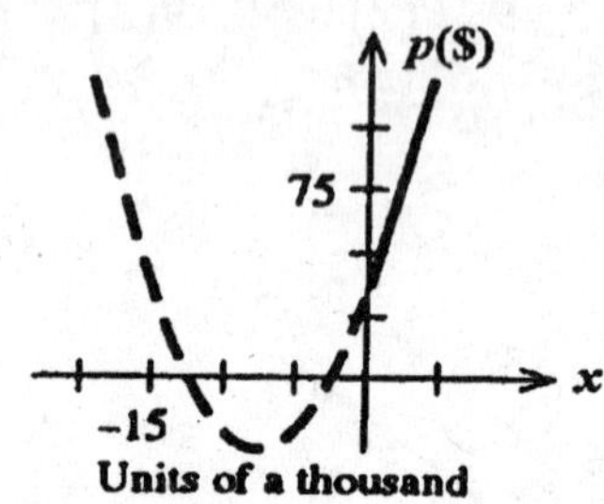

b. If $x = 2$, then $p = 2^2 + 16(2) + 40 = 76$, or \$76.

39. a.

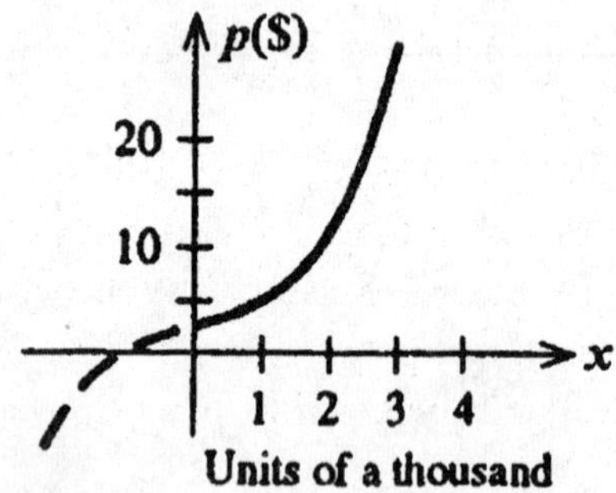

b. $p = 2^3 + 2(2) + 3 = 15$, or \$15.

41. a.

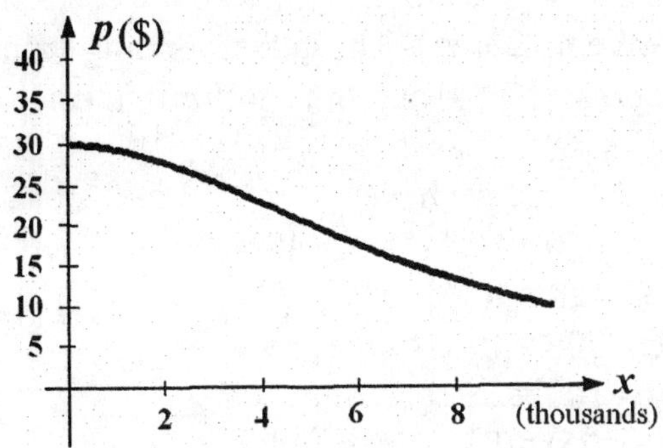

b. Substituting $x = 10$ into the demand function, we have

$$p = \frac{30}{0.02(10)^2 + 1} = \frac{30}{3} = 10, \text{ or } p = \$10.$$

43.

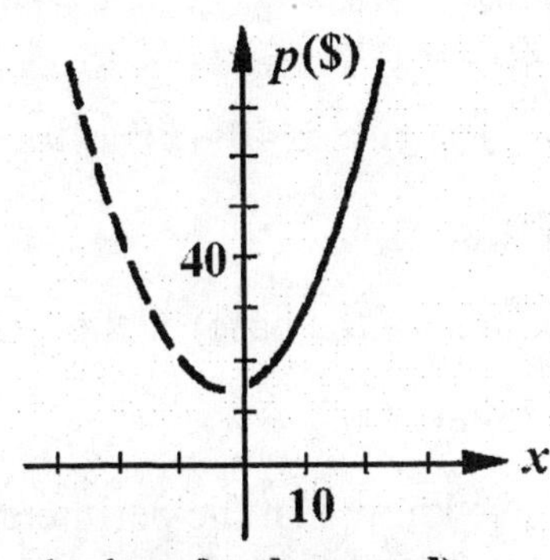

If $x = 5$, then $p = 0.1(5)^2 + 0.5(5) + 15 = 20$, or $20.

45. We solve the system of equations $p = -x^2 - 2x + 100$ and $p = 8x + 25$. Thus, $-x^2 - 2x + 100 = 8x + 25$, or $x^2 + 10x - 75 = 0$. Factoring this equation, we have $(x + 15)(x - 5) = 0$, or $x = -15$ and $x = 5$. Rejecting the negative root, we have $x = 5$ and the corresponding value of p is $p = 8(5) + 25 = 65$. We conclude that the equilibrium quantity is 5000 and the equilibrium price is $65.

47. We solve the system $p = 60 - 2x^2$
and $p = x^2 + 9x + 30$.
Equating these two equations, we have

$$x^2 + 9x + 30 = 60 - 2x^2$$
$$3x^2 + 9x - 30 = 0$$
$$x^2 + 3x - 10 = 0$$

$$(x + 5)(x - 2) = 0$$

and $x = -5$ or $x = 2$. We take $x = 2$. The corresponding value of p is 5. Therefore, the equilibrium quantity is 2000 and the equilibrium price is \$52.

49. Equating the two equations, we have

$$0.1x^2 + 2x + 20 = -0.1x^2 - x + 40$$
$$0.2x^2 + 3x - 20 = 0$$
$$2x^2 + 30x - 200 = 0$$
$$x^2 + 15x - 100 = 0$$
$$(x + 20)(x - 5) = 0,$$

and $x = -20$ or 5. Substituting $x = 5$ into the first equation gives

$$p = -0.1(25) - 5 + 40 = 32.5.$$

Therefore, the equilibrium quantity is 500 tents (x is measured in hundreds) and the equilibrium price is \$32.50.

51. Since there is 80 feet of fencing available,

$$2x + 2y = 80;\ x + y = 40 \text{ and } y = 40 - x.$$

Then the area of the garden is given by $f = xy = x(40 - x) = 40x - x^2$.
The domain of f is $[0, 40]$.

53. The volume of the box is given by the (area of the base) × the height of the box.
Thus, $V = f(x) = (15 - 2x)(8 - 2x)x$.

55. Since the perimeter of a circle is $2\pi r$, we know that the perimeter of the semicircle is πx. Next, the perimeter of the rectangular portion of the window is given by $2y + 2x$, so the perimeter of the Norman window is $\pi x + 2y + 2x$ and

$$\pi x + 2y + 2x = 28, \text{ or } y = \tfrac{1}{2}(28 - \pi x - 2x)$$

Since the area of the window is given by $2xy + \frac{1}{2}\pi x^2$, we see that

$$A = 2xy + \tfrac{1}{2}\pi x^2.$$

Substituting the value of y found earlier, we see that

$$A = f(x) = x(28 - \pi x - 2x) + \tfrac{1}{2}\pi x^2$$
$$= \tfrac{1}{2}\pi x^2 + 28x - \pi x^2 - 2x^2$$
$$= 28x - \frac{\pi}{2}x^2 - 2x^2$$
$$= 28x - \left(\frac{\pi}{2} + 2\right)x^2$$

57. $xy = 50$ and so $y = \dfrac{50}{x}$. The area of the printed page is

$$A = (x-1)(y-2) = (x-1)(\tfrac{50}{x} - 2) = -2x + 52 - \tfrac{50}{x}.$$

So the required function is $f(x) = -2x + 52 - \frac{50}{x}$. We must have $x > 0,\ x - 1 \geq 0,$ and $\frac{50}{x} - 2 \geq 2$; The last inequality is solved as follows:

$\dfrac{50}{x} \geq 4;\ \dfrac{x}{50} \leq \dfrac{1}{4};\ x \leq \dfrac{50}{4} = \dfrac{25}{2}$. So the domain is $[1, \frac{25}{2}]$.

59. a. Let x denote the number of people beyond 20 who sign up for the cruise. Then the revenue is

$$R(x) = (20 + x)(600 - 4x) = -4x^2 + 520x + 12{,}000$$

b. $R(40) = -4(40^2) + 520(40) + 12000 = 26{,}400$, or \$26,400.

$R(60) = -4(60^2) + 520(60) + 12{,}000 = 28{,}800$ or \$28,800.

61. True. If $P(x)$ is a polynomial function, then $P(x) = \dfrac{P(x)}{1}$ and so it is a rational function. The converse if false. For example, $R(x) = \dfrac{x+1}{x-1}$ is a rational function that is not a polynomial.

63. False. A power function has the form x^r, where r is a real number.

USING TECHNOLOGY EXERCISES 11.3, page 619

1. (-3.0414, 0.1503); (3.0414, 7.4497)

3. (-2.3371, 2.4117); (6.0514, -2.5015)

5. (-1.0219, -6.3461); (1.2414, -1.5931), and (5.7805, 7.9391)

7. a. 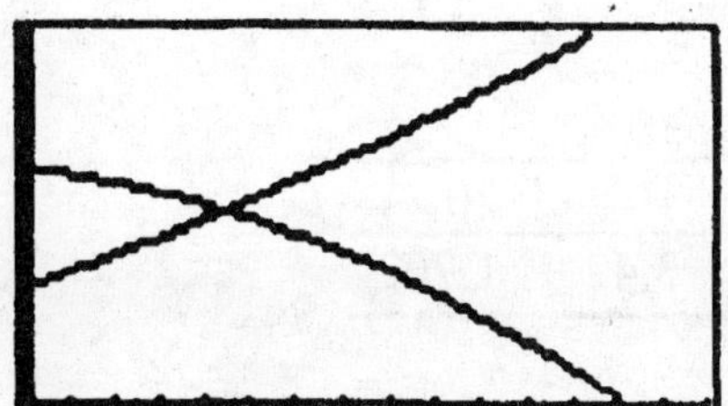

b. 438 wall clocks; \$40.92

9. a. $y = 0.1554t^2 + 0.5861t + 3.1607$

b.

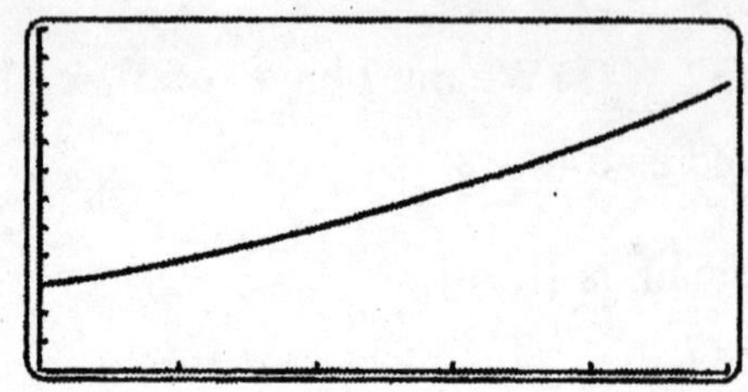

c. 3.16; 3.9; 4.95; 6.32; 7.99; 9.98

11. a. $y = -0.02028t^3 + 0.31393t^2 + 0.40873t + 0.66024$

b.

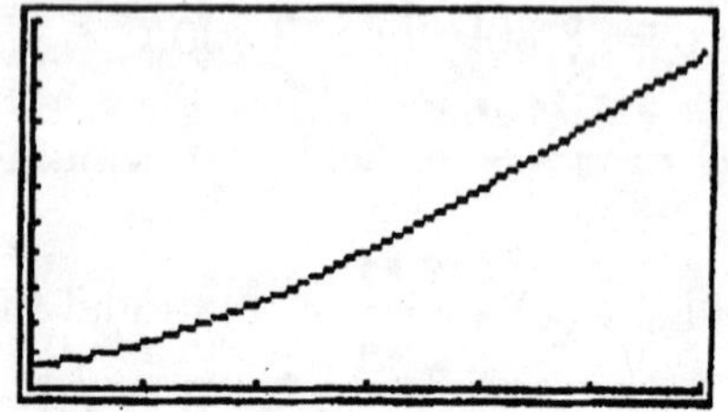

c. 0.66; 1.36; 2.57; 4.16; 6.02; 8.02; 10.03

13. a. $f(t) = 0.001532t^3 - 0.0588t^2 + 0.5208 + 2.55$

b.

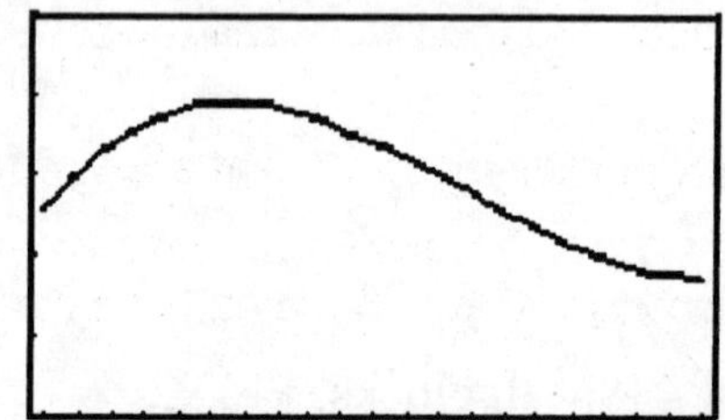

c.

t	0	7	10	20
$f(t)$ trillion dollars	2.55	3.84	3.42	1.702

15. a. $y = 0.05833t^3 - 0.325t^2 + 1.8881t + 5.07143$

b.

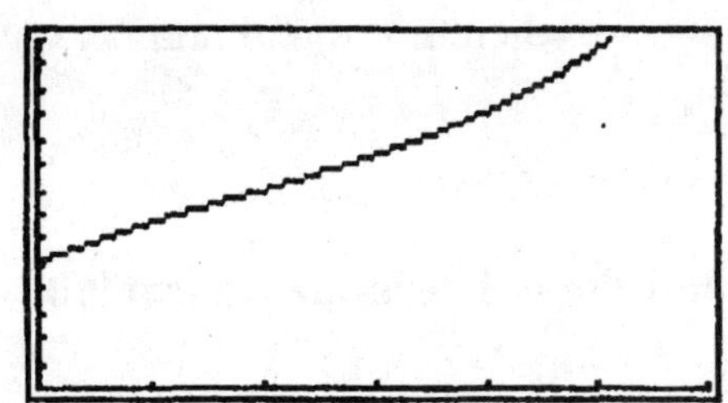

c. 6.7; 8.0; 9.4; 11.2; 13.7

17. a. $y = 0.0125t^4 - 0.01389t^3 + 0.55417t^2 + 0.53294t + 4.95238 \quad (0 \le t \le 5)$

b.

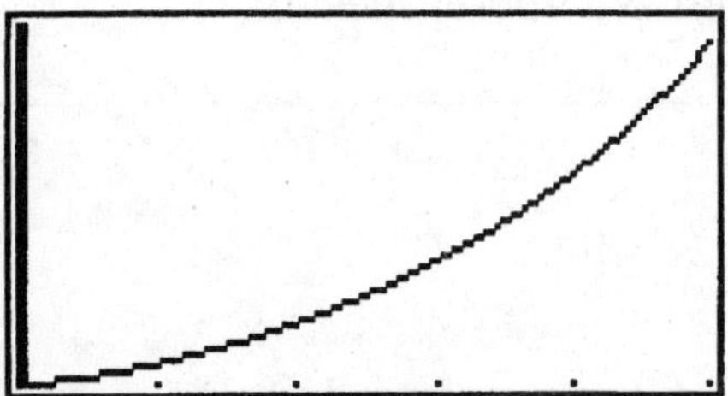

c. 5.0; 6.0; 8.3; 12.2; 18.3; 27.5

11.4 Problem Solving Tips

In this section, an important theorem was introduced (page 102). After you read Theorem 1, try to express the theorem in your own words. While you will not usually be required to prove these theorems in this course, you will be asked to understand the results of the theorem. For example, Theorem 1 gives us the properties of limits that allow us to evaluate the sum, difference, product, quotient, powers of functions and constant multiple of functions at a specified value with certain restrictions. You should be able to use the limit notation to write out each of these properties. You should also be able to use these properties to evaluate the limits of functions.

Here are some hints for solving the problems in the exercises that follow:

1. **To find the limit of a function $f(x)$ as $x \to a$,** where a is a real number, substitute a for x in the rule for f and simplify the result.

2. **To evaluate the limit of a quotient that has the indeterminate form 0/0:**

   a. First, replace the given function with an appropriate one that takes on the same values as the original function everywhere except at $x = a$.

   b. Next, evaluate the limit of this function as x approaches a.

11.4 Concept Questions, page 635

1. The values of $f(x)$ can be made as close to 3 as we please by taking x sufficiently close to $x = 2$.

3. a. $\lim\limits_{x\to 4}\sqrt{x}(2x^2+1) = \lim\limits_{x\to 4}(\sqrt{x})\lim\limits_{x\to 4}(2x^2+1)$ (Rule 4)

$= \sqrt{4}\left[2(4)^2+1\right]$ (Rules 1 and 3)

$= 66$

b. $\lim\limits_{x\to 1}\left(\dfrac{2x^2+x+5}{x^4+1}\right)^{3/2} = \left(\lim\limits_{x\to 1}\dfrac{2x^2+x+5}{x^4+1}\right)^{3/2}$ (Rule 1)

$= \left(\dfrac{2+1+5}{1+1}\right)^{3/2}$ (Rules 2, 3, and 5)

$= 4^{3/2} = 8$

5. $\lim\limits_{x\to\infty} f(x) = L$ means $f(x)$ can be made as close to L as we please by taking x sufficiently large. $\lim\limits_{x\to -\infty} f(x) = M$ means $f(x)$ can be made as close to M as we please by taking x as large as please in absolute value but negative.

EXERCISES 11.4, page 635

1. $\lim_{x \to -2} f(x) = 3.$

3. $\lim_{x \to 3} f(x) = 3.$

5. $\lim_{x \to -2} f(x) = 3.$

7. The limit does not exist. If we consider any value of x to the right of $x = -2$, $f(x) \le 2$. If we consider values of x to the left of $x = -2$, $f(x) \ge -2$. Since $f(x)$ does not approach any one number as x approaches $x = -2$, we conclude that the limit does not exist.

9. $\lim_{x \to 2} (x^2 + 1) = 5.$

x	1.9	1.99	1.999	2.001	2.01	2.1
$f(x)$	4.61	4.9601	4.9960	5.004	5.0401	5.41

11.

x	-0.1	-0.01	-0.001	0.001	0.01	0.1
$f(x)$	-1	-1	-1	1	1	1

The limit does not exist.

13.

x	0.9	0.99	0.999	1.001	1.01	1.1
$f(x)$	100	10,000	1,000,000	1,000,000	10,000	100

The limit does not exist.

15.

x	0.9	0.99	0.999	1.001	1.01	1.1
$f(x)$	2.9	2.99	2.999	3.001	3.01	3.1

$$\lim_{x \to 1} \frac{x^2 + x - 2}{x - 1} = 3.$$

17.

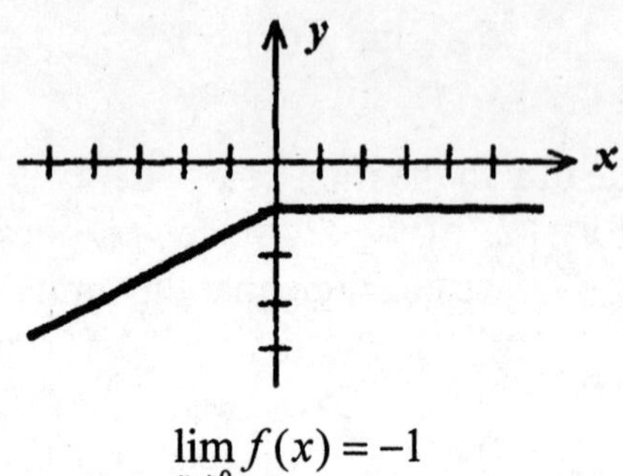

$\lim_{x\to 0} f(x) = -1$

19.

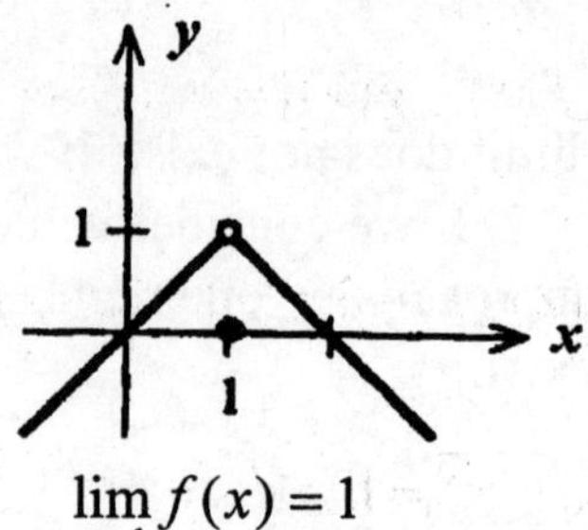

$\lim_{x\to 1} f(x) = 1$

21.

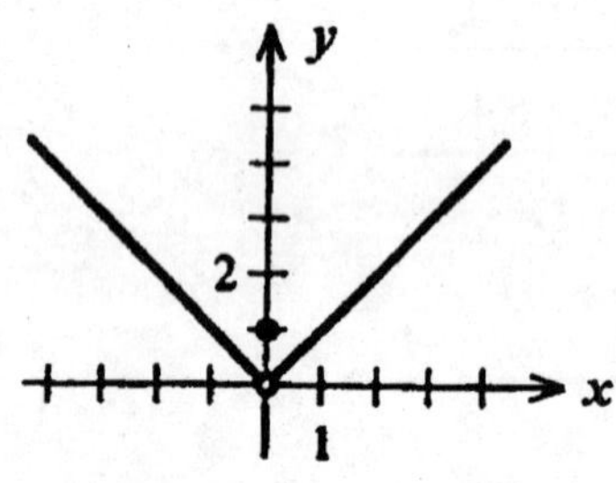

$\lim_{x\to 0} f(x) = 0$

23. $\lim_{x\to 2} 3 = 3$

25. $\lim_{x\to 3} x = 3$

27. $\lim_{x\to 1}(1-2x^2) = 1-2(1)^2 = -1$

29. $\lim_{x\to 1}(2x^3 - 3x^2 + x + 2) = 2(1)^3 - 3(1)^2 + 1 + 2 = 2.$

31. $\lim_{s\to 0}(2s^2 - 1)(2s+4) = (-1)(4) = -4.$

33. $\lim_{x\to 2}\frac{2x+1}{x+2} = \frac{2(2)+1}{2+2} = \frac{5}{4}.$

35. $\lim_{x\to 2}\sqrt{x+2} = \sqrt{2+2} = 2.$

37. $\lim_{x\to -3}\sqrt{2x^4 + x^2} = \sqrt{2(-3)^4 + (-3)^2} = \sqrt{162+9} = \sqrt{171} = 3\sqrt{19}.$

39. $\lim_{x\to -1}\frac{\sqrt{x^2+8}}{2x+4}=\frac{\sqrt{(-1)^2+8}}{2(-1)+4}=\frac{\sqrt{9}}{2}=\frac{3}{2}.$

41. $\lim_{x\to a}[f(x)-g(x)]=\lim_{x\to a}f(x)-\lim_{x\to a}g(x)=3-4=-1.$

43. $\lim_{x\to a}[2f(x)-3g(x)]=\lim_{x\to a}2f(x)-\lim_{x\to a}3g(x)=2(3)-3(4)=-6.$

45. $\lim_{x\to a}\sqrt{g(x)}=\lim_{x\to a}\sqrt{4}=2.$

47. $\lim_{x\to a}\frac{2f(x)-g(x)}{f(x)g(x)}=\frac{2(3)-(4)}{(3)(4)}=\frac{2}{12}=\frac{1}{6}.$

49. $\lim_{x\to 1}\frac{x^2-1}{x-1}=\lim_{x\to 1}\frac{(x-1)(x+1)}{x-1}=\lim_{x\to 1}(x+1)=1+1=2.$

51. $\lim_{x\to 0}\frac{x^2-x}{x}=\lim_{x\to 0}\frac{x(x-1)}{x}=\lim_{x\to 0}(x-1)=0-1=-1.$

53. $\lim_{x\to -5}\frac{x^2-25}{x+5}=\lim_{x\to -5}\frac{(x+5)(x-5)}{x+5}=\lim_{x\to -5}(x-5)=-10.$

55. $\lim_{x\to 1}\frac{x}{x-1}$ does not exist.

57. $\lim_{x\to -2}\frac{x^2-x-6}{x^2+x-2}=\lim_{x\to -2}\frac{(x-3)(x+2)}{(x+2)(x-1)}=\lim_{x\to -2}\frac{x-3}{x-1}=\frac{-2-3}{-2-1}=\frac{5}{3}.$

59. $\lim_{x\to 1}\frac{\sqrt{x}-1}{x-1}=\lim_{x\to 1}\frac{\sqrt{x}-1}{x-1}\cdot\frac{\sqrt{x}+1}{\sqrt{x}+1}=\lim_{x\to 1}\frac{x-1}{(x-1)(\sqrt{x}+1)}=\lim_{x\to 1}\frac{1}{\sqrt{x}+1}=\frac{1}{2}.$

61. $\lim_{x\to 1}\frac{x-1}{x^3+x^2-2x}=\lim_{x\to 1}\frac{x-1}{x(x-1)(x+2)}=\lim_{x\to 1}\frac{1}{x(x+2)}=\frac{1}{3}.$

63. $\lim_{x\to\infty}f(x)=\infty$ (does not exist) and $\lim_{x\to -\infty}f(x)=\infty$ (does not exist).

65. $\lim_{x\to\infty} f(x) = 0$ and $\lim_{x\to-\infty} f(x) = 0.$

67. $\lim_{x\to\infty} f(x) = -\infty$ (does not exist) and $\lim_{x\to-\infty} f(x) = -\infty$ (does not exist).

69.

x	1	10	100	1000
$f(x)$	0.5	0.009901	0.0001	0.000001

x	-1	-10	-100	-1000
$f(x)$	0.5	0.009901	0.0001	0.000001

$\lim_{x\to\infty} f(x) = 0$ and $\lim_{x\to-\infty} f(x) = 0$

71.

x	1	5	10	100	1000
$f(x)$	12	360	2910	2.99×10^6	2.999×10^9

x	-1	-5	-10	-100	-1000
$f(x)$	6	-390	-3090	-3.01×10^6	-3.0×10^9

$\lim_{x\to\infty} f(x) = \infty$ (does not exist) and $\lim_{x\to-\infty} f(x) = -\infty$ (does not exist).

73. $\lim_{x\to\infty} \dfrac{3x+2}{x-5} = \lim_{x\to\infty} \dfrac{3+\dfrac{2}{x}}{1-\dfrac{5}{x}} = \dfrac{3}{1} = 3.$

75. $\lim_{x\to-\infty} \dfrac{3x^3+x^2+1}{x^3+1} = \lim_{x\to-\infty} \dfrac{3+\dfrac{1}{x}+\dfrac{1}{x^3}}{1+\dfrac{1}{x^3}} = 3.$

77. $\lim_{x\to-\infty} \dfrac{x^4+1}{x^3-1} = \lim_{x\to-\infty} \dfrac{x+\dfrac{1}{x^3}}{1-\dfrac{1}{x^3}} = -\infty$; that is, the limit does not exist.

79. $\lim_{x\to\infty} \frac{x^5 - x^3 + x - 1}{x^6 + 2x^2 + 1} = \lim_{x\to\infty} \frac{\frac{1}{x} - \frac{1}{x^3} + \frac{1}{x^5} - \frac{1}{x^6}}{1 + \frac{2}{x^4} + \frac{1}{x^6}} = 0.$

81. a. The cost of removing 50 percent of the pollutant is

$$C(50) = \frac{0.5(50)}{100 - 50} = 0.5\text{, or \$500,000.}$$

Similarly, we find that the cost of removing 60, 70, 80, 90, and 95 percent of the pollutants is \$750,000; \$1,166,667; \$2,000,000, \$4,500,000, and \$9,500,000, respectively.

b. $\lim_{x\to 100} \frac{0.5x}{100 - x} = \infty,$

which means that the cost of removing the pollutant increases astronomically if we wish to remove almost all of the pollutant.

83. $\lim_{x\to\infty} \overline{C}(x) = \lim_{x\to\infty} 2.2 + \frac{2500}{x} = 2.2$, or \$2.20 per DVD.

In the long-run, the average cost of producing x DVDs will approach \$2.20/disc.

85. a. $T(1) = \frac{120}{1+4} = 24$, or \$24 million. $T(2) = \frac{120(4)}{8} = 60$, \$60 million.

$T(3) = \frac{120(9)}{13} = 83.1$, or \$83.1 million.

b. In the long run, the movie will gross

$$\lim_{x\to\infty} \frac{120x^2}{x^2 + 4} = \lim_{x\to\infty} \frac{120}{1 + \frac{4}{x^2}} = 120\text{, or \$120 million.}$$

87. a. The average cost of driving 5000 miles per year is

$$C(5) = \frac{2010}{5^{2.2}} + 17.80 = 76.07,$$

or 76.1 cents per mile. Similarly, we see that the average cost of driving 10,000 miles per year; 15,000 miles per year; 20,000 miles per year; and 25,000 miles per year is 30.5, 23; 20.6, and 19.5 cents per mile, respectively.

b.

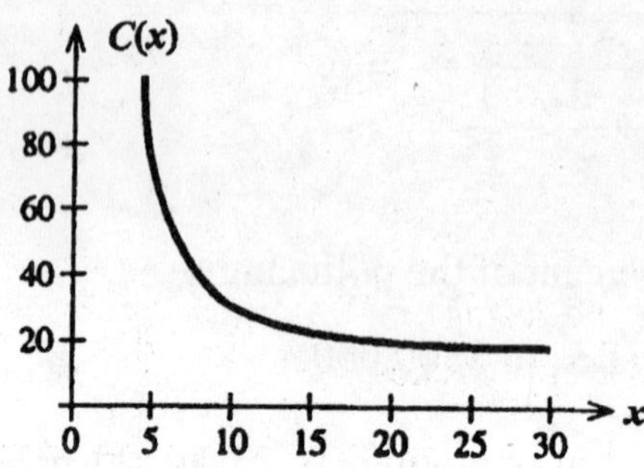

c. It approaches 17.80 cents per mile.

89. False. Let $f(x)=\begin{cases}-1 & \text{if } x<0\\ 1 & \text{if } x>0\end{cases}$. Then $\lim\limits_{x\to 0} f(x)=1$, but $f(1)$ is not defined.

91. True. Division by zero is not permitted.

93. True. Each limit in the sum exists. Therefore,

$$\lim_{x\to 2}\left(\frac{x}{x+1}+\frac{3}{x-1}\right)=\lim_{x\to 2}\frac{x}{x+1}+\lim_{x\to 2}\frac{3}{x-1}=\frac{2}{3}+\frac{3}{1}=\frac{11}{3}.$$

95. $\lim\limits_{x\to\infty}\dfrac{ax}{x+b}=\lim\limits_{x\to\infty}\dfrac{a}{1+\frac{b}{x}}=a.$ As the amount of substrate becomes very large, the initial speed approaches the constant a moles per liter per second.

97. Consider the functions $f(x)=\begin{cases}-1 & \text{if } x<0\\ 1 & \text{if } x\geq 0\end{cases}$ and $g(x)=\begin{cases}1 & \text{if } x<0\\ -1 & \text{if } x\geq 0\end{cases}$.

Then $\lim\limits_{x\to 0} f(x)$ and $\lim\limits_{x\to 0} g(x)$ do not exist, but $\lim\limits_{x\to 0}[f(x)g(x)]=\lim\limits_{x\to 0}(-1)=-1.$

This example does not contradict Theorem 1 because the hypothesis of Theorem 1 says that if $\lim\limits_{x\to 0} f(x)$ and $\lim\limits_{x\to 0} g(x)$ both exist, then the limit of the product of f and g also exists. It does not say that if the former do not exist, then the latter might not exist.

USING TECHNOLOGY EXERCISES 11.4, page 642

1. 5 3. 3 5. $\frac{2}{3}$ 7. $\frac{1}{2}$

9. e^2, or 7.38906

11. From the graph we see that $f(x)$ does not approach any finite number as x approaches 3.

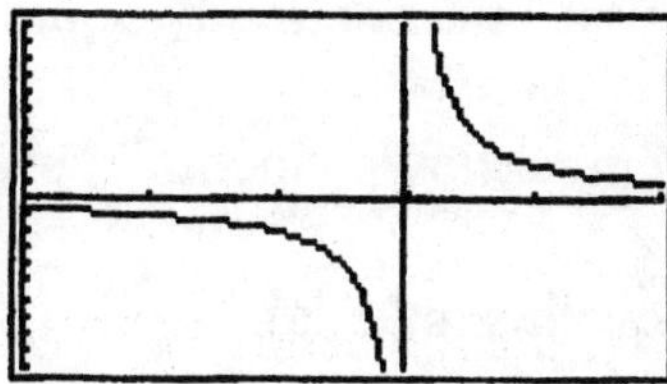

13. a.

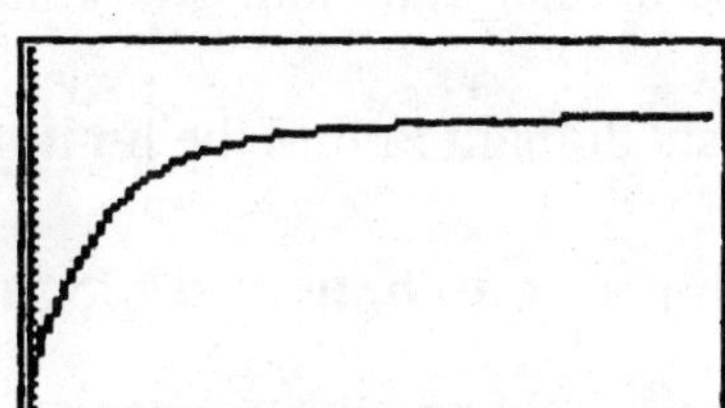

b. $\lim\limits_{t\to\infty}\frac{25t^2+125t+200}{t^2+5t+40}=25$, so in the long run the population will approach 25,000.

11.5 Problem Solving Tips

The problem-solving skills that you learned in earlier sections are building-blocks for the rest of the course. You can't skip a section or a concept and "hope" that you will understand the material in a new section. It just won't work. If you don't build a strong foundation, you won't be able to understand the later concepts. For example, in this section we discussed one-sided limits. You need to understand the definition of a limit before you can understand what is meant by one-sided limits. That means you should be able to express the definition of a limit in your own words. If you can't grasp a new concept, it may well be that you still don't understand a previous concept. If so, you need to go back and review the earlier section before you go on. As another example, we also discussed the continuity of polynomial and rational functions on page 123. If you don't remember how to identify polynomial and rational functions, go back to Section 2.3 and review this material.

Here are some hints for solving the problems in the exercises that follow:

1. **To evaluate the limit of a piecewise-defined function** at a real number a, follow the same procedure that you used to evaluate a piecewise-defined function. First find the subdomain that a lies in. Next, use the rule for that subdomain to find the limit of f at a.

2. **To determine the values of x at which a function is continuous** check to see if the function is a polynomial or rational function. A polynomial function $y = P(x)$ is continuous at every value of x and a rational function is continuous at every value of x where $q(x) \neq 0$.

11.5 CONCEPT QUESTIONS, page 651

1. $\lim_{x \to 3^-} f(x) = 2$ means $f(x)$ can be made as close to 2 as we please by taking x sufficiently close to but to the left of $x = 3$. $\lim_{x \to 3^+} f(x) = 4$ means $f(x)$ can be made as close to 4 as we please by taking x sufficiently close to but to the right of $x = 3$.
3. a. f is continuous at a if $\lim_{x \to a} f(x) = f(a)$.

   b. f is continuous on an interval I if f is continuous at each point in I.
5. Refer to page 648 in the text. (Answers will vary.)

EXERCISES 11.5, page 651

1. $\lim_{x\to 2^-} f(x) = 3$, $\lim_{x\to 2^+} f(x) = 2$, $\lim_{x\to 2} f(x)$ does not exist.

3. $\lim_{x\to -1^-} f(x) = \infty$, $\lim_{x\to -1^+} f(x) = 2$. Therefore $\lim_{x\to -1} f(x)$ does not exist.

5. $\lim_{x\to 1^-} f(x) = 0$, $\lim_{x\to 1^+} f(x) = 2$, $\lim_{x\to 1} f(x)$ does not exist.

7. $\lim_{x\to 0^-} f(x) = -2$, $\lim_{x\to 0^+} f(x) = 2$, $\lim_{x\to 0} f(x)$ does not exist.

9. True 11. True 13. False

15. True 17. False 19. True

21. $\lim_{x\to 1^+} (2x+4) = 6$.

23. $\lim_{x\to 2^-} \frac{x-3}{x+2} = \frac{2-3}{2+2} = -\frac{1}{4}$.

25. $\lim_{x\to 0^+} \frac{1}{x}$ does not exist because $1/x \to \infty$ as $x \to 0$ from the right.

27. $\lim_{x\to 0^+} \frac{x-1}{x^2+1} = \frac{-1}{1} = -1$.

29. $\lim_{x\to 0^+} \sqrt{x} = \sqrt{\lim_{x\to 0^+} x} = 0$.

31. $\lim_{x\to -2^+} (2x + \sqrt{2+x}) = \lim_{x\to -2^+} 2x + \lim_{x\to -2^+} \sqrt{2+x} = -4 + 0 = -4$.

33. $\lim_{x\to 1^-} \frac{1+x}{1-x} = \infty$, that is, the limit does not exist.

35. $\lim_{x\to 2^-} \frac{x^2-4}{x-2} = \lim_{x\to 2^-} \frac{(x+2)(x-2)}{x-2} = \lim_{x\to 2^-} (x+2) = 4$.

37. $\lim_{x\to 0^+} f(x) = \lim_{x\to 0^+} x^2 = 0$, $\lim_{x\to 0^-} f(x) = \lim_{x\to 0^-} 2x = 0$

39. The function is discontinuous at $x = 0$. Conditions 2 and 3 are violated.

41. The function is continuous everywhere.

43. The function is discontinuous at $x = 0$. Condition 3 is violated.

45. f is continuous for all values of x.

47. f is continuous for all values of x. Note that $x^2 + 1 \geq 1 > 0$.

49. f is discontinuous at $x = 1/2$, where the denominator is 0.

51. Observe that $x^2 + x - 2 = (x + 2)(x - 1) = 0$ if $x = -2$ or $x = 1$. So, f is discontinuous at these values of x.

53. f is continuous everywhere since all three conditions are satisfied.

55. f is continuous everywhere since all three conditions are satisfied.

57. Since the denominator $x^2 - 1 = (x - 1)(x + 1) = 0$ if $x = -1$ or 1, we see that f is discontinuous at -1 and 1.

59. Since $x^2 - 3x + 2 = (x - 2)(x - 1) = 0$ if $x = 1$ or 2, we see that the denominator is zero at these points and so f is discontinuous at these numbers.

61. The function f is discontinuous at $x = 1, 2, 3, ..., 11$ because the limit of f does not exist at these points.

63. Having made steady progress up to $x = x_1$, Michael's progress came to a standstill. Then at $x = x_2$ a sudden break-through occurs and he then continues to successfully complete the solution to the problem.

65. Conditions 2 and 3 are not satisfied at each of these points.

67. The graph of f follows.

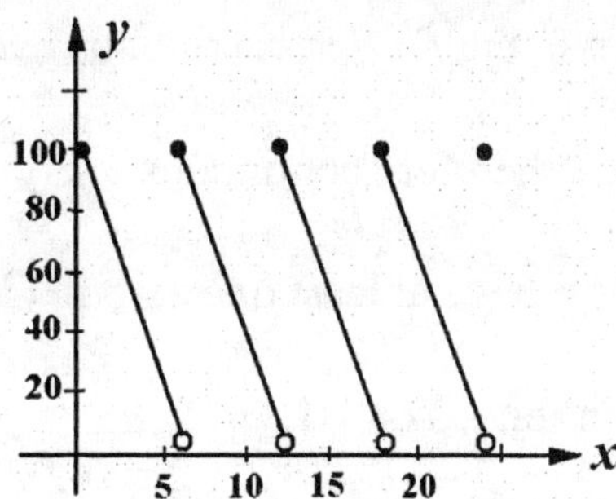

f is discontinuous at $x = 6, 12, 18, 24$.

69.

$$f(x) = \begin{cases} 2 & \text{if } 0 < x \le \frac{1}{2} \\ 3 & \text{if } \frac{1}{2} < x \le 1 \\ \vdots & \\ 10 & \text{if } 4\frac{1}{2} < x \le 5 \end{cases}$$

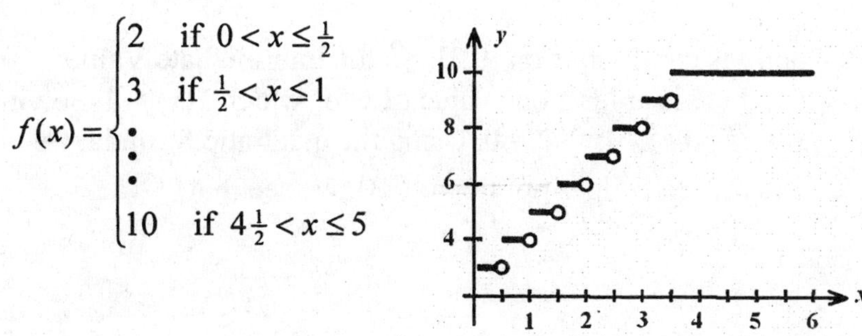

f is discontinuous at $x = \frac{1}{2}, 1, 1\frac{1}{2}, \ldots, 4$.

71. a. $\lim_{t \to 0^+} S(t) = \lim_{t \to 0^+} \frac{a}{t} + b = \infty$. As the time taken to excite the tissue is made smaller and smaller, the strength of the electric current gets stronger and stronger.

b. $\lim_{t \to \infty} \frac{a}{t} + b = b$. As the time taken to excite the tissue is made larger and larger, the strength of the electric current gets smaller and smaller and approaches b.

73. We require that $f(1) = 1 + 2 = 3 = \lim_{x \to 1^+} kx^2 = k$, or $k = 3$.

75. a. Yes, because if $f + g$ were continuous at a, then $g = (f + g) - f$ would be continuous (the difference of two continuous functions is continuous), and this would imply that g is continuous, a contradiction.

b. No. Consider the functions f and g defined by

$$f(x) = \begin{cases} -1 & \text{if } x < 0 \\ 1 & \text{if } x \ge 0 \end{cases} \quad \text{and} \quad g(x) = \begin{cases} 1 & \text{if } x < 0 \\ -1 & \text{if } x \ge 0 \end{cases}.$$

Both f and g are discontinuous at $x = 0$, but $f + g$ is continuous everywhere.

77. a. f is a polynomial of degree 2 and is therefore continuous everywhere, and in particular in [1,3].
b. $f(1) = 3$ and $f(3) = -1$ and so f must have at least one zero in (1,3).

79. f is a polynomial and is therefore continuous on [-1,1].
$$f(-1) = (-1)^3 - 2(-1)^2 + 3(-1) + 2 = -1 - 2 - 3 + 2 = -4.$$
$$f(1) = 1 - 2 + 3 + 2 = 4.$$
Since $f(-1)$ and $f(1)$ have opposite signs, we see that f has at least one zero in (-1,1).

81. $f(0) = 6$ and $f(3) = 3$ and f is continuous on [0,3]. So the Intermediate Value Theorem guarantees that there is at least one value of x for which $f(x) = 4$. Solving $f(x) = x^2 - 4x + 6 = 4$, we find $x^2 - 4x + 2 = 0$. Using the quadratic formula, we find that $x = 2 \pm \sqrt{2}$. Since $2 \pm \sqrt{2}$ does not lie in [0,3], we see that $x = 2 - \sqrt{2} \approx 0.59$.

83. $x^5 + 2x - 7 = 0$

Step	Root of f(x) = 0 lies in
1	(1,2)
2	(1,1.5)
3	(1.25,1.5)
4	(1.25,1.375)
5	(1.3125,1.375)
6	(1.3125,1.34375)
7	(1.328125,1.34375)
8	(1.3359375,1.34375)
9	(1.33984375,1.34375)

We see that the required root is approximately 1.34.

85. a. $h(t) = 4 + 64(0) - 16(0) = 4$, and $h(2) = 4 + 64(2) - 16(4) = 68$.
b. The function h is continuous on [0,2]. Furthermore, the number 32 lies between 4 and 68. Therefore, the Intermediate Value Theorem guarantees that there is at least one value of t in (0, 2] such that $h(t) = 32$, that is, Joan must see the ball at least once during the time the ball is in the air.

c. We solve

$$h(t) = 4 + 64t - 16t^2 = 32$$

or

$$16t^2 - 64t + 28 = 0$$
$$4t^2 - 16t + 7 = 0$$
$$(2t - 1)(2t - 7) = 0$$

giving $t = \frac{1}{2}$ or $t = \frac{7}{2}$. Joan sees the ball on its way up half a second after it was thrown and again $3\frac{1}{2}$ seconds later when it is on its way down.

87. False. Take

$$f(x) = \begin{cases} -1 & \text{if } x < 2 \\ 4 & \text{if } x = 2 \\ 1 & \text{if } x > 2 \end{cases}$$

Then $f(2) = 4$ but $\lim_{x \to 2} f(x)$ does not exist.

89. False. Consider the function $f(x) = x^2 - 1$ on the interval [-2, 2]. Here, $f(-2) = f(2) = 3$, but f has zeros at $x = -1$ and $x = 1$.

91. False. Let $f(x) = \begin{cases} x & \text{if } x \neq 0 \\ 1 & \text{if } x = 0 \end{cases}$. Then $\lim_{x \to 0^+} f(x) = \lim_{x \to 0^-} f(x)$, but $f(0) = 1$.

93. False. Take $f(x) = \begin{cases} \frac{1}{x} & \text{if } x \neq 0 \\ 0 & \text{if } x = 0 \end{cases}$. Then f is continuous for all $x \neq 0$ but $\lim_{x \to 0} f(x)$ does not exist.

95. a. Both $g(x) = x$ and $h(x) = \sqrt{1 - x^2}$ are continuous on [-1,1] and so $f(x) = x - \sqrt{1 - x^2}$ is continuous on [-1,1].
b. $f(-1) = -1$ and $f(1) = 1$ and so f has at least one zero in (-1,1).
c. Solving $f(x) = 0$, we have $x = \sqrt{1 - x^2}$, $x^2 = 1 - x^2$, $2x^2 = 1$, or $x = \frac{\pm\sqrt{2}}{2}$.

97. a. (i). Repeated use of Property 3 shows that $g(x) = x^n = x \cdot x \cdots x$ (n times) is a continuous function since $f(x) = x$ is continuous by Property 1.
(ii). Properties 1 and 5 combine to show that $c \cdot x^n$ is continuous using the results of (a).
(iii). Each of the terms of $p(x) = a_0x^n + a_1x^{n-1} + \cdots + a_n$ is continuous and so

Property 4 implies that p is continuous.

b. Property 6 now shows that $R(x) = \dfrac{p(x)}{q(x)}$ is continuous if $q(a) \neq 0$ since p and q are continuous at $x = a$.

USING TECHNOLOGY EXERCISES 11.5, page 658

1. $x = 0, 1$ 3. $x = 2$ 5. $x = 0, \frac{1}{2}$

7. $x = -\frac{1}{2}, 2$ 9. $x = -2, 1$

11.

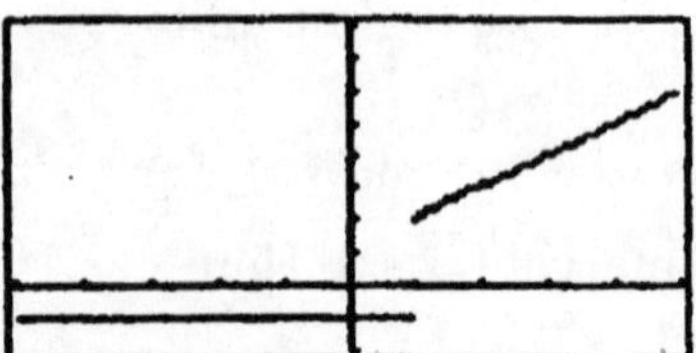

13.

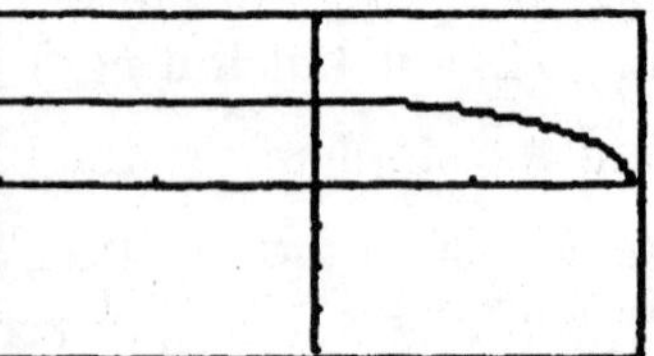

15.

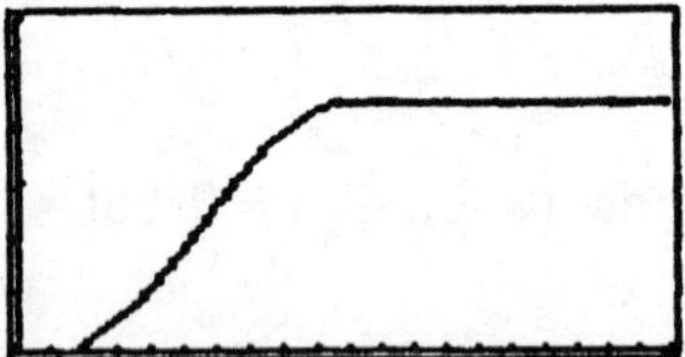

11.6 Problem Solving Tips

When you solve an applied problem, it is important to understand the question in mathematical terms. For example, if you are given a function $f(t)$ describing the size of a country's population at time t and asked to find the *rate of change* of that country's population at any time t, this means that you need to find the derivative of the given function; that is, find $f'(t)$. If you are then asked to find the population of the country at a specified time, say $t = 2$, you simply need to evaluate the function at that value of t. On the other hand, if you are asked to find the *rate of change* of the population at time $t = 2$, then you need to evaluate the derivative of the function at the value $t = 2$, that is find $f'(2)$. Here again, the key is to be familiar with the terminology and notation introduced in the chapter.

Here are some hints for solving the problems in the exercises that follow:

1. **To find the slope of the tangent line to the graph of a function at *any* point** on the graph of that function, find the derivative of f.

2. **To find the slope of the tangent line to the graph of a function at a *given* point** (x_0, y_0) on the graph of that function, find f' and then evaluate f' at x_0.

11.6 CONCEPT QUESTIONS, page 671

1. a. $m = \dfrac{f(2+h) - f(2)}{h}$

   b. Slope of the tangent line is $\lim\limits_{h \to 0} \dfrac{f(2+h) - f(2)}{h}$

3. a. It gives (i) the slope of the secant line passing through the points $(x, f(x))$ and $(x+h, f(x+h))$ and (ii) the average rate of change of f over the interval $[x, x+h]$.

   b. It gives (i) the slope of the tangent line to the graph of f at the point $(x, f(x))$ and (ii) the instantaneous rate of change of f at x.

EXERCISES 11.6, page 671

1. The rate of change of the average infant's weight when $t = 3$ is (7.5)/5, or 1.5 lb/month. The rate of change of the average infant's weight when $t = 18$ is (3.5)/6,

or approximately 0.6 lb/month. The average rate of change over the infant's first year of life is (22.5 – 7.5)/(12), or 1.25 lb/month.

3. The rate of change of the percentage of households watching television at 4 P.M. is (12.3)/4, or approximately 3.1 percent per hour. The rate at 11 P.M. is $(-42.3)/2 = -21.15$; that is, it is dropping off at the rate of 21.15 percent per hour.

5. a. Car *A* is travelling faster than Car *B* at t_1 because the slope of the tangent line to the graph of *f* is greater than the slope of the tangent line to the graph of *g* at t_1.

   b. Their speed is the same because the slope of the tangent lines are the same at t_2.

   c. Car *B* is travelling faster than Car *A*.

   d. They have both covered the same distance and are once again side by side at t_3.

7. a. P_2 is decreasing faster at t_1 because the slope of the tangent line to the graph of *g* at t_1 is greater than the slope of the tangent line to the graph of *f* at t_1.
   b. P_1 is decreasing faster than P_2 at t_2.
   c. Bactericide *B* is more effective in the short run, but bactericide *A* is more effective in the long run.

9. $f(x) = 13$

   Step 1 $f(x+h) = 13$

   Step 2 $f(x+h) - f(x) = 13 - 13 = 0$

   Step 3 $\dfrac{f(x+h) - f(x)}{h} = \dfrac{0}{h} = 0$

   Step 4 $f'(x) = \lim\limits_{h \to 0} \dfrac{f(x+h) - f(x)}{h} = \lim\limits_{h \to 0} 0 = 0$

11. $f(x) = 2x + 7$

    Step 1 $f(x+h) = 2(x+h) + 7$

    Step 2 $f(x+h) - f(x) = 2(x+h) + 7 - (2x+7) = 2h$

    Step 3 $\dfrac{f(x+h) - f(x)}{h} = \dfrac{2h}{h} = 2$

    Step 4 $f'(x) = \lim\limits_{h \to 0} \dfrac{f(x+h) - f(x)}{h} = \lim\limits_{h \to 0} 2 = 2$

13. $f(x)=3x^2$

Step 1 $f(x+h)= 3(x+h)^2 = 3x^2 +6xh+3h^2$

Step 2 $f(x+h)-f(x)= (3x^2 +6xh+3h^2)-3x^2 = 6xh+3h^2 = h(6x+3h)$

Step 3 $\dfrac{f(x+h)-f(x)}{h} = \dfrac{h(6x+3h)}{h} = 6x+3h$

Step 4 $f'(x)=\lim\limits_{h\to 0}\dfrac{f(x+h)-f(x)}{h} = \lim\limits_{h\to 0}(6x+3h) = 6x.$

15. $f(x)=-x^2+3x$

Step 1 $f(x+h)= -(x+h)^2+3(x+h) = -x^2-2xh-h^2+3x+3h$

Step 2 $f(x+h)-f(x) = (-x^2-2xh-h^2+3x+3h)-(-x^2+3x)$

$= -2xh-h^2+3h = h(-2x-h+3)$

Step 3 $\dfrac{f(x+h)-f(x)}{h} = \dfrac{h(-2x-h+3)}{h} = -2x-h+3$

Step 4 $f'(x)=\lim\limits_{h\to 0}\dfrac{f(x+h)-f(x)}{h} = \lim\limits_{h\to 0}(-2x-h+3) = -2x+3.$

17. $f(x)=2x+7$. Using the four-step process,

Step 1 $f(x+h)=2(x+h)+7 = 2x+2h+7$

Step 2 $f(x+h)-f(x)= 2x+2h+7-2x-7=2h$

Step 3 $\dfrac{f(x+h)-f(x)}{h} = \dfrac{2h}{h} = 2$

Step 4 $f'(x)=\lim\limits_{h\to 0}\dfrac{f(x+h)-f(x)}{h} = \lim\limits_{h\to 0} 2 = 2$

we find that $f'(x)=2$. In particular, the slope at $x=2$ is also 2. Therefore, a required equation is $y-11=2(x-2)$ or $y=2x+7$.

19. $f(x)=3x^2$. We first compute $f'(x)=6x$ (see Problem 13). Since the slope of the tangent line is $f'(1)=6$, we use the point-slope form of the equation of a line and find that a required equation is $y-3=6(x-1)$, or $y=6x-3$.

21. $f(x)=-1/x$. We first compute $f'(x)$ using the four-step process.

Step 1 $f(x+h)= -\dfrac{1}{x+h}$

Step 2 $f(x+h)-f(x) = -\frac{1}{x+h}+\frac{1}{x} = \frac{-x+(x+h)}{x(x+h)} = \frac{h}{x(x+h)}$

Step 3 $\frac{f(x+h)-f(x)}{h} = \frac{\frac{h}{x(x+h)}}{h} = \frac{1}{x(x+h)}$

Step 4 $f'(x) = \lim_{h\to 0}\frac{f(x+h)-f(x)}{h} = \lim_{h\to 0}\frac{1}{x(x+h)} = \frac{1}{x^2}.$

The slope of the tangent line is $f'(3)=1/9$. Therefore, a required equation is

$$y-(-\tfrac{1}{3}) = \tfrac{1}{9}(x-3) \quad \text{or} \quad y=\tfrac{1}{9}x-\tfrac{2}{3}.$$

23. a. $f(x)=2x^2+1$. We use the four-step process.

Step 1 $f(x+h) = 2(x+h)^2+1 = 2x^2+4xh+2h^2+1$

Step 2 $f(x+h)-f(x) = (2x^2+4xh+2h^2+1)-(2x^2+1) = 4xh+2h^2$
$= h(4x+2h)$

Step 3 $\frac{f(x+h)-f(x)}{h} = \frac{h(4x+2h)}{h} = 4x+2h$

Step 4 $f'(x) = \lim_{h\to 0}\frac{f(x+h)-f(x)}{h} = \lim_{h\to 0}(4x+2h) = 4x$

b. The slope of the tangent line is $f'(1)=4(1)=4$. Therefore, an equation is $y-3=4(x-1)$ or $y=4x-1$.

c.

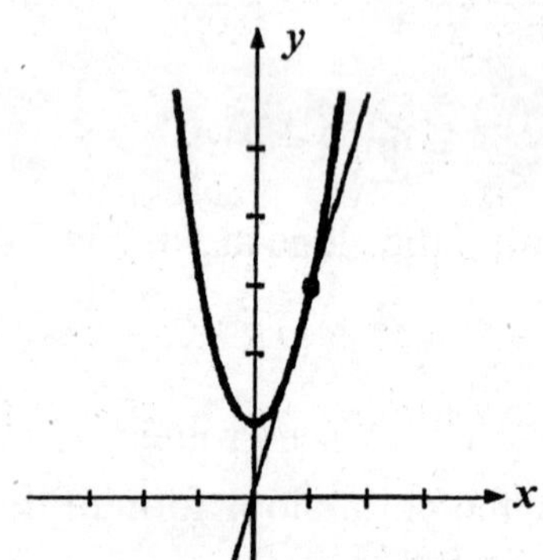

25. a. $f(x)=x^2-2x+1$. We use the four-step process:

Step 1 $f(x+h) = (x+h)^2-2(x+h)+1 = x^2+2xh+h^2-2x-2h+1$

Step 2 $f(x+h)-f(x) = (x^2+2xh+h^2-2x-2h+1)-(x^2-2x+1)\,]$
$= 2xh+h^2-2h = h(2x+h-2)$

Step 3 $\dfrac{f(x+h)-f(x)}{h}=\dfrac{h(2x+h-2)}{h}=2x+h-2$

Step 4 $f'(x)=\lim\limits_{h\to 0}\dfrac{f(x+h)-f(x)}{h}=\lim\limits_{h\to 0}(2x+h-2)=2x-2.$

b. At a point on the graph of f where the tangent line to the curve is horizontal, $f'(x)=0$. Then $2x-2=0$, or $x=1$. Since $f(1)=1-2+1=0$, we see that the required point is (1,0).

c.

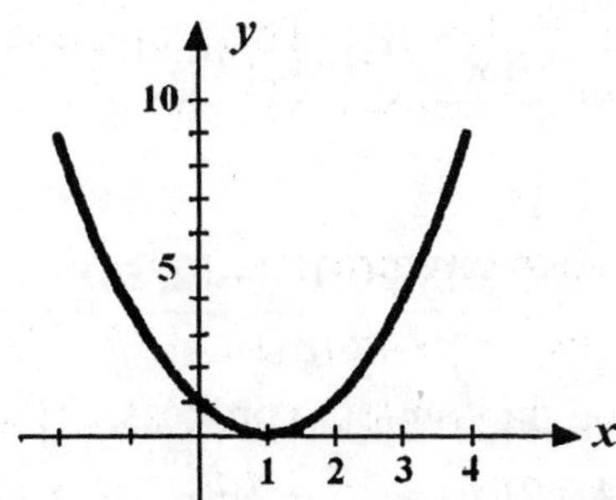

d. It is changing at the rate of 0 units per unit change in x.

27. a. $f(x)=x^2+x$

$$\frac{f(3)-f(2)}{3-2}=\frac{(3^2+3)-(2^2+2)}{1}=6$$

$$\frac{f(2.5)-f(2)}{2.5-2}=\frac{(2.5^2+2.5)-(2^2+2)}{0.5}=5.5$$

$$\frac{f(2.1)-f(2)}{2.1-2}=\frac{(2.1^2+2.1)-(2^2+2)}{0.1}=5.1$$

b. We first compute $f'(x)$ using the four-step process.

Step 1 $f(x+h)=(x+h)^2+(x+h)=x^2+2xh+h^2+x+h$

Step 2 $f(x+h)-f(x)=(x^2+2xh+h^2+x+h)-(x^2+x)]$

$$=2xh+h^2+h=h(2x+h+1)$$

Step 3 $\dfrac{f(x+h)-f(x)}{h}=\dfrac{h(2x+h+1)}{h}=2x+h+1$

Step 4 $f'(x)=\lim\limits_{h\to 0}\dfrac{f(x+h)-f(x)}{h}=\lim\limits_{h\to 0}(2x+h+1)=2x+1.$

The instantaneous rate of change of y at $x=2$ is $f'(2)=5$ or 5 units per unit change in x.

c. The results in (a) suggest that the average rates of change of f at $x = 2$ approach 5 as the interval $[2, 2+h]$ gets smaller and smaller ($h = 1$, 0.5, and 0.1). This number is the instantaneous rate of change of f at $x = 2$ as computed in (b).

29. a. $f(t) = 2t^2 + 48t$. The average velocity of the car over [20, 21] is

$$\frac{f(21) - f(20)}{21 - 20} = \frac{[2(21)^2 + 48(21)] - [2(20)^2 + 48(20)]}{1} = 130 \text{ ft / sec}$$

Its average velocity over [20,20.1] is

$$\frac{f(20.1) - f(20)}{20.1 - 20} = \frac{[2(20.1)^2 + 48(20.1)] - [2(20)^2 + 48(20)]}{0.1} = 128.2 \text{ ft / sec}$$

Its average velocity over [20,20.01]

$$\frac{f(20.01) - f(20)}{20.01 - 20} = \frac{[2(20.01)^2 + 48(20.01)] - [2(20)^2 + 48(20)]}{0.01} = 128.02 \text{ ft / sec}$$

b. We first compute $f'(t)$ using the four-step process.

Step 1 $f(t+h) = 2(t+h)^2 + 48(t+h) = 2t^2 + 4th + 2h^2 + 48t + 48h$

Step 2 $f(t+h) - f(t) = (2t^2 + 4th + 2h^2 + 48t + 48h) - (2t^2 + 48t)]$

$$= 4th + 2h^2 + 48h = h(4t + 2h + 48).$$

Step 3 $\dfrac{f(t+h) - f(t)}{h} = \dfrac{h(4t + 2h + 48)}{h} = 4t + 2h + 48$

Step 4 $f'(t) = \lim\limits_{t \to 0} \dfrac{f(t+h) - f(t)}{h} = \lim\limits_{t \to 0} 4t + 2h + 48 = 4t + 48$

The instantaneous velocity of the car at $t = 20$ is $f'(20) = 4(20) + 48$, or 128 ft/sec.

c. Our results shows that the average velocities do approach the instantaneous velocity as the intervals over which they are computed decreases.

31. a. We solve the equation $16t^2 = 400$ obtaining $t = 5$ which is the time it takes the screw driver to reach the ground.

b. The average velocity over the time [0, 5] is

$$\frac{f(5) - f(0)}{5 - 0} = \frac{16(25) - 0}{5} = 80, \text{ or } 80 \text{ ft/sec.} \quad [\text{Let } s = f(t) = 16t^2.]$$

c. The velocity of the screwdriver at time t is

$$v(t) = \lim_{h \to 0} \frac{f(t+h) - f(t)}{h} = \lim_{h \to 0} \frac{16(t+h)^2 - 16t^2}{h}$$

$$= \lim_{h \to 0} \frac{16t^2 + 32th + 16h^2 - 16t^2}{h} = \lim_{h \to 0} \frac{(32t + 16h)h}{h} = 32t.$$

In particular, the velocity of the screwdriver when it hits the ground (at $t = 5$) is
$v(5) = 32(5) = 160$, or 160 ft/sec.

33. a. $V = \dfrac{1}{p}$. The average rate of change of V is

$$\frac{f(3)-f(2)}{3-2} = \frac{\frac{1}{3}-\frac{1}{2}}{1} = -\frac{1}{6}, \qquad \left[\text{Write } V = f(p) = \frac{1}{p}.\right]$$

or a decrease of $\frac{1}{6}$ liter/atmosphere.

b.

$$V'(t) = \lim_{h\to 0}\frac{\frac{f(p+h)-f(p)}{h}}{h} = \lim_{h\to 0}\frac{\frac{1}{p+h}-\frac{1}{p}}{h}$$

$$= \lim_{h\to 0}\frac{p-(p+h)}{hp(p+h)} = \lim_{h\to 0} - \frac{1}{p(p+h)} = -\frac{1}{p^2}.$$

In particular, the rate of change of V when $p = 2$ is
$V'(2) = -\dfrac{1}{2^2}$, or a decrease of $\frac{1}{4}$ liter/atmosphere

35. a. Using the four-step process, we find that

$$P'(x) = \lim_{h\to 0}\frac{P(x+h)-P(x)}{h}$$

$$= \lim_{h\to 0}\frac{-\frac{1}{3}(x^2+2xh+h^2)+7x+7h+30-(-\frac{1}{3}x^2+7x+30)}{h}$$

$$P'(x) = \lim_{h\to 0}\frac{P(x+h)-P(x)}{h}$$

$$= \lim_{h\to 0}\frac{-\frac{2}{3}xh-\frac{1}{3}h+7h}{h} = \lim_{h\to 0}(-\tfrac{2}{3}x-\tfrac{1}{3}h+7) = -\tfrac{2}{3}x+7.$$

b. $P'(10) = -\frac{2}{3}(10)+7 \approx 0.333$, or \$333 per quarter.
$P'(30) = -\frac{2}{3}(30)+7 \approx -13$, or a decrease of \$13,000 per quarter.

37. $N(t) = t^2 + 2t + 50$. We first compute $N'(t)$ using the four–step process.

Step 1 $N(t+h) = (t+h)^2 + 2(t+h) + 50$
$= t^2 + 2th + h^2 + 2t + 2h + 50$

Step 2 $N(t+h) - N(t)$
$= (t^2 + 2th + h^2 + 2t + 2h + 50) - (t^2 + 2t + 50)$
$= 2th + h^2 + 2h = h(2t + h + 2)$.

Step 3 $\dfrac{N(t+h)-N(t)}{h}=2t+h+2.$

Step 4 $N'(t)=\lim_{h\to 0}(2t+h+2)=2t+2.$

The rate of change of the country's GNP two years from now will be $N'(2)=6$, or \$6 billion/yr. The rate of change four years from now will be $N'(4)=10$, or \$10 billion/yr.

39. $\dfrac{f(a+h)-f(a)}{h}$ gives the average rate of change of the seal population over the time interval $[a, a+h]$.

$\lim_{h\to 0}\dfrac{f(a+h)-f(a)}{h}$ gives the instantaneous rate of change of the seal population at $x=a$.

41. $\dfrac{f(a+h)-f(a)}{h}$ gives the average rate of change of the country's industrial production over the time interval $[a, a+h]$.

$\lim_{h\to 0}\dfrac{f(a+h)-f(a)}{h}$ gives the instantaneous rate of change of the country's industrial production at $x=a$.

43. $\dfrac{f(a+h)-f(a)}{h}$ gives the average rate of change of the atmospheric pressure over the altitudes $[a, a+h]$.

$\lim_{h\to 0}\dfrac{f(a+h)-f(a)}{h}$ gives the instantaneous rate of change of the atmospheric pressure at $x=a$.

45. a. f has a limit at $x=a$.
b. f is not continuous at $x=a$ because $f(a)$ is not defined.
c. f is not differentiable at $x=a$ because it is not continuous there.

47. a. f has a limit at $x=a$. b. f is continuous at $x=a$.
c. f is not differentiable at $x=a$ because f has a kink at the point $x=a$.

49. a. f does not have a limit at $x = a$ because it is unbounded in the neighborhood of a.
b. f is not continuous at $x = a$.
c. f is not differentiable at $x = a$ because it is not continuous there.

51. Our computations yield the following results:
32.1, 30.939, 30.814, 30.8014, 30.8001, 30.8000.
The motorcycle's instantaneous velocity at $t = 2$ is approximately 30.8 ft/sec.

53. False. Let $f(x) = |x|$. Then f is continuous at $x = 0$, but is not differentiable there.

55. Observe that the graph of f has a kink at $x = -1$.
We have

$$\frac{f(-1+h) - f(-1)}{h} = 1 \text{ if } h > 0, \text{ and } -1 \text{ if } h < 0,$$

so that $\lim\limits_{h \to 0} \frac{f(-1+h) - f(-1)}{h}$ does not exist.

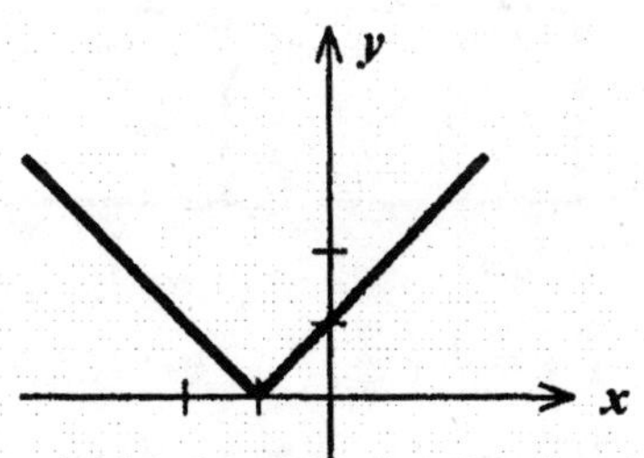

57. For continuity, we require that $f(1) = 1 = \lim\limits_{x \to 1^+}(ax + b) = a + b$, or $a + b = 1$.
In order that the derivative exist at $x = 1$, we require that $\lim\limits_{x \to 1^-} 2x = \lim\limits_{x \to 1^+} a$, or $2 = a$.
Therefore, $b = -1$ and so $f(x) = \begin{cases} x^2 & \text{if } x \leq 1 \\ 2x - 1 & \text{if } x > 1 \end{cases}$. The graph of f follows.

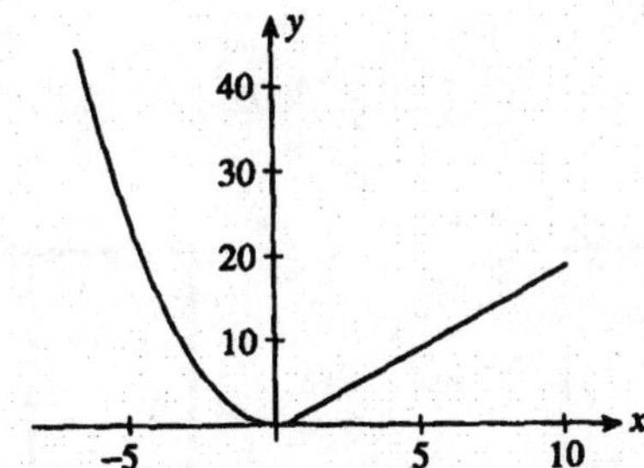

59. We have $f(x) = x$ if $x > 0$ and $f(x) = -x$ if $x < 0$. Therefore, when $x > 0$

$$f'(x) = \lim_{h \to 0} \frac{f(x+h) - f(x)}{h} = \lim_{h \to 0} \frac{x + h - x}{h} = \lim_{h \to 0} \frac{h}{h} = 1,$$

and when $x < 0$

$$f'(x) = \lim_{h \to 0} \frac{f(x+h) - f(x)}{h} = \lim_{h \to 0} \frac{-x - h - (-x)}{h} = \lim_{h \to 0} \frac{-h}{h} = -1.$$

Since the right–hand limit does not equal the left–hand limit, we conclude that $\lim_{h \to 0} f(x)$ does not exist.

USING TECHNOLOGY EXERCISES 11.6, page 678

1. a. $y = 4x - 3$

   b.

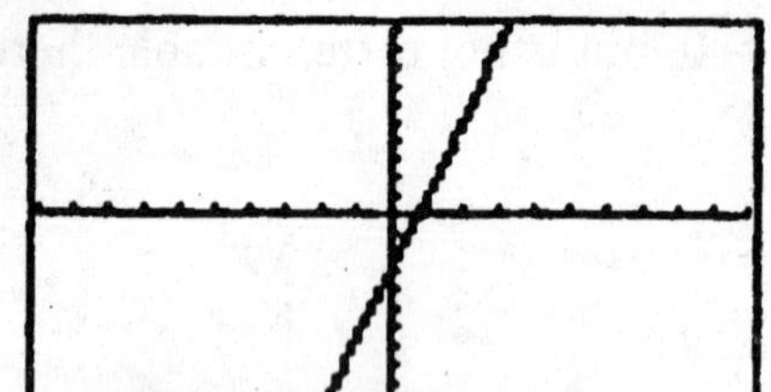

3. a. $y = 9x - 11$

   b.

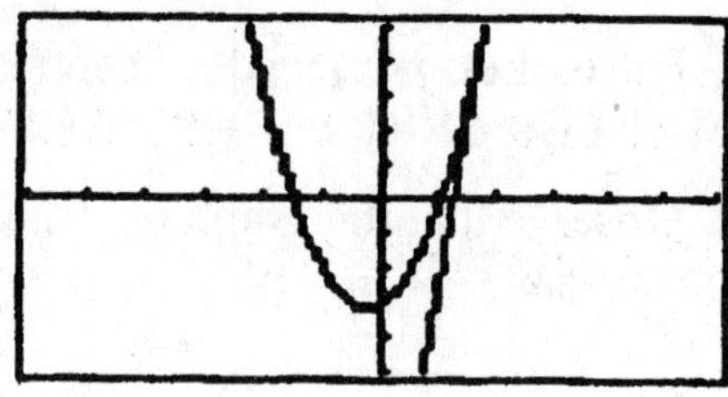

5. a. $y = \frac{1}{4}x + 1$

   b.

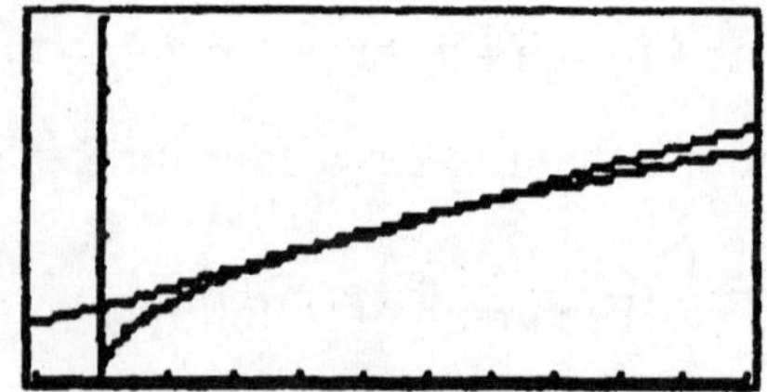

7. a. 4   b. $y = 4x - 1$

   c.

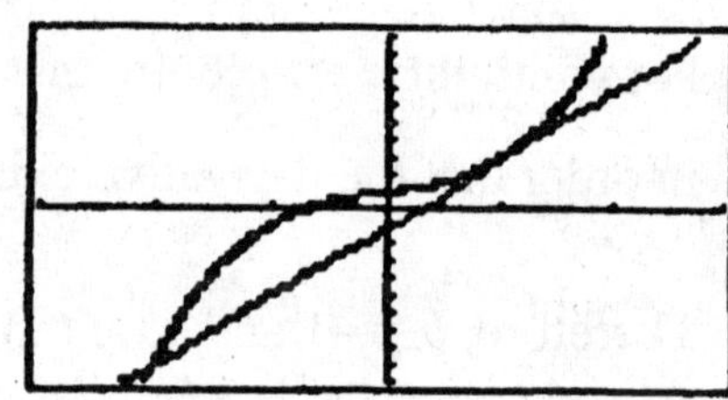

9. a. $\frac{3}{4}$ b. $y = \frac{3}{4}x - 1$

   c.

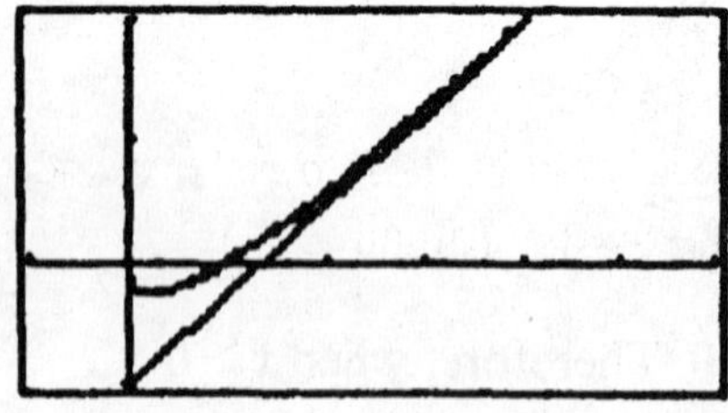

11. a. 4.02   b. $y = 4.02x - 3.57$

    c.

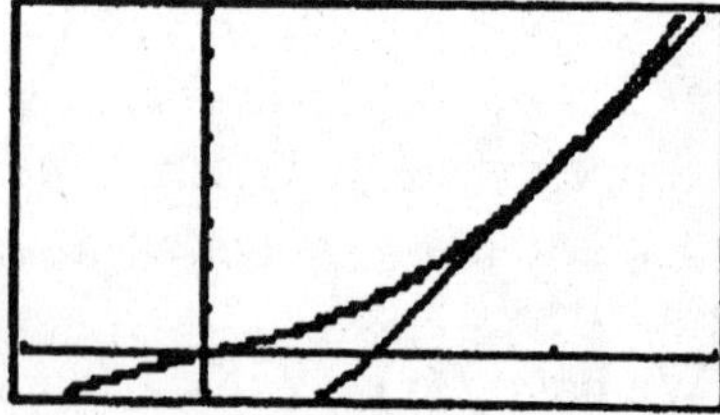

13. a. b. 41.22 cents/mile c. 1.22 cents/mile/yr

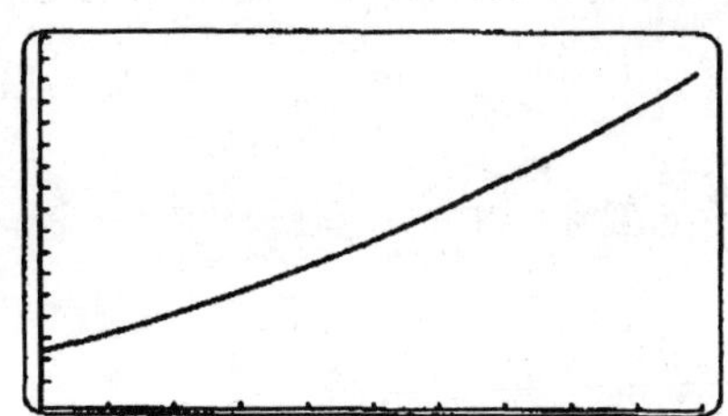

CHAPTER 11 CONCEPT REVIEW, page 679

1. Domain; range; B

3. $f(x) \pm g(x)$; $f(x)g(x)$; $\dfrac{f(x)}{g(x)}$; $A \cap B$; $A \cap B$; zero

5. a. $P(x) = a_0x^n + a_1x^{n-1} + \cdots + a_{n-1}x + a_n$ ($a_0 \neq 0$, n, a positive integer)
   b. Linear; quadratic; cubic c. Quotient; polynomials
   d. x^r, r, a real number

7. a. L^r b. $L \pm M$ c. LM d. $\dfrac{L}{M}$; $M \neq 0$

9. a. Right b. Left c. L; L 11. a. a; a: $g(a)$ b. Everywhere c. $Q(x)$

13. a. $f'(a)$ b. $y - f(a) = m(x - a)$

CHAPTER 11 REVIEW, page 680

1. a. $9 - x \geq 0$ gives $x \leq 9$ and the domain is $(-\infty, 9]$.
   b. $2x^2 - x - 3 = (2x - 3)(x + 1)$, and $x = 3/2$ or $x = -1$.
   Since the denominator of the given expression is zero at these points, we see that the domain of f cannot include these points and so the domain of f is
   $(-\infty, -1) \cup (-1, \frac{3}{2}) \cup (\frac{3}{2}, \infty)$.

3. a.

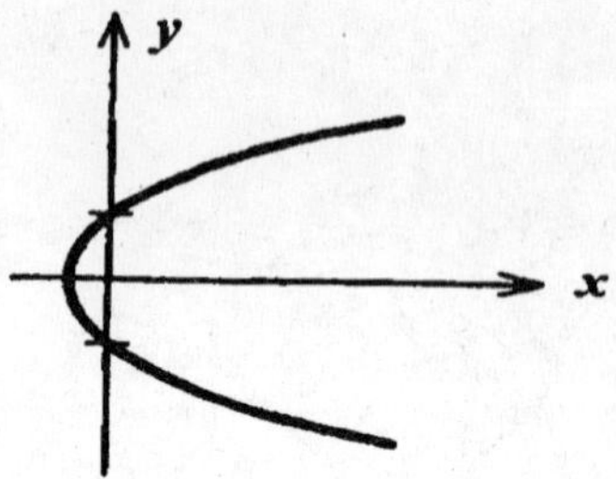

b. For each value of $x > 0$, there are two values of y. We conclude that y is not a function of x. Equivalently, the function fails the vertical line test.
c. Yes. For each value of y, there is only 1 value of x.

5. a. $f(x)g(x) = \dfrac{2x+3}{x}$ b. $\dfrac{f(x)}{g(x)} = \dfrac{1}{x(2x+3)}$

c. $f(g(x)) = \dfrac{1}{2x+3}$. d. $g(f(x)) = 2\left(\dfrac{1}{x}\right) + 3 = \dfrac{2}{x} + 3$.

7. $\lim\limits_{x\to 1}(x^2+1) = [(1)^2+1] = 1+1 = 2$. 9. $\lim\limits_{x\to 3}\dfrac{x-3}{x+4} = \dfrac{3-3}{3+4} = 0$.

11. $\lim\limits_{x\to -2}\dfrac{x^2-2x-3}{x^2+5x+6}$ does not exist. (The denominator is 0 at $x = -2$.)

13. $\lim\limits_{x\to 3}\dfrac{4x-3}{\sqrt{x+1}} = \dfrac{12-3}{\sqrt{4}} = \dfrac{9}{2}$.

15. $\lim\limits_{x\to 1^-}\dfrac{\sqrt{x}-1}{x-1} = \lim\limits_{x\to 1^-}\dfrac{(\sqrt{x}-1)(\sqrt{x}+1)}{(x-1)(\sqrt{x}+1)} = \lim\limits_{x\to 1^-}\dfrac{x-1}{(x-1)(\sqrt{x}+1)} = \lim\limits_{x\to 1^-}\dfrac{1}{\sqrt{x}+1} = \dfrac{1}{2}$.

17. $\lim\limits_{x\to -\infty}\dfrac{x+1}{x} = \lim\limits_{x\to -\infty}\left(1+\dfrac{1}{x}\right) = 1$.

19. $\lim_{x\to-\infty}\frac{x^2}{x+1}=\lim_{x\to-\infty}x\cdot\frac{1}{1+\frac{1}{x}}=-\infty$, so the limit does not exist.

21. $\lim_{x\to2^+}f(x)=\lim_{x\to2^+}(x+2)=4;\qquad \lim_{x\to2^-}f(x)=\lim_{x\to2^-}(4-x)=2.$

Therefore, $\lim_{x\to2}f(x)$ does not exist.

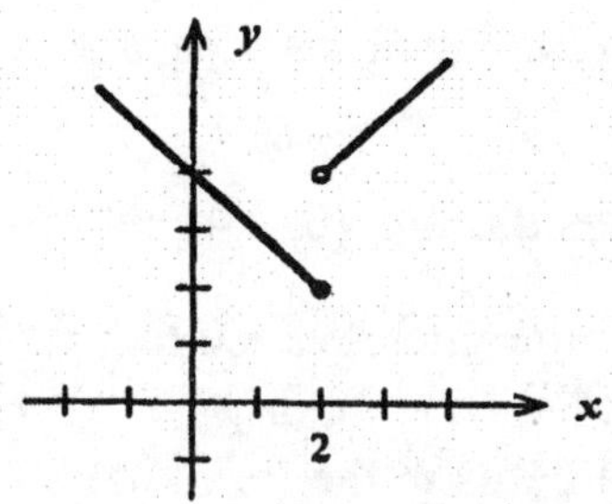

23. Since the denominator

$$4x^2-2x-2=2(2x^2-x-1)=2(2x+1)(x-1)=0$$

if $x=-1/2$ or 1, we see that f is discontinuous at these points.

25. The function is discontinuous at $x=0$.

27. $f(x)=3x+5$. Using the four-step process, we find

Step 1 $f(x+h)=3(x+h)+5=3x+3h+5$

Step 2 $f(x+h)-f(x)=3x+3h+5-3x-5=3h$

Step 3 $\frac{f(x+h)-f(x)}{h}=\frac{3h}{h}=3.$

Step 4 $f'(x)=\lim_{h\to0}\frac{f(x+h)-f(x)}{h}=\lim_{h\to0}(3)=3.$

29. $f(x)=\frac{3}{2}x+5$. We use the four-step process to obtain

Step 1 $f(x+h)=\frac{3}{2}(x+h)+5=\frac{3}{2}x+\frac{3}{2}h+5.$

Step 2 $f(x+h)-f(x)=\frac{3}{2}x+\frac{3}{2}h+5-\frac{3}{2}x-5=\frac{3}{2}h.$

Step 3 $\frac{f(x+h)-f(x)}{h}=\frac{3}{2}.$

Step 4 $f'(x)=\lim_{h\to0}\frac{f(x+h)-f(x)}{h}=\lim_{h\to0}\frac{3}{2}=\frac{3}{2}.$

Therefore, the slope of the tangent line to the graph of the function f at the point (-2,2) is 3/2. To find the equation of the tangent line to the curve at the point

(-2,2), we use the point–slope form of the equation of a line obtaining

$$y-2=\tfrac{3}{2}[x-(-2)] \quad \text{or} \quad y=\tfrac{3}{2}x+5.$$

31. a. f is continuous at $x = a$ because the three conditions for continuity are satisfied at $x = a$; that is, *i.* $f(x)$ is defined *ii.* $\lim_{x\to a} f(x)$ exists *iii.* $\lim_{x\to a} f(x) = f(a)$

b. f is not differentiable at $x = a$ because the graph of f has a kink at $x = a$.

33. a. The line passes through (0, 2.4) and (5, 7.4) and has slope $m=\dfrac{7.4-2.4}{5-0}=1.$

Letting y denote the sales, we see that an equation of the line is

$y - 2.4 = 1(t - 0)$, or $y = t + 2.4$.

We can also write this in the form $S(t) = t + 2.4$.

b. The sales in 2004 were $S(3) = 3 + 2.4 = 5.4$, or \$5.4 million.

35. Substituting the first equation into the second yields

$$3x-2(\tfrac{3}{4}x+6)+3=0 \quad \text{or} \quad \tfrac{3}{2}x-12+3=0$$

or $x = 6$. Substituting this value of x into the first equation then gives $y = 21/2$, so the point of intersection is $(6, \tfrac{21}{2})$.

37. We solve the system

$$3x + p - 40 = 0$$
$$2x - p + 10 = 0.$$

Adding these two equations, we obtain $5x - 30 = 0$, or $x = 6$. So,

$p = 2x + 10 = 12 + 10 = 22$.

Therefore, the equilibrium quantity is 6000 and the equilibrium price is \$22.

39. When 1000 units are produced,

$R(1000) = -0.1(1000)^2 + 500(1000) = 400{,}000$, or \$400,000.

41. $N(0) = 200(4 + 0)^{1/2} = 400$, and so there are 400 members initially.
$N(12) = 200(4 + 12)^{1/2} = 800$, and so there are 800 members a year later.

43. $T = f(n) = 4n\sqrt{n-4}$.

$f(4) = 0,\ f(5) = 20\sqrt{1} = 20,\ f(6) = 24\sqrt{2} \approx 33.9,\ f(7) = 28\sqrt{3} \approx 48.5,$

$f(8) = 32\sqrt{4} = 64,\ f(9) = 36\sqrt{5} \approx 80.5,\ f(10) = 40\sqrt{6} \approx 98,$

$f(11) = 44\sqrt{7} \approx 116$ and $f(12) = 48\sqrt{8} \approx 135.8.$

The graph of f follows:

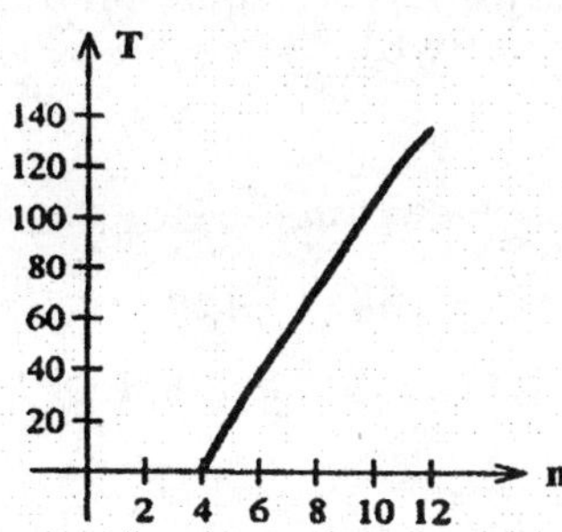

45. We solve

$$-1.1x^2+1.5x+40=0.1x^2+0.5x+15$$
$$1.2x^2-x-25=0$$
$$12x^2-10x-250=0$$
$$6x^2-5x-125=0;\quad (x-5)(6x+25)=0.$$

Therefore, $x=5$. Substituting this value of x into the second supply equation, we have $p=0.1(5)^2+0.5(5)+15=20$. So the equilibrium quantity is 5000 and the equilibrium price is \$20.

47.

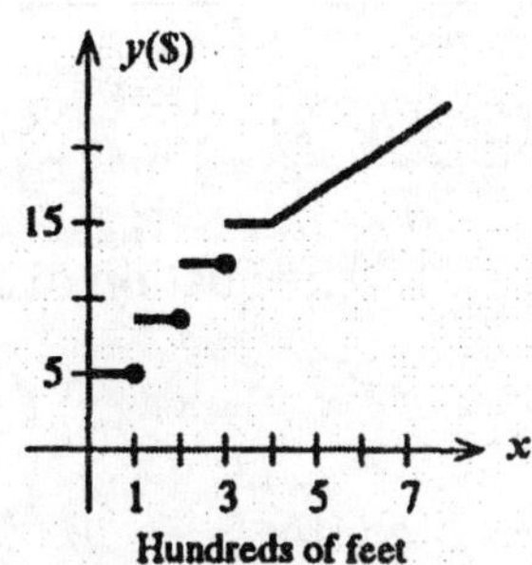

CHAPTER 11 BEFORE MOVING ON, page 682

1. a. $f(-1)=-2(-1)+1=3$ b. $f(0)=2$ c. $f(\frac{3}{2})=(\frac{3}{2})^2+2=\frac{17}{4}$.

2. a. $(f+g)(x)=f(x)+g(x)=\dfrac{1}{x+1}+x^2+1$ b. $(fg)(x)=f(x)g(x)=\dfrac{x^2+1}{x+1}$

c. $(f\circ g)(x)=f[g(x)]=\dfrac{1}{g(x)+1}=\dfrac{1}{x^2+2}$

d. $(g \circ f)(x) = g[f(x)]^2 + 1 = \dfrac{1}{(x+1)^2} + 1$

3. $4x + h = 108$ so $h = 108 - 4x$
Volume is $V = x^2 h = x^2(108 - 4x) = 108x^2 - 4x^3$

4. $\lim_{x \to -1} \dfrac{x^2 + 4x + 3}{x^2 + 3x + 2} = \lim_{x \to -1} \dfrac{(x+3)(x+1)}{(x+2)(x+1)} = 2$

5. a. $\lim_{x \to 1^-} f(x) = \lim_{x \to 1^-} (x^2 - 1) = 0$
b. $\lim_{x \to 1^+} f(x) = \lim_{x \to 1^+} x^3 = 1$.
Since $\lim_{x \to 1^-} f(x) \neq \lim_{x \to 1^+} f(x)$, f is not continuous at 1.

6. The slope of the tangent line at any point is

$$\lim_{h \to 0} \frac{f(x+h) - f(x)}{h} = \lim_{h \to 0} \frac{(x+h)^2 - 3(x+h) + 1 - (x^2 - 3x + 1)}{h}$$

$$= \lim_{h \to 0} \frac{x^2 + 2xh + h^2 - 3x - 3h - x^2 + 3x - 1}{h}$$

$$= \lim_{h \to 0} \frac{h(2x + h - 3)}{h} = \lim_{h \to 0} (2x + h - 3) = 2x - 3$$

Therefore, the slope at 1 is $2(1) - 3 = -1$. An equation of the tangent line is

$$y - (-1) = -1(x - 1)$$

$$y + 1 = -x + 1 \quad \text{, or} \quad y = -x$$

CHAPTER 12

12.1 Problem Solving Tips

In this section, you were given four basic rules for finding the derivative of a function. As you work through the exercises that follow, first decide which rule(s) you need to find the derivative. Then write out your solution. After doing this a few times, you should have the formulas memorized. The key here is to try not to look at the formula in the text, and to work the problem just as if you were taking a test. If you train yourself to work in this manner, test-taking will be a lot easier. Also, make sure that you distinguish between the notation *dy*/*dx* and *d*/*dx*. The first notation is used for *the derivative of a function y*, where as the second notation tells us *to find the derivative of the function that follows* with respect to *x*.

Here are some hints for solving the problems in the exercises that follow:

1. **To find the derivative of a function involving radicals**, first rewrite the expression in exponential form. For example, if $f(x) = 2x - 5\sqrt{x}$, rewrite the function in the form $f'(x) = 2x - 5x^{1/2}$.

2. **To find the point on the graph of *f* where the tangent line is horizontal,** simply set $f'(x) = 0$ and solve for *x*. (Here we are making use of the fact that the slope of a horizontal line is zero.) This yields the *x*-value of the point on the graph where the

tangent line is horizontal. To find the corresponding y-value, simply evaluate the function f at this value of x.

12.1 CONCEPT QUESTIONS, page 691

1. a. The derivative of a constant is zero.
   b. The derivative of $f(x)=x^n$ is the power times x raised to the power $n-1$.
   c. The derivative of a constant times a function is the constant times the derivative of the function.
   d. The derivative of the sum is the sum of the derivatives.

EXERCISES 12.1, page 691

1. $f'(x)=\dfrac{d}{dx}(-3)=0.$

3. $f'(x)=\dfrac{d}{dx}\left(x^5\right)=5x^4.$

5. $f'(x)=\dfrac{d}{dx}\left(x^{2.1}\right)=2.1x^{1.1}.$

7. $f'(x)=\dfrac{d}{dx}\left(3x^2\right)=6x.$

9. $f'(r)=\dfrac{d}{dr}\left(\pi r^2\right)=2\pi r.$

11. $f'(x)=\dfrac{d}{dx}\left(9x^{1/3}\right)=\dfrac{1}{3}(9)x^{(1/3-1)}=3x^{-2/3}.$

13. $f'(x)=\dfrac{d}{dx}\left(3\sqrt{x}\right)=\dfrac{d}{dx}\left(3x^{1/2}\right)=\dfrac{1}{2}(3)x^{-1/2}=\dfrac{3}{2}x^{-1/2}=\dfrac{3}{2\sqrt{x}}.$

15. $f'(x)=\dfrac{d}{dx}\left(7x^{-12}\right)=(-12)(7)x^{(-12-1)}=-84x^{-13}.$

17. $f'(x)=\dfrac{d}{dx}\left(5x^2-3x+7\right)=10x-3.$

19. $f'(x)=\dfrac{d}{dx}\left(-x^3+2x^2-6\right)=-3x^2+4x.$

21. $f'(x)=\dfrac{d}{dx}\left(0.03x^2-0.4x+10\right)=0.06x-0.4.$

23. If $f(x) = \dfrac{x^3 - 4x^2 + 3}{x} = x^2 - 4x + \dfrac{3}{x}$,

then $f'(x) = \dfrac{d}{dx}\left(x^2 - 4x + 3x^{-1}\right) = 2x - 4 - \dfrac{3}{x^2}$.

25. $f'(x) = \dfrac{d}{dx}\left(4x^4 - 3x^{5/2} + 2\right) = 16x^3 - \frac{15}{2}x^{3/2}$.

27. $f'(x) = \dfrac{d}{dx}\left(3x^{-1} + 4x^{-2}\right) = -3x^{-2} - 8x^{-3}$.

29. $f'(t) = \dfrac{d}{dt}\left(4t^{-4} - 3t^{-3} + 2t^{-1}\right) = -16t^{-5} + 9t^{-4} - 2t^{-2}$.

31. $f'(x) = \dfrac{d}{dx}\left(2x - 5x^{1/2}\right) = 2 - \dfrac{5}{2}x^{-1/2} = 2 - \dfrac{5}{2\sqrt{x}}$.

33. $f'(x) = \dfrac{d}{dx}\left(2x^{-2} - 3x^{-1/3}\right) = -4x^{-3} + x^{-4/3} = -\dfrac{4}{x^3} + \dfrac{1}{x^{4/3}}$.

35. a. $f'(x) = \dfrac{d}{dx}\left(2x^3 - 4x\right) = 6x^2 - 4$. $f'(-2) = 6(-2)^2 - 4 = 20$.

b. $f'(0) = 6(0) - 4 = -4$.

c. $f'(2) = 6(2)^2 - 4 = 20$.

37. The given limit is $f'(1)$ where $f(x) = x^3$. Since $f'(x) = 3x^2$, we have

$$\lim_{h\to 0}\frac{(1+h)^3 - 1}{h} = f'(1) = 3.$$

39. Let $f(x) = 3x^2 - x$. Then $\displaystyle\lim_{h\to 0}\frac{3(2+h)^2 - (2+h) - 10}{h} = \lim_{h\to 0}\frac{f(2+h) - f(2)}{h}$

because $f(2 + h) - f(2) = 3(2 + h)^2 - (2 + h) - [3(4) - 2]$

$= 3(2 + h)^2 - (2 + h) - 10$.

But the last limit is $f'(2)$. Since $f'(x) = 6x - 1$, we have $f'(2) = 11$.

Therefore, $\displaystyle\lim_{h\to 0}\frac{3(2+h)^2 - (2+h) - 10}{h} = 11$.

41. $f(x)=2x^2-3x+4$. The slope of the tangent line at any point $(x, f(x))$ on the graph of f is $f'(x)=4x-3$. In particular, the slope of the tangent line at the point (2,6) is $f'(2) = 4(2) - 3 = 5$. An equation of the required tangent line is

$$y - 6 = 5(x - 2) \quad \text{or} \quad y = 5x - 4.$$

43. $f(x)=x^4-3x^3+2x^2-x+1$. $f'(x)=4x^3-9x^2+4x-1$.
The slope is $f'(1) = 4 - 9 + 4 - 1 = -2$. An equation of the tangent line is $y - 0 = -2(x - 1)$ or $y = -2x + 2$.

45. a. $f'(x)=3x^2$. At a point where the tangent line is horizontal, $f'(x)=0$, or $3x^2=0$ giving $x=0$. Therefore, the point is (0,0).

b.

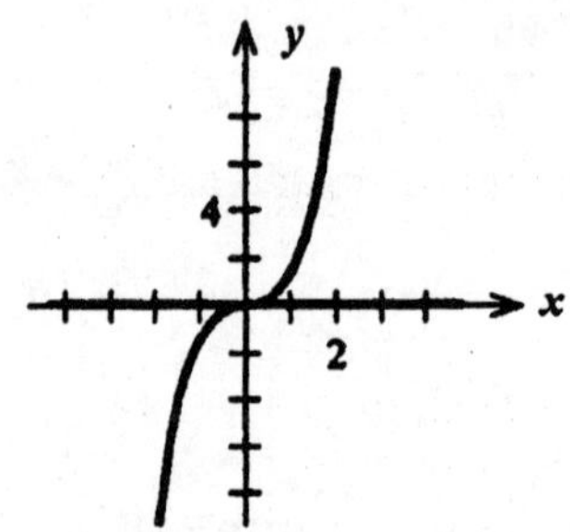

47. a. $f(x)=x^3+1$. The slope of the tangent line at any point $(x, f(x))$ on the graph of f is $f'(x)=3x^2$. At the point(s) where the slope is 12, we have $3x^2 = 12$, or $x = \pm 2$. The required points are (-2,-7) and (2,9).
b. The tangent line at (-2,-7) has equation

$$y - (-7) = 12[x - (-2)], \quad \text{or} \quad y = 12x + 17,$$

and the tangent line at (2,9) has equation

$$y - 9 = 12(x - 2), \quad \text{or} \quad y = 12x - 15.$$

c.

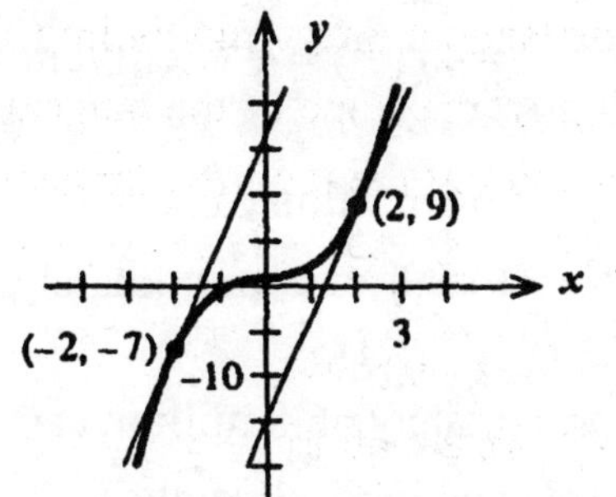

49. If $f(x)=\frac{1}{4}x^4-\frac{1}{3}x^3-x^2$, then $f'(x)=x^3-x^2-2x$.

a. $f'(x)=x^3-x^2-2x=-2x$

$$x^3-x^2=0$$

$$x^2(x-1)=0 \qquad \text{and} \qquad x=0 \text{ or } x=1.$$

$$f(1)=\tfrac{1}{4}(1)^4-\tfrac{1}{3}(1)^3-(1)^2=-\tfrac{13}{12}.$$

$$f(0)=\tfrac{1}{4}(0)^4-\tfrac{1}{3}(0)^3-(0)^2=0.$$

We conclude that the corresponding points on the graph are $(1,-\frac{13}{12})$ and $(0,0)$.

b. $f'(x)=x^3-x^2-2x=0$

$$x(x^2-x-2)=0$$

$$x(x-2)(x+1)=0 \quad \text{and} \quad x=0,\ 2, \text{ or } -1.$$

$$f(0)=0$$

$$f(2)=\frac{1}{4}(2)^4-\frac{1}{3}(2)^3-(2)^2=4-\frac{8}{3}-4=-\frac{8}{3}.$$

$$f(-1)=\frac{1}{4}(-1)^4-\frac{1}{3}(-1)^3-(-1)^2=\frac{1}{4}+\frac{1}{3}-1=-\frac{5}{12}.$$

We conclude that the corresponding points are $(0,0)$, $(2,-\frac{8}{3})$ and $(-1,-\frac{5}{12})$.

c. $f'(x)=x^3-x^2-2x=10x$

$$x^3-x^2-12x=0$$

$$x(x^2-x-12)=0$$

$$x(x-4)(x+3)=0$$

and $x=0$, 4, or -3.

$$f(0)=0$$

$$f(4)=\tfrac{1}{4}(4)^4-\tfrac{1}{3}(4)^3-(4)^2=48-\tfrac{64}{3}=\tfrac{80}{3}.$$

$$f(-3)=\tfrac{1}{4}(-3)^4-\tfrac{1}{3}(-3)^3-(-3)^2=\tfrac{81}{4}+9-9=\tfrac{81}{4}.$$

We conclude that the corresponding points are $(0,0)$, $(4,\frac{80}{3})$ and $(-3,\frac{81}{4})$.

51. $V(r)=\frac{4}{3}\pi r^3$. $V'(r)=4\pi r^2$.

a. $V'(\frac{2}{3})=4\pi(\frac{4}{9})=\frac{16}{9}\pi$ cm^3/cm. b. $V'(\frac{5}{4})=4\pi(\frac{25}{16})=\frac{25}{4}\pi$ cm^3/cm.

53. a. $N(1)=16.3(1^{0.8766})=16.3(1)=16.3$, or 16.3 million cameras.

b. $N'(t)=(16.3)90.8766)t^{-0.1234}$; $N'(1)\approx 14.29$, or approximately 14.3 million cameras/yr.

c. $N(5)=16.3(5^{0.8766})\approx 66.82$, or approximately 66.8 million cameras.

d. $N'(5)=(16.3)(0.8766)(5^{-0.1234})\approx 11.71$, or approximately 11.7 million cameras/year.

55. a.

1970 ($t=1$)	1980($t=2$)	1990 ($t=3$)	2000($t=4$)
49.6%	41.1%	36.9%	34.1%

b. $P'(t)=(49.6)(-0.27t^{-1.27})=-\dfrac{13.392}{t^{1.27}}$; In 1980, $P'(2)\approx -5.5$, or decreasing at 5.5%/decade. In 1990, $P'(3)\approx -3.3$, or decreasing at 3.3%/decade.

57. a. $P(9)=24.4(9)^{0.34}=51.5$, or 51.5%.

b. $P(t)=24.4t^{0.34}$; $P'(t)=\dfrac{d}{dx}(24.4t^{0.34})=(0.34)(24.4)t^{-0.66}=8.296t^{-0.66}$.

$P'(9)=8.296(9)^{-0.66}=1.946$, or approximately 1.95%/year.

59. a. $f(t)=120t-15t^2$. $v=f'(t)=120-30t$ b. $v(0)=120$ ft/sec

c. Setting $v=0$ gives $120-30t=0$, or $t=4$. Therefore, the stopping distance is $f(4)=120(4)-15(16)$ or 240 ft.

61. a. At the beginning of 1980, $P(0)=5\%$. At the beginning of 1990, $P(10)=-0.0105(10^2)+0.735(10)+5\approx 11.3\%$. At the beginning of 2000, $P(20)=-0.0105(20)^2+0.735(20)+5\approx 15.5\%$.

b. $P'(t)=-0.021t+0.735$; At the beginning of 1985, $P'(5)=-0.02(5)+0.735\approx 0.63\%/\text{yr}$. At the beginning of 1990, $P'(10)=-0.021(10)+0.735=0.525\%/\text{yr}$.

63. a $f(t) = 5.303t^2 - 53.977t + 253.8$. The rate of change of the groundfish population at any time t is given by $f'(t) = 10.606t - 53.977$. The rate of change at the beginning of 1994 is given by $f'(5) = 10.606(5) - 53.977 = -0.947$ and so the population is decreasing at the rate of 0.9 thousand metric tons/yr. At the beginning of 1996, the rate of change is $f'(7) = 10.606(7) - 53.977 = 20.265$ and so the population is increasing at the rate of 20.3 thousand metric tons/yr.
b. Yes.

65. $I'(t) = -0.6t^2 + 6t$.

a. In 2002, it was changing at a rate of $I'(5) = -0.6(25) + 6(5)$, or 15 points/yr. In 2004, it was $I'(7) = -0.6(49) + 6(7)$, or 12.6 pts/yr. In 2007, it was $I'(10) = -0.6(100) + 6(10)$, or 0 pts/yr.
b. The average rate of increase of the CPI over the period from 2002 to 2007 was

$$\frac{I(10) - I(5)}{5} = \frac{[-0.2(1000) + 3(100) + 100] - [-0.2(125) + 3(25) + 100]}{5}$$

$$= \frac{200 - 150}{5} = 10, \text{ or } 10 \text{ pts/yr.}$$

67. a. $f'(x) = \frac{d}{dx}\left[0.0001x^{5/4} + 10\right] = \frac{5}{4}(0.0001x^{1/4}) = 0.000125x^{1/4}$

b. $f'(10{,}000) = 0.000125(10{,}000)^{1/4} = 0.00125$, or \$0.00125/radio.

69. a. $f(t) = 20t - 40\sqrt{t} + 50$. $f'(t) = 20 - 40\left(\frac{1}{2}\right)t^{-1/2} = 20\left(1 - \frac{1}{\sqrt{t}}\right)$.

b. $f(0) = 20(0) - 40\sqrt{0} + 50 = 50$; $f(1) = 20(1) - 40\sqrt{1} + 50 = 30$
$f(2) = 20(2) - 40\sqrt{2} + 50 \approx 33.43$.
The average velocity at 6, 7, and 8 A.M. is 50 mph, 30 mph, and 33.43 mph, respectively.
c. $f'(\tfrac{1}{2}) = 20 - 20(\tfrac{1}{2})^{-1/2} \approx -8.28$. $f'(1) = 20 - 20(1)^{-1/2} \approx 0$.
$f'(2) = 20 - 20(2)^{-1/2} \approx 5.86$.
At 6:30 A.M. the average velocity is decreasing at the rate of 8.28 mph/hr; at 7 A.M., it is unchanged, and at 8 A.M., it is increasing at the rate of 5.86 mph.

71. $N(t) = 2t^3 + 3t^2 - 4t + 1000$. $N'(t) = 6t^2 + 6t - 4$.
$N'(2) = 6(4) + 6(2) - 4 = 32$, or 32 turtles/yr.
$N'(8) = 6(64) + 6(8) - 4 = 428$, or 428 turtles/yr.
The population ten years after implementation of the conservation measures will be $N(10) = 2(10^3) + 3(10^2) - 4(10) + 1000$, or 3260 turtles.

73. a. At the beginning of 1991, $P(0) = 12\%$. At the beginning of 2004, $P(13) = 0.0004(13^3) + 0.0036(13^2) + 0.8(13) + 12 \approx 23.9\%$.

b. $P'(t) = 0.0012t^2 + 0.0072t + 0.8$. At the beginning of 1991, $P'(0) = 0.8\%/\text{yr}$. At the beginning of 2004, $P'(13) = 0.0012(13^2) + 0.0072(3) + 0.8 \approx 1.1\%/\text{yr}$.

75. True. $\frac{d}{dx}[2f(x) - 5g(x)] = \frac{d}{dx}[2f(x)] - \frac{d}{dx}[5g(x)] = 2f'(x) - 5g'(x)$.

77. $$\frac{d}{dx}(x^3) = \lim_{h\to 0}\frac{(x+h)^3 - x^3}{h} = \lim_{h\to 0}\frac{x^3 + 3x^2h + 3xh^2 + h^3 - x^3}{h}$$
$$= \lim_{h\to 0}\frac{h(3x^2 + 3xh + h^2)}{h} = \lim_{h\to 0}(3x^2 + 3xh + h^2) = 3x^2.$$

USING TECHNOLOGY EXERCISES 12.1, page 696

1. 1 3. 0.4226 5. 0.1613

7. a.

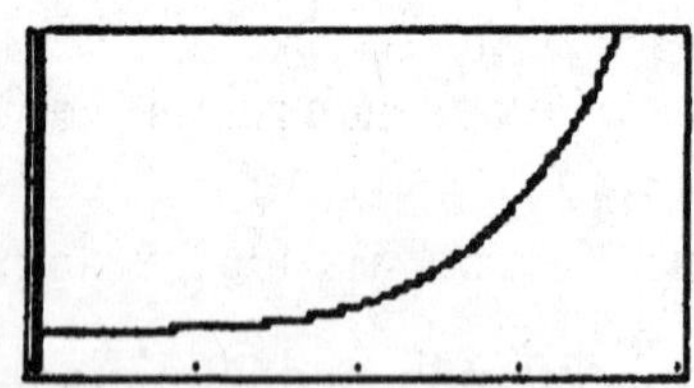

b. 3.4295 parts/million;
105.4332 parts/million

9. a.

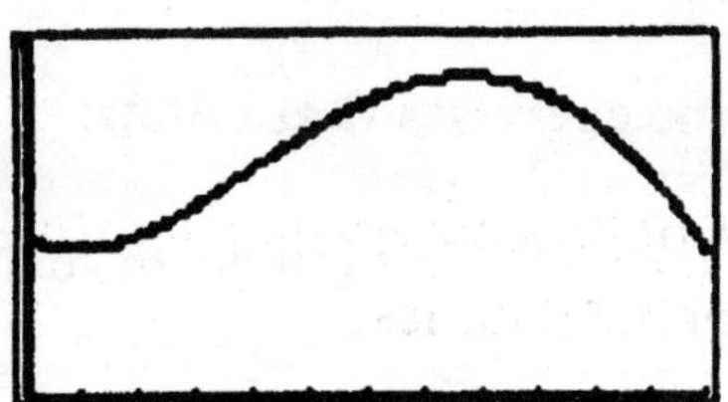

b. decreasing at the rate of 9 days/yr
increasing at the rate of 13 days/yr

11. a.

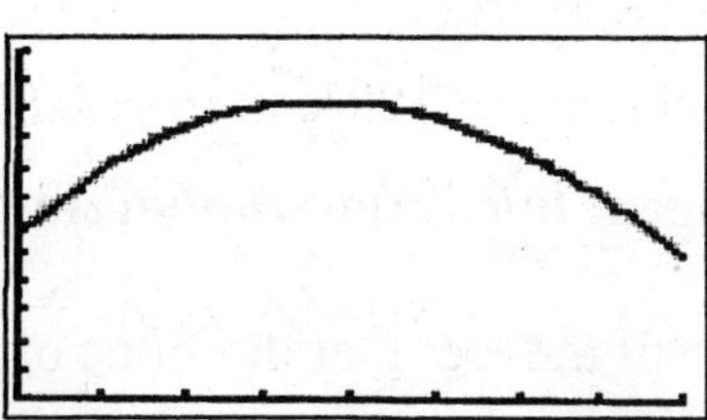

b. Increasing at the rate of 1.1557%/yr
decreasing at the rate of 0.2116%/yr

12.2 Problem Solving Tips

The answers at the back of the book for the exercises in this section are given in both simplified and unsimplified terms. Here, as with all of your homework, you should make it a practice to analyze your errors. If you do not get the right answer for the unsimplified form, it means that you are not applying the rules for differentiating correctly. In this case you need to review the rules, making sure that you can write out each rule. If you have the correct answer for the unsimplified form but the incorrect answer for the simplified form, it probably means that you have made an algebraic error. You may need to review the rules for simplifying algebraic expressions given on page 12 of the text and then work some of the exercises given in section 1.2 to get back into practice. In any case, you will need to simplify your answers when you work the problems on the applications of the derivative in the next chapter, so you should get in the habit of doing so now.

Here are some hints for solving the problems in the exercises that follow:

1. **To find the derivative of a function involving radicals,** first rewrite the expression in exponential form. For example, if $f(x) = 2x - 5\sqrt{x}$, rewrite the function in the form $f'(x) = 2x - 5x^{1/2}$.

2. **To find point on the graph of *f* where the tangent line is horizontal,** simply set $f'(x) = 0$ and solve for x. (Here we are making use of the fact that the slope of a horizontal line is zero.) This yields the x-value of the point on the graph where the tangent line is horizontal. To find the corresponding y-value, simply evaluate the function f at this value of x.

12.2 CONCEPT QUESTIONS, page 704

1. a. The derivative of the product of two functions is equal to the first function times the derivative of the second function plus the second function times the derivative of the first function.
   b. The derivative of the quotient of two functions is equal to the quotient whose numerator is given by the denominator times the derivative of the numerator minus the numerator times the derivative of the denominator and the denominator is the square of the denominator of the quotient.

EXERCISES 12.2, page 704

1. $f(x) = 2x(x^2 + 1)$.

$$f'(x) = 2x\frac{d}{dx}(x^2+1) + (x^2+1)\frac{d}{dx}(2x)$$
$$= 2x(2x) + (x^2+1)(2) = 6x^2 + 2.$$

3. $f(t) = (t - 1)(2t + 1)$

$$f'(t) = (t-1)\frac{d}{dt}\left(2t+1\right) + (2t+1)\frac{d}{dt}(t-1)$$
$$= (t-1)(2) + (2t+1)(1) \ = 4t - 1$$

5. $f(x) = (3x + 1)(x^2 - 2)$

$$f'(x) = (3x+1)\frac{d}{dx}\left(x^2-2\right) + (x^2-2)\frac{d}{dx}(3x+1)$$
$$= (3x+1)(2x) + (x^2-2)(3) = 9x^2 + 2x - 6.$$

7. $f(x) = (x^3 - 1)(x + 1).$

$$f'(x) = (x^3-1)\frac{d}{dx}\left(x+1\right) + (x+1)\frac{d}{dx}(x^3-1)$$
$$= (x^3-1)(1) + (x+1)(3x^2) = 4x^3 + 3x^2 - 1.$$

9. $f(w) = (w^3 - w^2 + w - 1)(w^2 + 2).$

$$f'(w) = (w^3 - w^2 + w - 1)\frac{d}{dw}\left(w^2+2\right) + (w^2+2)\frac{d}{dw}(w^3 - w^2 + w - 1)$$
$$= (w^3 - w^2 + w - 1)(2w) + (w^2+2)(3w^2 - 2w + 1)$$
$$= 2w^4 - 2w^3 + 2w^2 - 2w + 3w^4 - 2w^3 + w^2 + 6w^2 - 4w + 2$$
$$= 5w^4 - 4w^3 + 9w^2 - 6w + 2.$$

11. $f(x) = (5x^2 + 1)(2\sqrt{x} - 1)$

$$f'(x) = (5x^2+1)\frac{d}{dx}(2x^{1/2} - 1) + (2x^{1/2} - 1)\frac{d}{dx}(5x^2 + 1)$$
$$= (5x^2+1)(x^{-1/2}) + (2x^{1/2} - 1)(10x)$$
$$= 5x^{3/2} + x^{-1/2} + 20x^{3/2} - 10x \ = \frac{25x^2 - 10x\sqrt{x} + 1}{\sqrt{x}}.$$

13. $f(x) = (x^2 - 5x + 2)(x - \frac{2}{x})$

$$f'(x) = (x^2 - 5x + 2)\frac{d}{dx}(x - \frac{2}{x}) + (x - \frac{2}{x})\frac{d}{dx}(x^2 - 5x + 2)$$

$$= \frac{(x^2 - 5x + 2)(x^2 + 2)}{x^2} + \frac{(x^2 - 2)(2x - 5)}{x}$$

$$= \frac{(x^2 - 5x + 2)(x^2 + 2) + x(x^2 - 2)(2x - 5)}{x^2}$$

$$= \frac{x^4 + 2x^2 - 5x^3 - 10x + 2x^2 + 4 + 2x^4 - 5x^3 - 4x^2 + 10x}{x^2}$$

$$= \frac{3x^4 - 10x^3 + 4}{x^2}.$$

15. $f(x) = \frac{1}{x-2}$. $f'(x) = \frac{(x-2)\frac{d}{dx}(1) - (1)\frac{d}{dx}(x-2)}{(x-2)^2} = \frac{0 - 1(1)}{(x-2)^2} = -\frac{1}{(x-2)^2}.$

17. $f(x) = \frac{x-1}{2x+1}$.

$$f'(x) = \frac{(2x+1)\frac{d}{dx}(x-1) - (x-1)\frac{d}{dx}(2x+1)}{(2x+1)^2}$$

$$= \frac{2x + 1 - (x-1)(2)}{(2x+1)^2} = \frac{3}{(2x+1)^2}.$$

19. $f(x) = \frac{1}{x^2+1}$.

$$f'(x) = \frac{(x^2+1)\frac{d}{dx}(1) - (1)\frac{d}{dx}(x^2+1)}{(x^2+1)^2}$$

$$= \frac{(x^2+1)(0) - 1(2x)}{(x^2+1)^2} = -\frac{2x}{(x^2+1)^2}.$$

21. $f(s) = \frac{s^2 - 4}{s+1}$.

$$f'(s)=\frac{(s+1)\frac{d}{ds}(s^2-4)-(s^2-4)\frac{d}{ds}(s+1)}{(s+1)^2}$$

$$=\frac{(s+1)(2s)-(s^2-4)(1)}{(s+1)^2}=\frac{s^2+2s+4}{(s+1)^2}.$$

23. $f(x)=\dfrac{\sqrt{x}}{x^2+1}.$

$$f'(x)=\frac{(x^2+1)\frac{d}{dx}(x^{1/2})-(x^{1/2})\frac{d}{dx}(x^2+1)}{(x^2+1)^2}=\frac{(x^2+1)(\frac{1}{2}x^{-1/2})-(x^{1/2})(2x)}{(x^2+1)^2}$$

$$=\frac{(\frac{1}{2}x^{-1/2})[(x^2+1)-4x^2]}{(x^2+1)^2}=\frac{1-3x^2}{2\sqrt{x}(x^2+1)^2}.$$

25. $f(x)=\dfrac{x^2+2}{x^2+x+1}.$

$$f'(x)=\frac{(x^2+x+1)\frac{d}{dx}(x^2+2)-(x^2+2)\frac{d}{dx}(x^2+x+1)}{(x^2+x+1)^2}$$

$$=\frac{(x^2+x+1)(2x)-(x^2+2)(2x+1)}{(x^2+x+1)^2}$$

$$=\frac{2x^3+2x^2+2x-2x^3-x^2-4x-2}{(x^2+x+1)^2}=\frac{x^2-2x-2}{(x^2+x+1)^2}.$$

27. $f(x)=\dfrac{(x+1)(x^2+1)}{x-2}=\dfrac{(x^3+x^2+x+1)}{x-2}.$

$$f'(x)=\frac{(x-2)\frac{d}{dx}(x^3+x^2+x+1)-(x^3+x^2+x+1)\frac{d}{dx}(x-2)}{(x-2)^2}$$

$$=\frac{(x-2)(3x^2+2x+1)-(x^3+x^2+x+1)}{(x-2)^2}$$

$$=\frac{3x^3+2x^2+x-6x^2-4x-2-x^3-x^2-x-1}{(x-2)^2}=\frac{2x^3-5x^2-4x-3}{(x-2)^2}.$$

29. $f(x)=\dfrac{x}{x^2-4}-\dfrac{x-1}{x^2+4}=\dfrac{x(x^2+4)-(x-1)(x^2-4)}{(x^2-4)(x^2+4)}=\dfrac{x^2+8x-4}{(x^2-4)(x^2+4)}.$

$$f'(x)=\frac{(x^2-4)(x^2+4)\frac{d}{dx}(x^2+8x-4)-(x^2+8x-4)\frac{d}{dx}(x^4-16)}{(x^2-4)^2(x^2+4)^2}$$
$$=\frac{(x^2-4)(x^2+4)(2x+8)-(x^2+8x-4)(4x^3)}{(x^2-4)^2(x^2+4)^2}$$
$$=\frac{2x^5+8x^4-32x-128-4x^5-32x^4+16x^3}{(x^2-4)^2(x^2+4)^2}$$
$$=\frac{-2x^5-24x^4+16x^3-32x-128}{(x^2-4)^2(x^2+4)^2}.$$

31. $h'(x)=f(x)g'(x)+f'(x)g(x)$, by the Product Rule. Therefore, $h'(1)=f(1)g'(1)+f'(1)g(1)=(2)(3)+(-1)(-2)=8.$

33. Using the Quotient Rule followed by the Product Rule, we have

$$h'(x)=\frac{[x+g(x)]\frac{d}{dx}[xf(x)]-xf(x)\frac{d}{dx}[x+g(x)]}{[x+g(x)]^2}$$
$$=\frac{[x+g(x)][xf'(x)+f(x)]-xf(x)[1+g'(x)]}{[x+g(x)]^2}$$

Therefore, $h'(1)=\dfrac{[1+g(1)][f'(1)+f(1)]-f(1)[1+g'(1)]}{[1+g(1)]^2}$

$$=\frac{(1-2)(-1+2)-2(1+3)}{(1-2)^2}=\frac{-1-8}{1}=-9.$$

35. $f(x)=(2x-1)(x^2+3)$

$$f'(x)=(2x-1)\frac{d}{dx}(x^2+3)+(x^2+3)\frac{d}{dx}(2x-1)$$
$$=(2x-1)(2x)+(x^2+3)(2)=6x^2-2x+6=2(3x^2-x+3).$$

At $x=1, f'(1)=2[3(1)^2-(1)+3]=2(5)=10.$

37. $f(x) = \dfrac{x}{x^4 - 2x^2 - 1}$.

$$f'(x) = \frac{(x^4 - 2x^2 - 1)\dfrac{d}{dx}(x) - x\dfrac{d}{dx}(x^4 - 2x^2 - 1)}{(x^4 - 2x^2 - 1)^2}$$

$$= \frac{(x^4 - 2x^2 - 1)(1) - x(4x^3 - 4x)}{(x^4 - 2x^2 - 1)^2} = \frac{-3x^4 + 2x^2 - 1}{(x^4 - 2x^2 - 1)^2}.$$

Therefore, $f'(-1) = \dfrac{-3+2-1}{(1-2-1)^2} = -\dfrac{2}{4} = -\dfrac{1}{2}$.

39. $f(x) = (x^3 + 1)(x^2 - 2)$.

$$f'(x) = (x^3 + 1)\frac{d}{dx}(x^2 - 2) + (x^2 - 2)\frac{d}{dx}(x^3 + 1)$$

$$= (x^3 + 1)(2x) + (x^2 - 2)(3x^2).$$

The slope of the tangent line at (2,18) is $f'(2) = (8 + 1)(4) + (4 - 2)(12) = 60$.
An equation of the tangent line is $y - 18 = 60(x - 2)$, or $y = 60x - 102$.

41. $f(x) = \dfrac{x+1}{x^2+1}$.

$$f'(x) = \frac{(x^2 + 1)\dfrac{d}{dx}(x+1) - (x+1)\dfrac{d}{dx}(x^2 + 1)}{(x^2+1)^2}$$

$$= \frac{(x^2+1)(1) - (x+1)(2x)}{(x^2+1)^2} = \frac{-x^2 - 2x + 1}{(x^2+1)^2}.$$

At $x = 1$, $f'(1) = \dfrac{-1-2+1}{4} = -\dfrac{1}{2}$. Therefore, the slope of the tangent line at $x = 1$ is -1/2. Then an equation of the tangent line is

$$y - 1 = -\tfrac{1}{2}(x - 1) \quad \text{or} \quad y = -\tfrac{1}{2}x + \tfrac{3}{2}.$$

43. $f(x) = (x^3 + 1)(3x^2 - 4x + 2)$

$$f'(x) = (x^3 + 1)\frac{d}{dx}(3x^2 - 4x + 2) + (3x^2 - 4x + 2)\frac{d}{dx}(x^3 + 1)$$

$$= (x^3 + 1)(6x - 4) + (3x^2 - 4x + 2)(3x^2)$$

$$= 6x^4 + 6x - 4x^3 - 4 + 9x^4 - 12x^3 + 6x^2$$

$$= 15x^4 - 16x^3 + 6x^2 + 6x - 4.$$

At $x = 1$, $f'(1) = 15(1)^4 - 16(1)^3 + 6(1) + 6(1) - 4 = 7$.The slope of the tangent line at the point $x = 1$ is 7. The equation of the tangent line is

$$y - 2 = 7(x - 1), \quad \text{or} \quad y = 7x - 5.$$

45. $f(x) = (x^2+1)(2-x)$

$$f'(x) = (x^2+1)\frac{d}{dx}(2-x) + (2-x)\frac{d}{dx}(x^2+1)$$
$$= (x^2+1)(-1) + (2-x)(2x) = -3x^2 + 4x - 1.$$

At a point where the tangent line is horizontal, we have

$$f'(x) = -3x^2 + 4x - 1 = 0$$

or $\quad 3x^2 - 4x + 1 = (3x-1)(x-1) = 0$, giving $x = 1/3$ or $x = 1$.

Since $f(\frac{1}{3}) = (\frac{1}{9}+1)(2-\frac{1}{3}) = \frac{50}{27}$, and $f(1) = 2(2 - 1) = 2$, we see that the required points are $(\frac{1}{3}, \frac{50}{27})$ and $(1, 2)$.

47. $f(x) = (x^2+6)(x-5)$

$$f'(x) = (x^2+6)\frac{d}{dx}(x-5) + (x-5)\frac{d}{dx}(x^2+6)$$
$$= (x^2+6)(1) + (x-5)(2x) = x^2 + 6 + 2x^2 - 10x = 3x^2 - 10x + 6.$$

At a point where the slope of the tangent line is -2, we have

$f'(x) = 3x^2 - 10x + 6 = -2.$

This gives $3x^2 - 10x + 8 = (3x - 4)(x - 2) = 0$. So $x = \frac{4}{3}$ or $x = 2$.

Since $f(\frac{4}{3}) = (\frac{16}{9}+6)(\frac{4}{3}-5) = -\frac{770}{27}$ and $f(2) = (4 + 6)(2 - 5) = -30$, the required points are $(\frac{4}{3}, -\frac{770}{27})$ and $(2, -30)$.

49. $y = \dfrac{1}{1+x^2}$. $y' = \dfrac{(1+x^2)\frac{d}{dx}(1) - (1)\frac{d}{dx}(1+x^2)}{(1+x^2)^2} = \dfrac{-2x}{(1+x^2)^2}$.

So, the slope of the tangent line at $(1, \frac{1}{2})$ is

$$y'\big|_{x=1} = \frac{-2x}{(1+x^2)^2}\bigg|_{x=1} = \frac{-2}{4} = -\frac{1}{2}$$

and the equation of the tangent line is $y - \frac{1}{2} = -\frac{1}{2}(x-1)$, or $y = -\frac{1}{2}x + 1$.

Next, the slope of the required normal line is 2 and its equation is

$y-\frac{1}{2}=2(x-1)$, or $y=2x-\frac{3}{2}$.

51. $C(x)=\dfrac{0.5x}{100-x}$. $C'(x)=\dfrac{(100-x)(0.5)-0.5x(-1)}{(100-x)^2}=\dfrac{50}{(100-x)^2}$.

$C'(80)=\dfrac{50}{20^2}=0.125$; $\quad C'(90)=\dfrac{50}{10^2}=0.5$,

$C'(95)=\dfrac{50}{5^2}=2$; $\quad C'(99)=\dfrac{50}{1}=50$.

The rates of change of the cost in removing 80%, 90%, and 99% of the toxic waste are 0.125, 0.5, 2, and 50 million dollars per 1% more of the waste to be removed, respectively. It is too costly to remove *all* of the pollutant.

53. $N(t)=\dfrac{10{,}000}{1+t^2}+2000$

$N'(t)=\dfrac{d}{dt}[10{,}000(1+t^2)^{-1}+2000]=-\dfrac{10{,}000}{(1+t^2)^2}(2t)=-\dfrac{20{,}000t}{(1+t^2)^2}$.

The rate of change after 1 minute and after 2 minutes is

$N'(1)=-\dfrac{20{,}000}{(1+1^2)^2}=-5000$; $N'(2)=-\dfrac{20{,}000(2)}{(1+2^2)^2}=-1600$.

The population of bacteria after one minute is $N(1)=\dfrac{10{,}000}{1+1}+2000=7000$.

The population after two minutes is $N(2)=\dfrac{10{,}000}{1+4}+2000=4000$.

55. a. $N(t)=\dfrac{60t+180}{t+6}$.

$$N'(t)=\frac{(t+6)\frac{d}{dt}(60t+180)-(60t+180)\frac{d}{dt}(t+6)}{(t+6)^2}$$

$$=\frac{(t+6)(60)-(60t+180)(1)}{(t+6)^2}=\frac{180}{(t+6)^2}.$$

b. $N'(1)=\dfrac{180}{(1+6)^2}=3.7$, $N'(3)=\dfrac{180}{(3+6)^2}=2.2$, $N'(4)=\dfrac{180}{(4+6)^2}=1.8$,

$N'(7)=\dfrac{180}{(7+6)^2}=1.1$

We conclude that the rate at which the average student is increasing his or her speed one week, three weeks, four weeks, and seven weeks into the course is 3.7, 2.2, 1.8, and 1.1 words per minute, respectively.

c. Yes

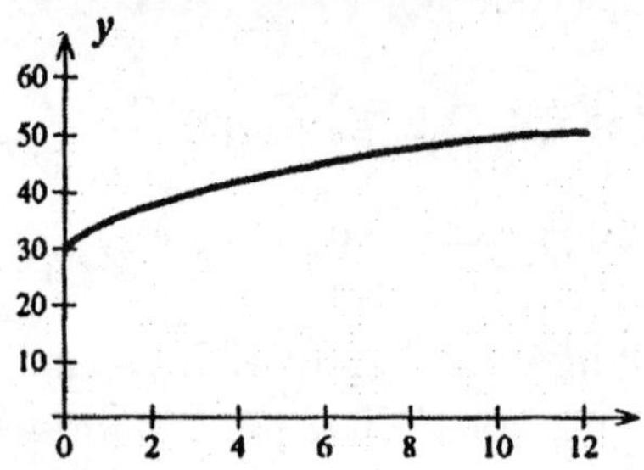

d. $N(12) = \dfrac{60(12)+180}{12+6} = 50$, or 50 words/minute.

57. $f(t) = \dfrac{0.055t + 0.26}{t+2}$; $f'(t) = \dfrac{(t+2)(0.055) - (0.055t + 0.26)(1)}{(t+2)^2} = -\dfrac{0.15}{(t+2)^2}$.

At the beginning, the formaldehyde level is changing at the rate of

$$f'(0) = -\frac{0.15}{4} = -0.0375;$$

that is, it is dropping at the rate of 0.0375 parts per million per year. Next,

$$f'(3) = -\frac{0.15}{5^2} = -0.006,$$

and so the level is dropping at the rate of 0.006 parts per million per year at the beginning of the fourth year ($t = 3$).

59. False. Take $f(x) = x$ and $g(x) = x$. Then $f(x)g(x) = x^2$. So

$$\frac{d}{dx}[f(x)g(x)] = \frac{d}{dx}(x^2) = 2x \neq f'(x)g'(x) = 1.$$

61. False. Let $f(x) = x^3$. Then

$$\frac{d}{dx}\left[\frac{f(x)}{x^2}\right] = \frac{d}{dx}\left(\frac{x^3}{x^2}\right) = \frac{d}{dx}(x) = 1 \neq \frac{f'(x)}{2x} = \frac{3x^2}{2x} = \frac{3}{2}x.$$

63. Let $f(x) = u(x)v(x)$ and $g(x) = w(x)$. Then $h(x) = f(x)g(x)$. Therefore,

$$h'(x) = f'(x)g(x) + f(x)g'(x).$$

But $\quad f'(x) = u(x)v'(x) + u'(x)v(x).$

Therefore, $h'(x) = [u(x)v'(x) + u'(x)v(x)]g(x) + u(x)v(x)w'(x)$

$$= u(x)v(x)w'(x) + u(x)v'(x)w(x) + u'(x)v(x)w(x).$$

USING TECHNOLOGY EXERCISES 12.2, page 708

1. 0.8750 3. 0.0774 5. -0.5000 7. 87,322 per year

12.3 Problem Solving Tips

Here are some hints for solving the problems in the exercises that follow:

1. It is often easier to find the derivative of a quotient when the numerator is a constant by using the General Power Rule, rather than the Quotient Rule. For example, to find the derivative of $f(x) = -\dfrac{1}{\sqrt{2x^2-1}}$ in self-check exercise #1 of this section, we first rewrite the function in the form $f(x) = -(2x^2-1)^{-1/2}$ and then use the General Power Rule to find the derivative.

2. **To simplify a function involving the powers of an expression**, factor out the lowest power of the expression. For example, to factor the expression

$$5(x+1)^{1/2} - 3(x+1)^{-1/2}$$

factor out $(x+1)^{-1/2}$ [which is the lowest power of $(x + 1)$ found in the expression] .

12.3 CONCEPT QUESTIONS, page 716

1. The derivative of $h(x) = g[f(x)]$ is equal to the derivative of g evaluated at $f(x)$ times the derivative of f.

EXERCISES 12.3, page 716

1. $f(x) = (2x-1)^4$. $f'(x) = 4(2x-1)^3 \dfrac{d}{dx}(2x-1) = 4(2x-1)^3(2) = 8(2x-1)^3$.

3. $f(x) = (x^2+2)^5$. $f'(x) = 5(x^2+2)^4(2x) = 10x(x^2+2)^4$.

5. $f(x) = (2x-x^2)^3$.

$$f'(x) = 3(2x-x^2)^2 \frac{d}{dx}(2x-x^2) = 3(2x-x^2)^2(2-2x) = 6x^2(1-x)(2-x)^2.$$

7. $f(x) = (2x+1)^{-2}$.

$$f'(x) = -2(2x+1)^{-3} \frac{d}{dx}(2x+1) = -2(2x+1)^{-3}(2) = -4(2x+1)^{-3}.$$

9. $f(x) = (x^2-4)^{3/2}$.

$$f'(x) = \tfrac{3}{2}(x^2-4)^{1/2} \frac{d}{dx}(x^2-4) = \tfrac{3}{2}(x^2-4)^{1/2}(2x) = 3x(x^2-4)^{1/2}.$$

11. $f(x) = \sqrt{3x-2} = (3x-2)^{1/2}$.

$$f'(x) = \frac{1}{2}(3x-2)^{-1/2}(3) = \frac{3}{2}(3x-2)^{-1/2} = \frac{3}{2\sqrt{3x-2}}.$$

13. $f(x) = \sqrt[3]{1-x^2}$.

$$f'(x) = \frac{d}{dx}(1-x^2)^{1/3} = \frac{1}{3}(1-x^2)^{-2/3} \frac{d}{dx}(1-x^2)$$

$$= \frac{1}{3}(1-x^2)^{-2/3}(-2x) = -\frac{2}{3}x(1-x^2)^{-2/3} = \frac{-2x}{3(1-x^2)^{2/3}}.$$

15. $f(x)=\dfrac{1}{(2x+3)^3}=(2x+3)^{-3}$.

$$f'(x)=-3(2x+3)^{-4}(2)=-6(2x+3)^{-4}=-\frac{6}{(2x+3)^4}.$$

17. $f(t)=\dfrac{1}{\sqrt{2t-3}}$.

$$f'(t)=\frac{d}{dt}(2t-3)^{-1/2}=-\frac{1}{2}(2t-3)^{-3/2}(2)=-(2t-3)^{-3/2}=-\frac{1}{(2t-3)^{3/2}}.$$

19. $y=\dfrac{1}{(4x^4+x)^{3/2}}$.

$$\frac{dy}{dx}=\frac{d}{dx}(4x^4+x)^{-3/2}=-\frac{3}{2}(4x^4+x)^{-5/2}(16x^3+1)=-\frac{3}{2}(16x^3+1)(4x^4+x)^{-5/2}.$$

21. $f(x)=(3x^2+2x+1)^{-2}$.

$$\begin{aligned}f'(x)&=-2(3x^2+2x+1)^{-3}\frac{d}{dx}(3x^2+2x+1)\\&=-2(3x^2+2x+1)^{-3}(6x+2)=-4(3x+1)(3x^2+2x+1)^{-3}.\end{aligned}$$

23. $f(x)=(x^2+1)^3-(x^3+1)^2$.

$$\begin{aligned}f'(x)&=3(x^2+1)^2\frac{d}{dx}(x^2+1)-2(x^3+1)\frac{d}{dx}(x^3+1)\\&=3(x^2+1)^2(2x)-2(x^3+1)(3x^2)=6x[(x^2+1)^2-x(x^3+1)]=6x(2x^2-x+1).\end{aligned}$$

25. $f(t)=(t^{-1}-t^{-2})^3$. $f'(t)=3(t^{-1}-t^{-2})^2\dfrac{d}{dt}(t^{-1}-t^{-2})=3(t^{-1}-t^{-2})^2(-t^{-2}+2t^{-3})$.

27. $f(x)=\sqrt{x+1}+\sqrt{x-1}=(x+1)^{1/2}+(x-1)^{1/2}$.

$$f'(x)=\tfrac{1}{2}(x+1)^{-1/2}(1)+\tfrac{1}{2}(x-1)^{-1/2}(1)=\tfrac{1}{2}[(x+1)^{-1/2}+(x-1)^{-1/2}].$$

29. $f(x)=2x^2(3-4x)^4$.

$$\begin{aligned}f'(x)&=2x^2(4)(3-4x)^3(-4)+(3-4x)^4(4x)=4x(3-4x)^3(-8x+3-4x)\\&=4x(3-4x)^3(-12x+3)=(-12x)(4x-1)(3-4x)^3.\end{aligned}$$

31. $f(x)=(x-1)^2(2x+1)^4$.

$$f'(x)=(x-1)^2\frac{d}{dx}(2x+1)^4+(2x+1)^4\frac{d}{dx}(x-1)^2 \quad \text{[Product Rule]}$$

$$=(x-1)^2(4)(2x+1)^3\frac{d}{dx}(2x+1)+(2x+1)^4(2)(x-1)\frac{d}{dx}(x-1)$$

$$=8(x-1)^2(2x+1)^3+2(x-1)(2x+1)^4$$

$$=2(x-1)(2x+1)^3(4x-4+2x+1)=6(x-1)(2x-1)(2x+1)^3.$$

33. $f(x)=\left(\frac{x+3}{x-2}\right)^3$.

$$f'(x)=3\left(\frac{x+3}{x-2}\right)^2\frac{d}{dx}\left(\frac{x-3}{x-2}\right)=3\left(\frac{x+3}{x-2}\right)^2\left[\frac{(x-2)(1)-(x+3)(1)}{(x-2)^2}\right]$$

$$=3\left(\frac{x+3}{x-2}\right)^2\left[-\frac{5}{(x-2)^2}\right]=-\frac{15(x+3)^2}{(x-2)^4}.$$

35. $s(t)=\left(\frac{t}{2t+1}\right)^{3/2}$.

$$s'(t)=\frac{3}{2}\left(\frac{t}{2t+1}\right)^{1/2}\frac{d}{dt}\left(\frac{t}{2t+1}\right)=\frac{3}{2}\left(\frac{t}{2t+1}\right)^{1/2}\left[\frac{(2t+1)(1)-t(2)}{(2t+1)^2}\right]$$

$$=\frac{3}{2}\left(\frac{t}{2t+1}\right)^{1/2}\left[\frac{1}{(2t+1)^2}\right]=\frac{3t^{1/2}}{2(2t+1)^{5/2}}.$$

37. $g(u)=\left(\frac{u+1}{3u+2}\right)^{1/2}$.

$$g'(u)=\frac{1}{2}\left(\frac{u+1}{3u+2}\right)^{-1/2}\frac{d}{du}\left(\frac{u+1}{3u+2}\right)$$

$$=\frac{1}{2}\left(\frac{u+1}{3u+2}\right)^{-1/2}\left[\frac{(3u+2)(1)-(u+1)(3)}{(3u+2)^2}\right]=-\frac{1}{2\sqrt{u+1})(3u+2)^{3/2}}.$$

39. $f(x)=\frac{x^2}{(x^2-1)^4}$.

$$f'(x)=\frac{(x^2-1)^4\frac{d}{dx}(x^2)-(x^2)\frac{d}{dx}(x^2-1)^4}{\left[(x^2-1)^4\right]^2}$$

$$=\frac{(x^2-1)^4(2x)-x^2(4)(x^2-1)^3(2x)}{(x^2-1)^8}$$

$$=\frac{(x^2-1)^3(2x)(x^2-1-4x^2)}{(x^2-1)^8}=\frac{(-2x)(3x^2+1)}{(x^2-1)^5}.$$

41. $h(x)=\dfrac{(3x^2+1)^3}{(x^2-1)^4}.$

$$h'(x)=\frac{(x^2-1)^4(3)(3x^2+1)^2(6x)-(3x^2+1)^3(4)(x^2-1)^3(2x)}{(x^2-1)^8}$$

$$=\frac{2x(x^2-1)^3(3x^2+1)^2[9(x^2-1)-4(3x^2+1)]}{(x^2-1)^8}$$

$$=-\frac{2x(3x^2+13)(3x^2+1)^2}{(x^2-1)^5}.$$

43. $f(x)=\dfrac{\sqrt{2x+1}}{x^2-1}.$

$$f'(x)=\frac{(x^2-1)(\frac{1}{2})(2x+1)^{-1/2}(2)-(2x+1)^{1/2}(2x)}{(x^2-1)^2}$$

$$=\frac{(2x+1)^{-1/2}[(x^2-1)-(2x+1)(2x)]}{(x^2-1)^2}=-\frac{3x^2+2x+1}{\sqrt{2x+1}(x^2-1)^2}.$$

45. $g(t)=\dfrac{(t+1)^{1/2}}{(t^2+1)^{1/2}}.$

$$g'(t)=\frac{(t^2+1)^{1/2}\frac{d}{dt}(t+1)^{1/2}-(t+1)^{1/2}\frac{d}{dt}(t^2+1)^{1/2}}{t^2+1}$$

$$=\frac{(t^2+1)^{1/2}(\frac{1}{2})(t+1)^{-1/2}(1)-(t+1)^{1/2}(\frac{1}{2})(t^2+1)^{-1/2}(2t)}{t^2+1}$$

$$= \frac{\frac{1}{2}(t+1)^{-1/2}(t^2+1)^{-1/2}[(t^2+1)-2t(t+1)]}{t^2+1} = -\frac{t^2+2t-1}{2\sqrt{t+1}(t^2+1)^{3/2}}.$$

47. $f(x)=(3x+1)^4(x^2-x+1)^3$

$$f'(x)=(3x+1)^4\cdot\frac{d}{dx}(x^2-x+1)^3+(x^2-x+1)^3\frac{d}{dx}(3x+1)^4$$
$$=(3x+1)^4\cdot 3(x^2-x+1)^2(2x-1)+(x^2-x+1)^3\cdot 4(3x+1)^3\cdot 3$$
$$=3(3x+1)^3(x^2-x+1)^2[(3x+1)(2x-1)+4(x^2-x+1)]$$
$$=3(3x+1)^3(x^2-x+1)^2(6x^2-3x+2x-1+4x^2-4x+4)$$
$$=3(3x+1)^3(x^2-x+1)^2(10x^2-5x+3)$$

49. $y=g(u)=u^{4/3}$ and $\dfrac{dy}{du}=\dfrac{4}{3}u^{1/3}$, $u=f(x)=3x^2-1$, and $\dfrac{du}{dx}=6x$.

So $\dfrac{dy}{dx}=\dfrac{dy}{du}\cdot\dfrac{du}{dx}=\frac{4}{3}u^{1/3}(6x)=\frac{4}{3}(3x^2-1)^{1/3}6x=8x(3x^2-1)^{1/3}$.

51. $\dfrac{dy}{du}=-\dfrac{2}{3}u^{-5/3}=-\dfrac{2}{3u^{5/3}}$, $\dfrac{du}{dx}=6x^2-1$.

$$\frac{dy}{dx}=\frac{dy}{du}\cdot\frac{du}{dx}=-\frac{2(6x^2-1)}{3u^{5/3}}=-\frac{2(6x^2-1)}{3(2x^3-x+1)^{5/3}}.$$

53. $\dfrac{dy}{du}=\frac{1}{2}u^{-1/2}-\frac{1}{2}u^{-3/2}$, $\dfrac{du}{dx}=3x^2-1$.

$$\frac{dy}{dx}=\frac{dy}{du}\cdot\frac{du}{dx}=\left[\frac{1}{2\sqrt{x^3-x}}-\frac{1}{2(x^3-x)^{3/2}}\right](3x^2-1)$$
$$=\frac{(3x^2-1)(x^3-x-1)}{2(x^3-x)^{3/2}}.$$

55. $F(x)=g(f(x))$; $F'(x)=g'(f(x))f'(x)$ and $F'(2)=g'(3)(-3)=(4)(-3)=-12$

57. Let $g(x)=x^2+1$, then $F(x)=f(g(x))$. Next, $F'(x)=f'(g(x))g'(x)$ and $F'(1)=f'(2)(2x)=(3)(2)=6$.

59. No. Suppose $h = g(f(x))$. Let $f(x) = x$ and $g(x) = x^2$. Then
$h = g(f(x)) = g(x) = x^2$ and $h'(x) = 2x \neq g'(f'(x)) = g'(1) = 2(1) = 2$.

61. $f(x) = (1-x)(x^2-1)^2$.
$f'(x) = (1-x)2(x^2-1)(2x) + (-1)(x^2-1)^2$
$= (x^2-1)(4x-4x^2-x^2+1) = (x^2-1)(-5x^2+4x+1)$.
Therefore, the slope of the tangent line at (2,–9) is
$$f'(2) = [(2)^2-1][-5(2)^2+4(2)+1] = -33.$$
Then the required equation is $y+9 = -33(x-2)$, or $y = -33x + 57$.

63. $f(x) = x\sqrt{2x^2+7}$. $f'(x) = \sqrt{2x^2+7} + x(\frac{1}{2})(2x^2+7)^{-1/2}(4x)$.
The slope of the tangent line is $f'(3) = \sqrt{25} + (\frac{3}{2})(25)^{-1/2}(12) = \frac{43}{5}$.
An equation of the tangent line is $y - 15 = \frac{43}{5}(x-3)$ or $y = \frac{43}{5}x - \frac{54}{5}$.

65. $N(t) = (60+2t)^{2/3}$. $N'(t) = \frac{2}{3}(60+2t)^{-1/3}\frac{d}{dx}(60+2t) = \frac{4}{3}(60+2t)^{-1/3}$.
The rate of increase at the end of the second week is
$N'(2) = \frac{4}{3}(64)^{-1/3} = \frac{1}{3}$, or $\frac{1}{3}$ million/week
At the end of the 12th week, $N'(12) = \frac{4}{3}(84)^{-1/3} \approx 0.3$ million/wk. The number of viewers in the 2nd and 24th week are $N(2) = (60 + 4)^{2/3} = 16$ million and $N(24) = (60 + 48)^{2/3} = 22.7$ million, respectively.

67. $P(t) = 33.55(t+5)^{0.205}$. $P'(t) = 33.55(0.205)(t+5)^{-0.795}(1) = 6.87775(t+5)^{-0.795}$
The rate of change at the beginning of 2000 is
$$P'(20) = 6.87775(25)^{-0.795} \approx 0.5322 \text{ or } 0.53\%/\text{yr.}$$
The percent of these mothers was $P(20) = 33.55(25)^{0.205} \approx 64.90$, or 64.9%.

69. a. $f(t) = 23.7(0.2t+1)^{1.32}$. The rate of change at any time t is given by
$$f'(t) = (23.7)(1.32)(0.2t+1)^{0.32}(0.2) = 6.2568(0.2t+1)^{0.32}$$
At the beginning of 2000, the rate of change is
$$f'(9) = 6.2568[0.2(9)+1]^{0.32} \approx 8.6985,$$ or approximately \$8.7 billion/yr.
b. The assets were $f(9) = 92.256$, or \$92.3 billion.

71. $C(t) = 0.01(0.2t^2 + 4t + 64)^{2/3}$.

a. $C'(t) = 0.01(\tfrac{2}{3})(0.2t^2 + 4t + 64)^{-1/3}\dfrac{d}{dt}(0.2t^2 + 4t + 64)$

$= (0.01)(0.667)(0.4t + 4)(0.2t^2 + 4t + 4)^{-1/3}$

$= 0.027(0.1t + 1)(0.2t^2 + 4t + 64)^{-1/3}$.

b. $C'(5) = 0.007[0.4(5) + 4][0.2(25) + 4(5) + 64]^{-1/3} \approx 0.009$,
or 0.009 parts per million per year.

73. a. $A(t) = 0.03t^3(t - 7)^4 + 60.2$

$A'(t) = 0.03[3t^2(t - 7)^4 + t^3(4)(t - 7)^3] = 0.03t^2(t - 7)^3[3(t - 7) + 4t]$

$= 0.21t^2(t - 3)(t - 7)^3$.

b. $A'(1) = 0.21(-2)(-6)^3 = 90.72$; $A'(3) = 0$. $A'(4) = 0.21(16)(1)(-3)^3 = -90.72$.
The amount of pollutant is increasing at the rate of 90.72 units/hr at 8 A.M. Its rate of change is 0 units/hr at 10 A.M.; its rate of change is –90.72 units/hr at 11 A.M.

75. $P(t) = \dfrac{300\sqrt{\frac{1}{2}t^2 + 2t + 25}}{t + 25} = \dfrac{300(\frac{1}{2}t^2 + 2t + 25)^{1/2}}{t + 25}$.

$$P'(t) = 300\left[\frac{(t + 25)\frac{1}{2}(\frac{1}{2}t^2 + 2t + 25)^{-1/2}(t + 2) - (\frac{1}{2}t^2 + 2t + 25)^{1/2}(1)}{(t + 25)^2}\right]$$

$$= 300\left[\frac{(\frac{1}{2}t^2 + 2t + 25)^{-1/2}[(t + 25)(t + 2) - 2(\frac{1}{2}t^2 + 2t + 25)}{(t + 25)^2}\right]$$

$$= \frac{3450t}{(t + 25)^2\sqrt{\frac{1}{2}t^2 + 2t + 25}}.$$

Ten seconds into the run, the athlete's pulse rate is increasing at

$P'(10) = \dfrac{3450(10)}{(35)^2\sqrt{50 + 20 + 25}} \approx 2.9$, or approximately 2.9 beats per minute per minute. Sixty seconds into the run, it is increasing at

$P'(60) = \dfrac{3450(60)}{(85)^2\sqrt{1800 + 120 + 25}} \approx 0.65$, or approximately 0.7 beats per minute per minute. Two minutes into the run, it is increasing at

$P'(120) = \dfrac{3450(120)}{(145)^2\sqrt{7200 + 240 + 25}} \approx 0.23$, or approximately 0.2 beats per minute per minute. The pulse rate two minutes into the run is given by

$$P(120) = \frac{300\sqrt{7200+240+25}}{120+25} \approx 178.8, \text{ or approximately 179 beats per minute.}$$

77. The area is given by $A = \pi r^2$. The rate at which the area is increasing is given by dA/dt, that is, $\frac{dA}{dt} = \frac{d}{dt}(\pi r^2) = \frac{d}{dt}(\pi r^2)\frac{dr}{dt} = 2\pi r\frac{dr}{dt}$.

If $r = 40$ and $dr/dt = 2$, then $\frac{dA}{dt} = 2\pi(40)(2) = 160\pi$, that is, it is increasing at the rate of 160π, or approximately 503, sq ft/sec.

79. $f(t) = 6.25t^2 + 19.75t + 74.75$. $g(x) = -0.00075x^2 + 67.5$.

$$\frac{dS}{dt} = g'(x)f'(t) = (-0.0015x)(12.5t + 19.75).$$

When $t = 4$, we have $x = f(4) = 6.25(16) + 19.75(4) + 74.75 = 253.75$

and $\left.\frac{dS}{dt}\right|_{t=4} = (-0.0015)(253.75)[12.5(4) + 19.75] \approx -26.55$;

that is, the average speed will be dropping at the rate of approximately 27 mph per decade. The average speed of traffic flow at that time will be

$$S = g(f(4)) = -0.00075(253.75^2) + 67.5 = 19.2,$$

or approximately 19 mph.

81. $N(x) = 1.42x$ and $x(t) = \frac{7t^2 + 140t + 700}{3t^2 + 80t + 550}$. The number of construction jobs as a function of time is $n(t) = N[x(t)]$. Using the Chain Rule,

$$n'(t) = \frac{dN}{dx}\cdot\frac{dx}{dt} = 1.42\frac{dx}{dt}$$

$$= (1.42)\left[\frac{(3t^2 + 80t + 550)(14t + 140) - (7t^2 + 140t + 700)(6t + 80)}{(3t^2 + 80t + 550)^2}\right]$$

$$= \frac{1.42(140t^2 + 3500t + 21000)}{(3t^2 + 80t + 550)^2}.$$

$$n'(12) = \frac{1.42[140(12)^2 + 3500(12) + 21000]}{[3(12)^2 + 80(12) + 550]^2} \approx 0.0313115, \text{ or approximately}$$

31,312 jobs/year.

83. $x = f(p) = 10\sqrt{\dfrac{50-p}{p}}$;

$$\frac{dx}{dp} = \frac{d}{dp}\left[10\left(\frac{50-p}{p}\right)^{1/2}\right] = (10)(\tfrac{1}{2})\left(\frac{50-p}{p}\right)^{-1/2}\frac{d}{dp}\left(\frac{50-p}{p}\right)$$

$$= 5\left(\frac{50-p}{p}\right)^{-1/2}\cdot\frac{d}{dp}\left(\frac{50}{p}-1\right) = 5\left(\frac{50-p}{p}\right)^{-1/2}\left(-\frac{50}{p^2}\right) = -\frac{250}{p^2\left(\dfrac{50-p}{p}\right)^{1/2}}$$

$$\left.\frac{dx}{dp}\right|_{p=25} = -\frac{250}{p^2\left(\dfrac{50-p}{p}\right)^{1/2}} = -\frac{250}{(625)\left(\dfrac{25}{25}\right)^{1/2}} = -0.4$$

So the quantity demanded is falling at the rate of 0.4(1000) or 400 wristwatches per dollar increase in price.

85. True. This is just the statement of the Chain Rule.

87. True. $\dfrac{d}{dx}\sqrt{f(x)} = \dfrac{d}{dx}[f(x)]^{1/2} = \dfrac{1}{2}[f(x)]^{-1/2}f'(x) = \dfrac{f'(x)}{2\sqrt{f(x)}}$.

89. Let $f(x) = x^{1/n}$ so that $[f(x)]^n = x$.
Differentiating both sides with respect to x, we get

$$n[f(x)]^{n-1}f'(x) = 1$$

$$f'(x) = \frac{1}{n[f(x)]^{n-1}} = \frac{1}{n[x^{1/n}]^{n-1}} = \frac{1}{nx^{1-(1/n)}} = \frac{1}{n}x^{(1/n)-1}.$$

as was to be shown.

USING TECHNOLOGY EXERCISES 12.3, page 721

1. 0.5774 3. 0.9390 5. –4.9498

7. a. 10,146,200/decade b. 7,810,520/decade

12.4 Problem Solving Tips

Here are some hints for solving the problems in the exercises that follow:

1. The *marginal cost function* is the derivative of the cost function. Similarly, the *marginal profit function* and the marginal *revenue function* are the derivatives of the profit function and the revenue function, respectively. The key word here is "marginal" as it indicates that we are dealing with the derivative of the function that follows.

2. The *average cost function* is given by $\overline{C}(x) = C(x)/x$ and the *marginal average cost* function is given by $\overline{C}'(x)$.

3. Remember that the revenue is *increasing* on an interval where the demand is *inelastic*, *decreasing* on an interval where the demand is *elastic*, and *stationary* at the point where the demand is *unitary*.

12.4 CONCEPT QUESTIONS, page 732

1. a. The marginal cost function is the derivative of the cost function.
   b. The average cost function is equal to the total cost function divided by the total number of the commodity produced.
   c. The marginal average cost function is the derivative of the average cost function.
   d. The marginal revenue function is the derivative of the revenue function.
   e. The marginal profit function is the derivative of the profit function.

EXERCISES 12.4, page 732

1. a. $C(x)$ is always increasing because as x, the number of units produced, increases, the greater the amount of money that must be spent on production.
   b. This occurs at $x = 4$, or a production level of 4000. You can see this by looking

at the slopes of the tangent lines for x less than, equal to, and a little larger then $x = 4$.

3. a. The actual cost incurred in the production of the 1001st disc is given by

$$C(1001) - C(1000) = [2000 + 2(1001) - 0.0001(1001)^2] - [2000 + 2(1000) - 0.0001(1000)^2] = 3901.7999 - 3900 = 1.7999,$$

or $1.80. The actual cost incurred in the production of the 2001st disc is given by $C(2001) - C(2000) = [2000 + 2(2001) - 0.0001(2001)^2] - [2000 + 2(2000) - 0.0001(2000)^2] = 5601.5999 - 5600 = 1.5999$, or $1.60.

b. The marginal cost is $C'(x) = 2 - 0.0002x$. In particular

$$C'(1000) = 2 - 0.0002(1000) = 1.80$$

and $$C'(2000) = 2 - 0.0002(2000) = 1.60.$$

5. a. $\overline{C}(x) = \dfrac{C(x)}{x} = \dfrac{100x + 200{,}000}{x} = 100 + \dfrac{200{,}000}{x}$.

b. $\overline{C}'(x) = \dfrac{d}{dx}(100) + \dfrac{d}{dx}(200{,}000x^{-1}) = -200{,}000x^{-2} = -\dfrac{200{,}000}{x^2}$.

c. $\lim\limits_{x\to\infty} \overline{C}(x) = \lim\limits_{x\to\infty}\left[100 + \dfrac{200{,}000}{x}\right] = 100$

and this says that the average cost approaches $100 per unit if the production level is very high.

7. $\overline{C}(x) = \dfrac{C(x)}{x} = \dfrac{2000 + 2x - 0.0001x^2}{x} = \dfrac{2000}{x} + 2 - 0.0001x.$

$\overline{C}'(x) = -\dfrac{2000}{x^2} + 0 - 0.0001 = -\dfrac{2000}{x^2} - 0.0001.$

9. a. $R'(x) = \dfrac{d}{dx}(8000x - 100x^2) = 8000 - 200x.$

b. $R'(39) = 8000 - 200(39) = 200$. $R'(40) = 8000 - 200(40) = 0$
$R'(41) = 8000 - 200(41) = -200$

c. This suggests the total revenue is maximized if the price charged/ passenger is $40.

11. a. $P(x) = R(x) - C(x) = (-0.04x^2 + 800x) - (200x + 300{,}000)$

$= -0.04x^2 + 600x - 300{,}000.$

b. $P'(x) = -0.08x + 600$

c. $P'(5000) = -0.08(5000) + 600 = 200 \quad P'(8000) = -0.08(8000) + 600 = -40.$

d.

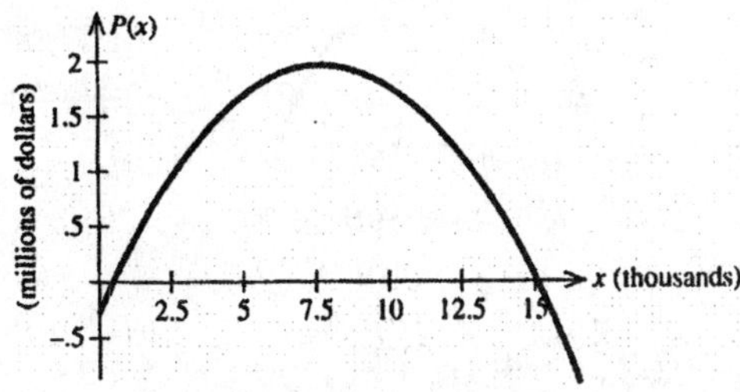

The profit realized by the company increases as production increases, peaking at a level of production of 7500 units. Beyond this level of production, the profit begins to fall.

13. a. The revenue function is $R(x) = px = (600 - 0.05x)x = 600x - 0.05x^2$ and the profit function is

$$\begin{aligned} P(x) &= R(x) - C(x) \\ &= (600x - 0.05x^2) - (0.000002x^3 - 0.03x^2 + 400x + 80{,}000) \\ &= -0.000002x^3 - 0.02x^2 + 200x - 80{,}000. \end{aligned}$$

b. $C'(x) = \dfrac{d}{dx}(0.000002x^3 - 0.03x^2 + 400x + 80{,}000) = 0.000006x^2 - 0.06x + 400.$

$$R'(x) = \frac{d}{dx}(600x - 0.05x^2) = 600 - 0.1x.$$

$$\begin{aligned} P'(x) &= \frac{d}{dx}(-0.000002x^3 - 0.02x^2 + 200x - 80{,}000) \\ &= -0.000006x^2 - 0.04x + 200. \end{aligned}$$

c. $C'(2000) = 0.000006(2000)^2 - 0.06(2000) + 400 = 304$, and this says that at a level of production of 2000 units, the cost for producing the 2001st unit is \$304. $R'(2000) = 600 - 0.1(2000) = 400$ and this says that the revenue realized in selling the 2001st unit is \$400. $P'(2000) = R'(2000) - C'(2000) = 400 - 304 = 96$, and this says that the revenue realized in selling the 2001st unit is \$96.

d.

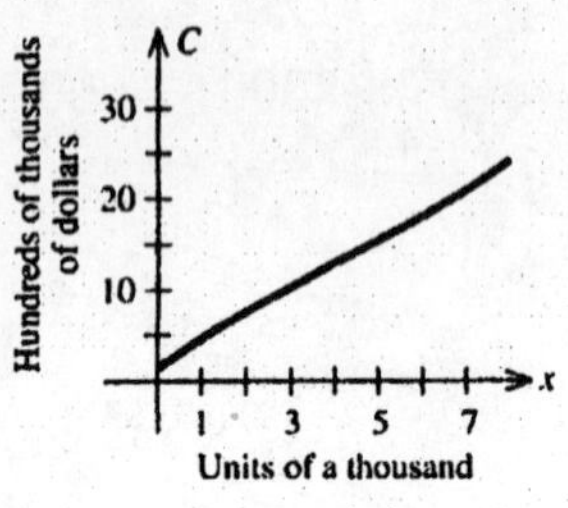

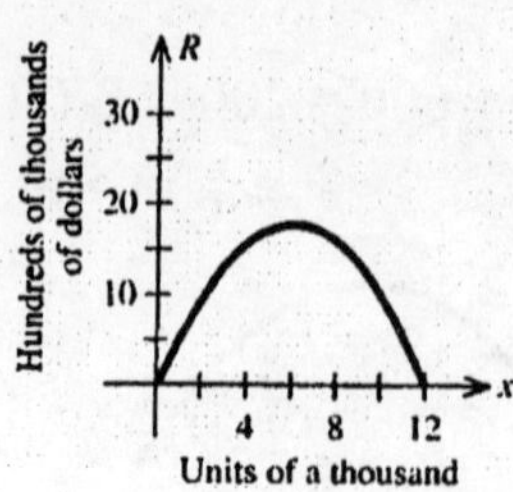

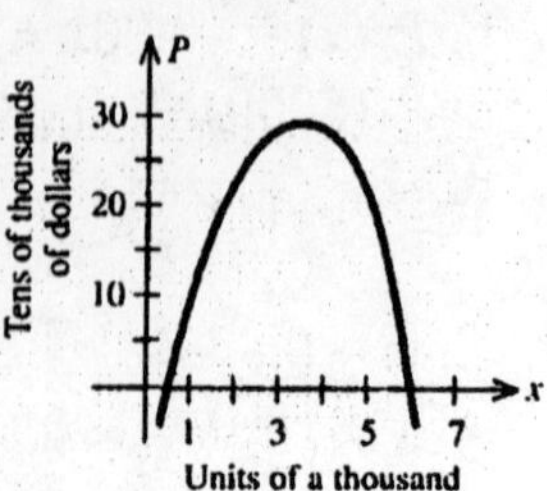

15. $\overline{C}(x) = \dfrac{C(x)}{x} = \dfrac{0.000002x^3 - 0.03x^2 + 400x + 80{,}000}{x}$

$= 0.000002x^2 - 0.03x + 400 + \dfrac{80{,}000}{x}.$

a. $\overline{C}'(x) = 0.000004x - 0.03 - \dfrac{80{,}000}{x^2}.$

b. $\overline{C}'(5000) = 0.000004(5000) - 0.03 - \dfrac{80{,}000}{5000^2} \approx -0.0132,$

and this says that, at a level of production of 5000 units, the average cost of production is dropping at the rate of approximately a penny per unit.

$\overline{C}'(10{,}000) = 0.000004(10000) - 0.03 - \dfrac{80{,}000}{10{,}000^2} \approx 0.0092,$

and this says that, at a level of production of 10,000 units, the average cost of production is increasing at the rate of approximately a penny per unit.

c.

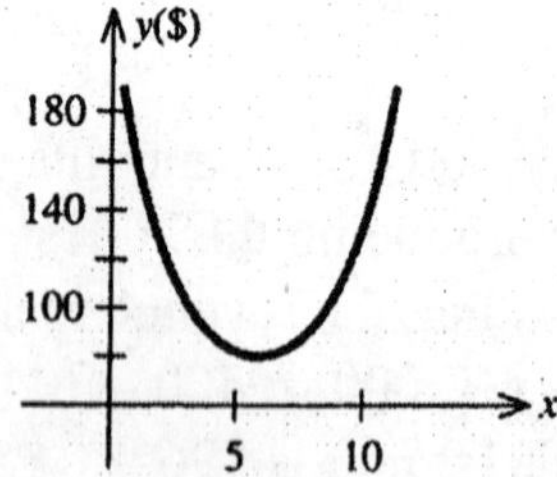

17. a. $R(x) = px = \dfrac{50x}{0.01x^2 + 1}.$

b. $R'(x) = \dfrac{(0.01x^2 + 1)50 - 50x(0.02x)}{(0.01x^2 + 1)^2} = \dfrac{50 - 0.5x^2}{(0.01x^2 + 1)^2}$

c. $R'(2) = \dfrac{50 - 0.5(4)}{[0.01(4)+1]^2} \approx 44.379.$

This result says that at a level of sales of 2000 units, the revenue increases at the rate of approximately \$44,379 per sales of 1000 units.

19. $C(x) = 0.873x^{1.1} + 20.34;\ \ C'(x) = 0.873(1.1)x^{0.1}$
$C'(10) = 0.873(1.1)(10)^{0.1} = 1.21$, or \$1.21 billion per billion dollars.

21. The consumption function is given by $C(x) = 0.712x + 95.05$. The marginal propensity to consume is given by $\dfrac{dC}{dx} = 0.712$. The marginal propensity to save is given by $\dfrac{dS}{dx} = 1 - \dfrac{dC}{dx} = 1 - 0.712 = 0.288$, or \$0.288 billion per billion dollars.

23. Here $x = f(p) = -\frac{5}{4}p + 20$ and so $f'(p) = -\frac{5}{4}$. Therefore,
$E(p) = -\dfrac{pf'(p)}{f(p)} = -\dfrac{p(-\frac{5}{4})}{-\frac{5}{4}p+20} = \dfrac{5p}{80-5p}.$
$E(10) = \dfrac{5(10)}{80-5(10)} = \dfrac{50}{30} = \dfrac{5}{3} > 1$, and so the demand is elastic.

25. $f(p) = -\frac{1}{3}p + 20;\ f'(p) = -\frac{1}{3}.$

Then the elasticity of demand is given by $E(p) = -\dfrac{p(-\frac{1}{3})}{-\frac{1}{3}p+20}$,

and $E(30) = -\dfrac{30(-\frac{1}{3})}{-\frac{1}{3}(30)+20} = 1,$

and we conclude that the demand is unitary at this price.

27. $x^2 = 169 - p$ and $f(p) = (169-p)^{1/2}$.
Next, $f'(p) = \frac{1}{2}(169-p)^{-1/2}(-1) = -\frac{1}{2}(169-p)^{-1/2}$.
Then the elasticity of demand is given by

$$E(p) = -\frac{pf'(p)}{f(p)} = -\frac{p(-\frac{1}{2})(169-p)^{-1/2}}{(169-p)^{1/2}} = \frac{\frac{1}{2}p}{169-p}.$$

Therefore, when $p = 29$, $E(p) = \dfrac{\frac{1}{2}(29)}{169-29} = \dfrac{14.5}{140} = 0.104.$

Since $E(p) < 1$, we conclude that demand is inelastic at this price.

29. $f(p) = \frac{1}{5}(225 - p^2)$; $f'(p) = \frac{1}{5}(-2p) = -\frac{2}{5}p$.
Then the elasticity of demand is given by

$$E(p) = -\frac{pf'(p)}{f(p)} = -\frac{p(-\frac{2}{5}p)}{\frac{1}{5}(225-p^2)} = \frac{2p^2}{225-p^2}.$$

a. When $p = 8$, $E(8) = \dfrac{2(64)}{225-64} = 0.8 < 1$ and the demand is inelastic. When $p = 10$,

$$E(10) = \frac{2(100)}{225-100} = 1.6 > 1$$

and the demand is elastic.

b. The demand is unitary when $E = 1$. Solving $\dfrac{2p^2}{225-p^2} = 1$ we find $2p^2 = 225 - p^2$, $3p^2 = 225$, and $p = 8.66$. So the demand is unitary when $p = 8.66$.
c. Since demand is elastic when $p = 10$, lowering the unit price will cause the revenue to increase.
d. Since the demand is inelastic at $p = 8$, a slight increase in the unit price will cause the revenue to increase.

31. $f(p) = \frac{2}{3}(36 - p^2)^{1/2}$
$f'(p) = \frac{2}{3}(\frac{1}{2})(36-p^2)^{-1/2}(-2p) = -\frac{2}{3}p(36-p^2)^{-1/2}$.
Then the elasticity of demand is given by

$$E(p) = -\frac{pf'(p)}{f(p)} = -\frac{-\frac{2}{3}p(36-p^2)^{-1/2}p}{\frac{2}{3}(36-p^2)^{1/2}} = \frac{p^2}{36-p^2}.$$

When $p = 2$, $E(2) = \dfrac{4}{36-4} = \dfrac{1}{8} < 1$, and we conclude that the demand is inelastic.
b. Since the demand is inelastic, the revenue will increase when the rental price is increased.

33. We first solve the demand equation for x in terms of p. Thus,

$$p = \sqrt{9 - 0.02x}$$
$$p^2 = 9 - 0.02x$$

or $x = -50p^2 + 450$. With $f(p) = -50p^2 + 450$, we find

$$E(p) = -\frac{pf'(p)}{f(p)} = -\frac{p(-100p)}{-50p^2 + 450} = \frac{2p^2}{9 - p^2}.$$

Setting $E(p) = 1$ gives $2p^2 = 9 - p^2$, so $p = \sqrt{3}$. So the demand is inelastic in $[0, \sqrt{3}]$, unitary when $p = \sqrt{3}$, and elastic in $(\sqrt{3}, 3)$.

35. True. $\overline{C}'(x) = \frac{d}{dx}\left[\frac{C(x)}{x}\right] = \frac{xC'(x) - C(x)\frac{d}{dx}(x)}{x^2} = \frac{xC'(x) - C(x)}{x^2}.$

12.5 Problem Solving Tips

When you work applied problems make sure that you keep track of the units of measure used. For example, if velocity is measured in ft/sec, then the units of acceleration will be ft/sec^2. If you are working an applied problem, and the units in your answer are not correct, it may indicate that you have made an error in your calculations or in the formulation of the problem.

Here are some hints for solving the problems in the exercises that follow:

1. Make sure that you simplify an expression before differentiating it to find the next order derivative.

2. The velocity of an object moving in a straight path is given by the derivative of the position function for that object. The acceleration of the object is given by the derivative of the velocity function.

12.5 CONCEPT QUESTIONS, page 740

1. a. The second derivative of f is the derivative of f'.

b. To find the second derivative of f, we differentiate f'.

3. The relative rate of change is $I'(c)/I(c)$.

EXERCISES 12.5, page 740

1. $f(x) = 4x^2 - 2x + 1$; $f'(x) = 8x - 2$; $f''(x) = 8$.

3. $f(x) = 2x^3 - 3x^2 + 1$; $f'(x) = 6x^2 - 6x$; $f''(x) = 12x - 6 = 6(2x - 1)$.

5. $h(t) = t^4 - 2t^3 + 6t^2 - 3t + 10$; $h'(t) = 4t^3 - 6t^2 + 12t - 3$
$h''(t) = 12t^2 - 12t + 12 = 12(t^2 - t + 1)$.

7. $f(x) = (x^2 + 2)^5$; $f'(x) = 5(x^2 + 2)^4(2x) = 10x(x^2 + 2)^4$ and
$f''(x) = 10(x^2 + 2)^4 + 10x(x^2 + 2)^3(2x)$
$= 10(x^2 + 2)^3[(x^2 + 2) + 8x^2] = 10(9x^2 + 2)(x^2 + 2)^3$.

9. $g(t) = (2t^2 - 1)^2(3t^2)$;
$g'(t) = 2(2t^2 - 1)(4t)(3t^2) - (2t^2 - 1)^2(6t)$
$= 6t(2t^2 - 1)[4t^2 + (2t^2 - 1)] = 6t(2t^2 - 1)(6t^2 - 1)$
$= 6t(12t^4 - 8t^2 + 1) = 72t^5 - 48t^3 + 6t$.
$g''(t) = 360t^4 - 144t^2 + 6 = 6(60t^4 - 24t^2 + 1)$

11. $f(x) = (2x^2 + 2)^{7/2}$; $f'(x) = \frac{7}{2}(2x^2 + 2)^{5/2}(4x) = 14x(2x^2 + 2)^{5/2}$;
$f''(x) = 14(2x^2 + 2)^{5/2} + 14x(\frac{5}{2})(2x^2 + 2)^{3/2}(4x)$
$= 14(2x^2 + 2)^{3/2}[(2x^2 + 2) + 10x^2] = 28(6x^2 + 1)(2x^2 + 2)^{3/2}$.

13. $f(x) = x(x^2 + 1)^2$;
$f'(x) = (x^2 + 1)^2 + x(2)(x^2 + 1)(2x)$
$= (x^2 + 1)[(x^2 + 1) + 4x^2] = (x^2 + 1)(5x^2 + 1)$;
$f''(x) = 2x(5x^2 + 1) + (x^2 + 1)(10x) = 2x(5x^2 + 1 + 5x^2 + 5) = 4x(5x^2 + 3)$.

15. $f(x) = \dfrac{x}{2x + 1}$; $f'(x) = \dfrac{(2x + 1)(1) - x(2)}{(2x + 1)^2} = \dfrac{1}{(2x + 1)^2}$;

$$f''(x) = \frac{d}{dx}(2x+1)^{-2} = -2(2x+1)^{-3}(2) = -\frac{4}{(2x+1)^3}.$$

17. $f(s) = \frac{s-1}{s+1}$; $f'(s) = \frac{(s+1)(1)-(s-1)(1)}{(s+1)^2} = \frac{2}{(s+1)^2}$.

$$f''(s) = 2\frac{d}{ds}(s+1)^{-2} = -4(s+1)^{-3} = -\frac{4}{(s+1)^3}.$$

19. $f(u) = \sqrt{4-3u} = (4-3u)^{1/2}$. $f'(u) = \frac{1}{2}(4-3u)^{-1/2}(-3) = -\frac{3}{2\sqrt{4-3u}}$.

$$f''(u) = -\frac{3}{2}\cdot\frac{d}{du}(4-3u)^{-1/2} = -\frac{3}{2}\left(-\frac{1}{2}\right)(4-3u)^{-3/2}(-3) = -\frac{9}{4(4-3u)^{3/2}}.$$

21. $f(x) = 3x^4 - 4x^3$; $f'(x) = 12x^3 - 12x^2$; $f''(x) = 36x^2 - 24x$; $f'''(x) = 72x - 24$.

23. $f(x) = \frac{1}{x}$; $f'(x) = \frac{d}{dx}(x^{-1}) = -x^{-2}$; $f''(x) = 2x^{-3}$; $f'''(x) = -6x^{-4} = -\frac{6}{x^4}$.

25. $g(s) = (3s-2)^{1/2}$; $g'(s) = \frac{1}{2}(3s-2)^{-1/2}(3) = \frac{3}{2(3s-2)^{1/2}}$;

$$g''(s) = \frac{3}{2}\left(-\frac{1}{2}\right)(3s-2)^{-3/2}(3) = -\frac{9}{4}(3s-2)^{-3/2} = -\frac{9}{4(3s-2)^{3/2}};$$

$$g'''(s) = \frac{27}{8}(3s-2)^{-5/2}(3) = \frac{81}{8}(3s-2)^{-5/2} = \frac{81}{8(3s-2)^{5/2}}.$$

27. $f(x) = (2x-3)^4$; $f'(x) = 4(2x-3)^3(2) = 8(2x-3)^3$
$f''(x) = 24(2x-3)^2(2) = 48(2x-3)^2$; $f'''(x) = 96(2x-3)(2) = 192(2x-3)$.

29. Its velocity at any time t is $v(t) = \frac{d}{dt}(16t^2) = 32t$. The hammer strikes the ground when $16t^2 = 256$ or $t = 4$ (we reject the negative root). Therefore, its velocity at the instant it strikes the ground is $v(4) = 32(4) = 128$ ft/sec. Its acceleration at time t is $a(t) = \frac{d}{dt}(32t) = 32$. In particular, its acceleration at $t = 4$ is 32 ft/sec^2.

31. $N(t) = -0.1t^3 + 1.5t^2 + 100$.

a. $N'(t) = -0.3t^2 + 3t = 0.3t(10 - t)$. Since $N'(t) > 0$ for $t = 0, 1, 2, ..., 7$, it is evident that $N(t)$ (and therefore the crime rate) was increasing from 1988 through 1995.

b. $N''(t) = -0.6t + 3 = 0.6(5 - t)$. Now $N''(4) = 0.6 > 0$, $N''(5) = 0$, $N''(6) = -0.6 < 0$ and $N''(7) = -1.2 < 0$. This shows that the rate of the rate of change was decreasing beyond $t = 5$ (1990). This shows that the program was working.

33. $N(t) = 0.00037t^3 - 0.0242t^2 + 0.52t + 5.3 \quad (0 \le t \le 10)$

$N'(t) = 0.00111t^2 - 0.0484t + 0.52$

$N''(t) = 0.00222t - 0.0484$

So $N(8) = 0.00037(8)^3 - 0.0242(8)^2 + 5.3 = 8.1$

$N'(8) = 0.00111(8)^2 - 0.0484(8) + 0.52 \approx 0.204.$

$N''(8) = 0.00222(8) - 0.0484 = -0.031.$

We conclude that at the beginning of 1998, there were 8.1 million persons receiving disability benefits, the number is increasing at the rate of 0.2 million/yr, and the rate of the rate of change of the number of persons is decreasing at the rate of 0.03 million persons/yr^2.

35. a. $h(t) = \frac{1}{16}t^4 - t^3 + 4t^2$. $h'(t) = \frac{1}{4}t^3 - 3t^2 + 8t$

b. $h'(0) = 0$ or zero feet per second.

$h'(4) = \frac{1}{4}(64) - 3(16) + 8(4) = 0$, or zero feet per second.

$h'(8) = \frac{1}{4}(8)^3 - 3(64) + 8(8) = 0$, or zero feet per second.

c. $h''(t) = \frac{3}{4}t^2 - 6t + 8$

d. $h''(0) = 8$ ft/sec^2; $h''(4) = \frac{3}{4}(16) - 6(4) + 8 = -4$ ft/sec^2.

$h''(8) = \frac{3}{4}(64) - 6(8) + 8 = 8$ ft/sec^2.

e. $h(0) = 0$ feet; $h(4) = \frac{1}{16}(4)^4 - (4)^3 + 4(4)^2 = 16$ feet.

$h(8) = \frac{1}{16}(8)^4 - (8)^3 + 4(8)^2 = 0$ feet.

37. $f(t) = 10.72(0.9t + 10)^{0.3}$.

$f'(t) = 10.72(0.3)(0.9t + 10)^{-0.7}(0.9) = 2.8944(0.9t + 10)^{-0.7}$

$f''(t) = 2.8944(-0.7)(0.9t + 10)^{-1.7}(0.9) = -1.823472(0.9t + 10)^{-1.7}$

So $f''(10) = -1.823472(19)^{-1.7} \approx -0.01222$. And this says that the rate of the rate of change of the population is decreasing at the rate of $0.01\%/\text{yr}^2$.

39. False. If f has derivatives of order two at $x = a$, then $f''(a) = [f'(a)]^2$.

41. True. If $f(x)$ is a polynomial function of degree n, then $f^{(n+1)}(x) = 0$.

43. True. Using the chain rule, $h'(x) = f'(2x) \cdot \frac{d}{dx}(2x) = f'(x) \cdot 2 = 2f'(2x)$
Using the chain rule again, $h''(x) = 2f''(2x) \cdot 2 = 4f''(2x)$.

45. Consider the function $f(x) = x^{(2n+1)/2} = x^{n+(1/2)}$.
Then $f'(x) = (n+\frac{1}{2})x^{n-(1/2)}$
$f''(x) = (n+\frac{1}{2})(n-\frac{1}{2})x^{n-(3/2)}$
....
$f^{(n)}(x) = (n+\frac{1}{2})(n-\frac{1}{2}) \cdots \frac{3}{2}x^{1/2}$
$f^{(n+1)}(x) = (n+\frac{1}{2})(n-\frac{1}{2}) \cdots \frac{1}{2}x^{-1/2}$.
The first n derivatives exist at $x = 0$, but the $(n + 1)$st derivative fails to be defined there.

USING TECHNOLOGY EXERCISES 12.5, page 744

1. –18 3. 15.2762 5. –0.6255 7. 0.1973

9. $f''(6) = -68.46214$ and it tells us that at the beginning of 1988, the rate of the rate of the rate at which banks were failing was 68 banks per year per year per year.

12.6 Problem Solving Tips

Here are some hints for solving the problems in the exercises that follow:

1. If an equation expresses y implicitly as a function of x, then we can use implicit differentiation to find its derivative. We apply the Chain rule to find the derivative of any

term involving y. (Note that the derivative of any term involving y will include the factor dy/dx.) The terms involving only x are differentiated in the usual manner.

2. Step 3 of the guidelines for solving related rates problems, page 227 in the text, asks you to find an equation giving the relationship between the variables in the related rates problem.. Make sure that you differentiate this equation implicitly with respect to t before you substitute the values of the variables back into the equation (Step 5).

12.6 CONCEPT QUESTIONS, page 753

1. a. We differentiate both sides of $F(x,y)=0$ with respect to x. Then solve for dy/dx.
   b. The chain rule is used to differentiate any expression involving the dependent variable y.
3. Suppose x and y are two variables that are related by an equation. Furthermore, suppose x and y are both functions of a third variable t. (Normally, t represents time). Then a related rates problem involving finding dx/dt or dy/dt.

EXERCISES 12.6, page 753

1. a. Solving for y in terms of x, we have $y=-\frac{1}{2}x+\frac{5}{2}$. Therefore, $y'=-\frac{1}{2}$.
   b. Next, differentiating $x+2y=5$ implicitly, we have $1+2y'=0$, or $y'=-\frac{1}{2}$.

3. a. $xy=1$, $y=\dfrac{1}{x}$, and $\dfrac{dy}{dx}=-\dfrac{1}{x^2}$.

   b.
$$x\frac{dy}{dx}+y=0$$
$$x\frac{dy}{dx}=-y$$
$$\frac{dy}{dx}=-\frac{y}{x}=\frac{-\frac{1}{x}}{x}=-\frac{1}{x^2}.$$

5. $x^3 - x^2 - xy = 4.$

a. $-xy = 4 - x^3 + x^2$

$$y = -\frac{4}{x} + x^2 - x \text{ and } y' = \frac{4}{x^2} + 2x - 1.$$

b. $x^3 - x^2 - xy = 4$

$$-x\frac{dy}{dx} = -3x^2 + 2x + y$$

$$\frac{dy}{dx} = 3x - 2 - \frac{y}{x}$$

$$= 3x - 2 - \frac{1}{x}(-\frac{4}{x} + x^2 - x) = 3x - 2 + \frac{4}{x^2} - x + 1$$

$$= \frac{4}{x^2} + 2x - 1.$$

7. a. $\frac{x}{y} - x^2 = 1$ is equivalent to $\frac{x}{y} = x^2 + 1$, or $y = \frac{x}{x^2 + 1}$. Therefore,

$$y' = \frac{(x^2 + 1) - x(2x)}{(x^2 + 1)^2} = \frac{1 - x^2}{(x^2 + 1)^2}.$$

b. Next, differentiating the equation $x - x^2y = y$ implicitly, we obtain

$$1 - 2xy - x^2y' = y',\ y'(1 + x^2) = 1 - 2xy, \text{ or } y' = \frac{1 - 2xy}{(1 + x^2)}.$$

(This may also be written in the form $-2y^2 + \frac{y}{x}$.) To show that this is equivalent to the results obtained earlier, use the value of y obtained before, to get

$$y' = \frac{1 - 2x\left(\frac{x}{x^2 + 1}\right)}{1 + x^2} = \frac{x^2 + 1 - 2x^2}{(1 + x^2)^2} = \frac{1 - x^2}{(1 + x^2)^2}.$$

9. $x^2 + y^2 = 16$. Differentiating both sides of the equation implicitly, we obtain

$$2x + 2yy' = 0 \text{ and so } y' = -\frac{x}{y}.$$

11. $x^2 - 2y^2 = 16$. Differentiating implicitly with respect to x, we have

$$2x - 4y\frac{dy}{dx} = 0 \text{ and } \frac{dy}{dx} = \frac{x}{2y}.$$

13. $x^2 - 2xy = 6$. Differentiating both sides of the equation implicitly, we obtain

$2x - 2y - 2xy' = 0$ and so $y' = \dfrac{x-y}{x} = 1 - \dfrac{y}{x}$.

15. $x^2y^2 - xy = 8$. Differentiating both sides of the equation implicitly, we obtain

$$2xy^2 + 2x^2yy' - y - xy' = 0,\ 2xy^2 - y + y'(2x^2y - x) = 0$$

and so
$$y' = \frac{y(1-2xy)}{x(2xy-1)} = -\frac{y}{x}.$$

17. $x^{1/2} + y^{1/2} = 1$. Differentiating implicitly with respect to x, we have

$\frac{1}{2}x^{-1/2} + \frac{1}{2}y^{-1/2}\dfrac{dy}{dx} = 0$. Therefore, $\dfrac{dy}{dx} = -\dfrac{x^{-1/2}}{y^{-1/2}} = -\dfrac{\sqrt{y}}{\sqrt{x}}$.

19. $\sqrt{x+y} = x$. Differentiating both sides of the equation implicitly, we obtain

$$\tfrac{1}{2}(x+y)^{-1/2}(1+y') = 1,\ 1 + y' = 2(x+y)^{1/2},$$

or
$$y' = 2\sqrt{x+y} - 1.$$

21. $\dfrac{1}{x^2} + \dfrac{1}{y^2} = 1$. Differentiating both sides of the equation implicitly, we obtain

$$-\frac{2}{x^3} - \frac{2}{y^3}y' = 0, \text{ or } y' = -\frac{y^3}{x^3}.$$

23. $\sqrt{xy} = x + y$. Differentiating both sides of the equation implicitly, we obtain

$$\tfrac{1}{2}(xy)^{-1/2}(xy' + y) = 1 + y'$$
$$xy' + y = 2\sqrt{xy}(1 + y')$$
$$y'(x - 2\sqrt{xy}) = 2\sqrt{xy} - y$$

or
$$y' = -\frac{(2\sqrt{xy} - y)}{(2\sqrt{xy} - x)} = \frac{2\sqrt{xy} - y}{x - 2\sqrt{xy}}.$$

25. $\dfrac{x+y}{x-y} = 3x$, or $x + y = 3x^2 - 3xy$. Differentiating both sides of the equation

implicitly, we obtain $1 + y' = 6x - 3xy' - 3y$ or $y' = \dfrac{6x - 3y - 1}{3x + 1}$.

27. $xy^{3/2} = x^2 + y^2$. Differentiating implicitly with respect to x, we obtain

$$y^{3/2} + x\left(\tfrac{3}{2}\right)y^{1/2}\frac{dy}{dx} = 2x + 2y\frac{dy}{dx}$$

$$2y^{3/2} + 3xy^{1/2}\frac{dy}{dx} = 4x + 4y\frac{dy}{dx} \qquad \text{(Multiplying by 2.)}$$

$$(3xy^{1/2} - 4y)\frac{dy}{dx} = 4x - 2y^{3/2}$$

$$\frac{dy}{dx} = \frac{2(2x - y^{3/2})}{3xy^{1/2} - 4y}.$$

29. $(x + y)^3 + x^3 + y^3 = 0$. Differentiating implicitly with respect to x, we obtain

$$3(x+y)^2\left(1 + \frac{dy}{dx}\right) + 3x^2 + 3y^2\frac{dy}{dx} = 0$$

$$(x+y)^2 + (x+y)^2\frac{dy}{dx} + x^2 + y^2\frac{dy}{dx} = 0$$

$$[(x+y)^2 + y^2]\frac{dy}{dx} = -[(x+y)^2 + x^2]$$

$$\frac{dy}{dx} = -\frac{2x^2 + 2xy + y^2}{x^2 + 2xy + 2y^2}.$$

31. $4x^2 + 9y^2 = 36$. Differentiating the equation implicitly, we obtain

$$8x + 18yy' = 0.$$

At the point (0,2), we have $0 + 36y' = 0$ and the slope of the tangent line is 0. Therefore, an equation of the tangent line is $y = 2$.

33. $x^2y^3 - y^2 + xy - 1 = 0$. Differentiating implicitly with respect to x, we have

$$2xy^3 + 3x^2y^2\frac{dy}{dx} - 2y\frac{dy}{dx} + y + x\frac{dy}{dx} = 0.$$

At (1,1), $2 + 3\frac{dy}{dx} - 2\frac{dy}{dx} + 1 + \frac{dy}{dx} = 0$, and

$$2\frac{dy}{dx} = -3 \text{ and } \frac{dy}{dx} = -\frac{3}{2}.$$

Using the point-slope form of an equation of a line, we have $y - 1 = -\frac{3}{2}(x - 1)$, and the equation of the tangent line to the graph of the function f at (1,1) is $y = -\frac{3}{2}x + \frac{5}{2}$.

35. $xy = 1$. Differentiating implicitly, we have $xy' + y = 0$, or $y' = -\frac{y}{x}$.

Differentiating implicitly once again, we have $xy'' + y' + y' = 0$.

Therefore, $y'' = -\frac{2y'}{x} = \frac{2\left(\frac{y}{x}\right)}{x} = \frac{2y}{x^2}$.

37. $y^2 - xy = 8$. Differentiating implicitly we have $2yy' - y - xy' = 0$

and $y' = \frac{y}{2y-x}$. Differentiating implicitly again, we have

$$2(y')^2 + 2yy'' - y' - y' - xy'' = 0, \quad \text{or} \quad y'' = \frac{2y' - 2(y')^2}{2y-x}.$$

Then $y'' = \frac{2\left(\frac{y}{2y-x}\right)\left(1 - \frac{y}{2y-x}\right)}{2y-x} = \frac{2y(2y-x-y)}{(2y-x)^3} = \frac{2y(y-x)}{(2y-x)^3}$.

39. a. Differentiating the given equation with respect to t, we obtain

$$\frac{dV}{dt} = \pi r^2 \frac{dh}{dt} + 2\pi rh\frac{dr}{dt} = \pi r\left(r\frac{dh}{dt} + 2h\frac{dr}{dt}\right).$$

b. Substituting $r = 2$, $h = 6$, $\frac{dr}{dt} = 0.1$ *and* $\frac{dh}{dt} = 0.3$ into the expression for $\frac{dV}{dt}$

we obtain $\frac{dV}{dt} = \pi(2)[2(0.3) + 2(6)(0.1)] = 3.6\pi$, and so the volume is increasing at the rate of 3.6π cu in/sec.

41. We are given $\frac{dp}{dt} = 2$ and are required to find $\frac{dx}{dt}$ when $x = 9$ and $p = 63$.

Differentiating the equation $p + x^2 = 144$ with respect to t, we obtain

$$\frac{dp}{dt} + 2x\frac{dx}{dt} = 0.$$

When $x = 9$, $p = 63$, and $\frac{dp}{dt} = 2$,

$$2 + 2(9)\frac{dx}{dt} = 0$$

and

$$\frac{dx}{dt} = -\frac{1}{9} \approx -0.111,$$

or the quantity demanded is decreasing at the rate of 111 tires per week.

43. $100x^2 + 9p^2 = 3600$. Differentiating the given equation implicitly with respect to t, we have $200x\frac{dx}{dt} + 18p\frac{dp}{dt} = 0$. Next, when $p = 14$, the given equation yields

$$100x^2 + 9(14)^2 = 3600$$
$$100x^2 = 1836,$$

or $x = 4.2849$. When $p = 14$, $\frac{dp}{dt} = -0.15$, and $x = 4.2849$, we have

$$200(4.2849)\frac{dx}{dt} + 18(14)(-0.15) = 0$$
$$\frac{dx}{dt} = 0.0441.$$

So the quantity demanded is increasing at the rate of 44 ten–packs per week.

45. From the results of Problem 44, we have

$$1250p\frac{dp}{dt} - 2x\frac{dx}{dt} = 0.$$

When $p = 1.0770$, $x = 25$, and $\frac{dx}{dt} = -1$, we find that

$$1250(1.077)\frac{dp}{dt} - 2(25)(-1) = 0,$$

and $$\frac{dp}{dt} = -\frac{50}{1250(1.077)} = -0.037.$$

We conclude that the price is decreasing at the rate of 3.7 cents per carton.

47. $p = -0.01x^2 - 0.2x + 8$. Differentiating the given equation implicitly with respect to p, we have

$$1 = -0.02x\frac{dx}{dp} - 0.2\frac{dx}{dp} = [0.02x + 0.2]\frac{dx}{dp}$$

or $$\frac{dx}{dp} = -\frac{1}{0.02x + 0.2}.$$

When $x = 15$, $p = -0.01(15)^2 - 0.2(15) + 8 = 2.75$

and $$\frac{dx}{dp} = -\frac{1}{0.02(15) + 0.2} = -2.$$

Therefore, $E(p) = -\dfrac{pf'(p)}{f(p)} = -\dfrac{(2.75)(-2)}{15} = 0.37 < 1$,

and the demand is inelastic.

49. $A = \pi r^2$. Differentiating with respect to t, we obtain

$$\frac{dA}{dt} = 2\pi r \frac{dr}{dt}.$$

When the radius of the circle is 40 ft and increasing at the rate of 2 ft/sec,

$$\frac{dA}{dt} = 2\pi(40)(2) = 160\pi \text{ ft}^2/\text{sec}.$$

51. Let D denote the distance between the two cars, x the distance traveled by the car heading east, and y the distance traveled by the car heading north as shown in the diagram at the right. Then

$D^2 = x^2 + y^2$. Differentiating with respect to t, we have

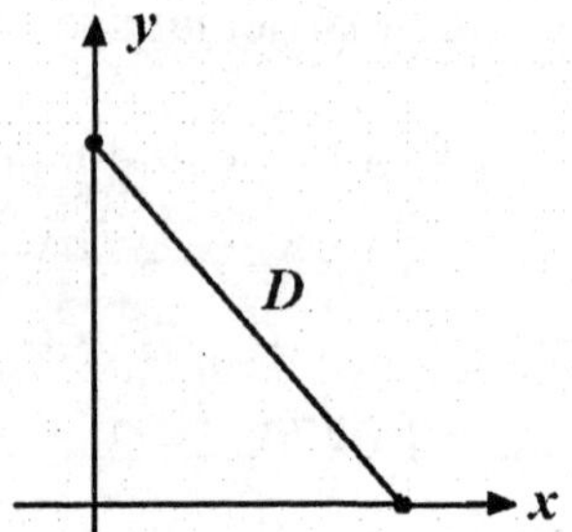

$$2D\frac{dD}{dt} = 2x\frac{dx}{dt} + 2y\frac{dy}{dt},$$

$$\text{or } \frac{dD}{dt} = \frac{x\frac{dx}{dt} + y\frac{dy}{dt}}{D}$$

When $t = 5$, $x = 30, y = 40, \dfrac{dx}{dt} = 2(5) + 1 = 11$, and $\dfrac{dy}{dt} = 2(5) + 3 = 13$.

Therefore, $\dfrac{dD}{dt} = \dfrac{(30)(11) + (40)(13)}{\sqrt{900 + 1600}} = 17$ ft/sec.

53. Referring to the diagram at the right, we see that

$D^2 = 120^2 + x^2$.

Differentiating this last equation with r respect to t, we have

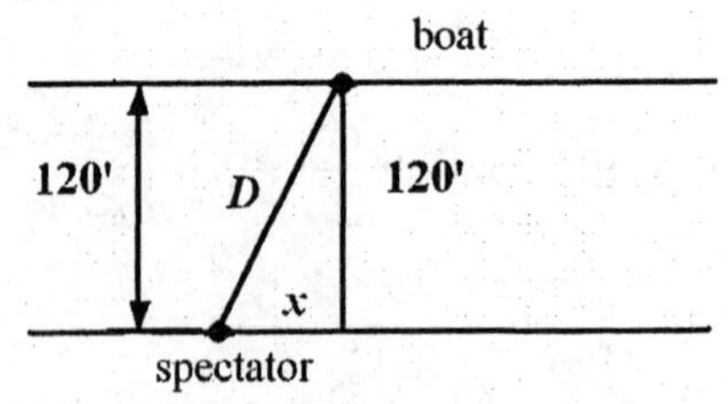

$$2D\frac{dD}{dt} = 2x\frac{dx}{dt} \quad \text{and} \quad \frac{dD}{dt} = \frac{x\frac{dx}{dt}}{D}.$$

When $x = 50$, $D = \sqrt{120^2 + 50^2} = 130$ and $\dfrac{dD}{dt} = \dfrac{(20)(50)}{130} \approx 7.69$, or 7.69 ft/sec.

55. Let V and S denote its volume and surface area. Then we are given that $\frac{dV}{dt} = kS$, where k is the constant of proportionality. But from $V = \left(\frac{4}{3}\right)\pi r^3$, we find, upon differentiating both sides with respect to t, that

$$\frac{dV}{dt} = \frac{d}{dt}\left(\frac{4}{3}\pi r^3\right) = 4\pi r^2 \frac{dr}{dt} = kS = k(4\pi r^2)$$

Therefore, $\frac{dr}{dt} = k$ a constant.

57. Refer to the figure at the right.

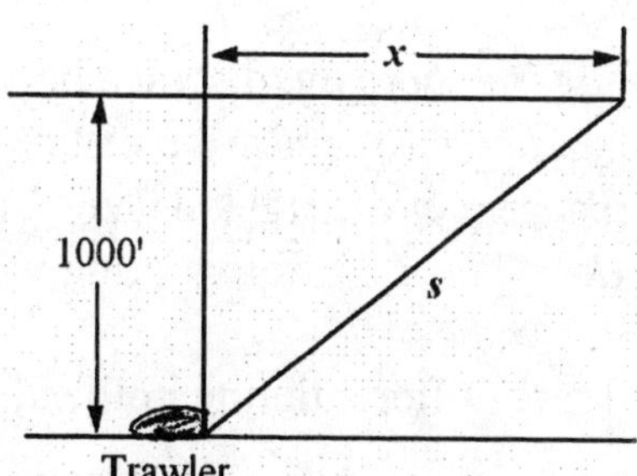

We are given that $\frac{dx}{dt} = 264$. Using the Pythagorean Theorem,

$$s^2 = x^2 + 1000^2 = x^2 + 1000000.$$

We want to find $\frac{ds}{dt}$ when $s = 1500$.

Differentiating both sides of the equation with respect to t, we have

$$2s\frac{ds}{dt} = 2x\frac{dx}{dt} \quad \text{and so} \quad \frac{ds}{dt} = \frac{x\frac{dx}{dt}}{s}.$$

Now, when $s = 1500$, we have

$$1500^2 = x^2 + 10000 \quad \text{or} \quad x = \sqrt{1250000}.$$

Therefore, $$\frac{ds}{dt} = \frac{\sqrt{1250000}\bullet(264)}{1500} \approx 196.8,$$

that is, the aircraft is receding from the trawler at the speed of approximately 196.8 ft/sec.

59. Refer to the diagram that follows.

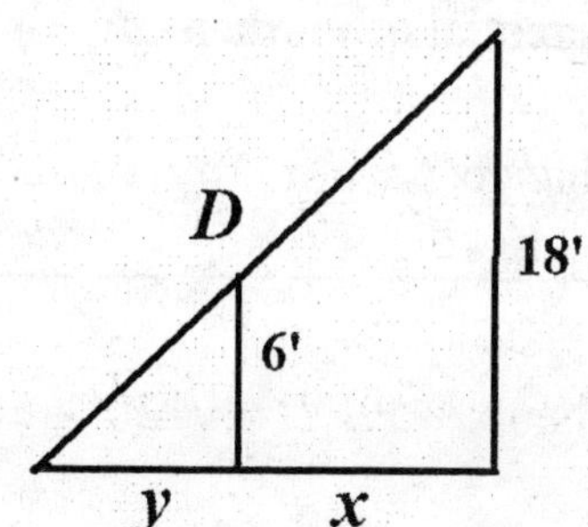

$$\frac{y}{6} = \frac{y+x}{18}, \quad 18y = 6(y+x)$$

$$3y = y + x, \quad 2y = x, \quad y = \tfrac{1}{2}x.$$

Then $D = y + x = \frac{3}{2}x$. Differentiating implicitly, we have $\frac{dD}{dt} = \frac{3}{2}\bullet\frac{dx}{dt}$

61. Differentiating $x^2 + y^2 = 13^2 = 169$ with respect to t gives

$$2x\frac{dx}{dt} + 2y\frac{dy}{dt} = 0.$$

When $x = 12$, we have

$144 + y^2 = 169$ or $y = 5$.

Therefore, with $x = 12$, $y = 5$, and $\frac{dx}{dt} = 8$, we find $2(12)(8) + 2(5)\frac{dy}{dt} = 0$, or

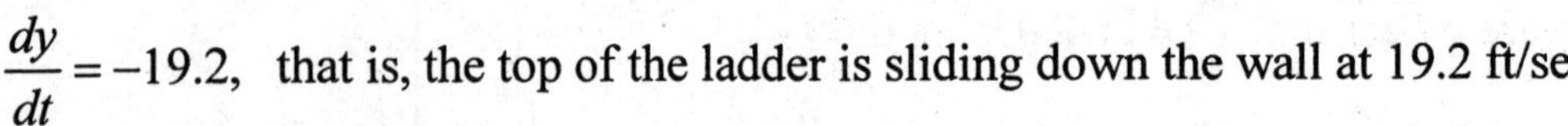

$\frac{dy}{dt} = -19.2$, that is, the top of the ladder is sliding down the wall at 19.2 ft/se

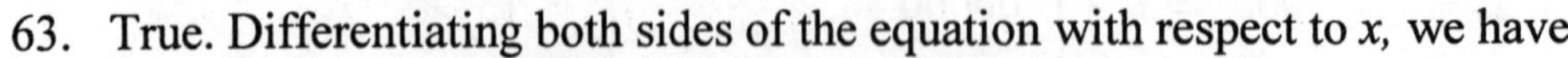

63. True. Differentiating both sides of the equation with respect to x, we have

$$\frac{d}{dx}[f(x)g(y)] = \frac{d}{dx}(0)$$

$$f(x)g'(y)\frac{dy}{dx} + f'(x)g(y) = 0$$

$$\frac{dy}{dx} = -\frac{f'(x)g(y)}{f(x)g'(y)}$$

provided $f(x) \neq 0$ and $g'(y) \neq 0$.

12.7 Problem Solving Tips

1. Δy is the *actual change* in y due to a small change Δx in x. It is approximated by dy, the differential of y, where $dy = f'(x)\Delta x = f'(x)dx$. Note that $dx = \Delta x$, *but* $dy \approx \Delta y$ for a small change Δx in x.

12.7 CONCEPT QUESTIONS, page 761

1. The differential of x is dx. The differential of y is $dy = f'(x)\,dx$.

EXERCISES 12.7, page 762

1. $f(x) = 2x^2$ and $dy = 4x\,dx$.

3. $f(x) = x^3 - x$ and $dy = (3x^2 - 1)\,dx$.

5. $f(x) = \sqrt{x+1} = (x+1)^{1/2}$ and $dy = \frac{1}{2}(x+1)^{-1/2}\,dx = \frac{dx}{2\sqrt{x+1}}$.

7. $f(x) = 2x^{3/2} + x^{1/2}$ and $dy = (3x^{1/2} + \frac{1}{2}x^{-1/2})\,dx = \frac{1}{2}x^{-1/2}(6x+1)dx = \frac{6x+1}{2\sqrt{x}}\,dx$.

9. $f(x) = x + \frac{2}{x}$ and $dy = \left(1 - \frac{2}{x^2}\right)dx = \frac{x^2-2}{x^2}\,dx$.

11. $f(x) = \frac{x-1}{x^2+1}$ and $dy = \frac{x^2+1-(x-1)2x}{(x^2+1)^2}\,dx = \frac{-x^2+2x+1}{(x^2+1)^2}\,dx$.

13. $f(x) = \sqrt{3x^2 - x} = (3x^2 - x)^{1/2}$ and

$$dy = \frac{1}{2}(3x^2 - x)^{-1/2}(6x-1)dx = \frac{6x-1}{2\sqrt{3x^2-x}}\,dx.$$

15. $f(x) = x^2 - 1$.
   a. $dy = 2x\,dx$.
   b. $dy \approx 2(1)(0.02) = 0.04$.
   c. $\Delta y = [(1.02)^2 - 1] - [1 - 1] = 0.0404$.

17. $f(x) = \frac{1}{x}$.
   a. $dy = -\frac{dx}{x^2}$.
   b. $dy \approx -0.05$
   c. $\Delta y = \frac{1}{-0.95} - \frac{1}{-1} = -0.05263$.

19. $y=\sqrt{x}$ and $dy=\dfrac{dx}{2\sqrt{x}}$. Therefore, $\sqrt{10}\approx 3+\dfrac{1}{2\cdot\sqrt{9}}=3.167$.

21. $y=\sqrt{x}$ and $dy=\dfrac{dx}{2\sqrt{x}}$. Therefore, $\sqrt{49.5}\approx 7+\dfrac{0.5}{2\cdot 7}=7.0357$.

23. $y=x^{1/3}$ and $dy=\frac{1}{3}x^{-2/3}\,dx$. Therefore, $\sqrt[3]{7.8}\approx 2-\dfrac{0.2}{3\cdot 4}=1.983$.

25. $y=\sqrt{x}$ and $dy=\dfrac{dx}{2\sqrt{x}}$. Therefore, $\sqrt{0.089}=\frac{1}{10}\sqrt{8.9}\approx\frac{1}{10}\left[3-\dfrac{0.1}{2\cdot 3}\right]\approx 0.298$.

27. $y=f(x)=\sqrt{x}+\dfrac{1}{\sqrt{x}}=x^{1/2}+x^{-1/2}$. Therefore,

$$\frac{dy}{dx}=\frac{1}{2}x^{-1/2}-\frac{1}{2}x^{-3/2}$$

$$dy=\left(\frac{1}{2x^{1/2}}-\frac{1}{2x^{3/2}}\right)dx.$$

Letting $x=4$ and $dx=0.02$, we find

$$\sqrt{4.02}+\frac{1}{\sqrt{4.02}}-f(4)=f(4.02)-f(4)=\Delta y\approx dy$$

$$\sqrt{4.02}+\frac{1}{\sqrt{4.02}}\approx f(4)+dy\Big|_{\substack{x=4\\ dx=0.02}}$$

$$\approx 2+\frac{1}{2}+\left(\frac{1}{2\cdot 2}-\frac{1}{16}\right)(0.02)\approx\ 2.50375.$$

29. The volume of the cube is given by $V=x^3$. Then $dV=3x^2\,dx$ and when $x=12$ and $dx=0.02$, $dV=3(144)(\pm 0.02)=\pm 8.64$, and the possible error that might occur in calculating the volume is ± 8.64 cm^3.

31. The volume of the hemisphere is given by $V=\frac{2}{3}\pi r^3$. The amount of rust-proofer needed is

$$\Delta V = \frac{2}{3}\pi(r+\Delta r)^3 - \frac{2}{3}\pi r^3$$

$$\approx dV = \left(\frac{2}{3}\right)(3\pi r^2)dr.$$

So, with $r = 60$, and $dr = \frac{1}{12}(0.01)$, we have

$$\Delta V \approx 2\pi(60^2)\left(\frac{1}{12}\right)(0.01) \approx 18.85.$$

So we need approximately 18.85 ft^3 of rust-proofer.

33. $dR = \frac{d}{dr}(k\ell r^{-4})dr = -4k\ell r^{-5}\,dr$. With $\frac{dr}{r} = 0.1$, we find

$$\frac{dR}{R} = -\frac{4k\ell r^{-5}}{k\ell r^{-4}}dr = -4\frac{dr}{r} = -4(0.1) = -0.4.$$

In other words, the resistance will drop by 40%.

35. $f(n) = 4n\sqrt{n-4} = 4n(n-4)^{1/2}$.

Then $df = 4[(n-4)^{1/2} + \frac{1}{2}n(n-4)^{-1/2}]dn$

When $n = 85$ and $dn = 5$, $df = 4[9 + \frac{85}{2\cdot 9}]5 \approx 274$ seconds.

37. $N(r) = \frac{7}{1+0.02r^2}$ and $dN = -\frac{0.28r}{(1+0.02r^2)^2}dr$. To estimate the decrease in the number of housing starts when the mortgage rate is increased from 12 to 12.5 percent, we compute

$$dN = -\frac{(0.28)(12)(0.5)}{(3.88)^2} \approx -0.111595 \quad (r = 12,\ dr = 0.5)$$

or 111,595 fewer housing starts.

39. $p = \frac{30}{0.02x^2+1}$ and $dp = -\frac{(1.2x)}{(0.02x^2+1)^2}dx$. To estimate the change in the price p when the quantity demanded changed from 5000 to 5500 units ($x = 5$ to $x = 5.5$) per week, we compute $dp = \frac{(-1.2)(5)(0.5)}{[0.02(25)+1]^2} \approx -1.33$, or a decrease of \$1.33.

41. $P(x) = -0.000032x^3 + 6x - 100$ and $dP = (-0.000096x^2 + 6)\,dx$. To determine the error in the estimate of Trappee's profits corresponding to a maximum error in the forecast of 15 percent [$dx = \pm 0.15(200)$], we compute

$$dP = [(-0.000096)(200)^2 + 6]\,(\pm 30) = (2.16)(30) = \pm 64.80$$

or ± 64,800.

43. $N(x) = \dfrac{500(400+20x)^{1/2}}{(5+0.2x)^2}$ and

$$N'(x) = \frac{(5+0.2x)^2 250(400+20x)^{-1/2}(20) - 500(400+20x)^{1/2}(2)(5+0.2x)(0.2)}{(5+0.2x)^4}dx.$$

To estimate the change in the number of crimes if the level of reinvestment changes from 20 cents per dollars deposited to 22 cents per dollar deposited, we compute

$$dN = \frac{(5+4)^2(250)(800)^{-1/2}(20) - 500(400+400)^{1/2}(2)(9)(0.2)}{(5+4)^4}(2)$$

$$= \frac{(14318.91 - 50911.69)}{9^4}(2) \approx -11$$

or a decrease of approximately 11 crimes per year.

45. $A = 10{,}000\left(1+\dfrac{r}{12}\right)^{120}$.

a. $dA = 10{,}000(120)\left(1+\dfrac{r}{12}\right)^{119}\left(\dfrac{1}{12}\right)dr = 100{,}000\left(1+\dfrac{r}{12}\right)^{119} dr.$

b. At 8.1%, it will be worth $100{,}000\left(1+\dfrac{0.08}{12}\right)^{119}(0.001)$, or \$220.49 more.

At 8.2%, it will be worth $100{,}000\left(1+\dfrac{0.08}{12}\right)^{119}(0.002)$, or \$440.99 more.

At 8.3%, it will be worth $100{,}000\left(1+\dfrac{0.08}{12}\right)^{119}(0.003)$, or \$661.48 more.

47. True. $dy = f'(x)\,dx = \dfrac{d}{dx}(ax+b)dx = a\,dx$. On the other hand,

$$\Delta y = f(x+\Delta x) - f(x) = [a(x+\Delta x)+b] - (ax+b) = a\Delta x = a\,dx.$$

USING TECHNOLOGY EXERCISES 12.7, page 766

1. $dy = f'(3)\,dx = 757.87(0.01) \approx 7.5787.$

3. $dy = f'(1)\,dx = 1.04067285926(0.03) \approx 0.031220185778.$

5. $dy = f'(4)(0.1) = -0.198761598(0.1) = -0.0198761598.$

7. If the interest rate changes from 10% to 10.3% per year, the monthly payment will increase by $dP = f'(0.1)(0.003) \approx 26.60279$, or approximately \$26.60 per month. If the interest rate changes from 10% to 10.4% per year, it will be \$35.47 per month. If the interest rate changes from 10% to 10.5% per year, it will be \$44.34 per month.

9. $dx = f'(40)(2) \approx -0.625$. That is, the quantity demanded will decrease by 625 watches per week.

CHAPTER 12, CONCEPT REVIEW QUESTIONS, page 767

1. a. 0 b. nx^{n-1} c. $cf'(x)$ d. $f'(x) \pm g'(x)$

3. a. $g'[f(x)]f'(x)$ b. $n[f(x)]^{n-1}f'(x)$

5. a. $-\dfrac{pf'(p)}{f(p)}$ b. Elastic; unitary; inelastic

7. y; dy/dt; a 9. a. $x_2 - x_1$ b. $f(x+\Delta x) - f(x)$ 10. Δx; Δx; x; $f'(x)\,dx$

CHAPTER 12 REVIEW, page 767

1. $f'(x) = \dfrac{d}{dx}(3x^5 - 2x^4 + 3x^2 - 2x + 1) = 15x^4 - 8x^3 + 6x - 2.$

3. $g'(x) = \dfrac{d}{dx}(-2x^{-3} + 3x^{-1} + 2) = 6x^{-4} - 3x^{-2}.$

5. $g'(t) = \dfrac{d}{dt}\,(2t^{-1/2} + 4t^{-3/2} + 2) = -t^{-3/2} - 6t^{-5/2}.$

7. $f'(t) = \dfrac{d}{dt}(t + 2t^{-1} + 3t^{-2}) = 1 - 2t^{-2} - 6t^{-3} = 1 - \dfrac{2}{t^2} - \dfrac{6}{t^3}.$

9. $h'(x) = \dfrac{d}{dx}(x^2 - 2x^{-3/2}) = 2x + 3x^{-5/2} = 2x + \dfrac{3}{x^{5/2}}.$

11. $g(t) = \dfrac{t^2}{2t^2 + 1}.$

$$g'(t) = \frac{(2t^2+1)\dfrac{d}{dt}(t^2) - t^2\dfrac{d}{dt}(2t^2+1)}{(2t^2+1)^2}$$

$$= \frac{(2t^2+1)(2t) - t^2(4t)}{(2t^2+1)^2} = \frac{2t}{(2t^2+1)^2}.$$

13. $f(x) = \dfrac{\sqrt{x} - 1}{\sqrt{x} + 1} = \dfrac{x^{1/2} - 1}{x^{1/2} + 1}.$

$$f'(x) = \frac{(x^{1/2}+1)(\frac{1}{2}x^{-1/2}) - (x^{1/2}-1)(\frac{1}{2}x^{-1/2})}{(x^{1/2}+1)^2}$$

$$= \frac{\frac{1}{2} + \frac{1}{2}x^{-1/2} - \frac{1}{2} + \frac{1}{2}x^{-1/2}}{(x^{1/2}+1)^2} = \frac{x^{-1/2}}{(x^{1/2}+1)^2} = \frac{1}{\sqrt{x}(\sqrt{x}+1)^2}.$$

15. $f(x) = \dfrac{x^2(x^2+1)}{x^2 - 1}.$

$$f'(x) = \frac{(x^2-1)\dfrac{d}{dx}(x^4+x^2) - (x^4+x^2)\dfrac{d}{dx}(x^2-1)}{(x^2-1)^2}$$

$$= \frac{(x^2-1)(4x^3+2x) - (x^4+x^2)(2x)}{(x^2-1)^2}$$

$$= \frac{4x^5 + 2x^3 - 4x^3 - 2x - 2x^5 - 2x^3}{(x^2 - 1)^2}$$

$$= \frac{2x^5 - 4x^3 - 2x}{(x^2 - 1)^2} = \frac{2x(x^4 - 2x^2 - 1)}{(x^2 - 1)^2}.$$

17. $f(x) = (3x^3 - 2)^8; f'(x) = 8(3x^3 - 2)^7(9x^2) = 72x^2(3x^3 - 2)^7.$

19. $f'(t) = \frac{d}{dt}(2t^2 + 1)^{1/2} = \frac{1}{2}(2t^2 + 1)^{-1/2}\frac{d}{dt}(2t^2 + 1)$

$$= \frac{1}{2}(2t^2 + 1)^{-1/2}(4t) = \frac{2t}{\sqrt{2t^2 + 1}}.$$

21. $s(t) = (3t^2 - 2t + 5)^{-2}$

$$s'(t) = -2(3t^2 - 2t + 5)^{-3}(6t - 2) = -4(3t^2 - 2t + 5)^{-3}(3t - 1)$$

$$= -\frac{4(3t - 1)}{(3t^2 - 2t + 5)^3}.$$

23. $h(x) = \left(x + \frac{1}{x}\right)^2 = (x + x^{-1})^2.$

$$h'(x) = 2(x + x^{-1})(1 - x^{-2}) = 2\left(x + \frac{1}{x}\right)\left(1 - \frac{1}{x^2}\right)$$

$$= 2\left(\frac{x^2 + 1}{x}\right)\left(\frac{x^2 - 1}{x^2}\right) = \frac{2(x^2 + 1)(x^2 - 1)}{x^3}.$$

25. $h'(t) = (t^2 + t)^4 \frac{d}{dt}(2t^2) + 2t^2 \frac{d}{dt}(t^2 + t)^4$

$$= (t^2 + t)^4(4t) + 2t^2 \cdot 4(t^2 + t)^3(2t + 1)$$

$$= 4t(t^2 + t)^3[(t^2 + t) + 4t^2 + 2t] = 4t^2(5t + 3)(t^2 + t)^3.$$

27. $g(x) = x^{1/2}(x^2 - 1)^3.$

$$g'(x) = \frac{d}{dx}[x^{1/2}(x^2-1)^3] = x^{1/2}\cdot 3(x^2-1)^2(2x) + (x^2-1)^3 \cdot \tfrac{1}{2}x^{-1/2}$$
$$= \tfrac{1}{2}x^{-1/2}(x^2-1)^2[12x^2 + (x^2-1)]$$
$$= \frac{(13x^2-1)(x^2-1)^2}{2\sqrt{x}}.$$

29. $h(x) = \dfrac{(3x+2)^{1/2}}{4x-3}.$

$$h'(x) = \frac{(4x-3)\frac{1}{2}(3x+2)^{-1/2}(3) - (3x+2)^{1/2}(4)}{(4x-3)^2}$$
$$= \frac{\frac{1}{2}(3x+2)^{-1/2}[3(4x-3) - 8(3x+2)]}{(4x-3)^2} = -\frac{12x+25}{2\sqrt{3x+2}(4x-3)^2}.$$

31. $f(x) = 2x^4 - 3x^3 + 2x^2 + x + 4.$

$$f'(x) = \frac{d}{dx}(2x^4 - 3x^3 + 2x^2 + x + 4) = 8x^3 - 9x^2 + 4x + 1.$$
$$f''(x) = \frac{d}{dx}(8x^3 - 9x^2 + 4x + 1) = 24x^2 - 18x + 4 = 2(12x^2 - 9x + 2).$$

33. $h(t) = \dfrac{t}{t^2+4}.$ $h'(t) = \dfrac{(t^2+4)(1) - t(2t)}{(t^2+4)^2} = \dfrac{4-t^2}{(t^2+4)^2}.$

$$h''(t) = \frac{(t^2+4)^2(-2t) - (4-t^2)2(t^2+4)(2t)}{(t^2+4)^4}$$
$$= \frac{-2t(t^2+4)[(t^2+4) + 2(4-t^2)]}{(t^2+4)^4} = \frac{2t(t^2-12)}{(t^2+4)^3}.$$

35. $f'(x) = \dfrac{d}{dx}(2x^2+1)^{1/2} = \dfrac{1}{2}(2x^2+1)^{-1/2}(4x) = 2x(2x^2+1)^{-1/2}.$

$$f''(x) = 2(2x^2+1)^{-1/2} + 2x\cdot(-\tfrac{1}{2})(2x^2+1)^{-3/2}(4x)$$
$$= 2(2x^2+1)^{-3/2}[(2x^2+1) - 2x^2] = \frac{2}{(2x^2+1)^{3/2}}.$$

37. $6x^2 - 3y^2 = 9$ so $12x - 6y\dfrac{dy}{dx} = 0$ and $-6y\dfrac{dy}{dx} = -12x.$

Therefore, $\dfrac{dy}{dx} = \dfrac{-12x}{-6y} = \dfrac{2x}{y}$.

39. $y^3 + 3x^2 = 3y$, so $3y^2y' + 6x = 3y'$, $3y^2y' - 3y' = -6x$,

and $y'(3y^2 - 3) = -6x$. Therefore, $y' = -\dfrac{6x}{3(y^2-1)} = -\dfrac{2x}{y^2-1}$.

41. $x^2 - 4xy - y^2 = 12$ so $2x - 4xy' - 4y - 2yy' = 0$ and $y'(-4x - 2y) = -2x + 4y$.

So $y' = \dfrac{-2(x-2y)}{-2(2x+y)} = \dfrac{x-2y}{2x+y}$.

43. $df = f'(x)dx = (2x - 2x^{-3})dx = \left(2x - \dfrac{2}{x^3}\right)dx = \dfrac{2(x^4-1)}{x^3}dx$

45. a. $df = f'(x)dx = \dfrac{d}{dx}(2x^2+4)^{1/2}\,dx = \dfrac{1}{2}(2x^2+4)^{-1/2}(4x) = \dfrac{2x}{\sqrt{2x^2+4}}dx$

b. $\Delta f \approx df\Big|_{\substack{x=4\\ dx=0.1}} = \dfrac{2(4)(0.1)}{\sqrt{2(16)+4}} = \dfrac{0.8}{6} = \dfrac{8}{60} = \dfrac{2}{15}$.

c. $\Delta f = f(4.1) - f(4) = \sqrt{2(4.1)^2+4} - \sqrt{2(16)+4} = 0.1335$

From (b), $\Delta f \approx \dfrac{2}{15} \approx 0.1333$.

47. $f(x) = 2x^3 - 3x^2 - 16x + 3$ and $f'(x) = 6x^2 - 6x - 16$.

a. To find the point(s) on the graph of f where the slope of the tangent line is equal to –4, we solve

$$6x^2 - 6x - 16 = -4,\ 6x^2 - 6x - 12 = 0,\ 6(x^2 - x - 2) = 0$$
$$6(x-2)(x+1) = 0$$

and $x = 2$ or $x = -1$. Then $f(2) = 2(2)^3 - 3(2)^2 - 16(2) + 3 = -25$ and $f(-1) = 2(-1)^3 - 3(-1)^2 - 16(-1) + 3 = 14$ and the points are (2,–25) and (–1,14).

b. Using the point-slope form of the equation of a line, we find

that $y - (-25) = -4(x-2)$, $y + 25 = -4x + 8$, or $y = -4x - 17$

and $y - 14 = -4(x+1)$, or $y = -4x + 10$

are the equations of the tangent lines at (2,–25) and (–1,14).

49. $y=(4-x^2)^{1/2}$. $y'=\frac{1}{2}(4-x^2)^{-1/2}(-2x)=-\dfrac{x}{\sqrt{4-x^2}}$.

The slope of the tangent line is obtained by letting $x = 1$, giving

$$m=-\frac{1}{\sqrt{3}}=-\frac{\sqrt{3}}{3}.$$

Therefore, an equation of the tangent line is

$$y-\sqrt{3}=-\frac{\sqrt{3}}{3}(x-1), \text{ or } y=-\frac{\sqrt{3}}{3}x+\frac{4\sqrt{3}}{3}.$$

51. $f(x)=(2x-1)^{-1}$; $f'(x)=-2(2x-1)^{-2}$, $f''(x)=8(2x-1)^{-3}=\dfrac{8}{(2x-1)^3}$.

$$f'''(x)=-48(2x-1)^4=-\frac{48}{(2x-1)^4}.$$

Since $(2x-1)^4=0$ when $x = 1/2$, we see that the domain of f''' is $(-\infty,\frac{1}{2})\cup(\frac{1}{2},\infty)$.

53. $x=\dfrac{25}{\sqrt{p}}-1$; $f'(p)=-\dfrac{25}{2p^{3/2}}$; $E(p)=-\dfrac{p\left(-\frac{25}{2p^{3/2}}\right)}{\frac{25}{p^{1/2}}-1}=\dfrac{\frac{25}{2p^{1/2}}}{\frac{25-p^{1/2}}{p^{1/2}}}=\dfrac{25}{2(25-p^{1/2})}$

Since $E(p)=1$,

$$2(25-p^{1/2})=25,$$

$$25-p^{1/2}=\tfrac{25}{2},\quad p^{1/2}=\tfrac{25}{2}, \text{ and } p=\tfrac{625}{4}.$$

$E(p)>1$ and demand is elastic if $p>156.25$; $E(p)=1$ and demand is unitary if $p=156.25$; and $E(p)<1$ and demand is inelastic, if $p<156.25$.

55. $p=9\sqrt[3]{1000-x}$; $\sqrt[3]{1000-x}=\dfrac{p}{9}$; $1000-x=\dfrac{p^3}{729}$; $x=1000-\dfrac{p^3}{729}$

Therefore, $x=f(p)=\dfrac{729{,}000-p^3}{729}$ and $f'(x)=-\dfrac{3p^2}{729}=-\dfrac{p^2}{243}$.

Then $E(p)=-\dfrac{p(-\frac{p^2}{243})}{\frac{729{,}000-p^3}{729}}=\dfrac{3p^3}{729{,}000-p^3}$.

So $E(60)=\dfrac{3(60)^3}{729{,}000-60^3}=\dfrac{648{,}000}{513{,}000}=\dfrac{648}{513}>1$, and so demand is elastic.

Therefore, raising the price slightly will cause the revenue to decrease.

57. $N(x) = 1000(1 + 2x)^{1/2}$. $N'(x) = 1000(\frac{1}{2})(1+2x)^{-1/2}(2) = \dfrac{1000}{\sqrt{1+2x}}$.

The rate of increase at the end of the twelfth week is $N'(12) = \dfrac{1000}{\sqrt{25}} = 200$,

or 200 subscribers/week.

59. He can expect to live $f(100) = 46.9[1+1.09(100)]^{0.1} \approx 75.0433$, or approximately 75.04 years. $f'(t) = 46.9(0.1)(1+1.09t)^{-0.9}(1.09) = 5.1121(1+1.09t)^{-0.9}$
So the required rate of change is $f'(100) = 5.1121(1+1.09)^{-0.9} = 0.074$, or approximately 0.07 yr/yr.

61. a. $R(x) = px = (-0.02x + 600)x = -0.02x^2 + 600x$
b. $R'(x) = -0.04x + 600$
c. $R'(10{,}000) = -0.04(10{,}000) + 600 = 200$ and this says that the sale of the 10,001st phone will bring a revenue of \$200.

CHAPTER 12, BEFORE MOVING ON, page 769

1. $f'(x) = 2(3x^2) - 3(\frac{1}{3}x^{-2/3}) + 5(-\frac{2}{3}x^{-5/3}) = 6x^2 - x^{-2/3} - \frac{10}{3}x^{-5/3}$

2. $g'(x) = \dfrac{d}{dx}[x(2x^2-1)^{1/2}] = (2x^2-1)^{1/2} + x(\frac{1}{2})(2x^2-1)^{-1/2}\dfrac{d}{dx}(2x^2-1)$

$$= (2x^2-1)^{1/2} + \tfrac{1}{2}x(2x^2-1)^{-1/2}(4x) = (2x^2-1)^{-1/2}[(2x^2-1)+2x^2) = \frac{4x^2-1}{\sqrt{2x^2-1}}$$

3. $\dfrac{dy}{dx} = \dfrac{(x^2+x+1)(2)-(2x+1)(2x+1)}{(x^2+x+1)^2} = \dfrac{2x^2+2x+2-(4x^2+4x+1)}{(x^2+x+1)^2}$

$$= -\frac{2x^2+2x-1}{(x^2+x+1)^2}.$$

4. $f'(x) = \dfrac{d}{dx}(x+1)^{-1/2} = -\dfrac{1}{2}(x+1)^{-3/2} = -\dfrac{1}{2(x+1)^{3/2}}$

$$f''(x) = -\frac{1}{2}\left(-\frac{3}{2}\right)(x+1)^{-5/2} = \frac{3}{4}(x+1)^{-5/2} = \frac{3}{4(x+1)^{5/2}}$$

$$f'''(x) = \frac{3}{4}\left(-\frac{5}{2}\right)(x+1)^{-7/2} = -\frac{15}{8}(x+1)^{-7/2} = -\frac{15}{8(x+1)^{7/2}}$$

5. Differentiating both sides of the equation with respect to x gives

$$y^2 + x(2yy') - 2xy - x^2y' + 3x^2 = 0$$

$$(2xy - x^2)y' + (y^2 - 2xy + 3x^2) = 0$$

$$y' = \frac{-y^2 + 2xy - 3x^2}{2xy - x^2} = \frac{-y^2 + 2xy - 3x^2}{x(2y - x)}$$

6. a. $dy = \frac{d}{dx}[x(x^2+5)^{1/2}]dx = [x(\tfrac{1}{2})(x^2+5)^{-1/2}(2x)]dx + [(x^2+5)^{1/2}(1)]dx$

$$= (x^2+5)^{-1/2}[(x^2+5) + x^2]dx = \frac{2x^2+5}{\sqrt{x^2+5}}dx$$

Here $dx = \Delta x = 2.01 - 2 = 0.01$. Therefore,

$$\Delta y \approx dy\Big|_{\substack{x=2 \\ dx=0.01}} = \frac{2(4)+5}{\sqrt{4+5}}(0.01) = \frac{0.13}{3} \approx 0.043$$

CHAPTER 13

13.1 Problem Solving Tips

1. The critical number of a function f is any number x in the domain of f such that $f'(x) = 0$, or $f'(x)$ does not exist. Note that the definition requires that x be in the domain of f. Consider the function $f(x) = x + 1/x$ in Example 8, Section 12.1, in the text. Even though f' is discontinuous at $x = 0$, $x = 0$ did not qualify as a critical number because it does not lie in the domain of f.

2. Note that when you use test values to find the sign of a derivative over an interval, you don't need to evaluate the derivative function at a test value. You only need to find the *sign* of the derivative function at that test value. For example, to find the sign of $f'(x) = \dfrac{(x+1)(x-1)}{x^2}$ in the interval (0,1) using the test value $x = ½$, we simply note that sign of the numerator is (+) × (-) = (-). Since x^2 is always positive, the denominator is always positive. Therefore, the sign of f' over the interval (0,1) is negative.

13.1 CONCEPT QUESTIONS, page 782

1. a. f is increasing on I if whenever x_1 and x_2 are in I with $x_1 < x_2$, then $f(x_1) < f(x_2)$.
   b. f is decreasing on I if whenever x_1 and x_2 are in I with $x_1 < x_2$, then $f(x_1) > f(x_2)$.

3. a. f has a relative maximum at $x = a$ if there is an open interval I containing a such that $f(x) \le f(a)$ for all x in I.

   b. f has a relative minimum at $x = a$ if there is an open interval I containing a such that $f(x) \ge f(a)$ for all x in I.

5. See text page 779.

EXERCISES 13.1, page 782

1. f is decreasing on $(-\infty, 0)$ and increasing on $(0, \infty)$.

3. f is increasing on $(-\infty,-1) \cup (1,\infty)$, and decreasing on $(-1,1)$.

5. f is increasing on $(0,2)$ and decreasing on $(-\infty,0) \cup (2,\infty)$.

7. f is decreasing on $(-\infty,-1) \cup (1,\infty)$ and increasing on $(-1,1)$.

9. Increasing on $(20.2, 20.6) \cup (21.7, 21.8)$, constant on $(19.6, 20.2) \cup (20.6, 21.1)$, and decreasing on $(21.1, 21.7) \cup (21.8, 22.7)$.

11. $f(x) = 3x + 5$; $f'(x) = 3 > 0$ for all x and so f is increasing on $(-\infty,\infty)$.

13. $f(x) = x^2 - 3x$. $f'(x) = 2x - 3$ is continuous everywhere and is equal to zero when $x = 3/2$. From the following sign diagram

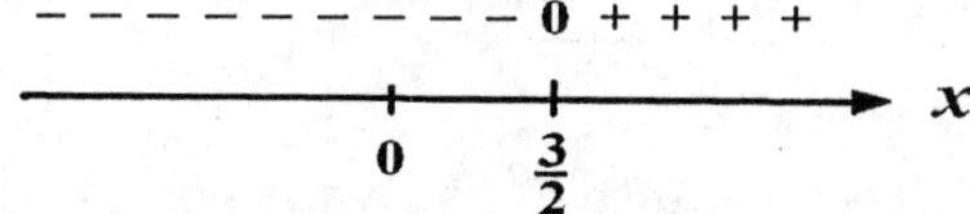

we see that f is decreasing on $(-\infty, \frac{3}{2})$ and increasing on $(\frac{3}{2}, \infty)$.

15. $g(x) = x - x^3$. $g'(x) = 1 - 3x^2$ is continuous everywhere and is equal to zero when $1 - 3x^2 = 0$, or $x = \pm\frac{\sqrt{3}}{3}$. From the following sign diagram

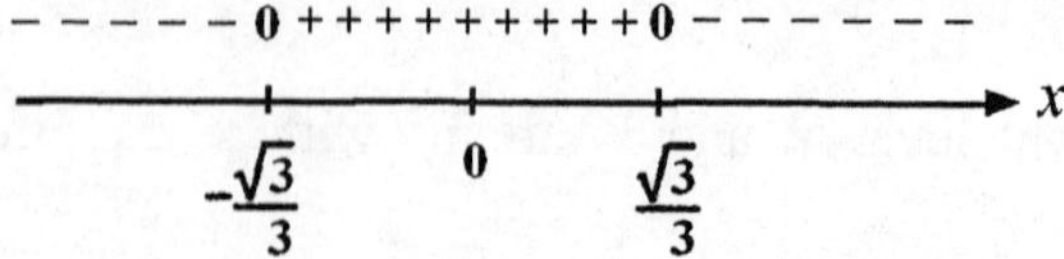

we see that f is decreasing on $(-\infty, -\frac{\sqrt{3}}{3}) \cup (\frac{\sqrt{3}}{3}, \infty)$ and increasing on $(-\frac{\sqrt{3}}{3}, \frac{\sqrt{3}}{3})$.

17. $g(x) = x^3 + 3x^2 + 1$; $g'(x) = 3x^2 + 6x = 3x(x + 2)$.
From the following sign diagram

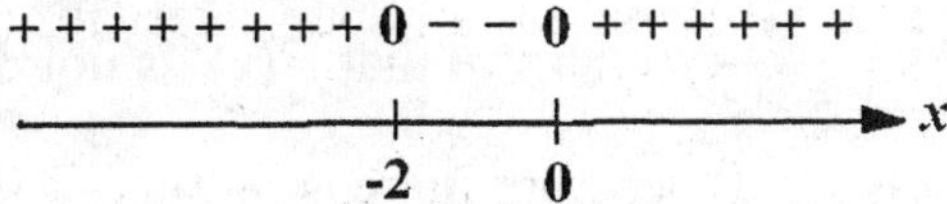

we see that g is increasing on $(-\infty,-2) \cup (0,\infty)$ and decreasing on $(-2,0)$.

19. $f(x) = \frac{1}{3}x^3 - 3x^2 + 9x + 20$; $f'(x) = x^2 - 6x + 9 = (x - 3)^2 > 0$ for all x except $x = 3$, at which point $f'(3) = 0$. Therefore, f is increasing on $(-\infty,3) \cup (3,\infty)$.

21. $h(x) = x^4 - 4x^3 + 10$; $h'(x) = 4x^3 - 12x^2 = 4x^2(x - 3)$ if $x = 0$ or 3. From the sign diagram of h',

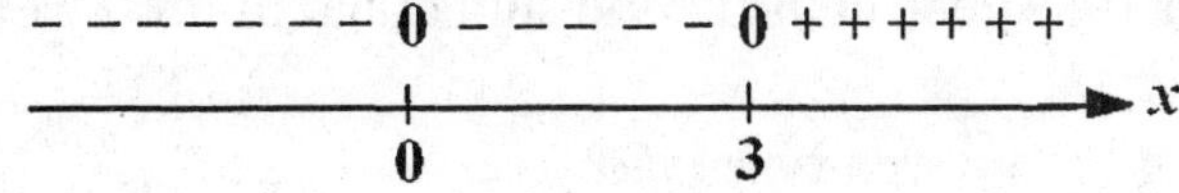

we see that h is increasing on $(3,\infty)$ and decreasing on $(-\infty,0) \cup (0,3)$.

23. $f(x) = \dfrac{1}{x-2} = (x-2)^{-1}$. $f'(x) = -1(x-2)^{-2}(1) = -\dfrac{1}{(x-2)^2}$ is discontinuous at $x = 2$ and is continuous everywhere else. From the sign diagram

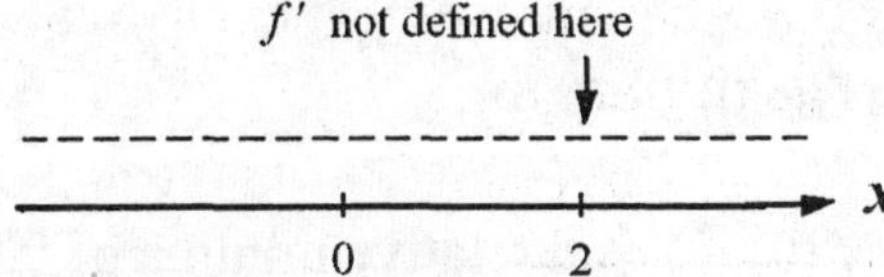

we see that f is decreasing on $(-\infty,2) \cup (2,\infty)$.

25. $h(t) = \dfrac{t}{t-1}$. $h'(t) = \dfrac{(t-1)(1) - t(1)}{(t-1)^2} = -\dfrac{1}{(t-1)^2}$.

From the following sign diagram,

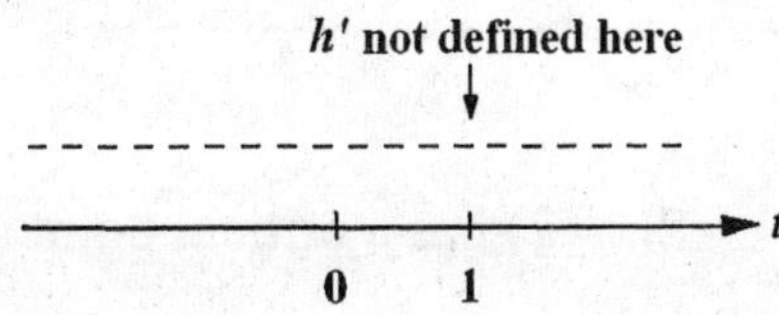

we see that $h'(t) < 0$ whenever it is defined. We conclude that h is decreasing on $(-\infty,1) \cup (1,\infty)$.

27. $f(x) = x^{3/5}$. $f'(x) = \frac{3}{5}x^{-2/5} = \frac{3}{5x^{2/5}}$. Observe that $f'(x)$ is not defined at $x = 0$, but is positive everywhere else and therefore increasing on $(-\infty,0) \cup (0,\infty)$.

29. $f(x) = \sqrt{x+1}$. $f'(x) = \frac{d}{dx}(x+1)^{1/2} = \frac{1}{2}(x+1)^{-1/2} = \frac{1}{2\sqrt{x+1}}$ and we see that $f'(x) > 0$ if $x > -1$. Therefore, f is increasing on $(-1, \infty)$.

31. $f(x) = \sqrt{16-x^2} = (16-x^2)^{1/2}$. $f'(x) = \frac{1}{2}(16-x^2)^{-1/2}(-2x) = -\frac{x}{\sqrt{16-x^2}}$.
Since the domain of f is $[-4,4]$, we consider the sign diagram for f' on this interval. Thus,

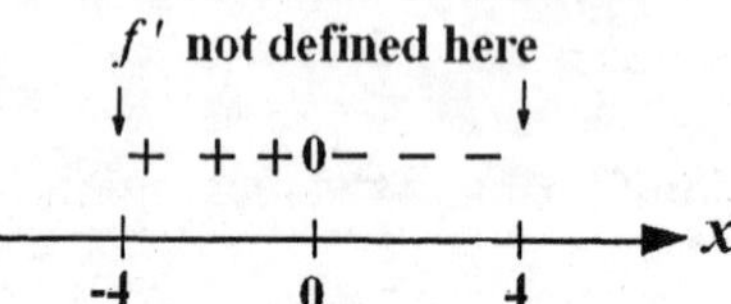

and we see that f is increasing on $(-4,0)$ and decreasing on $(0,4)$.

33. $f'(x) = \frac{d}{dx}(x - x^{-1}) = 1 + \frac{1}{x^2} = \frac{x^2+1}{x^2}$ and so $f'(x) > 0$ for all $x \neq 0$.
Therefore, f is increasing on $(-\infty,0) \cup (0,\infty)$.

35. f has a relative maximum of $f(0) = 1$ and relative minima of $f(-1) = 0$ and $f(1) = 0$.

37. f has a relative maximum of $f(-1) = 2$ and a relative minimum of $f(1) = -2$.

39. f has a relative maximum of $f(1) = 3$ and a relative minimum of $f(2) = 2$.

41. f has a relative minimum at $(0,2)$.

43. a

45. d

47. $f(x) = x^2 - 4x$. $f'(x) = 2x - 4 = 2(x - 2)$ has a critical point at $x = 2$. From the following sign diagram

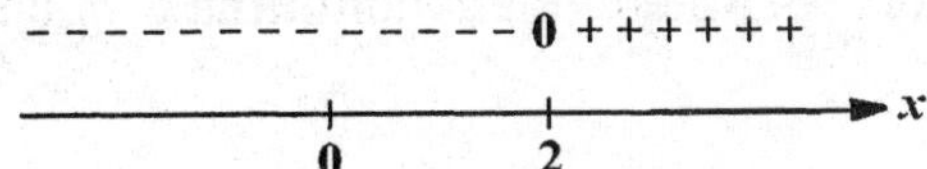

we see that $f(2) = -4$ is a relative minimum by the First Derivative Test.

49. $h(t) = -t^2 + 6t + 6$; $h'(t) = -2t + 6 = -2(t - 3) = 0$ if $t = 3$, a critical point. The sign

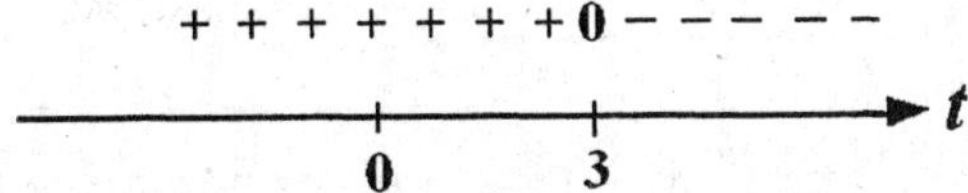

diagram and the First Derivative Test imply that h has a relative maximum at 3 with value $f(3) = -9 + 18 + 6 = 15$.

51. $f(x) = x^{5/3}$. $f'(x) = \frac{5}{3}x^{2/3}$ giving $x = 0$ as the critical point of f. From the sign diagram

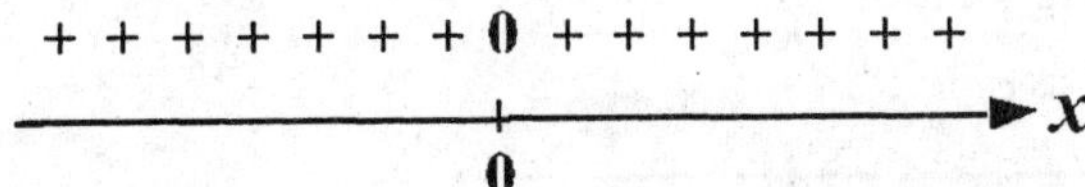

we see that f' does not change sign as we move across $x = 0$, and conclude that f has no relative extremum.

53. $g(x) = x^3 - 3x^2 + 4$. $g'(x) = 3x^2 - 6x = 3x(x - 2) = 0$ if $x = 0$ or 2. From the sign diagram, we see that the critical point $x = 0$ gives a relative maximum, whereas,

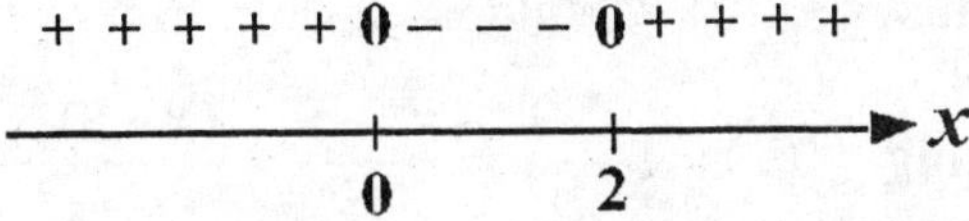

$x = 2$ gives a relative minimum. The values are $g(0) = 4$ and $g(2) = 8 - 12 + 4 = 0$.

55. $f(x) = \frac{1}{2}x^4 - x^2$. $f'(x) = 2x^3 - 2x = 2x(x^2 - 1) = 2x(x + 1)(x - 1)$ is continuous

everywhere and has zeros as $x = -1$, $x = 0$, and $x = 1$, the critical points of f. Using the First Derivative Test and the following sign diagram of f'

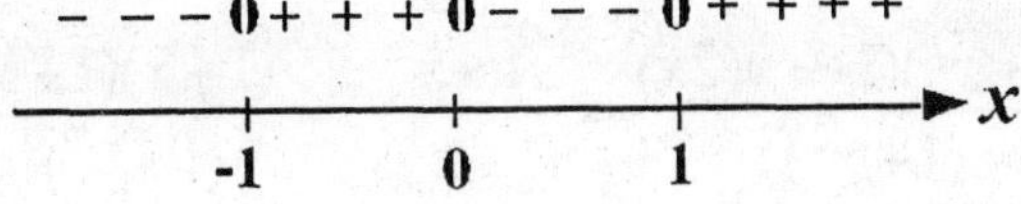

we see that $f(-1) = -1/2$ and $f(1) = -1/2$ are relative minima of f and $f(0) = 0$ is a relative maximum of f.

57. $F(x) = \frac{1}{3}x^3 - x^2 - 3x + 4$. Setting $F'(x) = x^2 - 2x - 3 = (x-3)(x+1) = 0$ gives $x = -1$ and $x = 3$ as critical points. From the sign diagram

```
+ + + + 0 - - - - - - 0 + + + + +
----------+---+--------+---------> x
         -1   0        3
```

we see that $x = -1$ gives a relative maximum and $x = 3$ gives a relative minimum. The values are

$$F(-1) = -\tfrac{1}{3} - 1 + 3 + 4 = \tfrac{17}{3} \quad \text{and} \quad F(3) = 9 - 9 - 9 + 4 = -5,$$

respectively.

59. $g(x) = x^4 - 4x^3 + 8$. Setting $g'(x) = 4x^3 - 12x^2 = 4x^2(x - 3) = 0$ gives $x = 0$ and $x = 3$ as critical points. From the sign diagram

```
- - - - - 0 - - -  0 + + + + +
----------+--------+----------> x
          0        3
```

we see that $x = 3$ gives a relative minimum. Its value is $g(3) = 3^4 - 4(3)^3 + 8 = -19$.

61. $g'(x) = \dfrac{d}{dx}\left(1 + \dfrac{1}{x}\right) = -\dfrac{1}{x^2}$. Observe that g' is never zero for all values of x. Furthermore, g' is undefined at $x = 0$, but $x = 0$ is not in the domain of g. Therefore, g has no critical points and so g has no relative extrema.

63. $f(x) = x + \dfrac{9}{x} + 2$. Setting $f'(x) = 1 - \dfrac{9}{x^2} = \dfrac{x^2 - 9}{x^2} = \dfrac{(x+3)(x-3)}{x^2} = 0$ gives $x = -3$ and $x = 3$ as critical points. From the sign diagram

```
                f' not defined here
                        ↓
+ + + + 0 - - -         0 - - - 0 + + + + +
--------+---------------+-------+----------> x
       -3               0       3
```

we see that (-3,-4) is a relative maximum and (3,8) is a relative minimum.

65. $f(x) = \dfrac{x}{1+x^2}$. $f'(x) = \dfrac{(1+x^2)(1) - x(2x)}{(1+x^2)^2} = \dfrac{1-x^2}{(1+x^2)^2} = \dfrac{(1-x)(1+x)}{(1+x^2)^2} = 0$ if $x = \pm 1$,

and these are critical points of f. From the sign diagram of f'

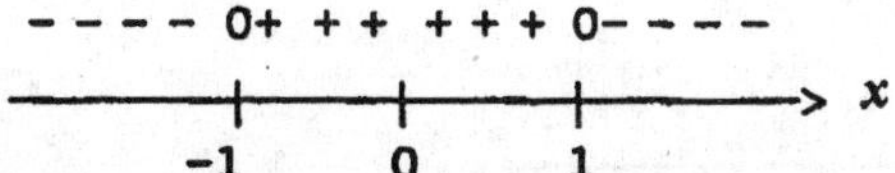

we see that f has a relative minimum at $(-1, -\frac{1}{2})$ and a relative maximum at $(1, \frac{1}{2})$.

67. $f(x) = (x - 1)^{2/3}$. $f'(x) = \frac{2}{3}(x-1)^{-1/3} = \frac{2}{3(x-1)^{1/3}}$.

$f'(x)$ is discontinuous at $x = 1$. The sign diagram for f' is

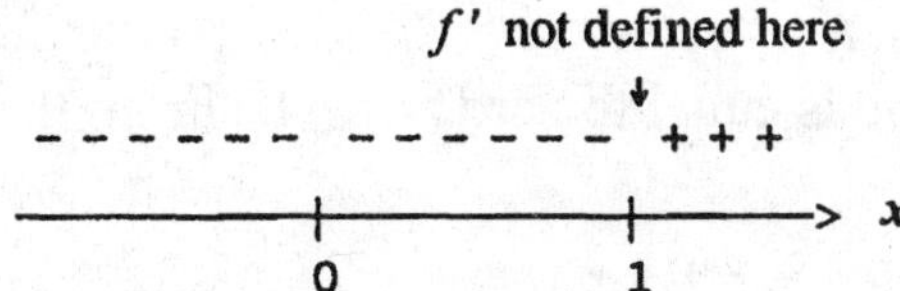

We conclude that $f(1) = 0$ is a relative minimum.

69. $h(t) = -16t^2 + 64t + 80$. $h'(t) = -32t + 64 = -32(t - 2)$ and has sign diagram

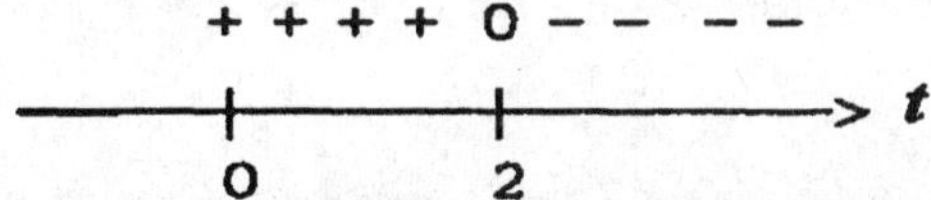

This tells us that the stone is rising on the time interval (0,2) and falling when $t > 2$. It hits the ground when $h(t) = -16t^2 + 64t + 80 = 0$ or $t^2 - 4t - 5 = (t - 5)(t + 1) = 0$ or $t = 5$ (we reject the root $t = -1$.)

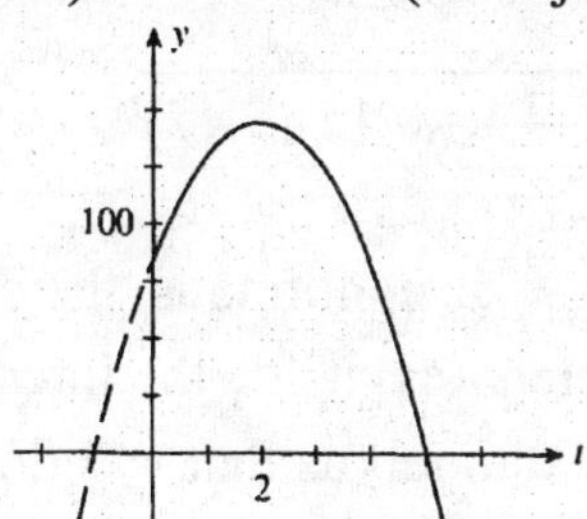

71. $P'(x) = \frac{d}{dx}(0.0726x^2 + 0.7902x + 4.9623) = 0.1452x + 0.7902.$

Since $P'(x) > 0$ on (0, 25), we see that P is increasing on the interval in question. Our result tells us that the percent of the population afflicted with Alzheimer's disease increases with age for those that are 65 and over.

73. $h(t) = -\frac{1}{3}t^3 + 16t^2 + 33t + 10$; $h'(t) = -t^2 + 32t + 33 = -(t + 1)(t - 33)$.

The sign diagram for h' is

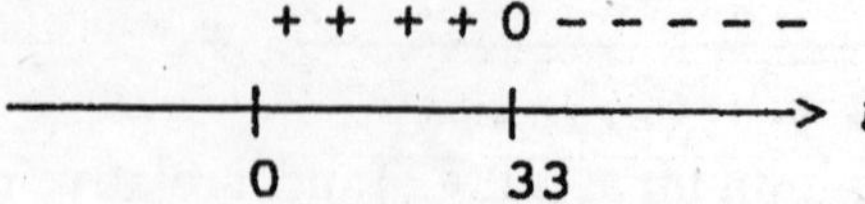

The rocket is rising on the time interval (0,33) and descending on (33,T) for some positive number T. The parachute is deployed 33 seconds after liftoff.

75. $f(t) = 20t - 40\sqrt{t} + 50 = 20t - 40t^{1/2} + 50$.

$$f'(t) = 20 - 40\left(\frac{1}{2}t^{-1/2}\right) = 20\left(1 - \frac{1}{\sqrt{t}}\right) = \frac{20(\sqrt{t} - 1)}{\sqrt{t}}.$$

Then f' is continuous on (0,4) and is equal to zero at $t = 1$. From the sign diagram

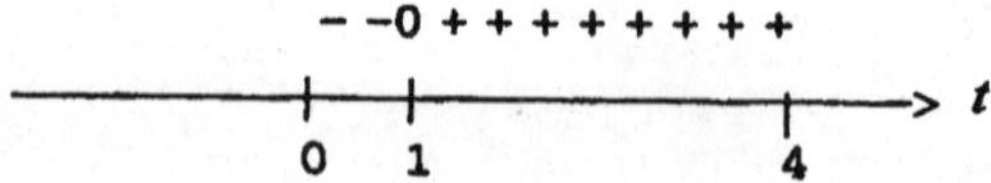

we see that f is decreasing on (0,1) and increasing on (1,4). We conclude that the average speed decreases from 6 A.M. to 7 A.M. and then picks up from 7 A.M. to 10 A. M.

77. a. $f'(t) = \frac{d}{dt}(-0.05t^3 + 0.56t^2 + 5.47t + 7.5) = -0.15t^2 + 1.12t + 5.47$.

Setting $f'(t) = 0$ gives $-0.15t^2 + 1.12t + 5.47 = 0$. Using the quadratic formula, we find

$$t = \frac{-1.12 \pm \sqrt{(1.12)^2 - 4(-0.15)(5.47)}}{-0.3}$$

that is, t = -3.37, or 10.83. Since f' is continuous, the only critical points of f are $t = -3.4$ and $t = 10.8$, both of which lie outside the interval of interest. Nevertheless this result can be used to tell us that f' does not change sign in the interval (-3.4, 10.8). Using $t = 0$ as the test point, we see that $f'(0) = 5.47 > 0$ and so we see that f is increasing on (-3.4, 10.8), and , in particular, in the interval (0, 6). Thus, we conclude that f is increasing on (0, 6).

b. The result of part (a) tells us that sales in the Web-hosting industry will be increasing throughout the years from 1999 through 2005.

79. a. $P'(t) = 0.00279t^2 - 0.036t - 0.51$. Setting $P'(t) = 0$ and solving the resulting

equation, we have

$$t = \frac{0.036 \pm \sqrt{(-0.036)^2 - 4(0.00279)(-0.51)}}{2(0.00279)}$$

$$\approx -8.53 \text{ or } 21.43$$

The sign diagram for P' follows. [Take t = -10, 0, 25 as test points.]

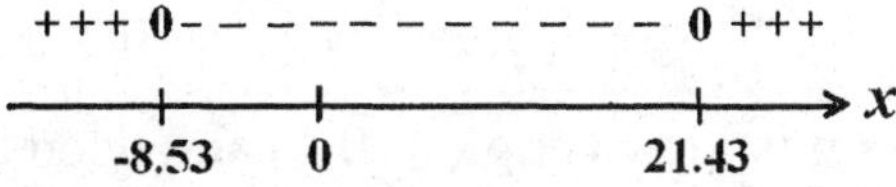

From the sign diagram, we see that P is decreasing on [0, 21.43] and increasing on [21.43, 30].

b. The percent of men 65 years and older in the workforce was decreasing from 1970 through about the middle of 1991, then starting increasing from then through the year 2000.

81. b. $S'(t) = \frac{d}{dt}(0.46t^3 - 2.22t^2 + 6.21t + 17.25) = 1.38t^2 - 4.44t + 6.21.$

Observe that S' is continuous everywhere. Setting $S'(t) = 0$ and solving, we find

$$t = \frac{4.44\sqrt{4.44^2 - 4(1.38)(6.21)}}{2(1.38)}$$

Now, the discriminant is -14.5656 < 0 which shows that the equation has no real roots. Since $S'(0) = 6.21 > 0$, we conclude that $S'(t) > 0$ for all t, in particular, for t in the interval (0,4). This shows that S is increasing on (0,4).

83. $S'(t) = -6.945t^2 + 68.65t + 1.32$. Setting $S'(t) = 0$ and solving the resulting equation, we obtain

$$t = \frac{-68.65 \pm \sqrt{(68.65)^2 - 4(-6.945)(1.32)}}{2(-6.945)} \approx -0.02 \text{ or } 9.90.$$

The sign diagram for S' follows.

- - - - - - - +0 + + + + + + + +++0 - - -

x

-0.02 0 9.9

From the sign diagram, we see that S' is increasing on the interval [0, 5]. We conclude that U.S. telephone company spending was projected to be increasing from 2001 through 2006.

85. $C(t)=\dfrac{t^2}{2t^3+1}$; $C'(t)=\dfrac{(2t^3+1)(2t)-t^2(6t^2)}{(2t^3+1)^2}=\dfrac{2t-2t^4}{(2t^3+1)^2}=\dfrac{2t(1-t^3)}{(2t^3+1)^2}$.

From the sign diagram of C' on $(0,\infty)$,

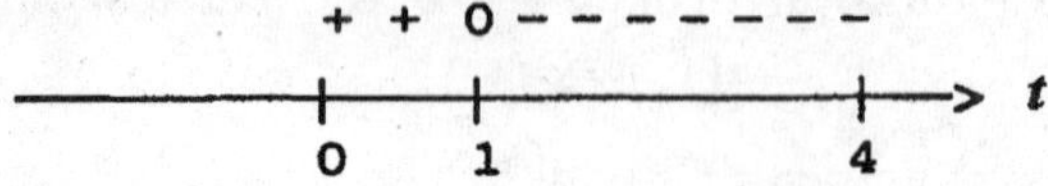

We see that the drug concentration is increasing on (0,1) and decreasing on (1,4).

87. $A(t)=\dfrac{136}{1+0.25(t-4.5)^2}+28$.

$$A'(t)=136\frac{d}{dt}[1+0.25(t-4.5)^2]^{-1}=-136[1+0.25(t-4.5)^2]^{-2}2(0.25)(t-4.5)$$

$$=-\frac{68(t-4.5)}{[1+0.25(t-4.5)^2]^2}.$$

Observe that $A'(t)>0$ if $t<4.5$ and $A'(t)<0$ if $t>4.5$, so the pollution is increasing from 7 A.M. to 11:30 A.M. and decreasing from 11:30 A.M. to 6 P.M.

89. a. $G(t)=(D-S)(t)=D(t)-S(t)$

$$=(0.0007t^2+0.0265t+2)-(-0.0014t^2+0.0326t+1.9)$$

$$=0.0021t^2-0.0061t+0.1$$

b. $G'(t)=0.0042t-0.0061=0$ implies $t\approx 1.45$. The sign diagram of G' follows:

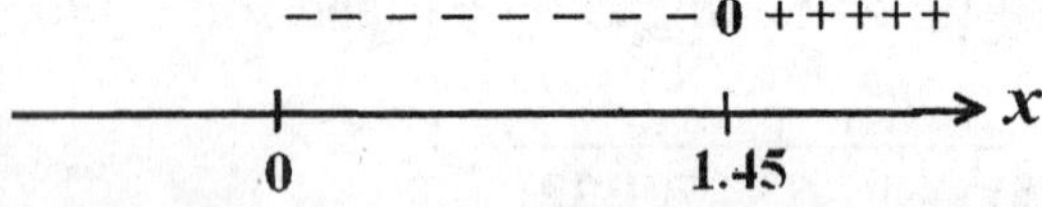

We see that G is decreasing on (0, 1.5) and increasing on (1.5, 15). This shows that the gap between the demand and supply of nurses was increasing from 2000 through the middle of 2001 but starts widening from the middle of 2001 through 2015.

c. The relative minimum of G occurs at $t=1.5$ and is $f(1.45)\approx 0.0956$. This says that at its best there is a shortage of approximately 96,000.

91. True. Let $a < x_1 < x_2 < b$. Then $f(x_2) > f(x_1)$ and $g(x_2) > g(x_1)$. Therefore,

$$(f+g)(x_2) = f(x_2) + g(x_2) > f(x_1) + g(x_1) = (f+g)(x_1)$$

and so $f+g$ is increasing on $(a,\ b)$.

93. True. Let $a < x_1 < x_2 < b$, then $f(x_1) < f(x_2)$ and $g(x_1) < g(x_2)$. We find

$$\begin{aligned}(fg)(x_2) - (fg)(x_1) &= f(x_2)g(x_2) - f(x_1)g(x_1) \\ &= f(x_2)g(x_2) - f(x_2)g(x_1) + f(x_2)g(x_1) - f(x_1)g(x_1) \\ &= f(x_2)[g(x_2) - g(x_1)] + g(x_1)[f(x_2) - f(x_1)] \\ &> 0\end{aligned}$$

So $(fg)(x_2) > (fg)(x_1)$ and fg is increasing on $(a,\ b)$.

95. False. Let $f(x) = |x|$. Then f has a relative minimum at $x = 0$, but $f'(0)$ does not exist.

97. $f'(x) = 3x^2 + 1$ is continuous on $(-\infty, \infty)$ and is always greater than or equal to 1. So f has no critical points in $(-\infty, \infty)$. Therefore f has no relative extrema on $(-\infty, \infty)$.

99. a. $f'(x) = -2x$ if $x \neq 0$. $f'(-1) = 2$ and $f'(1) = -2$ so $f'(x)$ changes sign from positive to negative as we move across $x = 0$.
b. f does not have a relative maximum at $x = 0$ because $f(0) = 2$ but a neighborhood of $x = 0$, for example $(-\frac{1}{2}, \frac{1}{2})$, contains points with values larger than 2. This does not contradict the First Derivative Test because f is not continuous at $x = 0$.

101. $f(x) = ax^2 + bx + c$. Setting $f'(x) = 2ax + b = 2a\left(x + \frac{b}{2a}\right) = 0$ gives $x = -\frac{b}{2a}$ as the only critical point of f. If $a < 0$, we have the sign diagram

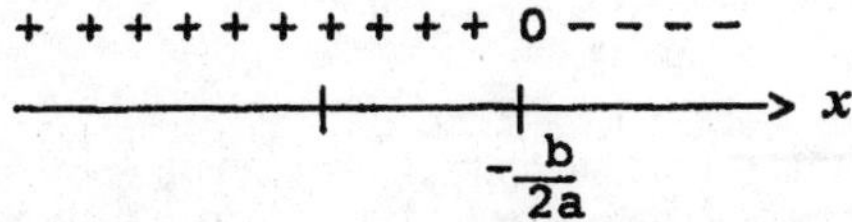

from which we see that $x = -b/2a$ gives a relative maximum. Similarly, you can show that if $a > 0$, then $x = -b/2a$ gives a relative minimum.

103. a. $f'(x) = 3x^2 + 1$ and so $f'(x) > 1$ on the interval (0,1). Therefore, f is increasing on (0,1).

b. $f(0) = -1$ and $f(1) = 1 + 1 - 1 = 1$. So the Intermediate Value Theorem guarantees that there is at least one root of $f(x) = 0$ in (0,1). Since f is increasing on (0,1), the graph of f can cross the x-axis at only one point in (0,1). So $f(x) = 0$ has exactly one root.

USING TECHNOLOGY EXERCISES 13.1, page 790

1. a. f is decreasing on $(-\infty,-0.2934)$ and increasing on $(-0.2934,\infty)$.
   b. Relative minimum: $f(-0.2934) = -2.5435$

3. a. f is increasing on $(-\infty,-1.6144) \cup (0.2390,\infty)$ and decreasing on $(-1.6144, 0.2390)$
   b. Relative maximum: $f(-1.6144) = 26.7991$; relative minimum: $f(0.2390) = 1.6733$

5. a. f is decreasing on $(-\infty,-1) \cup (0.33,\infty)$ and increasing on $(-1,0.33)$
   b Relative maximum: $f(0.33) = 1.11$; relative minimum: $f(-1) = -0.63$.

7. a. f is decreasing on $(-1,-0.71)$ and increasing on $(-0.71,1)$.
   b. f has a relative minimum at $(-0.71,-1.41)$.

9. a.

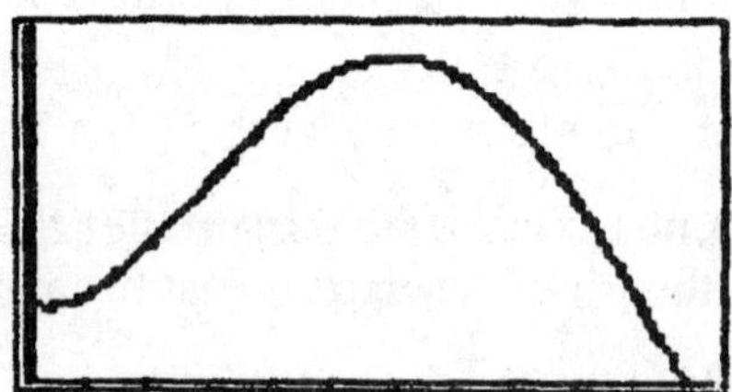

   b. f is decreasing on $(0,0.2398) \cup (6.8758,12)$ and increasing on $(0.2398,6.8758)$
   c. (6.8758, 200.14); The rate at which the number of banks were failing reached a peak of 200/yr during the latter part of 1988 ($t = 6.8758$).

11. a.

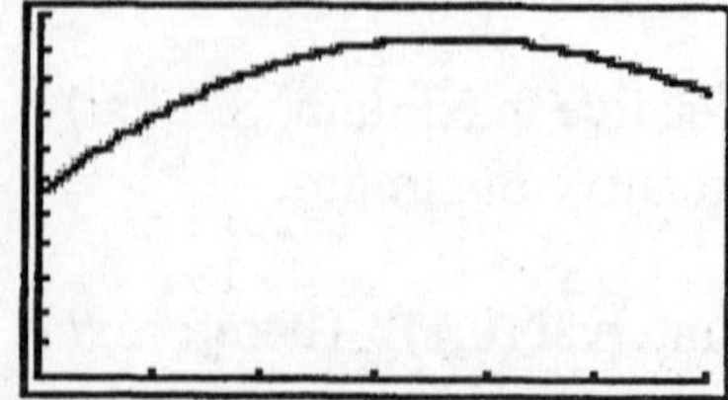

   b. increasing on (0, 3.6676) and decreasing on (3.6676, 6).

13. f is decreasing on the interval (0,1) and increasing on (1,4). The relative minimum occurs at the point (1, 32). These results indicate that the speed of traffic flow drops between 6 A.M. and 7 A.M. reaching a low of 32 mph. Thereafter, it increases till 10 A.M.

13.2 Problem Solving Tips

1. As an aid for remembering the shape of a graph,

   If $f''(x) > 0$, then the graph of f "holds water" and we say the graph of f is concave upward.

   If $f''(x) < 0$, then the graph of f "loses water" and we say the graph of f is concave downward.

2. To find the inflection points of a function f, we determine the number(s) in the domain of f for which $f''(x) = 0$ or $f''(x)$ does not exist. Note that each of these numbers c provides us with a *candidate* $(c, f(c))$ for an inflection point of f.

3. Note that the second derivative test is not valid when $f''(c) = 0$ or $f''(c)$ does not exist. In these cases you need to use the first derivative test to determine the relative extrema.

13.2 CONCEPT QUESTIONS, page 800

1. a. f is concave upward on (a,b) if f' is increasing on (a,b).
   f is concave downward on (a,b) if f' is decreasing on (a,b).

b. To determine where f is concave upward and concave downward, see page 793.

3. The second derivative test is stated in the text on page 798. In general, if f'' is easy to compute, then use the second derivative test. However, keep in mind that (1) in order to use this test f'' must exist, (2) the test is inconclusive if $f''(c) = 0$, and (3) the test is inconvenient to use if f'' is difficult to compute.

EXERCISES 13.2, page 800

1. f is concave downward on $(-\infty,0)$ and concave upward on $(0,\infty)$. f has an inflection point at (0,0).

3. f is concave downward on $(-\infty,0) \cup (0,\infty)$.

5. f is concave upward on $(-\infty,0) \cup (1,\infty)$ and concave downward on (0,1). (0,0) and (1,-1) are inflection points of f.

7. f is concave downward on $(-\infty,-2) \cup (-2,2) \cup (2,\infty)$.

9. a

11. b

13. a. $D_1'(t) > 0$, $D_2'(t) > 0$, $D_1''(t) > 0$, and $D_2''(t) < 0$ on (0,12).
    b. With or without the proposed promotional campaign, the deposits will increase, but with the promotion, the deposits will increase at an increasing rate whereas without the promotion, the deposits will increase at a decreasing rate.

15. The significance of the inflection point Q is that the restoration process is working at its peak at the time t_0 corresponding to its t-coordinate.

17. $f(x) = 4x^2 - 12x + 7$. $f'(x) = 8x - 12$ and $f''(x) = 8$. So, $f''(x) > 0$ everywhere and therefore f is concave upward everywhere.

19. $f(x) = \dfrac{1}{x^4} = x^{-4}$; $f'(x) = -\dfrac{4}{x^5}$ and $f''(x) = \dfrac{20}{x^6} > 0$ for all values of x in $(-\infty,0) \cup (0,\infty)$ and so f is concave upward everywhere.

21. $f(x) = 2x^2 - 3x + 4$; $f'(x) = 4x - 3$ and $f''(x) = 4x > 0$ for all values of x. So f is concave upward on $(-\infty,\infty)$.

23. $f(x) = x^3 - 1$. $f'(x) = 3x^2$ and $f''(x) = 6x$. The sign diagram of f'' follows.

- - - - - - - - - - 0 + + + + + + + + + +

x

0

We see that f is concave downward on $(-\infty,0)$ and concave upward on $(0,\infty)$.

25. $f(x) = x^4 - 6x^3 + 2x + 8$; $f'(x) = 4x^3 - 18x^2 + 2$ and $f''(x) = 12x^2 - 36x = 12x(x - 3)$. The sign diagram of f''

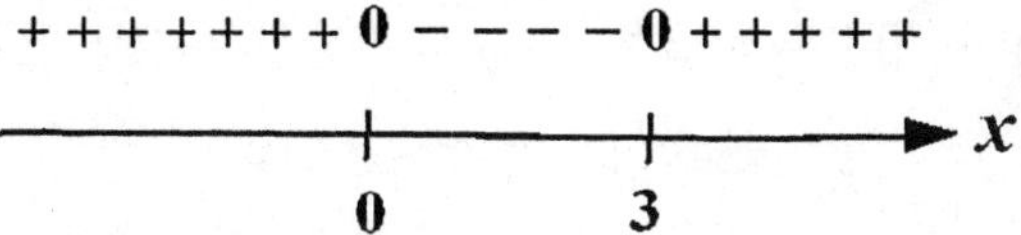

shows that f is concave upward on $(-\infty,0) \cup (3,\infty)$ and concave downward on $(0,3)$.

27. $f(x) = x^{4/7}$. $f'(x) = \dfrac{4}{7}x^{-3/7}$ and $f''(x) = -\dfrac{12}{49}x^{-10/7} = -\dfrac{12}{49x^{10/7}}$.
Observe that $f''(x) < 0$ for all x different from zero. So f is concave downward on $(-\infty,0) \cup (0,\infty)$.

29. $f(x) = (4-x)^{1/2}$. $f'(x) = \dfrac{1}{2}(4-x)^{-1/2}(-1) = -\dfrac{1}{2}(4-x)^{-1/2}$;

$$f''(x) = \frac{1}{4}(4-x)^{-3/2}(-1) = -\frac{1}{4(4-x)^{3/2}} < 0.$$

whenever it is defined. So f is concave downward on $(-\infty,4)$.

31. $f'(x) = \dfrac{d}{dx}(x-2)^{-1} = -(x-2)^{-2}$ and $f''(x) = 2(x-2)^{-3} = \dfrac{2}{(x-2)^3}$.
The sign diagram of f'' shows that f is concave downward on $(-\infty,2)$ and concave upward on $(2,\infty)$.

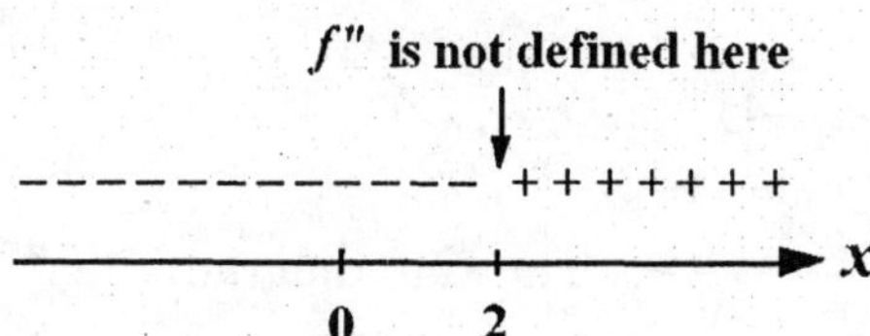

33. $f'(x) = \dfrac{d}{dx}\left(2+x^2\right)^{-1} = -(2+x^2)^{-2}(2x) = -2x(2+x^2)^{-2}$ and

$f''(x) = -2(2+x^2)^{-2} - 2x(-2)(2+x^2)^{-3}(2x)$

$= 2(2+x^2)^{-3}[-(2+x^2)+4x^2] = \dfrac{2(3x^2-2)}{(2+x^2)^3} = 0$ if $x = \pm\sqrt{2/3}$.

From the sign diagram of f''

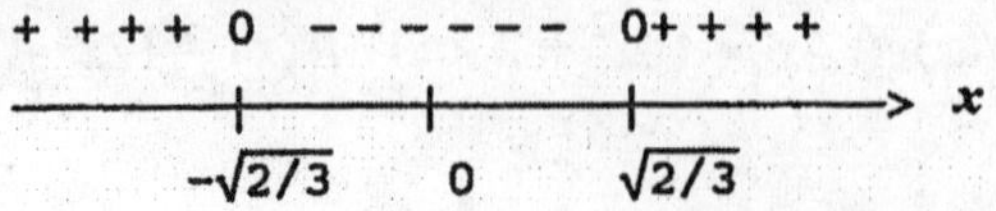

```
 + + + + 0  - - - - - -  0+ + + +
 ----+----------+----------+------> x
   -√2/3        0         √2/3
```

we see that f is concave upward on $(-\infty, -\sqrt{2/3}) \cup (\sqrt{2/3}, \infty)$ and concave downward on $(-\sqrt{2/3}, \sqrt{2/3})$.

35. $h(t) = \dfrac{t^2}{t-1}$; $h'(t) = \dfrac{(t-1)(2t)-t^2(1)}{(t-1)^2} = \dfrac{t^2-2t}{(t-1)^2}$;

$h''(t) = \dfrac{(t-1)^2(2t-2)-(t^2-2t)2(t-1)}{(t-1)^4}$

$= \dfrac{(t-1)(2t^2-4t+2-2t^2+4t)}{(t-1)^4} = \dfrac{2}{(t-1)^3}$.

The sign diagram of h'' is

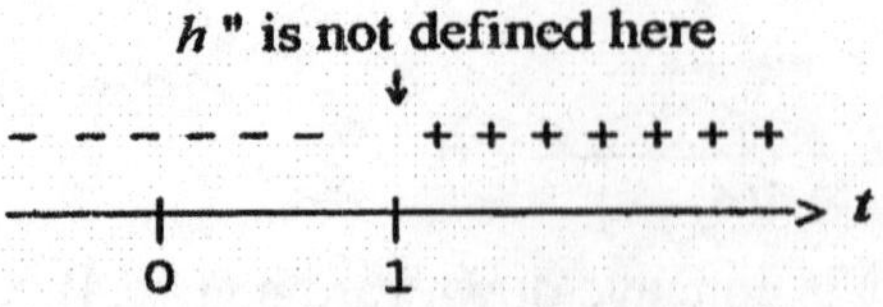

```
              h " is not defined here
                        ↓
 - - - - - - -            + + + + + + + +
 -------+---------------+-----------------> t
        0               1
```

and tells us that h is concave downward on $(-\infty,1)$ and concave upward on $(1,\infty)$.

37. $g(x) = x + \dfrac{1}{x^2}$. $g'(x) = 1 - 2x^{-3}$ and $g''(x) = 6x^{-4} = \dfrac{6}{x^4} > 0$ whenever $x \neq 0$.

Therefore, g is concave upward on $(-\infty,0) \cup (0,\infty)$.

39. $g(t) = (2t-4)^{1/3}$. $g'(t) = = \dfrac{1}{3}(2t-4)^{-2/3}(2) = \dfrac{2}{3}(2t-4)^{-2/3}$.

$g''(t) = -\dfrac{4}{9}(2t-4)^{-5/3} = -\dfrac{4}{9(2t-4)^{5/3}}$. The sign diagram of g''

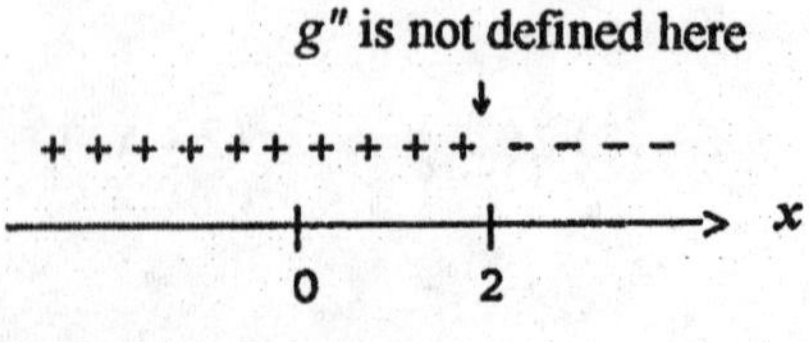

```
              g" is not defined here
                        ↓
 + + + + + + + + + + +   - - - -
 ----------+------------+----------> x
           0            2
```

tells us that g is concave upward on $(-\infty,2)$ and concave downward on $(2,\infty)$.

41. $f(x) = x^3 - 2$. $f'(x) = 3x^2$ and $f''(x) = 6x$. $f''(x)$ is continuous everywhere and has a zero at $x = 0$. From the sign diagram of f''

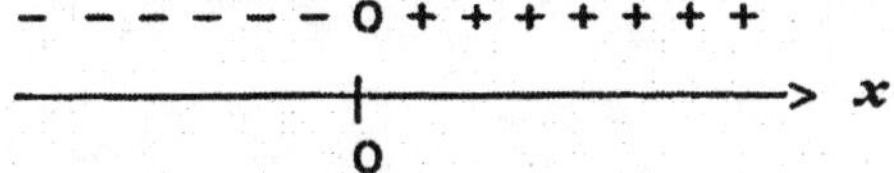

we conclude that $(0,-2)$ is an inflection point of f.

43. $f(x) = 6x^3 - 18x^2 + 12x - 15$; $f'(x) = 18x^2 - 36x + 12$ and $f''(x) = 36x - 36 = 36(x - 1) = 0$ if $x = 1$. The sign diagram of f'' tells us that f has an inflection point at $(1, -15)$.

45. $f(x) = 3x^4 - 4x^3 + 1$. $f'(x) = 12x^3 - 12x^2$ and $f''(x) = 36x^2 - 24x = 12x(3x - 2) = 0$ if $x = 0$ or $2/3$. These are candidates for inflection points. The sign diagram of f''

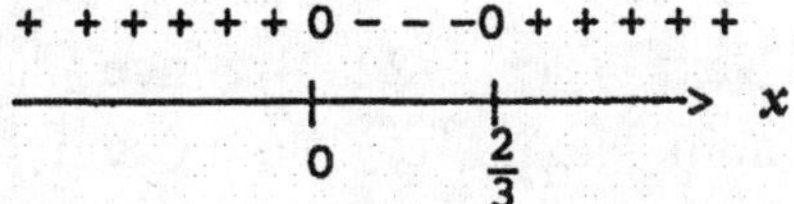

shows that $(0,1)$ and $(\frac{2}{3}, \frac{11}{27})$ are inflection points of f.

47. $g(t) = t^{1/3}$, $g'(t) = \frac{1}{3}t^{-2/3}$ and $g''(t) = -\frac{2}{9}t^{-5/3} = -\dfrac{2}{9t^{5/3}}$. Observe that $t = 0$ is in the domain of g. Next, since $g''(t) > 0$ if $t < 0$ and $g''(t) < 0$, if $t > 0$, we see that $(0,0)$ is an inflection point of g.

49. $f(x) = (x - 1)^3 + 2$. $f'(x) = 3(x - 1)^2$ and $f''(x) = 6(x - 1)$. Observe that $f''(x) < 0$ if $x < 1$ and $f''(x) > 0$ if $x > 1$ and so $(1, 2)$ is an inflection point of f.

51. $f(x) = \dfrac{2}{1+x^2} = 2(1+x^2)^{-1}$. $f'(x) = -2(1+x^2)^{-2}(2x) = -4x(1+x^2)^{-2}$.

$$f''(x) = -4(1+x^2)^{-2} - 4x(-2)(1+x^2)^{-3}(2x)$$

$$= 4(1+x^2)^{-3}[-(1+x^2)+4x^2] = \frac{4(3x^2-1)}{(1+x^2)^3},$$

is continuous everywhere and has zeros at $x = \pm\frac{\sqrt{3}}{3}$. From the sign diagram of f''

we conclude that $(-\frac{\sqrt{3}}{3},\frac{3}{2})$ and $(\frac{\sqrt{3}}{3},\frac{3}{2})$ are inflection points of f.

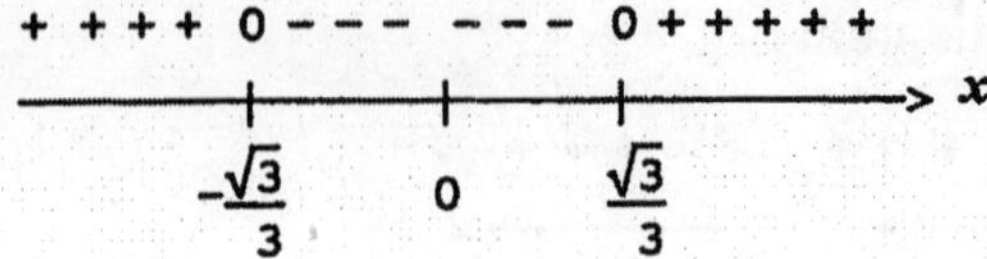

53. $f(x) = -x^2 + 2x + 4$ and $f'(x) = -2x + 2$. The critical point of f is $x = 1$. Since $f''(x) = -2$ and $f''(1) = -2 < 0$, we conclude that $f(1) = 5$ is a relative maximum of f.

55. $f(x) = 2x^3 + 1$; $f'(x) = 6x^2 = 0$ if $x = 0$ and this is a critical point of f. Next, $f''(x) = 12x$ and so $f''(0) = 0$. Thus, the Second Derivative Test fails. But the First Derivative Test shows that (0,0) is not a relative extremum.

57. $f(x) = \frac{1}{3}x^3 - 2x^2 - 5x - 10$. $f'(x) = x^2 - 4x - 5 = (x - 5)(x + 1)$ and this gives $x = -1$ and $x = 5$ as critical points of f. Next, $f''(x) = 2x - 4$. Since $f''(-1) = -6 < 0$, we see that $(-1,-\frac{22}{3})$ is a relative maximum. Next, $f''(5) = 6 > 0$ and this shows that $(5,-\frac{130}{3})$ is a relative minimum.

59. $g(t) = t + \dfrac{9}{t}$. $g'(t) = 1 - \dfrac{9}{t^2} = \dfrac{t^2 - 9}{t^2} = \dfrac{(t+3)(t-3)}{t^2}$ and this shows that $t = \pm 3$ are critical points of g. Now, $g''(t) = 18t^{-3} = \dfrac{18}{t^3}$. Since $g''(-3) = -\dfrac{18}{27} < 0$ the Second Derivative Test implies that g has a relative maximum at (-3,-6). Also, $g''(3) = \dfrac{18}{27} > 0$ and so g has a relative minimum at (3,6).

61. $f(x) = \dfrac{x}{1-x}$. $f'(x) = \dfrac{(1-x)(1) - x(-1)}{(1-x)^2} = \dfrac{1}{(1-x)^2}$ is never zero. So there are no critical points and f has no relative extrema.

63. $f(t) = t^2 - \dfrac{16}{t}$. $f'(t) = 2t + \dfrac{16}{t^2} = \dfrac{2t^3 + 16}{t^2} = \dfrac{2(t^3 + 8)}{t^2}$. Setting $f'(t) = 0$ gives $t = -2$ as a critical point. Next, we compute

$f''(t)=\frac{d}{dt}(2t+16t^{-2})=2-32t^{-3}=2-\frac{32}{t^3}$. Since $f''(-2)=2-\frac{32}{(-8)}=6>0$, we see that (-2,12) is a relative minimum.

65. $g(s)=\frac{s}{1+s^2}$; $g'(s)=\frac{(1+s^2)(1)-s(2s)}{(1+s^2)^2}=\frac{1-s^2}{(1+s^2)^2}=0$ gives $s=$ -1 and $s=1$ as critical points of g. Next, we compute

$$g''(s)=\frac{(1+s^2)^2(-2s)-(1-s^2)2(1+s^2)(2s)}{(1+s^2)^4}$$

$$=\frac{2s(1+s^2)(-1-s^2-2+2s^2)}{(1+s^2)^4}=\frac{2s(s^2-3)}{(1+s^2)^3}.$$

Now, $g''(-1)=\frac{1}{2}>0$ and so $g(-1)=-\frac{1}{2}$ is a relative minimum of g. Next, $g''(1)=-\frac{1}{2}<0$ and so $g(1)=\frac{1}{2}$ is a relative maximum of g.

67. $f(x)=\frac{x^4}{x-1}$.

$$f'(x)=\frac{(x-1)(4x^3)-x^4(1)}{(x-1)^2}=\frac{4x^4-4x^3-x^4}{(x-1)^2}=\frac{3x^4-4x^3}{(x-1)^2}=\frac{x^3(3x-4)}{(x-1)^2}$$

and so $x=0$ and $x=4/3$ are critical points of f. Next,

$$f''(x)=\frac{(x-1)^2(12x^3-12x^2)-(3x^4-4x^3)(2)(x-1)}{(x-1)^4}$$

$$=\frac{(x-1)(12x^4-12x^3-12x^3+12x^2-6x^4+8x^3)}{(x-1)^4}$$

$$=\frac{6x^4-16x^3+12x^2}{(x-1)^3}=\frac{2x^2(3x^2-8x+6)}{(x-1)^3}.$$

Since $f''(\frac{4}{3})>0$, we see that $f(\frac{4}{3})=\frac{256}{27}$ is a relative minimum. Since $f''(0)=0$, the Second Derivative Test fails. Using the sign diagram for f',

f' is not defined here

+ + + + + 0 – – – – – – 0 + + + + + +

0, 1, $\frac{4}{3}$, x

and the First Derivative Test, we see that $f(0)=0$ is a relative maximum.

69.

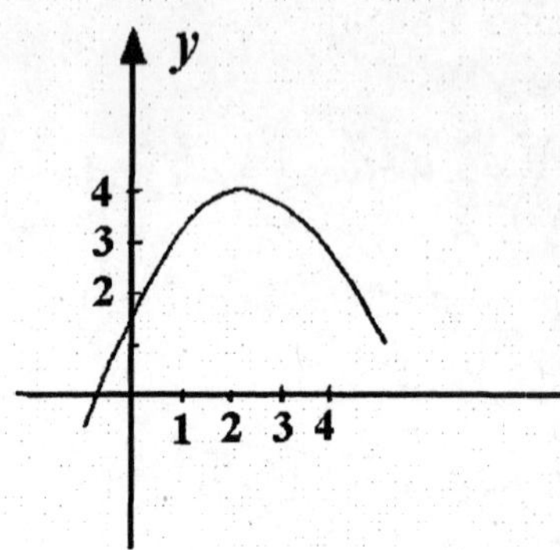

71.

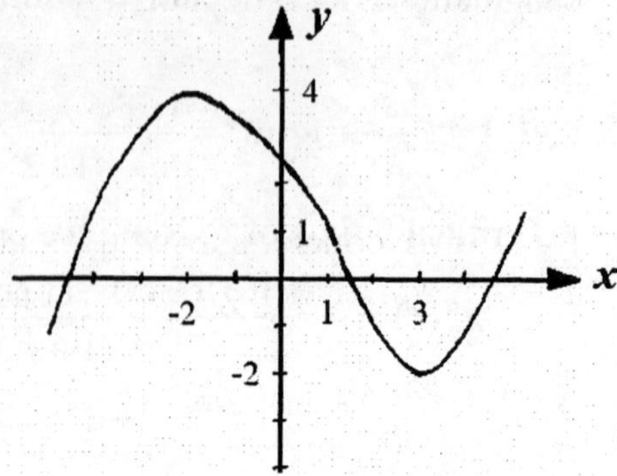

73.

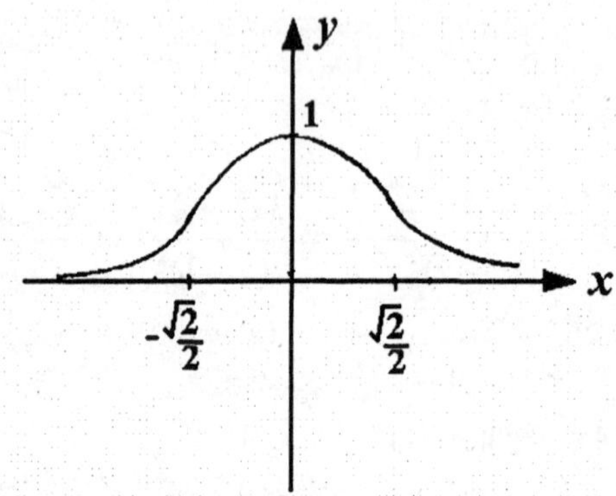

75. a. $N'(t)$ is positive because N is increasing on (0,12).
b. $N''(t) < 0$ on (0,6) and $N''(t) > 0$ on (6,12).
c. The rate of growth of the number of help-wanted advertisements was decreasing over the first six months of the year and increasing over the last six months.

77. $f(t)$ increases at an increasing rate until the water level reaches the middle of the vase at which time (and this corresponds to the inflection point of f), $f(t)$ is increasing at the fastest rate. Though $f(t)$ still increases until the vase is filled, it does so at a decreasing rate.

79. a. $S'(t) = 0.39t + 0.32 > 0$ on $[0, 7]$. So sales were increasing through the years in question.
b. $S''(t) = 0.39 > 0$ on $[0,7]$. So sales continued to accelerate through the years.

81. We wish to find the inflection point of the function $N(t) = -t^3 + 6t^2 + 15t$. Now, $N'(t) = -3t^2 + 12t + 15$ and $N''(t) = -6t + 12 = -6(t - 2)$ giving $t = 2$ as the only candidate for an inflection point of N. From the sign diagram

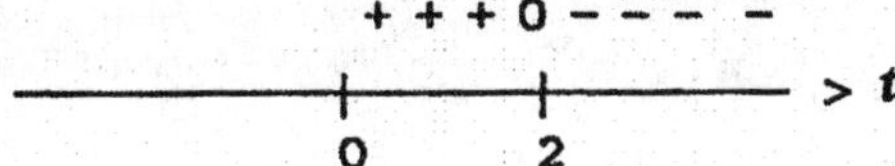

for N'', we conclude that $t = 2$ gives an inflection point of N. Therefore, the average worker is performing at peak efficiency at 10 A.M.

83. $S'(t) = -5.64t^2 + 60.66t - 76.14$; $S''(t) = -11.28t + 60.66 = 0$ if $t \approx 5.38$
The sign diagram of S'' follows:

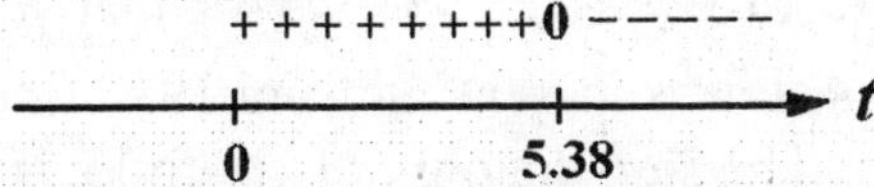

From the sign diagram for S'', we see that the graph of S is concave upward on (0,5). This says that the rate of business spending on technology is increasing from 2000 through 2005.

85. a. $R'(x) = -0.009x^2 + 2.7x + 2$; $R''(x) = -0.018x + 2.7$. Setting $R''(x) = 0$ gives $x = 150$. Since $R''(x) > 0$ if $x < 150$ and $R''(x) < 0$ if $x > 150$, we see that the graph of R is concave upward on (0, 150) and concave downward on (150, 400). So $x = 150$ gives rise to an inflection point of R. $R(150) = 28{,}550$. So the inflection point is (150, 28,550).
b. $R''(140) = 0.18$; $R''(160) = -0.18$ This shows that at $x = 140$, a slight increase in x (spending) would result in the revenue increasing. At $x = 160$, the opposite conclusion holds. So it would be more beneficial to increase the expenditure when it is \$140,000 than when it's at \$160,000.

87. a. $A'(t) = 0.92(0.61)(t+1)^{-0.39} = \dfrac{0.5612}{(t+1)^{0.39}} > 0$ on (0,4), so A is increasing on (0,4).
his tells us that the spending is increasing over the years in question.
b. $A''(t) = (0.5612)(-0.39)(t+1)^{-1.39} = -\dfrac{0.218868}{(t+1)^{1.39}} < 0$ on (0,4). And so A'' is concave downward on (0,4). This tells us that the spending is increasing but at a decreasing rate.

89. $S'(t) = -5.418t^2 + 20.476t + 93.35$; $S''(t) = -10.836t + 20.476 = 0$ if $t \approx 1.9$. The sign diagram for S'' follows.

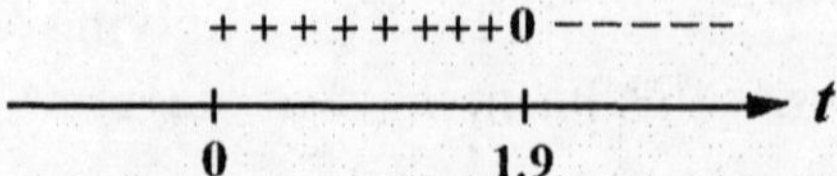

Since $S(1.9) = 784.9$, the inflection point is approximately (1.9, 784.9). The rate of annual spending slows down near the end of 2000.

91. a. $R'(t) = 74.925t^2 - 99.62t + 41.25$

$R''(t) = 149.85t - 99.62$

b. In solving the equation $R'(t) = 0$, we see that the discriminant is

$$(-99.62)^2 - 4(74.925)(41.25) = -2438.4806 < 0$$

and so R' has no zeros. Since $R'(0) = 41.25 > 0$, we see that $R'(t) > 0$ in (0,4). This shows that the revenue is always increasing from 1999 through 2003.

c. $R''(t) = 0$ implies $t = 0.66$. The sign diagram of R'' follows.

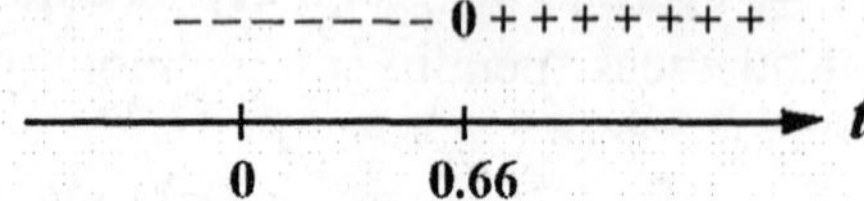

We conclude that the rate was increasing least rapidly around August 1999.

93. $A(t) = 1.0974t^3 - 0.0915t^4$. $A'(t) = 3.2922t^2 - 0.366t^3$ and $A''(t) = 6.5844t - 1.098t^2$. Setting $A'(t) = 0$, we obtain $t^2(3.2922 - 0.366t) = 0$, and this gives $t = 0$ or $t \approx 8.995 \approx 9$. Using the Second Derivative Test, we find $A''(9) = 6.5844(9) - 1.098(81) = -29.6784 < 0$, and this tells us that $t \approx 9$ gives rise to a relative maximum of A. Our analysis tells us that on that May day, the level of ozone peaked at approximately 4 P.M. in the afternoon.

95. a. $N'(t) = \dfrac{d}{dt}(-0.9307t^3 + 74.04t^2 + 46.8667t + 3967)$

$= -2.7921t^2 + 148.08t + 46.8667$

N' is continuous everywhere and has no zeros at

$$t = \frac{-146.08 \pm \sqrt{(148.08)^2 - 4(-0.9307)(46.86667)}}{2(-2.7921)}$$

that is, at $t = -0.1053$ or 53.1406. Both these points lie outside the interval of interest. Picking $t = 0$ for a test point, we see that $N'(0) = 46.86667 > 0$ and conclude that N is increasing on (0, 16). This shows that the number of participants is increasing over the years in question.

b. $N''(t) = \frac{d}{dt}(-2.7921t^2 + 148.08t + 46.86667) = -5.5842t + 148.08 = 0$ if $t = 26.518$. So $N''(t)$ does not change sign in the interval (0, 16).Since $N''(0) = 148.08 > 0$, we see that $N'(t)$ is increasing on (0, 16) and the desired conclusion follows.

97. True. If f' is increasing on (a,b), then $-f'$ is decreasing on (a,b), and so if the graph of f is concave upward on (a,b), the graph of $-f$ must be concave downward on (a,b).

99. True. The given conditions imply that $f''(0) < 0$ and the Second Derivative Test gives the desired conclusion.

101. $f(x) = ax^2 + bx + c$. $f'(x) = 2ax + b$ and $f''(x) = 2a$. So $f''(x) > 0$ if $a > 0$, and the parabola opens upward. If $a < 0$, then $f''(x) < 0$ and the parabola opens downward.

USING TECHNOLOGY EXERCISES 13.2, page 808

1. a. f is concave upward on $(-\infty, 0) \cup (1.1667, \infty)$ and concave downward on (0, 1.1667).
   b. (1.1667, 1.1153); (0,2)

3. a. f is concave downward on $(-\infty, 0)$ and concave upward on $(0, \infty)$.
   b. (0,2)

5. a. f is concave downward on $(-\infty, 0)$ and concave upward on $(0, \infty)$.
   b. (0,0)

7. a. f is concave downward on $(-\infty, -2.4495) \cup (0, 2.4495)$; f is concave upward on $(-2.4495, 0) \cup (2.4495, \infty)$. b. (-2.4495, -0.3402); (2.4495, 0.3402)

9. a.

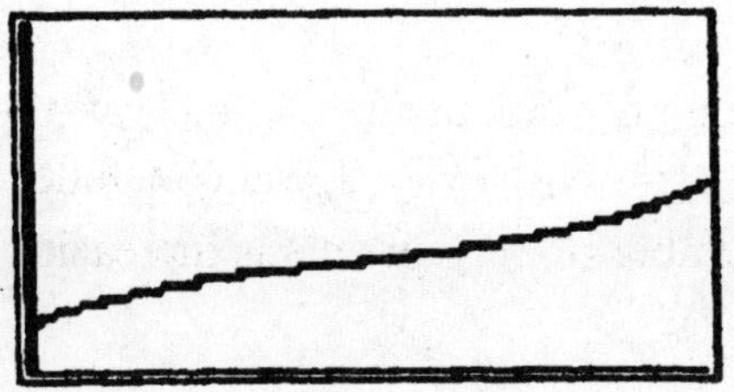

11. a.

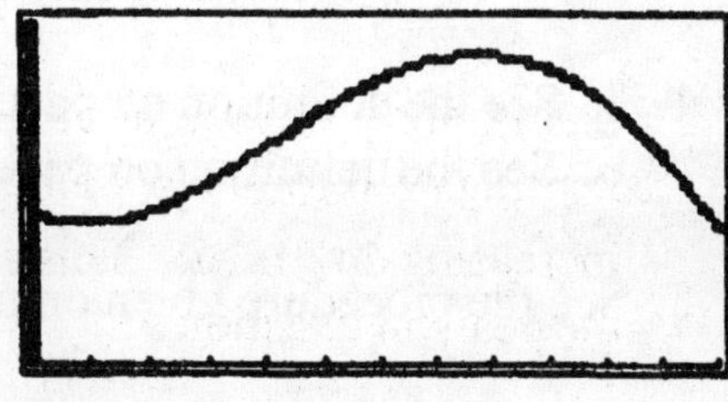

b. (5.5318, 35.9483)
c. $t = 5.5318$

b. (3.9024, 77.0919); sales of houses were increasing at the fastest rate in late 1988.

13. a.

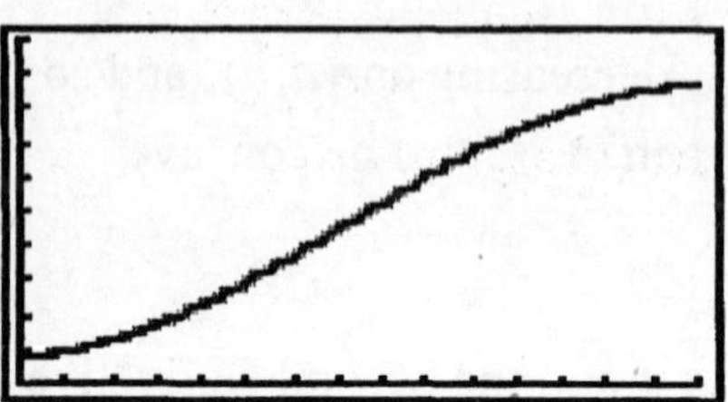

b. April 1993 ($t = 7.36$)

13.3 Problem Solving Tips

1. To find the horizontal asymptotes of a function f, find the limit of f as $x \to \infty$ and as $x \to -\infty$. If the limit is equal to a real number b, then $y = b$ is a horizontal asymptote of f.

2. To find the vertical asymptotes of a rational function $f(x) = P(x)/Q(x)$, determine the values a for which $Q(x) = 0$. If $Q(a) = 0$ but $P(a) \neq 0$, then the line $x = a$ is a vertical asymptote of f.

3. If a line $x = a$ is a vertical asymptote of the graph of a rational function f, then the denominator of $f(x)$ is equal to zero at $x = a$. However, if both the numerator and the denominator of $f(x)$ are equal to zero, then $x = a$ *need not* be a vertical asymptote.

13.3 CONCEPT QUESTIONS, page 817

1. a. See the definition on page 811 of the text.
   b. See the definition on page 812 of the text.

3. See the procedure given on page 811 of the text.

EXERCISES 13.3, page 817

1. $y = 0$ is a horizontal asymptote.

3. $y = 0$ is a horizontal asymptote and $x = 0$ is a vertical asymptote.

5. $y = 0$ is a horizontal asymptote and $x = -1$ and $x = 1$ are vertical asymptotes.

7. $y = 3$ is a horizontal asymptote and $x = 0$ is a vertical asymptote.

9. $y = 1$ and $y = -1$ are horizontal asymptotes.

11. $\lim_{x\to\infty} \frac{1}{x} = 0$ and so $y = 0$ is a horizontal asymptote. Next, since the numerator of the rational expression is not equal to zero and the denominator is zero at $x = 0$, we see that $x = 0$ is a vertical asymptote.

13. $f(x) = -\frac{2}{x^2}$. $\lim_{x\to\infty} -\frac{2}{x^2} = 0$, so $y = 0$ is a horizontal asymptote. Next, the denominator of $f(x)$ is equal to zero at $x = 0$. Since the numerator of $f(x)$ is not equal to zero at $x = 0$, we see that $x = 0$ is a vertical asymptote.

15. $\lim_{x\to\infty} \frac{x-1}{x+1} = \lim_{x\to\infty} \frac{1-\frac{1}{x}}{1+\frac{1}{x}} = 1$, and so $y = 1$ is a horizontal asymptote. Next, the denominator is equal to zero at $x = -1$ and the numerator is not equal to zero at this point, so $x = -1$ is a vertical asymptote.

17. $h(x) = x^3 - 3x^2 + x + 1$. $h(x)$ is a polynomial function and, therefore, it does not have any horizontal or vertical asymptotes.

19. $\lim_{t\to\infty} \frac{t^2}{t^2 - 9} = \lim_{t\to\infty} \frac{1}{1-\frac{9}{t^2}} = 1$, and so $y = 1$ is a horizontal asymptote. Next, observe that the denominator of the rational expression $t^2 - 9 = (t + 3)(t - 3) = 0$ if $t = -3$ and $t = 3$. But the numerator is not equal to zero at these points. Therefore, $t = -3$ and $t = 3$ are vertical asymptotes.

21. $\lim_{x\to\infty}\frac{3x}{x^2-x-6}=\lim_{x\to\infty}\frac{\frac{3}{x}}{1-\frac{1}{x}-\frac{6}{x^2}}=0$ and so $y=0$ is a horizontal asymptote. Next, observe that the denominator $x^2-x-6=(x-3)(x+2)=0$ if $x=-2$ or $x=3$. But the numerator $3x$ is not equal to zero at these points. Therefore, $x=-2$ and $x=3$ are vertical asymptotes.

23. $\lim_{t\to\infty}\left[2+\frac{5}{(t-2)^2}\right]=2$, and so $y=2$ is a horizontal asymptote. Next observe that

$\lim_{t\to 2^+}g(t)=\lim_{t\to 2^-}\left[2+\frac{5}{(t-2)^2}\right]=\infty$, and so $t=2$ is a vertical asymptote.

25. $\lim_{x\to\infty}\frac{x^2-2}{x^2-4}=\lim_{x\to\infty}\frac{1-\frac{2}{x^2}}{1-\frac{4}{x^2}}=1$ and so $y=1$ is a horizontal asymptote. Next, observe that the denominator $x^2-4=(x+2)(x-2)=0$ if $x=-2$ or 2. Since the numerator x^2-2 is not equal to zero at these points, the lines $x=-2$ and $x=2$ are vertical asymptotes.

27. $g(x)=\frac{x^3-x}{x(x+1)}$; Rewrite $g(x)$ as $g(x)=\frac{x^2-1}{x+1}$ $(x\neq 0)$ and note that

$\lim_{x\to-\infty}g(x)=\lim_{x\to-\infty}\frac{x-\frac{1}{x}}{1+\frac{1}{x}}=-\infty$ and $\lim_{x\to\infty}g(x)=\infty$. Therefore, there are no horizontal asymptotes. Next, note that the denominator of $g(x)$ is equal to zero at $x=0$ and $x=-1$. However, since the numerator of $g(x)$ is also equal to zero when $x=0$, we see that $x=0$ is not a vertical asymptote. Also, the numerator of $g(x)$ is equal to zero when $x=-1$, so $x=-1$ is not a vertical asymptote.

29. f is the derivative function of the function g. Observe that at a relative maximum (relative minimum) of g, $f(x)=0$.

31.

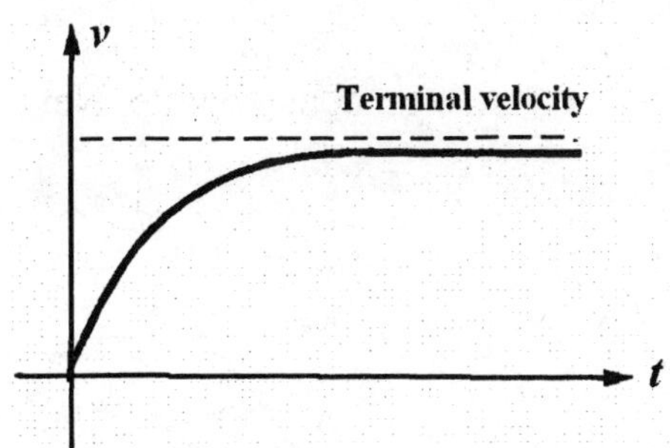

33.

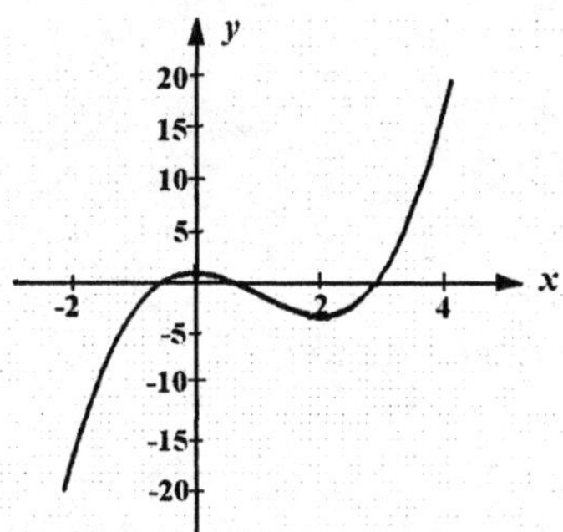

35.

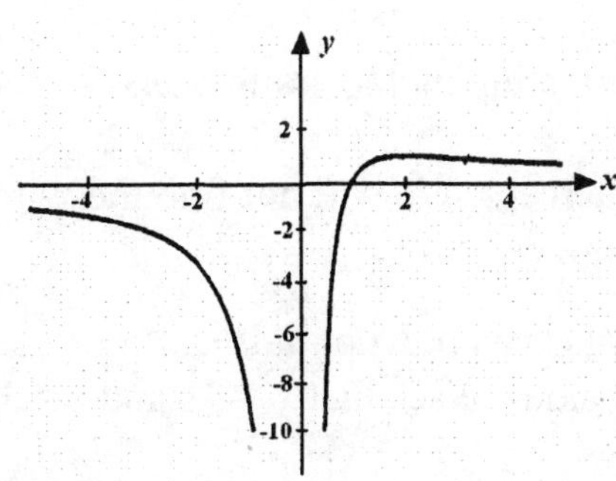

37. $g(x) = 4 - 3x - 2x^3$.

We first gather the following information on the graph of f.

1. The domain of f is $(-\infty, \infty)$.
2. Setting $x = 0$ gives $y = 4$ as the y-intercept. Setting $y = g(x) = 0$ gives a cubic equation which is not easily solved and we will not attempt to find the x-intercepts.
3. $\lim_{x\to-\infty} g(x) = \infty$ and $\lim_{x\to\infty} g(x) = -\infty$. 4. There are no asymptotes of g.
5. $g'(x) = -3 - 6x^2 = -3(2x^2 + 1) < 0$ for all values of x and so g is decreasing on $(-\infty, \infty)$.
6. The results of 5 show that g has no critical points and hence has no relative extrema.

7. $g''(x) = -12x$. Since $g''(x) > 0$ for $x < 0$ and $g''(x) < 0$ for $x > 0$, we see that g is concave upward on $(-\infty,0)$ and concave downward on $(0, \infty)$.
8. From the results of (7), we see that $(0,4)$ is an inflection point of g.

The graph of g follows.

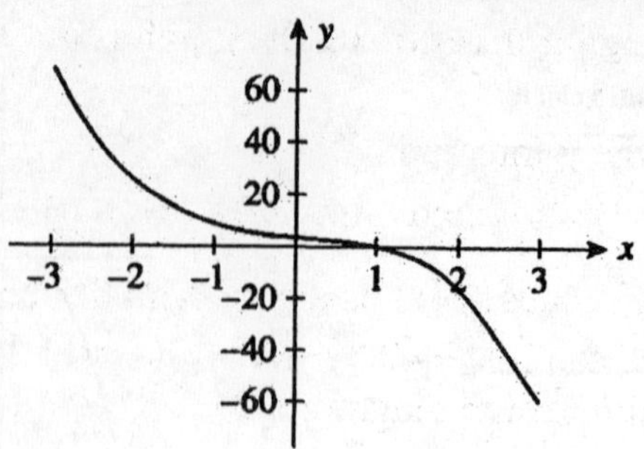

39. $h(x) = x^3 - 3x + 1$

We first gather the following information on the graph of h.

1. The domain of h is $(-\infty, \infty)$.
2. Setting $x = 0$ gives 1 as the y-intercept. We will not find the x-intercept.
3. $\lim_{x\to-\infty} (x^3 - 3x + 1) = -\infty$ and $\lim_{x\to\infty} (x^3 - 3x + 1) = \infty$
4. There are no asymptotes since $h(x)$ is a polynomial.
5. $h'(x) = 3x^2 - 3 = 3(x + 1)(x - 1)$, and we see that $x = -1$ and $x = 1$ are critical points. From the sign diagram

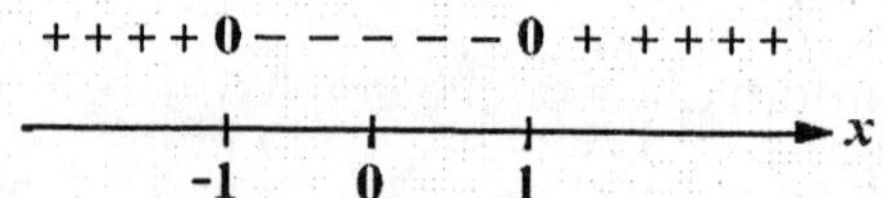

we see that h is increasing on $(-\infty,-1) \cup (1, \infty)$ and decreasing on $(-1,1)$.

6. The results of (5) shows that $(-1,3)$ is a relative maximum and $(1,-1)$ is a relative minimum.
7. $h''(x) = 6x$ and $h''(x) < 0$ if $x < 0$ and $h''(x) > 0$ if $x > 0$. So the graph of h is concave downward on $(-\infty,0)$ and concave upward on $(0, \infty)$.
8. The results of (7) show that $(0,1)$ is an inflection point of h.

The graph of h follows.

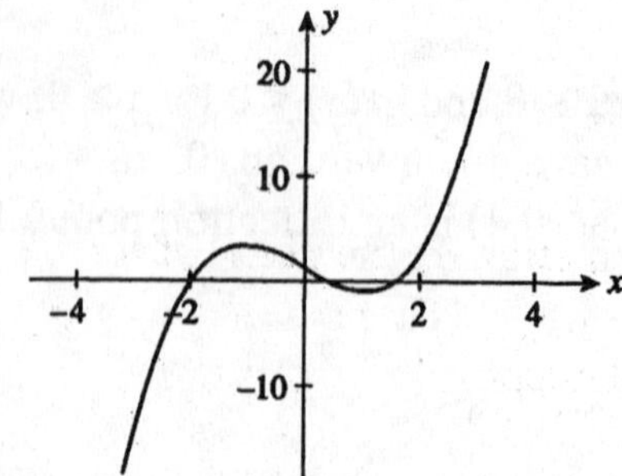

41. $f(x) = -2x^3 + 3x^2 + 12x + 2$

We first gather the following information on the graph of f.

1. The domain of f is $(-\infty, \infty)$.
2. Setting $x = 0$ gives 2 as the y-intercept.
3. $\lim_{x\to-\infty}(-2x^3 + 3x^2 + 12x + 2) = \infty$ and $\lim_{x\to\infty}(-2x^3 + 3x^2 + 12x + 2) = -\infty$
4. There are no asymptotes because $f(x)$ is a polynomial function.
5. $f'(x) = -6x^2 + 6x + 12 = -6(x^2 - x - 2) = -6(x-2)(x+1) = 0$ if $x = -1$ or $x = 2$, the critical points of f. From the sign diagram

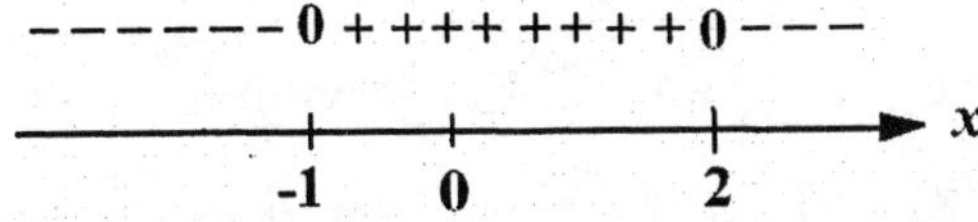

we see that f is decreasing on $(-\infty,-1) \cup (2, \infty)$ and increasing on $(-1,2)$.

6. The results of (5) show that $(-1,-5)$ is a relative minimum and $(2,22)$ is a relative maximum.
7. $f''(x) = -12x + 6 = 0$ if $x = 1/2$. The sign diagram of f''

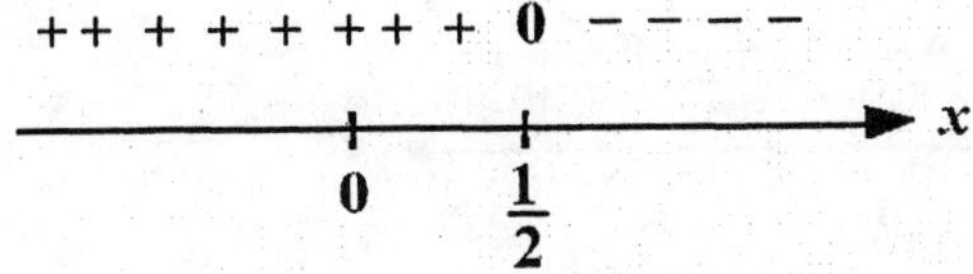

shows that the graph of f is concave upward on $(-\infty,1/2)$ and concave downward on $(1/2, \infty)$.

8. The results of (7) show that $(\frac{1}{2}, \frac{17}{2})$ is an inflection point.

The graph of f follows.

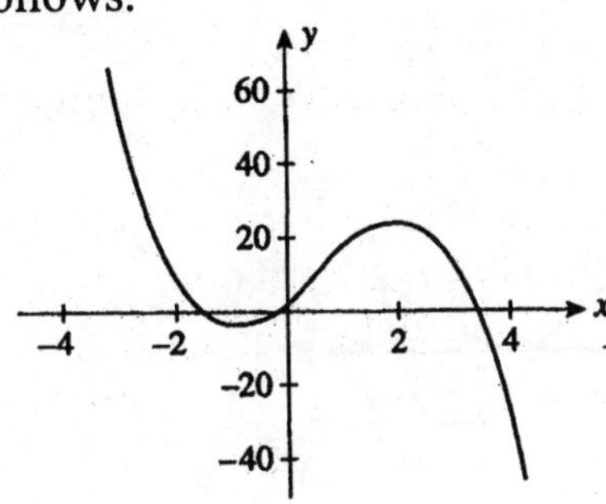

43. $h(x) = \frac{3}{2}x^4 - 2x^3 - 6x^2 + 8$

We first gather the following information on the graph of h.

1. The domain of h is $(-\infty, \infty)$.
2. Setting $x = 0$ gives 8 as the y-intercept.
3. $\lim_{x\to-\infty} h(x) = \lim_{x\to\infty} h(x) = \infty$

4. There are no asymptotes.

5. $h'(x) = 6x^3 - 6x^2 - 12x = 6x(x^2 - x - 2) = 6x(x-2)(x+1) = 0$ if $x = -1, 0,$ or 2, and these are the critical points of h. The sign diagram of h' is

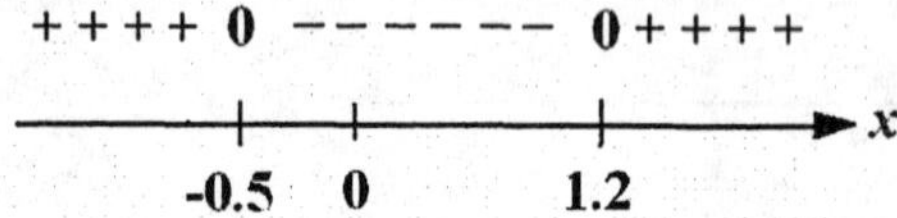

and this tells us that h is increasing on $(-1, 0) \cup (2, \infty)$ and decreasing on $(-\infty,-1) \cup (0,2)$.

6. The results of (5) show that $(-1, \frac{11}{2})$ and $(2,-8)$ are relative minima of h and $(0,8)$ is a relative maximum of h.

7. $h''(x) = 18x^2 - 12x - 12 = 6(3x^2 - 2x - 2)$. The zeros of h'' are

$$x = \frac{2 \pm \sqrt{4+24}}{6} \approx -0.5 \text{ or } 1.2.$$

The sign diagram of h'' is

+ + + + 0 - - - - - - 0 + + + +

x

-0.5 0 1.2

and tells us that the graph of h is concave upward on $(-\infty,-0.5) \cup (1.2, \infty)$ and is concave downward on $(0.5,1.2)$.

8. The results of (7) also show that $(-0.5,6.8)$ and $(1.2,-1)$ are inflection points. The graph of h follows.

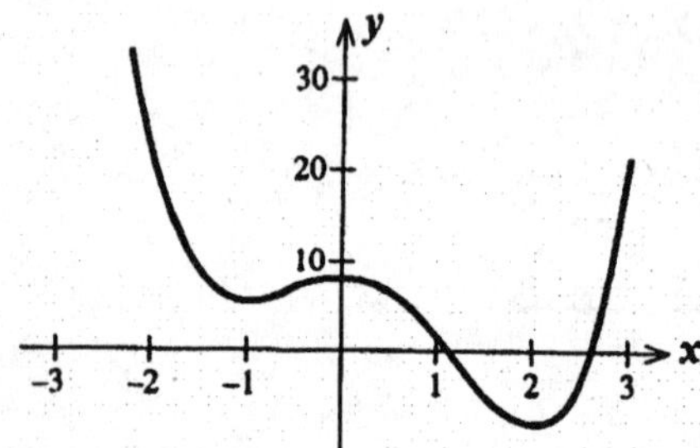

45. $f(t) = \sqrt{t^2 - 4}$.

We first gather the following information on f.

1. The domain of f is found by solving $t^2 - 4 \geq 0$ giving it as $(-\infty,-2] \cup [2,\infty)$.
2. Since $t \neq 0$, there is no y-intercept. Next, setting $y = f(t) = 0$ gives the t-intercepts as -2 and 2.
3. $\lim_{t \to -\infty} f(t) = \lim_{t \to \infty} f(t) = \infty$
4. There are no asymptotes.

5. $f'(t) = \frac{1}{2}(t^2 - 4)^{-1/2}(2t) = t(t^2 - 4)^{-1/2} = \frac{t}{\sqrt{t^2 - 4}}$.

Setting $f'(t) = 0$ gives $t = 0$. But $t = 0$ is not in the domain of f and so there are no critical points. The sign diagram for f' is

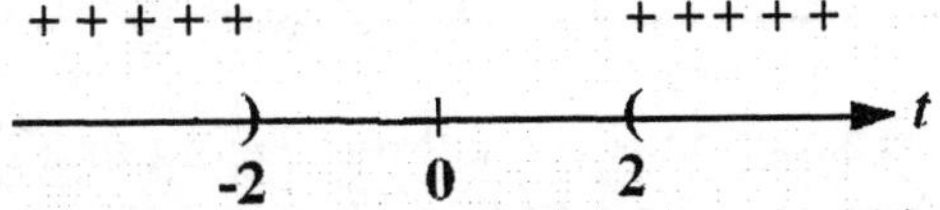

We see that f is increasing on $(2,\infty)$ and decreasing on $(-\infty,-2)$.

6. From the results of (5) we see that there are no relative extrema.

7. $f''(t) = (t^2 - 4)^{-1/2} + t(-\tfrac{1}{2})(t^2 - 4)^{-3/2}(2t) = (t^2 - 4)^{-3/2}(t^2 - 4 - t^2)$

$$= -\frac{4}{(t^2 - 4)^{3/2}}.$$

8. Since $f''(t) < 0$ for all t in the domain of f, we see that f is concave downward everywhere. From the results of (7), we see that there are no inflection points. The graph of f follows.

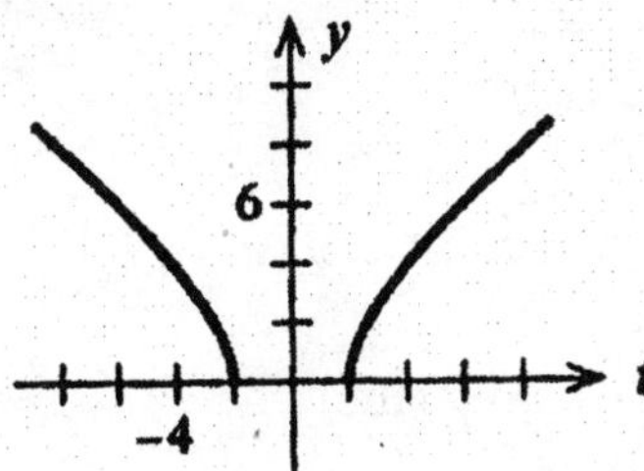

47. $g(x) = \tfrac{1}{2}x - \sqrt{x}$.

We first gather the following information on g.

1. The domain of g is $[0,\infty)$.
2. The y-intercept is 0. To find the x-intercept, set $y = 0$, giving

$$\tfrac{1}{2}x - \sqrt{x} = 0$$

$$x = 2\sqrt{x}$$

$$x^2 = 4x$$

$$x(x-4) = 0, \text{ and } x = 0 \text{ or } x = 4$$

3. $\lim_{x\to\infty} (\tfrac{1}{2}x - \sqrt{x}) = \lim_{x\to\infty} \tfrac{1}{2}x(1 - \tfrac{2}{\sqrt{x}}) = \infty$.
4. There are no asymptotes.

5. $g'(x) = \frac{1}{2} - \frac{1}{2}x^{-1/2} = \frac{1}{2}x^{-1/2}(x^{1/2} - 1) = \dfrac{\sqrt{x} - 1}{2\sqrt{x}}$

which is zero when $x = 1$. From the sign diagram for g'

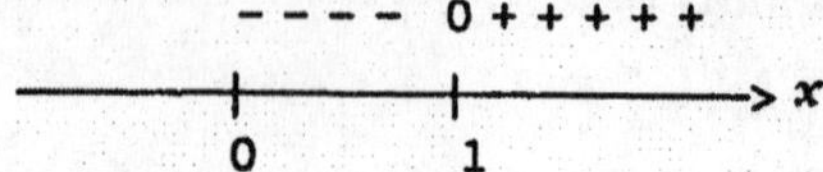

we see that g is decreasing on (0,1) and increasing on (1,∞).

6. From the sign diagram of g', we see that $g(1) = -1/2$ is a relative minimum.

7. $g''(x) = (-\frac{1}{2})(-\frac{1}{2})x^{-3/2} = \dfrac{1}{4x^{3/2}} > 0$ for $x > 0$, and so g is concave upward on (0,∞).

8. There are no inflection points.

The graph of g follows.

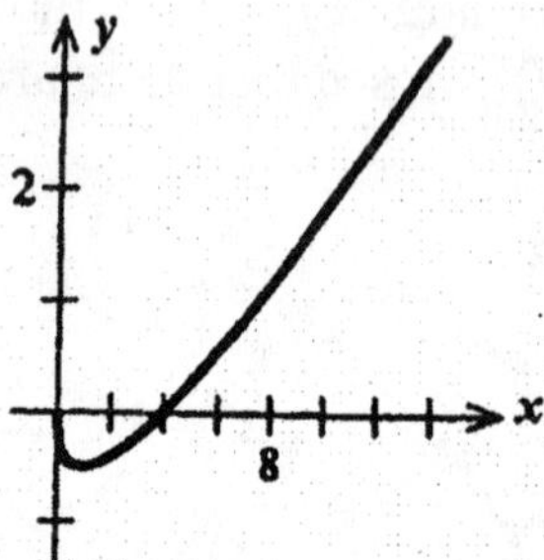

49. $g(x) = \dfrac{2}{x-1}$. We first gather the following information on g.

1. The domain of g is $(-\infty,1) \cup (1,\infty)$.
2. Setting $x = 0$ gives -2 as the y-intercept. There are no x-intercepts since $\dfrac{2}{x-1} \neq 0$ for all values of x.
3. $\displaystyle\lim_{x\to-\infty} \frac{2}{x-1} = 0$ and $\displaystyle\lim_{x\to\infty} \frac{2}{x-1} = 0$.

4. The results of (3) show that $y = 0$ is a horizontal asymptote. Furthermore, the denominator of $g(x)$ is equal to zero at $x = 1$ but the numerator is not equal to zero there. Therefore, $x = 1$ is a vertical asymptote.

5. $g'(x) = -2(x-1)^{-2} = -\dfrac{2}{(x-1)^2} < 0$ for all $x \neq 1$ and so g is decreasing on (-∞,1) and (1,∞).

6. Since g has no critical points, there are no relative extrema.

7. $g''(x) = \dfrac{4}{(x-1)^3}$ and so $g''(x) < 0$ if $x < 1$ and $g''(x) > 0$ if $x > 1$. Therefore, the graph of g is concave downward on $(-\infty,1)$ and concave upward on $(1,\infty)$.

8. Since $g''(x) \neq 0$, there are no inflection points.

The graph of g follows.

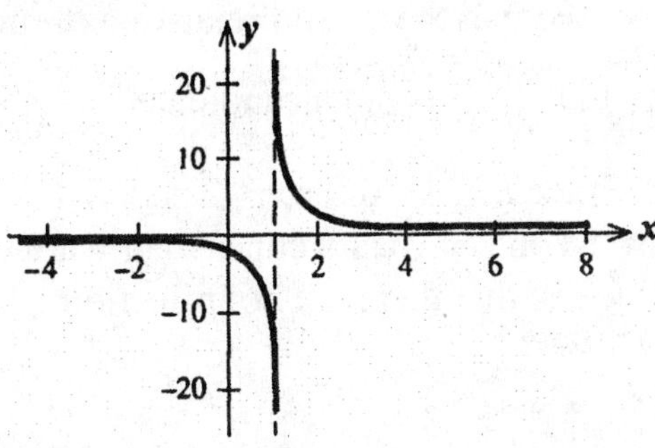

51. $h(x) = \dfrac{x+2}{x-2}$.

We first gather the following information on the graph of h.

1. The domain of h is $(-\infty,2) \cup (2,\infty)$.
2. Setting $x = 0$ gives $y = -1$ as the y-intercept. Next, setting $y = 0$ gives $x = -2$ as the x-intercept.
3. $\lim\limits_{x\to\infty} h(x) = \lim\limits_{x\to-\infty} \dfrac{1+\frac{2}{x}}{1-\frac{2}{x}} = \lim\limits_{x\to-\infty} h(x) = 1.$
4. Setting $x - 2 = 0$ gives $x = 2$. Furthermore,

$$\lim_{x\to2^+} \frac{x+2}{x-2} = \infty \quad\text{and}\quad \lim_{x\to2^+} \frac{x+2}{x-2} = -\infty$$

So $x = 2$ is a vertical asymptote of h. Also, from the resultsof (3), we see that $y = 1$ is a horizontal asymptote of h.

5. $h'(x) = \dfrac{(x-2)(1)-(x+2)(1)}{(x-2)^2} = -\dfrac{4}{(x-2)^2}$.

We see that there are no critical points of h. (Note $x = 2$ does not belong to the domain of h.) The sign diagram of h' follows.

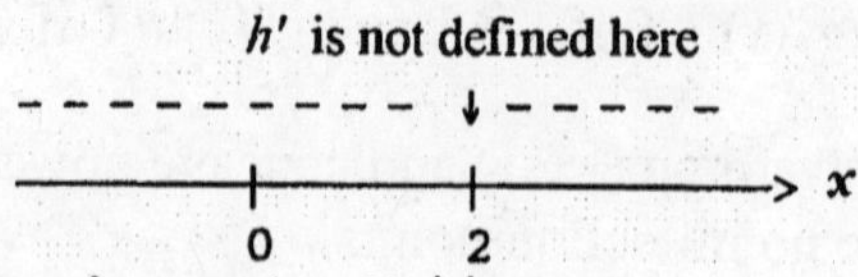

We see that h is decreasing on $(-\infty,2) \cup (2,\infty)$.
6. From the results of (5), we see that there is no relative extremum.

7. $h''(x) = \dfrac{8}{(x-2)^3}$. Note that $x = 2$ is not a candidate for an inflection point because $h(2)$ is not defined. Since $h''(x) < 0$ for $x < 2$ and $h''(x) > 0$ for $x > 2$, we see that h is concave downward on $(-\infty,2)$ and concave upward on $(2,\infty)$.
8. From the results of (7), we see that there are no inflection points.
The graph of h follows.

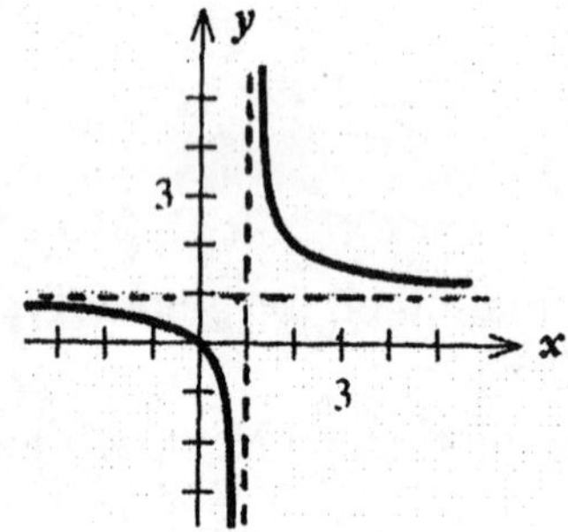

53. $f(t) = \dfrac{t^2}{1+t^2}$.

We first gather the following information on the graph of f.
1. The domain of f is $(-\infty, \infty)$.
2. Setting $t = 0$ gives the y-intercept as 0. Similarly, setting $y = 0$ gives the t-intercept as 0.
3. $\lim\limits_{t\to-\infty} \dfrac{t^2}{1+t^2} = \lim\limits_{t\to\infty} \dfrac{t^2}{1+t^2} = 1.$
4. The results of (3) show that $y = 1$ is a horizontal asymptote. There are no vertical asymptotes since the denominator is not equal to zero.
5. $f'(t) = \dfrac{(1+t^2)(2t) - t^2(2t)}{(1+t^2)^2} = \dfrac{2t}{(1+t^2)^2} = 0$, if $t = 0$, the only critical point of f.
Since $f'(t) < 0$ if $t < 0$ and $f'(t) > 0$ if $t > 0$, we see that f is decreasing on $(-\infty,0)$ and increasing on $(0,\infty)$.
6. The results of (5) show that $(0,0)$ is a relative minimum.

7. $f''(t) = \dfrac{(1+t^2)^2(2) - 2t(2)(1+t^2)(2t)}{(1+t^2)^4} = \dfrac{2(1+t^2)[(1+t^2) - 4t^2]}{(1+t^2)^4}$

$= \dfrac{2(1-3t^2)}{(1+t^2)^3} = 0$ if $t = \pm\dfrac{\sqrt{3}}{3}$.

The sign diagram of f'' is

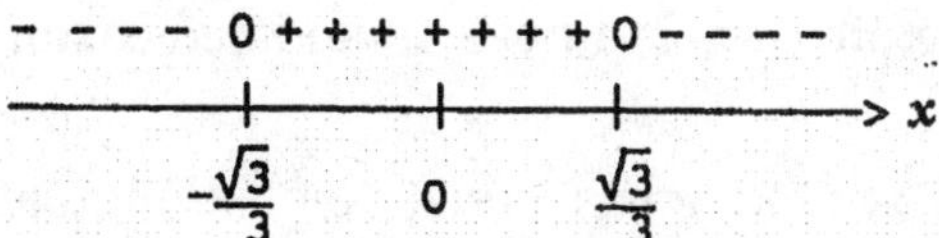

and shows that f is concave downward on $(-\infty, -\frac{\sqrt{3}}{3}) \cup (\frac{\sqrt{3}}{3}, \infty)$ and concave upward on $(-\frac{\sqrt{3}}{3}, \frac{\sqrt{3}}{3})$.

8. The results of (7) show that $(-\frac{\sqrt{3}}{3}, \frac{1}{4})$ and $(\frac{\sqrt{3}}{3}, \frac{1}{4})$ are inflection points.

The graph of f follows.

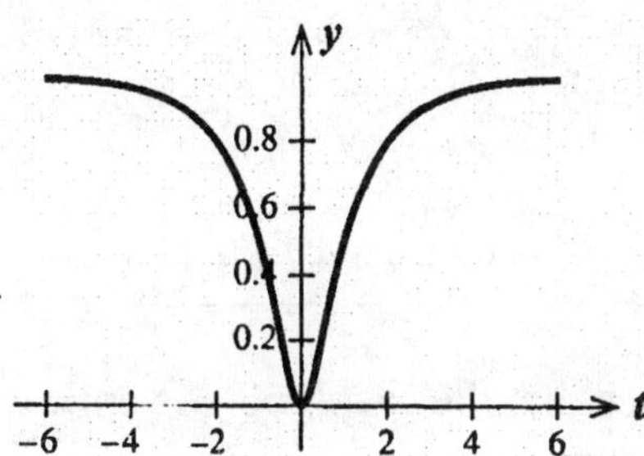

57. $g(t) = \dfrac{t^2}{t^2 - 1}$.

We first gather some information on the graph of g.

1. Since $t^2 - 1 = 0$ if $t = \pm 1$, we see that the domain of g is $(-\infty, -1) \cup (-1, 1) \cup (1, \infty)$.
2. Setting $t = 0$ gives 0 as the y-intercept. Setting $y = 0$ gives 0 as the t-intercept.
3. $\lim_{t\to-\infty} g(t) = \lim_{t\to\infty} g(t) = 1$.
4. The results of (3) show that $y = 1$ is a horizontal asymptote. Since the denominator (but not the numerator) is zero at $t = \pm 1$, we see that $t = \pm 1$ are vertical asymptotes.
5. $g'(t) = \dfrac{(t^2-1)(2t) - (t^2)(2t)}{(t^2-1)^2} = -\dfrac{2t}{(t^2-1)^2} = 0$, if $t = 0$.

The sign diagram of g' is

The sign diagram of g' is

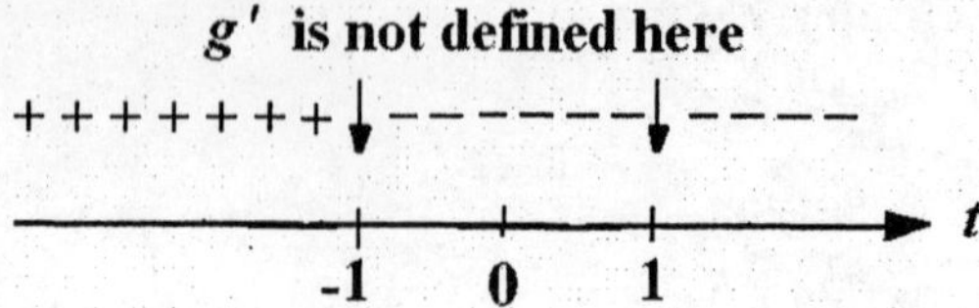

We see that g is increasing on $(-\infty,-1) \cup (-1, 0)$ and decreasing on $(0, 1) \cup (1, \infty)$.

6. From the results of (5), we see that g has a relative maximum at $t = 0$.
7. $g''(t) = \dfrac{(t^2-1)^2(-2)-(-2t)(2)(t^2-1)(2t)}{(t^2-1)^4}$

$$= \frac{2(t^2-1)^2[-(t^2-1)+4t^2]}{(t^2-1)^3}$$

$$= \frac{2(-t^2+1+4t^2)}{(t^2-1)^3} = \frac{2(3t^2+1)}{(t^2-1)^3}$$

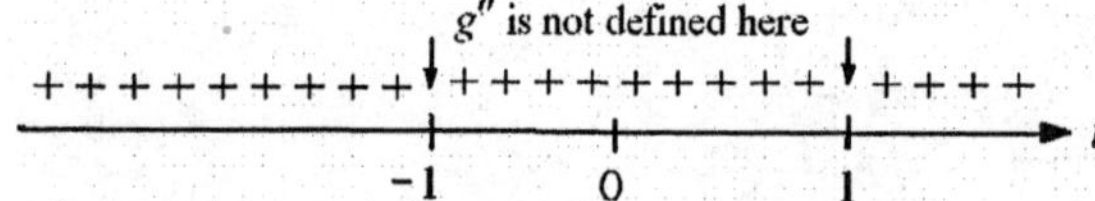

From the sign diagram we see that the graph of g is concave up on $(-\infty, -1)\cup(-1,1)\cup(1,\infty)$.

8. From (7), we see that the graph of g has no inflection points.

The graph of g follows.

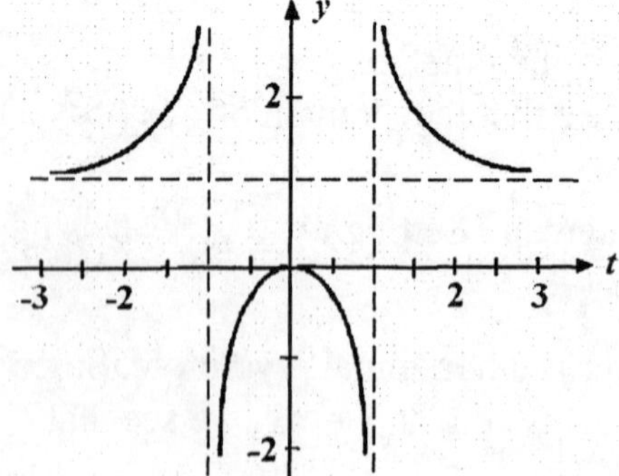

59. $h(x) = (x-1)^{2/3} + 1.$

1. The domain of h is $(-\infty, \infty)$.
2. Setting $x = 0$ gives 2 as the y-intercept; since $h(x) \neq 0$ there is no x-intercept.
3. $\lim_{x\to\infty} [(x-1)^{2/3}+1] = \infty$. Similarly, $\lim_{x\to-\infty} [(x-1)^{2/3}+1] = \infty$.
4. There are no asymptotes.
5. $h'(x) = \frac{2}{3}(x-1)^{-1/3}$ and is positive if $x > 1$ and negative if $x < 1$. So h is increasing on $(1,\infty)$, and decreasing on $(-\infty,1)$.
6. From (5), we see that h has a relative minimum at $(1,1)$.
7. $h''(x) = \frac{2}{3}(-\frac{1}{3})(x-1)^{-4/3} = -\frac{2}{9}(x-1)^{-4/3} = -\frac{2}{(x-1)^{4/3}}$. Since $h''(x) < 0$ on $(-\infty,1) \cup (1,\infty)$, we see that h is concave downward on $(-\infty,1) \cup (1,\infty)$. Note that $h''(x)$ is not defined at $x = 1$.
8. From the results of (7), we see h has no inflection points.

The graph of h follows.

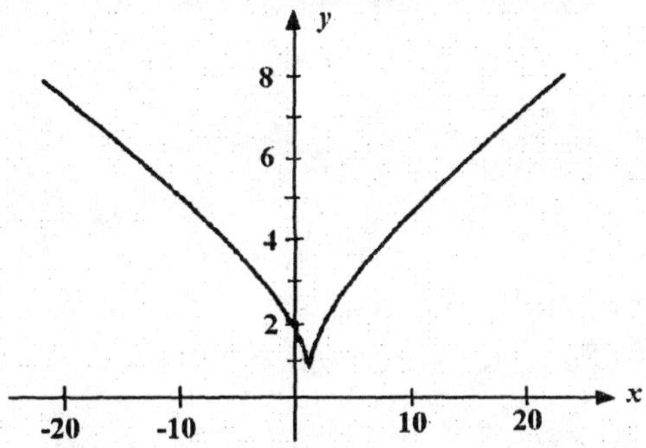

61. a. The denominator of $C(x)$ is equal to zero if $x = 100$. Also,

$$\lim_{x\to100^-} \frac{0.5x}{100-x} = \infty \quad \text{and} \quad \lim_{x\to100^+} \frac{0.5x}{100-x} = -\infty$$

Therefore, $x = 100$ is a vertical asymptote of C.

b. No, because the denominator will be equal to zero in that case.

63. a. Since $\lim_{t\to\infty} C(t) = \lim_{t\to\infty} \frac{0.2t}{t^2+1} = \lim_{t\to\infty} \left[\frac{0.2}{t+\frac{1}{t^2}}\right] = 0$, $y = 0$ is a horizontal asymptote.

b. Our results reveal that as time passes, the concentration of the drug decreases and approaches zero.

65. $G(t) = -0.2t^3 + 2.4t^2 + 60$.

We first gather the following information on the graph of G.

We first gather the following information on the graph of G.
1. The domain of G is $(0,\infty)$.
2. Setting $t = 0$ gives 60 as the y-intercept.
Note that Step 3 is not necessary in this case because of the restricted domain.
4. There are no asymptotes since G is a polynomial function.
5. $G'(t) = -0.6t^2 + 4.8t = -0.6t(t - 8) = 0$, if $t = 0$ or $t = 8$. But these points do not lie in the interval $(0,8)$, so they are not critical points. The sign diagram of G'

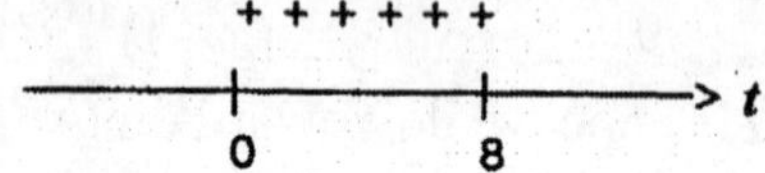

shows that G is increasing on $(0,8)$.
6. The results of (5) tell us that there are no relative extrema.
7. $G''(t) = -1.2t + 4.8 = -1.2(t - 4)$. The sign diagram of G'' is

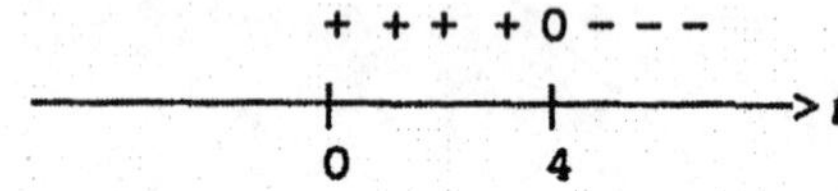

and shows that G is concave upward on $(0,4)$ and concave downward on $(4,8)$.
6. The results of (7) shows that $(4,85.6)$ is an inflection point.
The graph of G follows.

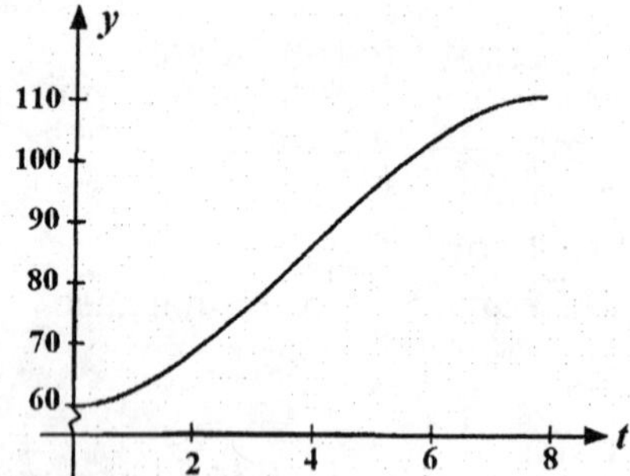

67. $C(t) = \dfrac{0.2t}{t^2 + 1}$.
We first gather the following information on the function C.
1. The domain of C is $[0,\infty)$.
2. If $t = 0$, then $y = 0$. Also, if $y = 0$, then $t = 0$.

3. $\lim_{t\to\infty}\frac{0.2t}{t^2+1}=0.$

4. The results of (3) imply that $y=0$ is a horizontal asymptote.

5. $$C'(t)=\frac{(t^2+1)(0.2)-0.2t(2t)}{(t^2+1)^2}=\frac{0.2(t^2+1-2t^2)}{(t^2+1)^2}=\frac{0.2(1-t^2)}{(t^2+1)^2}$$

and this is equal to zero at $t=\pm1$, so $t=1$ is a critical point of C. The sign diagram of C' is

```
         + + + 0 - - - -
    ---------|------|---------> t
             0      1
```

and tells us that C is decreasing on $(1,\infty)$ and increasing on $(0,1)$.

6. The results of (5) tell us that $(1,0.1)$ is a relative maximum.

7. $$C''(t)=0.2\left[\frac{(t^2+1)^2(-2t)-(1-t^2)2(t^2+1)(2t)}{(t^2+1)^4}\right]$$

$$=\frac{0.2(t^2+1)(2t)(-t^2-1-2+2t^2)}{(t^2+1)^4}=\frac{0.4t(t^2-3)}{(t^2+1)^3}.$$

The sign diagram of C'' is

```
                 0 - - 0 + + +
    -------------|------|--------> t
     -√3         0      √3
```

and so the graph of C is concave downward on $(0,\sqrt{3})$ and concave upward on $(\sqrt{3},\infty)$.

8. The results of (7) show that $(\sqrt{3},0.05\sqrt{3})$ is an inflection point.

The graph of C follows.

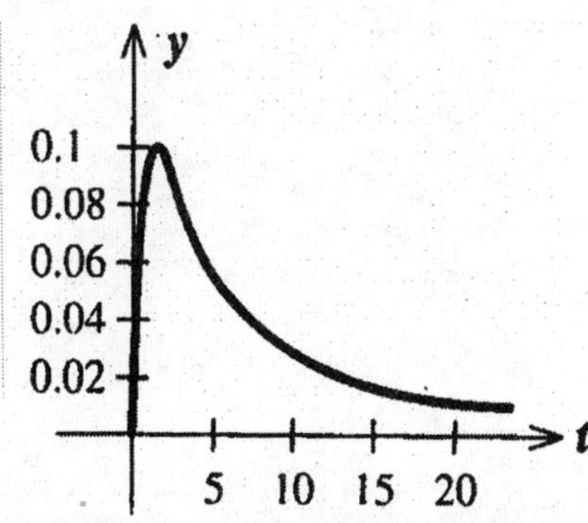

69. $T(x) = \dfrac{120x^2}{x^2+4}$.

We first gather the following information on the function T.

1. The domain of T is $[0,\infty)$.
2. Setting $x = 0$ gives 0 as the y-intercept.
3. $\lim\limits_{x\to\infty} \dfrac{120x^2}{x^2+4} = 120.$
4. The results of (3) show that $y = 120$ is a horizontal asymptote.
5. $T'(x) = 120\left[\dfrac{(x^2+4)2x - x^2(2x)}{(x^2+4)^2}\right] = \dfrac{960x}{(x^2+4)^2}$. Since $T'(x) > 0$

if $x > 0$, we see that T is increasing on $(0,\infty)$.

6. There are no relative extrema in $(0,\infty)$.
7. $T''(x) = 960\left[\dfrac{(x^2+4)^2 - x(2)(x^2+4)(2x)}{(x^2+4)^4}\right]$

$$= \frac{960(x^2+4)[(x^2+4) - 4x^2]}{(x^2+4)^4} = \frac{960(4-3x^2)}{(x^2+4)^3}.$$

The sign diagram for T'' is

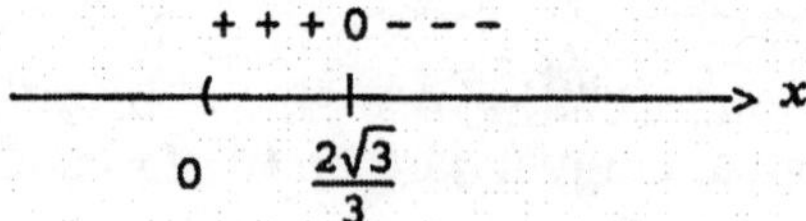

We see that T is concave downward on $(\frac{2\sqrt{3}}{3},\infty)$ and concave upward on $(0, \frac{2\sqrt{3}}{3})$.

8. We see from the results of (7) that $(\frac{2\sqrt{3}}{3}, 30)$ is an inflection point.

The graph of T follows.

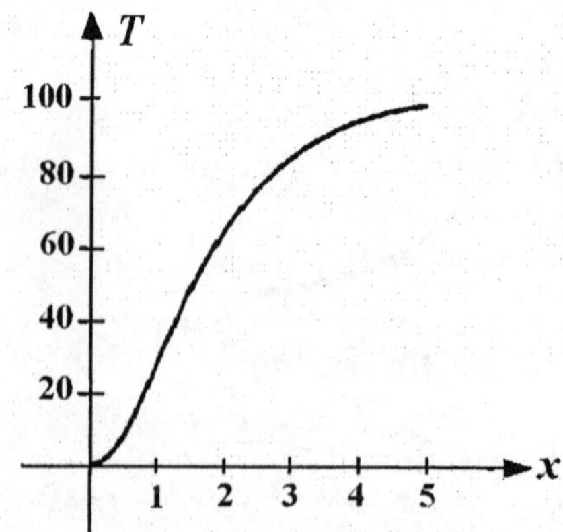

USING TECHNOLOGY EXERCISES 13.3, page 823

1.

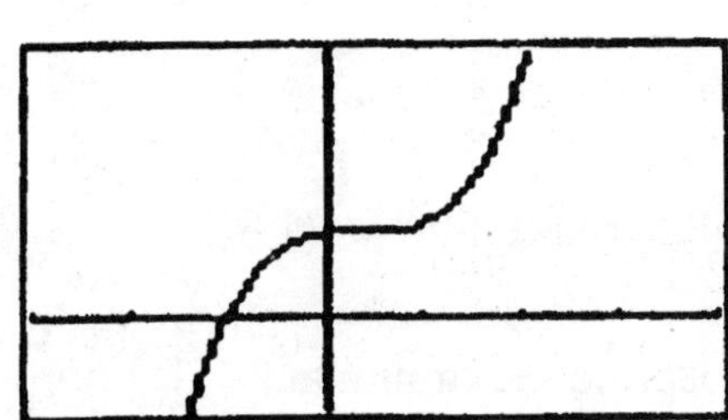

3.

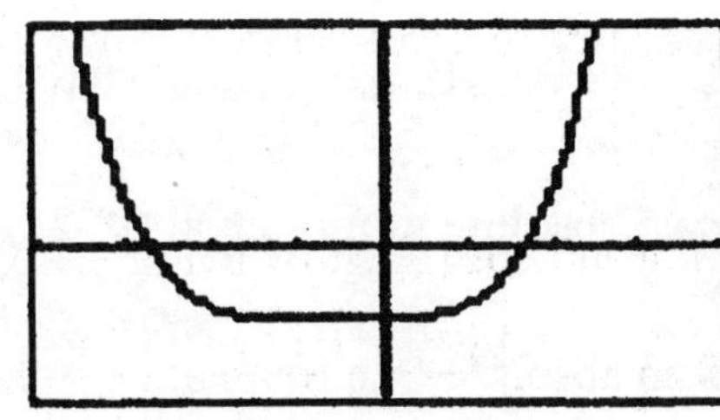

5. -0.9733; 2.3165, 4.6569

7 -1.1301; 2.9267

9. 1.5142

13.4 Problem Solving Tips

1. To determine the absolute maximum and absolute minimum of a continuous function f on a closed interval $[a,b]$, find the critical numbers of f that lie in (a,b). Then compute the value of f at each critical number of f and compute $f(a)$ and $f(b)$. The largest and smallest of these values will be the absolute maximum value and absolute minimum value of f, respectively.

2. Note that the procedure in (1) only holds for a continuous function f over a closed interval $[a,b]$.

13.4 CONCEPT QUESTIONS, page 831

1. a. A function f has an absolute maximum at a if $f(x) \le f(a)$ for all x in the domain of f.
   b. A function f has an absolute minimum at a if $f(x) \ge f(a)$ for all x in the domain of f.

EXERCISES 13.4, page 832

1. f has no absolute extrema.

3. f has an absolute minimum at (0,0).

5. f has an absolute minimum at (0,-2) and an absolute maximum at (1,3).

7. f has an absolute minimum at $(\frac{3}{2}, -\frac{27}{16})$ and an absolute maximum at (-1,3).

9. The graph of $f(x) = 2x^2 + 3x - 4$ is a parabola that opens upward. Therefore, the vertex of the parabola is the absolute minimum of f. To find the vertex, we solve the equation $f'(x) = 4x + 3 = 0$ giving $x = -3/4$. We conclude that the absolute minimum value is $f(-\frac{3}{4}) = -\frac{41}{8}$.

11. Since $\lim_{x \to -\infty} x^{1/3} = -\infty$ and $\lim_{x \to \infty} x^{1/3} = \infty$, we see that h is unbounded. Therefore it has no absolute extrema.

13. $f(x) = \dfrac{1}{1+x^2}$.
Using the techniques of graphing, we sketch the graph of f (see Figure 40, page 796, in the text). The absolute maximum of f is $f(0) = 1$. Alternatively, observe that $1 + x^2 \geq 1$ for all real values of x. Therefore, $f(x) \leq 1$ for all x, and we see that the absolute maximum is attained when $x = 0$.

15. $f(x) = x^2 - 2x - 3$ and $f'(x) = 2x - 2 = 0$, so $x = 1$ is a critical point. From the table,

x	-2	1	3
$f(x)$	5	-4	0

we conclude that the absolute maximum value is $f(-2) = 5$ and the absolute minimum value is $f(1) = -4$.

17. $f(x) = -x^2 + 4x + 6$; The function f is continuous and defined on the closed interval [0,5]. $f'(x) = -2x + 4$ and $x = 2$ is a critical point. From the table

x	0	2	5
$f(x)$	6	10	1

we conclude that $f(2) = 10$ is the absolute maximum value and $f(5) = 1$ is the absolute minimum value.

19. The function $f(x) = x^3 + 3x^2 - 1$ is continuous and defined on the closed interval [-3,2] and differentiable in (-3,2). The critical points of f are found by solving
$$f'(x) = 3x^2 + 6x = 3x(x + 2)$$
giving $x = -2$ and $x = 0$. Next, we compute the values of f given in the following table.

x	-3	-2	0	2
$f(x)$	-1	3	-1	19

From the table, we see that the absolute maximum value of f is $f(2) = 19$ and the absolute minimum value is $f(-3) = -1$ and $f(0) = -1$.

21. The function $g(x) = 3x^4 + 4x^3$ is continuous and differentiable on the closed interval [-2,1] and differentiable in (-2,1). The critical points of g are found by solving
$$g'(x) = 12x^3 + 12x^2 = 12x^2(x + 1)$$
giving $x = 0$ and $x = -1$. We next compute the values of g shown in the following table.

x	-2	-1	0	1
$g(x)$	16	-1	0	7

From the table we see that $g(-2) = 16$ is the absolute maximum value of g and $g(-1) = -1$ is the absolute minimum value of g.

23. $f(x) = \dfrac{x+1}{x-1}$ on [2,4]. Next, we compute,
$$f'(x) = \frac{(x-1)(1)-(x+1)(1)}{(x-1)^2} = -\frac{2}{(x-1)^2}.$$
Since there are no critical points, ($x = 1$ is not in the domain of f), we need only test the endpoints. From the table

x	2	4
$g(x)$	3	5/3

we conclude that $f(4) = 5/3$ is the absolute minimum value and $f(2) = 3$ is the absolute maximum value.

25. $f(x) = 4x + \dfrac{1}{x}$ is continuous on [1,3] and differentiable in (1,3). To find the critical points of f, we solve $f'(x) = 4 - \frac{1}{x^2} = 0$, obtaining $x = \pm\frac{1}{2}$. Since these critical points lie outside the interval [1, 3], they are not candidates for the absolute extrema of f. Evaluating f at the endpoints of the interval [1, 3], we find that the absolute maximum value of f is $f(3) = \frac{37}{3}$, and the absolute minimum value of f is $f(1) = 5$.

27. $f(x) = \frac{1}{2}x^2 - 2\sqrt{x} = \frac{1}{2}x^2 - 2x^{1/2}$. To find the critical points of f, we solve

$$f'(x) = x - x^{-1/2} = 0, \quad \text{or} \qquad x^{3/2} - 1 = 0,$$

obtaining $x = 1$. From the table

x	0	1	3
$f(x)$	0	$-\frac{3}{2}$	$\frac{9}{2} - 2\sqrt{3} \approx 1.04$

we conclude that $f(3) \approx 1.04$ is the absolute maximum value and $f(1) = -3/2$ is the absolute minimum value.

29. The graph of $f(x) = 1/x$ over the interval $(0,\infty)$ follows.

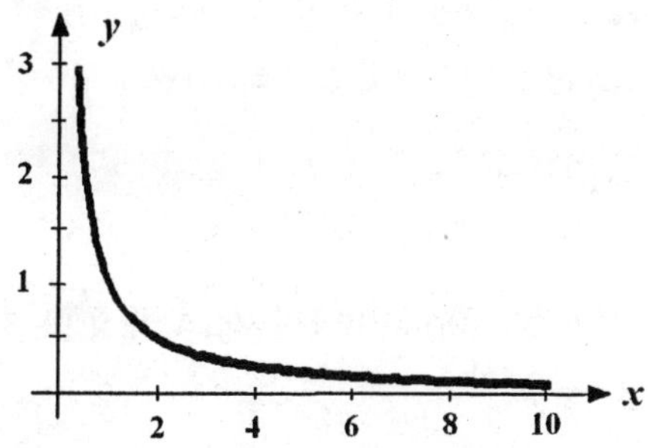

From the graph of f, we conclude that f has no absolute extrema.

31. $f(x) = 3x^{2/3} - 2x$. The function f is continuous on [0,3] and differentiable on (0,3). To find the critical points of f, we solve

$$f'(x) = 2x^{-1/3} - 2 = 0$$

obtaining $x = 1$ as the critical point. From the table,

x	0	1	3
$f(x)$	0	1	$3^{5/3} - 6 \approx 0.24$

we conclude that the absolute maximum value is $f(1) = 1$ and the absolute minimum value is $f(0) = 0$.

33. $f(x) = x^{2/3}(x^2 - 4)$.

$$f'(x) = x^{2/3}(2x) + \tfrac{2}{3}x^{-1/3}(x^2 - 4) = \tfrac{2}{3}x^{-1/3}[3x^2 + (x^2 - 4)]$$

$$= \frac{8(x^2 - 1)}{3x^{1/3}} = 0.$$

Observe that f' is not defined at $x = 0$. Furthermore, $f'(x) = 0$ at $x \pm 1$. So the critical points of f are -1, 0, 1. From the following table,

x	-1	0	1	2
$f(x)$	-3	0	-3	0

we see that f has an absolute minimum at (-1,-3) and (1,-3) and absolute maxima at (0,0) and (2,0).

35. $f(x)=\dfrac{x}{x^2+2}$. To find the critical points of f, we solve

$$f'(x)=\frac{(x^2+2)-x(2x)}{(x^2+2)^2}=\frac{2-x^2}{(x^2+2)^2}=0$$

obtaining $x=\pm\sqrt{2}$. Since $x=-\sqrt{2}$ lies outside [-1,2], $x=\sqrt{2}$ is the only critical point in the given interval. From the table

x	-1	$\sqrt{2}$	2
$f(x)$	$-\frac{1}{3}$	$\sqrt{2}/4\approx 0.35$	$\frac{1}{3}$

we conclude that $f(\sqrt{2}))=\sqrt{2}/4\approx 0.35$ is the absolute maximum value and $f(-1)=-1/3$ is the absolute minimum value.

37. The function $f(x)=\dfrac{x}{\sqrt{x^2+1}}=\dfrac{x}{(x^2+1)^{1/2}}$ is continuous and defined on the closed interval [-1,1] and differentiable on (-1,1). To find the critical points of f, we first compute

$$f'(x)=\frac{(x^2+1)^{1/2}(1)-x(\frac{1}{2})(x^2+1)^{-1/2}(2x)}{[(x^2+1)^{1/2}]^2}$$

$$=\frac{(x^2+1)^{-1/2}[x^2+1-x^2]}{x^2+1}=\frac{1}{(x^2+1)^{3/2}}$$

which is never equal to zero. Next, we compute the values of f shown in the following table.

x	-1	1
$f(x)$	$-\sqrt{2}/2$	$\sqrt{2}/2$

We conclude that $f(-1)=-\sqrt{2}/2$ is the absolute minimum value and $f(1)=\sqrt{2}/2$ is the absolute maximum value.

39. $h(t)=-16t^2+64t+80$. To find the maximum value of h, we solve

$$h'(t)=-32t+64=-32(t-2)=0$$

giving $t = 2$ as the critical point of h. Furthermore, this value of t gives rise to the absolute maximum value of h since the graph of h is a parabola that opens downward. The maximum height is given by

$$h(2) = -16(4) + 64(2) + 80 = 144, \text{ or } 144 \text{ feet.}$$

41. $P'(t) = 0.027t - 1.126 = 0$ implies $t \approx 41.7$, a critical number for P. $P''(t) = 0.027$ and $P''(41.7) = 0.027 > 0$. Therefore, $t = 41.7$ gives a minimum of P. This is around September of 1991. The percent is approximately $P(41.7) \approx 17.7$.

43. $N(t) = 0.81t - 1.14\sqrt{t} + 1.53$. $N'(t) = 0.81 - 1.14(\frac{1}{2}t^{-1/2}) = 0.81 - \dfrac{0.57}{t^{1/2}}$. Setting $N'(t) = 0$ gives $t^{1/2} = \dfrac{0.57}{0.81}$, or $t = 0.4952$ as a critical point of N. Evaluating $N(t)$ at the endpoints $t = 0$ and $t = 6$ as well as at the critical point, we have

t	0	0.4952	6
$N(t)$	1.53	1.13	3.60

From the table, we see that the absolute maximum of N occurs at $t = 6$ and the absolute minimum occurs at $t \approx 0.5$. Our results tell us that the number of nonfarm full-time self- employed women over the time interval from 1963 to 1993 was the highest in 1993 and stood at approximately 3.6 million.

45. $P(x) = -0.000002x^3 + 6x - 400$. $P'(x) = -0.000006x^2 + 6 = 0$ if $x = \pm 1000$. We reject the negative root. Next, we compute $P''(x) = -0.000012x$. Since $P''(1000) = -0.012 < 0$, the Second Derivative Test shows that $x = 1000$ affords a relative maximum of f. From physical considerations, or from a sketch of the graph of f, we see that the maximum profit is realized if 1000 cases are produced per day. The profit is $P(1000) = -0.000002(1000)^3 + 6(1000) - 400$, or \$3600/day.

47. The revenue is $R(x) = px = -0.0004x^2 + 10x$, and the profit is

$$P(x) = R(x) - C(x) = -0.0004x^2 + 10x - (400 + 4x + 0.0001x^2)$$
$$= -0.0005x^2 + 6x - 400.$$
$$P'(x) = -0.001x + 6 = 0$$

if $x = 6000$, a critical point. Since $P''(x) = -0.001 < 0$ for all x, we see that the graph of P is a parabola that opens downward. Therefore, a level of production of 6000 rackets/day will yield a maximum profit.

49. The total cost function is given by

$$C(x) = V(x) + 20,000$$
$$= 0.000001x^3 - 0.01x^2 + 50x + 20,000$$

The profit function is

$$P(x) = R(x) - C(x)$$
$$= -0.02x^2 + 150x - 0.000001x^3 + 0.01x^2 - 50x + 20,000$$
$$= -0.000001x^3 - 0.01x^2 + 100x - 20,000$$

We want to maximize P on [0, 7000].

$$P'(x) = -0.000003x^2 - 0.02x + 100$$

Setting $P'(x) = 0$ gives $3x^2 + 20,000x - 100,000,000 = 0$

or $x = \dfrac{-20,000 \pm \sqrt{20,000^2 + 1,200,000,000}}{6} = -10,000$ or 3,333.3.

So x = 3,333.30 is a critical point in the interval [0, 7000].

x	0	3,333	7,000
$P(x)$	-20,000	165,185	-519,700

From the table, we see that a level of production of 3,333 pagers per week will yield a maximum profit of $165,185 per week.

51. a. $\overline{C}(x) = \dfrac{C(x)}{x} = 0.0025x + 80 + \dfrac{10,000}{x}$.

b. $\overline{C}'(x) = 0.0025 - \dfrac{10,000}{x^2} = 0$ if $0.0025x^2 = 10,000$, or $x = 2000$.

Since $\overline{C}''(x) = \dfrac{20,000}{x^3}$, we see that $\overline{C}''(x) > 0$ for $x > 0$ and so $\overline{C}$ is concave upward on $(0,\infty)$. Therefore, $x = 2000$ yields a minimum.

c. We solve $\overline{C}(x) = C'(x)$. $0.0025x + 80 + \dfrac{10,000}{x} = 0.005x + 80$,

$0.0025x^2 = 10,000$, or $x = 2000$.

d. It appears that we can solve the problem in two ways.

NOTE This can be proved.

53. The demand equation is $p = \sqrt{800 - x} = (800 - x)^{1/2}$. The revenue function is $R(x) = xp = x(800 - x)^{1/2}$. To find the maximum of R, we compute

$$R'(x) = \tfrac{1}{2}(800-x)^{-1/2}(-1)(x) + (800-x)^{1/2}$$
$$= \tfrac{1}{2}(800-x)^{-1/2}[-x + 2(800-x)]$$
$$= \tfrac{1}{2}(800-x)^{-1/2}(1600-3x).$$

Next, $R'(x) = 0$ implies $x = 800$ or $x = 1600/3$ are critical points of R. Next, we compute the values of R given in the following table.

x	0	800	1600/3
$R(x)$	0	0	8709

We conclude that $R(\frac{1600}{3}) = 8709$ is the absolute maximum value. Therefore, the revenue is maximized by producing $1600/3 \approx 533$ dresses.

55. $f(t) = 100\left[\dfrac{t^2 - 4t + 4}{t^2 + 4}\right].$

a. $f'(t) = 100\left[\dfrac{(t^2+4)(2t-4) - (t^2 - 4t + 4)(2t)}{(t^2+4)^2}\right] = \dfrac{400(t^2-4)}{(t^2+4)^2}$

$$= \frac{400(t-2)(t+2)}{(t^2+4)^2}.$$

From the sign diagram for f'

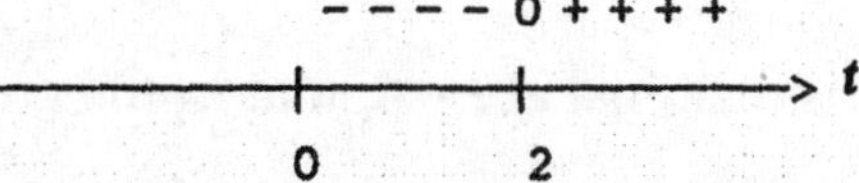

we see that $t = 2$ gives a relative minimum, and we conclude that the oxygen content is the lowest 2 days after the organic waste has been dumped into the pond.

b.

$$f''(t) = 400\left[\frac{(t^2+4)^2(2t) - (t^2-4)2(t^2+4)(2t)}{(t+4)^4}\right] = 400\left[\frac{(2t)(t^2+4)(t^2+4-2t^2+8)}{(t^2+4)^4}\right]$$

$$= -\frac{800t(t^2-12)}{(t^2+4)^3}$$

and $f''(t) = 0$ when $t = 0$ and $t = \pm 2\sqrt{3}$. We reject $t = 0$ and $t = -2\sqrt{3}$. From the sign diagram for f'',

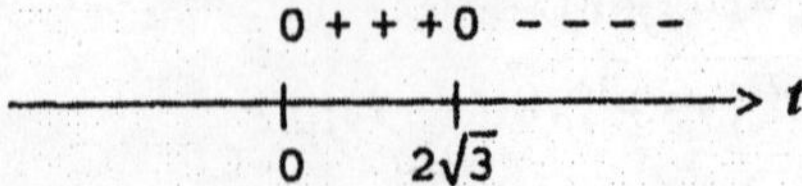

we see that $f'(2\sqrt{3})$ gives an inflection point of f and we conclude that this is an absolute maximum. Therefore, the rate of oxygen regeneration is greatest 3.5 days after the organic waste has been dumped into the pond.

57. We compute $\overline{R}'(x) = \dfrac{xR'(x) - R(x)}{x^2}$. Setting $\overline{R}'(x) = 0$ gives $xR'(x) - R(x) = 0$ or $R'(x) = \dfrac{R(x)}{x} = \overline{R}(x)$, so a critical point of $\overline{R}$ occurs when $\overline{R}(x) = R'(x)$.

Next, we compute

$$\overline{R}''(x) = \frac{x^2[R'(x) + xR''(x) - R'(x)] - [xR'(x) - R(x)](2x)}{x^4} = \frac{R''(x)}{x} < 0.$$

So, by the Second Derivative Test, the critical point does give a maximum revenue.

59. The growth rate is $G'(t) = -0.6t^2 + 4.8t$. To find the maximum growth rate, we compute $G''(t) = -1.2t + 4.8$. Setting $G''(t) = 0$ gives $t = 4$ as a critical point.

t	0	4	8
$G'(t)$	0	9.6	0

From the table, we see that G is maximal at $t = 4$; that is, the growth rate is greatest in 2001.

61. $P'(t) = 0.13089t^2 - 0.534t - 1.59 = 0$ gives

$$t = \frac{0.534 \pm \sqrt{(0.534)^2 - 4(0.13089)(-1.59)}}{2(0.13089)} \approx -2 \text{ or } 6.08$$

We reject the negative root. Since

$$P''(t) = 0.26178t - 0.534 \text{ and } P''(6.08) \approx 1.06 > 0$$

we conclude that t = 6.08 gives a minimum of P, and this number corresponds to approximately early 1970.

63. $R'(t) = -2.133t^2 + 7.52t + 0.2 = 0$ implies

$$t = \frac{-7.52 \pm \sqrt{7.52^2 - 4(-2.133)(0.2)}}{2(-2.133)} \approx -0.026, \text{ or } 3.55.$$

The root –0.026 lies outside the interval [0, 5], and is rejected.

$R''(t) = -4.266t + 7.52$ and $R''(3.55) \approx -7.62 > 0$

and so $t = 3.55$ gives a relative maximum. This is around the middle of 2000. The highest office space rent was given by $R(3.55) \approx 52.79$, or approximately \$52.79/sq ft.

65. a. On [0,3]: $f(t) = 0.6t^2 + 2.4t + 7.6$; $f'(t) = 1.2t + 2.4 = 0$ implies t = -2 which lies outside the interval [0, 3].

t	0	3
$f(t)$	7.6	20.2

On [3, 5]: $f(t) = 3t^2 + 18.8t - 63.2$, $f'(t) = 6t + 18.8 = 0$ implies t = -3.13 which lies outside the interval [3, 5].

t	3	5
$f(t)$	20.2	105.8

On [5, 8]: $f(t) = -3.3167t^3 + 80.1t^2 - 642.583t + 1730.8025$

$f'(t) = -9.9501t^2 + 160.2t - 642.583$

Solving the equation $f'(t) = 0$, we find

$$t = \frac{-160.2 \pm \sqrt{160.2^2 - 4(-9.9501)(642.583)}}{2(-9.9501)} \approx 7.58, \text{ or } 8.52.$$

Only the critical number t = 7.58 lies inside the interval [5, 8].

t	5	7.58	8
$f(t)$	105.8	17.8	18.4

From the tables, we see that the investment peaked when t = 5, that is, in 2000. The amount was \$105.8 billion.

b. The investment (\$7.6 billion) was lowest when t = 0.

67. $R = D^2\left(\frac{k}{2} - \frac{D}{3}\right) = \frac{kD^2}{2} - \frac{D^3}{3}$. $\frac{dR}{dD} = \frac{2kD}{2} - \frac{3D^2}{3} = kD - D^2 = D(k - D)$

Setting $\frac{dR}{dD} = 0$, we have $D = 0$ or $k = D$. We only consider $k = D$ (since $D > 0$). If $k > 0$, $\frac{dR}{dD} > 0$ and if $k < 0$, $\frac{dR}{dD} < 0$. Therefore $k = D$ provides a relative maximum. The nature of the problem suggests that $k = D$ gives the absolute maximum of R. We can also verify this by graphing R.

69. False. Let $f(x) = \begin{cases} |x| & if\ x \neq 0 \\ 1 & if\ x = 0 \end{cases}$ on [-1, 1].

71. False. Let $f(x) = \begin{cases} -x & \text{if } -1 \leq x < 0 \\ \frac{1}{2} & \text{if } 0 \leq x < 1 \end{cases}$. Then f is discontinuous at $x = 0$. But f has an absolute maximum value of 1 attained at $x = -1$.

73. Since $f(x) = c$ for all x, the function f satisfies $f(x) \leq c$ for all x and so f has an absolute maximum at all points of x. Similarly, f has an absolute minimum at all points of x.

75. a. f is not continuous at $x = 0$ because $\lim_{x \to 0} f(x)$ does not exist.

b. $\lim_{x \to 0} f(x) = \lim_{x \to 0^-} \frac{1}{x} = -\infty$ and $\lim_{x \to 0^+} f(x) = \lim_{x \to 0^+} \frac{1}{x} = \infty$.

c.

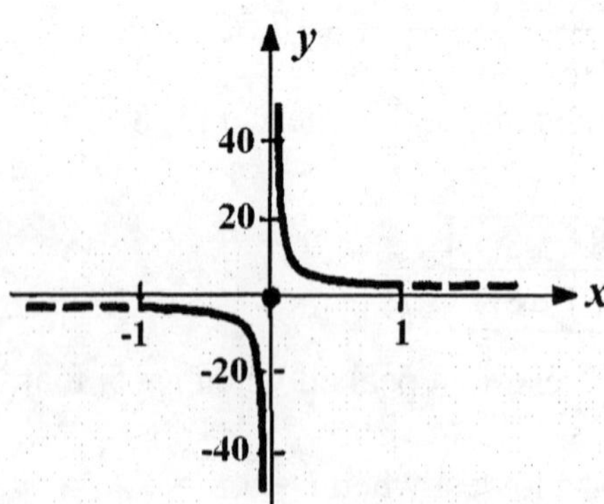

USING TECHNOLOGY EXERCISES 13.4, page 838

1. Absolute maximum value: 145.8985; absolute minimum value: -4.3834

3. Absolute maximum value: 16; absolute minimum value: -0.1257

5. Absolute maximum value: 2.8889; absolute minimum value: 0

7. a.

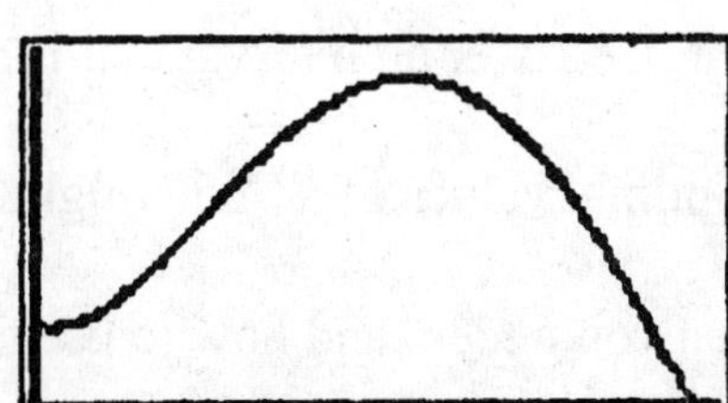

b. 200.1410 banks/yr

9. a.

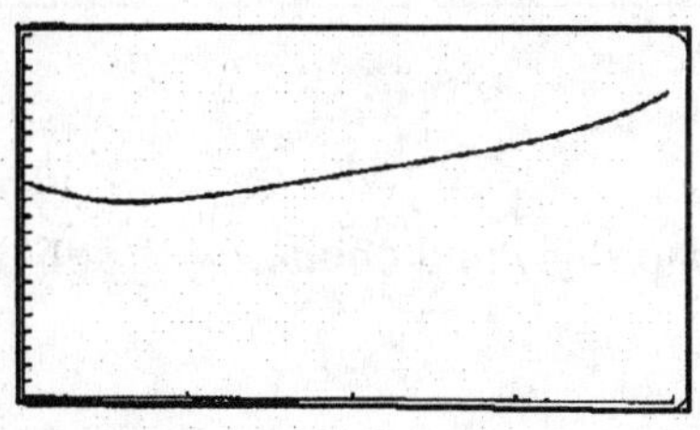

b. 21.51%

. 11.b. 1145

13. a.

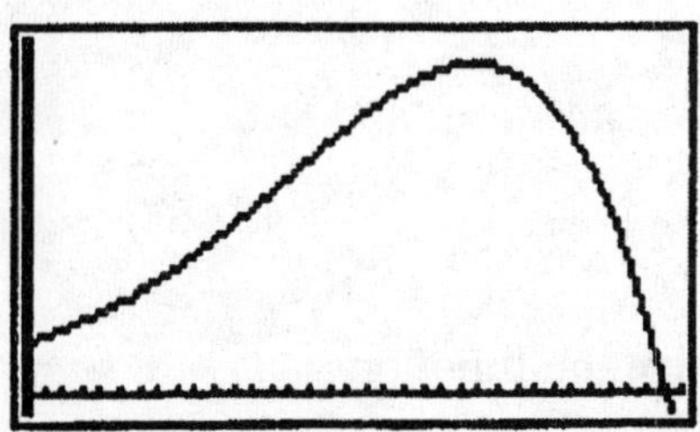

13.5 Problem Solving Tips

1. Follow the guidelines given on page 839 of the text to solve the optimization problems that follow. Remember, Theorem 3 in Section 13.4 provides us with a method of computing the absolute extrema of a continuous function over a closed interval $[a,b]$. If the problem involves a function that is to be optimized over an interval that is not closed, then use the graphical method to find the optimal values of f. You might review Example 4 on page 844 in the text to make sure that you understand how to use the graphical method.

13.5 CONCEPT QUESTIONS, page 846

1. We could solve the problem by sketching the graph of f and checking to see if there is (are) an absolute extrema.

EXERCISES 13.5, page 846

1. Refer to the following figure.

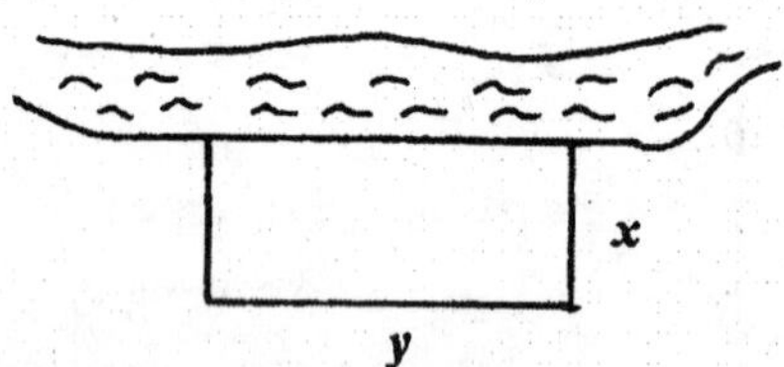

We have $2x + y = 3000$ and we want to maximize the function

$$A = f(x) = xy = x(3000 - 2x) = 3000x - 2x^2$$

on the interval $[0,1500]$. The critical point of A is obtained by solving $f'(x) = 3000 - 4x = 0$, giving $x = 750$. From the table of values

x	0	750	1500
$f(x)$	0	1,125,000	0

we conclude that $x = 750$ yields the absolute maximum value of A.Thus, the required dimensions are 750 × 1500 yards. The maximum area is 1,125,000 sq yd.

3. Let x denote the length of the side made of wood and y the length of the side made of steel. The cost of construction will be $C = 6(2x) + 3y$. But $xy = 800$. So $y = 800/x$ and therefore $C = f(x) = 12x + 3\left(\frac{800}{x}\right) = 12x + \frac{2400}{x}$. To minimize C, we compute

$$f'(x) = 12 - \frac{2400}{x^2} = \frac{12x^2 - 2400}{x^2} = \frac{12(x^2 - 200)}{x^2}.$$

Setting $f'(x) = 0$ gives $x = \pm\sqrt{200}$ as critical points of f. The sign diagram of f'

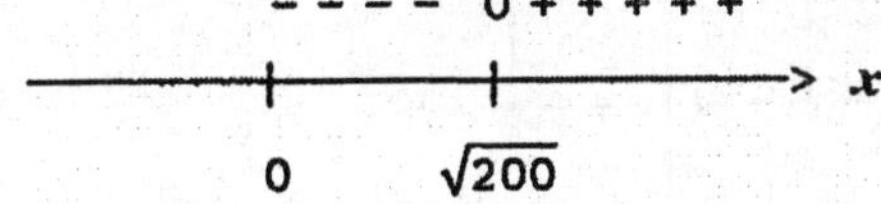

shows that $x = \pm\sqrt{200}$ gives a relative minimum of f. $f''(x) = \frac{4800}{x^3} > 0$ if $x > 0$ and so f is concave upward for $x > 0$. Therefore $x = \sqrt{200} = 10\sqrt{2}$ actually yields the absolute minimum. So the dimensions of the enclosure should be

$$10\sqrt{2}\text{ ft} \times \frac{800}{10\sqrt{2}}\text{ ft, or } 14.1\text{ ft} \times 56.6\text{ ft.}$$

5. Let the dimensions of each square that is cut out be x" × x". Refer to the following diagram.

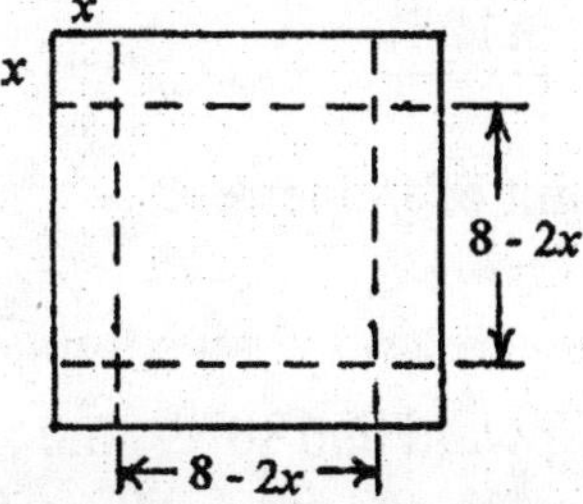

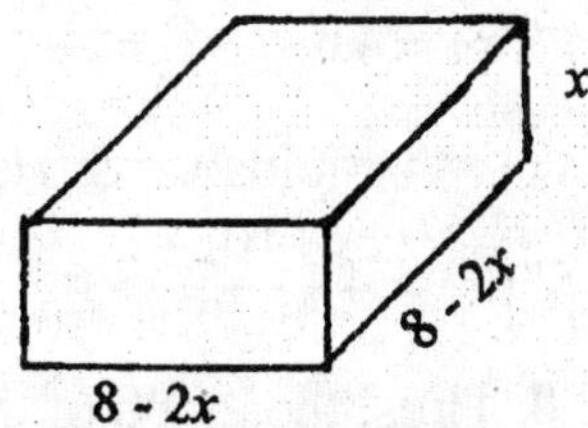

Then the dimensions of the box will be $(8 - 2x)$" by $(8 - 2x)$" by x". Its volume will be $V = f(x) = x(8 - 2x)^2$. We want to maximize f on [0,4].

$$f'(x) = (8 - 2x)^2 + x(2)(8 - 2x)(-2) \qquad \text{[Using the Product Rule.]}$$
$$= (8 - 2x)[(8 - 2x) - 4x] = (8 - 2x)(8 - 6x) = 0$$

if $x = 4$ or 4/3. The latter is a critical point in (0,4).

x	0	4/3	4
$f(x)$	0	1024/27	0

We see that $x = 4/3$ yields an absolute maximum for f. So the dimensions of the box should be $\frac{16}{3}'' \times \frac{16}{3}'' \times \frac{4}{3}''$.

7. Let x denote the length of the sides of the box and y denote its height. Referring to the following figure, we see that the volume of the box is given by $x^2y = 128$. The

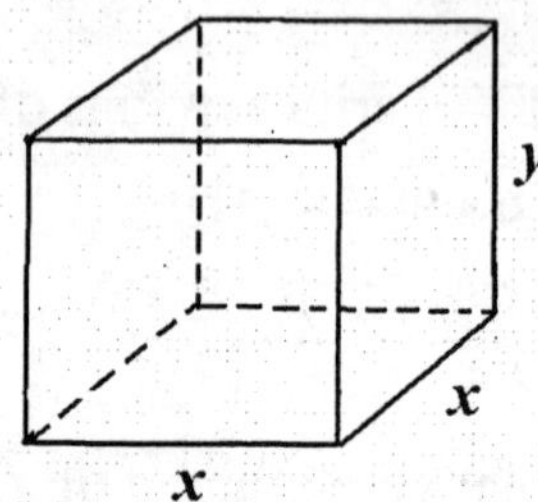

amount of material used is given by

$$S = f(x) = 2x^2 + 4xy$$
$$= 2x^2 + 4x\left(\frac{128}{x^2}\right)$$
$$= 2x^2 + \frac{512}{x} \text{ square inches.}$$

We want to minimize f subject to the condition that $x > 0$. Now

$$f'(x) = 4x - \frac{512}{x^2} = \frac{4x^3 - 512}{x^2} = \frac{4(x^3 - 128)}{x^2}.$$

Setting $f'(x) = 0$ yields $x = 5.04$, a critical point of f. Next,

$$f''(x) = 4 + \frac{1024}{x^3} > 0$$

for all $x > 0$. Thus, the graph of f is concave upward and so $x = 5.04$ yields an absolute minimum of f. Thus, the required dimensions are 5.04" × 5.04" × 5.04".

9. The length plus the girth of the box is $4x + h = 108$ and $h = 108 - 4x$. Then

$$V = x^2h = x^2(108 - 4x) = 108x^2 - 4x^3$$

and $V' = 216x - 12x^2$. We want to maximize V on the interval $[0,27]$. Setting $V'(x) = 0$ and solving for x, we obtain $x = 18$ and $x = 0$. Evaluating $V(x)$ at $x = 0$, $x = 18$, and $x = 27$, we obtain

$$V(0) = 0,\ V(18) = 11{,}664, \text{ and } V(27) = 0$$

Thus, the dimensions of the box are 18" × 18" × 36" and its maximum volume is approximately 11,664 cu in.

11. We take $2\pi r + \ell = 108$. We want to maximize

$$V = \pi r^2 \ell = \pi r^2(-2\pi r + 108) = -2\pi^2 r^3 + 108\pi r^2$$

subject to the condition that $0 \le r \le \frac{54}{\pi}$. Now

$$V'(r) = -6\pi^2 r^2 + 216\pi r = -6\pi r(\pi r - 36).$$

Since $V' = 0$, we find $r = 0$ or $r = 36/\pi$, the critical points of V. From the table

r	0	$36/\pi$	$54/\pi$
V	0	$46{,}656/\pi$	0

we conclude that the maximum volume occurs when $r = 36/\pi \approx 11.5$ inches and $\ell = 108 - 2\pi\left(\frac{36}{\pi}\right) = 36$ inches and its volume is $46{,}656/\pi$ cu in .

13. Let y denote the height and x the width of the cabinet. Then $y = (3/2)x$. Since the volume is to be 2.4 cu ft, we have $xyd = 2.4$, where d *is* the depth of the cabinet.

We have $\quad x\left(\frac{3}{2}x\right)d = 2.4$ or $d = \dfrac{2.4(2)}{3x^2} = \dfrac{1.6}{x^2}$.

The cost for constructing the cabinet is

$$C = 40(2xd + 2yd) + 20(2xy) = 80\left[\frac{1.6}{x} + \left(\frac{3}{2}x\right)\left(\frac{1.6}{x^2}\right)\right] + 40x\left(\frac{3}{2}x\right)$$

$$= \frac{320}{x} + 60x^2.$$

$$C'(x) = -\frac{320}{x^2} + 120x = \frac{120x^3 - 320}{x^2} = 0 \text{ if } x = \sqrt[3]{\frac{8}{3}} = \frac{2}{\sqrt[3]{3}} = \frac{2}{3}\sqrt[3]{9}$$

Therefore, $x = \frac{2}{3}\sqrt[3]{9}$ is a critical point of C. The sign diagram

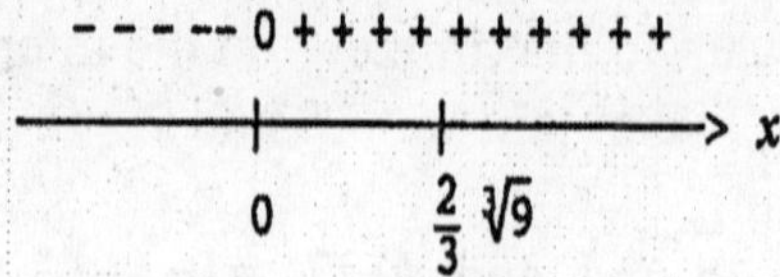

shows that $x = \frac{2}{3}\sqrt[3]{9}$ gives a relative minimum. Next, $C''(x) = \dfrac{640}{x^3} + 120 > 0$ for all $x > 0$ tells us that the graph of C is concave upward. So $x = \frac{2}{3}\sqrt[3]{9}$ yields an absolute minimum. The required dimensions are $\frac{2}{3}\sqrt[3]{9}' \times \sqrt[3]{9}' \times \frac{2}{5}\sqrt[3]{9}'$.

15. We want to maximize the function
$$R(x) = (200 + x)(300 - x) = -x^2 + 100x + 60000.$$
Then $$R'(x) = -2x + 100 = 0$$
gives $x = 50$ and this is a critical point of R. Since $R''(x) = -2 < 0$, we see that $x = 50$ gives an absolute maximum of R. Therefore, the number of passengers should be 250. The fare will then be \$250/passenger and the revenue will be \$62,500.

17. Let x denote the number of people beyond 20 who sign up for the cruise. Then the revenue is $R(x) = (20 + x)(600 - 4x) = -4x^2 + 520x + 12{,}000$. We want to maximize R on the closed bounded interval [0, 70].
$$R'(x) = -8x + 520 = 0 \text{ implies } x = 65,$$
a critical point of R. Evaluating R at this critical point and the endpoints, we have

x	0	65	70
$R(x)$	12,000	28,900	28,800

From this table, we see that R is maximized if $x = 65$. Therefore, 85 passengers will result in a maximum revenue of \$28,900. The fare would be \$340/passenger.

19. We want to maximize $S = kh^2w$. But $h^2 + w^2 = 24^2$ or $h^2 = 576 - w^2$. So $S = f(w) = kw(576 - w^2) = k(576w - w^3)$. Now, setting
$$f'(w) = k(576 - 3w^2) = 0$$
gives $w = \pm\sqrt{192} \approx \pm 13.86$. Only the positive root is a critical point of interest. Next, we find $f''(w) = -6kw$, and in particular,
$$f''(\sqrt{192}) = -6\sqrt{192}\,k < 0,$$

so that $w = \pm\sqrt{192} \approx \pm 13.86$ gives a relative maximum of f. Since $f''(w) < 0$ for $w > 0$, we see that the graph of f is concave downward on $(0,\infty)$ and so, $w = \sqrt{192}$ gives an absolute maximum of f. We find $h^2 = 576 - 192 = 384$ or $h \approx 19.60$. So the width and height of the log should be approximately 13.86 inches and 19.60 inches, respectively.

21. We want to minimize $C(x) = 1.50(10{,}000 - x) + 2.50\sqrt{3000^2 + x^2}$ subject to $0 \le x \le 10{,}000$. Now

$$C'(x) = -1.50 + 2.5(\tfrac{1}{2})(9{,}000{,}000 + x^2)^{-1/2}(2x) = -1.50 + \frac{2.50x}{\sqrt{9{,}000{,}000 + x^2}}$$

$$C'(x) = 0 \Rightarrow 2.5x = 1.50\sqrt{9{,}000{,}000 + x^2}$$

$$6.25x^2 = 2.25(9{,}000{,}000 + x^2) \text{ or } 4x^2 = 20250000,\ x = 2250.$$

x	0	2250	10000
$f(x)$	22500	21000	26101

From the table, we see that $x = 2250$ gives the absolute minimum.

23. The time taken for the flight is

$$T = f(x) = \frac{12 - x}{6} + \frac{\sqrt{x^2 + 9}}{4}.$$

$$f'(x) = -\frac{1}{6} + \frac{1}{4}\left(\frac{1}{2}\right)(x^2 + 9)^{-1/2}(2x) = -\frac{1}{6} + \frac{x}{4\sqrt{x^2 + 9}}$$

$$= \frac{3x - 2\sqrt{x^2 + 9}}{12\sqrt{x^2 + 9}}.$$

Setting $f'(x) = 0$ gives $3x = 2\sqrt{x^2 + 9}$, $9x^2 = 4(x^2 + 9)$ or $5x^2 = 36$. Therefore, $x = \pm 6/\sqrt{5} = \pm 6\sqrt{5}/5$. Only the critical point $x = 6\sqrt{5}/5$ is of interest. The nature of the problem suggests $x \approx 2.68$ gives an absolute minimum for T.

25. The area enclosed by the rectangular region of the racetrack is $A = (\ell)(2r) = 2r\ell$. The length of the racetrack is $2\pi r + 2\ell$, and is equal to 1760. That is,

$$2(\pi r + \ell) = 1760;\ \pi r + \ell = 880, \text{ or } \ell = 880 - \pi r.$$

Therefore, we want to maximize $A = f(r) = 2r(880 - \pi r) = 1760r - 2\pi r^2$.
The restriction on r is $0 \le r \le \frac{880}{\pi}$. To maximize A, we compute

$f'(r) = 1760 - 4\pi r$. Setting $f'(r) = 0$ gives $r = \dfrac{1760}{4\pi} = \dfrac{440}{\pi} \approx 140$. Since

$f(0) = f\left(\frac{880}{\pi}\right) = 0$, we see that the maximum rectangular area is enclosed if we

take $r = \frac{440}{\pi}$ and $\ell = 880 - \pi\left(\frac{440}{\pi}\right) = 440$. So $r = 140$ and $\ell = 440$. The total area

enclosed is $2r\ell + \pi r^2 = 2\left(\frac{440}{\pi}\right)(440) + \pi\left(\frac{440}{\pi}\right)^2 = \frac{2(440)^2}{\pi} + \frac{440^2}{\pi} = \frac{580,800}{\pi} \approx 184,874$ sq ft.

27. Let x denote the number of bottles in each order. We want to minimize

$$C(x) = 200\left(\frac{2,000,000}{x}\right) + \frac{x}{2}(0.40) = \frac{400,000,000}{x} + 0.2x.$$

We compute $C'(x) = -\dfrac{400,000,000}{x^2} + 0.2$. Setting $C'(x) = 0$ gives

$x^2 = \dfrac{400,000,000}{0.2} = 2,000,000,000$, or $x = 44,721$, a critical point of C.

$C''(x) = \dfrac{800,000,000}{x^3} > 0$ for all $x > 0$, and we see that the graph of C is concave upward and so $x = 44,721$ gives an absolute minimum of C. Therefore, there should be $2,000,000/x \approx 45$ orders per year (since we can not have fractions of an order.) Then each order should be for $2,000,000/45 \approx 44,445$ bottles.

29. a. Since the sales are assumed to be at a steady rate and D units are expected to be sold per year, the number of orders/yr is D/x. Since is costs \$$K$ per order, the ordering cost is KD/x. The purchasing cost is PD (cost per item times number purchased). Finally, the holding cost is $(x/2)h$ (the average number on hand times holding cost per item). Therefore

$$C(x) = \frac{KD}{x} + pD + \frac{hx}{2}$$

b. $C'(x) = -\dfrac{KD}{x^2} + \dfrac{h}{2} = 0$

implies $\dfrac{KD}{x^2} = \dfrac{h}{2}$

$$x^2 = \frac{2KD}{h}$$

$$x = \pm\sqrt{\frac{2KD}{h}}$$

We reject the negative root. So $x = \sqrt{\frac{2KD}{h}}$ is the only critical number. Next,

$$C''(x) = \frac{2KD}{x^3} > 0 \text{ for } x > 0$$

So $C''\left(\sqrt{\frac{2KD}{h}}\right) > 0$ and the second derivative test shows that $x = \sqrt{\frac{2KD}{h}}$ does give a relative minimum and because C is concave upward, the absolute minimum.

CHAPTER 13 CONCEPT REVIEW, page 851

1. a. $f(x_1) < f(x_2)$ b. $f(x_1) > f(x_2)$
3. a. $f(x) \leq f(c)$ b. $f(x) \geq f(c)$
5. a. $f'(x)$ b. > 0 c. Concavity d. Relative maximum; relative extremum
7. 0; 0
9. a. $f(x) \leq f(c)$; absolute maximum value b. $f(x) \geq f(c)$; open interval

CHAPTER 13 REVIEW, page 852

1. a. $f(x) = \frac{1}{3}x^3 - x^2 + x - 6$. $f'(x) = x^2 - 2x + 1 = (x - 1)^2$. $f'(x) = 0$ gives $x = 1$, the critical point of f. Now, $f'(x) > 0$ for all $x \neq 1$. Thus, f is increasing on $(-\infty,1) \cup (1,\infty)$.
   b. Since $f'(x)$ does not change sign as we move across the critical point $x = 1$, the First Derivative Test implies that $x = 1$ does not give rise to a relative extremum of f.
   c. $f''(x) = 2(x - 1)$. Setting $f''(x) = 0$ gives $x = 1$ as a candidate for an inflection point of f. Since $f''(x) < 0$ for $x < 1$, and $f''(x) > 0$ for $x > 1$, we see that f is concave downward on $(-\infty,1)$ and concave upward on $(1,\infty)$.
   d. The results of (c) imply that $(1, -\frac{17}{3})$ is an inflection point.

3. a. $f(x) = x^4 - 2x^2$. $f'(x) = 4x^3 - 4x = 4x(x^2 - 1) = 4x(x + 1)(x - 1)$. The sign diagram of f' shows that f is decreasing on $(-\infty,-1) \cup (0,1)$ and increasing on $(-1,0) \cup (1,\infty)$.

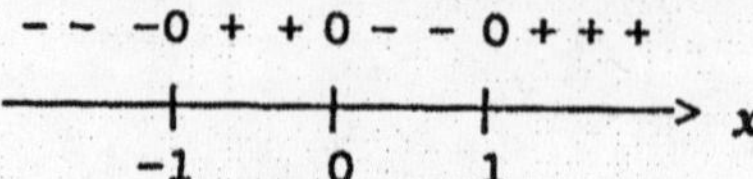

b. The results of (a) and the First Derivative Test show that (-1,-1) and (1,-1) are relative minima and (0,0) is a relative maximum.

c. $f''(x) = 12x^2 - 4 = 4(3x^2 - 1) = 0$ if $x = \pm\sqrt{3}/3$. The sign diagram

+ + + + 0 - - - - - - - - 0 + + + +

$-\frac{\sqrt{3}}{3}$ 0 $\frac{\sqrt{3}}{3}$ x

shows that f is concave upward on $(-\infty, -\sqrt{3}/3) \cup (\sqrt{3}/3, \infty)$ and concave downward on $(-\sqrt{3}/3, \sqrt{3}/3)$.

d. The results of (c) show that $(-\sqrt{3}/3, -5/9)$ and $(\sqrt{3}/3, -5/9)$ are inflection points.

5. a. $f(x) = \dfrac{x^2}{x-1}$. $f'(x) = \dfrac{(x-1)(2x) - x^2(1)}{(x-1)^2} = \dfrac{x^2 - 2x}{(x-1)^2} = \dfrac{x(x-2)}{(x-1)^2}$.

The sign diagram of f'

f' is not defined here

+ + + + + 0 - - - ↓ - - - 0 + + + + + +

0   1   2 x

shows that f is increasing on $(-\infty,0) \cup (2,\infty)$ and decreasing on $(0,1) \cup (1,2)$.

b. The results of (a) show that (0,0) is a relative maximum and (2,4) is a relative minimum.

c. $$f''(x) = \frac{(x-1)^2(2x-2) - x(x-2)2(x-1)}{(x-1)^4} = \frac{2(x-1)[(x-1)^2 - x(x-2)]}{(x-1)^4}$$

$$= \frac{2}{(x-1)^3}.$$

Since $f''(x) < 0$ if $x < 1$ and $f''(x) > 0$ if $x > 1$, we see that f is concave downward on $(-\infty,1)$ and concave upward on $(1,\infty)$.

d. Since $x = 1$ is not in the domain of f, there are no inflection points.

7. $f(x) = (1 - x)^{1/3}$. $f'(x) = -\dfrac{1}{3}(1-x)^{-2/3} = -\dfrac{1}{3(1-x)^{2/3}}$.

The sign diagram for f' is

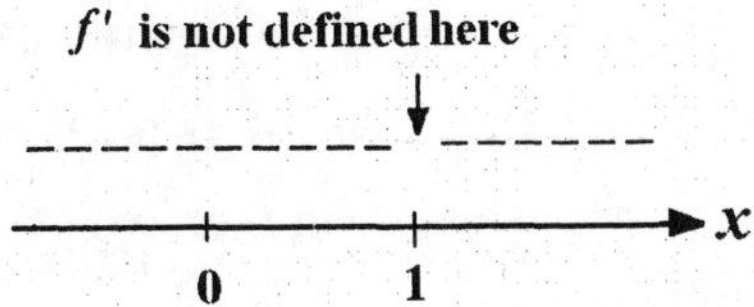

a. f is decreasing on $(-\infty,1) \cup (1,\infty)$.
b. There are no relative extrema.
c. Next, we compute $f''(x) = -\frac{2}{9}(1-x)^{-5/3} = -\frac{2}{9(1-x)^{5/3}}$.

The sign diagram for f'' is

f'' is not defined here

- - - - - - - - - - ↓ + + + + + +

0   1 x

We find f is concave downward on $(-\infty,1)$ and concave upward on $(1,\infty)$.
d. $x = 1$ is a candidate for an inflection point of f. Referring to the sign diagram for f'', we see that $(1,0)$ is an inflection point.

9. a. $f(x) = \frac{2x}{x+1}$. $f'(x) = \frac{(x+1)(2)-2x(1)}{(x+1)^2} = \frac{2}{(x+1)^2} > 0$ if $x \neq -1$.
Therefore f is increasing on $(-\infty,-1) \cup (-1,\infty)$.
b. Since there are no critical points, f has no relative extrema.
c. $f''(x) = -4(x+1)^{-3} = -\frac{4}{(x+1)^3}$. Since $f''(x) > 0$ if $x < -1$ and $f''(x) < 0$ if $x > -1$, we see that f is concave upward on $(-\infty,-1)$ and concave downward on $(-1,\infty)$.
d. There are no inflection points since $f''(x) \neq 0$ for all x in the domain of f.

11. $f(x) = x^2 - 5x + 5$
1. The domain of f is $(-\infty, \infty)$.
2. Setting $x = 0$ gives 5 as the y-intercept.
3. $\lim_{x\to-\infty}(x^2 - 5x + 5) = \lim_{x\to\infty}(x^2 - 5x + 5) = \infty$.
4. There are no asymptotes because f is a quadratic function.
5. $f'(x) = 2x - 5 = 0$ if $x = 5/2$. The sign diagram

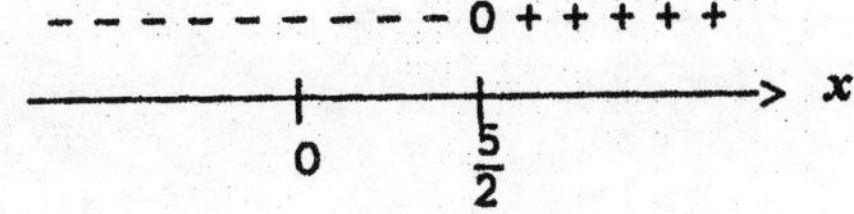

shows that f is increasing on $(\frac{5}{2},\infty)$ and decreasing on $(-\infty,\frac{5}{2})$.

6. The First Derivative Test implies that $(\frac{5}{2},-\frac{5}{4})$ is a relative minimum.
7. $f''(x) = 2 > 0$ and so f is concave upward on $(-\infty, \infty)$.
8. There are no inflection points.

The graph of f follows.

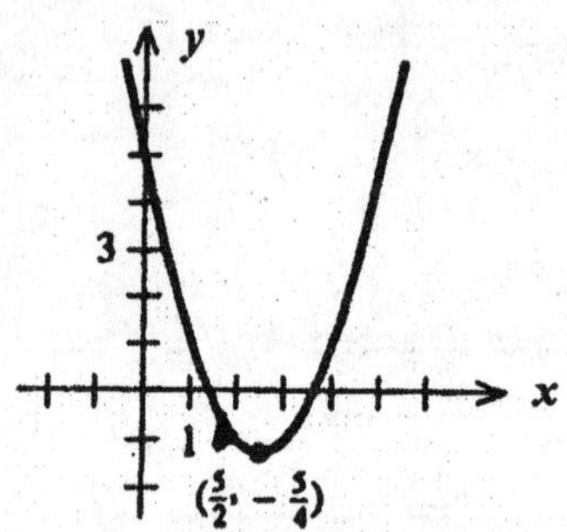

13. $g(x) = 2x^3 - 6x^2 + 6x + 1$.
    1. The domain of g is $(-\infty, \infty)$.
    2. Setting $x = 0$ gives 1 as the y-intercept.
    3. $\lim_{x\to-\infty} g(x) = -\infty$, $\lim_{x\to\infty} g(x) = \infty$.
    4. There are no vertical or horizontal asymptotes.
    5. $g'(x) = 6x^2 - 12x + 6 = 6(x^2 - 2x + 1) = 6(x - 1)^2$. Since $g'(x) > 0$ for all $x \neq 1$, we see that g is increasing on $(-\infty,1) \cup (1,\infty)$.
    6. $g'(x)$ does not change sign as we move across the critical point $x = 1$, so there is no extremum.
    7. $g''(x) = 12x - 12 = 12(x - 1)$. Since $g''(x) < 0$ if $x < 1$ and $g''(x) > 0$ if $x > 1$, we see that g is concave upward on $(1,\infty)$ and concave downward on $(-\infty,1)$.
    8. The point $x = 1$ gives rise to the inflection point $(1,3)$.
    9. The graph of g follows.

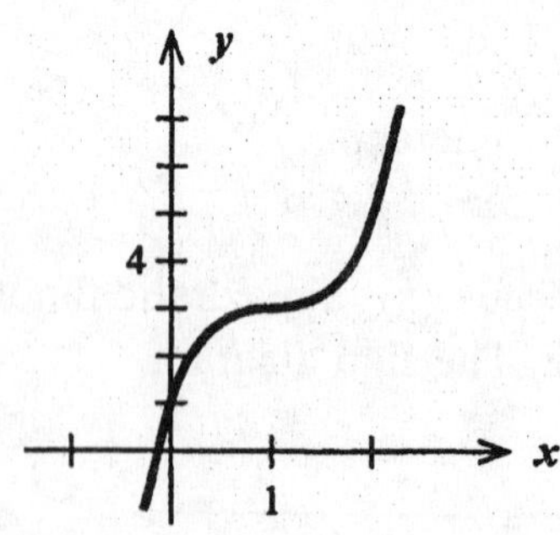

15. $h(x) = x\sqrt{x-2}$.

1. The domain of h is $[2,\infty)$.
2. There are no y-intercepts. Next, setting $y = 0$ gives 2 as the x-intercept.
3. $\lim_{x\to\infty} x\sqrt{x-2} = \infty$.
4. There are no asymptotes.
5. $h'(x) = (x-2)^{1/2} + x(\frac{1}{2})(x-2)^{-1/2} = \frac{1}{2}(x-2)^{-1/2}[2(x-2)+x]$

$$= \frac{3x-4}{2\sqrt{x-2}} > 0 \text{ on } [2,\infty)$$

and so h is increasing on $[2,\infty)$.

6. Since h has no critical points in $(2,\infty)$, there are no relative extrema.
7. $h''(x) = \frac{1}{2}\left[\frac{(x-2)^{1/2}(3) - (3x-4)\frac{1}{2}(x-2)^{-1/2}}{x-2}\right]$

$$= \frac{(x-2)^{-1/2}[6(x-2)-(3x-4)]}{4(x-2)} = \frac{3x-8}{4(x-2)^{3/2}}.$$

The sign diagram for h''

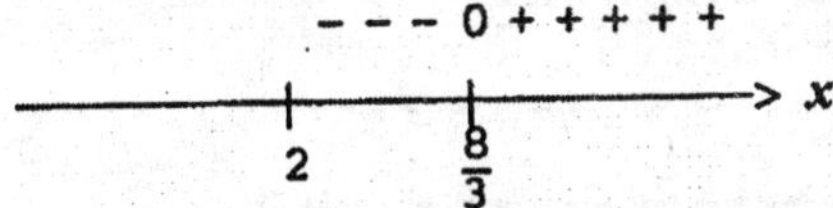

shows that h is concave downward on $(2,\frac{8}{3})$ and concave upward on $(\frac{8}{3},\infty)$.

8. The results of (7) tell us that $(\frac{8}{3},\frac{8\sqrt{6}}{9})$ is an inflection point.

The graph of h follows.

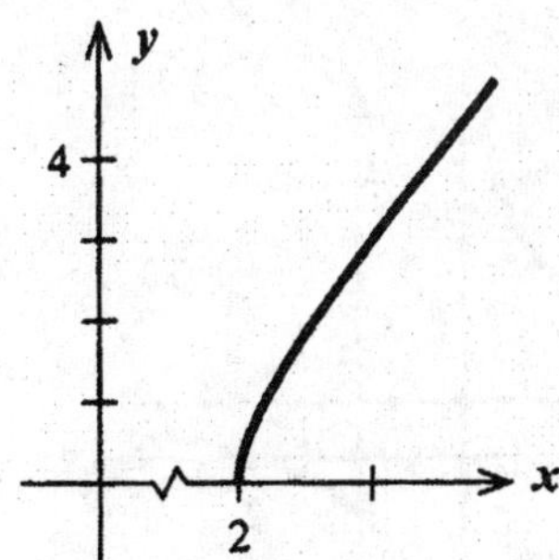

17. $f(x) = \dfrac{x-2}{x+2}$.

1. The domain of f is $(-\infty,-2) \cup (-2,\infty)$.

2. Setting $x = 0$ gives -1 as the y-intercept. Setting $y = 0$ gives 2 as the x-intercept.

3. $\lim_{x\to-\infty} \frac{x-2}{x+2} = \lim_{x\to\infty} \frac{x-2}{x+2} = 1.$

4. The results of (3) tell us that $y = 1$ is a horizontal asymptote. Next, observe that the denominator of $f(x)$ is equal to zero at $x = -2$, but its numerator is not equal to zero there. Therefore, $x = -2$ is a vertical asymptote.

5. $$f'(x) = \frac{(x+2)(1)-(x-2)(1)}{(x+2)^2} = \frac{4}{(x+2)^2}.$$

The sign diagram of f'

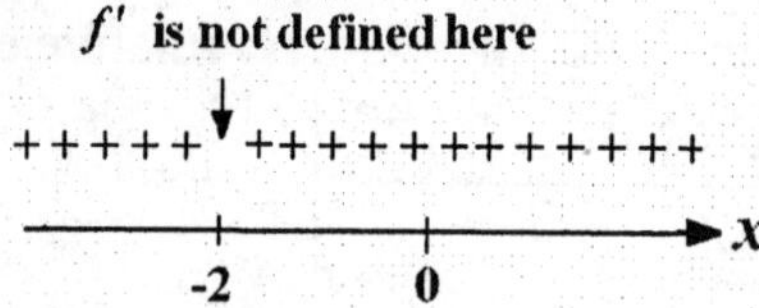

tells us that f is increasing on $(-\infty,-2) \cup (-2,\infty)$.

6. The results of (5) tells us that there are no relative extrema.

7. $f''(x) = -\frac{8}{(x+2)^3}$. The sign diagram of f'' follows

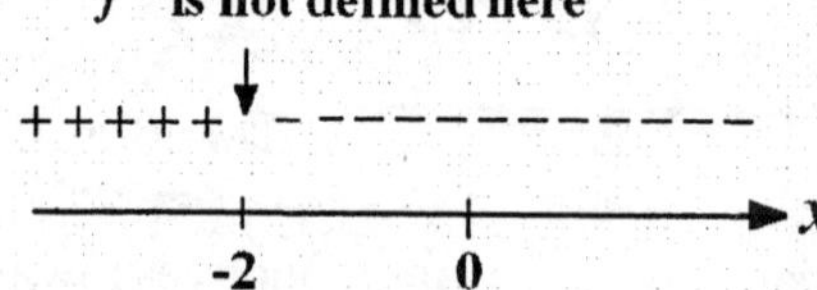

and it shows that f is concave upward on $(-\infty,-2)$ and concave downward on $(-2,\infty)$.

8. There are no inflection points.

The graph of f follows.

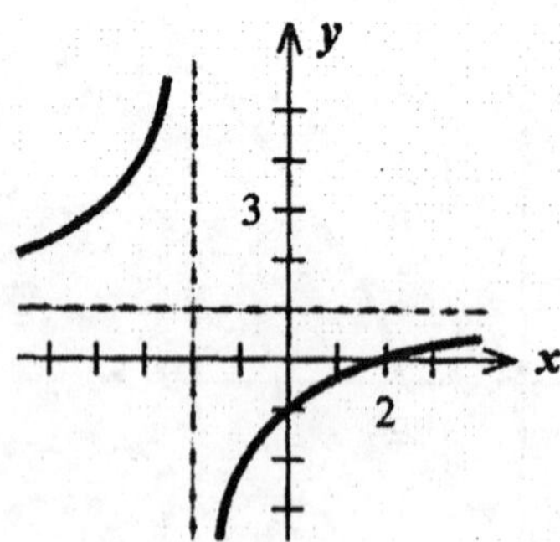

19. $\lim_{x\to-\infty} \dfrac{1}{2x+3} = \lim_{x\to\infty} \dfrac{1}{2x+3} = 0$ and so $y = 0$ is a horizontal asymptote. Since the denominator is equal to zero at $x = -3/2$, but the numerator is not equal to zero there, we see that $x = -3/2$ is a vertical asymptote.

21. $\lim_{x\to-\infty} \dfrac{5x}{x^2-2x-8} = \lim_{x\to\infty} \dfrac{5x}{x^2-2x-8} = 0$ and so $y = 0$ is a horizontal asymptote. Next, note that the denominator is zero if $x^2 - 2x - 8 = (x - 4)(x + 2) = 0$, or $x = -2$ or $x = 4$. Since the numerator is not equal to zero at these points, we see that $x = -2$ and $x = 4$ are vertical asymptotes.

23. $f(x) = 2x^2 + 3x - 2$; $f'(x) = 4x + 3$. Setting $f'(x) = 0$ gives $x = -3/4$ as a critical point of f. Next, $f''(x) = 4 > 0$ for all x, so f is concave upward on $(-\infty, \infty)$. Therefore, $f(-\frac{3}{4}) = -\frac{25}{8}$ is an absolute minimum of f. There is no absolute maximum.

25. $g(t) = \sqrt{25-t^2} = (25-t^2)^{1/2}$. Differentiating $g(t)$, we have

$$g'(t) = \tfrac{1}{2}(25-t^2)^{-1/2}(-2t) = -\frac{t}{\sqrt{25-t^2}}.$$

Setting $g'(t) = 0$ gives $t = 0$ as a critical point of g. The domain of g is given by solving the inequality $25 - t^2 \geq 0$ or $(5 - t)(5 + t) \geq 0$ which implies that $t \in [-5,5]$. From the table

t	-5	0	5
$g(t)$	0	5	0

we conclude that $g(0) = 5$ is the absolute maximum of g and $g(-5) = 0$ and $g(5) = 0$ is the absolute minimum value of g.

27. $h(t) = t^3 - 6t^2$. $h'(t) = 3t^2 - 12t = 3t(t-4) = 0$ if $t = 0$ or $t = 4$, critical points of h. But only $t = 4$ lies in (2,5).

t	2	4	5
$h(t)$	-16	-32	-25

From the table, we see that there is an absolute minimum at (4,-32) and an absolute maximum at (2,-16).

29. $f(x) = x - \frac{1}{x}$ on [1,3]. $f'(x) = 1 + \frac{1}{x^2}$. Since $f'(x)$ is never zero, f has no critical point.

x	1	3
$f(x)$	0	$\frac{8}{3}$

We see that $f(1) = 0$ is the absolute minimum value and $f(3) = 8/3$ is the absolute maximum value.

31. $f(s) = s\sqrt{1-s^2}$ on [-1,1]. The function f is continuous on [-1,1] and differentiable on (-1,1). Next,

$$f'(s) = (1-s^2)^{1/2} + s(\tfrac{1}{2})(1-s^2)^{-1/2}(-2s) = \frac{1-2s^2}{\sqrt{1-s^2}}.$$

Setting $f'(s) = 0$, we have $s = \pm\sqrt{2}/2$, giving the critical points of f. From the table

x	-1	$-\sqrt{2}/2$	$\sqrt{2}/2$	1
$f(x)$	0	-1/2	1/2	0

we see that $f(-\sqrt{2}/2) = -1/2$ is the absolute minimum value and $f(\sqrt{2}/2) = 1/2$ is the absolute maximum value of f.

33. We want to maximize $P(x) = -x^2 + 8x + 20$. Now, $P'(x) = -2x + 8 = 0$ if $x = 4$, a critical point of P. Since $P''(x) = -2 < 0$, the graph of P is concave downward. Therefore, the critical point $x = 4$ yields an absolute maximum. So, to maximize profit, the company should spend $4000 on advertising per month.

35. a. $N'(t) = 16.25t + 24.625$; Since $N'(t) > 0$ for $0 < t < 3$, we see that sales of camera phones is always increasing between 2002 and 2005.
b. $N''(t) = 16.25$; Since $N''(t) > 0$ for $0 < t < 3$, we see that the rate of sales is increasing between 2002 and 2005.

37. $C(x) = 0.0001x^3 - 0.08x^2 + 40x + 5000$; $C'(x) = 0.0003x^2 - 0.16x + 40$; $C''(x) = 0.0006x - 0.16$. Thus, $x = 266.67$ is a candidate for an inflection point of C. The sign diagram for C'' is

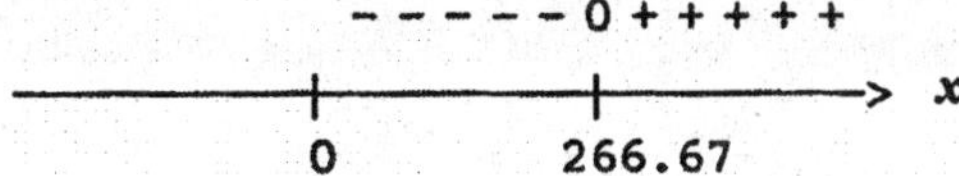

We see that the inflection point of C is (266.67, 11,874.08).

39. The revenue is $R(x) = px = x(-0.0005x^2 + 60) = -0.0005x^3 + 60x$. Therefore, the total profit is $P(x) = R(x) - C(x) = -0.0005x^3 + 0.001x^2 + 42x - 4000$. $P'(x) = -0.0015x^2 + 0.002x + 42$. Setting $P'(x) = 0$, we have $3x^2 - 4x - 84{,}000 = 0$. Solving for x, we find

$$x = \frac{4 \pm \sqrt{16 - 4(3)(84{,}000)}}{2(3)} = \frac{4 \pm 1004}{6} = 168, \text{ or } -167.$$

We reject the negative root. Next, $P''(x) = -0.003x + 0.002$ and $P''(168) = -0.003(168) + 0.002 = -0.502 < 0$. By the Second Derivative Test, $x = 168$ gives a relative maximum. Therefore, the required level of production is 168 DVDs.

41. a. $C(x) = 0.001x^2 + 100x + 4000$.

$$\overline{C}(x) = \frac{C(x)}{x} = \frac{0.001x^2 + 100x + 4000}{x} = 0.001x + 100 + \frac{4000}{x}.$$

b. $\overline{C}'(x) = 0.001 - \dfrac{4000}{x^2} = \dfrac{0.001x^2 - 4000}{x^2} = \dfrac{0.001(x^2 - 4{,}000{,}000)}{x^2}$.

Setting $\overline{C}'(x) = 0$ gives $x = \pm\, 2000$. We reject the negative root.

The sign diagram of $\overline{C}'$ shows that $x = 2000$ gives rise to a relative minimum of $\overline{C}$. Since $\overline{C}''(x) = \dfrac{8000}{x^3} > 0$ if $x > 0$, we see that $\overline{C}$ is concave upward on $(0,\infty)$. So $x = 2000$ yields an absolute minimum. So the required production level is 2000 units.

43. a. $P'(t) = -0.0006t^2 + 0.036t - 0.36$; Setting $P'(t) = 0$ gives $-0.0006t^2 + 0.036t - 0.36 = 0$, or $t^2 - 60t + 600 = 0$

So $t = \dfrac{60 \pm \sqrt{60^2 - 4(1)(600)}}{2} \approx 12.7$ or 47.3

We reject the root 47.3 because it lies outside [0, 30]. The sign diagram for P' follows.

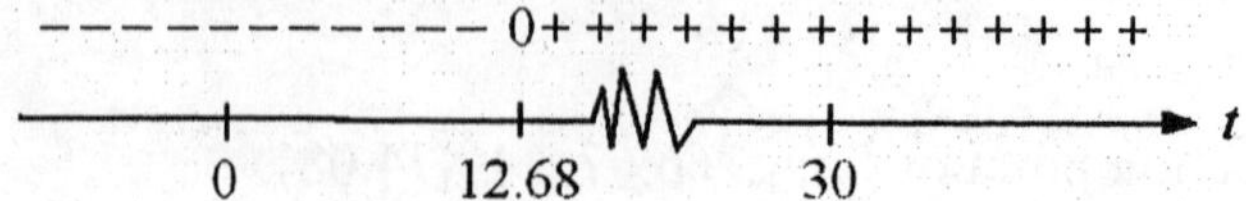

So P is decreasing on (0, 12.7) and increasing on (12.7, 30).

b. The absolute minimum of P occurs at $t = 12.7$ and $P(12.7) \approx 7.9$.

c. The percent of women 65 years and older in the workforce was decreasing from 1970 to Sept 1982 and increasing from Sept. 1982 to 2000. It reached a minimum value of 7.9% in Sept. 1982.

45. The volume is $V = f(x) = x(10-2x)^2$ cubic units for $0 \le x \le 5$.

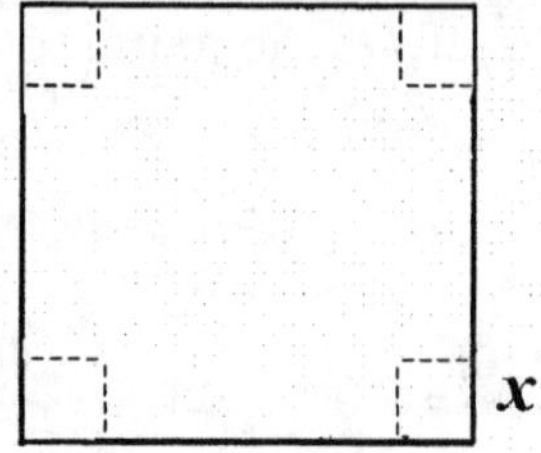

To maximize V, we compute

$$f'(x) = 12x^2 - 80x + 100 = 4(3x^2 - 20x + 25) = 4(3x-5)(x-5).$$

Setting $f'(x) = 0$ gives $x = 5/3$, or 5 as critical points of f. From the table

x	0	5/3	5
$f(x)$	0	$2000/27 \approx 74.07$	0

We see that the box has a maximum volume of 74.07 cu in.

47. Refer to the following picture.

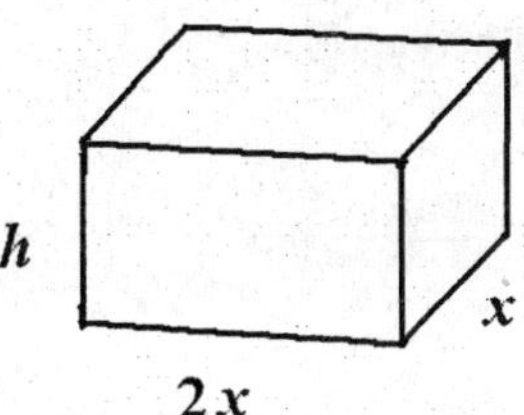

$C(x) = 30(2)(2x)(x) + 20(2)(2xh + xh) = 120x^2 + 120xh$. But $x(2x)h = 4$, or $h = \frac{2}{x^2}$.

Therefore, $C(x) = 120x^2 + 120x\left(\frac{2}{x^2}\right) = 120x^2 + \frac{240}{x}$

$$C'(x) = 240x - \frac{240}{x^2}.$$

Setting $C'(x) = 0$ gives $240x - \frac{240}{x^2} = 0$, or $x^3 = 1$. Therefore, $x = 1$.

$C''(x) = 240 + \frac{480}{x^3}$. In particular, $C''(1) > 0$. Therefore, the cost is minimized by taking $x = 1$. The required dimensions are 1 ft × 2 ft × 2 ft.

49. $f(x) = ax^2 + bx + c$; $f'(x) = 2ax + b = 2a\left(x + \frac{b}{2a}\right)$. Then f' is continuous everywhere and has a zero at $x = -\frac{b}{2a}$. The sign diagram of f' is

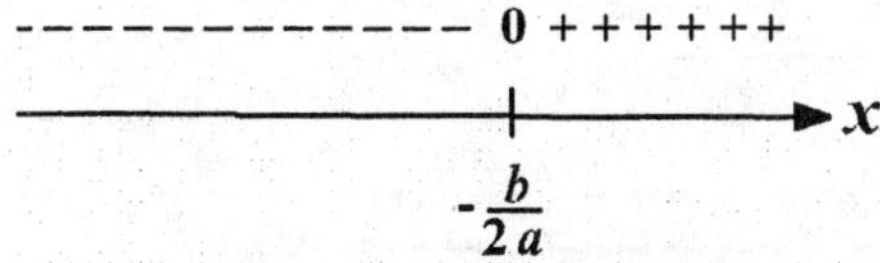

when $a > 0$, or

+ + + + + + + + + + + 0 - - - - - -

x

$-\frac{b}{2a}$

when $a < 0$. Therefore, if $a > 0$, f is decreasing on $(-\infty, -\frac{b}{2a})$ and increasing on $(-\frac{b}{2a}, \infty)$, and if $a < 0$, f is increasing on $(-\infty, -\frac{b}{2a})$ and decreasing on $(-\frac{b}{2a}, \infty)$.

CHAPTER 13, BEFORE MOVING ON, page 854

1. $f'(x) = \frac{(1-x)(2x) - x^2(-1)}{(1-x)^2} = \frac{2x - 2x^2 + x^2}{(1-x)^2} = \frac{x(2-x)}{(1-x)^2}$; f' is not defined at = 1 and has zeros at $x = 0$ and $x = 2$. The sign diagram of f follows:

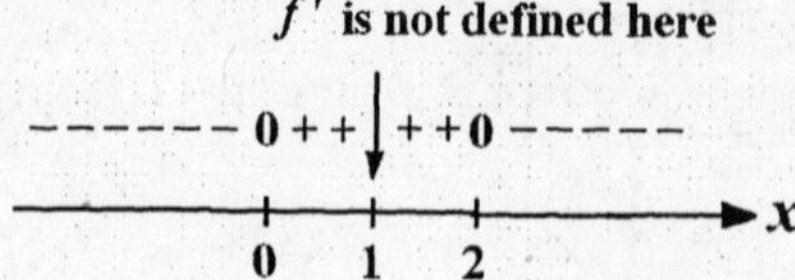

We see that f is decreasing on $(-\infty, 0) \cup (2, \infty)$ and increasing on $(0,1) \cup (1,2)$.

2. $f'(x) = 4x - 4x^{-2/3} = 4x^{-2/3}(x^{5/3} - 1) = \dfrac{4(x^{5/3} - 1)}{x^{2/3}}$; f' is discontinuous at $x = 0$ and has a zero where $x^{5/3} = 1$ or $x = 1$. Therefore, f has critical numbers at 0 and 1. The sign diagram for f' follows:

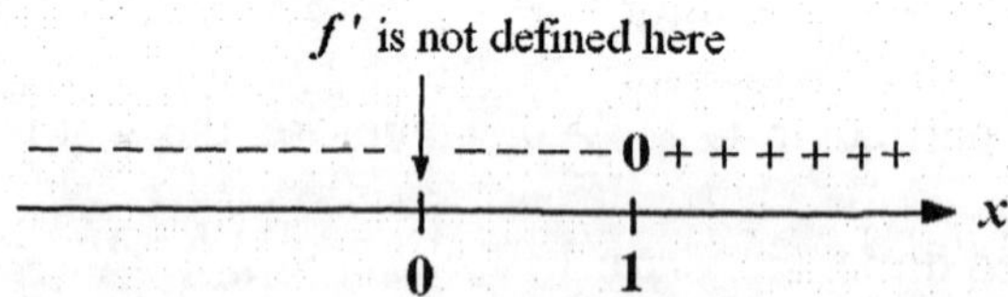

We see that $x = 1$ gives a relative minimum. Since $f(1) = 2 - 12 = -10$, the relative minimum is (1,-10). There are no relative maxima.

3. $f'(x) = x^2 - \frac{1}{2}x - \frac{1}{2}$; $f''(x) = 2x - \frac{1}{2}$; $f''(x) = 0$ gives $x = \frac{1}{4}$. The sign diagram of f'' follows:

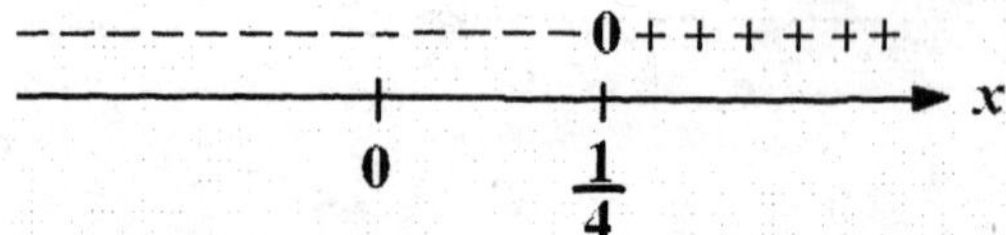

We see that f is concave downward on $(-\infty, \frac{1}{4})$ and concave upward on $(\frac{1}{4}, \infty)$. Since

$$f(\tfrac{1}{4}) = \tfrac{1}{3}(\tfrac{1}{4})^3 - \tfrac{1}{4}(\tfrac{1}{4})^2 - \tfrac{1}{2}(\tfrac{1}{4}) + 1 = \frac{83}{96}$$

the inflection point is $(\frac{1}{4}, \frac{83}{96})$.

4. $f(x) = 2x^3 - 9x^2 + 12x - 1$
   1. Domain of f is $(-\infty, \infty)$.
   2. Setting $y = f(x) = 0$ gives -1 as the y-intercept of f.
   3. $\lim\limits_{x \to -\infty} f(x) = -\infty$ and $\lim\limits_{x \to \infty} f(x) = \infty$.

4. There are no asymptotes.
5. $f'(x) = 6x^2 - 18x + 12 = 6(x^2 - 3x + 2) = 6(x-2)(x-1)$. The sign diagram of f'

shows that f is increasing on $(-\infty, 1) \cup (2, \infty)$ and decreasing on (1, 2).
6. We see that (1, 4) is a relative maximum and (2, 3) is a relative minimum.
7. $f''(x) = 12x - 18 = 6(2x-3)$. The sign diagram of f'' is

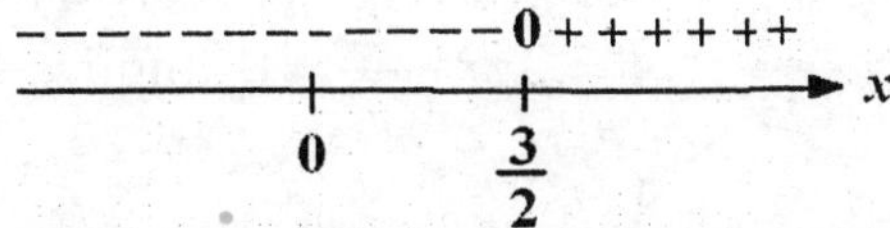

and shows that f is concave downward on $(-\infty, \frac{3}{2})$ and concave upward on $(\frac{3}{2}, \infty)$.
8. $f(\frac{3}{2}) = 2(\frac{3}{2})^3 - 9(\frac{3}{2})^2 + 12(\frac{3}{2}) - 1 = \frac{7}{2}$; So $(\frac{3}{2}, \frac{7}{2})$ is an inflection point of f.

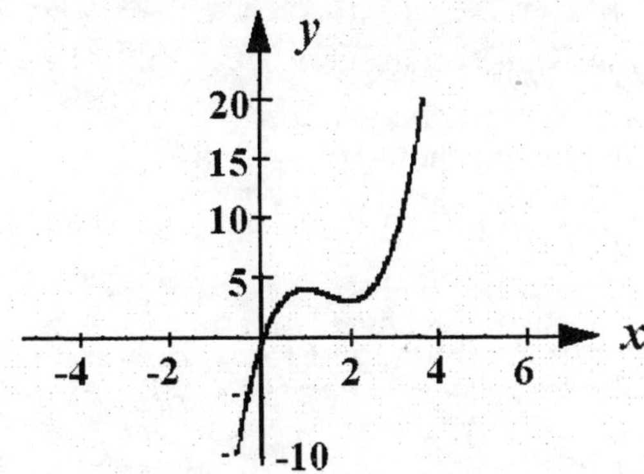

5. f is continuous on a closed interval [-2, 3]. $f'(x) = 6x^2 + 6x = 6x(x+1)$. The critical numbers of f are –1 and 0.

x	-2	-1	0	3
y	-5	0	-1	80

The absolute maximum value of f is 80; the absolute minimum value is –5.

6. The amount of material used (the surface area) is

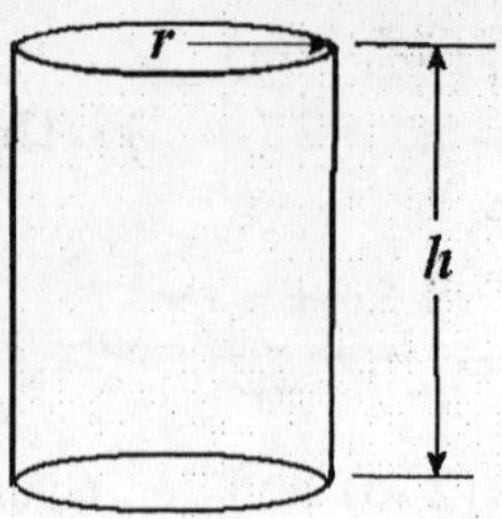

$A = \pi r^2 + 2\pi rh$. But $V = \pi r^2 h = 1$ and so $h = \dfrac{1}{\pi r^2}$. Therefore,

$A = \pi r^2 + 2\pi rh\left(\dfrac{1}{\pi r^2}\right) = \pi r^2 + \dfrac{2}{r}$; $A' = 2\pi r - \dfrac{2}{r^2} = 0$ implies

$2\pi r = \dfrac{2}{r^2}$, $r^3 = \dfrac{2}{r^2}$, $r^3 = \dfrac{1}{\pi}$, or $r = \dfrac{1}{\sqrt[3]{\pi}}$; Since $A'' = 2\pi + \dfrac{4}{r^3} > 0$ for $r > 0$, we see

that $r = \dfrac{1}{\sqrt[3]{\pi}}$ does give an absolute minimum. Also

$$h = \frac{1}{\pi r^2} = \frac{1}{\pi} \cdot \pi^{2/3} = \frac{1}{\pi^{1/3}} = \frac{1}{\sqrt[3]{\pi}}$$

Therefore both the radius and height should be $\dfrac{1}{\sqrt[3]{\pi}}$ ft.

CHAPTER 14

14.1 Problem Solving Tips

1. Be careful to remember the order of operations when you work with exponents. Note that $-5^2 \neq 25$, but rather $-5^2 = -(5)^2 = -25$. On the other hand $(-5)^2 = 25$.

2. $b^{-x} = \dfrac{1}{b^x} = \left(\dfrac{1}{b}\right)^x$. If $b > 1$, then $0 < \frac{1}{b} < 1$. So the graph of b^{-x} for $b > 1$ is similar to the graph of $y = (1/2)^x$

14.1 CONCEPT QUESTIONS, page 859

1. $f(x) = b^x$; $b > 0$, $b \neq 1$. Its domain is $(-\infty, \infty)$.

EXERCISES 14.1, page 859

1. a. $4^{-3} \times 4^5 = 4^{-3+5} = 4^2 = 16$ b. $3^{-3} \times 3^6 = 3^{6-3} = 3^3 = 27$.

3. a. $9(9)^{-1/2} = \dfrac{9}{9^{1/2}} = \dfrac{9}{3} = 3$. b. $5(5)^{-1/2} = 5^{1/2} = \sqrt{5}$.

5. a. $\dfrac{(-3)^4(-3)^5}{(-3)^8} = (-3)^{4+5-8} = (-3)^1 = -3$. b. $\dfrac{(2^{-4})(2^6)}{2^{-1}} = 2^{-4+6+1} = 2^3 = 8$.

7. a. $\dfrac{5^{3.3} \cdot 5^{-1.6}}{5^{-0.3}} = \dfrac{5^{3.3-1.6}}{5^{-0.3}} = 5^{1.7+(0.3)} = 5^2 = 25$.

   b. $\dfrac{4^{2.7} \cdot 4^{-1.3}}{4^{-0.4}} = 4^{2.7-1.3+0.4} = 4^{1.8} \approx 12.1257$.

9. a. $(64x^9)^{1/3} = 64^{1/3}(x^{9/3}) = 4x^3$.
   b. $(25x^3y^4)^{1/2} = 25^{1/2}(x^{3/2})(y^{4/2}) = 5x^{3/2}y^2 = 5xy^2\sqrt{x}$.

11. a. $\dfrac{6a^{-5}}{3a^{-3}} = 2a^{-5+3} = 2a^{-2} = \dfrac{2}{a^2}$. b. $\dfrac{4b^{-4}}{12b^{-6}} = \dfrac{1}{3}b^{-4+6} = \dfrac{1}{3}b^2$.

13. a. $(2x^3y^2)^3 = 2^3 \times x^{3(3)} \times y^{2(3)} = 8x^9y^6$.
b. $(4x^2y^2z^3)^2 = 4^2 \times x^{2(2)} \times y^{2(2)} \times z^{3(2)} = 16x^4y^4z^6$.

15. a. $\dfrac{5^0}{(2^{-3}x^{-3}y^2)^2} = \dfrac{1}{2^{-3(2)}x^{-3(2)}y^{2(2)}} = \dfrac{2^6x^6}{y^4} = \dfrac{64x^6}{y^4}$.
b. $\dfrac{(x+y)(x-y)}{(x-y)^0} = (x+y)(x-y)$.

17. $6^{2x} = 6^4$ if and only if $2x = 4$ or $x = 2$.

19. $3^{3x-4} = 3^5$ if and only if $3x - 4 = 5$, $3x = 9$, or $x = 3$.

21. $(2.1)^{x+2} = (2.1)^5$ if and only if $x + 2 = 5$, or $x = 3$.

23. $8^x = (\frac{1}{32})^{x-2}$, $(2^3)^x = (32)^{2-x} = (2^5)^{2-x}$, so $2^{3x} = 2^{5(2-x)}$, $3x = 10 - 5x$, $8x = 10$, or $x = 5/4$.

25. Let $y = 3^x$, then the given equation is equivalent to

$$y^2 - 12y + 27 = 0$$
$$(y-9)(y-3) = 0$$

giving $y = 3$ or 9. So $3^x = 3$ or $3^x = 9$, and therefore, $x = 1$ or $x = 2$.

27. $y = 2^x, y = 3^x$, and $y = 4^x$

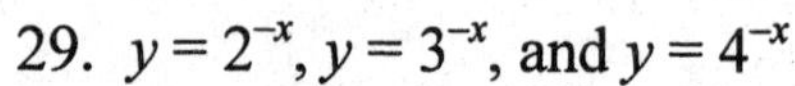
29. $y = 2^{-x}, y = 3^{-x}$, and $y = 4^{-x}$

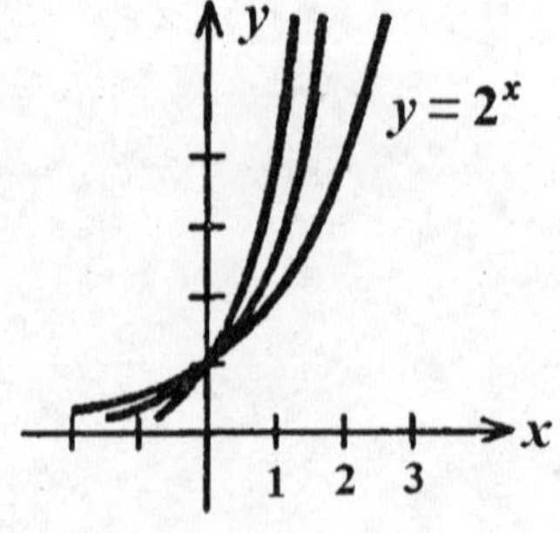

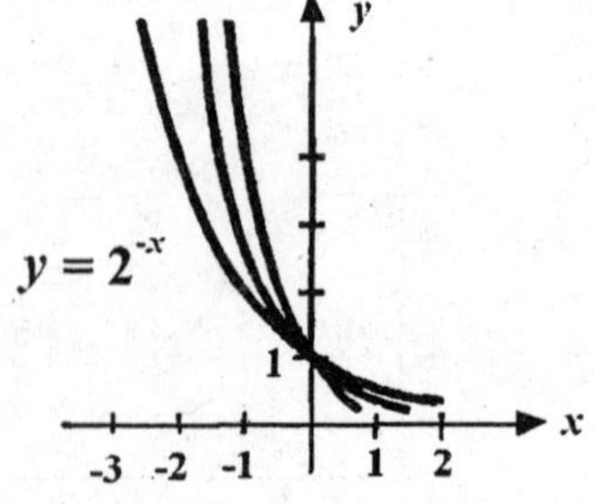

31. $y = 4^{0.5x}, y = 4x$, and $y = 4^{2x}$

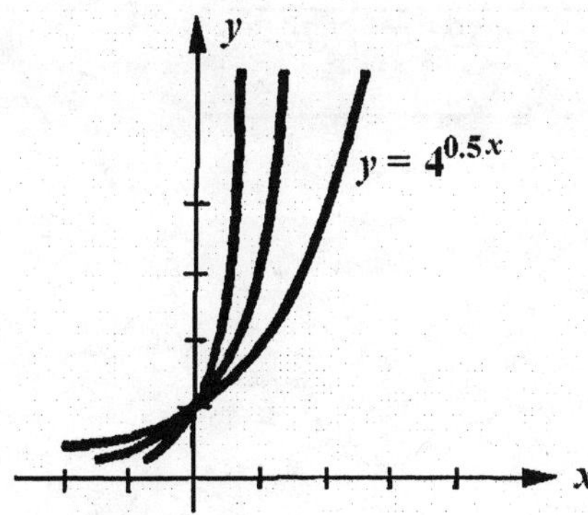

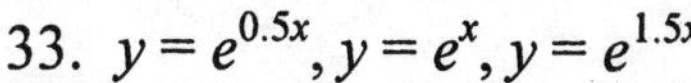
33. $y = e^{0.5x}, y = e^{x}, y = e^{1.5x}$

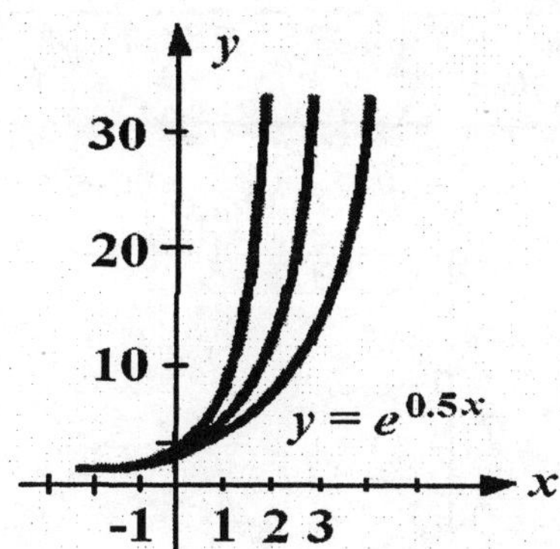

35. $y = 0.5e^{-x}, y = e^{-x}$, and $y = 2e^{-x}$

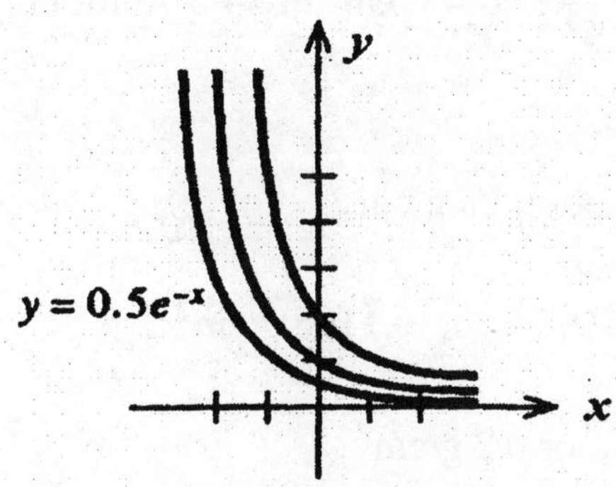

37. a. $R(t) = 26.3e^{-0.016t}$; In 1982, the rate was $R(0) = 26.3\%$. In 1986, the rate was $R(4) = 24.7\%$. In 1994, the rate was $R(12) = 21.7\%$. In 2000, the rate was $R(18) = 19.7\%$.

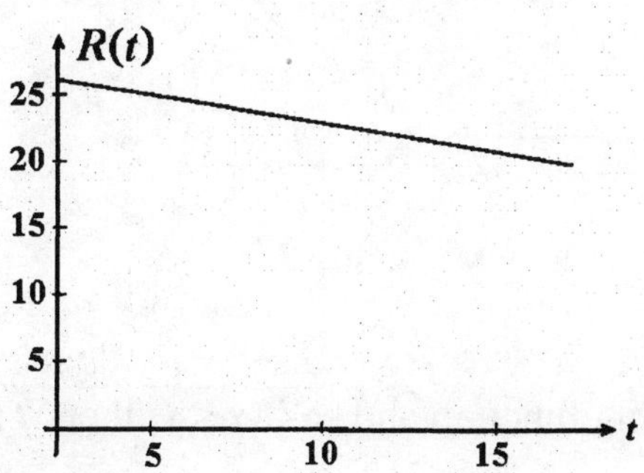

39. a.

Year	0	1	2	3	4	5
Number (billions)	0.45	0.80	1.41	2.49	4.39	7.76

b.

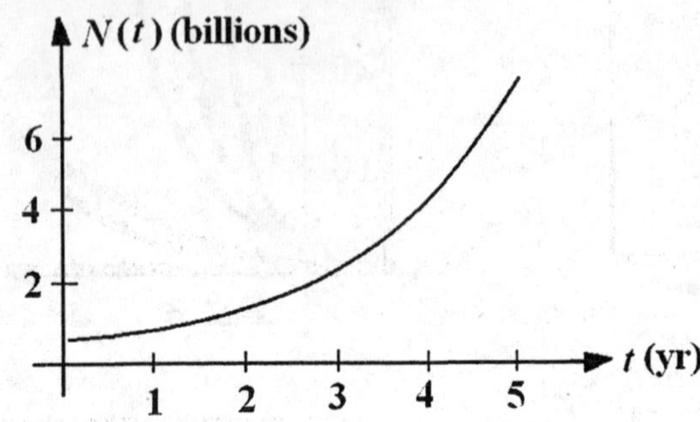

41. $N(t)=\dfrac{35.5}{1+6.89e^{-0.8674t}}$; $N(6)=\dfrac{35.5}{1+6.89e^{-0.8674(6)}}\approx 34.2056$, or 34.21 million.

43. a. The concentration initially is given by

$$N(0)=0.08+0.12(1-e^{-0.02(0)})=0.08\text{ , or }0.08\text{ g/cm}^3.$$

b. The concentration after 20 seconds is given by

$$N(20)=0.08+0.12(1-e^{-0.02(20)})=0.11956,\text{ or }0.1196\text{ g/cm}^3.$$

c. The concentration in the long run is given by

$$\lim_{t\to\infty}x(t)=\lim_{t\to\infty}[0.08+0.12(1-e^{-0.02t})]=0.2,\text{ or }0.2\text{ g/cm}^3.$$

d.

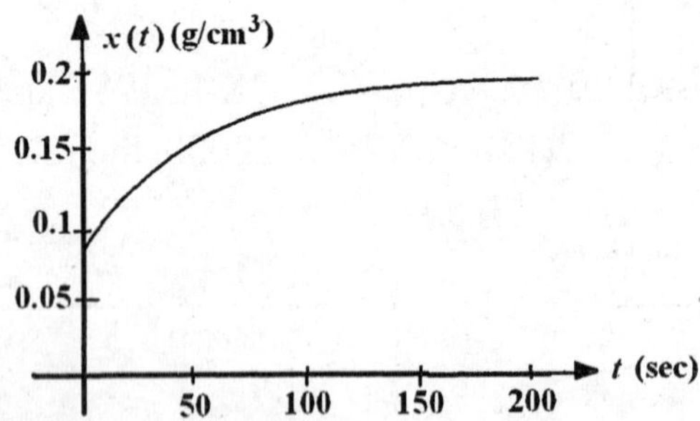

45. False. $(x^2+1)^3=x^6+3x^4+3x^2+1$.

47. True. $f(x)=e^x$ is an increasing function and so if $x<y$, then $f(x)<f(y)$, or $e^x<e^y$.

USING TECHNOLOGY EXERCISES 14.1, page 862

1.

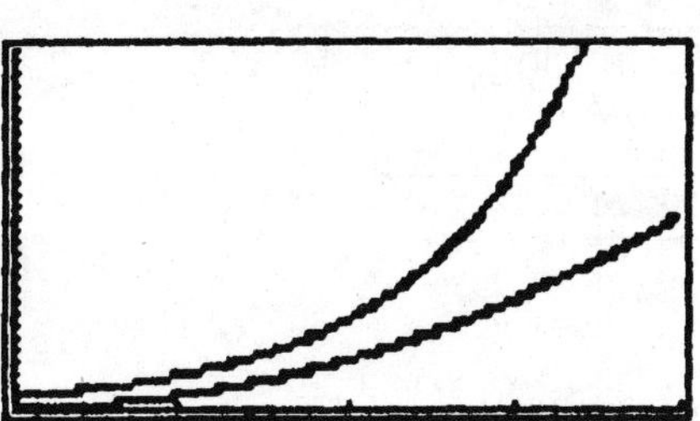

3.

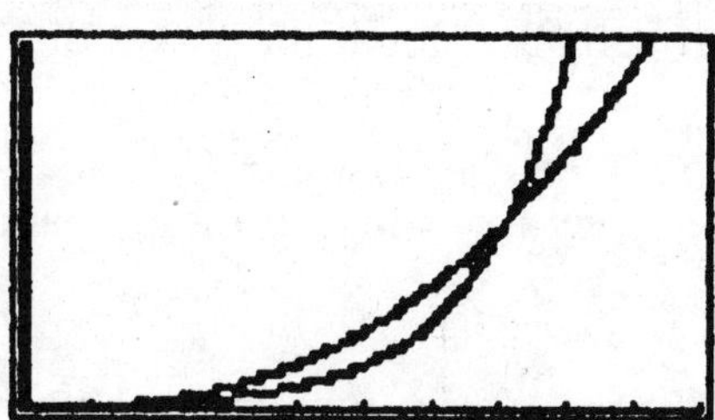

5.

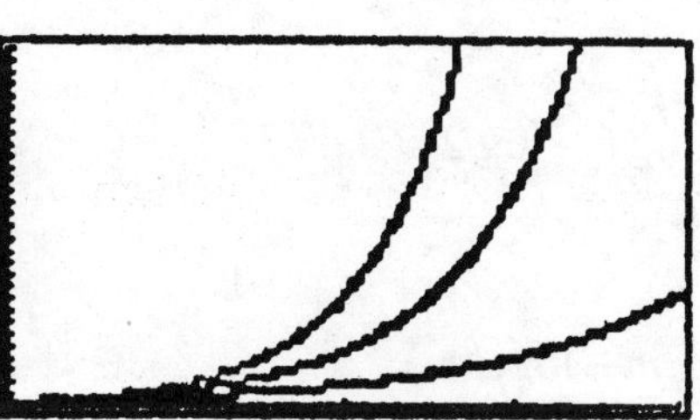

7.

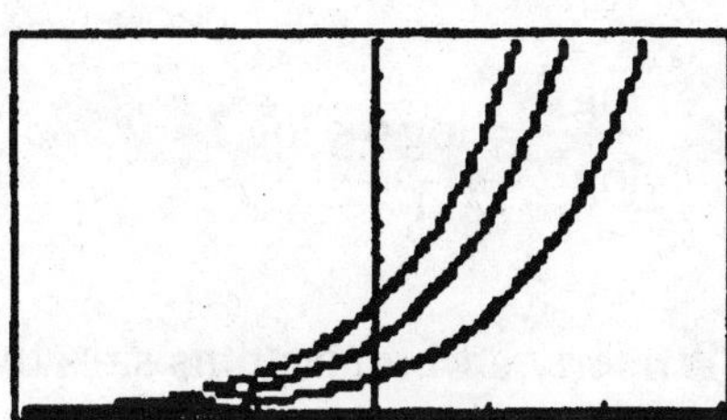

9.

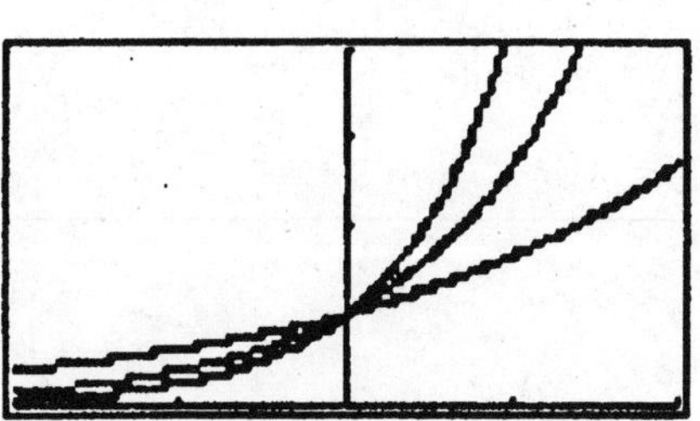

11. a.

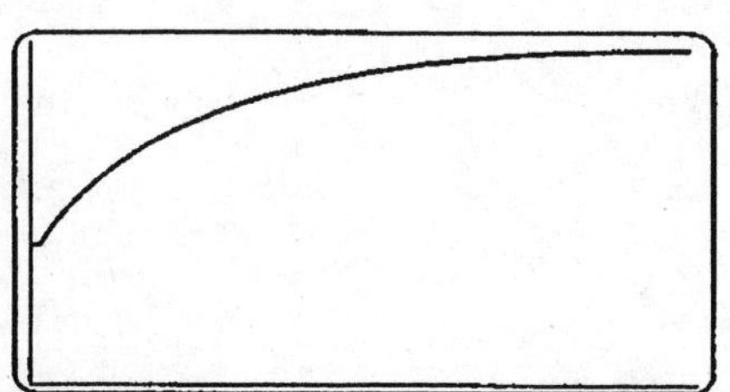

b. 0.08 g/cm^3 c. 0.12 g/cm^3
d. 0.2 g/cm^3

13. a.

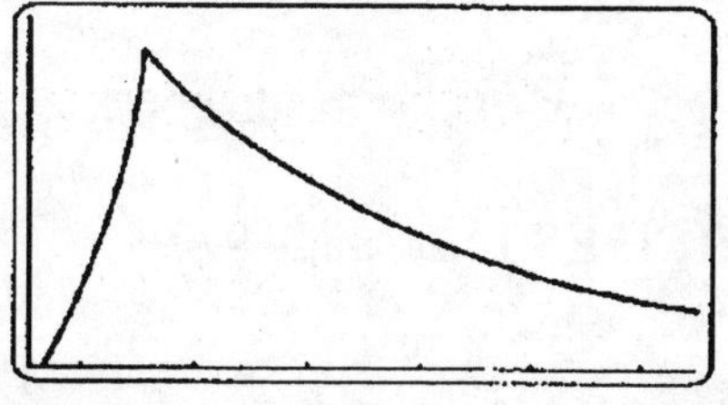

b. 20 sec. c. 35.1 sec

14.2 Problem Solving Tips

1. Property 1 of logarithms says that $\log_b mn = \log_b m + \log_b n$. However,

$$\log_b(m+n) \neq \log_b m + \log_b n \text{ and } \log_b \frac{m}{n} \neq \frac{\log_b m}{\log_b n}$$

2. When you work with logarithms be sure that you can distinguish between the following two operations:

$$\frac{\log 6}{\log 2} = \log 6 \div \log 2 \approx 2.585 \text{ and } \log\frac{6}{2} = \log 6 - \log 2 \approx 0.477$$

(Property 2 of logarithms says that $\log_b \frac{m}{n} = \log_b m - \log_b n$.)

3. The domain of the logarithmic function is $(0, \infty)$. So the logarithm of 0 and the logarithm of negative numbers are not defined.

14.2 CONCEPT QUESTIONS, page 870

1. a. $y = \log_b x$ if and only if $x = b^y$.
   b. $f(x) = \log_b x$; $b > 0$, $b \neq 1$. Its domain is $(0, \infty)$.

3. a. $e^{\ln x} = x$ b. $\ln e^x = x$

EXERCISES 14.2 , page 870

1. $\log_2 64 = 6$ 3. $\log_3 \frac{1}{9} = -2$ 5. $\log_{1/3} \frac{1}{3} = 1$ 7. $\log_{32} 8 = \frac{3}{5}$

9. $\log_{10} 0.001 = -3$

11. $\log 12 = \log 4 \times 3 = \log 4 + \log 3 = 0.6021 + 0.4771 = 1.0792.$

13. $\log 16 = \log 4^2 = 2 \log 4 = 2(0.6021) = 1.2042.$

15. $\log 48 = \log 3 \times 4^2 = \log 3 + 2 \log 4 = 0.4771 + 2(0.6021) = 1.6813.$

17. $2\ln a + 3\ln b = \ln a^2b^3.$

19. $\ln 3 + \frac{1}{2}\ln x + \ln y - \frac{1}{3}\ln z = \ln \frac{3\sqrt{x}y}{\sqrt[3]{z}}$

21. $\log x(x+1)^4 = \log x + \log (x+1)^4 = \log x + 4 \log (x+1).$

23. $\log \frac{\sqrt{x+1}}{x^2+1} = \log (x+1)^{1/2} - \log(x^2+1) = \frac{1}{2} \log (x+1) - \log (x^2+1)$

25. $\ln xe^{-x^2} = \ln x - x^2.$

27. $\ln \left(\frac{x^{1/2}}{x^2\sqrt{1+x^2}} \right) = \ln x^{1/2} - \ln x^2 - \ln (1+x^2)^{1/2}$

$= \frac{1}{2} \ln x - 2 \ln x - \frac{1}{2} \ln (1+x^2) = -\frac{3}{2} \ln x - \frac{1}{2} \ln (1+x^2).$

29. $y = \log_3 x$

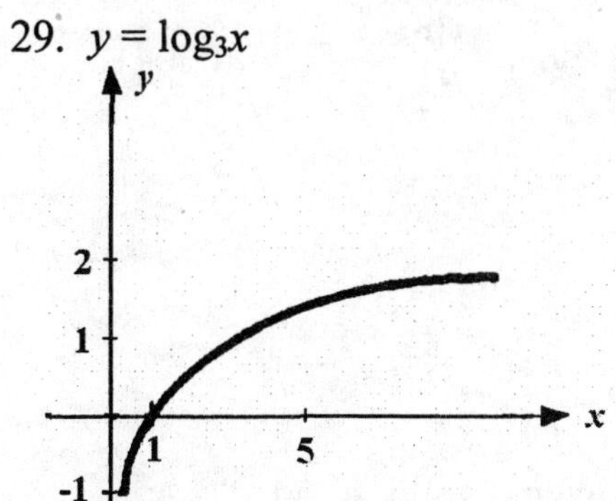

31. $y = \ln 2x$

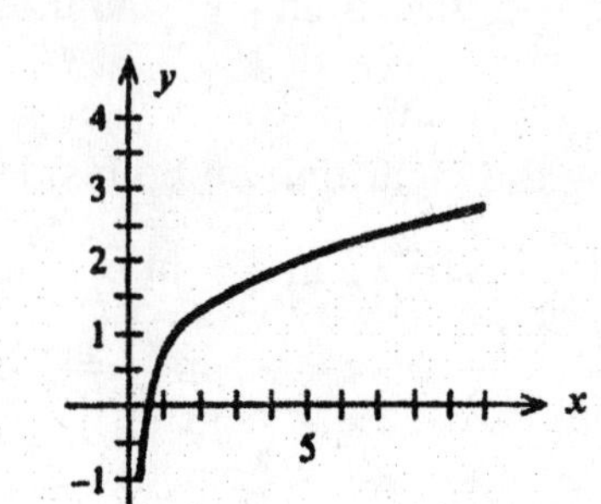

33. $y = 2^x$ and $y = \log_2 x$

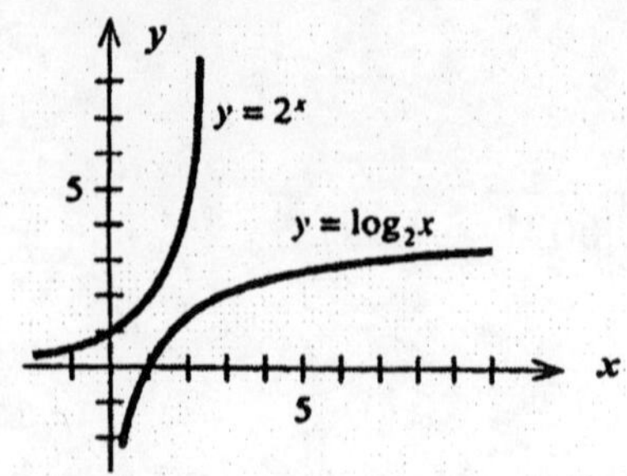

35. $e^{0.4t} = 8$, $0.4t \ln e = \ln 8$, and $0.4t = \ln 8$ ($\ln e = 1$.) So, $t = \dfrac{\ln 8}{0.4} = 5.1986$.

37. $5e^{-2t} = 6$, $e^{-2t} = \frac{6}{5} = 1.2$. Taking the logarithm, we have

$$-2t \ln e = \ln 1.2, \text{ or } t = -\frac{\ln 1.2}{2} \approx -0.0912.$$

39. $2e^{-0.2t} - 4 = 6$, $2e^{-0.2t} = 10$. Taking the logarithm on both sides of this last equation, we have $\ln e^{-0.2t} = \ln 5$; $-0.2t \ln e = \ln 5$; $-0.2t = \ln 5$;

and $$t = -\frac{\ln 5}{0.2} \approx -8.0472.$$

41. $$\frac{50}{1+4e^{0.2t}} = 20,\quad 1+4e^{0.2t} = \frac{50}{20} = 2.5,\quad 4e^{0.2t} = 1.5,$$

$$e^{0.2t} = \frac{1.5}{4} = 0.375,\ \ln e^{0.2t} = \ln 0.375,\ 0.2t = \ln 0.375. \text{ So } t = \frac{\ln 0.375}{0.2} \approx -4.9041.$$

43. Taking the logarithm on both sides, we obtain

$$\ln A = \ln Be^{-t/2},\ \ln A = \ln B + \ln e^{-t/2},\ \ln A - \ln B = -t/2 \ln e,$$

$$\ln\frac{A}{B} = -\frac{t}{2} \text{ or } t = -2\ln\frac{A}{B} = 2\ln\frac{B}{A}.$$

45. We use the compound interest formula with $A = 7500$, $P = 5000$, $m = 12$, and $t = 3$. Thus

$$7500 = 5000\left(1+\tfrac{r}{12}\right)^{36};$$

$\left(1+\frac{r}{12}\right)^{36} = \frac{7500}{5000} = \frac{3}{2}$, $\ln(1+\frac{r}{12})^{36} = \ln 1.5$;

$36\left(1+\frac{r}{12}\right) = \ln 1.5$

$\left(1+\frac{r}{12}\right) = \frac{\ln 1.5}{36} = 0.0112629$

$1+\frac{r}{12} = e^{0.0112629} = 1.011327; \frac{r}{12} = 0.011327;$

$r = 0.13592$

So the interest rate is 13.59% per year.

47. We use the compound interest formula) with $A = 8000$, $P = 5000$, $m = 2$, and $t = 4$. Thus $8000 = 5000\left(1+\frac{r}{2}\right)^{8}$

$\left(1+\frac{r}{2}\right)^{8} = \frac{8000}{5000} = 1.6$, $\ln(1+\frac{r}{2})^{8} = \ln 1.6$;

$8\ln\left(1+\frac{r}{2}\right) = \ln 1.6$

$\ln\left(1+\frac{r}{2}\right) = \frac{\ln 1.6}{8} = 0.05875$

$1+\frac{r}{2} = e^{0.05875} = 1.06051$; $\frac{r}{2} = 0.06051$

$r = 0.1210$

So the required interest rate is 12.1% per year.

49. We use the compound interest formula with $A = 4000$; $P = 2000$, $m = 1$, and $t = 5$. Thus

$4000 = 2000\left(1+r\right)^{5}$; $(1+r)^{5} = 2$; $5\ln(1+r) = \ln 2$; $\ln(1+r) = \frac{\ln 2}{5} = 0.138629$

$1+r = e^{0.138629} = 1.148698$; $r = 0.1487$

So the required interst rate is 14.87% per year.

51. We use the compound interest formula with $A = 6500$, $P = 5000$, $m = 12$, and $r = 0.12$. Thus

$$6500 = 5000\left(1+\frac{0.12}{12}\right)^{12t};\ (1.01)^{12t} = \frac{6500}{5000} = 1.3;\ 12t\ln(1.01) = \ln 1.3$$

$$t = \frac{\ln 1.3}{12\ln 1.01} \approx 2.197$$

So, it will take approximately 2.2 years.

53. We use the compound interest formula with $A = 4000$, $P = 2000$, $m = 12$, and $r = 0.09$. Thus,

$$4000 = 2000\left(1+\frac{0.09}{12}\right)^{12t}$$

$$\left(1+\tfrac{0.09}{12}\right)^{12t}=2$$

$$12t\ln\left(1+\tfrac{0.09}{12}\right)=\ln 2 \text{ and } t=\frac{\ln 2}{12\ln\left(1+\tfrac{0.09}{12}\right)}\approx 7.73.$$

So it will take approximately 7.7 years.

55. $p(x) = 19.4 \ln x + 18$. For a child weighing 92 lb, we find
$p(92) = 19.4 \ln 92 + 18 = 105.72$ millimeters of mercury.

57. a. $30 = 10\log\dfrac{I}{I_0}$; $3=\log\dfrac{I}{I_0}$; $\dfrac{I}{I_0}=10^3=1000$. So $I = 1000\,I_0$.
b. When $D = 80$, $I = 10^8 I_0$ and when $D = 30$, $I = 10^3 I_0$. Therefore, an 80–decibel sound is $10^8/10^3$ or $10^5 = 100{,}000$ times louder than a 30–decibel sound.
c. It is $10^{15}/10^8 = 10^7$, or 10,000,000, times louder.

59. We solve the following equation for t. Thus,

$$\frac{160}{1+240e^{-0.2t}}=80;\quad 1+240e^{-0.2t}=\frac{160}{80},$$

$$240e^{-0.2t}=2-1=1;\ e^{-0.2t}=\frac{1}{240};\ -0.2t=\ln\frac{1}{240}$$

$$t=-\frac{1}{0.2}\ln\frac{1}{240}\approx 27.40,$$ or approximately 27.4 years old.

61. We solve the following equation for t:

$$200(1-0.956e^{-0.18t})=140$$

$$1-0.956e^{-0.18t}=\frac{140}{200}=0.7$$

$$-0.956e^{-0.18t}=0.7-1=-0.3$$

$$e^{-0.18t}=\tfrac{0.3}{0.956}$$

$$-0.18t=\ln\left(\tfrac{0.3}{0.956}\right)$$

$$t=-\frac{\ln\left(\tfrac{0.3}{0.956}\right)}{0.18}\approx 6.43875.$$

So, its approximate age is 6.44 years.

63. a. We solve the equation $0.08+0.12e^{-0.02t}=0.18$.

$$0.12e^{-0.02t} = 0.1;\ e^{-0.02t} = \frac{0.1}{0.12} = \frac{1}{1.2}$$

$$\ln e^{-0.02t} = \ln\frac{1}{1.2} = \ln 1 - \ln 1.2 = -\ln 1.2$$

$$-0.02t = -\ln 1.2$$

$$t = \frac{\ln 1.2}{0.02} \approx 9.116, \quad \text{or } 9.12 \text{ sec.}$$

b. We solve the equation $0.08 + 0.12e^{-0.02t} = 0.16$.

$$0.12e^{-0.02t} = 0.08;\ e^{-0.02t} = \frac{0.08}{0.12} = \frac{2}{3};\ -0.02t = \ln\frac{2}{3}$$

$$t = -\frac{\ln\left(\frac{2}{3}\right)}{0.02} \approx 20.2733, \text{ or } 20.27 \text{ sec.}$$

65. False. Take $x = e$. Then $(\ln e)^3 = 1^3 = 1 \neq 3\ln e = 3$.

67. True. $g(x) = \ln x$ is continuous and greater than zero on $(1,\infty)$. Therefore, $f(x) = \dfrac{1}{\ln x}$ is continuous on $(1,\infty)$.

69. a. Taking the logarithm on both sides gives $\ln 2^x = \ln e^{kx}$, $x \ln 2 = kx(\ln e) = kx$. So, $x(\ln 2 - k) = 0$ for all x and this implies that $k = \ln 2$.
b. Tracing the same steps as done in (a), we find that $k = \ln b$.

71. Let $\log_b m = p$, then $m = b^p$. Therefore, $m^n = (b^p)^n = b^{np}$. Therefore,

$$\log_b m^n = \log_b b^{np} = np \log_b b = np \qquad \text{(Since } \log_b b = 1.)$$
$$= n \log_b m,$$

as was to be shown.

14.3 CONCEPT QUESTIONS, page 879

1. a. $f'(x) = e^x$ b. $g'(x) = e^{f(x)} \cdot f'(x)$

EXERCISES 14.3, page 879

1. $f(x) = e^{3x};\ f'(x) = 3e^{3x}$

3. $g(t) = e^{-t};\ g'(t) = -e^{-t}$

5. $f(x) = e^x + x;\ f'(x) = e^x + 1$

7. $f(x) = x^3e^x$, $f'(x) = x^3e^x + e^x(3x^2) = x^2e^x(x+3)$.

9. $f(x) = \dfrac{2e^x}{x}$, $f'(x) = \dfrac{x(2e^x) - 2e^x(1)}{x^2} = \dfrac{2e^x(x-1)}{x^2}$.

11. $f(x) = 3(e^x + e^{-x})$; $f'(x) = 3(e^x - e^{-x})$.

13. $f(w) = \dfrac{e^w + 1}{e^w} = 1 + \dfrac{1}{e^w} = 1 + e^{-w}$. $f'(w) = -e^{-w} = -\dfrac{1}{e^w}$.

15. $f(x) = 2e^{3x-1}$, $f'(x) = 2e^{3x-1}(3) = 6e^{3x-1}$.

17. $h(x) = e^{-x^2}$; $h'(x) = e^{-x^2}(-2x) = -2xe^{-x^2}$.

19. $f(x) = 3e^{-1/x}$; $f'(x) = 3e^{-1/x} \cdot \dfrac{d}{dx}\left(-\dfrac{1}{x}\right) = 3e^{-1/x}\left(\dfrac{1}{x^2}\right) = \dfrac{3e^{-1/x}}{x^2}$.

21. $f(x) = (e^x + 1)^{25}$, $f'(x) = 25(e^x + 1)^{24}e^x = 25e^x(e^x + 1)^{24}$.

23. $f(x) = e^{\sqrt{x}}$; $f'(x) = e^{\sqrt{x}}\dfrac{d}{dx}x^{1/2} = e^{\sqrt{x}}\dfrac{1}{2}x^{-1/2} = \dfrac{e^{\sqrt{x}}}{2\sqrt{x}}$.

25. $f(x) = (x-1)e^{3x+2}$; $f'(x) = (x-1)(3)e^{3x+2} + e^{3x+2} = e^{3x+2}(3x - 3 + 1) = e^{3x+2}(3x-2)$.

27. $f(x) = \dfrac{e^x - 1}{e^x + 1}$; $f'(x) = \dfrac{(e^x+1)(e^x) - (e^x-1)(e^x)}{(e^x+1)^2} = \dfrac{e^x(e^x + 1 - e^x + 1)}{(e^x+1)^2} = \dfrac{2e^x}{(e^x+1)^2}$.

29. $f(x) = e^{-4x} + 2e^{3x}$; $f'(x) = -4e^{-4x} + 6e^{3x}$ and
$f''(x) = 16e^{-4x} + 18e^{3x} = 2(8e^{-4x} + 9e^{3x})$.

31. $f(x) = 2xe^{3x}$; $f'(x) = 2e^{3x} + 2xe^{3x}(3) = 2(3x+1)e^{3x}$.
$f''(x) = 6e^{3x} + 2(3x+1)e^{3x}(3) = 6(3x+2)e^{3x}$.

33. $y = f(x) = e^{2x-3}$. $f'(x) = 2e^{2x-3}$. To find the slope of the tangent line to the graph

of f at $x = 3/2$, we compute $f'(\frac{3}{2}) = 2e^{3-3} = 2$. Next, using the point–slope form of the equation of a line, we find that

$$y - 1 = 2(x - \tfrac{3}{2})$$
$$= 2x - 3, \quad \text{or} \quad y = 2x - 2.$$

35. $f(x) = e^{-x^2/2}$, $f'(x) = e^{-x^2/2}(-x) = -xe^{-x^2/2}$. Setting $f'(x) = 0$, gives $x = 0$ as the only critical point of f. From the sign diagram,

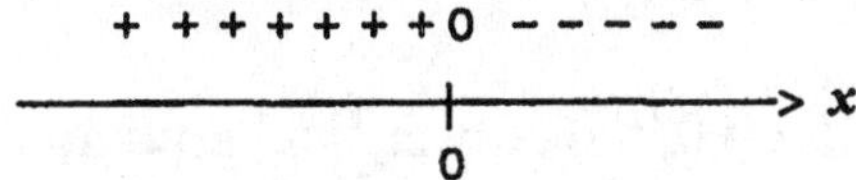

we conclude that f is increasing on $(-\infty,0)$ and decreasing on $(0,\infty)$.

37. $f(x) = \frac{1}{2}e^x - \frac{1}{2}e^{-x}$, $f'(x) = \frac{1}{2}(e^x + e^{-x})$, $f''(x) = \frac{1}{2}(e^x - e^{-x})$. Setting $f''(x) = 0$, gives $e^x = e^{-x}$ or $e^{2x} = 1$, and $x = 0$. From the sign diagram for f'',

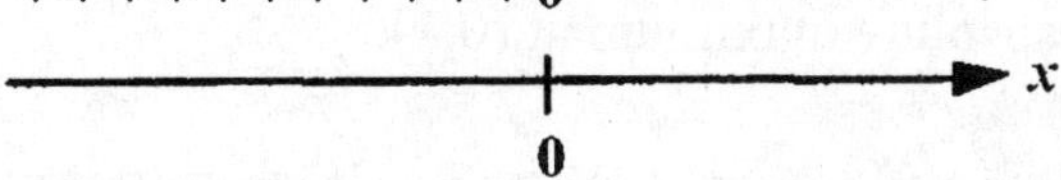

we conclude that f is concave upward on $(0,\infty)$ and concave downward on $(-\infty,0)$.

39. $f(x) = xe^{-2x}$. $f'(x) = e^{-2x} + xe^{-2x}(-2) = (1 - 2x)e^{-2x}$.
$f''(x) = -2e^{-2x} + (1 - 2x)e^{-2x}(-2) = 4(x - 1)e^{-2x}$.
Observe that $f''(x) = 0$ if $x = 1$. The sign diagram of f''

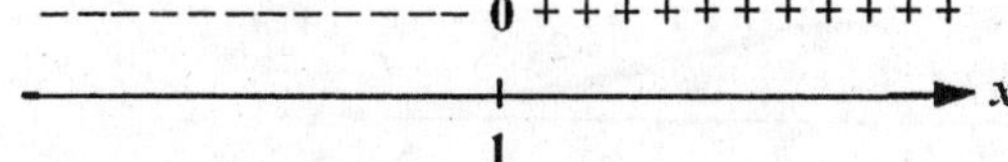

shows that $(1, e^{-2})$ is an inflection point.

41. $f(x) = e^{-x^2}$, $f'(x) = -2xe^{-x^2}$;
$f''(x) = -2e^{-x^2} - 2xe^{-x^2} \cdot (-2x) = -2e^{-x^2}(1-2x^2) = 0$ implies $x = \pm\frac{\sqrt{2}}{2}$. The sign diagram of f'' follows:

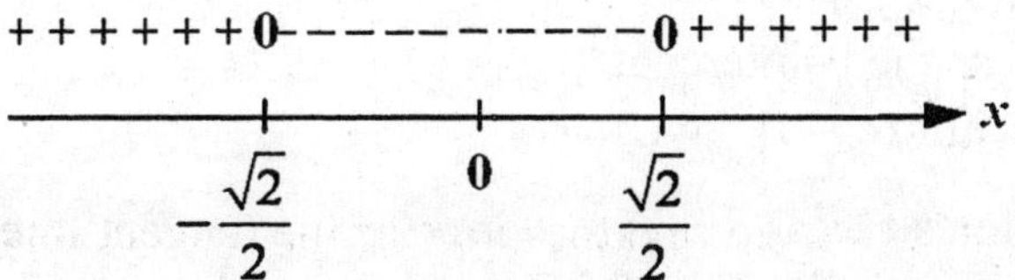

We see that the graph of f has inflection points at $(-\frac{\sqrt{2}}{2}, e^{-1/2})$ and $(\frac{\sqrt{2}}{2}, e^{-1/2})$. The slope of the tangent line at $(-\frac{\sqrt{2}}{2}, e^{-1/2})$ is $f'(-\frac{\sqrt{2}}{2}) = \sqrt{2}e^{-1/2}$. The tangent line has equation

$$y - e^{-1/2} = \sqrt{2}e^{-1/2}(x + \tfrac{\sqrt{2}}{2}) \text{ or } y = \sqrt{\tfrac{2}{e}}x + \tfrac{2}{\sqrt{e}} \text{ or } e^{-1/2}(\sqrt{2}x + 2)$$

The slope of the tangent line at $(\frac{\sqrt{2}}{2}, e^{-1/2})$ is $f'(\frac{\sqrt{2}}{2}) = -\sqrt{2}e^{-1/2}$. The tangent line has equation

$$y - e^{-1/2} = -\sqrt{2}e^{-1/2}(x - \tfrac{\sqrt{2}}{2}) \text{ or } y = e^{-1/2}(-\sqrt{2}x + 2)$$

43. $f(x) = e^{-x^2}$. $f'(x) = -2xe^{-x^2} = 0$ if $x = 0$, the only critical point of f.

x	-1	0	1
$f(x)$	e^{-1}	1	e^{-1}

From the table, we see that f has an absolute minimum value of e^{-1} attained at $x = -1$ and $x = 1$. It has an absolute maximum at $(0,1)$.

45. $g(x) = (2x - 1)e^{-x}$; $g'(x) = 2e^{-x} + (2x - 1)e^{-x}(-1) = (3 - 2x)e^{-x} = 0$, if $x = 3/2$. The graph of g shows that $(\frac{3}{2}, 2e^{-3/2})$ is an absolute maximum, and $(0,-1)$ is an absolute minimum.

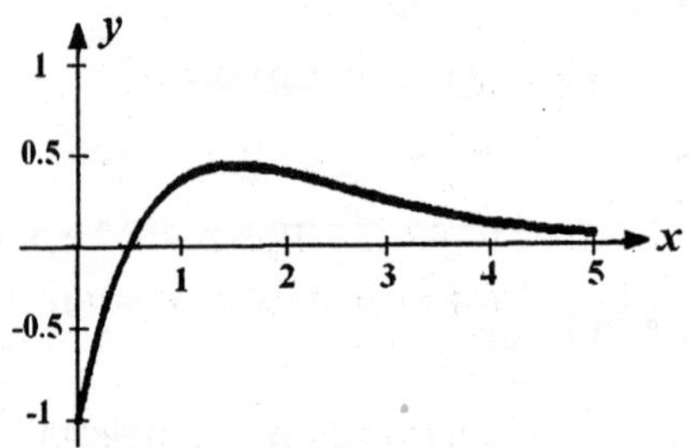

47. $f(t) = e^t - t$;

We first gather the following information on f.

1. The domain of f is $(-\infty, \infty)$.
2. Setting $t = 0$ gives 1 as the y–intercept.
3. $\lim_{t \to -\infty} (e^t - t) = \infty$ and $\lim_{t \to \infty} (e^t - t) = \infty$.
4. There are no asymptotes.
5. $f'(t) = e^t - 1$ if $t = 0$, a critical point of f. From the sign diagram for f'

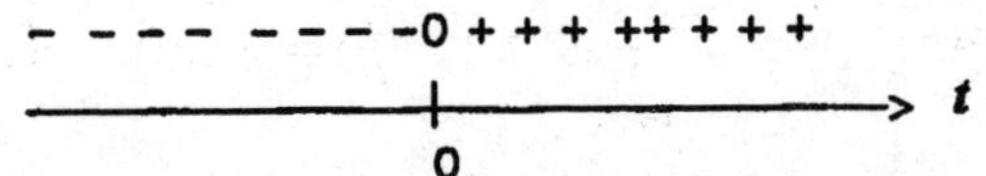

we see that f is decreasing on $(-\infty,0)$ and increasing on $(0,\infty)$.

6. From the results of f(5), we see that (0,1) is a relative minimum of f.
7. $f''(t) = e^t > 0$ for all t in $(-\infty,\infty)$. So the graph of f is concave upward on $(-\infty,\infty)$.
8. There are no inflection points.

The graph of f follows.

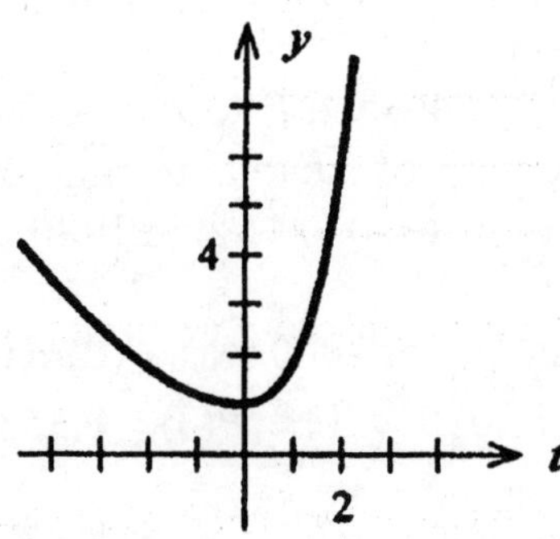

49. $f(x) = 2 - e^{-x}$.

We first gather the following information on f.

1. The domain of f is $(-\infty,\infty)$.
2. Setting $x = 0$ gives 1 as the y-intercept.
3. $\lim_{x\to-\infty} (2 - e^{-x}) = -\infty$ and $\lim_{x\to\infty} (2 - e^{-x}) = 2$,
4. From the results of (3), we see that $y = 2$ is a horizontal asymptote of f.
5. $f'(x) = e^{-x}$. Observe that $f'(x) > 0$ for all x in $(-\infty,\infty)$ and so f is increasing on $(-\infty,\infty)$.
6. Since there are no critical points, f has no relative extrema.
7. $f''(x) = -e^{-x} < 0$ for all x in $(-\infty,\infty)$ and so the graph of f is concave downward on $(-\infty,\infty)$.
8. There are no inflection points

The graph of f follows.

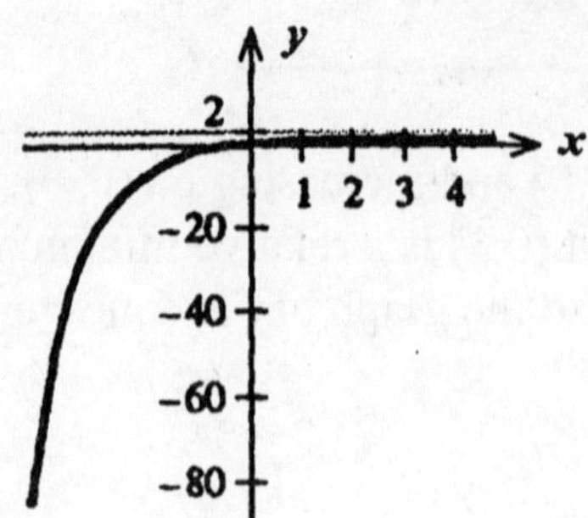

51. $P'(t) = 20.6(-0.009)e^{-0.009t} = -0.1854e^{-0.009t}$
$P'(10) = -0.1694$, $P'(20) = -0.1549$, and $P'(30) = -0.1415$,
and this tells us that the percentage of the total population relocating was decreasing at the rate of 0.17% in 1970, 0.15% in 1980, and 0.14% in 1990.

53. a. The population at the beginning of 2000 was $P(0) = 0.07$, or 70,000. The population at the beginning of 2030 will be $P(3) = 0.3537$, or approximately 353,700.
b. $P'(t) = 0.0378e^{0.54t}$; The population was changing at the rate of $P'(0) = 0.0378$ or 37,800/decade at the beginning of 2000. At the beginning of 2030, it was changing at the rate of $P'(3) \approx 0.191$, or approximately 191,000/decade.

55. a. The total loans outstanding in 1998 were $L(0) = 4.6$, or \$4.6 trillion. The total loans outstanding in 2004 were $L(6) = 3.6$, or \$3.6 trillion.
b. $L'(t) = -0.184e^{-0.04t}$; The total loans outstanding were changing at the rate of $L'(0) = -0.184$, that is, they were declining at the rate of \$0.18 trillion/yr in 1998. In 2004, they were changing at $L'(16) \approx -0.145$ or declining at the rate of \$0.145 trillion/yr.
c. $L''(t) = 0.00736e^{-0.04t}$; Since $L''(t) > 0$ on the interval (0, 6), we see L is decreasing but at a slower rate and this proves the assertion.

57. a. $S(t) = 20{,}000(1 + e^{-0.5t})$
$S'(t) = 20{,}000(-0.5e^{-0.5t}) = -10{,}000e^{-0.5t}$;
$S'(1) = -10{,}000e^{-0.5} = -6065$, or –\$6065/day.
$S'(2) = -10{,}000e^{-1} = -3679$, or –\$3679/day.
$S'(3) = -10{,}000(e^{-1.5}) = -2231$, or –\$2231/day.
$S'(4) = -10{,}000e^{-2} = -1353$, or –\$1353/day.

b. $S(t) = 20{,}000(1 + e^{-0.5t}) = 27{,}400$

$$1 + e^{-0.5t} = \frac{27{,}400}{20{,}000}$$

$$e^{-0.5t} = \frac{274}{200} - 1$$

$$-0.5t = \ln\left(\frac{274}{200} - 1\right)$$

$$t = \frac{\ln\left(\frac{274}{200} - 1\right)}{-0.5} \approx 2, \text{ or 2 days}$$

59. $N(t) = 5.3e^{0.095t^2 - 0.85t}$.

a. $N'(t) = 5.3e^{0.095t^2 - 0.85t}(0.19t - 0.85)$. Since $N'(t)$ is negative for $(0 \leq t \leq 4)$, we see that $N(t)$ is decreasing over that interval.

b. To find the rate at which the number of polio cases was decreasing at the beginning of 1959, we compute

$$N'(0) = 5.3e^{0.095(0^2) - 0.85(0)}(0.85) = 5.3(-0.85) = -4.505$$

(t is measured in thousands), or 4,505 cases per year. To find the rate at which the number of polio cases was decreasing at the beginning of 1962, we compute

$$N'(3) = 5.3e^{0.095(9) - 0.85(3)}(0.57 - 0.85)$$
$$= (-0.28)(0.9731) \approx -0.273, \text{ or 273 cases per year.}$$

61. From the results of Exercise 60, we see that $R'(x) = 100(1 - 0.0001x)e^{-0.0001x}$. Setting $R'(x) = 0$ gives $x = 10{,}000$, a critical point of R. From the graph of R

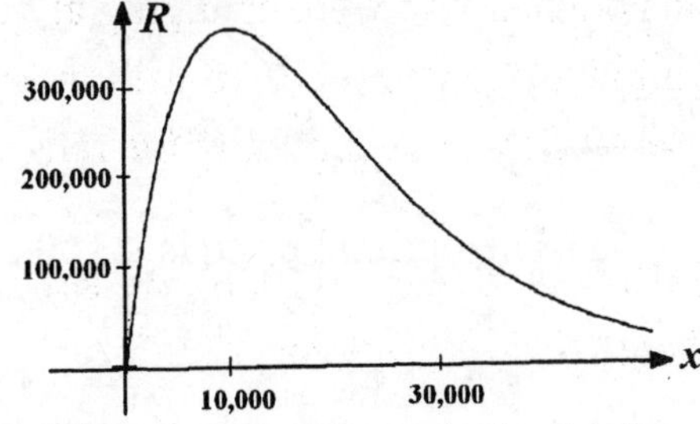

we see that the revenue is maximized when $x = 10{,}000$. So 10,000 pairs must be sold, yielding a maximum revenue of $R(10{,}000) = 367{,}879.44$, or \$367,879.

63. $p = 240\left(1 - \dfrac{3}{3 + e^{-0.0005x}}\right) = 240[1 - 3(3 + e^{-0.0005x})^{-1}]$.

$p' = 720(3 + e^{-0.0005x})^{-2}(-0.0005e^{-0.0005x})$

$$p'(1000) = 720(3 + e^{-0.0005(1000)})^{-2}(-0.0005e^{-0.0005(1000)})$$

$$= -\frac{0.36(0.606531)}{(3+0.606531)^2} \approx -0.0168, \quad \text{or } -1.68 \text{ cents per case.}$$

$$p(1000) = 240(1 - \frac{3}{3.606531}) \approx 40.36, \quad \text{or } \$40.36/\text{case.}$$

65. a. $W = 2.4e^{1.84h}$; $W = 2.4e^{1.84(16)} \approx 45.58$, or approximately 45.6 kg.

b. $\Delta W \approx dW = (2.4)(1.84)e^{1.84h}dh$. With $h = 1.6$ and $dh = \Delta h = 1.65 - 1.6 = 0.05$, we find

$$\Delta W \approx (2.4)(1.84)e^{1.84(1.6)} \cdot (0.05) \approx 4.19, \text{ or approximately 4.2 kg.}$$

67. $P(t) = 80{,}000\, e^{\sqrt{t}/2 - 0.09t} = 80{,}000\, e^{\frac{1}{2}t^{1/2} - 0.09t}$.

$P'(t) = 80{,}000(\frac{1}{4}t^{-1/2} - 0.09)e^{\frac{1}{2}t^{1/2} - 0.09t}$.

Setting $P'(t) = 0$, we have

$$\tfrac{1}{4}t^{-1/2} = 0.09, \; t^{-1/2} = 0.36, \; \frac{1}{\sqrt{t}} = 0.36, \; t = \left(\frac{1}{0.36}\right)^2 \approx 7.72.$$

Evaluating $P(t)$ at each of its endpoints and at the point $t = 7.72$, we find

t	$P(t)$
0	80,000
7.72	160,207.69
8	160,170.71

We conclude that P is optimized at $t = 7.72$. The optimal price is \$160,207.69.

69. $f(t) = 1.5 + 1.8te^{-1.2t}$

$f'(t) = 1.8\frac{d}{dt}(te^{-1.2t}) = 1.8[e^{-1.2t} + te^{-1.2t}(-1.2)] = 1.8e^{-1.2t}(1 - 1.2t)$.

$f'(0) = 1.8$, $f'(1) = -0.11$, $f'(2) = -0.23$, and $f'(3) = -0.13$,
and this tells us that the rate of change of the amount of oil used is 1.8 barrels per \$1000 of output per decade in 1965; it is decreasing at the rate of 0.11 barrels per \$1000 of output per decade in 1966, and so on.

71. a. The price at $t = 0$ is 8 + 4, or 12, dollars per unit.

b. $\dfrac{dp}{dt} = -8e^{-2t} + e^{-2t} - 2te^{-2t}$.

$$\left.\frac{dp}{dt}\right|_{t=0} = -8e^{-2t} + e^{-2t} - 2te^{-2t}\Big|_{t=0} = -8 + 1 = -7.$$

That is, the price is decreasing at the rate of \$7/week.

c. The equilibrium price is $\lim\limits_{t\to\infty}(8 + 4e^{-2t} + te^{-2t}) = 8 + 0 + 0$, or \$8 per unit.

73. We are given that

$$c(1 - e^{-at/V}) < m$$

$$1 - e^{-at/V} < \frac{m}{c}$$

$$-e^{-at/V} < \frac{m}{c} - 1 \text{ and } \quad e^{-at/V} > 1 - \frac{m}{c}.$$

Taking the log of both sides of the inequality, we have

$$-\frac{at}{V}\ln e > \ln\frac{c-m}{c}$$

$$-\frac{at}{V} > \ln\frac{c-m}{c}$$

$$-t > \frac{V}{a}\ln\frac{c-m}{c} \quad \text{or} \quad t < \frac{V}{a}\left(-\ln\frac{c-m}{c}\right) = \frac{V}{a}\ln\left(\frac{c}{c-m}\right).$$

Therefore the liquid must not be allowed to enter the organ for a time longer than $t = \dfrac{V}{a}\ln\left(\dfrac{c}{c-m}\right)$ minutes.

75. $$C'(t) = \begin{cases} 0.3 + 18e^{-t/60}\left(-\frac{1}{60}\right) & 0 \le t \le 20 \\ -\frac{18}{60}e^{-t/60} + \frac{12}{60}e^{-(t-20)/60} & t > 20 \end{cases} = \begin{cases} 0.3(1 - e^{-t/60}) & 0 \le t \le 20 \\ -0.3e^{-t/60} + 0.2e^{-(t-20)/60} & t > 20 \end{cases}$$

a. $C'(10) = 0.3\left(1 - e^{-10/60}\right) \approx 0.05$ or 0.05 g/cm^3/sec.

b. $C'(30) = -0.3e^{-30/60} + 0.2e^{-10/60} \approx -0.01$, or decreasing at the rate of 0.01 g/ cm^3/sec.

c. On the interval (0, 20), $C'(t) = 0$ implies $1 - e^{-t/60} = 0$, or $t = 0$. Therefore, C attains its absolute maximum value at an endpoint. In this case, at $t = 20$. On the interval $[20, \infty)$, $C'(t) = 0$ implies

$$-0.3e^{-t/60} = -0.2e^{-(t-20)/60}$$

$$\frac{e^{-\left(\frac{t-20}{60}\right)}}{e^{-t/60}} = \frac{3}{2}; \text{ or } e^{1/3} = \frac{3}{2},$$

which is not possible. Therefore $C'(t) \neq 0$ on $[20, \infty)$. Since $C(t) \to 0$ as $t \to \infty$, the absolute maximum of c occurs at $t = 20$. Thus, the concentration of the drug reaches a maximum at $t = 20$.

d. The maximum concentration is $C(20) = 0.90$ g/cm^3.

77. False. $f(x) = 3^x = e^{x\ln 3}$ and so $f'(x) = e^{x\ln 3} \cdot \frac{d}{dx}(x \ln 3) = (\ln 3)e^{x\ln 3} = (\ln 3)3^x$.

79. False. $f'(x) = (\ln \pi)\pi^x$. See Exercise 74.

USING TECHNOLOGY EXERCISES 14.3, page 884

1. 5.4366 3. 12.3929 5. 0.1861

7. a. The initial population of crocodiles is $P(0) = \frac{300}{6} = 50$.

b. $\lim_{t\to 0} P(t) = \lim_{t\to 0} \frac{300e^{-0.024t}}{5e^{-0.024t} + 1} = \frac{0}{0+1} = 0.$

c.

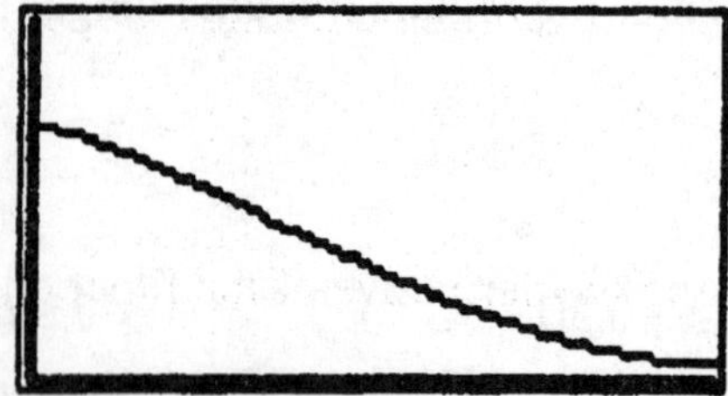

9. a.

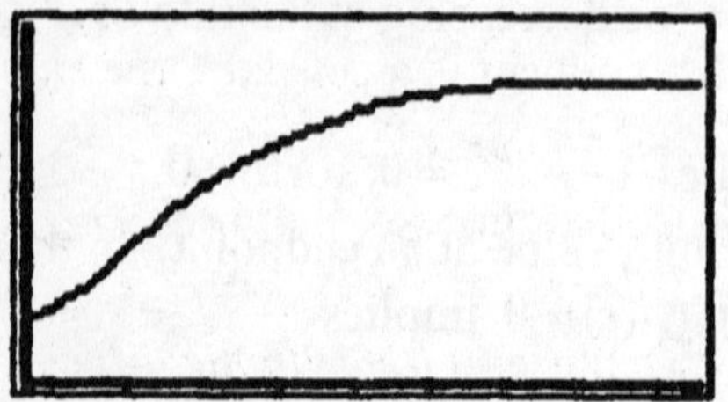

b. 4.2720 billion/half century

11. a. Using the function evaluation capabilities of a graphing utility, we find

$$f(11) = 153.024 \text{ and } g(11) = 235.180977624$$

and this tells us that the number of violent-crime arrests will be 153,024 at the beginning of the year 2000, but if trends like inner-city drug use and wider availability of guns continue, then the number of arrests will be 235,181.

b. Using the differentiation capability of a graphing utility, we find

$$f'(11) = -0.634 \text{ and } g'(11) = 18.4005596893$$

and this tells us that the number of violent-crime arrests will be decreasing at the rate of 634 per year at the beginning of the year 2000. But if the trends like inner-city drug use and wider availability of guns continues, then the number of arrests will be increasing at the rate of 18,401 per year at the beginning of the year 2000.

13. a. $P(10) = \dfrac{74}{1+2.6e^{-0.166(10)+0.04536(10)^2-0.0066(10)^3}} \approx 69.63$ percent.

b. $P'(10) = 5.09361$, or 5.09361%/decade

14.4 Problem Solving Tips

1. If you trying to find the derivative of a complicated function involving products, quotients, or powers, check to see if you can use logarithmic differentiation to simplify the process. Look at problems 37-46, and try to become familiar with the type of problems this method is especially suitable for.

2. Example 7 provided us with a method for finding the derivative of the function $y = x^x$ $(x > 0)$. Note that we use the *Power rule* ($\frac{d}{dx}[x^n] = nx^{n-1}$) to differentiate functions of the form $y = x^n$, where the base is a variable and the exponent is a constant, the *rule for differentiating exponential functions* to differentiate functions of the form $y = e^x$, where the base is the constant *e* and the exponent is a variable, and *logarithmic differentiation*

to differentiate functions of the form $y = x^x$ where both the base and the exponent of the function are variables. Be sure that you can distinguish between these functions and the rule to be applied in each of these cases.

14.4 CONCEPT QUESTIONS, page 890

1. a. $f'(x) = \frac{1}{x}$; $g'(x) = \frac{f'(x)}{f(x)}$

EXERCISES 14.4, page 890

1. $f(x) = 5 \ln x; f'(x) = 5\left(\frac{1}{x}\right) = \frac{5}{x}$.

3. $f(x) = \ln (x + 1); f'(x) = \frac{1}{x+1}$.

5. $f(x) = \ln x^8; f'(x) = \frac{8x^7}{x^8} = \frac{8}{x}$.

7. $f(x) = \ln x^{1/2}$; $f'(x) = \frac{\frac{1}{2}x^{-1/2}}{x^{1/2}} = \frac{1}{2x}$.

9. $f(x) = \ln \left(\frac{1}{x^2}\right) = \ln x^{-2} = -2 \ln x$; $f'(x) = -\frac{2}{x}$.

11. $f(x) = \ln (4x^2 - 6x + 3)$; $f'(x) = \frac{8x-6}{4x^2-6x+3} = \frac{2(4x-3)}{4x^2-6x+3}$.

13. $f(x) = \ln \left(\frac{2x}{x+1}\right) = \ln 2x - \ln (x + 1)$.

$$f'(x) = \frac{2}{2x} - \frac{1}{x+1} = \frac{2(x+1)-2x}{2x(x+1)} = \frac{2x+2-2x}{2x(x+1)} = \frac{2}{2x(x+1)} = \frac{1}{x(x+1)}.$$

15. $f(x) = x^2 \ln x$; $f'(x) = x^2\left(\frac{1}{x}\right) + (\ln x)(2x) = x + 2x \ln x = x(1 + 2 \ln x)$

17. $f(x) = \frac{2 \ln x}{x}$. $f'(x) = \frac{x\left(\frac{2}{x}\right) - 2 \ln x}{x^2} = \frac{2(1-\ln x)}{x^2}$.

19. $f(u) = \ln (u-2)^3$; $f'(u) = \frac{3(u-2)^2}{(u-2)^3} = \frac{3}{u-2}$.

21. $f(x) = (\ln x)^{1/2}$ and $f'(x) = \frac{1}{2}(\ln x)^{-1/2}\left(\frac{1}{x}\right) = \frac{1}{2x\sqrt{\ln x}}$.

23. $f(x) = (\ln x)^3$; $f'(x) = 3(\ln x)^2\left(\frac{1}{x}\right) = \frac{3(\ln x)^2}{x}$.

25. $f(x) = \ln(x^3 + 1)$; $f'(x) = \frac{3x^2}{x^3+1}$.

27. $f(x) = e^x \ln x$. $f'(x) = e^x \ln x + e^x\left(\frac{1}{x}\right) = \frac{e^x(x \ln x + 1)}{x}$.

29. $f(t) = e^{2t} \ln(t + 1)$

$$f'(t) = e^{2t}\left(\frac{1}{t+1}\right) + \ln(t+1)\cdot(2e^{2t}) = \frac{[2(t+1)\ln(t+1)+1]e^{2t}}{t+1}.$$

31. $f(x)$ $\frac{\ln x}{x}$. $f'(x) = \frac{x(\frac{1}{x}) - \ln x}{x^2} = \frac{1 - \ln x}{x^2}$.

33. $f(x) = \ln 2 + \ln x$; So $f'(x) = \frac{1}{x}$ and $f''(x) = -\frac{1}{x^2}$.

35. $f(x) = \ln(x^2 + 2)$; $f'(x) = \frac{2x}{(x^2+2)}$ and

$$f''(x) = \frac{(x^2+2)(2) - 2x(2x)}{(x^2+2)^2} = \frac{2(2-x^2)}{(x^2+2)^2}.$$

37. $y = (x + 1)^2(x + 2)^3$

$\ln y = \ln(x + 1)^2(x + 2)^3 = \ln(x + 1)^2 + \ln(x + 2)^3$

$= 2\ln(x + 1) + 3\ln(x + 2)$.

$$\frac{y'}{y} = \frac{2}{x+1} + \frac{3}{x+2} = \frac{2(x+2)+3(x+1)}{(x+1)(x+2)} = \frac{5x+7}{(x+1)(x+2)}$$

$$y' = \frac{(5x+7)(x+1)^2(x+2)^3}{(x+1)(x+2)} = (5x+7)(x+1)(x+2)^2.$$

39. $y = (x - 1)^2(x + 1)^3(x + 3)^4$

$\ln y = 2\ln(x - 1) + 3\ln(x + 1) + 4\ln(x + 3)$

$$\frac{y'}{y}=\frac{2}{x-1}+\frac{3}{x+1}+\frac{4}{x+3}$$
$$=\frac{2(x+1)(x+3)+3(x-1)(x+3)+4(x-1)(x+1)}{(x-1)(x+1)(x+3)}$$
$$=\frac{2x^2+8x+6+3x^2+6x-9+4x^2-4}{(x-1)(x+1)(x+3)}=\frac{9x^2+14x-7}{(x-1)(x+1)(x+3)}.$$

Therefore,

$$y'=\frac{9x^2+14x-7}{(x-1)(x+1)(x+3)}\cdot y$$
$$=\frac{(9x^2+14x-7)(x-1)^2(x+1)^3(x+3)^4}{(x-1)(x+1)(x+3)}$$
$$=(9x^2+14x-7)(x-1)(x+1)^2(x+3)^3.$$

41. $y=\dfrac{(2x^2-1)^5}{\sqrt{x+1}}.$

$$\ln y=\ln\frac{(2x^2-1)^5}{(x+1)^{1/2}}=5\ln(2x^2-1)-\frac{1}{2}\ln(x+1)$$

So $\dfrac{y'}{y}=\dfrac{20x}{2x^2-1}-\dfrac{1}{2(x+1)}=\dfrac{40x(x+1)-(2x^2-1)}{2(2x^2-1)(x+1)}$

$$=\frac{38x^2+40x+1}{2(2x^2-1)(x+1)}.$$

$$y'=\frac{38x^2+40x+1}{2(2x^2-1)(x+1)}\cdot\frac{(2x^2-1)^5}{\sqrt{x+1}}=\frac{(38x^2+40x+1)(2x^2-1)^4}{2(x+1)^{3/2}}.$$

43. $y=3^x$; $\quad \ln y=x\ln 3$; $\quad \dfrac{1}{y}\cdot\dfrac{dy}{dx}=\ln 3$; $\quad \dfrac{dy}{dx}=y\ln 3=3^x\ln 3.$

45. $y=(x^2+1)^x$; $\ln y=\ln(x^2+1)^x=x\ln(x^2+1)$. So

$$\frac{y'}{y}=\ln(x^2+1)+x\left(\frac{2x}{x^2+1}\right)=\frac{(x^2+1)\ln(x^2+1)+2x^2}{x^2+1}.$$

$$y'=\frac{[(x^2+1)\ln(x^2+1)+2x^2](x^2+1)^x}{x^2+1}$$

47. $y = x \ln x$. The slope of the tangent line at any point is

$$y' = \ln x + x\left(\tfrac{1}{x}\right) = \ln x + 1.$$

In particular, the slope of the tangent line at (1,0) where $x = 1$ is $m = \ln 1 + 1 = 1$. So, an equation of the tangent line is $y - 0 = 1(x - 1)$ or $y = x - 1$.

49. $f(x) = \ln x^2 = 2 \ln x$ and so $f'(x) = 2/x$. Since $f'(x) < 0$ if $x < 0$, and $f'(x) > 0$ if $x > 0$, we see that f is decreasing on $(-\infty,0)$ and increasing on $(0,\infty)$.

51. $f(x) = x^2 + \ln x^2$; $f'(x) = 2x + \dfrac{2x}{x^2} = 2x + \dfrac{2}{x}$; $f''(x) = 2 - \dfrac{2}{x^2}$.

To find the intervals of concavity for f, we first set $f''(x) = 0$ giving

$$2 - \frac{2}{x^2} = 0, \quad 2 = \frac{2}{x^2}, \quad 2x^2 = 2$$

or $\qquad x^2 = 1$ and $x = \pm 1$.

Next, we construct the sign diagram for f''

f" is not defined here

↓

+ + + + + + 0 - - - - - - - - - - 0 + + + +

-1 0 1 x

and conclude that f is concave upward on $(-\infty,-1) \cup (1,\infty)$ and concave downward on $(-1,0) \cup (0,1)$.

53. $f(x) = \ln(x^2 + 1)$. $f'(x) = \dfrac{2x}{x^2+1}$; $f''(x) = \dfrac{(x^2+1)(2) - (2x)(2x)}{(x^2+1)^2} = -\dfrac{2(x^2-1)}{(x^2+1)^2}$.

Setting $f''(x) = 0$ gives $x = \pm 1$ as candidates for inflection points of f.

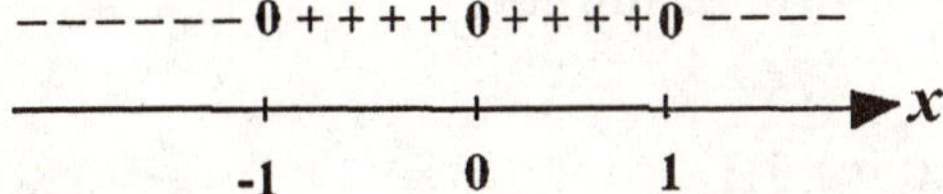

From the sign diagram for f'', we see that $(-1, \ln 2)$ and $(1, \ln 2)$ are inflection points of f.

55. $f(x) = x^2 + 2\ln x$, $f'(x) = 2x + \dfrac{2}{x}$, $f''(x) = 2 - \dfrac{2}{x^2} = 0$ implies

$2 - \dfrac{2}{x^2} = 0$, $x^2 = 1$, or $x = \pm 1$. We reject the negative root because the domain of f is $(0, \infty)$. The sign diagram of f'' follows:

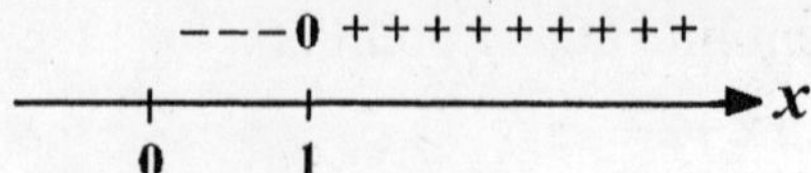

We see that (1,1) is an inflection point of the graph of *f*. $f'(1) = 4$. So, an equation of the required tangent line is

$$y - 1 = 4(x - 1) \quad \text{or} \quad y = 4x - 3$$

57. $f(x) = x - \ln x; \quad f'(x) = 1 - \dfrac{1}{x} = \dfrac{x-1}{x} = 0$ if $x = 1$, a critical point of *f*.

x	1/2	1	3
$f(x)$	1/2 + ln 2	1	3 - ln 3

From the table, we see that *f* has an absolute minimum at (1,1) and an absolute maximum at (3, 3 – ln 3).

59. $f(x) = 7.2956 \ln(0.0645012x^{0.95} + 1)$

$$f'(x) = 7.2956 \cdot \frac{\frac{d}{dx}(0.0645012x^{0.95} + 1)}{0.0645012x^{0.95} + 1} = \frac{7.2956(0.0645012)(0.95x^{-0.05})}{0.0645012x^{0.95} + 1}$$

$$= \frac{0.4470462}{x^{0.05}(0.0645012x^{0.95} + 1)}$$

So $\quad f'(100) = 0.05799$, or approximately 0.0580 percent/kg and
$f'(500) = 0.01329$, or approximately 0.0133 percent/kg.

61. a. The projected number at the beginning of 2005 will be

$$N(1) = 34.68 + 23.88 \ln(6.35) \approx 78.82 \text{ million}$$

b. $N'(t) = 23.88 \dfrac{1.05}{1.05t + 5.3} = \dfrac{25.074}{1.05t + 5.3}$

The projected number will be changing at the rate of

$$N'(1) = \frac{25.074}{1.05 + 5.3} \approx 3.95 \text{ (million/yr)}$$

63. a. If $0 < r < 100$, then $c = 1 - \frac{r}{100}$ sastisfies $0 < c < 1$. It suffices to show that $A_1(n) = -(1 - \frac{r}{100})^n$ is increasing, (why?), or equivalently $A_2(n) = -A_1(n) = \left(1 - \frac{r}{100}\right)^n$ is decreasing. Let $y = \left(1 - \frac{r}{100}\right)^n$. Then $\ln y = \ln\left(1 - \frac{r}{100}\right)^n = \ln c^n = n \ln c$. Differentiating both sides with respect to n, we find

$$\frac{y'}{y} = \ln c \text{ and so } y' = (\ln c)\left(1 - \tfrac{r}{100}\right)^n < 0$$

since $\ln c < 0$ and $\left(1 - \frac{r}{m}\right)^n > 0$ for $0 < r < 100$. Therefore, A is an increasing function of n on $(0, \infty)$.

b.

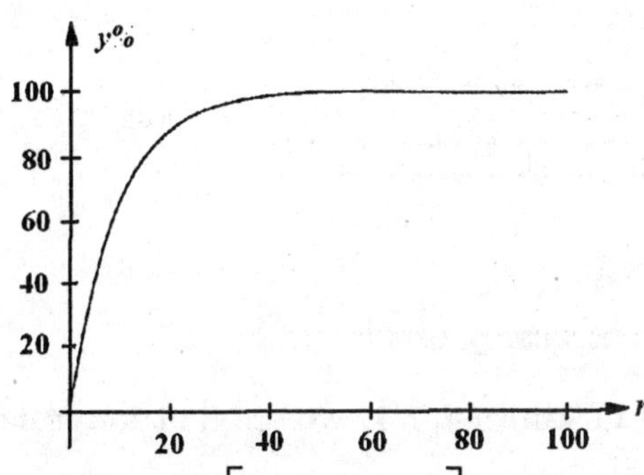

c. $\lim\limits_{n \to \infty} A(n) = \lim\limits_{n \to \infty} 100\left[1 - \left(1 - \tfrac{r}{100}\right)^n\right] = 100$

65. a. $R = \log \dfrac{10^6 I_0}{I_0} = \log 10^6 = 6.$

b. $I = I_0 10^R$ by definition. Taking the natural logarithm on both sides, we find

$$\ln I = \ln I_0 10^R = \ln I_0 + \ln 10^R = \ln I_0 + R \ln 10.$$

Differentiating implicitly with respect to R, we obtain

$$\frac{I'}{I} = \ln 10 \text{ or } \frac{dI}{dR} = (\ln 10) I.$$

Therefore, $\Delta I \approx dI = \dfrac{dI}{dR} \Delta R = (\ln 10) I \Delta R$. With $|\Delta R| \le (0.02)(6) = 0.12$ and $I = 1{,}000{,}000 I_0$, (see part a), we have

$$|\Delta I| \le (\ln 10)(1{,}000{,}000 I_0)(0.12) = 276310.21 I_o$$

So the error is at most 276,310 times the standard reference intensity.

67. $f(x) = \ln (x - 1)$.

1. The domain of f is obtained by requiring that $x - 1 > 0$. We find the domain to be $(1, \infty)$.
2. Since $x \neq 0$, there are no y-intercepts. Next, setting $y = 0$ gives $x - 1 = 1$ or $x = 2$ as the x-intercept.
3. $\lim\limits_{x \to 1^+} \ln (x - 1) = -\infty$.

4. There are no horizontal asymptotes. Observe that $\lim_{x \to 1^+} \ln(x-1) = -\infty$ so $x = 1$ is a vertical asymptote.

5. $f'(x) = \dfrac{1}{x-1}$.

The sign diagram for f' is

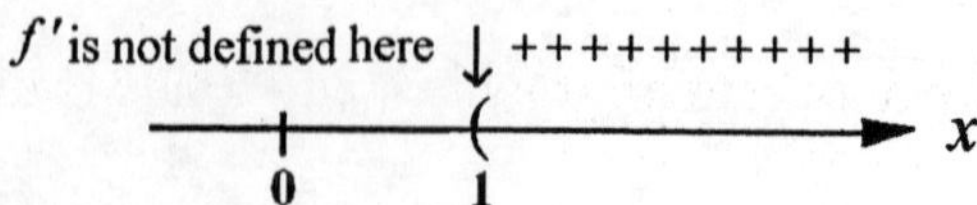

We conclude that f is increasing on $(1,\infty)$.

6. The results of (5) show that f is increasing on $(1,\infty)$.

7. $f''(x) = -\dfrac{1}{(x-1)^2}$. Since $f''(x) < 0$ for $x > 1$, we see that f is concave downward on $(1,\infty)$.

8. From the results of (7), we see that f has no inflection points.

The graph of f follows.

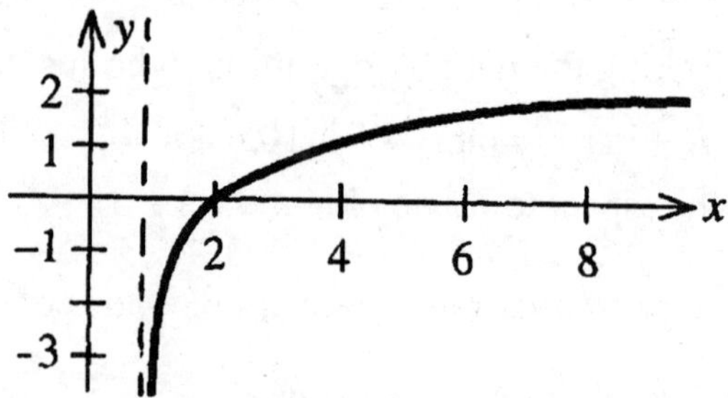

69. False. $\ln 5$ is a constant function and $f'(x) = 0$.

71. If $x \le 0$, then $|x| = -x$. Therefore, $\ln|x| = \ln(-x)$. Writing $f(x) = \ln|x|$ we have $|x| = -x = e^{f(x)}$. Differentiating both sides with respect to x and using the Chain Rule, we have $-1 = e^{f(x)} \cdot f'(x)$ or $f'(x) = -\dfrac{1}{e^{f(x)}} = -\dfrac{1}{-x} = \dfrac{1}{x}$.

14.5 Problem Solving Tips

Four mathematical models were introduced in this section:

Exponential growth: $Q(t) = Q_0e^{kt}$ that describes a quantity $Q(t)$ that is initially present in the amount of $Q(0) = Q_0$ and whose rate of growth at any time t is directly proportional to the amount of the quantity present at time t.

Exponential decay: $Q(t) = Q_0e^{-kt}$ that describes a quantity $Q(t)$ that is initially present in the amount of $Q(0) = Q_0$ and decreases at a rate that is directly proportional to its size.

Learning curves: $Q(t) = C - Ae^{-kt}$ that describes a quantity $Q(t)$, where $Q(0) = C - A$, and $Q(t)$ increases and approaches the number C as t increases without bound.

Logistic growth functions: $Q(t) = \dfrac{A}{1+Be^{-kt}}$ that describes a quantity $Q(t)$, where $Q(0) = \dfrac{A}{1+B}$. Note that $Q(t)$ increases rather rapidly for small values of t but the rate of growth of $Q(t)$ decreases quite rapidly as t increases and $Q(t)$ approaches the number A as t increases without bound.

Try to familiarize yourself with the examples and graphs for each of these models before you work through the applied problems in this section.

CONCEPT QUESTIONS, 14.5 CONCEPT QUESTIONS, page 900

1. $Q(t) = Q_0 e^{kt}$ where $k > 0$ is exponential growth and $k < 0$ is exponential decay. The larger the magnitude of k the faster the former grows and the faster the latter decays.

3. $Q(t) = \dfrac{A}{1+Be^{-kt}}$, where A, B, and k are positive constants. Q increases rapidly for small values of t but the rate of increase slows down as Q (always increasing) approaches the number A.

EXERCISES 14.5 , page 900

1. a. The growth constant is $k = 0.05$. b. Initially, the quantity present is 400 units.

c.

t	0	10	20	100	1000
Q	400	660	1087	59,365	2.07×10^{24}

3. a. $Q(t) = Q_0 e^{kt}$. Here $Q_0 = 100$ and so $Q(t) = 100e^{kt}$. Since the number of cells doubles in 20 minutes, we have

$$Q(20) = 100e^{20k} = 200,\ e^{20k} = 2,\ 20k = \ln 2, \text{ or } k = \tfrac{1}{20} \ln 2 \approx 0.03466.$$

$$Q(t) = 100e^{0.03466t}$$

b. We solve the equation $100e^{0.03466t} = 1{,}000{,}000$. We obtain

$$e^{0.03466t} = 10000 \text{ or } 0.03466t = \ln 10000,$$

$$t = \frac{\ln 10{,}000}{0.03466} \approx 266, \text{ or 266 minutes.}$$

c. $Q(t) = 1000e^{0.03466t}$.

5. a. We solve the equation $5.3e^{0.02t} = 3(5.3)$ or $e^{0.02t} = 3$,

or $\quad 0.02t = \ln 3$ and $t = \dfrac{\ln 3}{0.02} \approx 54.93$.

So the world population will triple in approximately 54.93 years.

b. If the growth rate is 1.8 percent, then proceeding as before, we find $N(t) = 5.3e^{0.018t}$. If $t = 54.93$, the population would be

$$N(54.93) = 5.3e^{0.018(54.93)} \approx 14.25, \text{ or approximately 14.25 billion.}$$

7. $P(h) = p_0 e^{-kh}$, $P(0) = 15$, therefore, $p_0 = 15$.

$$P(4000) = 15e^{-4000k} = 12.5; \quad e^{-4000k} = \frac{12.5}{15},$$

$$-4000k = \ln\left(\frac{12.5}{15}\right) \quad \text{and } k = 0.00004558.$$

Therefore, $P(12{,}000) = 15e^{-0.00004558(12{,}000)} = 8.68$, or 8.7 lb/sq in.

The rate of change of the atmospheric pressure with respect to altitude is given by

$$P'(h) = \frac{d}{dh}(15e^{-0.00004558h}) = -0.0006837e^{-0.00004558h}.$$

So, the rate of change of the atmospheric pressure with respect to altitude when the altitude is 12,000 feet is $P'(12{,}000) = -0.0006837e^{-0.00004558(12{,}000)} \approx -0.00039566$. That is, it is dropping at the rate of approximately 0.0004 lbs per square inch/foot.

9. Suppose the amount of phosphorus 32 at time t is given by

$$Q(t) = Q_0 e^{-kt}$$

where Q_0 is the amount present initially and k is the decay constant. Since this element has a half–life of 14.2 days, we have

$$\tfrac{1}{2}Q_0 = Q_0 e^{-14.2k}, \quad e^{-14.2k} = \tfrac{1}{2}, \ -14.2k = \ln\tfrac{1}{2}, \ k = -\frac{\ln\frac{1}{2}}{14.2} \approx 0.0488.$$

Therefore, the amount of phosphorus 32 present at any time t is given by

$$Q(t) = 100e^{-0.0488t}$$

The amount left after 7.1 days is given by

$$Q(7.1) = 100e^{-0.0488(7.1)} = 100e^{-0.3465}$$
$$= 70.617, \text{ or } 70.617 \text{ grams.}$$

The rate at which the phosphorus 32 is decaying when $t = 7.1$ is given by

$$Q'(t) = \frac{d}{dt}[100e^{-0.0488t}] = 100(-0.0488)e^{-0.0488t} = -4.88e^{-0.0488t}.$$

Therefore, $Q'(7.1) = -4.88e^{-0.0488(7.1)} \approx -3.451$; that is, it is changing at the rate of 3.451 gms/day.

11. We solve the equation $0.2Q_0 = Q_0 e^{-0.00012t}$

obtaining $t = \dfrac{\ln 0.2}{-0.00012} \approx 13{,}412$, or approximately 13,412 years.

13. The graph of $Q(t)$ follows.

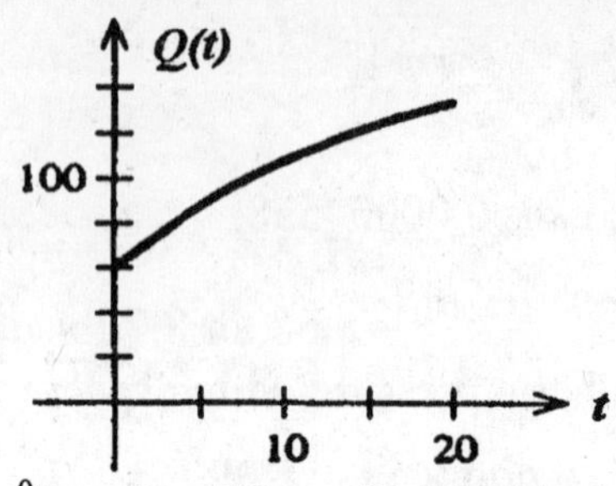

a. $Q(0) = 120(1 - e^{0}) + 60 = 60$, or 60 w.p.m.
b. $Q(10) = 120(1 - e^{-0.5}) + 60 = 107.22$, or approximately 107 w.p.m.
c. $Q(20) = 120(1 - e^{-1}) + 60 = 135.65$, or approximately 136 w.p.m.

15. The graph of $D(t)$ follows.

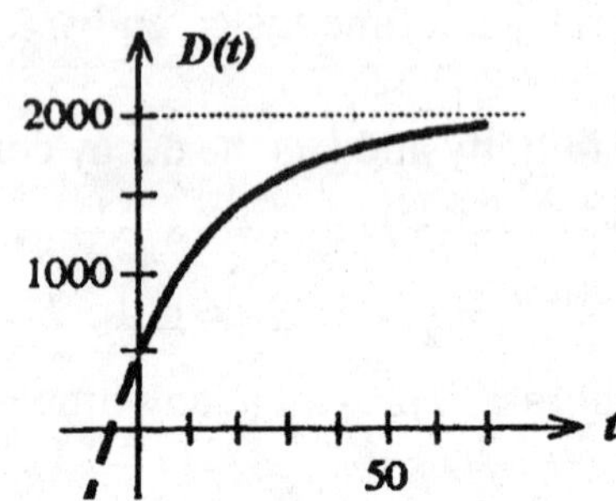

a. After one month, the demand is $D(1) = 2000 - 1500e^{-0.05} \approx 573$.
After twelve months, the demand is $D(12) = 2000 - 1500e^{-0.6} \approx 1177$.
After twenty-four months the demand is $D(24) = 2000 - 1500e^{-1.2} \approx 1548$.
After sixty months, the demand is $D(60) = 2000 - 1500e^{-3} \approx 1925$.
b. $$\lim_{t\to\infty} D(t) = \lim_{t\to\infty} 2000 - 1500e^{-0.05t} = 2000$$
and we conclude that the demand is expected to stabilize at 2000 computers per month.
c. $D'(t) = -1500e^{-0.05t}(-0.05) = 75e^{-0.05t}$. Therefore, the rate of growth after ten months is given by $D'(10) = 75e^{-0.5} \approx 45.49$, or approximately 45 computers per month.

17. a. The length is given by $f(5) = 200(1 - 0.956e^{-0.18(5)}) \approx 122.26$, or approximately 122.3 cm.
b. $f'(t) = 200(-0.956)e^{-0.18t}(-0.18) = 34.416e^{-0.18t}$. So, a 5-yr old is growing at the rate of $f'(5) = 34.416e^{-0.18(5)} \approx 13.9925$, or approximately 14 cm/yr.

c. The maximum length is given by $\lim_{t\to\infty} 200(1-0.956e^{-0.18t}) = 200$, or 200 cm.

19. a. The percent of lay teachers is $f(3) = \dfrac{98}{1+2.77e^{-3}} \approx 86.1228$, or 86.12%.

b. $f'(t) = \dfrac{d}{dt}[98(1+2.77e^{-t})^{-1}] = 98(-1)(1+2.77e^{-t})^{-2}(2.77e^{-t})(-1)$

$$= \frac{271.46e^{-t}}{(1+2.77e^{-t})^2}$$

$$f'(3) = \frac{271.46e^{-3}}{(1+2.77e^{-3})^2} \approx 10.4377.$$

So it is increasing at the rate of 10.44%/yr.

c. $f''(t) = 271.46\left[\dfrac{(1+2.77e^{-t})^2(-e^{-t}) - e^{-t}\cdot 2(1+2.77e^{-t})(-2.77e^{-t})}{(1+2.77e^{-t})^4}\right]$

$$= \frac{271.46[-(1+2.77e^{-t}+5.54e^{-t}]}{e^t(1+2.77e^{-t})^3} = \frac{271.46(2.77e^{-t}-1)}{e^t(1+2.77e^{-t})^3}.$$

Setting $f''(t) = 0$ gives $2.77e^{-t} = 1$

$$e^{-t} = \frac{1}{2.77}; \quad -t = \ln\left(\tfrac{1}{2.77}\right), \text{ and } t = 1.0188.$$

The sign diagram of f'' shows that $t = 1.02$ gives an inflection point of P. So, the

+ + + + 0 − − − − − − −

0 1.02 t

percent of lay teachers was increasing most rapidly in 1970.

21. The projected population of citizens 45-64 in 2010 is

$$P(20) = \frac{197.9}{1+3.274e^{-0.0361(20)}} \approx 76.3962$$

or 76.4 million.

23. The first of the given conditions implies that $f(0) = 300$, that is,

$$300 = \frac{3000}{1+Be^0} = \frac{3000}{1+B}.$$

So $1+B = 10$, or $B = 9$. Therefore, $f(t) = \dfrac{3000}{1+9e^{-kt}}$. Next, the condition

$f(2) = 600$ gives the equation

$$600 = \frac{3000}{1+9e^{-2k}},\ 1+9e^{-2k} = 5,\ e^{-2k} = \frac{4}{9},\ \text{or } k = -\frac{1}{2}\ln\left(\frac{4}{9}\right).$$

Therefore, $f(t) = \dfrac{3000}{1+9e^{(1/2)t\cdot\ln(4/9)}} = \dfrac{3000}{1+9(\frac{4}{9})^{t/2}}$.

The number of students who had heard about the policy four hours later is given by

$$f(4) = \frac{3000}{1+9(\frac{4}{9})^2} = 1080, \quad \text{or 1080 students.}$$

To find the rate at which the rumor was spreading at any time time, we compute

$$f'(t) = \frac{d}{dt}\left[3000(1+9e^{-0.405465t})^{-1}\right]$$

$$= (3000)(-1)(1+9e^{-0.405465})^{-2}\frac{d}{dt}(9e^{-0.405465t})$$

$$= -3000(9)(-0.405465)e^{-0.405465t}(1+9e^{-0.405465t})^{-2}$$

$$= \frac{10947.555e^{-0.405465t}}{(1+9e^{-0.405465t})^2}$$

In particular, the rate at which the rumor was spreading 4 hours after the ceremony is given by $f'(4) = \dfrac{10947.555e^{-0.405465(4)}}{(1+9e^{-0.405465(4)})^2} \approx 280.25737.$

So , the rumor is spreading at the rate of 280 students per hour.

25. $x(t) = \dfrac{15\left(1-(\frac{2}{3})^{3t}\right)}{1-\frac{1}{4}(\frac{2}{3})^{3t}}$; $\lim\limits_{t\to\infty} x(t) = \lim\limits_{t\to\infty}\dfrac{15\left(1-(\frac{2}{3})^{3t}\right)}{1-\frac{1}{4}(\frac{2}{3})^{3t}} = \dfrac{15(1-0)}{1-0} = 15$

or 15 lbs.

27. a. $C(t) = \dfrac{k}{b-a}\left(e^{-at} - e^{-bt}\right);$

$$C'(t) = \frac{k}{b-a}(-ae^{-at} + be^{-bt}) = \frac{kb}{b-a}\left[e^{-bt} - \left(\frac{a}{b}\right)e^{-at}\right]$$

$$= \frac{kb}{b-a}e^{-bt}\left[1-\frac{a}{b}e^{(b-a)t}\right]$$

$C'(t) = 0$ implies that $1 = \dfrac{a}{b}e^{(b-a)t}$ or $t = \dfrac{\ln\left(\frac{b}{a}\right)}{b-a}$.

The sign diagram of C'

$$\begin{array}{c} 0+\ +\ +\ +\ +-\ ---\\ \longrightarrow t \\ 0 \qquad \frac{\ln(\frac{b}{a})}{b-a} \end{array}$$

shows that this value of t gives a minimum.

b. $\lim_{t\to\infty} C(t) = \dfrac{k}{b-a}$.

29. a. We solve $Q_0 e^{-kt} = \frac{1}{2}Q_0$ for t. Proceeding, we have

$$e^{-kt} = \frac{1}{2},\ \ln e^{-kt} = \ln\frac{1}{2} = \ln 1 = \ln 2 = -\ln 2;$$

$$-kt = -\ln 2;$$

So $$\bar{t} = \frac{\ln 2}{k}$$

b. $$\bar{t} = \frac{\ln 2}{0.0001238} \approx 5598.927,$$ or approximately 5599 years.

USING TECHNOLOGY EXERCISES 14.5, page 904

1. a.

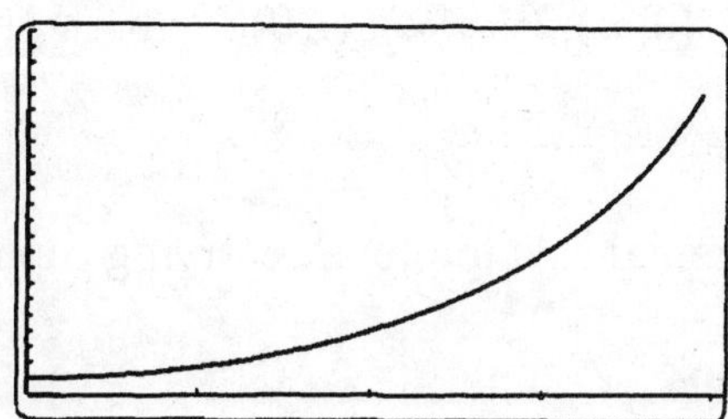

b. 12.146%/yr; 9.474%/yr

3. a.

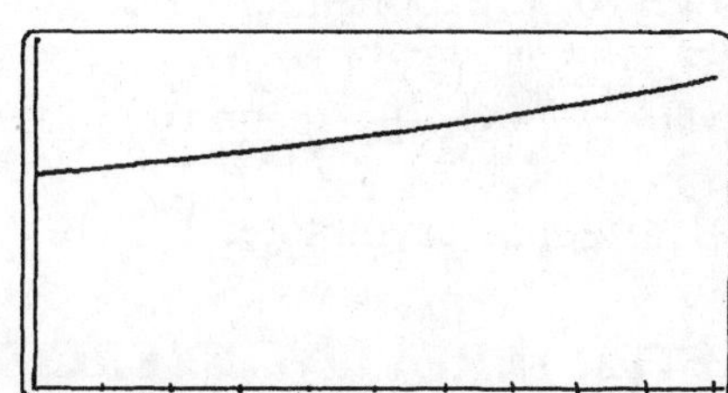

b. 666 million, 818.8 million

c. 33.8 million/yr

5. a.

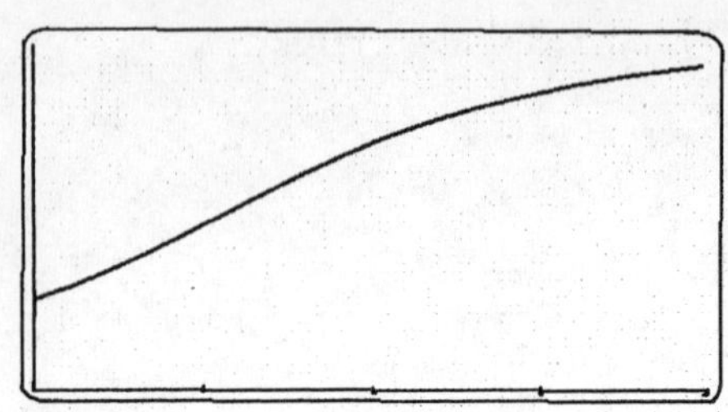

b. 86.12%/yr c. 10.44%/yr
d. 1970

7. a.

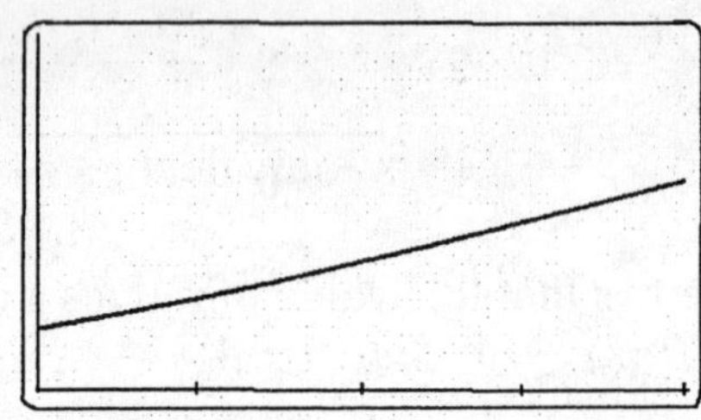

b. 325 million
c. 76.84 million/decade

9. a.

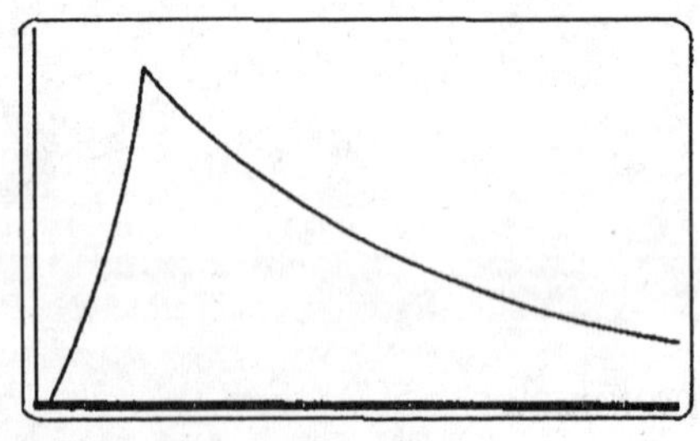

b. 0 c. 0.237 g/cm^3
d. 0.760 g/cm^3 e. 0

CHAPTER 14 CONCEPT REVIEW, page 906

1. Power; 0; 1; exponential

3. a. $(0,\infty)$; $(-\infty,\infty)$; $(1,0)$ b. <1; >1

5. a. $e^{f(x)} \cdot f'(x)$ b. $\dfrac{f'(x)}{f(x)}$

7. a. Horizontal asymptote; C b. Horizontal asymptote; A; carrying capacity

CHAPTER 14 REVIEW EXERCISES, page 906

1. a-b

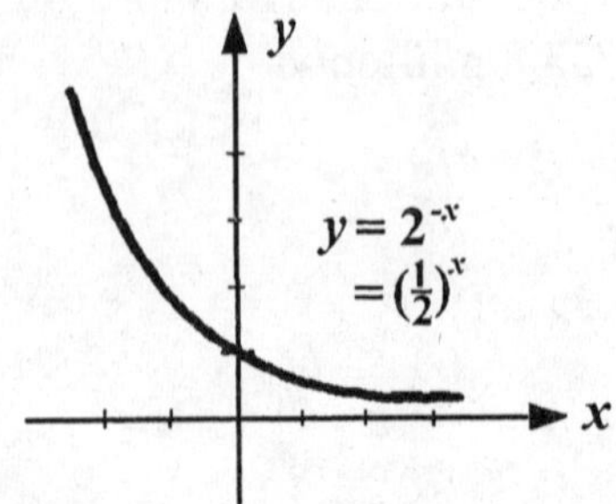

Since $y = \left(\frac{1}{2}\right)^x = \frac{1}{2^x} = 2^{-x}$, it has the same graph as that of $y = 2^{-x}$.

3. $16^{-3/4} = 0.125$ is equivalent to $-\frac{3}{4} = \log_{16} 0.125$.

5.
$$\ln(x-1) + \ln 4 = \ln(2x+4) - \ln 2$$
$$\ln(x-1) - \ln(2x+4) = -\ln 2 - \ln 4 = -(\ln 2 + \ln 4)$$
$$\ln\left(\frac{x-1}{2x+4}\right) = -\ln 8 = \ln \tfrac{1}{8}.$$

$$\left(\frac{x-1}{2x+4}\right) = \frac{1}{8}$$
$$8x - 8 = 2x + 4$$
$$6x = 12, \text{ or } x = 2.$$

CHECK: l.h.s. $\ln(2-1) + \ln 4 = \ln 4$

r.h.s $\ln(4+4) - \ln 2 = \ln 8 - \ln 2 = \ln \frac{8}{2} = \ln 4.$

7. $\ln 3.6 = \ln \frac{36}{10} = \ln 36 - \ln 10 = \ln 6^2 - \ln 2\cdot 5 = 2\ln 6 - \ln 2 - \ln 5$
$= 2(\ln 2 + \ln 3) - \ln 2 - \ln 5 = 2(x+y) - x - z = x + 2y - z.$

9. We first sketch the graph of $y = 2^{x-3}$. Then we take the reflection of this graph with respect to the line $y = x$.

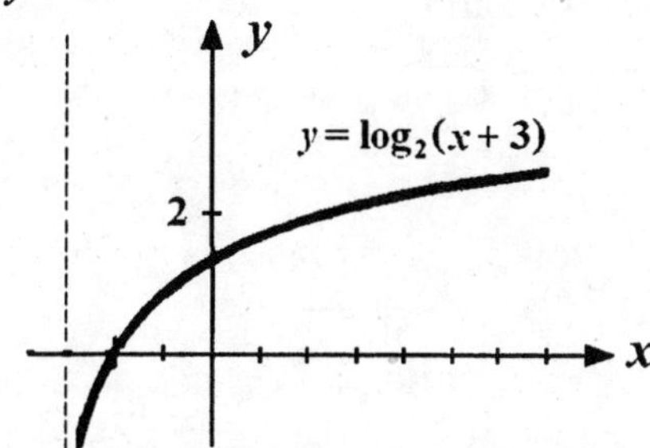

11. $f(x) = xe^{2x}$; $f'(x) = e^{2x} + xe^{2x}(2) = (1+2x)e^{2x}$.

13. $g(t) = \sqrt{t}e^{-2t}$; $g'(t) = \frac{1}{2}t^{-1/2}e^{-2t} + \sqrt{t}e^{-2t}(-2) = \dfrac{1-4t}{2\sqrt{t}e^{2t}}$.

15. $y = \dfrac{e^{2x}}{1+e^{-2x}}$; $y' = \dfrac{(1+e^{-2x})e^{2x}(2) - e^{2x}\cdot e^{-2x}(-2)}{(1+e^{-2x})^2} = \dfrac{2(e^{2x}+2)}{(1+e^{-2x})^2}$.

17. $f(x) = xe^{-x^2}; f'(x) = e^{-x^2} + xe^{-x^2}(-2x) = (1 - 2x^2)e^{-x^2}.$

19. $f(x) = x^2e^x + e^x;$
$f'(x) = 2xe^x + x^2e^x + e^x = (x^2 + 2x + 1)e^x = (x + 1)^2e^x.$

21. $f(x) = \ln(e^{x^2} + 1); f'(x) = \dfrac{e^{x^2}(2x)}{e^{x^2} + 1} = \dfrac{2xe^{x^2}}{e^{x^2} + 1}.$

23. $f(x) = \dfrac{\ln x}{x+1}.\quad f'(x) = \dfrac{(x+1)\left(\frac{1}{x}\right) - \ln x}{(x+1)^2} = \dfrac{1 + \frac{1}{x} - \ln x}{(x+1)^2} = \dfrac{x - x\ln x + 1}{x(x+1)^2}.$

25. $y = \ln(e^{4x} + 3); y' = \dfrac{e^{4x}(4)}{e^{4x} + 3} = \dfrac{4e^{4x}}{e^{4x} + 3}.$

27. $f(x) = \dfrac{\ln x}{1 + e^x};$

$$f'(x) = \frac{(1+e^x)\frac{d}{dx}\ln x - \ln x\frac{d}{dx}(1+e^x)}{(1+e^x)^2} = \frac{(1+e^x)\left(\frac{1}{x}\right) - (\ln x)e^x}{(1+e^x)^2}$$
$$= \frac{1 + e^x - xe^x\ln x}{x(1+e^x)^2} = \frac{1 + e^x(1 - x\ln x)}{x(1+e^x)^2}.$$

29. $y = \ln(3x + 1); y' = \dfrac{3}{3x+1};$
$$y'' = 3\frac{d}{dx}(3x+1)^{-1} = -3(3x+1)^{-2}(3) = -\frac{9}{(3x+1)^2}.$$

31. $h'(x) = g'(f(x))f'(x)$. But $g'(x) = 1 - \dfrac{1}{x^2}$ and $f'(x) = e^x$.
So $f(0) = e^0 = 1$ and $f'(0) = e^0 = 1$. Therefore,

$h'(0) = g'(f(0))f'(0) = g'(1)f'(0) = 0 \cdot 1 = 0.$

33. $y = (2x^3 + 1)(x^2 + 2)^3.\quad \ln y = \ln(2x^3 + 1) + 3\ln(x^2 + 2).$
$$\frac{y'}{y} = \frac{6x^2}{2x^3+1} + \frac{3(2x)}{x^2+2} = \frac{6x^2(x^2+2) + 6x(2x^3+1)}{(2x^3+1)(x^2+2)}$$

$$= \frac{6x^4 + 12x^2 + 12x^4 + 6x}{(2x^3+1)(x^2+2)} = \frac{18x^4 + 12x^2 + 6x}{(2x^3+1)(x^2+2)}.$$

Therefore, $y' = 6x(3x^3 + 2x + 1)(x^2 + 2)^2$.

35. $y = e^{-2x}$. $y' = -2e^{-2x}$ and this gives the slope of the tangent line to the graph of $y = e^{-2x}$ at any point (x, y). In particular, the slope of the tangent line at $(1, e^{-2})$ is $y'(1) = -2e^{-2}$. The required equation is $y - e^{-2} = -2e^{-2}(x-1)$ or $y = \frac{1}{e^2}(-2x+3)$.

37. $f(x) = xe^{-2x}$.

We first gather the following information on f.

1. The domain of f is $(-\infty,\infty)$.
2. Setting $x = 0$ gives 0 as the y-intercept.
3. $\lim_{x\to-\infty} xe^{-2x} = -\infty$ and $\lim_{x\to\infty} xe^{-2x} = 0$.
4. The results of (3) show that $y = 0$ is a horizontal asymptote.
5. $f'(x) = e^{-2x} + xe^{-2x}(-2) = (1-2x)e^{-2x}$. Observe that $f'(x) = 0$ if $x = 1/2$, a critical point of f. The sign diagram of f'

shows that f is increasing on $(-\infty, \frac{1}{2})$ and decreasing on $(\frac{1}{2}, \infty)$.

6. The results of (5) show that $(\frac{1}{2}, \frac{1}{2}e^{-1})$ is a relative maximum.

7. $f''(x) = -2e^{-2x} + (1-2x)e^{-2x}(-2) = 4(x-1)e^{-2x}$ and is equal to zero if $x = 1$. The sign diagram of f''

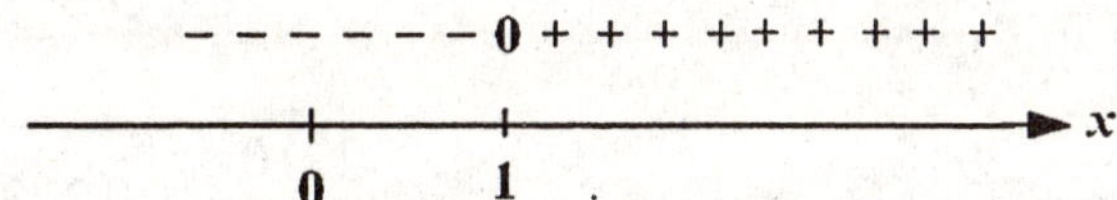

shows that the graph of f is concave downward on $(-\infty,1)$ and concave upward on $(1,\infty)$.

The graph of f follows.

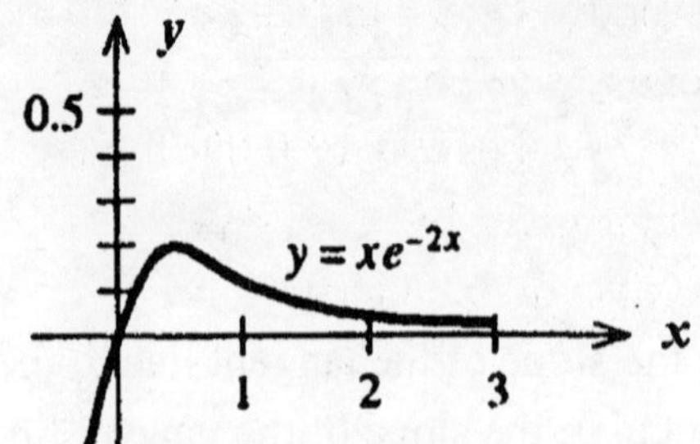

39. $f(t) = te^{-t}$. $f'(t) = e^{-t} + t(-e^{-t}) = e^{-t}(1-t)$. Setting $f'(t) = 0$ gives $t = 1$ as the only critical point of f. From the sign diagram of f'

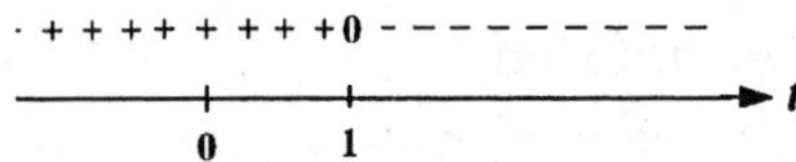

we see that $f(1) = e^{-1} = 1/e$ is the absolute maximum value of f.

41. We solve the equation $2 = 1(1+0.075)^t$ for t. Taking the logarithm on both sides, we have $\ln 2 = \ln(1.075)^t \approx t\ln 1.075$. So $t = \dfrac{\ln 2}{\ln 1.075} \approx 9.58$, or 9.6 years.

43. We have $Q(t) = Q_0e^{-kt}$, where Q_0 is the amount of radium present initially. Since the half-life of radium is 1600 years, we have $\frac{1}{2}Q_0 = Q_0e^{-1600k}$, $\quad e^{-1600k} = \frac{1}{2}$, $-1600k = \ln \frac{1}{2} = -\ln 2$, and $k = \dfrac{\ln 2}{1600} \approx 0.0004332$.

45. We have $Q(10) = 90$ and this gives $\dfrac{3000}{1+499e^{-10k}} = 90$, $1 + 499e^{-10k} = \dfrac{3000}{90}$,

$499e^{-10k} = \dfrac{2910}{90}$, $e^{-10k} = \dfrac{2910}{90(499)}$, and $k = -\dfrac{1}{10}\ln\dfrac{2910}{90(499)} \approx 0.2737$.

So $N(t) = \dfrac{3000}{1+499e^{-0.2737t}}$. The number of students who have contracted the flu by the 20th day is $N(20) = \dfrac{3000}{1+499e^{-0.2737(20)}} \approx 969.92$, or approximately 970 students.

47. a. The concentration initially is given by $C(0) = 0.08(1 - e^{-0.02(0)}) = 0$, or 0 g/cm^3.

b. The concentration after 30 seconds is given by

$$C(30) = 0.08(1 - e^{-0.02(30)}) = 0.03609, \text{ or } 0.0361 \text{ g/cm}^3.$$

c. The concentration in the long run is given by

$$\lim_{t\to\infty} 0.08(1 - e^{-0.02t}) = 0.08 \text{ or } 0.08 \text{ g/cm}^3.$$

d.

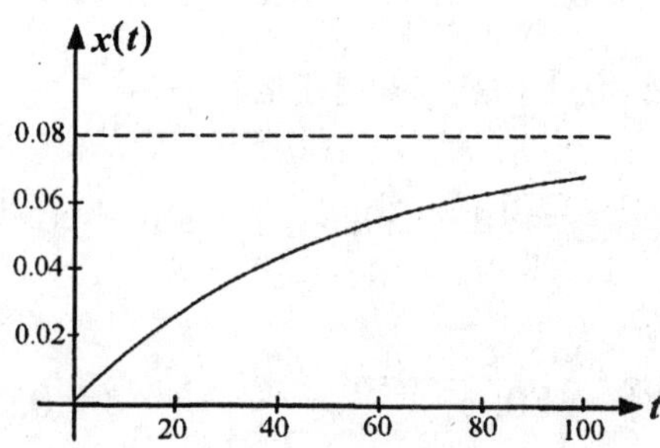

CHAPTER 14 BEFORE MOVING ON, page 908

1. $\dfrac{100}{1+2e^{0.3t}} = 40,\ 1+2e^{0.3t} = \dfrac{100}{40} = 2.5;$

$$2e^{0.3t} = 1.5,\ e^{0.3t} = \frac{1.5}{2} = 0.75,\ 0.3t = \ln 0.75,\ t = \frac{\ln 0.75}{0.3} \approx -0.959$$

2. $f'(x) = \dfrac{d}{dx} e^{x^{1/2}} = e^{x^{1/2}} \dfrac{d}{dx}(x^{1/2}) = e^{x^{1/2}} (\dfrac{1}{2} x^{-1/2}) = \dfrac{e^{\sqrt{x}}}{2\sqrt{x}}$

3. $\dfrac{dy}{dx} = x\dfrac{d}{dx}\ln(x^2+1) + \ln(x^2+1)\cdot\dfrac{d}{dx}(x) = x\cdot\dfrac{2x}{x^2+1} + \ln(x^2+1) = \dfrac{2x^2}{x^2+1} + \ln(x^2+1)$

$$\left.\frac{dy}{dx}\right|_{x=1} = \frac{1}{1+1} + \ln 2 = 1 + \ln 2$$

4. $y' = e^{2x}\dfrac{d}{dx}\ln 3x + \ln 3x \cdot \dfrac{d}{dx} e^{2x} = \dfrac{e^{2x}}{x} + 2e^{2x}\ln 3x$

$$y'' = \frac{d}{dx}(x^{-1}e^{2x}) + 2e^{2x}\frac{d}{dx}\ln 3x + (\ln 3x)\frac{d}{dx}(2e^{2x})$$

$$= -x^{-2}e^{2x} + 2x^{-1}e^{2x} + 2e^{2x}\left(\frac{1}{x}\right) + 4e^{2x} \cdot \ln 3x$$

$$= -\frac{1}{x^2}e^{2x} + \frac{4e^{2x}}{x} + 4(\ln 3x)e^{2x} = e^{2x}\left(\frac{4x^2 \ln 3x + 4x - 1}{x^2}\right)$$

5. $T(0) = 200$ gives $70 + ce^0 = 70 + C = 200$, so $C = 130$. So $T(t) = 70 + 130e^{-kt}$.

$T(3) = 180$ implies $70 + 130e^{-3k} = 180, 130e^{-3k} = 110, e^{-3k} = \frac{110}{130}, \; -3k = \ln\frac{11}{13}$,

$k = -\frac{1}{3}\ln\frac{11}{13} \approx 0.0557$. Therefore, $T(t) = 70 + 130e^{-0.0557t}$. So when $T(t) = 150$, we have

$$70 + 130e^{-0.0557t} = 150; \; 130e^{-0.0557t} = 80; \; e^{-0.0557t} = \frac{80}{130} = \frac{8}{13}; -0.0557t = \ln\frac{8}{13}$$

Thus, $t = -\frac{\ln\frac{8}{13}}{0.0557} \approx 8.716$, or approximately 8.7 minutes.

CHAPTER 15

15.1 Problem Solving Tips

1. Get into the habit of using the correct notation for integration. The indefinite integral of $f(x)$ with respect to x is written $\int x\,dx$. It is incorrect to write $\int x$ without indicating that you are integrating with respect to x. You will appreciate how important the correct notation is if you use CAS or a graphic calculator with the capability to do symbolic integration. If you don't enter this information (the variable with respect to which you are performing the integration) into your calculator or computer, the integration cannot be completed.

2. If you are finding an indefinite integral, be sure to include a constant of integration in your answer. Remember $\int f(x)\,dx$ is the family of functions given by $F(x)+C$, where $F'(x)=f(x)$.

3. It's very easy to check your answer if you are finding an indefinite integral. Just take the derivative of your answer and you should get the integrand. (You are just verifying that $F'(x)=f(x)$.) If not, you know immediately that you have made an error.

15.1 CONCEPT QUESTIONS, page 918

1. An antiderivative of a continuous function f on an interval I is a function F such that $F'(x) = f(x)$ for every x in I. For example, an antiderivative of $f(x) = x^2$ on $(-\infty, \infty)$ is the function $F(x) = \frac{1}{3}x^3$ on $(-\infty, \infty)$.
3. The indefinite integral of f is the family of functions $F(x) + C$, where F is an antiderivative of f, and C is an arbitrary constant.

EXERCISES 15.1, page 918

1. $F(x) = \frac{1}{3}x^3 + 2x^2 - x + 2$; $F'(x) = x^2 + 4x - 1 = f(x)$.

3. $F(x) = (2x^2 - 1)^{1/2}$; $F'(x) = \frac{1}{2}(2x^2 - 1)^{-1/2}(4x) = 2x(2x^2 - 1)^{-1/2} = f(x)$.

5. a. $G'(x) = \dfrac{d}{dx}(2x) = 2 = f(x)$ b. $F(x) = G(x) + C = 2x + C$

c.

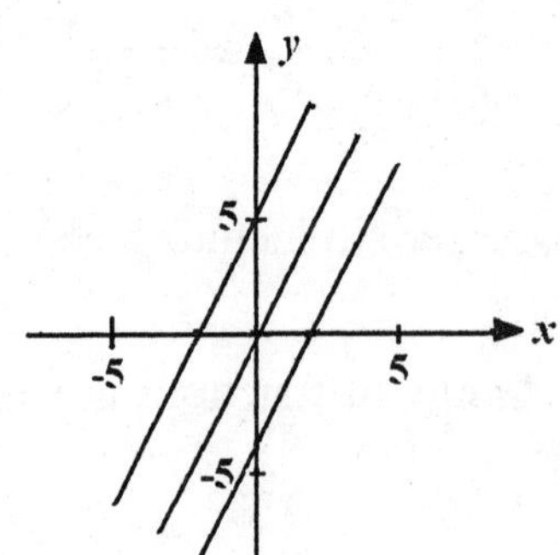

7. a. $G'(x) = \dfrac{d}{dx}(\frac{1}{3}x^3) = x^2 = f(x)$ b. $F(x) = G(x) + C = \frac{1}{3}x^3 + C$

c.

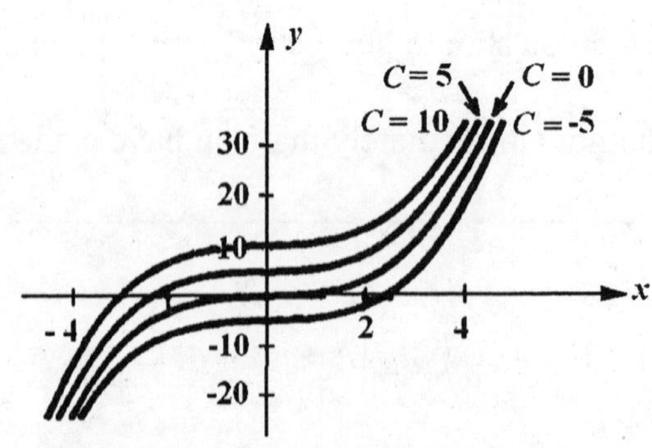

9. $\int 6\,dx = 6x + C.$

11. $\int x^3 dx = \frac{1}{4}x^4 + C$

13. $\int x^{-4} dx = -\frac{1}{3}x^{-3} + C$

15. $\int x^{2/3} dx = \frac{3}{5}x^{5/3} + C$

17. $\int x^{-5/4} dx = -4x^{-1/4} + C$

19. $\int \frac{2}{x^2}\,dx = 2\int x^{-2} dx = 2(-1x^{-1}) + C = -\frac{2}{x} + C$

21. $\int \pi\sqrt{t}\,dt = \pi\int t^{1/2} dt = \pi(\frac{2}{3}t^{3/2}) + C = \frac{2\pi}{3}t^{3/2} + C$

23. $\int (3-2x)\,dx = \int 3\,dx - 2\int x\,dx = 3x - x^2 + C$

25. $\int (x^2 + x + x^{-3})\,dx = \int x^2\,dx + \int x\,dx + \int x^{-3}\,dx = \frac{1}{3}x^3 + \frac{1}{2}x^2 - \frac{1}{2}x^{-2} + C$

27. $\int 4e^x\,dx = 4e^x + C$

29. $\int (1 + x + e^x)\,dx = x + \frac{1}{2}x^2 + e^x + C$

31. $\int (4x^3 - \frac{2}{x^2} - 1)\,dx = \int (4x^3 - 2x^{-2} - 1)\,dx = x^4 + 2x^{-1} - x + C = x^4 + \frac{2}{x} - x + C$

33. $\int (x^{5/2} + 2x^{3/2} - x)\,dx = \frac{2}{7}x^{7/2} + \frac{4}{5}x^{5/2} - \frac{1}{2}x^2 + C$

35. $\int (x^{1/2} + 3x^{-1/2})\,dx = \frac{2}{3}x^{3/2} + 6x^{1/2} + C$

37. $\int \left(\frac{u^3 + 2u^2 - u}{3u}\right) du = \frac{1}{3}\int (u^2 + 2u - 1)\,du = \frac{1}{9}u^3 + \frac{1}{3}u^2 - \frac{1}{3}u + C$

39. $\int (2t+1)(t-2)\,dt = \int (2t^2-3t-2)\,dt = \frac{2}{3}t^3 - \frac{3}{2}t^2 - 2t + C$

41. $\int \frac{1}{x^2}(x^4-2x^2+1)\,dx = \int (x^2-2+x^{-2})\,dx = \frac{1}{3}x^3 - 2x - x^{-1} + C$

$$= \frac{1}{3}x^3 - 2x - \frac{1}{x} + C$$

43. $\int \frac{ds}{(s+1)^{-2}} = \int (s+1)^2\,ds = \int (s^2+2s+1)\,ds = \frac{1}{3}s^3 + s^2 + s + C$

45. $\int (e^t + t^e)\,dt = e^t + \frac{1}{e+1}t^{e+1} + C$

47. $\int \left(\frac{x^3+x^2-x+1}{x^2}\right)dx = \int \left(x+1-\frac{1}{x}+\frac{1}{x^2}\right)dx = \frac{1}{2}x^2 + x - \ln|x| - x^{-1} + C$

49. $\int \left(\frac{(x^{1/2}-1)^2}{x^2}\right)dx = \int \left(\frac{x-2x^{1/2}+1}{x^2}\right)dx = \int (x^{-1}-2x^{-3/2}+x^{-2})\,dx$

$$= \ln|x| + 4x^{-1/2} - x^{-1} + C = \ln|x| + \frac{4}{\sqrt{x}} - \frac{1}{x} + C$$

51. $\int f'(x)\,dx = \int (2x+1)\,dx = x^2 + x + C$. The condition $f(1) = 3$ gives $f(1) = 1 + 1 + C = 3$, or $C = 1$. Therefore, $f(x) = x^2 + x + 1$.

53. $f'(x) = 3x^2 + 4x - 1$; $f(x) = x^3 + 2x^2 - x + C$. Using the given initial condition, we have $f(2) = 8 + 2(4) - 2 + C = 9$, so $16 - 2 + C = 9$, or $C = -5$. Therefore, $f(x) = x^3 + 2x^2 - x - 5$.

55. $f(x) = \int f'(x)\,dx = \int \left(1+\frac{1}{x^2}\right)dx = \int (1+x^{-2})\,dx = x - \frac{1}{x} + C.$

Using the given initial condition, we have $f(1) = 1 - 1 + C = 2$, or $C = 2$.

Therefore, $f(x) = x - \frac{1}{x} + 2.$

57. $f(x)=\int \frac{x+1}{x}\,dx=\int \left(1+\frac{1}{x}\right)dx=x+\ln|x|+C$. Using the initial condition, we have $f(1) = 1 + \ln 1 + C = 1 + C = 1$, or $C = 0$. So $f(x)=x+\ln|x|$.

59. $f(x)=\int f'(x)\,dx=\int \frac{1}{2}x^{-1/2}\,dx=\frac{1}{2}(2x^{1/2})+C=x^{1/2}+C$; $f(2)=\sqrt{2}+C=\sqrt{2}$ implies $C = 0$. So $f(x)=\sqrt{x}$.

61. $f'(x)=e^x+x$; $f(x)=e^x+\frac{1}{2}x^2+C$; $f(0)=e^0+\frac{1}{2}(0)+C=1+C$
So $3 = 1 + C$ or $2 = C$. Therefore, $f(x)=e^x+\frac{1}{2}x^2+2$.

63. The net amount on deposit in Branch A is given by the area under the graph of f from $t = 0$ to $t = 180$. On the other hand, the net amount on deposit in Branch B is given by the area under the graph of g over the same interval. Evidently the first area is larger than the second. Therefore, we see that Branch A has the larger net deposit.

65. The position of the car is
$s(t)=\int f(t)\,dt=\int 2\sqrt{t}\,dt=\int 2t^{1/2}\,dt=2(\frac{2}{3}t^{3/2})+C=\frac{4}{3}t^{3/2}+C$.
$s(0) = 0$ implies $s(0) = C = 0$. So $s(t)=\frac{4}{3}t^{3/2}$.

67. $C(x)=\int C'(x)\,dx=\int (0.000009x^2-0.009x+8)\,dx$
$=0.000003x^3-0.0045x^2+8x+k$.
$C(0)=k=120$ and so $C(x)=0.000003x^3-0.0045x^2+8x+120$.
$C(500)=0.000003(500)^3-0.0045(500)^2+8(500)+120$, or \$3370.

69. $P'(x) = -0.004x + 20$, $P(x) = -0.002x^2 + 20x + C$. Since $C = -16{,}000$, we find that $P(x) = -0.002x^2 + 20x - 16{,}000$. The company realizes a maximum profit when $P'(x) = 0$, that is, when $x = 5000$ units. Next,
$P(5000) = -0.002(5000)^2 + 20(5000) - 16{,}000 = 34{,}000$.
Thus, a maximum profit of \$34,000 is realized at a production level of 5000 units.

71. a. $N(t)=\int N'(t)\,dt=\int (-3t^2+12t+45)\,dt=-t^3+6t^2+45t+C$. But $N(0) = C = 0$

and so $N(t) = -t^3 + 6t^2 + 45t$.
b. The number is $N(4) = -4^3 + 6(4)^2 + 45(4) = 212$.

73. a. The number of subscribers in year t is given by

$$N(t) = \int r(t)\,dt = \int (-0.375t^2 + 2.1t + 2.45)dt$$
$$= -0.125t^3 + 1.05t^2 + 2.45t + C$$

To find C, note that $N(0) = 1.5$. This gives

$$N(0) = C = 1.5.$$

Therefore,

$$N(t) = -0.125t^3 + 1.05t^2 + 2.45t + 1.5$$

b. $N(5) = -0.125(5^3) + 1.05(5^2) + 2.45(5) + 1.5 = 24.375$, or 24.375 million subscribers.

75. a. The approximate average credit card debt per U.S. household in year t is

$$A(t) = \int D(t)\,dt = \int (-4.479t^2 + 69.8t + 279.5)\,dt$$
$$= -1.493t^3 + 34.9t^2 + 279.5t + C$$

Using the condition $A(0) = 2917$, we find

$A(0) = C = 2917$. Therefore,

$$A(t) = -1.493t^3 + 34.9t^2 + 279.5t + 2917$$

b. The average credit card debt per U.S. household in 2003 was

$$A(13) = -1.493(13^3) + 34.9(13^2) + 279.5(13) + 2917 \approx 9168.479,$$

or approximately \$9168.

77. a. The number of gastric bypass surgeries performed in year t is

$$N(t) = \int R(t)\,dt = \int (9.399t^2 - 13.4t + 14.07)dt$$
$$= 3.133t^3 - 6.7t^2 + 14.07t + C$$

Using the condition $N(0) = 36.7$, we find

$N(0) = C = 36.7$. Therefore, $N(t) = 3.133t^3 - 6.7t^2 + 14.07t + 36.7$.

b. The number of bypass surgeries performed in 2003 was

$$N(3) = 3.133(3^3) - 6.7(3^2) + 14.07(3) + 36.7 = 103,201,$$

or approximately 103,201.

79. a. We have the initial-value problem:

$$C'(t) = 12.288t^2 - 150.5594t + 695.23$$
$$C(0) = 3142$$

Integrating, we find

$$C(t) = \int C'(t)dt = \int (12.288t^2 - 150.5594t + 695.23)dt$$
$$= 4.096t^3 - 75.2797t^2 + 695.23t + k$$

Using the initial condition, we find

$C(0) = 0 + k = 3142$, and so $k = 3142$.

Therefore, $C(t) = 4.096t^3 - 75.2797t^2 + 695.23t + 3142$.

b. The projected average out-of-pocket costs for beneficiaries is 2010 is

$$C(2) = 4.096(2^3) - 75.2797(2^2) + 695.23(2) + 3142 = 4264.1092$$

or \$4264.11.

81. The number of new subscribers at any time is

$$N(t) = \int (100 + 210t^{3/4})\,dt = 100t + 120t^{7/4} + C.$$

The given condition implies that $N(0) = 5000$. Using this condition, we find $C = 5000$. Therefore, $N(t) = 100t + 120t^{7/4} + 5000$. The number of subscribers 16 months from now is

$N(16) = 100(16) + 120(16)^{7/4} + 5000$, or 21,960.

83. $A(t) = \int A'(t)\,dt = \int (3.2922t^2 - 0.366t^3)\,dt = 1.0974t^3 - 0.0915t^4 + C$

Now, $A(0) = C = 0$. So, $A(t) = 1.0974t^3 - 0.0915t^4$.

85. $h(t) = \int h'(t)\,dt = \int (-3t^2 + 192t + 120)\,dt = -t^3 + 96t^2 + 120t + C$

$= -t^3 + 96t^2 + 120t + C$.

$h(0) = C = 0$ implies $h(t) = -t^3 + 96t^2 + 120t$.

The altitude 30 seconds after lift-off is

$h(30) = -30^3 + 96(30)^2 + 120(30) = 63{,}000$ ft.

87. a. The number of health-care agencies in year t is

$$N(t) = \int -0.186te^{-0.02t}\,dt = 9.3e^{-0.02t} + C$$

Using the condition $N(0) = 9.3$, we find

$N(0) = 9.3 + C = 9.3$ or $C = 0$.

Therefore, $N(t) = 9.3e^{-0.02t}$

b. The number of health-care agencies in 2002 was

$$N(14) = 9.3e^{-0.02(14)} \approx 7.03, \text{ or } 7.03 \text{ thousand units.}$$

c. The number of health-care agencies in 2005 would be

$$N(17) = 9.3e^{-0.02(17)} \approx 6.62, \text{ or } 6.62 \text{ thousand units.}$$

89. $v(r) = \int v'(r)\,dr = \int -kr\,dr = -\frac{1}{2}kr^2 + C.$

But $v(R) = 0$ and so $v(R) = -\frac{1}{2}kR^2 + C = 0$, or $C = \frac{1}{2}kR^2$. Therefore, $v(R) = -\frac{1}{2}kr^2 + \frac{1}{2}kR^2 = \frac{1}{2}k(R^2 - r^2).$

91. Denote the constant deceleration by k (ft/sec^2). Then $f''(t) = -k$, so $f'(t) = v(t) = -kt + C_1$. Next, the given condition implies that $v(0) = 88$. This gives $C_1 = 88$,or $f'(t) = -kt + 88$.

$$s = f(t) = \int f'(t)\,dt = \int (-kt + 88)\,dt = -\frac{1}{2}kt^2 + 88t + C_2.$$

Also, $f(0) = 0$ gives $s = f(t) = -\frac{1}{2}kt^2 + 88t$. Since the car is brought to rest in 9 seconds, we have $v(9) = -9k + 88 = 0$, or $k = \frac{88}{9}$, or $9\frac{7}{9}$. So the deceleration is $9\frac{7}{9}$ ft/sec^2. The distance covered is

$$s = f(9) = -\frac{1}{2}\left(\frac{88}{9}\right)(81) + 88(9) = 396.$$

So the stopping distance is 396 ft.

93. The time taken by runner A to cross the finish line is

$$t = \frac{200}{22} = \frac{100}{11} \text{ sec}$$

Let a be the constant acceleration of runner B as he begins to spurt. Then

$$\frac{dv}{dt} = a \text{ or } v = \int a\,dt = at + c,$$

the velocity of runner B as he runs towards the finish line. At $t = 0$, $v = 20$ and so $v = at = 20$. Now, $\frac{ds}{dt} = v = at + 20$, so $s = \int (at + 20)dt = \frac{1}{2}at^2 + 20t + k$ (k, constant of integration) . Next, $s(0) = 0$ gives $s = \frac{1}{2}at^2 + 20t = (\frac{1}{2}at + 20)t$. In order for runner B to cover 220 ft in $\frac{100}{11}$ sec, we must have

$$[\tfrac{1}{2}a(\tfrac{100}{11}) + 20](\tfrac{100}{11}) = 220$$

$$\tfrac{50}{11}a + 20 = \tfrac{(220(11)}{100} = \tfrac{121}{5};\ \tfrac{50}{11}a = \tfrac{121}{5} - 20 = \tfrac{21}{5} \text{ or } a = \tfrac{21}{5}\cdot\tfrac{11}{50} = 0.924 \text{ (ft/sec}^2)$$

So B must have a minimum acceleration of 0.924 ft/sec^2.

95. a. We have the initial-value problem $R'(t) = \frac{8}{(t+4)^2}$ and $R(0) = 0$.

Integrating, we find $R(t) = \int \frac{8}{(t+4)^2}\,dt = 8\int (t+4)^{-2}\,dt = -\frac{8}{t+4} + C$

$R(0) = 0$ implies $-\frac{8}{4} + C = 0$ or $C = 2$.

Therefore, $R(t) = -\frac{8}{t+4} + 2 = \frac{-8+2t+8}{t+4} = \frac{2t}{t+4}$.

b. After 1 hr, $R(1) = \frac{2}{5} = 0.4$, or 0.0.4" had fallen. After 2 hr, $R(2) = \frac{4}{6} = \frac{2}{3}$, or $\frac{2}{3}$" had fallen.

97. True. See proof in Section15.1 in the text.

99. True. Use the Sum Rule followed by the Constant Multiple Rule.

15.2 Problem Solving Tips

1. Here are some hints for using the method of substitution.

   a. The idea here is to replace the given integral by a simpler integral. So look for a substitution $u = g(x)$ that simplifies the integral.

   b. Check to see that $du = g'(x)\,dx$ appears in the integral.

2. Look through Problems 1-50, so that you familiarize yourself with the types of functions that can be integrated using the method of substitution. Even if you don't complete every problem, check to see if you can set up the given integral so that the method of substitution can be used to complete the integration.

15.2 CONCEPT QUESTIONS, page 931

1. To find $I = \int f(g(x))g'(x)\,dx$ by the Method of Substitution, let $u = g(x)$, so that $du = g'(x)dx$. Making the substitution, we obtain $I = \int f(u)\,du$, which can be integrated with respect to u. Finally, replace u by $u = g(x)$ to obtain the integral.

EXERCISES 15.2, page 931

1. Put $u = 4x + 3$ so that $du = 4\,dx$, or $dx = \frac{1}{4}du$. Then

$$\int 4(4x+3)^4\,dx = \int u^4\,du = \tfrac{1}{5}u^5 + C = \tfrac{1}{5}(4x+3)^5 + C.$$

3. Let $u = x^3 - 2x$ so that $du = (3x^2 - 2)\,dx$. Then

$$\int (x^3-2x)^2(3x^2-2)\,dx = \int u^2\,du = \tfrac{1}{3}u^3 + C = \tfrac{1}{3}(x^3-2x)^3 + C.$$

5. Let $u = 2x^2 + 3$ so that $du = 4x\,dx$. Then

$$\int \frac{4x}{(2x^2+3)^3}dx = \int \frac{1}{u^3}du = \int u^{-3}\,du = -\tfrac{1}{2}u^{-2} + C = -\frac{1}{2(2x^2+3)^2} + C.$$

7. Put $u = t^3 + 2$ so that $du = 3t^2\,dt$ or $t^2\,dt = \frac{1}{3}du$. Then

$$\int 3t^2\sqrt{t^3+2}\,dt = \int u^{1/2}\,du = \tfrac{2}{3}u^{3/2} + C = \tfrac{2}{3}(t^3+2)^{3/2} + C$$

9. Let $u = x^2 - 1$ so that $du = 2x\,dx$ and $x\,dx = \frac{1}{2}du$. Then,

$$\int (x^2-1)^9\,x\,dx = \int \tfrac{1}{2}u^9\,du = \tfrac{1}{20}u^{10} + C = \tfrac{1}{20}(x^2-1)^{10} + C.$$

11. Let $u = 1 - x^5$ so that $du = -5x^4\,dx$ or $x^4\,dx = -\frac{1}{5}du$. Then

$$\int \frac{x^4}{1-x^5}dx = -\frac{1}{5}\int \frac{du}{u} = -\frac{1}{5}\ln|u| + C = -\frac{1}{5}\ln\left|1-x^5\right| + C.$$

13. Let $u = x - 2$ so that $du = dx$. Then

15. Let $u = 0.3x^2 - 0.4x + 2$. Then $du = (0.6x - 0.4)\,dx = 2(0.3x - 0.2)\,dx$.

$$\int \frac{0.3x - 0.2}{0.3x^2 - 0.4x + 2}\,dx = \int \frac{1}{2u}\,du = \frac{1}{2}\ln|u| + C = \frac{1}{2}\ln(0.3x^2 - 0.4x + 2) + C.$$

17. Let $u = 3x^2 - 1$ so that $du = 6x\,dx$, or $x\,dx = \frac{1}{6}\,du$. Then

$$\int \frac{x}{3x^2 - 1}\,dx = \frac{1}{6}\int \frac{du}{u} = \frac{1}{6}\ln|u| + C = \frac{1}{6}\ln|3x^2 - 1| + C.$$

19. Let $u = -2x$ so that $du = -2\,dx$ or $dx = -\frac{1}{2}\,du$. Then

$$\int e^{-2x}\,dx = -\tfrac{1}{2}\int e^u\,du = -\tfrac{1}{2}e^u + C = -\tfrac{1}{2}e^{-2x} + C.$$

21. Let $u = 2 - x$ so that $du = -\,dx$ or $dx = -\,du$. Then

$$\int e^{2-x}dx = -\int e^u\,du = -e^u + C = -e^{2-x} + C.$$

23. Let $u = -x^2$, then $du = -2x\,dx$ or $x\,dx = -\frac{1}{2}\,du$.

$$\int xe^{-x^2}\,dx = \int -\tfrac{1}{2}e^u\,du = -\tfrac{1}{2}e^u + C = -\tfrac{1}{2}e^{-x^2} + C.$$

25. $\int (e^x - e^{-x})\,dx = \int e^x\,dx - \int e^{-x}\,dx = e^x - \int e^{-x}\,dx.$

To evaluate the second integral on the right, let $u = -x$ so that $du = -dx$ or

$dx = -du$. Therefore,

$$\int (e^x - e^{-x})\,dx = e^x + \int e^u\,du = e^x + e^u + C = e^x + e^{-x} + C.$$

27. Let $u = 1 + e^x$ so that $du = e^x\,dx$. Then

$$\int \frac{e^x}{1 + e^x}\,dx = \int \frac{du}{u} = \ln|u| + C = \ln(1 + e^x) + C.$$

29. Let $u = \sqrt{x} = x^{1/2}$. Then $du = \frac{1}{2}x^{-1/2}\,dx$ or $2\,du = x^{-1/2}\,dx$.

$$\int \frac{e^{\sqrt{x}}}{\sqrt{x}}\,dx = \int 2e^u\,du = 2e^u + C = 2e^{\sqrt{x}} + C.$$

31. Let $u = e^{3x} + x^3$ so that $du = (3e^{3x} + 3x^2)\,dx = 3(e^{3x} + x^2)\,dx$ or $(e^{3x} + x^2)\,dx = \frac{1}{3}du$. Then

$$\int \frac{e^{3x} + x^2}{(e^{3x} + x^3)^3}\,dx = \frac{1}{3}\int \frac{du}{u^3} = \frac{1}{3}\int u^{-3}\,du = -\frac{1}{6}u^{-2} + C = -\frac{1}{6(e^{3x} + x^3)^2} + C.$$

33. Let $u = e^{2x} + 1$, so that $du = 2e^{2x}\,dx$, or $\frac{1}{2}du = e^{2x}\,dx$.

$$\int e^{2x}(e^{2x} + 1)^3\,dx = \int \frac{1}{2}u^3\,du = \frac{1}{8}u^4 + C = \frac{1}{8}(e^{2x} + 1)^4 + C.$$

35. Let $u = \ln 5x$ so that $du = \dfrac{1}{x}\,dx$. Then

$$\int \frac{\ln 5x}{x}\,dx = \int u\,du = \frac{1}{2}u^2 + C = \frac{1}{2}(\ln 5x)^2 + C.$$

37. Let $u = \ln x$ so that $du = \frac{1}{x}\,dx$. Then

$$\int \frac{1}{x\ln x}\,dx = \int \frac{du}{u} = \ln|u| + C = \ln|\ln x| + C.$$

39. Let $u = \ln x$ so that $du = \frac{1}{x}\,dx$. Then

$$\int \frac{\sqrt{\ln x}}{x}\,dx = \int \sqrt{u}\,du = \frac{2}{3}u^{3/2} + C = \frac{2}{3}(\ln x)^{3/2} + C.$$

41. $$\int \left(xe^{x^2} - \frac{x}{x^2+2}\right)dx = \int xe^{x^2}\,dx - \int \frac{x}{x^2+2}\,dx.$$

To evaluate the first integral, let $u = x^2$ so that $du = 2x\,dx$, or $x\,dx = \frac{1}{2}du$. Then

$$\int xe^{x^2}\,dx = \frac{1}{2}\int e^u\,du + C_1 = \frac{1}{2}e^u + C_1 = \frac{1}{2}e^{x^2} + C_1.$$

To evaluate the second integral, let $u = x^2 + 2$ so that $du = 2x\,dx$, or $x\,dx = \frac{1}{2}\,du$.

Then $\displaystyle\int \frac{x}{x^2+2}\,dx = \frac{1}{2}\int \frac{du}{u} = \frac{1}{2}\ln|u| + C_2 = \frac{1}{2}\ln(x^2+2) + C_2.$

Therefore, $\displaystyle\int \left(xe^{x^2} - \frac{x}{x^2+2}\right) dx = \frac{1}{2}e^{x^2} - \frac{1}{2}\ln(x^2+2) + C.$

43. Let $u = \sqrt{x} - 1$ so that $du = \frac{1}{2}x^{-1/2}\,dx = \dfrac{1}{2\sqrt{x}}\,dx$ or $dx = 2\sqrt{x}\,du$.

Also, we have $\sqrt{x} = u + 1$, so that $x = (u+1)^2 = u^2 + 2u + 1$ and $dx = 2(u+1)\,du$.
So

$$\int \frac{x+1}{\sqrt{x}-1}\,dx = \int \frac{u^2+2u+2}{u}\cdot 2(u+1)\,du = 2\int \frac{(u^3+3u^2+4u+2)}{u}\,du$$

$$= 2\int \left(u^2+3u+4+\frac{2}{u}\right) du = 2\left(\frac{1}{3}u^3 + \frac{3}{2}u^2 + 4u + 2\ln|u|\right) + C$$

$$= 2\left[\frac{1}{3}(\sqrt{x}-1)^3 + \frac{3}{2}(\sqrt{x}-1)^2 + 4(\sqrt{x}-1) + 2\ln\left|\sqrt{x}-1\right|\right] + C.$$

45. Let $u = x - 1$ so that $du = dx$. Also, $x = u + 1$ and so

$$\int x(x-1)^5\,dx = \int (u+1)u^5\,du = \int (u^6+u^5)\,du$$

$$= \frac{1}{7}u^7 + \frac{1}{6}u^6 + C = \frac{1}{7}(x-1)^7 + \frac{1}{6}(x-1)^6 + C$$

$$= \frac{(6x+1)(x-1)^6}{42} + C.$$

47. Let $u = 1 + \sqrt{x}$ so that $du = \frac{1}{2}x^{-1/2}\,dx$ and $dx = 2\sqrt{x} = 2(u-1)\,du$

$$\int \frac{1-\sqrt{x}}{1+\sqrt{x}}\,dx = \int \left(\frac{1-(u-1)}{u}\right)\cdot 2(u-1)\,du = 2\int \frac{(2-u)(u-1)}{u}\,du$$

$$= 2\int \frac{-u^2+3u-2}{u}\,du = 2\int \left(-u+3-\frac{2}{u}\right) du = -u^2 + 6u - 4\ln|u| + C$$

$$= -(1+\sqrt{x})^2 + 6(1+\sqrt{x}) - 4\ln(1+\sqrt{x}) + C$$

$$= -1 - 2\sqrt{x} - x + 6 + 6\sqrt{x} - 4\ln(1+\sqrt{x}) + C$$

$$= -x + 4\sqrt{x} + 5 - 4\ln(1+\sqrt{x}) + C.$$

49. $I = \int v^2(1-v)^6\,dv$. Let $u = 1 - v$, then $du = -dv$. Also, $1 - u = v$, and $(1 - u)^2 = v^2$. Therefore,

$$I = \int -(1-2u+u^2)u^6\,du = \int -(u^6 - 2u^7 + u^8)\,du = -\left(\frac{u^7}{7} - \frac{2u^8}{8} + \frac{u^9}{9}\right) + C$$

$$= -u^7\left(\frac{1}{7} - \frac{1}{4}u + \frac{1}{9}u^2\right) + C = -\frac{1}{252}(1-v)^7[36 - 63(1-v) + 28(1-2v+v^2)]$$

$$= -\frac{1}{252}(1-v)^7[36 - 63 + 63v + 28 - 56v + 28v^2]$$

$$= -\frac{1}{252}(1-v)^7(28v^2 + 7v + 1) + C.$$

51. $f(x) = \int f'(x)\,dx = 5\int (2x-1)^4\,dx$. Let $u = 2x - 1$ so that $du = 2x-1$ so that $du = 2\ dx$, or $dx = \frac{1}{2}du$. Then $f(x) = \dfrac{5}{2}\int u^4\,du = \dfrac{1}{2}u^5 + C = \dfrac{1}{2}(2x-1)^5 + C.$

Next, $f(1) = 3$ implies $\dfrac{1}{2} + C = 3$ or $C = \dfrac{5}{2}$. Therefore,

$$f(x) = \frac{1}{2}(2x-1)^5 + \frac{5}{2}.$$

53. $f(x) = \int -2xe^{-x^2+1}\,dx$. Let $u = -x^2 + 1$ so that $du = -2x\,dx$. Then

$f(x) = \int e^u\,du = e^u + C = e^{-x^2+1} + C.$ The condition $f(1) = 0$ implies

$f(1) = 1 + C = 0$, or $C = -1$. Therefore, $f(x) = e^{-x^2+1} - 1.$

55. The number of subscribers at time t is

$$N(t) = \int R(t)\,dt = \int 3.36(t+1)^{0.05}\,dt$$

Let $u = t+1$, so that $du = dt$. Thus

$$N = 3.36\int u^{0.05}\,du = 3.2u^{1.05} + C = 3.2(t+1)^{1.05} + C$$

To find C, use the condition $N(0) = 3.2$ giving

$$N(0) = 3.2 + C = 3.2,$$

so $C = 0$. Therefore, $N(t) = 3.2(t+1)^{1.05}$. If the projection holds true, then the number of subscribers at the beginning of 2008 will be

$N(4) = 3.2(4+1)^{1.05} \approx 17.341$, or 17.341 million.

57. $N'(t) = 2000(1 + 0.2t)^{-3/2}$. Let $u = 1 + 0.2t$. Then $du = 0.2\, dt$ and $5\, du = dt$. Therefore, $N(t) = (5)(2000)$

$$\int u^{-3/2}\, du = -20{,}000u^{-1/2} + C = -20{,}000(1+0.2t)^{-1/2} + C.$$

Next, $N(0) = -20{,}000(1)^{-1/2} + C = 1000$. Therefore, $C = 21{,}000$ and $N(t) = -\dfrac{20{,}000}{\sqrt{1+0.2t}} + 21{,}000$. In particular, $N(5) = -\dfrac{20{,}000}{\sqrt{2}} + 21{,}000 \approx 6{,}858$.

59. $p(x) = \displaystyle\int -\frac{250x}{(16+x^2)^{3/2}}\, dx = -250\int \frac{x}{(16+x^2)^{3/2}}\, dx.$

Let $u = 16 + x^2$ so that $du = 2x\, dx$ and $x\, dx = \frac{1}{2}\, du$.

Then $p(x) = -\frac{250}{2}\displaystyle\int u^{-3/2}\, du = (-125)(-2)u^{-1/2} + C = \frac{250}{\sqrt{16+x^2}} + C.$

$p(3) = \dfrac{250}{\sqrt{16+9}} + C = 50$ implies $C = 0$ and $p(x) = \dfrac{250}{\sqrt{16+x^2}}$.

61. Let $u = 2t + 4$, so that $du = 2\, dt$. Then

$$r(t) = \int \frac{30}{\sqrt{2t+4}}\, dt = 30\int \frac{1}{2}u^{-1/2}\, du = 30u^{1/2} + C = 30\sqrt{2t+4} + C.$$

$r(0) = 60 + C = 0$, and $C = -60$. Therefore, $r(t) = 30\left(\sqrt{2t+4} - 2\right)$. Then

$r(16) = 30\left(\sqrt{36} - 2\right) = 120$ ft. Therefore, the polluted area is

$\pi r^2 = \pi(120)^2 = 14{,}400\pi$, or $14{,}400\pi$ sq ft.

63. Let $u = 1 + 2.449e^{-0.3277t}$ so that $du = -0.802537e^{-0.3277t}\, dt$ and $e^{-0.3277t}\, dt = -1.24605\, du$.

Then $h(t) = \displaystyle\int \frac{52.8706e^{-0.3277t}}{(1+2.449e^{-0.3277t})^2}\, dt = (52.8706)(-1.24605)\int \frac{du}{u^2}$

$$= 65.8794u^{-1} + C = \frac{65.8794}{1+2.449e^{-0.3277t}} + C.$$

$$h(0) = \frac{65.8794}{1+2.449} + C = 19.4, \text{ and } C = 0.3.$$

Therefore, $h(t) = \dfrac{65.8794}{1+2.449e^{-0.3277t}} + 0.3,$

and $h(8) = \dfrac{65.8794}{1+2.449e^{-0.3277(8)}} + 0.3 \approx 56.22$, or 56.22 inches.

65. $A(t) = \int A'(t)\,dt = r\int e^{-at}\,dt$. Let $u = -at$ so that $du = -a\,dt$, or $dt = -\frac{1}{a}du$.

$A(t) = r(-\frac{1}{a})\int e^{u}\,du = -\frac{r}{a}e^{u} + C = -\frac{r}{a}e^{-at} + C$

$A(0)$ implies $-\dfrac{r}{a} + C = 0$, or $C = \dfrac{r}{a}$. So, $A(t) = -\dfrac{r}{a}e^{-at} + \dfrac{r}{a} = \dfrac{r}{a}(1-e^{-at})$.

15.3 Problem Solving Tips

1. In Sections 15.1 and 15.2, we found the indefinite integral of a function , that is, $\int f(x)\,dx = F(x) + C$. Note that our answer ($F(x)+C$) was a *family of functions* where $F'(x) = f(x)$. In this section, we found the definite integral of a function, $\int_a^b f(x)\,dx = \lim_{n\to\infty}[f(x_1)\Delta x + f(x_2)\Delta x + \cdots + f(x_n)\Delta x]$). Note that the answer here is a *number*.

2. The geometric interpretation of a definite integral follows: If f is continuous on $[a,b]$, then $\int_a^b f(x)\,dx$ is equal to the area of the region above $[a, b]$ minus the area of the region below $[a,b]$.

15.3 CONCEPT QUESTIONS, page 941

1. See text page 938.

EXERCISES 15.3, page 941

1. $\frac{1}{3}(1.9 + 1.5 + 1.8 + 2.4 + 2.7 + 2.5) = \frac{12.8}{3} \approx 4.27.$

3. a. $A = \frac{1}{2}(2)(6) = 6$ sq units.

   b. $\Delta x = \frac{2}{4} = \frac{1}{2}$; $x_1 = 0$, $x_2 = \frac{1}{2}$, $x_3 = 1$, $x_4 = \frac{3}{2}$.

   $A \approx \frac{1}{2}[3(0) + 3(\frac{1}{2}) + 3(1) + 3(\frac{3}{2})] = \frac{9}{2}$
   $= 4.5$ sq units.

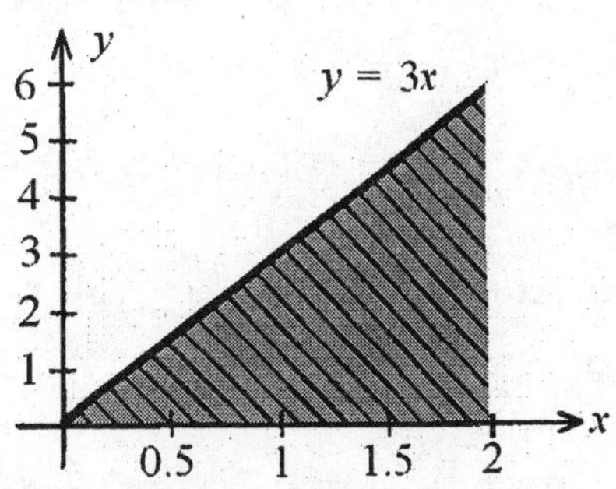

   c. $\Delta x = \frac{2}{8} = \frac{1}{4}$. $x_1 = 0, \ldots, x_8 = \frac{7}{4}$.

$$A \approx \tfrac{1}{4}\left[3(0) + 3(\tfrac{1}{4}) + 3(\tfrac{1}{2}) + 3(\tfrac{3}{4}) + 3(1) + 3(\tfrac{5}{4}) + 3(\tfrac{3}{2}) + 3(\tfrac{7}{4})\right]$$

$$= \tfrac{21}{4} = 5.25 \text{ sq units.}$$

   d. Yes.

5. a. $A = 4$

   b. $\Delta x = \frac{2}{5} = 0.4$; $x_1 = 0$, $x_2 = 0.4$, $x_3 = 0.8$,
   $x_4 = 1.2$, $x_5 = 1.6$

   $A \approx 0.4\{[4 - 2(0)] + [4 - 2(0.4)] + [4 - 2(0.8)]$
   $+ [4 - 2(1.2)] + [4 - 2(1.6)]\}$
   $= 4.8$

   c. $\Delta x = \frac{2}{10} = 0.2$, $x_1 = 0$, $x_2 = 0.2$, $x_3 = 0.4$, ...,
   $x_{10} = 1.8$.

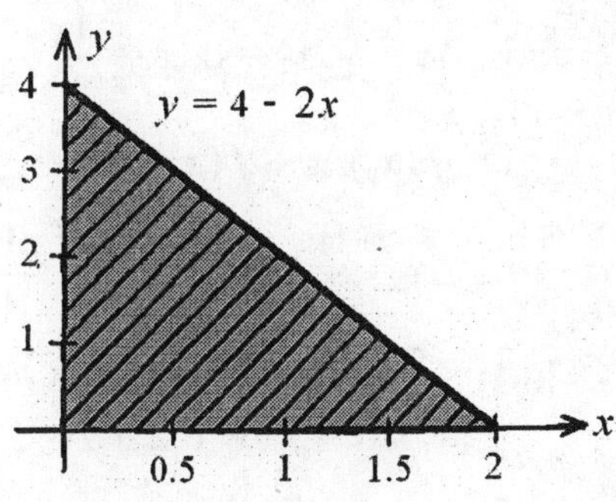

$$A \approx 0.2\{[4 - 2(0)] + [4 - 2(0.2)] + [4 - 2(0.4)] + \cdots + [4 - 2(1.8)]\} = 4.4$$

   d. Yes.

7. a. $\Delta x = \dfrac{4-2}{2} = 1$; $x_1 = 2.5, x_2 = 3.5$; The Riemann sum is $[(2.5)^2 + (3.5)^2] = 18.5$.

b. $\Delta x = \dfrac{4-2}{5} = 0.4$; $x_1 = 2.2, x_2 = 2.6, x_3 = 3.0, x_4 = 3.4, x_5 = 3.8$.

The Riemann sum is $0.4[2.2^2 + 2.6^2 + 3.0^2 + 3.4^2 + 3.8^2] = 18.64$.

c. $\Delta x = \dfrac{4-2}{10} = 0.2$; $x_1 = 2.1, x_2 = 2.3, x_2 = 2.5, \ldots, x_{10} = 3.9$

The Riemann sum is $0.2[2.1^2 + 2.3^2 + 2.5^2 + \cdots + 3.9^2] = 18.66$.

The area seems to be $18\frac{2}{3}$ sq units.

9. a. $\Delta x = \dfrac{4-2}{2} = 1$; $x_1 = 3, x_2 = 4$. The Riemann sum is $(1)[3^2 + 4^2] = 25$.

b. $\Delta x = \dfrac{4-2}{5} = 0.4$; $x_1 = 2.4, x_2 = 2.8, x_3 = 3.2, x_4 = 3.6, x_5 = 4$.

The Riemann sum is $0.4[2.4^2 + 2.8^2 + \cdots + 4^2] = 21.12$.

c. $\Delta x = \dfrac{4-2}{10} = 0.2$; $x_1 = 2.2, x_2 = 2.4, x_3 = 2.6, \ldots, x_{10} = 4$.

The Riemann sum is $0.2[2.2^2 + 2.4^2 + 2.6^2 + \cdots + 4^2] = 19.88$.

d. 19.9 sq units.

11. a. $\Delta x = \dfrac{1}{2}$, $x_1 = 0, x_2 = \dfrac{1}{2}$. The Riemann sum is

$f(x_1)\Delta x + f(x_2)\Delta x = \left[(0)^3 + (\tfrac{1}{2})^3\right]\tfrac{1}{2} = \tfrac{1}{16} = 0.0625$.

b. $\Delta x = \dfrac{1}{5}$, $x_1 = 0, x_2 = \dfrac{1}{5}, x_3 = \dfrac{2}{5}, x_4 = \dfrac{3}{5}, x_5 = \dfrac{4}{5}$. The Riemann sum

is $f(x_1)\Delta x + f(x_2)\Delta x + \cdots + f(x_5)\Delta x = \left[(\tfrac{1}{5})^3 + (\tfrac{2}{5})^3 + \cdots + (\tfrac{4}{5})^3\right]\tfrac{1}{5} = \tfrac{100}{625} = 0.16$.

c. $\Delta x = \dfrac{1}{10}$; $x_1 = 0, x_2 = \dfrac{1}{10}, x_3 = \dfrac{2}{10}, \ldots, x_{10} = \dfrac{9}{10}$.

The Riemann sum is

$$f(x_1)\Delta x + f(x_2)\Delta x + \cdots + f(x_{10})\Delta x = \left[(\tfrac{1}{10})^3 + (\tfrac{2}{10})^3 + \cdots + (\tfrac{9}{10})^3\right]\tfrac{1}{10}$$

$$= \tfrac{2025}{10{,}000} = 0.2025 \approx 0.2 \text{ sq units.}$$

The Riemann sum seems to approach 0.2.

13. $\Delta x = \dfrac{2-0}{5} = \dfrac{2}{5}$; $x_1 = \dfrac{1}{5}, x_2 = \dfrac{3}{5}, x_3 = \dfrac{5}{5}, x_4 = \dfrac{7}{5}, x_5 = \dfrac{9}{5}$.

$$A \approx \left\{\left[\left(\tfrac{1}{5}\right)^2+1\right]+\left[\left(\tfrac{3}{5}\right)^2+1\right]+\left[\left(\tfrac{5}{5}\right)^2+1\right]+\left[\left(\tfrac{7}{5}\right)^2+1\right]+\left[\left(\tfrac{9}{5}\right)^2+1\right]\left(\tfrac{2}{5}\right)\right\}$$

$$= \tfrac{580}{125} = 4.64 \text{ sq units.}$$

15. $\Delta x = \dfrac{3-1}{4} = \dfrac{1}{2}$; $x_1 = \dfrac{3}{2}$, $x_2 = \dfrac{4}{2}$, $x_3 = \dfrac{5}{2}$, $x_4 = 3$.

$$A \approx \left[\frac{1}{\frac{3}{2}} + \frac{1}{\frac{4}{2}} + \frac{1}{\frac{5}{2}} + \frac{1}{3}\right]\frac{1}{2} \approx 0.95 \text{ sq units.}$$

17. $A = 20[f(10) + f(30) + f(50) + f(70) + f(90)]$
$= 20(80 + 100 + 110 + 100 + 80) = 9400$ sq ft.

15.4 CONCEPT QUESTIONS, page 951

1. See the Fundamental of Calculus Theorem on page 943 in the text.

EXERCISES 15.4, page 951

1. $A = \int_1^4 2\,dx = 2x\Big|_1^4 = 2(4-1) = 6$, or 6 square units. The region is a rectangle whose area is

$3 \cdot 2$, or 6, square units.

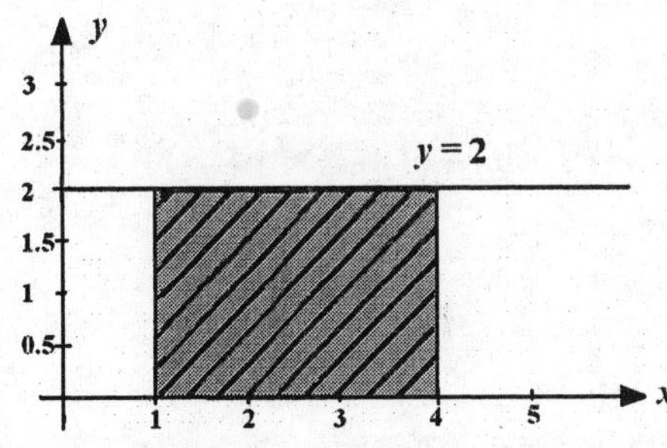

3. $A=\int_1^3 2x\,dx = x^2\Big|_1^3 = 9-1=8$, or 8 sq units.

The region is a parallelogram of area (1/2)(3 − 1)(2 + 6) = 8 sq units.

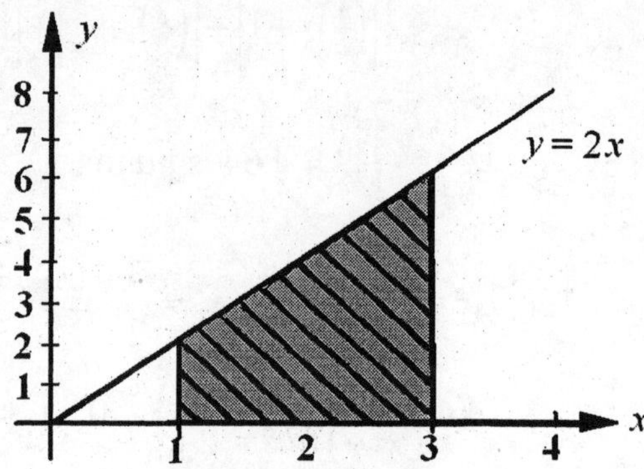

5. $A=\int_{-1}^2 (2x+3)\,dx = x^2+3x\Big|_{-1}^2 = (4+6)-(1-3)=12$, or 12 sq units.

7. $A=\int_{-1}^2 (-x^2+4)\,dx = -\frac{1}{3}x^3+4x\Big|_{-1}^2 = \left(-\frac{8}{3}+8\right)-\left(\frac{1}{3}-4\right)=9$, or 9 sq units.

9. $A=\int_1^2 \frac{1}{x}\,dx = \ln|x|\Big|_1^2 = \ln 2-\ln 1=\ln 2$, or ln 2 sq units.

11. $A=\int_1^9 \sqrt{x}\,dx = \frac{2}{3}x^{3/2}\Big|_1^9 = \frac{2}{3}(27-1)=\frac{52}{3}$, or $17\frac{1}{3}$ sq units.

13. $A=\int_{-8}^{-1} (1-x^{1/3})\,dx = x-\frac{3}{4}x^{4/3}\Big|_{-8}^{-1} = (-1-\frac{3}{4})-(-8-12)=18\frac{1}{4}$, or $18\frac{1}{4}$ sq units.

15. $A=\int_0^2 e^x\,dx = e^x\Big|_0^2 = (e^2-1)$, or approximately 6.39 sq units.

17. $\int_2^4 3\,dx = 3x\Big|_2^4 = 3(4-2)=6$.

19. $\int_1^3 (2x+3)\,dx = x^2+3x\Big|_1^3 = (9+9)-(1+3)=14$.

21. $\int_{-1}^3 2x^2\,dx = \frac{2}{3}x^3\Big|_{-1}^3 = \frac{2}{3}(27)-\frac{2}{3}(-1)=\frac{56}{3}$.

23. $\int_{-2}^2 (x^2-1)\,dx = \frac{1}{3}x^3-x\Big|_{-2}^2 = \left(\frac{8}{3}-2\right)-\left(-\frac{8}{3}+2\right)=\frac{4}{3}$.

25. $\int_1^8 4x^{1/3}\,dx = (4)(\tfrac{3}{4})x^{4/3}\Big|_1^8 = 3(16-1) = 45.$

27. $\int_0^1 (x^3 - 2x^2 + 1)\,dx = \tfrac{1}{4}x^4 - \tfrac{2}{3}x^3 + x\Big|_0^1 = \tfrac{1}{4} - \tfrac{2}{3} + 1 = \tfrac{7}{12}$

29. $\int_2^4 \frac{1}{x}\,dx = \ln|x|\Big|_2^4 = \ln 4 - \ln 2 = \ln(\tfrac{4}{2}) = \ln 2.$

31. $\int_0^4 x(x^2-1)\,dx = \int_0^4 (x^3 - x)\,dx = \tfrac{1}{4}x^4 - \tfrac{1}{2}x^2\Big|_0^4 = 64 - 8 = 56.$

33. $\int_1^3 (t^2 - t)^2\,dt = \int_1^3 t^4 - 2t^3 + t^2)\,dt = \frac{1}{5}t^5 - \frac{1}{2}t^4 + \frac{1}{3}t^3\Big|_1^3$

$$= \left(\frac{243}{5} - \frac{81}{2} + \frac{27}{3}\right) - \left(\frac{1}{5} - \frac{1}{2} + \frac{1}{3}\right) = \frac{512}{30} = \frac{256}{15}.$$

35. $\int_{-3}^{-1} x^{-2}\,dx = -\frac{1}{x}\Big|_{-3}^{-1} = 1 - \frac{1}{3} = \frac{2}{3}.$

37. $\int_1^4 \left(\sqrt{x} - \frac{1}{\sqrt{x}}\right)dx = \int_1^4 (x^{1/2} - x^{-1/2})\,dx = \frac{2}{3}x^{3/2} - 2x^{1/2}\Big|_1^4$

$$= \left(\frac{16}{3} - 4\right) - \left(\frac{2}{3} - 2\right) = \frac{8}{3}.$$

39. $\int_1^4 \frac{3x^3 - 2x^2 + 4}{x^2}\,dx = \int_1^4 (3x - 2 + 4x^{-2})\,dx = \frac{3}{2}x^2 - 2x - \frac{4}{x}\Big|_1^4$

$$= (24 - 8 - 1) - (\tfrac{3}{2} - 2 - 40 = \tfrac{39}{2}.$$

41. a. $C(300) - C(0) = \int_0^{300} (0.0003x^2 - 0.12x + 20)\,dx = 0.0001x^3 - 0.06x^2 + 20x\Big|_0^{300}$

$$= 0.0001(300)^3 - 0.06(300)^2 + 20(300) = 3300.$$

Therefore $C(300) = 3300 + C(0) = 3300 + 800 = 4100$, or \$4100.

b. $\int_{200}^{300} C'(x)\,dx = (0.0001x^3 - 0.06x^2 + 20x)\Big|_{200}^{300}$

$$= [0.0001(300)^3 - 0.06(300)^2 + 20(300)]$$
$$-[0.0001(200)^3 - 0.06(200)^2 + 20(200)]$$
$$= 900 \text{ or } \$900.$$

43. a. The profit is $\int_0^{200} (-0.0003x^2 + 0.02x + 20)\,dx + P(0)$

$$= -0.0001x^3 + 0.01x^2 + 20x\Big|_0^{200} + P(0)$$
$$= 3600 + P(0) = 3600 - 800, \text{ or } \$2800.$$

b. $\int_{200}^{220} P'(x)\,dx = P(220) - P(200) = -0.0001x^3 + 0.01x^2 + 20x\Big|_{200}^{220}$

$$= 219.20, \text{ or } \$219.20.$$

45. The distance is

$$\int_0^{20} v(t)\,dt = \int_0^{20} (-t^2 + 20t + 440)\,dt = -\tfrac{1}{3}t^3 + 10t^2 + 440t\Big|_0^{20} \approx 10{,}133\tfrac{1}{3} \text{ ft.}$$

47. a. The percent of these households in decade t is

$$P(t) = \int R(t)\,dt = \int (0.8499t^2 - 3.872t + 5)\,dt$$
$$= 0.2833t^3 - 1.936t^2 + 5t + C$$

The condition $P(0) = 5.6$ gives

$$P(0) = C = 5.6$$

Therefore,

$$P(t) = 0.2833t^3 - 1.936t^2 + 5t + 5.6$$

b. The percent of these households in 2010 will be

$$P(4) = 0.2833(4^3) - 1.936(4^2) + 5(4) + 5.6 = 12.7552$$

or approximately 12.8%.

c. The percent of these households in 2000 was

$$P(3) = 0.2833(3^3) - 1.936(3^2) + 5(3) + 5.6 = 10.8251.$$

Therefore, the net increase in the percent of these households from 1970 ($t = 0$) to 2000 ($t = 3$) is $P(3) - P(0) = 10.825 - 5.6 = 5.2$, or approximately 5.2%.

49. The average population over the period in question is

$$A = \tfrac{1}{3}\int \frac{85}{1 + 1.859e^{-0.66t}}\,dt$$

Multiplying the integrand by $e^{0.66t} / e^{0.66t}$ gives

$$A = \frac{85}{3}\int_0^3 \frac{e^{0.66t}}{e^{0.66t}+1.859}\,dt$$

Let $u = 1.859 + e^{0.66t}$, $du = 0.66e^{0.66t}dt$, or $e^{0.66t}dt = \dfrac{du}{0.66}$.

If $t = 0$, then $u = 2.859$. If $t = 3$, then $u = 9.1017$.

Substituting

$$A = \frac{85}{3}\int_{2.859}^{9.1017} \frac{du}{(0.66)u} = \frac{85}{3(0.66)}\ln u\Big|_{2.859}^{9.1017}$$

$$= \frac{85}{3(0.66)}(\ln 9.1017 - \ln 2.859) = \frac{85}{3(0.66)}\ln\frac{9.1017}{2.859} \approx 49.712,$$

or approximately 49.7 million people.

51. $f(x) = x^4 - 2x^2 + 2$. $f'(x) = 4x^3 - 4x = 4x(x^2-1) = 4x(x+1)(x-1)$
Setting $f'(x) = 0$ gives $x = -1, 0,$ and 1 as critical numbers.
$f''(x) = 12x^2 - 4 = 4(3x^2 - 1)$
Using the second derivative test, we find
$f''(-1) = 8 > 0$ and so (-1,1) is a relative minimum. $f''(0) = -4 < 0$ and so (0, 2) is a relative maximum. $f''(1) = 8 > 0$ and so (1, 1) is a relative minimum. The graph of f is symmetric with respect to the y-axis because

$$f(-x) = (-x)^4 - 2(-x)^2 + 2 = x^4 - 2x^2 + 2.$$

So, the required area is the area under the graph of f between $x = 0$ and $x = 1$. Thus,

$$A = \int_0^1 (x^4 - 2x^2 + 2)\,dx = \tfrac{1}{5}x^5 - \tfrac{2}{3}x^3 + 2x\Big|_0^1 = \tfrac{1}{5} - \tfrac{2}{3} + 2 = \tfrac{23}{15} \text{ sq units.}$$

53. False. The integrand $f(x) = \dfrac{1}{x^3}$ is discontinuous at $x = 0$.

55. False. $f(x)$ is not nonnegative on [0, 2].

USING TECHNOLOGY EXERCISES 15.4, page 954

1. 6.1787 3. 0.7873 5. −0.5888 7. 2.7044

9. 3.9973 11. 46 %; 24% 13. 333,209 15. 903,213

15.5 Problem Solving Tips

1. If you use the method of substitution to evaluate a definite integral, make sure that you change the corresponding limits of integration to reflect the fact that the integration is being performed with respect to the new variable *u*.

15.5 CONCEPT QUESTIONS, page 961

1. Approach I: We first find the indefinite integral. Let $u = x^3 + 1$ so that $du = 3x^2\,dx$ or $x^2 dx = \frac{1}{3} du$. Then

$$\int x^2(x^3+1)^2\,dx = \tfrac{1}{3}\int u^2\,du = \tfrac{1}{9}u^3 + C = \tfrac{1}{9}(x^3+1)^3 + C.$$

Therefore,

$$\int_0^1 x^2(x^3+1)^2\,dx = \tfrac{1}{9}(x^3+1)^3\Big|_0^1 = \tfrac{1}{9}(8-1) = \tfrac{7}{9}.$$

Approach II: Transform the definite integral in *x* into an integral in *u*: Let $u = x^3 + 1$, so that $du = 3x^2\,dx$ *or* $x^2\,dx = \frac{1}{3} du$. Next, find the limits of integration with respect to *u*: If $x = 0$, then $u = 0^3 + 1 = 1$ and if $x = 1$, then $u = 1^3 + 1 = 2$. Therefore,

$$\int_0^1 x^2(x^3+1)^2\,dx = \tfrac{1}{3}\int_1^2 u^2\,du = \tfrac{1}{9}u^3\Big|_1^2 = \tfrac{1}{9}(8-1) = \tfrac{7}{9}.$$

EXERCISES 15.5, page 961

1. Let $u = x^2 - 1$ so that $du = 2x\,dx$ or $x\,dx = \frac{1}{2}\,du$. Also, if $x = 0$, then $u = -1$ and if $x = 2$, then $u = 3$. So

$$\int_0^2 x(x^2-1)^3\,dx = \frac{1}{2}\int_{-1}^3 u^3\,du = \tfrac{1}{8}u^4\Big|_{-1}^3 = \tfrac{1}{8}(81) - \tfrac{1}{8}(1) = 10.$$

3. Let $u = 5x^2 + 4$ so that $du = 10x\,dx$ or $x\,dx = \frac{1}{10}\,du$. Also, if $x = 0$, then $u = 4$, and if $x = 1$, then $u = 9$. So

$$\int_0^1 x\sqrt{5x^2+4}\,dx = \frac{1}{10}\int_4^9 u^{1/2}\,du = \tfrac{1}{15}u^{3/2}\Big|_4^9 = \tfrac{1}{15}(27) - \tfrac{1}{15}(8) = \tfrac{19}{15}.$$

5. Let $u = x^3 + 1$ so that $du = 3x^2\,dx$ or $x^2\,dx = \frac{1}{3}\,du$. Also, if $x = 0$, then $u = 1$, and if $x = 2$, then $u = 9$. So,

$$\int_0^2 x^2(x^3+1)^{3/2}\,dx = \frac{1}{3}\int_1^9 u^{3/2}\,du = \frac{2}{15}u^{5/2}\Big|_1^9 = \frac{2}{15}(243) - \frac{2}{15}(1) = \frac{484}{15}.$$

7. Let $u = 2x + 1$ so that $du = 2\,dx$ or $dx = \frac{1}{2}\,du$. Also, if $x = 0$, then $u = 1$ and if $x = 1$ then $u = 3$. So

$$\int_0^1 \frac{1}{\sqrt{2x+1}}\,dx = \frac{1}{2}\int_1^3 \frac{1}{\sqrt{u}}\,du = \frac{1}{2}\int_1^3 u^{-1/2}\,du = u^{1/2}\Big|_1^3 = \sqrt{3} - 1.$$

9. $\int_1^2 (2x-1)^4\,dx$. Put $u = 2x - 1$ so that $du = 2\,dx$ or $dx = \frac{1}{2}\,du$. Then if $x = 1$, $u = 1$

and if $x = 2$, then $u = 3$. Then

$$\int_1^2 (2x-1)^4\,dx = \frac{1}{2}\int_1^3 u^4\,du = \frac{1}{10}u^5\Big|_1^3 = \frac{1}{10}(243-1) = \frac{121}{5} = 24\tfrac{1}{5}.$$

11. Let $u = x^3 + 1$ so that $du = 3x^2\,dx$ or $x^2\,dx = \frac{1}{3}\,du$. Also, if $x = -1$, then $u = 0$ and if $x = 1$, then $u = 2$. So

$$\int_{-1}^1 x^2(x^3+1)^4\,dx = \frac{1}{3}\int_0^2 u^4\,du = \frac{1}{15}u^5\Big|_0^2 = \frac{32}{15}.$$

13. Let $u = x - 1$ so that $du = dx$. Then if $x = 1$, $u = 0$, and if $x = 5$, then $u = 4$.

$$\int_1^5 x\sqrt{x-1}\,dx = \int_0^4 (u+1)u^{1/2}\,du = \int_0^4 (u^{3/2} + u^{1/2})du$$

$$= \frac{2}{5}u^{5/2} + \frac{2}{3}u^{3/2}\Big|_0^4 = \frac{2}{5}(32) + \frac{2}{3}(8) = 18\tfrac{2}{15}.$$

15. Let $u = x^2$ so that $du = 2x\,dx$ or $x\,dx = \frac{1}{2}\,du$. If $x = 0$, $u = 0$ and if $x = 2$, $u = 4$. So

$$\int_0^2 xe^{x^2}\,dx = \frac{1}{2}\int_0^4 e^u\,du = \frac{1}{2}e^u\Big|_0^4 = \frac{1}{2}(e^4 - 1).$$

17. $$\int_0^1 (e^{2x} + x^2 + 1)\,dx = \frac{1}{2}e^{2x} + \frac{1}{3}x^3 + x\Big|_0^1 = \left(\frac{1}{2}e^2 + \frac{1}{3} + 1\right) - \frac{1}{2}$$

$$= \frac{1}{2}e^2 + \frac{5}{6}.$$

19. Put $u = x^2 + 1$ so that $du = 2x\,dx$ or $x\,dx = \frac{1}{2}\,du$. Then

$$\int_{-1}^{1} xe^{x^2+1}dx = \frac{1}{2}\int_{2}^{2} e^u\,du = \frac{1}{2}e^u\Big|_2^2 = 0$$

(Since the upper and lower limits are equal.)

21. Let $u = x - 2$ so that $du = dx$. If $x = 3$, $u = 1$ and if $x = 6$, $u = 4$. So

$$\int_3^6 \frac{2}{x-2}dx = 2\int_1^4 \frac{du}{u} = 2\ln|u|\Big|_1^4 = 2\ln 4.$$

23. Let $u = x^3 + 3x^2 - 1$ so that $du = (3x^2 + 6x)dx = 3(x^2 + 2x)dx$. If $x = 1$, $u = 3$, and if $x = 2$, $u = 19$. So

$$\int_1^2 \frac{x^2+2x}{x^3+3x^2-1}dx = \frac{1}{3}\int_3^{19} \frac{du}{u} = \frac{1}{3}\ln u\Big|_3^{19} = \frac{1}{3}(\ln 19 - \ln 3).$$

25. $\displaystyle\int_1^2 \left(4e^{2u} - \frac{1}{u}\right)du = 2e^{2u} - \ln u\Big|_1^2 = (2e^4 - \ln 2) - (2e^2 - 0) = 2e^4 - 2e^2 - \ln 2.$

27. $\displaystyle\int_1^2 (2e^{-4x} - x^{-2})dx = -\frac{1}{2}e^{-4x} + \frac{1}{x}\Big|_1^2 = (-\frac{1}{2}e^{-8} + \frac{1}{2}) - (-\frac{1}{2}e^{-4} + 1)$

$$= -\frac{1}{2}e^{-8} + \frac{1}{2}e^{-4} - \frac{1}{2} = \frac{1}{2}(e^{-4} - e^{-8} - 1).$$

29. AV $\displaystyle = \frac{1}{2}\int_0^2 (2x+3)\,dx = \frac{1}{2}(x^2 + 3x)\Big|_0^2 = \frac{1}{2}(10) = 5.$

31. AV $\displaystyle = \frac{1}{2}\int_1^3 (2x^2 - 3)\,dx = \frac{1}{2}(\frac{2}{3}x^3 - 3x)\Big|_1^3 = \frac{1}{2}(9 + \frac{7}{3}) = \frac{17}{3}.$

33. AV $\displaystyle = \frac{1}{3}\int_{-1}^2 (x^2 + 2x - 3)\,dx = \frac{1}{3}(\frac{1}{3}x^3 + x^2 - 3x)\Big|_{-1}^2$

$$= \frac{1}{3}[(\frac{8}{3} + 4 - 6) - (-\frac{1}{3} + 1 + 3)] = \frac{1}{3}(\frac{8}{3} - 2 + \frac{1}{3} - 4) = -1.$$

35. AV $\displaystyle = \frac{1}{4}\int_0^4 (2x+1)^{1/2}\,dx = (\frac{1}{4})(\frac{1}{2})(\frac{2}{3})(2x+1)^{3/2}\Big|_0^4 = \frac{1}{12}(27-1) = \frac{13}{6}$

37. $\text{AV} = \frac{1}{2}\int_0^2 xe^{x^2}\,dx = \frac{1}{4}e^{x^2}\Big|_0^2 = \frac{1}{4}(e^4 - 1).$

39. The amount produced was

$$\int_0^{20} 3.5e^{0.05t}\,dt = \frac{3.5}{0.05}e^u\Big|_0^{20} \qquad \text{(Use the substitution } u = 0.05t.\text{)}$$

$$= 70(e-1) \approx 120.3, \quad \text{or } 120.3 \text{ billion metric tons.}$$

41. The amount is $\int_1^2 t(\frac{1}{2}t^2+1)^{1/2}\,dt$. Let $u = \frac{1}{2}t^2 + 1$, so that $du = t\,dt$. Therefore,

$$\int_1^2 t(\tfrac{1}{2}t^2+1)^{1/2}\,dt = \int_{3/2}^3 u^{1/2}\,du = \tfrac{2}{3}u^{3/2}\Big|_{3/2}^3 = \tfrac{2}{3}[(3)^{3/2} - (\tfrac{3}{2})^{3/2}]$$

$$\approx 2.24 \text{ million dollars.}$$

43. The tractor will depreciate

$$\int_0^5 13388.61e^{-0.22314t}\,dt = \frac{13388.61}{-0.22314}e^{-0.22314t}\Big|_0^5$$

$$= -60{,}000.94e^{-0.22314t}\Big|_0^5 = -60{,}000.94(-0.672314)$$

$$= 40{,}339.47, \quad \text{or } \$40{,}339.$$

45. $\bar{A} = \frac{1}{5}\int(\frac{1}{12}t^2 + 2t + 44)dt = \frac{1}{5}\left[\frac{1}{36}t^3 + t^2 + 44t\Big|_0^5\right]$

$$= \tfrac{1}{5}\left[\tfrac{125}{36} + 25 + 220\right] = \frac{125+900+7920}{5(36)} \approx 49.69, \text{ or } 49.7 \text{ ft/sec.}$$

47. The average whale population will be

$$\frac{1}{10}\int_0^{10}(3t^3 + 2t^2 - 10t + 600)\,dt = \tfrac{1}{10}(\tfrac{3}{4}t^4 + \tfrac{2}{3}t^3 - 5t^2 + 600t\Big|_0^{10}$$

$$\approx \tfrac{1}{10}(7500 + 666.67 - 500 + 6000) \approx 1367 \text{ whales.}$$

49. The average yearly sales of the company over its first 5 years of operation is given by $\frac{1}{5-0}\int_0^5 t(0.2t^2+4)^{1/2}\,dt = \frac{1}{5}[(\frac{5}{2})(\frac{2}{3})(0.2t^2+4)^{3/2}]\Big|_0^5$ [Let $u = -0.2t^2+4$.]

$$= \tfrac{1}{5}[\tfrac{5}{3}(5+4)^{3/2} - \tfrac{5}{3}(4)^{3/2}] = \tfrac{1}{3}(27-8) = \tfrac{19}{3}, \text{ or } 6\tfrac{1}{3} \text{ million dollars.}$$

51. The average velocity is $\frac{1}{4}\int_0^4 3t\sqrt{16-t^2}\,dt = \frac{1}{4}(64) = 16$, or 16 ft/sec.

(Using the results of Exercise 44.)

53. $\int_0^5 p\,dt = \int_0^5 (18-3e^{-2t}-6e^{-t/3})\,dt = \frac{1}{5}\left[18t+\frac{3}{2}e^{-2t}+18e^{-t/3}\right]_0^5$

$= \frac{1}{5}\left[18(5)+\frac{3}{2}e^{-10}+18e^{-5/3}-\frac{3}{2}-18\right] = 14.78$, or \$14.78.

55. The average content of oxygen in the pond over the first 10 days is

$$\frac{1}{10-0}\int_0^{10} 100\left(\frac{t^2+10t+100}{t^2+20t+100}\right)dt = \frac{100}{10}\int_0^{10}\left[1-\frac{10}{t+10}+\frac{100}{(t+10)^2}\right]dt$$

$$= 10\int_0^{10}\left[1-\frac{10}{t+10}+100(t+10)^{-2}\right]$$

$$= 10\left[t-10\ln(t+10)-\frac{100}{t+10}\right]\Bigg|_0^{10}$$ [Use the substitution $u = t + 10$ for the third integral.]

$$= 10\{[10-10\ln 20-\frac{100}{2p}]-[-10\ln 10-10]\}$$

$$= 10[10-10\ln 20-5+10\ln 10+10]$$

≈ 80.6853, or approximately 80.7%.

57. $\int_a^a f(x)\,dx = F(x)\Big|_a^a = F(a)-F(a) = 0$, where $F'(x) = f(x)$.

59. $\int_1^3 x^2\,dx = \frac{1}{3}x^3\Big|_1^3 = 9-\frac{1}{3} = \frac{26}{3} = -\int_3^1 x^2\,dx = -\frac{1}{3}x^3\Big|_3^1 = -\frac{1}{3}+9 = \frac{26}{3}$.

61. $\int_1^9 2\sqrt{x}\,dx = \frac{4}{3}x^{3/2}\Big|_1^9 = \frac{4}{3}(27-1) = \frac{104}{3} = 2\int_1^9 \sqrt{x}\,dx = (2)(\frac{2}{3}x^{3/2})\Big|_1^9 = \frac{104}{3}$.

63. $\int_0^3 (1+x^3)\,dx = (x+\frac{1}{4}x^4)\Big|_0^3 = 3+\frac{81}{4} = \frac{93}{4}$.

$\int_0^1 (1+x^3)\,dx + \int_1^2 (1+x^3)\,dx + \int_2^3 (1+x^3)\,dx$

$= (x+\frac{1}{4}x^4)\Big|_0^1 + (x+\frac{1}{4}x^4)\Big|_1^2 + (x+\frac{1}{4}x^4)\Big|_2^3$

$= (1+\frac{1}{4})+(2+4)-(1+\frac{1}{4})+(3+\frac{81}{4})-(2+4) = \frac{93}{4}$. [Property 5.]

65. $\int_3^3 (1+\sqrt{x})e^{-x}\,dx = 0$ by Property 1 of definite integrals.

67. a. $\int_{-1}^{2} [2f(x)+g(x)]\,dx = 2\int_{-1}^{2} f(x)\,dx + \int_{-1}^{2} g(x)\,dx = 2(-2) + 3 = -1.$

b. $\int_{-1}^{2} [g(x)-f(x)]\,dx = \int_{-1}^{2} g(x)\,dx - \int_{-1}^{2} f(x)\,dx = 3 - (-2) = 5.$

c. $\int_{-1}^{2} [2f(x)-3g(x)]\,dx = 2\int_{-1}^{2} f(x)\,dx - 3\int_{-1}^{2} g(x)\,dx = 2(-2) - 3(3) = -13.$

69. True. This follows from Property 1 of the definite integral.

71. False. Only a constant can be "moved out" of the integral sign.

73. True. This follows from Properties 3 and 4 of the definite integral.

USING TECHNOLOGY EXERCISES 15.5, page 965

1. 7.71667 3. 17.5649 5. 10,140 7. 60.5mg/day

15.6 Problem Solving Tips

1. Note that the formula for the area between two curves where f and g are continuous functions and $f(x) \geq g(x)$ on $[a,b]$ is given by $\int_a^b [f(x)-g(x)]\,dx$. The condition $f(x) \geq g(x)$ on $[a,b]$ tells us that we cannot interchange $f(x)$ and $g(x)$ in this formula, as that would yield a negative answer, and area cannot be negative

15.6 CONCEPT QUESTIONS, page 972

1. $\int_a^b [f(x)-g(x)]\,dx$

EXERCISES 15.6, page 972

1. $-\int_0^6 (x^3-6x^2)\,dx = -\frac{1}{4}x^4+2x^3\Big|_0^6 = -\frac{1}{4}(6^4)+2(6^3) = 108$ sq units.

3. $A = -\int_{-1}^0 x\sqrt{1-x^2}\,dx + \int_0^1 x\sqrt{1-x^2}\,dx = 2\int_0^1 x(1-x^2)^{1/2}\,dx$ (by symmetry). Let

   $u = 1-x^2$ so that $du = -2x\,dx$ or $x\,dx = -\frac{1}{2}\,du$. Also, if $x = 0$, then $u = 1$ and

   if $x = 1$, $u = 0$. So $A = (2)(-\frac{1}{2})\int_0^1 u^{1/2}\,du = -\frac{2}{3}u^{3/2}\Big|_1^0 = \frac{2}{3}$, or $\frac{2}{3}$sq unit.

5. $A = -\int_0^4 (x-2\sqrt{x})\,dx = \int_0^4 (-x+2x^{1/2})dx = -\frac{1}{2}x^2+\frac{4}{3}x^{3/2}\Big|_0^4$

   $= 8 + \frac{32}{3} = \frac{8}{3}$ sq units.

7. The required area is given by

   $\int_{-1}^0 (x^2-x^{1/3})\,dx + \int_0^1 (x^{1/3}-x^2)\,dx = \frac{1}{3}x^3-\frac{3}{4}x^{4/3}\Big|_{-1}^0 + \frac{3}{4}x^{4/3}-\frac{1}{3}x^3\Big|_0^1$

   $= -(-\frac{1}{3}-\frac{3}{4})+(\frac{3}{4}-\frac{1}{3}) = 1\frac{1}{2}$ sq units.

9. The required area is given by

   $-\int_{-1}^2 -x^2\,dx = \frac{1}{3}x^3\Big|_{-1}^2 = \frac{8}{3}+\frac{1}{3} = 3$ sq units.

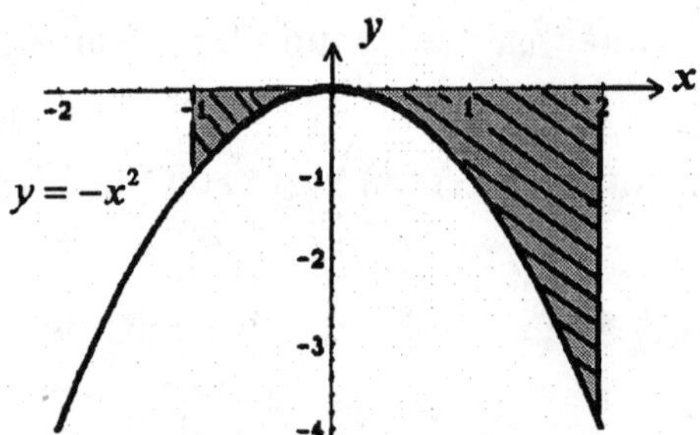

11. $y = x^2-5x+4 = (x-4)(x-1) = 0$

    if $x = 1$ or 4. These give the x-intercepts.

    $A = -\int_1^3 (x^2-5x+4)\,dx = -\frac{1}{3}x^3+\frac{5}{2}x^2-4x\Big|_1^3$

    $= (-9+\frac{45}{2}-12)-(-\frac{1}{3}+\frac{5}{2}-4) = \frac{10}{3} = 3\frac{1}{3}$.

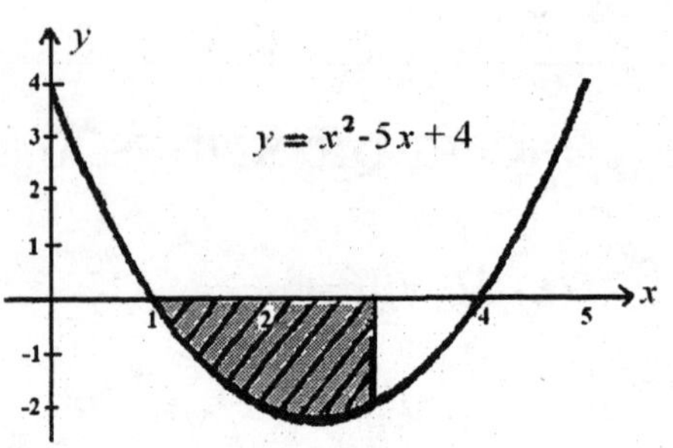

13. The required area is given by

$$-\int_0^9 -(1+\sqrt{x})\,dx = x+\tfrac{2}{3}x^{3/2}\Big|_0^9 = 9+18=27.$$

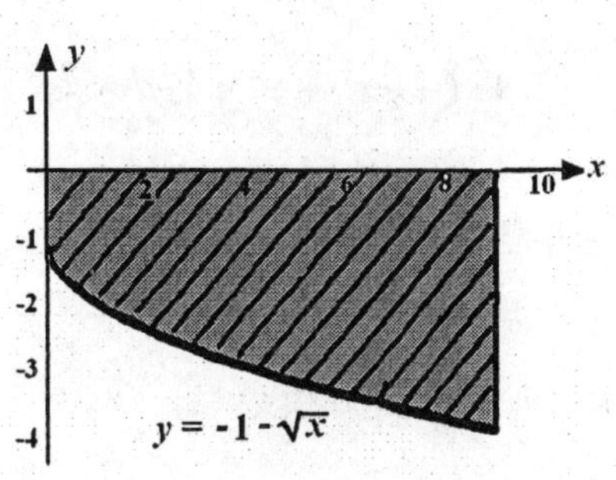

15. $-\int_{-2}^{4} -e^{(1/2)x}\,dx = 2e^{(1/2)x}\Big|_{-2}^{4}$

$= 2(e^2 - e^{-1})$ sq units.

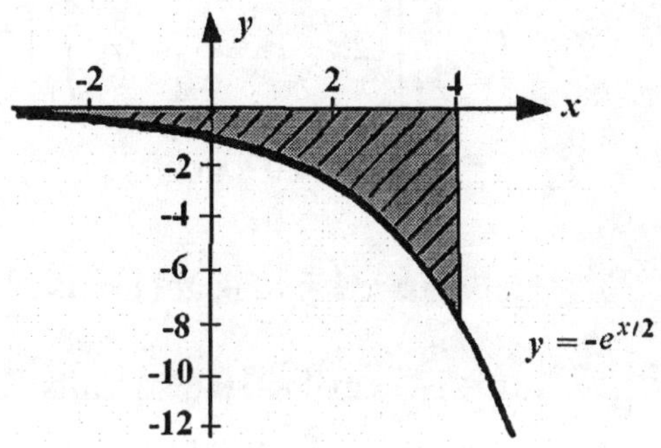

17. $A = \int_1^3 [(x^2+3)-1]\,dx$

$= \int_1^3 (x^2+2)\,dx = \tfrac{1}{3}x^3 + 2x\Big|_1^3$

$= (9+6) - (\tfrac{1}{3}+2) = \tfrac{38}{3}.$

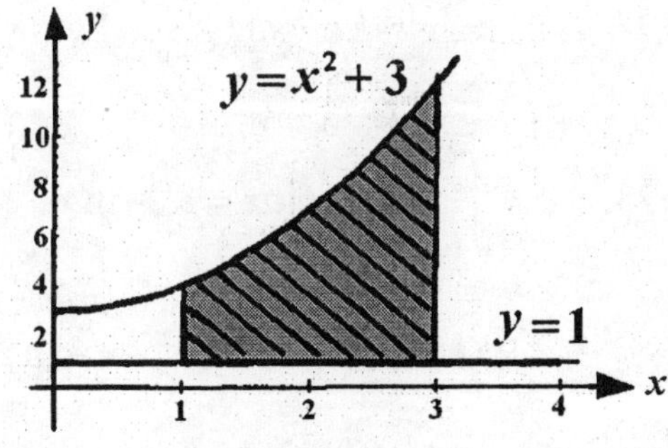

19. $A = \int_0^2 (-x^2+2x+3+x-3)\,dx$

$= -\tfrac{1}{3}x^3 + \tfrac{3}{2}x^2\Big|_0^2 = -\tfrac{1}{3}(8) + \tfrac{3}{2}(4)$

$= 6 - \tfrac{8}{3} = \tfrac{10}{3}$ sq units

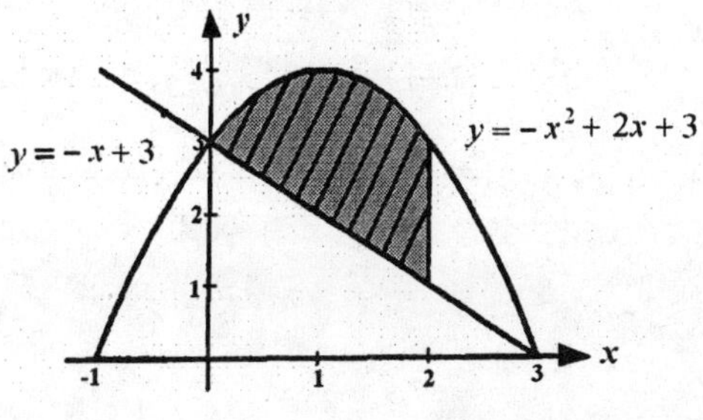

21. $A = \int_{-1}^{2} [(x^2+1) - \frac{1}{3}x^3]\,dx$

$= \int_{-1}^{2} (-\frac{1}{3}x^3 + x^2 + 1)\,dx$

$= -\frac{1}{12}x^4 + \frac{1}{3}x^3 + x\Big|_{-1}^{2}$

$= (-\frac{4}{3} + \frac{8}{3} + 2) - (-\frac{1}{12} - \frac{1}{3} - 1)$

$= 4\frac{3}{4}$ sq units.

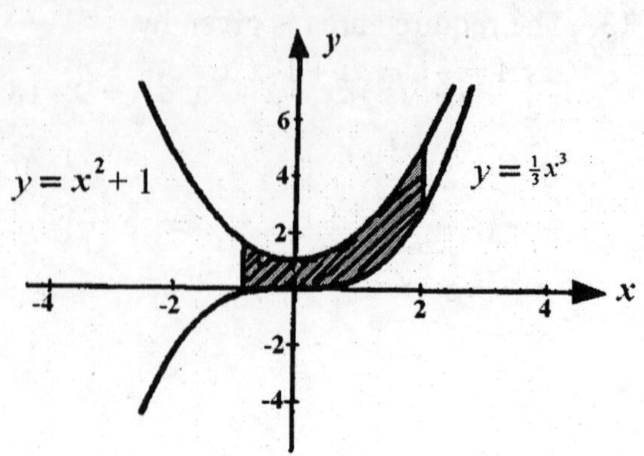

23. $A = \int_{1}^{4} \left[(2x-1) - \frac{1}{x}\right] dx = \int_{1}^{4} \left(2x - 1 - \frac{1}{x}\right) dx$

$= (x^2 - x - \ln x)\Big|_{1}^{4}$

$= (16 - 4 - \ln 4) - (1 - 1 - \ln 1)$

$= 12 - \ln 4 \approx 10.6$ sq units.

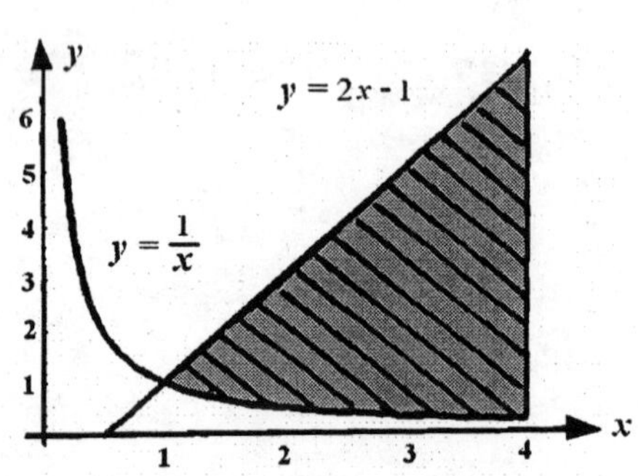

25. $A = \int_{1}^{2} \left(e^x - \frac{1}{x}\right) dx = e^x - \ln x\Big|_{1}^{2}$

$= (e^2 - \ln 2) - e = (e^2 - e - \ln 2)$ sq units.

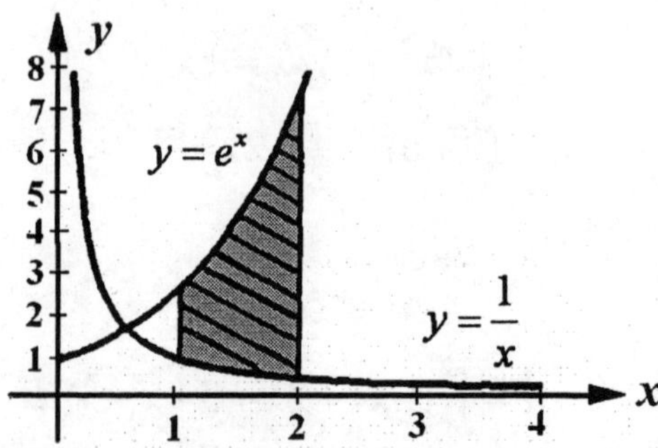

27. $$A = -\int_{-1}^{0} x\,dx + \int_{0}^{2} x\,dx$$

$$= -\tfrac{1}{2}x^2\Big|_{-1}^{0} + \tfrac{1}{2}x^2\Big|_{0}^{2}$$

$$= \tfrac{1}{2} + 2 = 2\tfrac{1}{2} \text{ sq units.}$$

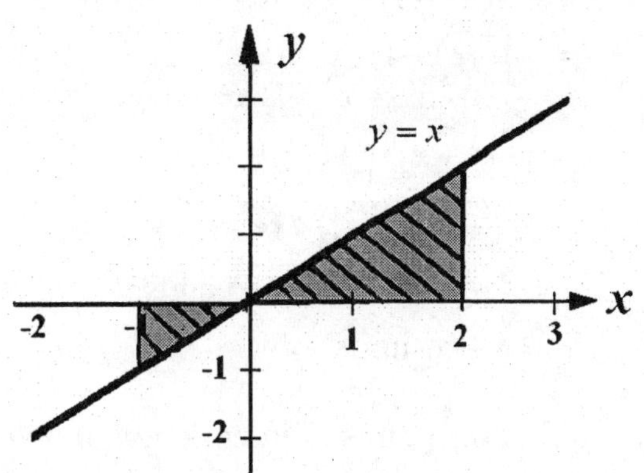

29. The x–intercepts are found by solving $x^2 - 4x + 3 = (x-3)(x-1) = 0$ giving $x = 1$ or 3. The region is shown in the figure.

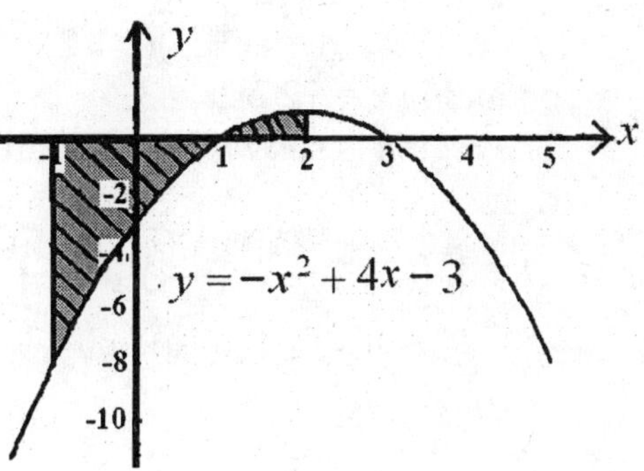

$$A = -\int_{-1}^{1} [(-x^2 + 4x - 3)\,dx + \int_{1}^{2} (-x^2 + 4x - 3)\,dx$$
$$= \tfrac{1}{3}x^3 - 2x^2 + 3x\Big|_{-1}^{1} + (-\tfrac{1}{3}x^3 + 2x^2 - 3x)\Big|_{1}^{2}$$
$$= (\tfrac{1}{3} - 2 + 3) - (-\tfrac{1}{3} - 2 - 3) + (-\tfrac{8}{3} + 8 - 6) - (-\tfrac{1}{3} + 2 - 3) = \tfrac{22}{3} \text{ sq units.}$$

31. The region is shown in the figure at the right.

$$A = \int_{0}^{1} (x^3 - 4x^2 + 3x)\,dx - \int_{1}^{2} (x^3 - 4x^2 + 3x)\,dx$$
$$= (\tfrac{1}{4}x^4 - \tfrac{4}{3}x^3 + \tfrac{3}{2}x^2)\Big|_{0}^{1}$$
$$-(\tfrac{1}{4}x^4 - \tfrac{4}{3}x^3 + \tfrac{3}{2}x^2)\Big|_{1}^{2} = \tfrac{3}{2} \text{ sq units.}$$

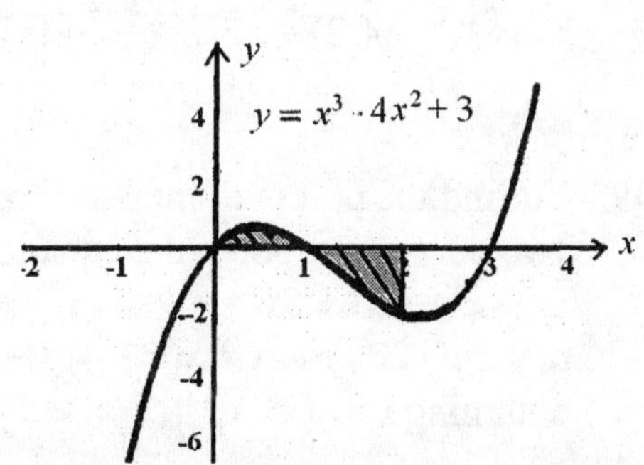

33. The region is shown in the figure at the right.

$$A = -\int_{-1}^{0}(e^x - 1)\,dx + \int_{0}^{3}(e^x - 1)\,dx$$

$$= (-e^x + x)\Big|_{-1}^{0} + (e^x - x)\Big|_{0}^{3}$$

$$= -1 - (-e^{-1} - 1) + (e^3 - 3) - 1$$

$$= e^3 - 4 + \tfrac{1}{e} \approx 16.5 \quad \text{sq units.}$$

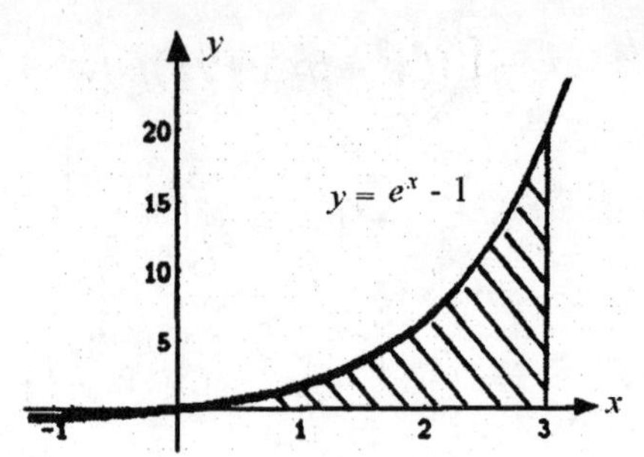

35. To find the points of intersection of the two curves, we solve the equation

$$x^2 - 4 = x + 2$$

$$x^2 - x - 6 = (x-3)(x+2) = 0,$$

obtaining $x = -2$ or $x = 3$.

The region is shown in the figure at the right.

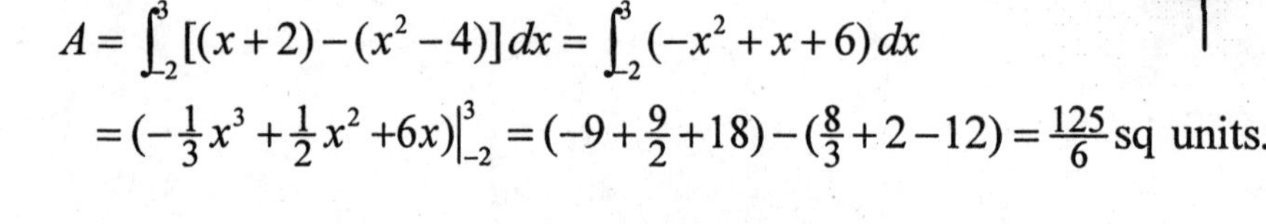

$$A = \int_{-2}^{3}[(x+2) - (x^2 - 4)]\,dx = \int_{-2}^{3}(-x^2 + x + 6)\,dx$$

$$= (-\tfrac{1}{3}x^3 + \tfrac{1}{2}x^2 + 6x)\Big|_{-2}^{3} = (-9 + \tfrac{9}{2} + 18) - (\tfrac{8}{3} + 2 - 12) = \tfrac{125}{6} \text{ sq units.}$$

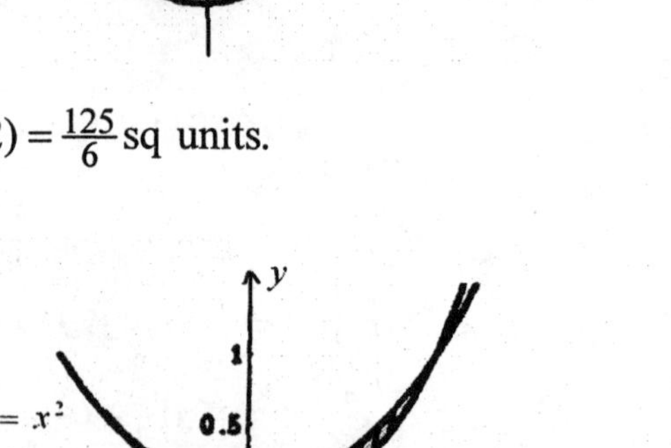

37. To find the points of intersection of the two curves, we solve the equation $x^3 = x^2$ or $x^3 - x^2 = x^2(x-1) = 0$ giving $x = 0$ or 1. The region is shown in the figure.

$$A = -\int_{0}^{1}(x^2 - x^3)\,dx$$

$$= (\tfrac{1}{3}x^3 - \tfrac{1}{4}x^4)\Big|_{0}^{1} = \tfrac{1}{3} - \tfrac{1}{4} = \tfrac{1}{12} \text{ sq units}$$

39. To find the points of intersection of the two curves, we solve the equation

$$x^3 - 6x^2 + 9x = x^2 - 3x,$$

or $x^3 - 7x^2 + 12x = x(x-4)(x-3) = 0$

obtaining $x = 0, 3,$ or 4.

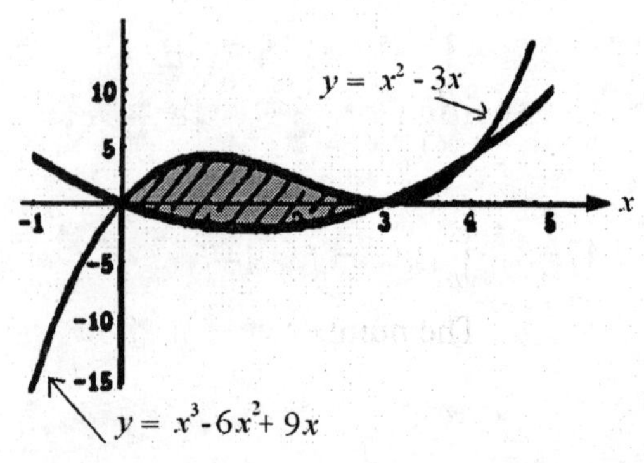

$$A = \int_0^3 [(x^3 - 6x^2 + 9x) - (x^2 + 3x)]\,dx$$
$$+ \int_3^4 [(x^2 - 3x) - (x^3 - 6x^2 + 9x)\,dx$$
$$= \int_0^3 (x^3 - 7x^2 + 12x)\,dx - \int_3^4 (x^3 - 7x^2 + 12x)\,dx$$
$$= (\tfrac{1}{4}x^4 - \tfrac{7}{3}x^3 + 6x^2)\Big|_0^3 - (\tfrac{1}{4}x^4 - \tfrac{7}{3}x^3 + 6x^2)\Big|_3^4$$
$$= (\tfrac{81}{4} - 63 + 54) - (64 - \tfrac{448}{3} + 96) + (\tfrac{81}{4} - 63 + 54) = \tfrac{71}{6}.$$

41. By symmetry, $A = 2\int_0^3 x(9 - x^2)^{1/2}\,dx$. We integrate by substitution with $u = 9 - x^2$, $du = -2x\,dx$. If $x = 0$, $u = 9$, and if $x = 3, u = 0$. So
$$A = 2\int_9^0 -\tfrac{1}{2}u^{1/2}\,du = -\int_9^0 u^{1/2}\,du = -\tfrac{2}{3}u^{3/2}\Big|_9^0 = \tfrac{2}{3}(9)^{3/2} = 18 \text{ sq units.}$$

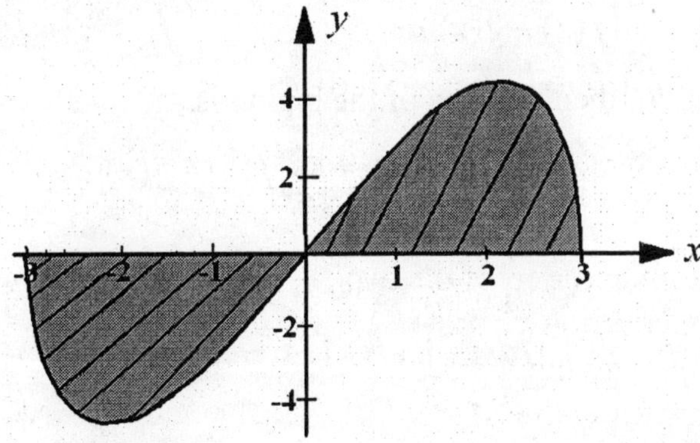

43. S gives the additional revenue that the company would realize if it used a different advertising agency. $S = \int_0^b [g(x) - f(x)]\,dx.$

45. Shortfall $= \int_{2010}^{2050} [f(t) - g(t)]\,dt$

47. a. $\int_{T_1}^{T} [g(t) - f(t)]\,dt - \int_0^{T_1} [f(t) - g(t)]\,dt = A_2 - A_1.$

b. The number $A_2 - A_1$ gives the distance car 2 is ahead of car 1 after t seconds.

49. The turbo-charged model is moving at

$$A = \int_0^{10} [(4+1.2t+0.03t^2)-(4+0.8t)]\,dt$$

$$= \int_0^{10} (0.4t+0.03t^2)\,dt = (0.2t^2+0.1t^3)\Big|_0^{10}$$

$= 20 + 10$, or 30 ft/sec faster than the standard model.

51. The additional number of cars will be given by

$$\int_0^5 (5e^{0.3t} - 5 - 0.5t^{3/2})dt = \frac{5}{0.3}e^{0.3t} - 5t - 0.2t^{5/2}\Big|_0^5$$

$$= \frac{5}{0.3}e^{1.5} - 25 - 0.2(5)^{5/2} - \frac{5}{0.3} = 74.695 - 25 - 0.2(5)^{5/2} - \frac{50}{3}$$

≈ 21.85, or 21,850 cars. (Remember t is measured in thousands.)

53. True. If $f(x) \geq g(x)$ on $[a, b]$, then the area of the said region is

$$\int_a^b [f(x)-g(x)]\,dx = \int_a^b |f(x)-g(x)|\,dx$$

If $f(x) \leq g(x)$ on $[a, b]$, then the area of the region is

$$\int_a^b [g(x)-f(x)]\,dx = \int_a^b -[f(x)-g(x)]\,dx = \int_a^b |f(x)-g(x)|\,dx$$

55. The area of R' is

$$A = \int_a^b \{[f(x)+C]-[g(x)+C]\}\,dx = \int_a^b [f(x)+C-g(x)-C]\,dx$$

$$= \int_a^b [f(x)-g(x)]\,dx$$

USING TECHNOLOGY EXERCISES 15.6, page 977

1. a.

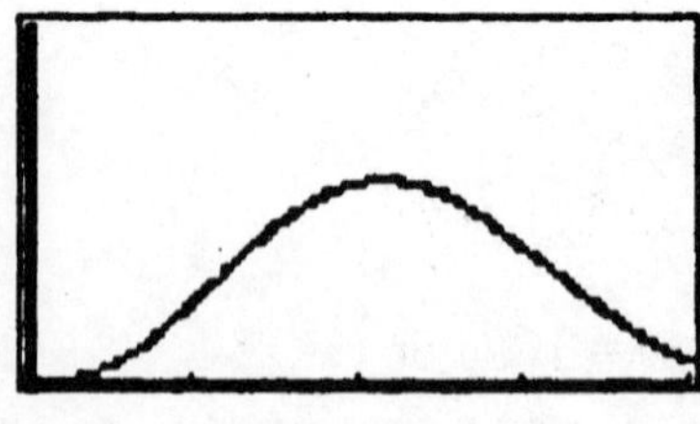

b. 1074.2857 sq units

3. a.

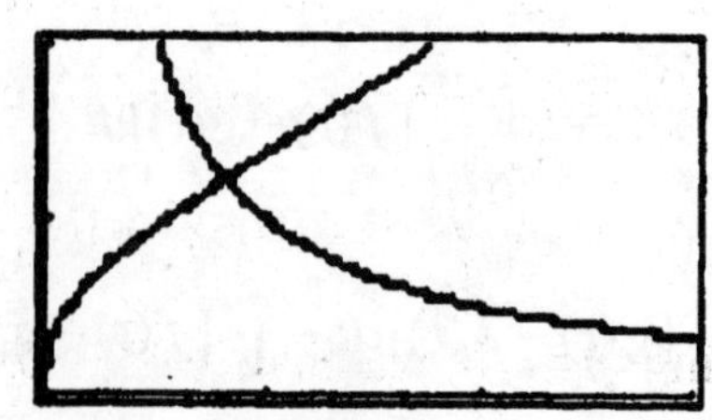

b. 0.9961 sq units

5. a.

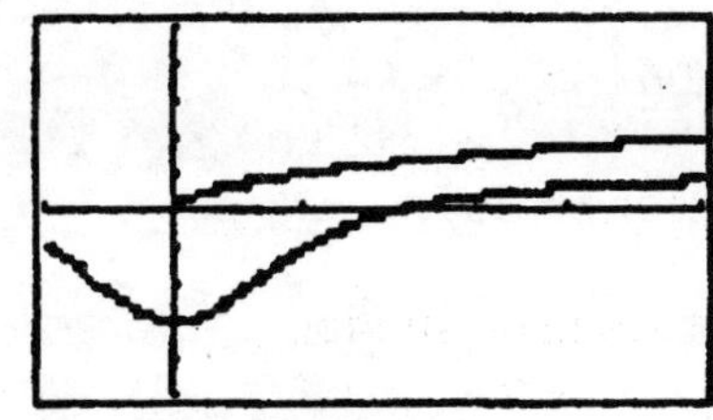

b. 5.4603 sq units

7. a.

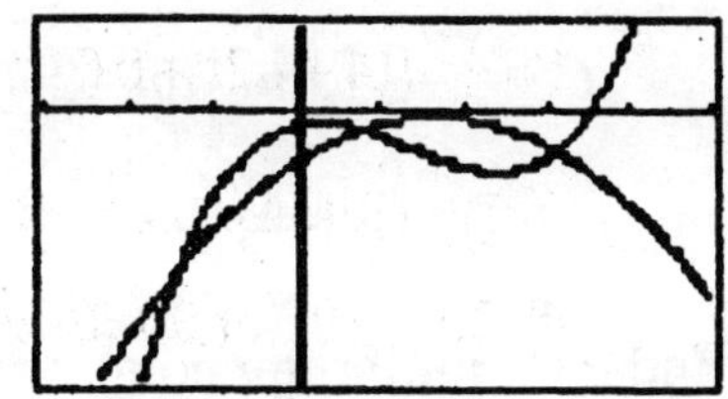

b. 25.8549 sq units

9. a.

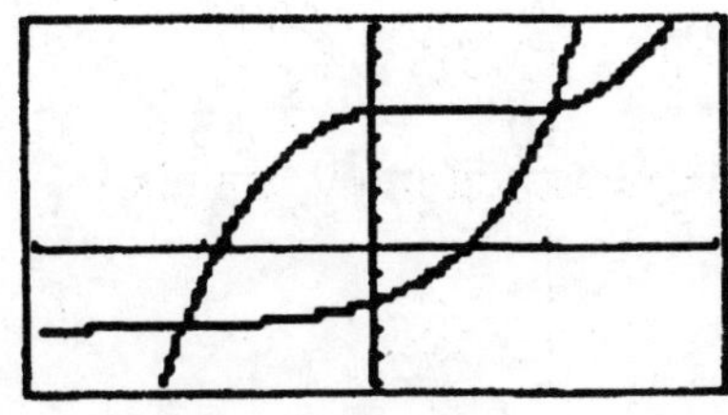

b. 10.5144 sq units

11. a.

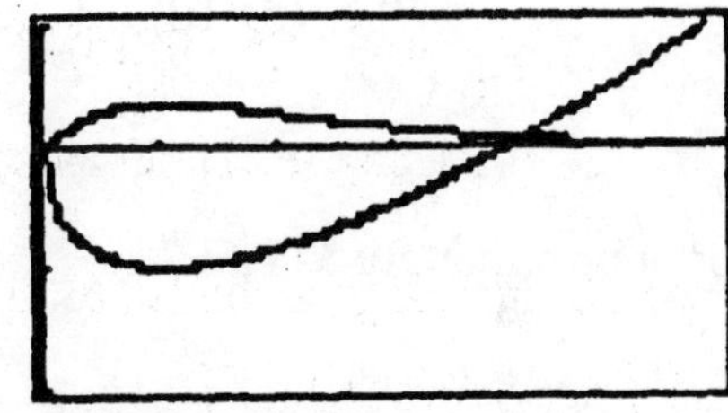

b. 3.5799 sq units

13. 207.43 sq units

15.7 CONCEPT QUESTIONS, page 988

1. a. See definition in text on page 979. b. See definition in text on page 979.
3. See definition in text on page 984.

EXERCISES 15.7, page 988

1. When $p = 4$, $-0.01x^2 - 0.1x + 6 = 4$ or $x^2 + 10x - 200 = 0$, $(x - 10)(x + 20) = 0$ and $x = 10$ or -20. We reject the root $x = -20$. The consumers' surplus is

$$CS = \int_0^{10} (-0.01x^2 - 0.1x + 6)\,dx - (4)(10)$$

$$= -\frac{0.01}{3}x^3 - 0.05x^2 + 6x\Big|_0^{10} - 40 \approx 11.667\text{, or \$11,667.}$$

3. Setting $p = 10$, we have $\sqrt{225 - 5x} = 10$, $225 - 5x = 100$, or $x = 25$.
Then $CS = \int_0^{25} \sqrt{225 - 5x}\,dx - (10)(25) = \int_0^{25} (225 - 5x)^{1/2}\,dx - 250.$
To evaluate the integral, let $u = 225 - 5x$ so that $du = -5\,dx$ or

$dx = -\frac{1}{5}\,du$. If $x = 0$, $u = 225$ and if $x = 25$, $u = 100$. So

$$CS = -\frac{1}{5}\int_{225}^{100} u^{1/2}\,du - 250 = -\frac{2}{15}u^{3/2}\Big|_{225}^{100} - 250$$
$$= -\frac{2}{15}(1000 - 3375) - 250 = 66.667, \text{ or } \$6{,}667.$$

5. To find the equilibrium point, we solve
$$0.01x^2 + 0.1x + 3 = -0.01x^2 - 0.2x + 8, \quad \text{or} \quad 0.02x^2 + 0.3x - 5 = 0,$$
$$2x^2 + 30x - 500 = (2x - 20)(x + 25) = 0$$
obtaining $x = -25$ or 10. So the equilibrium point is (10,5). Then
$$PS = (5)(10) - \int_0^{10} (0.01x^2 + 0.1x + 3)\,dx$$
$$= 50 - \left(\frac{0.01}{3}x^3 + 0.05x^2 + 3x\right)\Big|_0^{10} = 50 - \frac{10}{3} - 5 - 30 = \frac{35}{3},$$
or approximately \$11,667.

7. To find the market equilibrium, we solve
$$-0.2x^2 + 80 = 0.1x^2 + x + 40, \quad 0.3x^2 + x - 40 = 0,$$
$$3x^2 + 10x - 400 = 0, \ (3x + 40)(x - 10) = 0$$
giving $x = -\frac{40}{3}$ or $x = 10$. We reject the negative root. The corresponding equilibrium price is \$60. The consumers' surplus is
$$CS = \int_0^{10} (-0.2x^2 + 80)\,dx - (60)(10) = -\frac{0.2}{3}x^3 + 80x\Big|_0^{10} - 600 = 133\tfrac{1}{3},$$
or \$13,333. The producers' surplus is
$$PS = 600 - \int_0^{10} (0.1x^2 + x + 40)\,dx = 600 - \left[\tfrac{0.1}{3}x^3 + \tfrac{1}{2}x^2 + 40x\right]_0^{10}$$
$$= 116\tfrac{2}{3}, \text{ or } \$11{,}667.$$

9. Here $P = 200{,}000$, $r = 0.08$, and $T = 5$. So
$$PV = \int_0^5 200{,}000e^{-0.08t}\,dt = -\frac{200{,}000}{0.08}e^{-0.08t}\Big|_0^5 = -2{,}500{,}000(e^{-0.4} - 1)$$
$$\approx 824{,}199.85, \text{ or } \$824{,}200.$$

11. Here $P = 250$, $m = 12$, $T = 20$, and $r = 0.08$. So
$$A = \frac{mP}{r}(e^{rT} - 1) = \frac{12(250)}{0.08}(e^{1.6} - 1) \approx 148{,}238.70, \text{ or approximately } \$148{,}239.$$

13. Here $P = 150$, $m = 12$, $T = 15$, and $r = 0.08$. So

$$A = \frac{12(150)}{0.08}(e^{1.2} - 1) \approx 52{,}202.60, \text{ or approximately } \$52{,}203.$$

15. Here $P = 2000$, $m = 1$, $T = 15.75$, and $r = 0.1$. So

$$A = \frac{1(2000)}{0.1}(e^{1.575} - 1) \approx 76{,}615, \text{ or approximately } \$76{,}615.$$

17. Here $P = 1200$, $m = 12$, $T = 15$, and $r = 0.1$. So

$$PV = \frac{12(1200)}{0.1}(1 - e^{-1.5}) \approx 111{,}869, \text{ or approximately } \$111{,}869.$$

19. We want the present value of an annuity with $P = 300$, $m = 12$, $T = 10$, and $r = 0.12$. So

$$PV = \frac{12(300)}{0.12}(1 - e^{-1.2}) \approx 20{,}964, \text{ or approximately } \$20{,}964.$$

21. a.

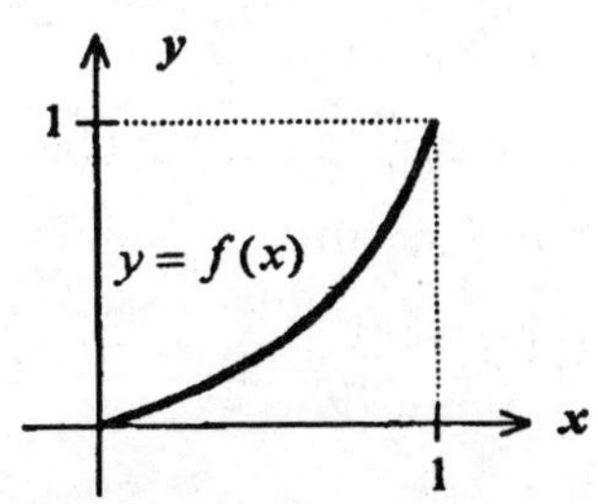

b. $f(0.4) = \frac{15}{16}(0.4)^2 + \frac{1}{16}(0.4) \approx 0.175$; $f(0.9) = \frac{15}{16}(0.9)^2 + \frac{1}{16}(0.9) \approx 0.816$.
So, the lowest 40 percent of the people receive 17.5 percent of the total income and the lowest 90 percent of the people receive 81.6 percent of the income.

23. a.

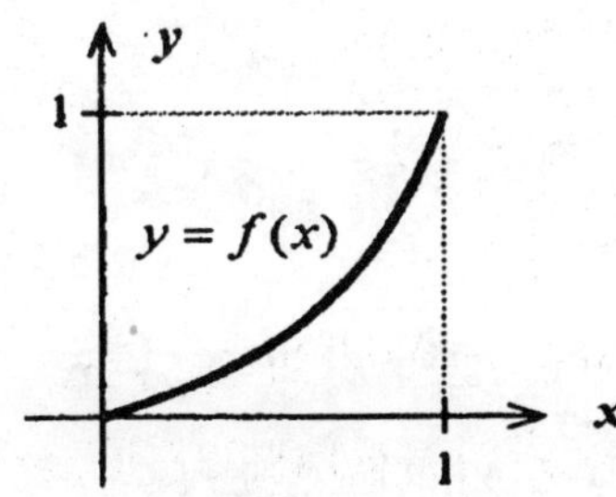

b. $f(0.3) = \frac{14}{15}(0.03)^2 + \frac{1}{15}(0.3) = 0.104$

$$f(0.7)=\tfrac{14}{15}(0.7)^2+\tfrac{1}{15}(0.7)\approx 0.504.$$

USING TECHNOLOGY EXERCISES 15.7, page 990

1. Consumer's surplus: $18,000,000; producer's surplus: $11,700,000.

3. Consumer's surplus: $33,120; producer's surplus: $2,880.

5. Investment A

CHAPTER 15, CONCEPT REVIEW, page 993

1. a. $F'(x)=f(x)$ b. $F(x)+C$ 3. a. Unknown b. Function

5. a. $\int_a^b f(x)\,dx$ b. Minus 7. a. $\dfrac{1}{b-a}\int_a^b f(x)\,dx$ b. Area; area

9. a. $\int_0^{\bar{x}} D(x)\,dx-\bar{p}\,\bar{x}$ b. $\bar{p}\bar{x}-\int_0^{\bar{x}} S(x)\,dx$

11. $\dfrac{mP}{r}(e^{rT}-1)$

CHAPTER 15 REVIEW EXERCISES, page 993

1. $\int (x^3+2x^2-x)\,dx=\frac{1}{4}x^4+\frac{2}{3}x^3-\frac{1}{2}x^2+C.$

3. $\int\left(x^4-2x^3+\dfrac{1}{x^2}\right)dx=\dfrac{x^5}{5}-\dfrac{1}{2}x^4-\dfrac{1}{x}+C.$

5. $\int x(2x^2+x^{1/2})dx=\int (2x^3+x^{3/2})\,dx=\frac{1}{2}x^4+\frac{2}{5}x^{5/2}+C.$

7. $$\begin{aligned}\int (x^2-x+\tfrac{2}{x}+5)\,dx&=\int x^2\,dx-\int x\,dx+2\int \tfrac{dx}{x}+5\int dx\\&=\tfrac{1}{3}x^3-\tfrac{1}{2}x^2+2\ln|x|+5x+C.\end{aligned}$$

9. Let $u = 3x^2 - 2x + 1$ so that $du = (6x - 2)\,dx = 2(3x - 1)\,dx$ or $(3x - 1)\,dx = \frac{1}{2}\,du$.

So $\int (3x-1)(3x^2-2x+1)^{1/3}\,dx = \frac{1}{2}\int u^{1/3}\,du = \frac{3}{8}u^{4/3} + C = \frac{3}{8}(3x^2-2x+1)^{4/3} + C.$

11. Let $u = x^2 - 2x + 5$ so that $du = 2(x - 1)\,dx$ or $(x - 1)\,dx = \frac{1}{2}\,du$.

$$\int \frac{x-1}{x^2-2x+5}\,dx = \frac{1}{2}\int \frac{du}{u} = \frac{1}{2}\ln|u| + C = \frac{1}{2}\ln(x^2-2x+5) + C.$$

13. Put $u = x^2 + x + 1$ so that $du = (2x + 1)\,dx = 2(x + \frac{1}{2})\,dx$ and $(x+\frac{1}{2})dx = \frac{1}{2}du$.

$\int (x+\frac{1}{2})e^{x^2+x+1}dx = \frac{1}{2}\int e^u\,du = \frac{1}{2}e^u + C = \frac{1}{2}e^{x^2+x+1} + C.$

15. Let $u = \ln x$ so that $du = \frac{1}{x}\,dx$. Then

$$\int \frac{(\ln x)^5}{x}\,dx = \int u^5\,du = \frac{1}{6}u^6 + C = \frac{1}{6}(\ln x)^6 + C.$$

17. Let $u = x^2 + 1$ so that $du = 2x\,dx$ or $x\,dx = \frac{1}{2}\,du$. Then

$$\int x^3(x^2+1)^{10}\,dx = \frac{1}{2}\int (u-1)u^{10}\,du \qquad (x^2 = u - 1)$$

$$= \frac{1}{2}\int (u^{11} - u^{10})\,du = \frac{1}{2}(\tfrac{1}{12}u^{12} - \tfrac{1}{11}u^{11}) + C$$

$$= \frac{1}{264}u^{11}(11u-12) + C = \frac{1}{264}(x^2+1)^{11}(11x^2-1) + C.$$

19. Put $u = x - 2$ so that $du = dx$. Then $x = u + 2$ and

$$\int \frac{x}{\sqrt{x-2}}\,dx = \int \frac{u+2}{\sqrt{u}}\,du = \int (u^{1/2} + 2u^{-1/2})\,du = \int u^{1/2}\,du + 2\int u^{-1/2}\,du$$

$$= \frac{2}{3}u^{3/2} + 4u^{1/2} + C = \frac{2}{3}u^{1/2}(u+6) + C = \frac{2}{3}\sqrt{x-2}(x-2+6) + C$$

$$= \frac{2}{3}(x+4)\sqrt{x-2} + C.$$

21. $\int_0^1 (2x^3 - 3x^2 + 1)\,dx = \frac{1}{2}x^4 - x^3 + x\Big|_0^1 = \frac{1}{2} - 1 + 1 = \frac{1}{2}.$

23. $\int_1^4 (x^{1/2}+x^{-3/2})\,dx = \frac{2}{3}x^{3/2} - 2x^{-1/2}\Big|_1^4 = \frac{2}{3}x^{3/2} - \frac{2}{\sqrt{x}}\Big|_1^4 = (\frac{16}{3}-1)-(\frac{2}{3}-2) = \frac{17}{3}.$

25. Put $u = x^3 - 3x^2 + 1$ so that $du = (3x^2 - 6x)\,dx = 3(x^2 - 2x)\,dx$ or $(x^2 - 2x)\,dx = \frac{1}{3}\,du$. Then if $x = -1$, $u = -3$, and if $x = 0$, $u = 1$,

$$\int_{-1}^{0} 12(x^2-2x)(x^3-3x^2+1)^3\,dx = (12)(\tfrac{1}{3})\int_{-3}^{1} u^3\,du = 4(\tfrac{1}{4})u^4\Big|_{-3}^{1}$$
$$= 1 - 81 = -80.$$

27. Let $u = x^2 + 1$ so that $du = 2x\,dx$ or $x\,dx = \frac{1}{2}\,du$. Then, if $x = 0$, $u = 1$, and if $x = 2$, $u = 5$, so

$$\int_0^2 \frac{x}{x^2+1}\,dx = \frac{1}{2}\int_1^5 \frac{du}{u} = \frac{1}{2}\ln u\Big|_1^5 = \frac{1}{2}\ln 5.$$

29. Let $u = 1 + 2x^2$ so that $du = 4x\,dx$ or $x\,dx = \frac{1}{4}\,du$. If $x = 0$, then $u = 1$ and if $x = 2$, then $u = 9$.

$$\int_0^2 \frac{4x}{\sqrt{1+2x^2}}\,dx = \int_1^9 \frac{du}{u^{1/2}} = 2u^{1/2}\Big|_1^9 = 2(3-1) = 4.$$

31. Let $u = 1 + e^{-x}$ so that $du = -e^{-x}\,dx$ and $e^{-x}\,dx = -\,du$. Then

$$\int_{-1}^{0} \frac{e^{-x}}{(1+e^{-x})^2}\,dx = -\int_{1+e}^{2} \frac{du}{u^2} = \frac{1}{u}\Big|_{1+e}^{2} = \frac{1}{2} - \frac{1}{1+e} = \frac{e-1}{2(1+e)}.$$

33. $f(x) = \int f'(x)\,dx = \int (3x^2 - 4x + 1)\,dx = 3\int x^2\,dx - 4\int x\,dx + \int dx$

$= x^3 - 2x^2 + x + C.$

The given condition implies that $f(1) = 1$ or $1 - 2 + 1 + C = 1$, and $C = 1$. Therefore, the required function is $f(x) = x^3 - 2x^2 + x + 1$.

35. $f(x) = \int f'(x)\,dx = \int (1 - e^{-x})\,dx = x + e^{-x} + C$, $f(0) = 2$ implies $0 + 1 + C = 2$ or $C = 1$. So $f(x) = x + e^{-x} + 1$.

37. $\Delta x = \frac{2-1}{5} = \frac{1}{5}$; $x_1 = \frac{6}{5}$, $x_2 = \frac{7}{5}$, $x_3 = \frac{8}{5}$, $x_4 = \frac{9}{5}$, $x_5 = \frac{10}{5}$. The Riemann sum is

$$f(x_1)\Delta x + \cdots + f(x_5)\Delta x = \left\{\left[-2(\tfrac{6}{5})^2 + 1\right] + \left[-2(\tfrac{7}{5})^2 + 1\right] + \cdots + \left[-2(\tfrac{10}{5})^2 + 1\right]\right\}(\tfrac{1}{5})$$

$$= \tfrac{1}{5}(-1.88 - 2.92 - 4.12 - 5.48 - 7) = -4.28.$$

39. a. $R(x) = \int R'(x)\,dx = \int (-0.03x + 60)\,dx = -0.015x^2 + 60x + C$.

$R(0) = 0$ implies that $C = 0$. So, $R(x) = -0.015x^2 + 60x$.

b. From $R(x) = px$, we have $-0.015x^2 + 60x = px$ or $p = -0.015x + 60$.

41. The total number of systems that Vista may expect to sell t months from the time they are put on the market is given by $f(t) = 3000t - 50{,}000(1 - e^{-0.04t})$.

The number is $\int_0^{12} (3000 - 2000e^{-0.04t})\,dt = \left(3000t - \frac{2000}{-0.04}e^{-0.04t}\right)\Big|_0^{12}$

$$= 3000(12) + 50{,}000e^{-0.48} - 50{,}000 = 16{,}939.$$

43. The number of speakers sold at the end of 5 years is

$f(t) = \int f'(t)\,dt = \int_0^5 2000(3 - 2e^{-t})\,dt = 2000[3(5) - 2e^{-5}] - 2000[3 - 2(1)]$

$= 26{,}027.$

45. The number will be

$\int_0^{10} (0.00933t^3 + 0.019t^2 - 0.10833t + 1.3467)\,dt$

$$= 0.0023325t^4 + 0.0063333t^3 - 0.054165t^2 + 1.3467t\Big|_0^{10} = 37.7,$$

or approximately 37.7 million Americans.

47. $A = \int_{-1}^{2} (3x^2 + 2x + 1)\,dx = x^3 + x^2 + x\Big|_{-1}^{2} = [2^3 + 2^2 + 2] - [(-1)^3 + 1 - 1]$

$= 14 - (-1) = 15$ sq units.

49. $A = \int_1^3 \frac{1}{x^2}\,dx = \int_1^3 x^{-2}\,dx = -\frac{1}{x}\Big|_1^3 = -\frac{1}{3} + 1 = \frac{2}{3}$ sq units.

51.

$$A = \int_a^b [f(x) - g(x)]\,dx$$
$$= \int_0^2 (e^x - x)\,dx$$
$$= \left(e^x - \frac{1}{2}x^2\right)\Big|_0^2$$
$$= (e^2 - 2) - (1 - 0) = e^2 - 3 \text{ sq units}$$

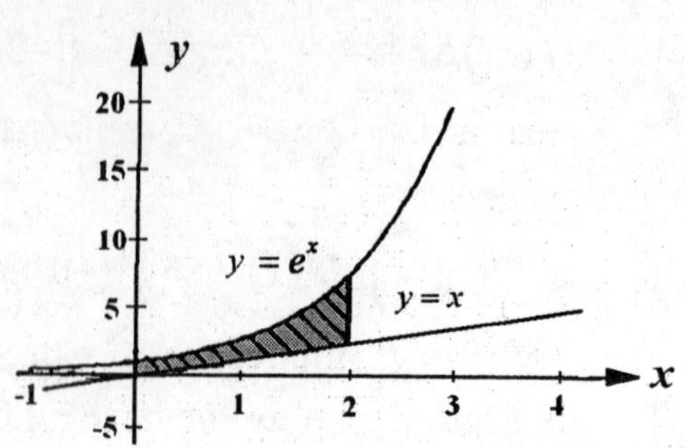

53. $A = \int_0^1 (x^3 - 3x^2 + 2x)\,dx - \int_1^2 (x^3 - 3x^2 + 2x)\,dx$

$$= \frac{x^4}{4} - x^3 + x^2\Big|_0^1 - \left(\frac{x^4}{4} - x^3 + x^2\right)\Big|_1^2$$
$$= \tfrac{1}{4} - 1 + 1 - [(4 - 8 + 4) - (\tfrac{1}{4} - 1 + 1)]$$
$$= \tfrac{1}{4} + \tfrac{1}{4} = \tfrac{1}{2}.$$

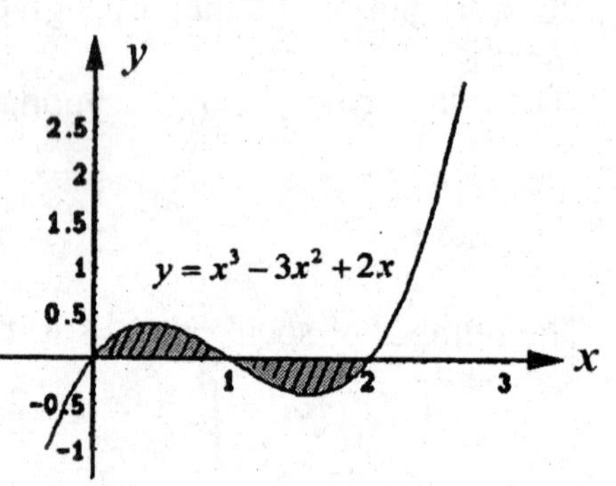

55.

$$A = \frac{1}{3}\int_0^3 \frac{x}{\sqrt{x^2 + 16}}\,dx = \frac{1}{3}\cdot\frac{1}{2}\cdot 2(x^2 + 16)^{1/2}\Big|_0^3$$
$$= \frac{1}{3}(x^2 + 16)^{1/2}\Big|_0^3 = \frac{1}{3}(5 - 4) = \frac{1}{3} \text{ sq units.}$$

57. Setting $p = 8$, we have $-0.01x^2 - 0.2x + 23 = 8$, $-0.01x^2 - 0.2x + 15 = 0$, or $x^2 + 20x - 1500 = (x - 30)(x + 50) = 0$, giving $x = -50$ or 30.

$$CS = \int_0^{30} (-0.01x^2 - 0.2x + 23)\,dx - 8(30) = -\frac{0.01}{3}x^3 - 0.1x^2 + 23x\Big|_0^{30} - 240$$
$$= -\frac{0.01(30)^3}{3} - 0.1(900) + 23(30) - 240 = 270, \text{ or } \$270{,}000.$$

59. Use Equation (17) with $P = 4000$, $r = 0.08$, $T = 20$, and $m = 1$, obtaining

$$A = \frac{(1)(4000)}{0.08}(e^{1.6} - 1) \approx 197{,}651.62$$

that is, Chi-Tai will have approximately \$197,652 in his account after 20 years.

61. Here $P = 80{,}000$, $m = 1$, $T = 10$, and $r = 0.1$, so

$$PV = \frac{(1)(80{,}000)}{0.1}(1 - e^{-1}) \approx 505{,}696,$$ or approximately \$505,696.

63. The average population will be

$$\tfrac{1}{5}\int 80{,}000e^{-0.05t}\,dt = \frac{80{,}000}{5}\cdot\left(-\frac{1}{0.05}\right)e^{-0.05t}\Big|_0^5 = -320{,}000(e^{-0.25} - 1) \approx 70{,}784.$$

CHAPTER 15, BEFORE MOVING ON, page 996

1. $$\int (2x^3 + \sqrt{x} + \frac{2}{x} - \frac{2}{\sqrt{x}})dx = 2\int x^3\,dx + \int x^{1/2}\,dx + 2\int \frac{1}{x}dx - 2\int x^{-1/2}\,dx$$
$$= \tfrac{1}{2}x^4 + \tfrac{2}{3}x^{3/2} + 2\ln|x| - 4x^{1/2} + C$$

2. $f(x) = \int f'(x)\,dx = \int (e^x + x)\,dx = e^x + \tfrac{1}{2}x^2 + C$

$f(0) = 2$ implies $f(0) = e^0 + 0 + C = 2$ or $C = 1$. Therefore, $f(x) = e^x + \tfrac{1}{2}x^2 + 1$.

3. Let $u = x^2 + 1$ so that $du = 2x\,dx$ or $x\,dx = \tfrac{1}{2}du$. Then

$$\int \frac{x}{\sqrt{x^2+1}}dx = \frac{1}{2}\int \frac{du}{\sqrt{u}} = \frac{1}{2}\int u^{-1/2}\,du = \frac{1}{2}\left(2u^{1/2}\right) + C = \sqrt{u} + C = \sqrt{x^2+1} + C.$$

4. Let $u = 2 - x^2$. Then $du = -2x\,dx$ or $x\,dx = -\tfrac{1}{2}du$. If $x = 0$, then $u = 2$ and if $x = 1$, then $u = 1$. Therefore,

$$\int_0^1 x\sqrt{2-x^2}\,dx = -\tfrac{1}{2}\int_2^1 u^{1/2}\,du = -\tfrac{1}{2}\cdot\tfrac{2}{3}u^{3/2}\Big|_2^1 = -\tfrac{1}{3}u^{3/2}\Big|_2^1$$
$$= -\tfrac{1}{3}(1 - 2^{3/2}) = \tfrac{1}{3}(2\sqrt{2} - 1).$$

5. To find the points of intersection, we solve

$$x^2-1=1-x,\ x^2+x-2=0,$$

$$(x+2)(x-1)=0$$

Giving $x=-2$ or $x=1$. The points of intersection are (-2,3) and (1,0). The required area is

$$A=\int_{-2}^{1}[(1-x)-(x^2-1)]\,dx=\int_{-2}^{1}(2-x-x^2)\,dx=(2x-\tfrac{1}{2}x^2-\tfrac{1}{3}x^3)\Big|_{-2}^{1}$$

$$=(2-\tfrac{1}{2}-\tfrac{1}{3})-(4-2+\tfrac{8}{3})=\tfrac{9}{2},\text{ or } \tfrac{9}{2}\text{ sq units.}$$

CHAPTER 16

16.1 Problem Solving Tips

1. When you use integration by parts, remember to choose u and dv so that du is simpler than u and dv is easy to integrate.

2. When you use the method of integration of parts, it is helpful follow the pattern used in the examples in this section.

$$u = \_\_\_\_\_ \qquad dv = \_\_\_\_\_$$

and $\qquad du = \_\_\_\_\_ \qquad v = \_\_\_\_\_$

Then $\qquad \int u\,dv = uv - \int v\,du$

16.1 CONCEPT QUESTIONS, page 1002

1. $\int u\,dv = uv - \int v\,du$

EXERCISES 16.1, page 1003

1. $I = \int xe^{2x}\,dx$. Let $u = x$ and $dv = e^{2x}\,dx$. Then $du = dx$ and $v = \frac{1}{2}e^{2x}$. Therefore,

$$I = uv - \int v\,du = \tfrac{1}{2}xe^{2x} - \int \tfrac{1}{2}e^{2x}\,dx = \tfrac{1}{2}xe^{2x} - \tfrac{1}{4}e^{2x} = \tfrac{1}{4}e^{2x}(2x-1) + C.$$

3. $I = \int xe^{x/4}dx$. Let $u = x$ and $dv = e^{x/4}\,dx$. Then $du = dx$ and $v = 4e^{x/4}$.

$$\int xe^{x/4}\,dx = uv - \int v\,du = 4xe^{x/4} - 4\int e^{x/4}\,dx = 4xe^{x/4} - 16e^{x/4} + C$$

$$= 4(x-4)e^{x/4} + C.$$

5. $\int (e^x - x)^2\,dx = \int (e^{2x} - 2xe^x + x^2)\,dx = \int e^{2x}\,dx - 2\int xe^x\,dx + \int x^2\,dx.$

Using the result $\int xe^x\,dx = (x-1)e^x + k$, from Example 1, we see that

$\int (e^x - x)^2\,dx = \frac{1}{2}e^{2x} - 2(x-1)e^x + \frac{1}{3}x^3 + C.$

7. $I = \int (x+1)e^x\,dx$. Let $u = x + 1$, $dv = e^x\,dx$. Then $du = dx$ and $v = e^x$. Therefore,

$I = (x+1)e^x - \int e^x\,dx = (x+1)e^x - e^x + C = xe^x + C.$

9. Let $u = x$ and $dv = (x + 1)^{-3/2}\,dx$. Then $du = dx$ and $v = -2(x + 1)^{-1/2}$.

$$\begin{aligned}\int x(x+1)^{-3/2}\,dx &= uv - \int v\,du = -2x(x+1)^{-1/2} + 2\int (x+1)^{-1/2}\,dx \\ &= -2x(x+1)^{-1/2} + 4(x+1)^{1/2} + C \\ &= 2(x+1)^{-1/2}[-x + 2(x+1)] + C = \frac{2(x+2)}{\sqrt{x+1}} + C.\end{aligned}$$

11. $I = \int x(x-5)^{1/2}\,dx$. Let $u = x$ and $dv = (x - 5)^{1/2}\,dx$. Then $du = dx$ and

$v = \frac{2}{3}(x - 5)^{3/2}$. Therefore,

$$\begin{aligned}I &= \tfrac{2}{3}x(x-5)^{3/2} - \int \tfrac{2}{3}(x-5)^{3/2}\,dx = \tfrac{2}{3}x(x-5)^{3/2} - \tfrac{2}{3}\cdot\tfrac{2}{5}(x-5)^{5/2} + C \\ &= \tfrac{2}{3}(x-5)^{3/2}[x - \tfrac{2}{5}(x-5)] + C = \tfrac{2}{15}(x-5)^{3/2}(5x - 2x + 10) + C \\ &= \tfrac{2}{15}(x-5)^{3/2}(3x+10) + C.\end{aligned}$$

13. $I = \int x \ln 2x\,dx$. Let $u = \ln 2x$ and $dv = x\,dx$. Then $du = \frac{1}{x}\,dx$ and $v = \frac{1}{2}x^2$.

Therefore, $I = \frac{1}{2}x^2 \ln 2x - \int \frac{1}{2}x\,dx = \frac{1}{2}x^2 \ln 2x - \frac{1}{4}x^2 + C = \frac{1}{4}x^2(2\ln 2x - 1) + C.$

15. Let $u = \ln x$ and $dv = x^3\,dx$, then $du = \frac{1}{x}\,dx$, and $v = \frac{1}{4}x^4$.

$$\begin{aligned}\int x^3 \ln x\,dx &= \tfrac{1}{4}x^4 \ln x - \tfrac{1}{4}\int x^3\,dx = \tfrac{1}{4}x^4 \ln x - \tfrac{1}{16}x^4 + C \\ &= \tfrac{1}{16}x^4(4\ln x - 1) + C.\end{aligned}$$

17. Let $u = \ln x^{1/2}$ and $dv = x^{1/2}\,dx$. Then $du = \frac{1}{2x}\,dx$ and $v = \frac{2}{3}x^{3/2}$,

and $\int \sqrt{x}\ln\sqrt{x}\,dx = uv - \int v\,du = \frac{2}{3}x^{3/2}\ln x^{1/2} - \frac{1}{3}\int x^{1/2}\,dx$

$$= \frac{2}{3}x^{3/2}\ln x^{1/2} - \frac{2}{9}x^{3/2} + C = \frac{2}{9}x\sqrt{x}(3\ln\sqrt{x} - 1) + C.$$

19. Let $u = \ln x$ and $dv = x^{-2}\,dx$. Then $du = \frac{1}{x}\,dx$ and $v = -x^{-1}$,

$$\int \frac{\ln x}{x^2}\,dx = uv - \int v\,du = -\frac{\ln x}{x} + \int x^{-2}\,dx = -\frac{\ln x}{x} - \frac{1}{x} + C$$

$$= -\frac{1}{x}(\ln x + 1) + C.$$

21. Let $u = \ln x$ and $dv = dx$. Then $du = \frac{1}{x}\,dx$ and $v = x$ and

$$\int \ln x\,dx = uv - \int v\,du = x\ln x - \int dx = x\ln x - x + C = x(\ln x - 1) + C.$$

23. Let $u = x^2$ and $dv = e^{-x}\,dx$. Then $du = 2x\,dx$ and $v = -e^{-x}$, and

$\int x^2e^{-x}\,dx = uv - \int v\,du = -x^2e^{-x} + 2\int xe^{-x}\,dx.$

We can integrate by parts again, or, using the result of Problem 2, we find

$$\int x^2e^{-x}\,dx = -x^2e^{-x} + 2[-(x+1)e^{-x}] + C = -x^2e^{-x} - 2(x+1)e^{-x} + C$$

$$= -(x^2 + 2x + 2)e^{-x} + C.$$

25. $I = \int x(\ln x)^2\,dx$. Let $u = (\ln x)^2$ and $dv = x\,dx$, so that

$du = 2(\ln x)\left(\frac{1}{x}\right) = \frac{2\ln x}{x}$ and $v = \frac{1}{2}x^2$. Then $I = \frac{1}{2}x^2(\ln x)^2 - \int x\ln x\,dx$.

Next, we evaluate $\int x\ln x\,dx$, by letting $u = \ln x$ and $dv = x\,dx$, so that $du = \frac{1}{x}\,dx$

and $v = \frac{1}{2}x^2$. Then $\int x\ln x\,dx = \frac{1}{2}x^2(\ln x) - \frac{1}{2}\int x\,dx = \frac{1}{2}x^2\ln x - \frac{1}{4}x^2 + C.$

Therefore, $\int x(\ln x)^2\,dx = \frac{1}{2}x^2(\ln x)^2 - \frac{1}{2}x^2\ln x + \frac{1}{4}x^2 + C$

$$= \frac{1}{4}x^2[2(\ln x)^2 - 2\ln x + 1] + C.$$

27. $\int_0^{\ln 2} xe^x\,dx = (x-1)e^x\Big|_0^{\ln 2}$ (Using the results of Example 1.)

$$= (\ln 2 - 1)e^{\ln 2} - (-e^0) = 2(\ln 2 - 1) + 1 = 2\ln 2 - 1.$$ (Recall $e^{\ln 2} = 2$.)

29. We first integrate $I = \int \ln x\,dx$. Integrating by parts with $u = \ln x$ and $dv = dx$ so that $du = \frac{1}{x}\,dx$ and $v = x$, we find

$$I = x\ln x - \int dx = x\ln x - x + C = x(\ln x - 1) + C.$$

Therefore, $\int_1^4 \ln x\,dx = x(\ln x - 1)\Big|_1^4 = 4(\ln 4 - 1) - 1(\ln 1 - 1) = 4\ln 4 - 3.$

31. Let $u = x$ and $dv = e^{2x}\,dx$. Then $u = dx$ and $v = \frac{1}{2}e^{2x}$ and

$$\int_0^2 xe^{2x}\,dx = \frac{1}{2}xe^{2x}\Big|_0^2 - \frac{1}{2}\int_0^2 e^{2x}\,dx = e^4 - \frac{1}{4}e^{2x}\Big|_0^2$$
$$= e^4 - \frac{1}{4}e^4 + \frac{1}{4} = \frac{1}{4}(3e^4 + 1).$$

33. Let $u = x$ and $dv = e^{-2x}\,dx$, so that $du = dx$ and $v = -\frac{1}{2}e^{-2x}$.

$f(x) = \int xe^{-2x}\,dx = -\frac{1}{2}xe^{-2x} - \frac{1}{4}e^{-2x} + C$; $f(0) = -\frac{1}{4} + C = 3$ and $C = \frac{13}{4}$.

Therefore, $y = -\frac{1}{2}xe^{-2x} - \frac{1}{4}e^{-2x} + \frac{13}{4}$.

35. The required area is given by $\int_1^5 \ln x\,dx$. We first find $\int \ln x\,dx$. Using the technique of integration by parts with $u = \ln x$ and $dv = dx$ so that $du = \frac{1}{x}\,dx$ and $v = x$, we have $\int \ln x\,dx = x\ln x - \int dx = x\ln x - x = x(\ln x - 1) + C.$

Therefore, $\int_1^5 \ln x\,dx = x(\ln x - 1)\Big|_1^5 = 5(\ln 5 - 1) - 1(\ln 1 - 1) = 5\ \ln 5 - 4$

and the required area is (5 ln 5 - 4) sq units.

37. The distance covered is given by $\int_0^{10} 100te^{-0.2t}\,dt = 100\int_0^{10} te^{-0.2t}\,dt$.

We integrate by parts, letting $u = t$ and $dv = e^{-0.2t}\,dt$ so that $du = dt$ and $v = -\frac{1}{0.2}e^{-0.2t} = -5e^{-0.2t}$. Therefore,

$$100\int_0^{10} te^{-0.2t}\,dt = 100\left[-5te^{-0.2t}\Big|_0^{10}\right] + 5\int_0^{10} e^{-0.2t}\,dt$$

$$= 100[-5te^{-0.2t} - 25e^{-0.2t}]\Big|_0^{10} = -500e^{-0.2t}(t+5)\Big|_0^{10}$$

$$= -500e^{-2}(15) + 500(5) = 1485, \text{ or } 1485 \text{ feet.}$$

39. The average concentration is $C = \frac{1}{12}\int_0^{12} 3te^{-t/3}\,dt = \frac{1}{4}\int_0^{12} te^{-t/3}\,dt.$

Let $u = t$ and $dv = e^{-t/3}\,dt$. So $du = dt$ and $v = -3e^{-t/3}$. Then

$$C = \frac{1}{4}\left[-3te^{-t/3}\Big|_0^{12} + 3\int_0^{12} e^{-t/3}\,dt\right] = \frac{1}{4}\left\{-36e^{-4} - \left[9e^{-t/3}\Big|_0^{12}\right]\right\}$$

$$= \tfrac{1}{4}(-36e^{-4} - 9e^{-4} + 9) \approx 2.04 \text{ mg/ml.}$$

41. $N = 2\int te^{-0.1t}\,dt$. Let $u = t$ and $dv = e^{-0.1t}$, so that $du = dt$ and $v = -10e^{-0.1t}$. Then

$v = -10e^{-0.1t}$. Then

$$N(t) = 2[-10te^{-0.1t} + 10\int e - 0.1t\,dt] = 2(-10te^{-0.1t} - 100e^{-0.1t}) + C$$

$$= -20e^{-0.1t}(t+10) + 200. \qquad [N(0) = 0]$$

43. $PV = \int_0^5 (30{,}000 + 800t)e^{-0.08t}\,dt = 30{,}000\int_0^5 e^{-0.08t}\,dt + 800\int_0^5 te^{-0.08t}\,dt$.

Let $I = \int te^{-0.08t}\,dt$. To evaluate I by parts, let $u = t$, $dv = e^{-0.08t}\,dt$

and $du = dt$, $v = -\frac{1}{0.08}e^{-0.08t} = -12.5e^{-0.08t}$.

Therefore, $I = -12.5te^{-0.08t} + 12.5\int e^{-0.08t}\,dt = -12.5te^{-0.08t} - 156.25e^{-0.08t} + C.$

$$PV = \left[-\frac{30{,}000}{0.08}e^{-0.08t} - 800(12.5)te^{-0.08t} - 800(156.25)e^{-0.08t}\right]_0^5$$

$$= -375{,}000\,e^{-0.4} + 375{,}000 - 50{,}000e^{-0.4} - 125{,}000e^{-0.4} + 125{,}000$$

$$= 500{,}000 - 550{,}000e^{-0.4} = 131{,}323.97, \text{ or approximately } \$131{,}324.$$

45. The membership will be

$$N(5) = N(0) + \int_0^5 9\sqrt{t+1}\ln\sqrt{t+1}\,dt = 50 + 9\int_0^5 \sqrt{t+1}\,\ln\sqrt{t+1}\,dt$$

To evaluate the integral, let $u = t + 1$ so that $du = dt$. Also, if $t = 0$, then $u = 1$ and if $t = 5$, then $u = 6$. So $9\int_0^5 \sqrt{t+1}\ \ln\sqrt{t+1}\,dt = 9\int_1^6 \sqrt{u}\ln\sqrt{u}\,du.$

Using the results of Problem 17, we find $9\int_1^6 \sqrt{u}\ln\sqrt{u}\,du = 2u\sqrt{u}(3\ln\sqrt{u}-1)\Big|_1^6.$

Therefore, $N = 50 + 51.606 \approx 101.606$ or 101,606 people.

47. The average concentration from $r = r_1$ to $r = r_2$ is

$$A = \frac{1}{r^2 - r_1}\int_{r_1}^{r_2} c(r)\,dr = \frac{1}{r^2 - r_1}\int_{r_1}^{r_2}\left[\left(\frac{c_1 - c_2}{\ln r_1 - \ln r_2}\right)(\ln r - \ln r_2) + c_2\right]dr$$

$$= \frac{1}{r_2 - r_1}\left(\frac{c_1 - c_2}{\ln r_1 - \ln r_2}\right)\left[\int_{r_1}^{r_2}\ln r\,dr - \int_{r_1}^{r_2}\ln r_2\,dr\right] + \frac{1}{r_2 - r_1}\int_{r_1}^{r_2} c_2\,dr$$

We integrate $\int \ln r\,dr$ by parts, letting $u = \ln r$, $dv = dr$, or $du = \frac{dr}{r}$, and $v = r$, so that $\int \ln r\,dr = r\ln r - \int r\frac{dr}{r} = r\ln r - r = r(\ln r - 1).$

Therefore,

$$A = \frac{1}{r_2 - r_1}\left(\frac{c_1 - c_2}{\ln r_1 - \ln r_2}\right)\left[r(\ln r - 1)\Big|_{r_1}^{r_2} - (\ln r_2)\Big|_{r_1}^{r_2}\right] + c_2$$

$$= \frac{1}{r_2 - r_1}\left(\frac{c_1 - c_2}{\ln r_1 - \ln r_2}\right)\{r_2(\ln r_2 - 1) - r_1(\ln r_1 - 1) - (r_2 - r_1)\ln r_2] + c_2$$

$$= \frac{1}{r_2 - r_1}\left(\frac{c_1 - c_2}{\ln r_1 - \ln r_2}\right)[r_1(\ln r_2 - \ln r_1) - (r_2 - r_1)] + c_2]$$

$$= (c_2 - c_1)\left[\frac{r_1}{r_2 - r_1} + \frac{1}{\ln r_1 - \ln r_2}\right] + c_2$$

49. True. This is just the integration by parts formula.

16.2 Problem Solving Tips

1. The integrals that follow in the exercise set may not be exactly in the same form as those in the table of integrals. Sometimes you may need to rewrite the integral (as in Example 2, page 1008 of the text) or you may need to apply one rule more than once (as

in Example 5, page 1009 in the text).

16.2 CONCEPT QUESTIONS, page 1010

1. a. We would chose Formula 19.

   b. Put $a=\sqrt{2}$ and $x=u$. Then, using Formula 19, we find

$$\int \frac{\sqrt{2-x^2}}{x}dx = \int \frac{\sqrt{(\sqrt{2})^2-x^2}}{x}dx = \sqrt{2-x^2}-\sqrt{2}\ln\left|\frac{\sqrt{a}+\sqrt{2-x^2}}{x}\right|+C$$

EXERCISES 16.2, page 1010

1. First we note that

$$\int \frac{2x}{2+3x}dx = 2\int \frac{x}{2+3x}dx.$$

Next, we use Formula 1 with $a=2$, $b=3$, and $u=x$. Then

$$\int \frac{2x}{2+3x}dx = \frac{2}{9}[2+3x-2\ln|2+3x|]+C.$$

3. $\displaystyle\int \frac{3x^2}{2+4x}dx = \frac{3}{2}\int \frac{x^2}{1+2x}dx.$

Use Formula 2 with $a=1$ and $b=2$ obtaining

$$\int \frac{3x^2}{2+4x}dx = \frac{3}{32}[(1+2x)^2-4(1+2x)+2\ln|1+2x|]+C.$$

5. $\displaystyle\int x^2\sqrt{9+4x^2}\,dx = \int x^2\sqrt{4(\tfrac{9}{4})+x^2)}\,dx = 2\int x^2\sqrt{(\tfrac{3}{2})^2+x^2}\,dx.$

Use Formula 8 with $a=3/2$, we find that

$$\int x^2\sqrt{9+4x^2}\,dx == 2[\tfrac{x}{8}(\tfrac{9}{4}+2x^2)\sqrt{\tfrac{9}{4}+x^2}-\tfrac{81}{128}\ln\left|x+\sqrt{\tfrac{9}{4}+x^2}\right|+C.$$

7. Use Formula 6 with $a=1$, $b=4$, and $u=x$, then

$$\int \frac{dx}{x\sqrt{1+4x}} = \ln\left|\frac{\sqrt{1+4x}-1}{\sqrt{1+4x}+1}\right|+C.$$

9. Use Formula 9 with $a=3$ and $u=2x$. Then $du=2\,dx$ and

$$\int_0^2 \frac{dx}{\sqrt{9+4x^2}} = \frac{1}{2}\int_0^4 \frac{du}{\sqrt{3^2+u^2}} = \frac{1}{2}\ln\left|u+\sqrt{9+u^2}\right|\Big|_0^4$$

$$= \frac{1}{2}(\ln 9 - \ln 3) = \frac{1}{2}\ln 3.$$

Note that the limits of integration have been changed from $x = 0$ to $x = 2$ and from $u = 0$ to $u = 4$.

11. Using Formula 22 with $a = 3$, we see that $\int \frac{dx}{(9-x^2)^{3/2}} = \frac{x}{9\sqrt{9-x^2}} + C.$

13. $\int x^2\sqrt{x^2-4}\,dx.$

Use Formula 14 with $a = 2$ and $u = x$, obtaining

$\int x^2\sqrt{x^2-4}\,dx = \frac{x}{8}(2x^2-4)\sqrt{x^2-4} - 2\ln\left|x+\sqrt{x^2-4}\right| + C.$

15. Using Formula 19 with $a = 2$ and $u = x$, we have

$$\int \frac{\sqrt{4-x^2}}{x}\,dx = \sqrt{4-x^2} - 2\ln\left|\frac{2+\sqrt{4-x^2}}{x}\right| + C.$$

17. $\int xe^{2x}\,dx.$

Use Formula 23 with $u = x$ and $a = 2$, obtaining

$\int xe^{2x}\,dx = \frac{1}{4}(2x-1)e^{2x} + C.$

19. $\int \frac{dx}{(x+1)\ln(x+1)}.$

Let $u = x + 1$ so that $du = dx$. Then $\int \frac{dx}{(x+1)\ln(x+1)} = \int \frac{du}{u\ln u}.$

Use Formula 28 with $u = x$, obtaining $\int \frac{du}{u\ln u} = \ln\left|\ln u\right| + C.$

Therefore, $\int \frac{dx}{(x+1)\ln(x+1)} = \ln\left|\ln(x+1)\right| + C$

21. $\int \frac{e^{2x}}{(1+3e^x)^2}dx.$

Put $u = e^x$ then $du = e^x dx$. Then we use Formula 3 with $a = 1, b = 3$. Then

$$I = \int \frac{u}{(1+3u)^2}du = \frac{1}{9}\left[\frac{1}{1+3u} + \ln|1+3u|\right] + C = \frac{1}{9}\left[\frac{1}{1+3e^x} + \ln(1+3e^x)\right] + C$$

23. $\int \frac{3e^x}{1+e^{x/2}}dx = 3\int \frac{e^{x/2}}{e^{-x/2}+1}dx$.

Let $v = e^{x/2}$ so that $dv = \frac{1}{2}e^{x/2}dx$ or $e^{x/2}\,dx = 2\,dv$. Then

$$\int \frac{3e^x}{1+e^{x/2}}dx = 6\int \frac{dv}{\frac{1}{v}+1} = 6\int \frac{v}{v+1}dv.$$

Use Formula 1 with $a = 1$, $b = 1$, and $u = v$, obtaining

$6\int \frac{v}{v+1}dv = 6[1+v-\ln|1+v|] + C.$ So $\int \frac{3e^x}{1+e^{x/2}}dx = 6[1+e^{x/2} - \ln(1+e^{x/2})] + C.$

This answer may be written in the form $6[e^{x/2} - \ln(1+e^{x/2})] + C$ since C is an arbitrary constant.

25. $\int \frac{\ln x}{x(2+3\ln x)}dx.$ Let $v = \ln x$ so that $dv = \frac{1}{x}dx.$ Then

$$\int \frac{\ln x}{x(2+3\ln x)}dx = \int \frac{v}{2+3v}dv.$$

Use Formula 1 with $a = 2$, $b = 3$, and $u = v$ to obtain

$\int \frac{v}{2+3v}dv = \frac{1}{9}[2+3\ln x - 2\ln|2+3\ln x|] + C$. So

$$\int \frac{\ln x}{x(2+3\ln x)}dx = \frac{1}{9}[2+3\ln x - 2\ln|2+3\ln x| + C.$$

27. Using Formula 24 with $a = 1$, $n = 2$, and $u = x$. Then

$$\int_0^1 x^2e^x\,dx = x^2e^x\Big|_0^1 - 2\int_0^1 xe^x\,dx = x^2e^x - 2(xe^x - e^x)\Big|_0^1$$
$$= x^2e^x - 2xe^x + 2e^x\Big|_0^1 = e - 2e + 2e - 2 = e - 2.$$

29. $\int x^2 \ln x\,dx.$ Use Formula 27 with $n = 2$ and $u = x$, obtaining

$\int x^2 \ln x\,dx = \frac{x^3}{9}(3\ln x - 1) + C.$

31. $\int (\ln x)^3 dx$. Use Formula 29 with $n = 3$ to write

$\int (\ln x)^3\,dx = x(\ln x)^3 - 3\int (\ln x)^2 dx.$ Using Formula 29 again with $n = 2$, we obtain

$$\int (\ln x)^3 dx = x(\ln x)^3 - 3[x(\ln x)^2 - 2\int \ln x\,dx].$$

Using Formula 29 one more time with $n = 1$ gives

$$\begin{aligned}\int (\ln x)^3 dx &= x(\ln x)^3 - 3x(\ln x)^2 + 6(x\ln x - x) + C \\ &= x(\ln x)^3 - 3x(\ln x)^2 + 6x\ln x - 6x + C.\end{aligned}$$

33. Letting $p = 50$ gives $50 = \frac{250}{\sqrt{16+x^2}}$

from which we deduce that $\sqrt{16+x^2} = 5$, $16+x^2 = 25$, and $x = 3$.

Using Formula 9 with $u = 3$, we see that

$$\begin{aligned}CS &= \int_0^3 \frac{250}{\sqrt{16+x^2}}\,dx = 50(3) = 250\int_0^3 \frac{1}{\sqrt{16+x^2}}\,dx - 150 \\ &= 250\ln\left|x+\sqrt{16+x^2}\right|_0^3 - 150 = 250[\ln 8 - \ln 4] - 150 \\ &= 23.286795, \text{ or approximately } \$2,329.\end{aligned}$$

35. The number of visitors admitted to the amusement park by noon is found by evaluating the integral

$$\int_0^3 \frac{60}{(2+t^2)^{3/2}}\,dt = 60\int_0^3 \frac{dt}{(2+t)^{3/2}}.$$

Using Formula 12 with $a = \sqrt{2}$ and $u = t$, we find

$$60\int_0^3 \frac{dt}{(2+t^2)^{3/2}} = 60\left[\frac{t}{2\sqrt{2+t^3}}\right]_0^3 = 60\left[\frac{3}{2\sqrt{11}-0}\right] = \frac{90}{\sqrt{11}} = 27.136, \text{ or } 27,136.$$

37. In the first 10 days

$$\frac{1}{10}\int_0^{10}\frac{1000}{1+24e^{-0.02t}}dt = 100\int_0^{10}\frac{1}{1+24e^{-0.02t}dt} = 100\left[t+\frac{1}{0.02}\ln(1+24e^{-0.02t}\right]_0^{10}$$

(Use Formula 25 with a = 0.02 and b = 24.)

$$= 100[10+50\ \ln\ 20.64953807-50\ \ln\ 25] = 44.0856,$$

or approximately 44 fruitflies. In the first 20 days:

$$\frac{1}{20}\int_0^{20}\frac{1000}{1+24e^{-0.02t}}dt = 50\int_0^{10}\frac{1}{1+24e^{-0.02t}}dt$$

$$= 500[t+\ln(1+24e^{-0.02t}]\Big|_0^{20}$$

$$= 50[20+50\ln 17.0876822-50\ln 25] = 48.71$$

or approximately 49 fruitflies.

39. $\frac{1}{5}\int_0^5\frac{100,000}{2(1+1.5e^{-0.2t})}dt = 10,000\int_0^5\frac{1}{1+1.5e^{-0.2t}}dt$

$$= 10,000[t+5\ln(1+1.5e^{-0.2t})]\Big|_0^5$$

(Use Formula 25 with a = -0.2 and b = 1.5.)

$$= 10,000[5+5\ln 1.551819162-5\ln 2.5] \approx 26157,$$

or approximately 26,157 people.

41. $\int_0^5 20,000te^{0.15t}dt = 20,000\int_0^5 te^{0.15t}dt = 20,000\left[\frac{1}{(0.15)^2}(0.15t-1)e^{0.15t}\right]_0^5$

(Use Formula 23 with a = 0.15.)

$$= 888,888.8889[-0.25e^{0.75}+1] = \$418,444.$$

16.3 CONCEPT QUESTIONS, page 1023

1. In the trapezoidal rule, each region under the graph of f (or over the graph of f) is approximated by the area of a trapezoid whose base consists of two consecutive points in the partition. Therefore, n can be odd or even. In Simpson's rule, the area of each subregion is approximated by part of a parabola passing through those points. Therefore, there are two subintervals involved in the approximations. This implies that n must be even.

3. If we use the trapezoidal rule and f is a linear function, then $f''(x) = 0$, and therefore, $M = 0$, and consequently the maximum error is 0. If we use Simpson's rule, then $f^4(x) = 0$ and therefore, $M = 0$ and consequently the maximum error is 0.

EXERCISES 16.3, page 1024

1. $\Delta x = \frac{b-a}{n} = \frac{2-0}{6} = \frac{1}{3}, x_0 = 0, x_1 = \frac{1}{3}, x_2 = \frac{2}{3}, x_3 = 1, x_4 = \frac{4}{3}, x_5 = \frac{5}{3}, x_6 = 2.$

Trapezoidal Rule:

$$\int_0^2 x^2\,dx \approx \tfrac{1}{6}\left[0 + 2(\tfrac{1}{3})^2 + 2(\tfrac{2}{3})^2 + 2(1)^2 + 2(\tfrac{4}{3})^2 + 2(\tfrac{5}{3})^2 + 2^2\right]$$

$$\approx \tfrac{1}{6}\ (0.22222 + 0.88889 + 2 + 3.55556 + 5.55556 + 4) \approx 2.7037.$$

Simpson's Rule:

$$\int x^2\,dx = \tfrac{1}{9}[0 + 4(\tfrac{1}{3})^2 + 2(\tfrac{2}{3})^2 + 4(1)^2 + 2(\tfrac{4}{3})^2 + 4(\tfrac{5}{3})^2 + 2^2]$$

$$\approx \tfrac{1}{9}\ (0.44444 + 0.88889 + 4 + 3.55556 + 11.11111 + 4) \approx 2.6667.$$

Exact Value: $\int_0^2 x^2\,dx = \frac{1}{3}x^3\Big|_0^2 = \frac{8}{3} = 2\frac{2}{3}.$

3. $\Delta x = \frac{b-a}{n} = \frac{1-0}{4} = \frac{1}{4}; x_0 = 0, x_1 = \frac{1}{4}, x_2 = \frac{1}{2}, x_3 = \frac{3}{4}, x_4 = 1.$

Trapezoidal Rule:

$$\int_0^1 x^3\,dx \approx \tfrac{1}{8}\left[0 + 2(\tfrac{1}{4})^3 + 2(\tfrac{1}{2})^3 + 2(\tfrac{3}{4})^3 + 1^3\right] \approx \tfrac{1}{8}(0 + 0.3125 + 0.25 + 0.8)$$

$$\approx 0.265625.$$

Simpson's Rule:

$$\int_0^1 x^3\,dx \approx \tfrac{1}{12}\left[0 + 4(\tfrac{1}{4})^3 + 2(\tfrac{1}{2})^3 + 4(\tfrac{3}{4})^3 + 1\right] \approx \tfrac{1}{12}[0 + 0.625 + 0.25 + 1.6875 + 1]$$

$$\approx 0.25.$$

Exact Value: $\int_0^1 x^3\,dx = \frac{1}{4}x^4\Big|_0^1 = \frac{1}{4} - 0 = \frac{1}{4}.$

5. a. Here $a = 1$, $b = 2$, and $n = 4$; so $\Delta x = \frac{2-1}{4} = \frac{1}{4} = 0.25$, and $x_0 = 1$, $x_1 = 1.25$, $x_2 = 1.5$, $x_3 = 1.75$, $x_4 = 2$.

Trapezoidal Rule:

$$\int_1^2 \frac{1}{x}\,dx \approx \frac{0.25}{2}\left[1 + 2\left(\frac{1}{1.25}\right) + 2\left(\frac{1}{1.5}\right) + 2\left(\frac{1}{1.75}\right) + \frac{1}{2}\right] \approx 0.697.$$

Simpson's Rule:

$$\int_1^2 \frac{1}{x}\,dx \approx \frac{0.25}{3}\left[1+4\left(\frac{1}{1.25}\right)+2\left(\frac{1}{1.5}\right)+4\left(\frac{1}{1.75}\right)+\frac{1}{2}\right] \approx 0.6933.$$

$$\int_1^2 \frac{1}{x}\,dx = \ln x\Big|_1^2 = \ln 2 - \ln 1 \approx 0.6931.$$

7. $\Delta x = \frac{1}{4},\ x_0 = 1,\ x_1 = \frac{5}{4},\ x_2 = \frac{3}{2},\ x_3 = \frac{7}{4},\ x_4 = 2.$

Trapezoidal Rule:

$$\int_1^2 \frac{1}{x^2}\,dx \approx \tfrac{1}{8}\left[1+2(\tfrac{4}{5})^2+2(\tfrac{2}{3})^2+2(\tfrac{4}{7})^2+(\tfrac{1}{2})^2\right] \approx 0.5090.$$

Simpson's Rule:

$$\int_1^2 \frac{1}{x^2}\,dx \approx \tfrac{1}{12}\left[1+4(\tfrac{4}{5})^2+2(\tfrac{2}{3})^2+4(\tfrac{4}{7})^2+(\tfrac{1}{2})^2\right] \approx 0.5004.$$

Exact Value: $\displaystyle\int_1^2 \frac{1}{x^2}\,dx = -\frac{1}{x}\Big|_1^2 = -\frac{1}{2}+1 = \frac{1}{2}.$

9. $\Delta x = \frac{b-a}{n} = \frac{4-0}{8} = \frac{1}{2};\ x_0 = 0, x_1 = \frac{1}{2}, x_2 = \frac{2}{2}, x_3 = \frac{3}{2}, \ldots, x_8 = \frac{8}{2}.$

Trapezoidal Rule:

$$\int_0^4 \sqrt{x}\,dx \approx \frac{\frac{1}{2}}{2}\left[0+2\sqrt{0.5}+2\sqrt{1}+2\sqrt{1.5}+\cdots+2\sqrt{3.5}+\sqrt{4}\right] \approx 5.26504.$$

Simpson's Rule:

$$\int_0^4 \sqrt{x}\,dx \approx \frac{\frac{1}{2}}{3}\left[0+4\sqrt{0.5}+2\sqrt{1}+4\sqrt{1.5}+\cdots+4\sqrt{3.5}+\sqrt{4}\right] \approx 5.30463.$$

The actual value is $\displaystyle\int_0^4 \sqrt{x}\,dx \approx \frac{2}{3}x^{3/2}\Big|_0^4 = \frac{2}{3}(8) = \frac{16}{3} \approx 5.333333.$

11. $\Delta x = \frac{1-0}{6} = \frac{1}{6};\ x_0 = 0, x_1 = \frac{1}{6}, x_2 = \frac{2}{6}, \ldots, x_6 = \frac{6}{6}.$

Trapezoidal Rule:

$$\int_0^1 e^{-x}\,dx \approx \frac{\frac{1}{6}}{2}[1+2e^{-1/6}+2e^{-2/6}+\cdots+2e^{-5/6}+e^{-1}] \approx 0.633583.$$

Simpson's Rule:

$$\int_0^1 e^{-x}\,dx \approx \frac{\frac{1}{6}}{3}[1+4e^{-1/6}+2e^{-2/6}+\cdots+4e^{-5/6}+e^{-1}] \approx 0.632123.$$

The actual value is $\displaystyle\int_0^1 e^{-x}\,dx = -e^{-x}\Big|_0^1 = -e^{-1}+1 \approx 0.632121.$

13. $\Delta x = \frac{1}{4}; x_0 = 0, x_1 = \frac{5}{4}, x_2 = \frac{3}{2}, x_3 = \frac{7}{4}, x_4 = 2.$

Trapezoidal Rule:

$$\int_1^2 \ln x\,dx \approx \tfrac{1}{8}[\ln 1 + 2\ln\tfrac{5}{4} + 2\ln\tfrac{3}{2} + 2\ln\tfrac{7}{4} + \ln 2] \approx 0.38370.$$

Simpson's Rule:

$$\int_1^2 \ln x\,dx \approx \tfrac{1}{12}[\ln 1 + 4\ln\tfrac{5}{4} + 2\ln\tfrac{3}{2} + 4\ln\tfrac{7}{4} + \ln 2] \approx 0.38626.$$

Exact Value: $\int_1^2 \ln x\,dx \approx x(\ln x - 1)\Big|_1^2 = 2(\ln 2 - 1) + 1 = 2\ln 2 - 1 \approx 0.3863.$

15. $\Delta x = \frac{1-0}{4} = \frac{1}{4}; x_0 = 0, x_1 = \frac{1}{4}, x_2 = \frac{2}{4}, x_3 = \frac{3}{4}, x_4 = \frac{4}{4}.$

Trapezoidal Rule:

$$\int_0^1 \sqrt{1+x^3}\,dx \approx \tfrac{\frac{1}{4}}{2}\left[\sqrt{1} + 2\sqrt{1+(\tfrac{1}{4})^3} + \cdots + 2\sqrt{1+(\tfrac{3}{4})^3} + \sqrt{2}\right] \approx 1.1170.$$

Simpson's Rule:

$$\int_0^1 \sqrt{1+x^3}\,dx \approx \tfrac{\frac{1}{4}}{3}\left[\sqrt{1} + 4\sqrt{1+(\tfrac{1}{4})^3} + 2\sqrt{1+(\tfrac{2}{4})^3}\ \cdots + 4\sqrt{1+(\tfrac{3}{4})^3} + \sqrt{2}\right] \approx 1.1114.$$

17. $\Delta x = \frac{2-0}{4} = \frac{1}{2}; x_0 = 0, x_1 = \frac{1}{2}, x_2 = \frac{2}{2}, x_3 = \frac{3}{2}, x_4 = \frac{4}{2}.$

Trapezoidal Rule:

$$\int_0^2 \frac{1}{\sqrt{x^3+1}}\,dx = \frac{\frac{1}{2}}{2}\left[1 + \frac{2}{\sqrt{(\frac{1}{2})^3+1}} + \frac{2}{\sqrt{(1)^3+1}} + \frac{2}{\sqrt{(\frac{3}{2})^3+1}} + \frac{1}{\sqrt{(2)^3+1}}\right] \approx 1.3973$$

Simpson's Rule:

$$\int_0^2 \frac{1}{\sqrt{x^3+1}}\,dx = \frac{\frac{1}{2}}{3}\left[1 + \frac{4}{\sqrt{(\frac{1}{2})^3+1}} + \frac{2}{\sqrt{(1)^3+1}} + \frac{4}{\sqrt{(\frac{3}{2})^3+1}} + \frac{1}{\sqrt{(2)^3+1}}\right] \approx 1.4052$$

19. $\Delta x = \frac{2}{4} = \frac{1}{2}; x_0 = 0, x_1 = \frac{1}{2}, x_2 = 1, x_3 = \frac{3}{2}, x_4 = 2.$

Trapezoidal Rule:

$$\int_0^2 e^{-x^2}\,dx = \tfrac{1}{4}[e^{-0} + 2e^{-(1/2)^2} + 2e^{-1} + 2e^{-(3/2)^2} + e^{-4}] \approx 0.8806.$$

Simpson's Rule:

$$\int_0^2 e^{-x^2}\,dx = \tfrac{1}{6}[e^{-0} + 4e^{-(1/2)^2} + 2e^{-1} + 4e^{-(3/2)^2} + e^{-4}] \approx 0.8818.$$

21. $\Delta x = \frac{2-1}{4} = \frac{1}{4}; x_0 = 1, x_1 = \frac{5}{4}, x_2 = \frac{6}{4}, x_3 = \frac{7}{4}, x_4 = \frac{8}{4}$.

Trapezoidal Rule:

$$\int_1^2 x^{-1/2} e^x \, dx = \frac{\frac{1}{4}}{2}\left[e + \frac{2e^{5/4}}{\sqrt{\frac{5}{4}}} + \cdots + \frac{2e^{7/4}}{\sqrt{\frac{7}{4}}} + \frac{e^2}{\sqrt{2}}\right] \approx 3.7757.$$

Simpson's Rule:

$$\int_1^2 x^{-1/2} e^x \, dx = \frac{\frac{1}{4}}{3}\left[e + \frac{4e^{5/4}}{\sqrt{\frac{5}{4}}} + \cdots + \frac{4e^{7/4}}{\sqrt{\frac{7}{4}}} + \frac{e^2}{\sqrt{2}}\right] \approx 3.7625.$$

23. a. Here $a = -1$, $b = 2$, $n = 10$, and $f(x) = x^5$. $f'(x) = 5x^4$ and $f''(x) = 20x^3$. Because $f'''(x) = 60x^2 > 0$ on $(-1,0) \cup (0,2)$, we see that $f''(x)$ is increasing on $(-1,0) \cup (0,2)$. So, we take $M = f''(2) = 20(2^3) = 160$.
Using (7), we see that the maximum error incurred is

$$\frac{M(b-a)^3}{12n^2} = \frac{160[2-(-1)]^3}{12(100)} = 3.6.$$

b. We compute $f''' = 60x^2$ and $f^{(iv)}(x) = 120x$. $f^{(iv)}(x)$ is clearly increasing on $(-1,2)$, so we can take $M = f^{(iv)}(2) = 240$. Therefore, using (8), we see that an error bound is $\frac{M(b-a)^3}{180n^4} = \frac{240(3)^5}{180(10^4)} \approx 0.0324$.

25. a. Here $a = 1$, $b = 3$, $n = 10$, and $f(x) = \frac{1}{x}$. We find $f'(x) = -\frac{1}{x^2}$, $f''(x) = \frac{2}{x^3}$.

Since $f'''(x) = -\frac{6}{x^4} < 0$ on $(1,3)$, we see that $f''(x)$ is decreasing there. We may take $M = f''(1) = 2$. Using (7), we find an error bound is

$$\frac{M(b-a)^3}{12n^2} = \frac{2(3-1)^3}{12(100)} \approx 0.013.$$

b. $f'''(x) = -\frac{6}{x^4}$ and $f^{(iv)}(x) = \frac{24}{x^5}$. $f^{(iv)}(x)$ is decreasing on $(1,3)$, so we can take $M = f^{(iv)}(1) = 24$. Using (8), we find an error bound is $\frac{24(3-1)^5}{180(10^4)} \approx 0.00043$.

27. a. Here $a = 0$, $b = 2$, $n = 8$, and $f(x) = (1+x)^{-1/2}$. We find

$$f'(x) = -\tfrac{1}{2}(1+x)^{-3/2}, \; f''(x) = \tfrac{3}{4}(1+x)^{-5/2}.$$

Since f'' is positive and decreasing on (0,2), we see that $|f''(x)| \le \frac{3}{4}$.

So the maximum error is $\dfrac{\frac{3}{4}(2-0)^3}{12(8)^2} = 0.0078125.$

b. $f''' = -\frac{15x}{8}(1+x)^{-7/2}$ and $f^{(4)}(x) = \dfrac{105}{16}(1+x)^{-9/2}$. Since $f^{(4)}$ is positive and decreasing on (0,2), we find $\left|f^{(4)}(x)\right| \le \frac{105}{16}$.

Therefore, the maximum error is $\dfrac{\frac{105}{16}(2-0)^5}{180(8)^4} = 0.000285.$

29. The distance covered is given by

$$d = \int_0^2 V(t)\,dt = \frac{\frac{1}{4}}{2}\left[V(0) + 2V(\tfrac{1}{4}) + \cdots + 2V(\tfrac{7}{4}) + V(2)\right]$$
$$= \tfrac{1}{8}[19.5 + 2(24.3) + 2(34.2) + 2(40.5) + 2(38.4) + 2(26.2)$$
$$+ 2(18) + 2(16) + 8] \approx 52.84, \text{ or } 52.84 \text{ miles.}$$

31. $$\frac{1}{13}\int_0^{13} f(t)\,dt = (\tfrac{1}{13})(\tfrac{1}{2})\{13.2 + 2[14.8 + 16.6 + 17.2 + 18.7 + 19.3 + 22.6 + 24.2 + 25$$
$$+24.6 + 25.6 + 26.4 + 26.6] + 26.6\} \approx 21.65, \text{ or } 21.65 \text{ mpg.}$$

33. The average daily consumption of oil is

$$A = \frac{1}{b-1}\int_a^b f(t)\,dt$$

where $f(t)$ has the values shown in the table where $t = 0$ corresponds to 1980.
Using Simpson's Rule with $n = 10$ and $\Delta t = 2$

$$A = \frac{1}{20-0}\int_0^{20} f(t)\,dt \approx \frac{1}{20}\cdot\frac{2}{3}[f(0) + 4f(2) + 2f(4) + 4f(6) + \cdots + 4f(18) + f(20)]$$
$$= \frac{1}{30}[17.1 + 4(15.3) + 2(15.7) + 4(16.3) + 2(17.3) + 4(17) + 2(17) + 4(17.7)$$
$$+ 2(18.3) + 4(18.9) + 19.7]$$
$$= 17.14, \qquad \text{or } 17.14 \text{ million barrels.}$$

35. The required rate of flow is

$$R = (\text{area of cross section of the river}) \times \text{rate of flow}$$
$$= (4)(\text{area of cross section})$$
$$= 4\int_0^{78} y(x)\,dx$$

Approximating the integral using the trapezoidal rule,

$$R \approx (4)(\tfrac{6}{2})[0.8 + 2(2.6) + 2(5.8) + 2(6.2) + 2(8.2) + 2(10.1) + 2(10.8) + 2(9.8)$$
$$+ 2(7.6) + 2(6.4) + 2(5.2) + 2(3.9) + 2(2.4) + 1.4]$$
$$= 1922.4$$

or 1922.4 cu ft/sec.

37. We solve the equation $8 = \sqrt{0.01x^2 + 0.11x + 38}$.
$64 = 0.01x^2 + 0.11x + 38$, $0.01x^2 + 0.11x - 26 = 0$, $x^2 + 11x - 2600 = 0$,
and $x = \dfrac{-11 \pm \sqrt{121 + 10{,}400}}{2} \approx 45.786$. Therefore

$$PS = (8)(45.786) - \int_0^{45.786} \sqrt{0.01x^2 + 0.11x + 38}\, dx.$$

a. $\Delta x = \dfrac{45.786}{8} = 5.72$; $x_0 = 0, x_1 = 5.72, x_2 = 11.44, \ldots, x_8 = 45.79$

$$PS = 366.288 - \frac{5.72}{2}\Big[\sqrt{38} + 2\sqrt{0.01(5.72)^2 + 0.11(5.72) + 38} + \cdots$$
$$+ \sqrt{0.01(45.79)^2 + 0.11(45.79) + 38}\Big] \approx 51{,}558, \text{ or } \$51{,}558.$$

$$PS = 366.288 - \frac{5.72}{2}\Big[\sqrt{38} + 4\sqrt{0.01(5.72)^2 + 0.11(5.72) + 38} + \cdots$$
$$+ \sqrt{0.01(45.79)^2 + 0.11(45.79) + 38}\Big] \approx 51{,}708, \text{ or } \$51{,}708.$$

39. The average petroleum reserves from 1981 through 1990 were

$$A = \frac{1}{9-0}\int_0^9 S(t)\, dt = \frac{1}{9}\int_0^9 \frac{613.7t^2 + 1449.1}{t^2 + 6.3}\, dt$$

Using the trapezoidal rule with $a = 0$, $b = 9$, and $n = 9$ so that $\Delta t = (9-0)/9 = 1$, we have $t_0 = 0, t = 1, \ldots, t_9 = 9$ so that

$$A = \frac{1}{9}\int_0^9 S(t)\, dt = \left(\frac{1}{9}\right)\left(\frac{1}{2}\right)[S(0) + 2S(1) + 2f(x) + 2f(x) + \cdots + S(9)$$
$$\approx \frac{1}{18}[130.02 + 2(282.58) + 2(379.02) + 2(455.71) + 2(505.30) + 2(536.47)$$
$$+ 2(556.56) + 2(569.99) + 2(579.32) + 586.01]$$
$$\approx 474.77$$

or approximately 474.77 million barrels.

41. $\Delta x = \dfrac{40{,}000-30{,}000}{10} = 1000$; $x_0 = 30{,}000$, $x_1 = 31{,}000$, x_2, ..., $x_{10} = 40{,}000$.

$$P = \frac{100}{2000\sqrt{2\pi}} \int_{30{,}000}^{40{,}000} e^{-0.5[x-40{,}000)/2000]^2}\, dx$$

$$P = \frac{100(1000)}{2000\sqrt{2\pi}}\left[e^{-0.5[30{,}000-40{,}000)/2000]^2} + 4e^{-0.5[(31{,}000-40{,}000)/2000]^2} + \cdots + 1]\right]$$
$$\approx 0.50, \text{ or } 50 \text{ percent.}$$

43. $R = \dfrac{60D}{\int_0^T C(t)\,dt} = \dfrac{480}{\int_0^{24} C(t)\,dt}$. Now,

$$\int_0^{24} C(t)\,dt \approx \tfrac{24}{12}\cdot\tfrac{1}{3}[0+4(0)+2(2.8)+4(6.1)+2(9.7)+4(7.6)+2(4.8)$$
$$+4(3.7)+2(1.9)+4(0.8)+2(0.3)+4(0.1)+0] \approx 74.8$$

and $R = \frac{480}{74.8} \approx 6.42$, or 6.42 liters/min.

45. False. The number n can be odd or even.

47. True.

49. Taking the limit and recalling the definition of the Riemann sum, we find

$$\lim_{\Delta t\to 0}[c(t_1)R\Delta t + c(t_2)R\Delta t + \cdots + c(t_n)R\Delta t]/60 = D$$
$$\frac{R}{60}\lim_{\Delta t\to 0}[c(t_1)\Delta t + c(t_2)\Delta t + \cdots + c(t_n)\Delta t] = D$$
$$\frac{R}{60}\int_0^T c(t)\,dt = D, \text{ or } R = \frac{60D}{\int_0^T c(t)\,dt}.$$

16.4 Problem Solving Tips

1. The improper integral on the left-hand side of the equation

   $\int_{-\infty}^{\infty} f(x)\,dx = \int_{-\infty}^{c} f(x)\,dx + \int_{c}^{\infty} f(x)\,dx$ is only convergent if both integrals on the right-hand side of the equation converge.

2. It is often convenient to choose $c = 0$ in the formula given in (1).

16.4 CONCEPT QUESTIONS, page 1034

1. a. $\int_{a}^{\infty} f(x)\,dx = \lim_{b\to\infty}\int_{a}^{b} f(x)\,dx$

   b. $\int_{-\infty}^{b} f(x)\,dx = \lim_{a\to-\infty}\int_{a}^{b} f(x)\,dx$

   c. $\int_{-\infty}^{\infty} f(x)\,dx = \int_{-\infty}^{c} f(x)\,dx + \int_{c}^{\infty} f(x)\,dx$ where c is any real number.

EXERCISES 16.4, page 1034

1. The required area is given by

$$\int_{3}^{\infty}\frac{2}{x^2}\,dx = \lim_{b\to\infty}\int_{3}^{b}\frac{2}{x^2}\,dx = \lim_{b\to\infty}\left(-\frac{2}{x}\right)\Bigg|_{3}^{b} = \lim_{b\to\infty}\left(-\frac{2}{b}+\frac{2}{3}\right) = \frac{2}{3} \text{ or } \frac{2}{3} \text{ sq units.}$$

3. $$A = \int_{3}^{\infty}\frac{1}{(x-2)^2}\,dx = \lim_{b\to\infty}\int_{3}^{b}(x-2)^{-2}\,dx = \lim_{b\to\infty} -\frac{1}{x-2}\Bigg|_{3}^{b} = \lim_{b\to\infty}\left(-\frac{1}{b-2}+1\right)$$
   $= 1$ sq unit.

5. $$A = \int_{1}^{\infty}\frac{1}{x^{3/2}}\,dx = \lim_{b\to\infty}\int_{1}^{b} x^{-3/2}\,dx = \lim_{b\to\infty} -\frac{2}{\sqrt{x}}\Bigg|_{1}^{b} = \lim_{b\to\infty}\left(-\frac{2}{\sqrt{b}}+2\right) = 2 \text{ sq units.}$$

7. $$A = \int_{0}^{\infty}\frac{1}{(x+1)^{5/2}}\,dx = \lim_{b\to\infty}\int_{1}^{b}(x+1)^{-5/2}\,dx = \lim_{b\to\infty} -\frac{2}{3}(x+1)^{-3/2}\Bigg|_{0}^{b}$$

$$= \lim_{b\to\infty}\left[-\frac{2}{3(b+1)^{3/2}}+\frac{2}{3}\right]=\frac{2}{3}\text{ sq units.}$$

9. $A=\int_{-\infty}^{2} e^{2x}\,dx=\lim_{a\to-\infty}\int_a^2 e^{2x}\,dx=\lim_{a\to-\infty}\tfrac{1}{2}e^{2x}\Big|_a^2=\lim_{a\to-\infty}\left(\tfrac{1}{2}e^4-\tfrac{1}{2}e^{2a}\right)=\tfrac{1}{2}e^4$ sq units.

11. Using symmetry, the required area is given by

$$2\int_0^{\infty}\frac{x}{(1+x^2)^2}\,dx=2\lim_{b\to\infty}\int_0^{\infty}\frac{x}{(1+x^2)^2}\,dx.$$

To evaluate the indefinite integral $\int\frac{x}{(1+x^2)^2}\,dx$, put $u=1+x^2$ so that

$du=2x\,dx$ or $x\,dx=\tfrac{1}{2}\,du$.

Then $\int\frac{x}{(1+x^2)^2}\,dx=\frac{1}{2}\int\frac{du}{u^2}=-\frac{1}{2u}+C=-\frac{1}{2(1+x^2)}+C.$

Therefore, $2\lim_{b\to\infty}\int_0^b\frac{x}{(1+x^2)}\,dx=\lim_{b\to\infty}-\frac{1}{(1+x^2)^2}\Big|_0^b=\lim_{b\to\infty}\left[-\frac{1}{(1+b^2)}+1\right]=1,$

or 1 sq unit.

13. a. $I(b)=\int_0^b\sqrt{x}\,dx=\tfrac{2}{3}x^{3/2}\Big|_0^b=\tfrac{2}{3}b^{3/2}$. b. $\lim_{b\to\infty}I(b)=\lim_{b\to\infty}\tfrac{2}{3}b^{3/2}=\infty.$

15. $\int_1^{\infty}\frac{3}{x^4}\,dx=\lim_{b\to\infty}\int_1^b 3x^{-4}\,dx=\lim_{b\to\infty}\left(-\frac{1}{x^3}\right)\Big|_1^b=\lim_{b\to\infty}\left(-\frac{1}{b^3}+1\right)=1.$

17. $A=\int_4^{\infty}\frac{2}{x^{3/2}}\,dx=\lim_{b\to\infty}\int_4^b 2x^{-3/2}\,dx=\lim_{b\to\infty}-4x^{-1/2}\Big|_4^b=\lim_{b\to\infty}\left(-\frac{4}{\sqrt{b}}+2\right)=2.$

19. $\int_1^{\infty}\frac{4}{x}\,dx=\lim_{b\to\infty}\int_1^b\frac{4}{x}\,dx=\lim_{b\to\infty}4\ln x\Big|_1^b=\lim_{b\to\infty}(4\ln b)=\infty.$

21. $\int_{-\infty}^{0}(x-2)^{-3}\,dx=\lim_{a\to-\infty}\int_a^0(x-2)^{-3}\,dx=\lim_{a\to-\infty}-\frac{1}{2(x-2)^2}\Big|_a^0=-\frac{1}{8}.$

23. $\displaystyle\int_1^{\infty} \frac{1}{(2x-1)^{3/2}}\,dx = \lim_{b\to\infty}\int_1^b (2x-1)^{-3/2}\,dx = \lim_{b\to\infty} -\frac{1}{(2x-1)^{1/2}}\Bigg|_1^b$

$$= \lim_{b\to\infty}\left(-\frac{1}{\sqrt{2b-1}}+1\right) = 1.$$

25. $\displaystyle\int_0^{\infty} e^{-x}\,dx = \lim_{b\to\infty}\int_0^b e^{-x}\,dx = \lim_{b\to\infty} -e^{-x}\Big|_0^b = \lim_{b\to\infty}(-e^{-b}+1) = 1.$

27. $\displaystyle\int_{-\infty}^0 e^{2x}\,dx = \lim_{a\to-\infty} \tfrac{1}{2}e^{2x}\Big|_a^0 = \lim_{a\to-\infty}\left(\tfrac{1}{2}-\tfrac{1}{2}e^{2a}\right) = \tfrac{1}{2}.$

29. $\displaystyle\int_1^{\infty} \frac{e^{\sqrt{x}}}{\sqrt{x}}\,dx = \lim_{b\to\infty}\int_1^b \frac{e^{\sqrt{x}}}{\sqrt{x}}\,dx = \lim_{b\to\infty} -2e^{\sqrt{x}}\Big|_1^b$ (Integrate by substitution: $u=\sqrt{x}$.)

$= \displaystyle\lim_{b\to\infty}(2e^{\sqrt{b}}-2e) = \infty$, and so it diverges.

31. $\displaystyle\int_{-\infty}^0 xe^x\,dx = \lim_{a\to-\infty}\int_a^0 xe^x\,dx = \lim_{a\to-\infty}(x-1)e^x\Big|_a^0 = \lim_{a\to-\infty}[-1+(a-1)e^a] = -1.$

Note: We have used integration by parts to evaluate the integral.

33. $\displaystyle\int_{-\infty}^{\infty} x\,dx = \lim_{a\to-\infty}\tfrac{1}{2}x^2\Big|_a^0 + \lim_{b\to\infty}\tfrac{1}{2}x^2\Big|_0^b$ both of which diverge and so the integral diverges.

35. $\displaystyle\int_{-\infty}^{\infty} x^3(1+x^4)^{-2}\,dx = \int_{-\infty}^0 x^3(1+x^4)^{-2}\,dx + \int_0^{\infty} x^3(1+x^4)^{-2}\,dx$

$$= \lim_{a\to-\infty}\int_a^0 x^3(1+x^4)^{-2}\,dx + \lim_{b\to\infty}\int_0^b x^3(1+x^4)^{-2}\,dx$$

$$= \lim_{a\to-\infty}\left[-\frac{1}{4}(1+x^4)^{-1}\Bigg|_a^0\right] + \lim_{b\to\infty}\left[-\frac{1}{4}(1+x^4)^{-1}\Bigg|_0^b\right]$$

$$= \lim_{a\to-\infty}\left[-\frac{1}{4}+\frac{1}{4(1+a^4)}\right] + \lim_{b\to\infty}\left[-\frac{1}{4(1+b^4)}+\frac{1}{4}\right]$$

$$= -\tfrac{1}{4}+\tfrac{1}{4} = 0.$$

37. $\displaystyle\int_{-\infty}^{\infty} xe^{1-x^2}\,dx = \lim_{a\to-\infty}\int_a^0 xe^{1-x^2}\,dx + \lim_{b\to\infty}\int_0^b xe^{1-x^2}\,dx = \lim_{a\to-\infty} -\tfrac{1}{2}e^{1-x^2}\Big|_a^0 + \lim_{b\to\infty} -\tfrac{1}{2}e^{1-x^2}\Big|_0^b$

$$= \lim_{a\to-\infty}\left(-\tfrac{1}{2}e+\tfrac{1}{2}e^{1-a^2}\right) + \lim_{b\to\infty}\left(-\tfrac{1}{2}e^{1-b^2}+\tfrac{1}{2}e\right) = 0.$$

39. $\int_{-\infty}^{\infty} \frac{e^{-x}}{1+e^{-x}}\,dx = \lim_{a\to-\infty} -\ln(1+e^{-x})\Big|_a^0 + \lim_{b\to\infty} -\ln(1+e^{-x})\Big|_0^b = \infty$, and it is divergent.

41. First, we find the indefinite integral $I = \int \frac{1}{x\ln^3 x}\,dx$. Let $u = \ln x$ so that

$du = \frac{1}{x}dx$. Therefore, $I = \int \frac{du}{u^3} = -\frac{1}{2u^2+C} = -\frac{1}{2\ln^2 x} + C.$ So

$$\int_e^{\infty} \frac{1}{x\ln^3 x}\,dx = \lim_{b\to\infty} \int_e^b \frac{1}{x\ln^3 x}\,dx$$

$$= \lim_{b\to\infty}\left[-\frac{1}{2\ln^2 x}\Big|_e^b\right] = \lim_{b\to\infty}\left(-\frac{1}{2(\ln b)^2}+\frac{1}{2}\right) = \frac{1}{2}$$

and so the given integral is convergent.

43. We want the present value PV of a perpetuity with $m = 1$, $P = 1500$, and $r = 0.08$.

We find $PV = \frac{(1)(1500)}{0.08} = 18{,}750$, or \$18,750.

45. $PV = \int_0^{\infty} (10{,}000 + 4000t)e^{-rt}\,dt = 10{,}000\int_0^{\infty} e^{-rt}\,dt + 4000\int_0^{\infty} te^{-rt}\,dt$

$$= \lim_{b\to\infty}\left(-\frac{10{,}000}{r}e^{-rt}\Big|_0^b\right) + 4000\left(\frac{1}{r^2}\right)\left(-rt-1)e^{-rt}\Big|_0^b\right)$$

(Integrating by parts.)

$$= \frac{10{,}000}{r} + \frac{4000}{r^2} = \frac{10{,}000r + 4000}{r^2} \text{ dollars.}$$

47. True. $\int_a^{\infty} f(x)\,dx = \int_a^b f(x)\,dx + \int_b^{\infty} f(x)\,dx$. So if $\int_a^{\infty} f(x)\,dx$ exists then $\int_b^{\infty} f(x)\,dx = \int_a^{\infty} f(x)\,dx - \int_a^b f(x)\,dx$.

49. False. Let $f(x) = \begin{cases} e^{2x} & \text{if } -\infty < x \le 0 \\ e^{-x} & \text{if } 0 < x < \infty \end{cases}$. Then

$\int_{-\infty}^{\infty} f(x)\,dx = \int_{-\infty}^{0} e^{2x}\,dx + \int_0^{\infty} e^{-x}\,dx = \frac{1}{2} + 1 = \frac{3}{2}$. But $2\int_0^{\infty} f(x)\,dx = 2\int_0^{\infty} e^{-x}\,dx = 2$.

51. a. $CV \approx \int_0^\infty Re^{-it}\,dt = \lim_{b\to\infty}\int_0^b Re^{-it}\,dt = \lim_{b\to\infty} -\frac{R}{i}e^{-it}\Big|_0^b = \lim_{b\to\infty}\left(-\frac{R}{i}e^{-ib} + \frac{R}{i}\right) = \frac{R}{i}.$

b. $CV \approx \dfrac{10{,}000}{0.12} \approx 83{,}333$, or \$83,333.

53. $\int_{-\infty}^b e^{px}\,dx = \lim_{a\to-\infty}\int_a^b e^{px}\,dx = \lim_{a\to-\infty}\left[\frac{1}{p}e^{px}\Big|_a^b\right] = \lim_{a\to-\infty}\left(\frac{1}{p}e^{pb} - \frac{1}{p}e^{pa}\right)$

$= -\dfrac{1}{p}e^{pa}$ if $p > 0$ and is divergent if $p < 0$.

16.5 CONCEPT QUESTIONS, page 1044

1. See the definition given on page 1036 of the test. For example, $f(x) = \dfrac{3x^2}{125}$ on the interval [0, 5].

EXERCISES 16.5, page 1044

1. $f(x) \geq 0$ on [2,6]. Next $\int_2^6 \frac{2}{32}x\,dx = \frac{1}{32}x^2\Big|_2^6 = \frac{1}{32}(36-4) = 1,$
and so f is a probability density function on [2,6].

3. $f(x) = \frac{3}{8}x^2$ is nonnegative on [0,2]. Next, we compute

$$\int_0^2 \tfrac{3}{8}x^2\,dx = \tfrac{1}{8}x^3\Big|_0^2 = 1$$

and so f is a probability density function.

5. $\int_0^1 20(x^3 - x^4)dx = 20(\frac{1}{4}x^4 - \frac{1}{5}x^5)\Big|_0^1 = 20(\frac{1}{4} - \frac{1}{5}) = 20(\frac{1}{20}) = 1.$
Furthermore, $f(x) = 20(x^3 - x^4) = 20x^3(1 - x) \geq 0$ on [0,1].
Therefore, f is a density function on [0,1] as asserted.

7. Clearly $f(x) \geq 0$ on [1,4]. Next,

$$\int_1^4 f(x)\,dx = \tfrac{3}{14}\int_1^4 x^{1/2}\,dx = (\tfrac{3}{14})(\tfrac{2}{3})x^{3/2}\Big|_1^4 = \tfrac{1}{7}(8-1) = 1,$$

and so f is a probability density function on [1,4].

9. First, $f(x) \geq 0$ on $[0,\infty)$. Next, we compute

$$I=\int_0^{\infty}\frac{x}{(x^2+1)^{3/2}}\,dx=\lim_{b\to\infty}\int_0^b x(x^2+1)^{-3/2}\,dx.$$

Letting $u=x^2+1$, so that $du=2x\,dx$, we find

$$I=\lim_{b\to\infty}\int_1^{b^2+1}u^{-3/2}du=\tfrac{1}{2}\lim_{b\to\infty}-2u^{-1/2}\Big|_1^{b^2+1}=\lim_{b\to\infty}\left(-\frac{1}{\sqrt{b^2+1}}+1\right)=1.$$

So the given function is a probability density function on $[0,\infty)$.

11. a. $\int_0^4 k(4-x)\,dx=k\int_0^4(4-x)\,dx=k(4x-\tfrac{1}{2}x^2)\Big|_0^4=k(16-8)=8k=1$ implies that $k=1/8$.

b. $P(1\le x\le 3)=\tfrac{1}{8}\int_1^3(4-x)\,dx=\tfrac{1}{8}(4x-\tfrac{1}{2}x^2)\Big|_1^3=\tfrac{1}{8}\left[(12-\tfrac{9}{2})-(4-\tfrac{1}{2})\right]=\tfrac{1}{2}.$

13. a. $\int_0^4 f(x)\,dx=\int_0^4 2ke^{-kx}\,dx=(2k)\left(-\frac{1}{k}\right)e^{-kx}\Big|_0^4$

$$=-2e^{-kx}\Big|_0^4=-2e^{-4k}+2=2(1-e^{-4k})=1$$

gives

$$1-e^{-4k}=\frac{1}{2};\quad e^{-4k}=\frac{1}{2}$$

$$-4k=\ln\frac{1}{2}=\ln 1-\ln 2=-\ln 2$$

So $\qquad k=\dfrac{\ln 2}{4}$

b. $P(1\le x\le 2)=\dfrac{2\ln 2}{4}\int_1^2 e^{-\left(\frac{\ln 2}{4}\right)x}\,dx=\dfrac{2\ln 2}{4}\cdot\left(-\dfrac{4}{\ln 2}\right)e^{-\left(\frac{\ln 2}{4}\right)x}\Big|_1^2$

$$=-2e^{-\left(\frac{\ln 4}{2}\right)2}+2e^{-\frac{\ln 4}{2}}\approx 0.2676.$$

15. a. Here $k=\frac{1}{15}$ and so $f(x)=\frac{1}{15}e^{(-1/15)x}$.

b. The probability is

$$\int_{10}^{12}\tfrac{1}{15}e^{(-1/15)x}\,dx=-e^{(-1/15)x}\Big|_{10}^{12}=-e^{-12/15}+e^{-10/15}\approx 0.06.$$

c. The probability is

$$\int_{15}^{\infty}\tfrac{1}{15}e^{(-1/15)x}\,dx=\lim_{b\to\infty}\int_{15}^{b}\frac{1}{15}e^{(-1/15)x}dx$$

$$= \lim_{b\to\infty} -e^{(-1/15)x}\Big|_{15}^{b} = \lim_{b\to\infty} -e^{(-1/15)b} + e^{-1} \approx 0.37.$$

17. $\mu = \int_0^5 t \cdot \frac{2}{25} t\, dt = \frac{2}{25}\int_0^5 t^2\, dt = \frac{2}{75} t^3\Big|_0^5 = \frac{2}{75}(125) = 3\frac{1}{3}.$

So a shopper is expected to spend $3\frac{1}{3}$ minutes in the magazine section.

19. $\mu = \int_0^5 x \cdot \frac{6}{125} x(5-x)\, dx = \frac{6}{125}\int_0^5 (5x^2 - x^3) dx = \frac{6}{125}\left(\frac{5}{3}x^3 - \frac{1}{4}x^4\right)\Big|_0^5$

$= \frac{6}{125}\left(\frac{625}{3} - \frac{625}{4}\right) = 2.5.$

So the expected demand is 2500 lb.

21. The required probability is given by

$$P(0 \le x \le \tfrac{1}{2}) = \int_0^{1/2} 12x^2(1-x)\, dx = 12\int_0^{1/2} (x^2 - x^3)\, dx$$

$$= 12\left(\tfrac{1}{3}x^3 - \tfrac{1}{4}x^4\right)\Big|_0^{1/2} = 12x^3\left[\tfrac{1}{3} - \tfrac{1}{4}(x)\right]\Big|_0^{1/2}$$

$$= 12\left(\tfrac{1}{2}\right)^3\left[\tfrac{1}{3} - \tfrac{1}{4}\left(\tfrac{1}{2}\right)\right] - 0 = \tfrac{5}{16}.$$

23. $\mu = \int_0^\infty t \cdot 9(9+t^2)^{-3/2}\, dt = \lim_{b\to\infty}\int_0^b 9t(9+t^2)^{-3/2}\, dt$

$= \lim_{b\to\infty}[(9)(\tfrac{1}{2})(-2)(9+t^2)^{-1/2}]\Big|_0^b = \lim_{b\to\infty}\left[-\frac{9}{\sqrt{9+t^2}} + 3\right] = 3.$

So the tubes are expected to last 3 years.

25. The probability function is $f(x) = \frac{1}{8}e^{-x/8}$. The required probability is

$$P(x \ge 8) = \tfrac{1}{8}\int_8^\infty e^{-x/8}\, dx = \lim_{b\to\infty} -e^{-x/8}\Big|_8^b = \lim_{b\to\infty}(-e^{-b/8} + e^{-1}) = e^{-1} \approx 0.37.$$

27. The probability function is $f(x) = 0.00001e^{-0.00001x}$. The required probability is

$$P(x \le 20{,}000) = 0.00001\int_0^{20{,}000} e^{-0.00001x}\, dx = -e^{-0.00001x}\Big|_0^{20{,}000} = -e^{0.2} + 1 \approx 0.18.$$

29. False. f must be nonnegative on $[a, b]$ as well.

31. False. Let $f(x) = 1$ for all x in the interval [0, 1]. Then f is a probability density

function on [0, 1], but f is not a probability density function on $\left[\frac{1}{2}, \frac{3}{4}\right]$ since

$$\int_{1/2}^{3/4} 1\, dx = \tfrac{1}{4} \neq 1.$$

CHAPTER 16 CONCEPT REVIEW, page 1048

1. Product; $uv - \int v\,du$; u; easy to integrate

3. $\frac{\Delta x}{2}[f(x_0)+2f(x_1)+2f(x_2)+\cdots 2f(x_{n-1})+f(x_n)]$; even; $\frac{M(b-a)^3}{12n^2}$

5. $\lim\limits_{a\to-\infty}\int_a^b f(x)\,dx$; $\lim\limits_{b\to\infty}\int_a^b f(x)\,dx$; $\int_{-\infty}^c f(x)\,dx + \int_c^\infty f(x)\,dx$

CHAPTER 16 REVIEW EXERCISES, page 1048

1. Let $u = 2x$ and $dv = e^{-x}\,dx$ so that $du = 2\,dx$ and $v = -e^{-x}$. Then

$$\int 2xe^{-x}\,dx = uv - \int v\,du = -2xe^{-x} + 2\int e^{-x}\,dx$$

$$= -2xe^{-x} - 2e^{-x} + C = -2(1+x)e^{-x} + C.$$

3. Let $u = \ln 5x$ and $dv = dx$, so that $du = \frac{1}{x}\,dx$ and $v = x$. Then

$$\int \ln 5x\,dx = x\ln 5x\,dx - \int dx = x\ln 5x - x + C = x(\ln 5x - 1) + C.$$

5. Let $u = x$ and $dv = e^{-2x}\,dx$ so that $du = dx$ and $v = -\frac{1}{2}e^{-2x}$. Then

$$\int_0^1 xe^{-2x}\,dx = -\tfrac{1}{2}xe^{-2x}\Big|_0^1 + \tfrac{1}{2}\int_0^1 e^{-2x}\,dx = -\tfrac{1}{2}e^{-2} - \tfrac{1}{4}e^{-2x}\Big|_0^1$$

$$= -\tfrac{1}{2}e^{-2} - \tfrac{1}{4}e^{-2} + \tfrac{1}{4} = \tfrac{1}{4}(1 - 3e^{-2}).$$

7. $f(x) = \int f'(x)\,dx = \int \frac{\ln x}{\sqrt{x}}\,dx$. To evaluate the integral, we integrate by parts with $u = \ln x$, $dv = x^{-1/2}\,dx$, $du = \frac{1}{x}\,dx$ and $v = 2x^{1/2}\,dx$. Then

$$\int \frac{\ln x}{x^{1/2}}\,dx = 2x^{1/2}\ln x - \int 2x^{-1/2}\,dx = 2x^{1/2}\ln x - 4x^{1/2} + C$$

$$= 2x^{1/2}(\ln x - 2) + C = 2\sqrt{x}(\ln x - 2) + C.$$

But $f(1) = -2$ and this gives $2\sqrt{1}(\ln 1 - 2) + C = -2$, or $C = 2$. Therefore,

$f(x) = 2\sqrt{x}(\ln x - 2) + 2.$

9. Using Formula 4 with $a = 3$ and $b = 2$, we obtain

$$\int \frac{x^2}{(3+2x)^2}\,dx = \frac{1}{8}\left[3 + 2x - \frac{9}{3+2x} - 6\ln|3+2x|\right] + C.$$

11. Use Formula 24 with $a = 4$ and $n = 2$, obtaining $\int x^2 e^{4x}\,dx = \frac{1}{4}x^2e^{4x} - \frac{1}{2}\int xe^{4x}\,dx.$

Use Formula 23 to obtain

$$\begin{aligned}\int x^2e^{4x}\,dx &= \tfrac{1}{4}x^2e^{4x} - \tfrac{1}{2}\left[\tfrac{1}{16}(4x-1)e^{4x}\right] + C\\ &= \tfrac{1}{32}(8x^2 - 4x + 1)e^{4x} + C.\end{aligned}$$

13. Use Formula 17 with $a = 2$ obtaining $\int \frac{dx}{x^2\sqrt{x^2-4}} = \frac{\sqrt{x^2-4}}{4x} + C.$

15. $\int_0^\infty e^{-2x}\,dx = \lim_{b\to\infty}\int_0^b e^{-2x}\,dx = \lim_{b\to\infty}(-\tfrac{1}{2}e^{-2x})\Big|_0^b = \lim_{b\to\infty}(-\tfrac{1}{2}e^{-2b} + \tfrac{1}{2}) = \tfrac{1}{2}.$

17. $\int_3^\infty \frac{2}{x}\,dx = \lim_{b\to\infty}\int_3^b \frac{2}{x}\,dx = \lim_{b\to\infty} 2\ \ln x\Big|_3^b = \lim_{b\to\infty}(2\ln b - 2\ln 3) = \infty.$

19.
$$\begin{aligned}\int_2^\infty \frac{dx}{(1+2x)^2} &= \lim_{b\to\infty}\int_2^b (1+2x)^{-2}\,dx = \lim_{b\to\infty}(\tfrac{1}{2})(-1)(1+2x)^{-1}\Big|_2^b\\ &= \lim_{b\to\infty}\left(-\frac{1}{2(1+2b)} + \frac{1}{2(5)}\right) = \frac{1}{10}.\end{aligned}$$

21. $\Delta x = \frac{b-a}{n} = \frac{3-1}{4} = \frac{1}{2}; x_0 = 1,\ x_1 = \frac{3}{2},\ x_2 = 2,\ x_3 = \frac{5}{2},\ x_4 = 3.$

Trapezoidal Rule:

$$\int_1^3 \frac{dx}{1+\sqrt{x}} \approx \frac{\frac{1}{2}}{2}\left[\frac{1}{2} + \frac{2}{1+\sqrt{1.5}} + \frac{2}{1+\sqrt{2}} + \frac{2}{1+\sqrt{2.5}} + \frac{1}{1+\sqrt{3}}\right] \approx 0.8421.$$

Simpson's Rule

$$\int_1^3 \frac{dx}{1+\sqrt{x}} \approx \frac{\frac{1}{2}}{3}\left[\frac{1}{2} + \frac{4}{1+\sqrt{1.5}} + \frac{2}{1+\sqrt{2}} + \frac{4}{1+\sqrt{2.5}} + \frac{1}{1+\sqrt{3}}\right] \approx 0.8404.$$

23. $\Delta x = \frac{1-(-1)}{4} = \frac{1}{2}$; $x_0 = -1$, $x_1 = -\frac{1}{2}$, $x_2 = 0$, $x_3 = \frac{1}{2}$, $x_4 = 1$.

Trapezoidal Rule:

$$\int_{-1}^{1} \sqrt{1+x^4}\,dx \approx \frac{0.5}{2}\left[\sqrt{2} + 2\sqrt{1+(-0.5)^4} + 2 + 2\sqrt{1+(0.5)^4} + \sqrt{2}\right]$$
$$\approx 2.2379.$$

Simpson's Rule:

$$\int_{-1}^{1} \sqrt{1+x^4}\,dx \approx \frac{0.5}{3}\left[\sqrt{2} + 4\sqrt{1+(-0.5)^4} + 2 + 4\sqrt{1+(0.5)^4} + \sqrt{2}\right]$$
$$\approx 2.1791.$$

25. a. Here $a = 0, b = 1$, and $f(x) = \dfrac{1}{x+1}$. We have

$$f'(x) = -\frac{1}{(x+1)^2} \quad \text{and} \quad f''(x) = \frac{2}{(x+1)^3}$$

Since f'' is positive and decreasing on (0,1), it attains its maximum value of 2 at $x = 0$. So we take $M = 2$. Using (7), we see that the maximum error incurred is

$$\frac{M(b-a)^3}{12n^2} = \frac{2(1^3)}{12(8^2)} = \frac{1}{384} \approx 0.002604$$

b. We compute

$$f'''(x) = -\frac{6}{(x+1)^4} \quad \text{and} \quad f^{(iv)}(x) = \frac{24}{(x+1)^5}$$

Since $f^{(4)}(x)$ is positive and decreasing on (0,1), we take $M = 24$. The maximum error is $\dfrac{24(1^5)}{180(8^4)} = \dfrac{1}{30720} \approx 0.000033$.

27. $f(x) \geq 0$ on [0,3]. Next,

$$\tfrac{1}{9}\int_0^3 x\sqrt{9-x^2}\,dx = \tfrac{1}{9}\int_0^3 x(9-x^2)^{1/2}\,dx = (\tfrac{1}{9})(-\tfrac{1}{2})(\tfrac{2}{3})(9-x^2)^{3/2}\Big|_0^3$$
$$= -\tfrac{1}{27}(9-x^2)^{3/2}\Big|_0^3 = 0 + \tfrac{1}{27}(9)^{3/2} = 1.$$

29. a. $\displaystyle\int_1^4 \frac{k}{\sqrt{x}}\,dx = k\int_1^4 x^{-1/2}\,dx = 2k\sqrt{x}\Big|_1^4 = 2k(2-1) = 2k = 1$, or $k = 1/2$.

b. $\displaystyle\int_2^3 \frac{\frac{1}{2}}{\sqrt{x}}\,dx = \frac{1}{2}\int_2^3 x^{-1/2}\,dx = 2\left(\frac{1}{2}\right)\sqrt{x}\Big|_2^3 = \sqrt{3} - \sqrt{2} = 0.3178,$

31. a. The probability that a woman entering the maternity wing stays in the hospital more than 6 days is given by

$$\int_6^\infty \tfrac{1}{4}e^{-0.25t}\,dt = \lim_{b\to\infty}\int_6^b \tfrac{1}{4}e^{-0.25t}dt = \lim_{b\to\infty} -e^{-0.25t}\Big|_6^b = \lim_{b\to\infty}(-e^{-0.25b} + e^{-1.5}) = 0.22$$

b. The probability that a woman entering the maternity wing at the hospital stays there less than 2 days is given by $\int_0^2 \tfrac{1}{4}e^{-0.25t}\,dt = -e^{-0.25t}\Big|_0^2 = -e^{-0.5} + 1 = 0.39.$

c. $E = \dfrac{1}{k} - \dfrac{1}{\frac{1}{4}} = 4$, or 4 days.

33. Let $u = t$ and $dv = e^{-0.05t}$ so that $du = 1$ and $v = -20e^{-0.05t}$, and integrate by parts obtaining $S(t) = -20te^{-0.05t} + \int 20e^{-0.05t}\,dt = -20te^{-0.05t} - 400e^{-0.05t} + C$

$$= -20te^{-0.05t} - 400e^{-0.05t} + C = -20e^{-0.05t}(t+20) + C.$$

The initial condition implies $S(0) = 0$ giving $-20(20) + C = 0$, or $C = 400$. Therefore, $S(t) = -20e^{-0.05t}(t + 20) + 400$. By the end of the first year, the number of units sold is given by $S(12) = -20e^{-0.6}(32) + 400 = 48.761$, or 48,761 cartridges.

35. Trapezoidal Rule:

$A = \frac{100}{2}[0 + 480 + 520 + 600 + 680 + 680 + 800 + 680 + 600 + 440 + 0]$

$= 274{,}000$, or 274,000 sq ft.

Simpson's Rule:

$A = \frac{100}{3}[0 + 960 + 520 + 1200 + 680 + 1360 + 800 + 1360 + 600 + 880 + 0]$

$= 278{,}667$, or 278,667 sq ft.

37. We want the present value of a perpetuity with $m = 1$, $P = 10{,}000$, and $r = 0.09$. We find $PV = \dfrac{(1)(10{,}000)}{0.09} \approx 111{,}111$ or approximately $111,111.

CHAPTER 16 BEFORE MOVING ON, page 1050

1. Let $u = \ln x$ and $dv = x^2\,dx$. Then $du = \dfrac{1}{x}dx$ and $v = \dfrac{1}{3}x^3$.

$$\int x^2 \ln x\,dx = \frac{1}{3}x^3 \ln x - \int \frac{1}{3}x^2\,dx = \frac{1}{3}x^3 \ln x - \frac{1}{9}x^3 + C$$

$$= \frac{1}{9}x^3(3\ln x - 1) + C$$

2. $I = \int \frac{du}{x^2\sqrt{8+2x^2}}$. Let $u = \sqrt{2}x$. Then $du = \sqrt{2}\,dx$ or $dx = \frac{du}{\sqrt{2}} = \frac{\sqrt{2}}{2}du.$

$$I = \frac{\sqrt{2}}{2}\int \frac{du}{\frac{u^2}{2}\sqrt{8+u^2}} = \sqrt{2}\int \frac{du}{u^2\sqrt{(2\sqrt{2})^2+u^2}}$$

With $a = 2\sqrt{2}$ and $x = u$,

$$I = \sqrt{2}\int \frac{du}{u^2\sqrt{(2\sqrt{2})^2+u^2}} = \sqrt{2}\left[-\frac{\sqrt{8+u^2}}{8u}\right] + C = -\frac{\sqrt{8+2x^2}}{8x} + C$$

3. $n = 5,\ \Delta x = \frac{4-2}{5} = 0.4;\ x_0 = 2,\ x_1 = 2.4,\ x_2 = 2.8,\ x_3 = 3.2,\ x_4 = 3.6,\ x_5 = 4$

$$\int_2^4 \sqrt{x^2+1}\,dx \approx \frac{0.4}{2}[f(2)+2f(2.4)+2f(2.8)+2f(3.2)+2f(3.2)+2f(3.6)+f(4)]$$
$$= 0.2[2.23607+2(2.6)+2(2.97321)+2(3.35261)+2(3.73631)+4.12311]$$
$$\approx 6.3367.$$

4. $n = 6,\ \Delta x = \frac{3-1}{6} = \frac{1}{3};\ x_0 = 1,\ x_1 = \frac{4}{3},\ x_2 = \frac{5}{3},\ x_3 = 2,\ x_4 = \frac{7}{3},\ x_5 = \frac{8}{3},\ x_6 = 3$

$$\int_1^3 e^{0.2x}\,dx \approx \frac{\frac{1}{3}}{3}[f(1)+4f(\tfrac{4}{3})+2f(\tfrac{5}{3})+4f(2)+2f(\tfrac{7}{3})+4f(\tfrac{8}{3})+f(3)]$$
$$\approx \frac{1}{9}[1.2214+4(1.30561)+2(1.39561)+4(1.49182)+2(1.59467)$$
$$+4(1.7046)+1.82212]$$
$$\approx 3.0036$$

5. $\int_1^\infty e^{-2x}\,dx = \lim_{b\to\infty}\int_1^b e^{-2x}\,dx = \lim_{b\to\infty}\left[-\tfrac{1}{2}e^{-2x}\Big|_1^b\right] = \lim_{b\to\infty}\left(-\frac{1}{2}e^{-2b}+\frac{1}{2}e^{-2}\right) = \frac{1}{2}e^{-2} = \frac{1}{2e^2}$

6. a. $f(x) \ge 0$ on $[0,8]$ and $\int_0^8 \frac{5}{96}x^{2/3}\,dx = \frac{5}{96}\left[\frac{3}{5}x^{5/3}\Big|_0^8\right] = \frac{5}{96}(\frac{3}{5})(8^{5/3}) = 1$

Therefore f is a probability density function on [0, 8].

b. $P(1 \le x \le 8) = \int_1^8 \frac{5}{96}x^{2/3}\,dx = \frac{5}{96}(\frac{3}{5})x^{5/3}\Big|_1^8 = \frac{1}{32}(32-1) = \frac{31}{32}.$

CHAPTER 17

17.1 CONCEPT QUESTIONS, page 1058

1. A function of two variables is a rule that assigns to each point (x, y) in a subset of the plane, a unique number $f(x, y)$. For example, $f(x, y) = x^2 + 2y^2$ has the whole *xy*-plane as its domain.
3. a. The graph of $f(x, y)$ is the set $S = \{(x, y, z) | z = f(x, y), (x, y) \in \text{ Domain of } f\}$.
   b. The level curve of *f* is the projection onto the *xy*-plane of the trace of $f(x, y)$ in the plane $z = k$, where *k* is a constant in the range of *f*.

EXERCISES 17.1, page 1058

1. $f(0, 0) = 2(0) + 3(0) - 4 = -4.$ $f(1, 0) = 2(1) + 3(0) - 4 = -2.$
   $f(0, 1) = 2(0) + 3(1) - 4 = -1.$ $f(1, 2) = 2(1) + 3(2) - 4 = 4.$
   $f(2, -1) = 2(2) + 3(-1) - 4 = -3.$

3. $f(1, 2) = 1^2 + 2(1)(2) - 1 + 3 = 7;$ $f(2, 1) = 2^2 + 2(2)(1) - 2 + 3 = 9$
   $f(-1, 2) = (-1)^2 + 2(-1)(2) - (-1) + 3 = 1;$ $f(2, -1) = 2^2 + 2(2)(-1) - 2 + 3 = 1.$

5. $g(s, t) = 3s\sqrt{t} + t\sqrt{s} + 2;$ $g(1, 2) = 3(1)\sqrt{2} + 2\sqrt{1} + 2 = 4 + 3\sqrt{2}$
   $g(2, 1) = 3(2)\sqrt{1} + \sqrt{2} + 2 = 8 + \sqrt{2};$
   $g(0, 4) = 0 + 0 + 2 = 2,$ $g(4, 9) = 3(4)\sqrt{9} + 9\sqrt{4} + 2 = 56.$

7. $h(1,e) = \ln e - e \ln 1 = \ln e = 1;$ $h(e,1) = e \ln 1 - \ln e = -1;$
   $h(e,e) = e \ln e - e \ln e = 0.$

9. $g(r, s, t) = re^{s/t};$ $g(1,1,1) = e,$ $g(1,0,1) = 1,$ $g(-1,-1,-1) = -e^{-1/(-1)} = -e.$

11. The domain of f is the set of all ordered pairs (x, y) where x and y are real numbers.

13. All real values of u and v except those satisfying the equation $u = v$.

15. The domain of g is the set of all ordered pairs (r,s) satisfying $rs \geq 0$, that is the set of all ordered pairs where both $r \geq 0$ and $s \geq 0$, or in which both $r \leq 0$ and $s \leq 0$.

17. The domain of h is the set of all ordered pairs (x, y) such that $x + y > 5$.

19. The level curves of $z = f(x, y) = 2x + 3y$ for $z = -2, -1, 0, 1, 2$, follow.

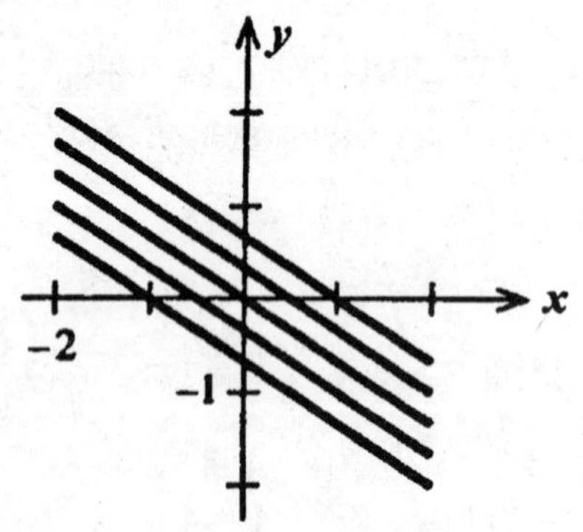

21. The level curves of $f(x, y) = 2x^2 + y$ for $z = -2, -1, 0, 1, 2$, are shown below.

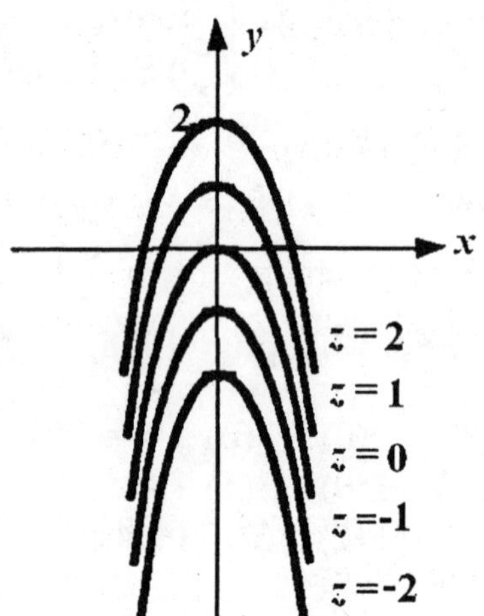

23. The level curves of $f(x,y) = \sqrt{16 - x^2 - y^2}$ for $z = 0, 1, 2, 3, 4$ follow.

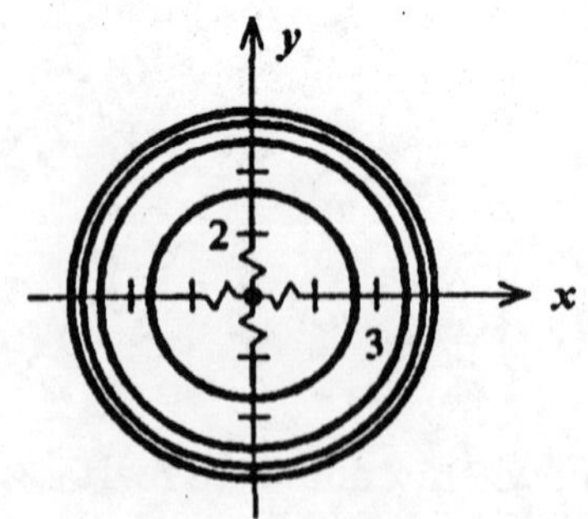

25. $V = f(1.5, 4) = \pi(1.5)^2(4) = 9\pi$, or 9π cu ft

27. a. $M = \dfrac{80}{(1.8)^2} = 24.69$.

b. $\dfrac{w}{(1.8)^2} < 25$; that is, $w < 25(1.8)^2 = 81$, that is less than 81 kg.

29. a. $R(x,y) = xp + yq = x(200 - \frac{1}{5}x - \frac{1}{10}y) + y(160 - \frac{1}{10}x - \frac{1}{4}y)$
$= -\frac{1}{5}x^2 - \frac{1}{4}y^2 - \frac{1}{5}xy + 200x + 160y.$

b. The domain of R is the set of all points (x,y) satisfying
$200 - \frac{1}{5}x - \frac{1}{10}y \geq 0,\ 160 - \frac{1}{10}x - \frac{1}{4}y \geq 0$

31. a. $R(x,y) = xp + yq = 20x - 0.005x^2 - 0.001xy + 15y - 0.001xy - 0.003y^2$
$= -0.005x^2 - 0.003y^2 - 0.002xy + 20x + 15y.$

b. Since p and q must both be nonnegative, the domain of R is the set of all ordered pairs (x, y) for which

$20 - 0.005x - 0.001y \geq 0$

and $15 - 0.001x - 0.003y \geq 0.$

33. a. The domain of V is the set of all ordered pairs (P,T) where P and T are positive real numbers.

b. $V = \dfrac{30.9(273)}{760} = 11.10$ liters.

35. The output is $f(32,243) = 100(32^{3/5})(243)^{2/5} = 100(8)(9) = 7200$ or \$7,200 billlion.

37. The number of suspicious fires is

$$N(100,20) = \frac{100[1000 + 0.03(100^2)(20)]^{1/2}}{[5 + 0.2(20)]^2} = 103.29, \text{ or approximately } 103.$$

39. a. $P = f(100{,}000,\ 0.08,\ 30) = \dfrac{100{,}000(0.08)}{12\left[1 - \left(1 + \dfrac{0.08}{12}\right)^{-360}\right]} \approx 733.76$, or \$733.76.

$$P = f(100{,}000,\ 0.1,\ 30) = \frac{100{,}000(0.1)}{12\left[1 - \left(1 + \dfrac{0.1}{12}\right)^{-360}\right]} \approx 877.57, \text{ or } \$877.57.$$

b. $P = f(100{,}000, 0.08, 20) = \dfrac{100{,}000(0{,}08)}{12\left[1-\left(1+\dfrac{0.08}{12}\right)^{-240}\right]} \approx 836.44$, or \$836.44.

41. $f(M,600,10) = \dfrac{\pi^2(360{,}000)M(10)}{900} \approx 39{,}478.42M$

or $\dfrac{39{,}478.42}{980} \approx 40.28$ times gravity.

43. False. Let $h(x, y) = xy$. Then there are no functions f and g such that $h(x, y) = f(x) + g(y)$.

45. False. Since $x^2 - y^2 = (x+y)(x-y)$, we see that $x^2 - y^2 = 0$ if $y = \pm x$. Therefore, the domain of f is $\{(x, y) | y \neq \pm x\}$.

17.2 Problem Solving Tips

1. The notation f_{xy} and f_{yx} is used to denote the second partial derivative of the function $f(x, y)$. Note that when this notation is used the differentiation is carried out in the order (left to right) in which x and y appear.

   The notation $\dfrac{\partial^2 f}{\partial y \partial x}$ and $\dfrac{\partial^2 f}{\partial x \partial y}$ is also used to denote the second partial derivatives of the function $f(x, y)$, but in this case the differentiation is carried out in the reverse order (right to left) in which x and y appear.

17.2 CONCEPT QUESTIONS, page 1070

1. a. $\frac{\partial f}{\partial x}(a,b) = \frac{\partial f}{\partial x}(x,y)\Big|_{(a,b)} = \left[\lim_{h\to 0}\frac{f(x+h,y)-f(x,y)}{h}\right]\Big|_{(a,b)}$

   b. See page 1067 of the text.

3. f_{xx}, f_{yy}, f_{xy}, and f_{yx}.

EXERCISES 17.2, page 1071

1. $f_x = 2,\ f_y = 3$

3. $g_x = 4x,\ g_y = 4$

5. $f_x = -\frac{4y}{x^3};\ f_y = \frac{2}{x^2}.$

7. $g(u,v) = \frac{u-v}{u+v};\ \frac{\partial g}{\partial u} = \frac{(u+v)(1)-(u-v)(1)}{(u+v)^2} = \frac{2v}{(u+v)^2}.$

   $\frac{\partial g}{\partial v} = \frac{(u+v)(-1)-(u-v)(1)}{(u+v)^2} = -\frac{2u}{(u+v)^2}.$

9. $f(s,t) = (s^2 - st + t^2)^3;\ f_s = 3(s^2 - st + t^2)^2(2s - t)$ and $f_t = 3(s^2 - st + t^2)^2(2t - s)$

11. $f(x,y) = (x^2 + y^2)^{2/3};\ f_x = \frac{2}{3}(x^2 + y^2)^{-1/3}(2x) = \frac{4}{3}x(x^2 + y^2)^{-1/3}$. Similarly, $f_y = \frac{4}{3}y(x^2 + y^2)^{-1/3}.$

13. $f(x,y) = e^{xy+1};\ f_x = ye^{xy+1},\ f_y = xe^{xy+1}.$

15. $f(x,y) = x\ln y + y\ln x;\ f_x = \ln y + \frac{y}{x},\ f_y = \frac{x}{y} + \ln x.$

17. $g(u,v) = e^u \ln v.\ g_u = e^u \ln v,\ g_v = \frac{e^u}{v}.$

19. $f(x,y,z) = xyz + xy^2 + yz^2 + zx^2;\ f_x = yz + y^2 + 2xz,$
    $f_y = xz + 2xy + z^2,\ f_z = xy + 2yz + x^2.$

21. $h(r,s,t) = e^{rst};\ h_r = ste^{rst},\ h_s = rte^{rst},\ h_t = rse^{rst}.$

23. $f(x,y)=x^2y+xy^2$; $f_x(1,2)=2xy+y^2\big|_{(1,2)}=8$; $f_y(1,2)=x^2+2xy\big|_{(1,2)}=5$.

25. $f(x,y)=x\sqrt{y}+y^2=xy^{1/2}+y^2$; $f_x(2,1)=\sqrt{y}\,\big|_{(2,1)}=1$,

$f_y(2,1)=\dfrac{x}{2\sqrt{y}}+2y\,\big|_{(2,1)}=3$.

27. $f(x,y)=\dfrac{x}{y}$; $f_x(1,2)=\dfrac{1}{y}\bigg|_{(1,2)}=\dfrac{1}{2}$, $f_y(1,2)=-\dfrac{x}{y^2}\bigg|_{(1,2)}=-\dfrac{1}{4}$.

29. $f(x,y)=e^{xy}$. $f_x(1,1)=ye^{xy}\big|_{(1,1)}=e$, $f_y(1,1)=xe^{xy}\big|_{(1,1)}=e$.

31. $f(x,y,z)=x^2yz^3$; $f_x(1,0,2)=2xyz^3\big|_{(1,0,2)}=0$; $f_y(1,0,2)=x^2z^3\big|_{(1,0,2)}=8$.

$f_z(1,0,2)=3x^2yz^2\big|_{(1,0,2)}=0$.

33. $f(x,y)=x^2y+xy^3$; $f_x=2xy+y^3$, $f_y=x^2+3xy^2$.

Therefore, $f_{xx}=2y$, $f_{xy}=2x+3y^2=f_{yx}$, $f_{yy}=6xy$.

35. $f(x,y)=x^2-2xy+2y^2+x-2y$; $f_x=2x-2y+1$, $f_y=-2x+4y-2$; $f_{xx}=2$,

$f_{xy}=-2=f_{yx}$, $f_{yy}=4$.

37. $f(x,y)=(x^2+y^2)^{1/2}$; $f_x=\frac{1}{2}(x^2+y^2)^{-1/2}(2x)=x(x^2+y^2)^{-1/2}$;

$f_y=y(x^2+y^2)^{-1/2}$.

$$f_{xx}=(x^2+y^2)^{-1/2}+x(-\tfrac{1}{2})(x^2+y^2)^{-3/2}(2x)=(x^2+y^2)^{-1/2}-x^2(x^2+y^2)^{-3/2}$$

$$=(x^2+y^2)^{-3/2}(x^2+y^2-x^2)=\frac{y^2}{(x^2+y^2)^{3/2}}.$$

$$f_{xy}=x(-\tfrac{1}{2})(x^2+y^2)^{-3/2}(2y)=-\frac{xy}{(x^2+y^2)^{3/2}}=f_{yx}.$$

$$f_{yy}=(x^2+y^2)^{-1/2}+y(-\tfrac{1}{2})(x^2+y^2)^{-3/2}(2y)=(x^2+y^2)^{-1/2}-y^2(x^2+y^2)^{-3/2}$$

$$=(x^2+y^2)^{-3/2}(x^2+y^2-y^2)=\frac{x^2}{(x^2+y^2)^{3/2}}.$$

39. $f(x,y)=e^{-x/y}$; $f_x=-\frac{1}{y}e^{-x/y}$; $f_y=\frac{x}{y^2}e^{-x/y}$; $f_{xx}=\frac{1}{y^2}e^{-x/y}$;

$$f_{xy}=-\frac{x}{y^3}e^{-x/y}+\frac{1}{y^2}e^{-x/y}=\left(\frac{-x+y}{y^3}\right)e^{-x/y}=f_{yx}.$$

$$f_{yy}=-\frac{2x}{y^3}e^{-x/y}+\frac{x^2}{y^4}e^{-x/y}=\frac{x}{y^3}\left(\frac{x}{y}-2\right)e^{-x/y}.$$

41. a. $f(x,y)=20x^{3/4}y^{1/4}$. $f_x(256,16)=15\left(\frac{y}{x}\right)^{1/4}\Big|_{(256,16)}$

$$=15\left(\frac{16}{256}\right)^{1/4}=15\left(\frac{2}{4}\right)=7.5.$$

$$f_y(256,16)=5\left(\frac{x}{y}\right)^{3/4}\Big|_{(256,16)}=5\left(\frac{256}{16}\right)^{3/4}=5(80)=40.$$

b. Yes.

43. $p(x,y)=200-10(x-\frac{1}{2})^2-15(y-1)^2$. $\frac{\partial p}{\partial x}(0,1)=-20(x-\frac{1}{2})\big|_{(0,1)}=10$;

At the location (0,1) in the figure, the price of land is changing at the rate of \$10 per sq ft per mile change to the right.

$$\frac{\partial p}{\partial y}(0,1)=-30(y-1)\big|_{(0,1)}=0;$$

At the location (0,1) in the figure, the price of land is constant per mile change upwards.

45. $f(p,q)=10{,}000-10p-e^{0.5q}$; $g(p,q)=50{,}000-4000q-10p$.

$$\frac{\partial f}{\partial q}=-0.5e^{0.5q}<0 \text{ and } \frac{\partial g}{\partial p}=-10<0$$

and so the two commodities are complementary commodities.

47. $R(x,y)=-0.2x^2-0.25y^2-0.2xy+200x+160y$.

$$\frac{\partial R}{\partial x}(300,250)=-0.4x-0.2y+200\big|_{(300,250)}$$

$$=-0.4(300)-0.2(250)+200=30$$

and this says that at a sales level of 300 finished and 250 unfinished units the revenue is increasing at the rate of \$30 per week per unit increase in the finished units.

$$\frac{\partial R}{\partial y}(300,250) = -0.5y - 0.2x + 160\Big|_{(300,250)}$$

$$= -0.5(250) - 0.2(300) + 160 = -25$$

and this says that at a level of 300 finished and 250 unfinished units the revenue is decreasing at the rate of \$25 per week per increase in the unfinished units.

49. a. $T = f(32,20) = 35.74 + 0.6215(32) - 35.75(20^{0.16}) + 0.4275(32)(20^{0.16})$
≈ 19.99, or approximately $20\,°F$.

b. $$\frac{\partial T}{\partial s} = -35.75(0.16S^{-0.84}) + 0.4275t(0.16S^{-0.84})$$

$$= 0.16(-35.75 + 0.4275t)s^{-0.84}$$

$$\left.\frac{\partial T}{\partial s}\right|_{(32,20)} = 0.16[-35.75 + 0.4275(32)]20^{-0.84} \approx -0.285;$$

that is, the wind chill will drop by 0.3 degrees for each 1 mph increase in wind speed.

51. $V = \dfrac{30.9T}{P}$. $\dfrac{\partial V}{\partial T} = \dfrac{30.9}{P}$ and $\dfrac{\partial V}{\partial P} = -\dfrac{30.9T}{P^2}$.

Therefore, $\left.\dfrac{\partial V}{\partial T}\right|_{T=300,\,P=800} = \dfrac{30.9}{800} = 0.039$, or 0.039 liters/degree.

$$\left.\frac{\partial V}{\partial P}\right|_{T=300,\,P=800} = -\frac{(30.9)(300)}{800^2} = -0.015$$

or -0.015 liters/mm of mercury.

53. $V = \dfrac{kT}{P}$ and $\dfrac{\partial V}{\partial T} = \dfrac{k}{P}$; $T = \dfrac{VP}{k}$ and $\dfrac{\partial T}{\partial P} = \dfrac{V}{k} = \dfrac{T}{P}$; and

$P = \dfrac{kT}{V}$ and $\dfrac{\partial P}{\partial V} = -\dfrac{kT}{V^2} = -kT \cdot \dfrac{P^2}{(kT)^2} = -\dfrac{P^2}{kT}$

Therefore $\dfrac{\partial V}{\partial T} \cdot \dfrac{\partial T}{\partial P} \cdot \dfrac{\partial P}{\partial V} = \dfrac{k}{P} \cdot \dfrac{T}{P} \cdot -\dfrac{P^2}{kT} = -1.$

55. False. Let $f(x,y) = xy^{1/2}$. Then $f_x = y^{1/2}$ is defined at (0, 0). But

$$f_y = \frac{1}{2}xy^{-1/2} = \frac{x}{2y^{1/2}}$$ is not defined at (0, 0).

57. True. See Section 17.2.

USING TECHNOLOGY EXERCISES 17.2, page 1074

1. 1.3124; 0.4038 3. −1.8889; 0.7778 5. −0.3863; −0.8497

17.3 Problem Solving Tips

1. To find the relative extrema of a function of several variables, first find the critical points of $f(x,y)$ by solving the simultaneous equations $f_x = 0$ and $f_y = 0$, and then use the second derivative test to classify those points.
2. To use the second derivative test, first evaluate the function $D(x,y) = f_{xx}f_{yy} - f_{xy}^2$ for each critical point found in (1). If $D(a,b) > 0$ and $f_{xx}(a,b) < 0$ then $f(x,y)$ has a relative maximum at the point (*a*,*b*). If $D(a,b) > 0$ and $f_{xx}(a,b) > 0$, then $f(x,y)$ has a relative minimum at the point (*a*,*b*). If $D(a,b) < 0$ then $f(x,y)$ has neither a relative maximum nor a relative minimum at the point (*a*,*b*). If $D(a,b) = 0$, then the test is inconclusive.

17.3 CONCEPT QUESTIONS, page 1083

1. a. A function $f(x,y)$ has a relative maximum at (a,b) if $f(a,b)$ is the largest number compared to $f(x,y)$ for all (x,y) near (a,b).
   b. $f(a,y)$ has an absolute maximum at (*a*,*b*) if $f(a,b)$ is the largest number compared to $f(x,y)$ for all (x,y) in the domain of *f*.

3. See the procedure given in the text on page 1078.

EXERCISES 17.3, page 1083

1. $f(x,y)=1-2x^2-3y^2$. To find the critical point(s) of f, we solve the system

$$\begin{cases} f_x = -4x = 0 \\ f_y = -6y = 0 \end{cases}$$

obtaining (0,0) as the only critical point of f. Next,

$f_{xx} = -4, f_{xy} = 0$, and $f_{yy} = -6$.

In particular, $f_{xx}(0, 0) = -4, f_{xy}(0, 0) = 0$, and $f_{yy}(0, 0) = -6$, giving

$D(0, 0) = (-4)(-6) - 0^2 = 24 > 0$.

Since $f_{xx}(0, 0) < 0$, the Second Derivative Test implies that (0, 0) gives rise to a relative maximum of f. Finally, the relative maximum of f is $f(0, 0) = 1$.

3. To find the critical points of f, we solve the system

$$\begin{cases} f_x = \ \ 2x-2=0 \\ f_y = -2y+4=0 \end{cases}$$

obtaining $x = 1$ and $y = 2$ so that (1, 2) is the only critical point.

$f_{xx} = 2, f_{xy} = 0$, and $f_{yy} = -2$.

So $D(x,y) = f_{xx}f_{yy} - f_{xy}^2 = -4$. In particular, $D(1, 2) = -4 < 0$ and so (1, 2) affords a saddle point of f and $f(1, 2) = 4$.

5. $f(x,y)=x^2+2xy+2y^2-4x+8y-1$. To find the critical point(s) of f, we solve the system

$$\begin{cases} f_x = 2x+2y-4=0 \\ f_y = 2x+4y+8=0 \end{cases}$$

obtaining (8,–6) as the critical point of f. Next, $f_{xx} = 2, f_{xy} = 2, f_{yy} = 4$. In particular, $f_{xx}(8,-6) = 2, f_{xy}(8,-6) = 2, f_{yy}(8,-6) = 4$, giving $D = 2(4) - 4 = 4 > 0$. Since $f_{xx}(8,-6) > 0$, (8,–6) gives rise to a relative minimum of f. Finally, the relative minimum value of f is $f(8,-6) = -41$.

7. $f(x,y)=2x^3+y^2-9x^2-4y+12x-2$.. To find the critical points of f, we solve the system

$$\begin{cases} f_x = 6x^2-18x+12=0 \\ f_y = 2y-4=0 \end{cases}$$

The first equation is equivalent to $x^2 - 3x + 2 = 0$, or $(x - 2)(x - 1) = 0$ which gives $x = 1$ or 2. The second equation of the system gives $y = 2$. Therefore, there are two critical points, (1,2) and (2,2). Next, we compute

$f_{xx} = 12x - 18 = 6(2x - 3), f_{xy} = 0, f_{yy} = 2.$

At the point (1,2):

$f_{xx}(1,2) = 6(2 - 3) = -6, f_{xy}(1,2) = 0,$ and $f_{yy}(1,2) = 2.$

Therefore, $D = (-6)(2) - 0 = -12 < 0$ and we conclude that (1,2) gives rise to a saddle point of f. At the point (2,2):

$f_{xx}(2,2) = 6(4 - 3) = 6, f_{xy}(2,2) = 0,$ and $f_{yy}(2,2) = 2.$

Therefore, $D = (6)(2) - 0 = 12 > 0$. Since $f_{xx}(2,2) > 0$, we see that (2,2) gives rise to a relative minimum with value $f(2,2) = -2$.

9. To find the critical points of f, we solve the system

$$\begin{cases} f_x = 3x^2 - 2y + 7 = 0 \\ f_y = 2y - 2x - 8 = 0 \end{cases}$$

Adding the two equations gives $3x^2 - 2x - 1 = 0$, or $(3x + 1)(x - 1) = 0$. Therefore, $x = -1/3$ or 1. Substituting each of these values of x into the second equation gives $y = 8/3$ and $y = 5$, respectively. Therefore, $(-\frac{1}{3}, \frac{11}{3})$ and (1,5) are critical points of f.

Next, $f_{xx} = 6x, f_{xy} = -2,$ and $f_{yy} = 2$. So $D(x,y) = 12x - 4 = 4(3x - 1)$. Then

$$D(-\tfrac{1}{3}, \tfrac{11}{3}) = 4(-1-1) = -8 < 0$$

and so $(-\frac{1}{3}, \frac{11}{3})$ gives a saddle point. Next, $D(1,5) = 4(3 - 1) = 8 > 0$ and since $f_{xx}(1,5) = 6 > 0$, we see that (1,5) gives rise to a relative minimum.

11. To find the critical points of f, we solve the system

$$\begin{cases} f_x = 3x^2 - 3y = 0 \\ f_y = -3x + 3y^2 = 0 \end{cases}$$

The first equation gives $y = x^2$ which when substituted into the second equation gives $-3x + 3x^4 = 3x(x^3 - 1) = 0$. Therefore, $x = 0$ or 1. Substituting these values of x into the first equation gives $y = 0$ and $y = 1$, respectively. Therefore, (0,0) and (1,1) are critical points of f. Next, we find $f_{xx} = 6x, f_{xy} = -3,$ and $f_{yy} = 6y$. So $D = f_{xx}f_{yy} - f_{xy}^2 = 36xy - 9$. Since $D(0,0) = -9 < 0$, we see that (0,0) gives a saddle point of f. Next, $D(1,1) = 36 - 9 = 27 > 0$ and since $f_{xx}(1,1) = 6 > 0$, we see that $f(1,1) = -3$ is a relative minimum value of f.

13. Solving the system of equations

$$\begin{cases} f_x = y - \frac{4}{x^2} = 0 \\ f_y = x - \frac{2}{y^2} = 0 \end{cases}$$

we obtain $y = \frac{4}{x^2}$. Therefore, $x - 2\left(\frac{x^4}{16}\right) = 0$ and $8x - x^4 = x(8 - x^3) = 0$, and $x = 0$, or $x = 2$. Since $x = 0$ is not in the domain of f, (2,1) is the only critical point of f. Next, $f_{xx} = \frac{8}{y^3}$, $f_{xy} = 1$, and $f_{yy} = \frac{4}{y^3}$. Therefore,

$D(2,1) = \frac{32}{x^3 y^3} - 1\Big|_{(2,1)} = 4 - 1 = 3 > 0$ and $f_{xx}(2,1) = 1 > 0$. Therefore, the relative minimum value of f is $f(2,1) = 2 + 4/2 + 2/1 = 6$.

15. Solving the system of equations $f_x = 2x = 0$ and $f_y = -2ye^{y^2} = 0$, we obtain $x = 0$ and $y = 0$. Therefore, (0,0) is the only critical point of f. Next,
$$f_{xx} = 2, f_{xy} = 0, f_{yy} = -2e^{y^2} - 4y^2e^{y^2}.$$
Therefore, $D(0,0) = -4e^{y^2}(1 + 2y^2)\Big|_{(0,0)} = -4(1) < 0$, and we conclude that (0,0) gives rise to a saddle point.

17. $f(x,y) = e^{x^2+y^2}$

Solving the system

$$\begin{cases} f_x = 2xe^{x^2+y^2} = 0 \\ f_y = 2ye^{x^2+y^2} = 0 \end{cases}$$

we see that $x = 0$ and $y = 0$ (recall that $e^{x^2+y^2} \neq 0$). Therefore, (0,0) is the only critical point of f. Next, we compute

$$f_{xx} = 2e^{x^2+y^2} + 2x(2x)e^{x^2+y^2} = 2(1 + 2x^2)e^{x^2+y^2}$$

$$f_{xy} = 2x(2y)e^{x^2+y^2} = 4xye^{x^2+y^2}; \; f_{yy} = 2(1 + 2y^2)e^{x^2+y^2}.$$

In particular, at the point (0,0), $f_{xx}(0,0) = 2$, $f_{xy}(0,0) = 0$, and $f_{yy}(0,0) = 2$. Therefore, $D = (2)(2) - 0 = 4 > 0$. Furthermore, since $f_{xx}(0,0) > 0$, we conclude that (0,0) gives rise to a relative minimum of f. The relative minimum value of f is $f(0,0) = 1$.

19. $f(x,y) = \ln(1 + x^2 + y^2)$. We solve the system of equations

$$f_x = \frac{2x}{1 + x^2 + y^2} = 0 \text{ and } f_y = \frac{2y}{1 + x^2 + y^2} = 0,$$

obtaining $x = 0$ and $y = 0$. Therefore, (0,0) is the only critical point of f. Next,

$$f_{xx} = \frac{(1+x^2+y^2)2-(2x)(2x)}{(1+x^2+y^2)^2} = \frac{2+2y^2-2x^2}{(1+x^2+y^2)^2}$$

$$f_{yy} = \frac{(1+x^2+y^2)2-(2y)(2y)}{(1+x^2+y^2)^2} = \frac{2+2x^2-2y^2}{(1+x^2+y^2)^2}$$

$$f_{xy} = -2x(1+x^2+y^2)^{-2}(2y) = -\frac{4xy}{(1+x^2+y^2)^2}.$$

Therefore, $D(x,y) = \dfrac{(2+2y^2-2x^2)(2+2x^2-2y^2)}{(1+x^2+y^2)^4} - \dfrac{16x^2y^2}{(1+x^2+y^2)^4}$.

Since $D(0,0) = \frac{4}{1} > 0$ and $f_{xx}(0,0) = 2 > 0$, $f(0,0) = 0$ is a relative minimum value.

21. $P(x) = -0.2x^2 - 0.25y^2 - 0.2xy + 200x + 160y - 100x - 70y - 4000$
$= -0.2x^2 - 0.25y^2 - 0.2xy + 100x + 90y - 4000.$

Then $\begin{cases} P_x = -0.4x - 0.2y + 100 = 0 \\ P_y = -0.5y - 0.2x + 90 = 0 \end{cases}$

implies that $\begin{cases} 4x + 2y = 1000 \\ 2x + 5y = \ \ 900 \end{cases}$. Solving, we find $x = 200$ and $y = 100$.

Next, $P_{xx} = -0.4$, $P_{yy} = -0.5$, $P_{xy} = -0.2$, and

$D(200,100) = (-0.4)(-0.5) - (-0.2)^2 > 0$. Since $P_{xx}(200, 100) < 0$, we conclude that (200,100) is a relative maximum of P. Thus, the company should manufacture 200 finished and 100 unfinished units per week. The maximum profit is

$P(200,100) = -0.2(200)^2 - 0.25(100)^2 - 0.2(100)(200) + 100(200) + 90(100) - 4000$
$= 10{,}500$, or \$10,500.

23. $p(x,y) = 200 - 10(x - \frac{1}{2})^2 - 15(y-1)^2$. Solving the system of equations

$$\begin{cases} p_x = -20(x - \frac{1}{2}) = 0 \\ p_y = -30(y-1) = 0 \end{cases}$$

we obtain $x = 1/2$, $y = 1$. We conclude that the only critical point of f is $(\frac{1}{2},1)$.

Next, $p_{xx} = -20$, $p_{xy} = 0$, $p_{yy} = -30$

so $D(\frac{1}{2},1) = (-20)(-30) = 600 > 0$.

Since $p_{xx} = -20 < 0$, we conclude that $f(\frac{1}{2},1)$ gives a relative maximum. So we conclude that the price of land is highest at $(\frac{1}{2},1)$.

25. We want to minimize

$f(x,y) = D^2 = (x-5)^2 + (y-2)^2 + (x+4)^2 + (y-4)^2 + (x+1)^2 + (y+3)^2.$

Next, $\begin{cases} f_x = 2(x-5)+2(x+4)+2(x+1) = 6x = 0, \\ f_y = 2(y-2)+2(y-4)+2(y+3) = 6y-6 = 0 \end{cases}$

and we conclude that $x = 0$ and $y = 1$. Also,

$$f_{xx} = 6, f_{xy} = 0, f_{yy} = 6 \text{ and } D(x,y) = (6)(6) = 36 > 0.$$

Since $f_{xx} > 0$, we conclude that the function is minimized at (0,1) and so (0,1) gives the desired location.

27. Refer to the figure in the text.

$$xy + 2xz + 2yz = 300; \; z(2x+2y) = 300 - xy; \text{ and } z = \frac{300-xy}{2(x+y)}.$$

Then the volume is given by

$$V = xyz = xy\left[\frac{300-xy}{2(x+y)}\right] = \frac{300xy - x^2y^2}{2(x+y)}.$$

We find

$$\frac{\partial V}{\partial x} = \frac{1}{2}\frac{(x+y)(300y-2xy^2)-(300xy-x^2y^2)}{(x+y)^2}$$

$$= \frac{300xy - 2x^2y^2 + 300y^2 - 2xy^3 - 300xy + x^2y^2}{2(x+y)^2}$$

$$= \frac{300y^2 - 2xy^3 - x^2y^2}{2(x+y)^2} = \frac{y^2(300-2xy-x^2)}{2(x+y)^2}.$$

Similarly, $\dfrac{\partial V}{\partial y} = \dfrac{x^2(300-2xy-y^2)}{2(x+y)^2}$. Setting both $\partial V/\partial x$ and $\partial V/\partial y$ equal to zero and observing that both $x > 0$ and $y > 0$, we have the system

$\begin{cases} 2yx + x^2 = 300 \\ 2yx + y^2 = 300 \end{cases}$. Subtracting, we find $y^2 - x^2 = 0$; that is $(y-x)(y+x) = 0$. So $y = x$ or $y = -x$. The latter is not possible since $x, y > 0$. Therefore, $y = x$. Substituting this value into the first equation in the system gives

$$2x^2 + x^2 = 300; \; x^2 = 100; \text{ and } x = 10.$$

Therefore, $y = 10$. Substituting this value into the expression for z

gives $z = \dfrac{300-10^2}{2(10+10)} = 5$. So the dimensions are $10" \times 10" \times 5"$. The volume is

500 cu in.

29. The heating cost is $C = 2xy + 8xz + 6yz$. But $xyz = 12{,}000$ or $z = 12{,}000/xy$. Therefore,

$$C = f(x,y) = 2xy + 8x\left(\frac{12{,}000}{xy}\right) + 6y\left(\frac{12{,}000}{xy}\right) = 2xy + \frac{96{,}000}{y} + \frac{72{,}000}{x}.$$

To find the minimum of f, we find the critical point of f by solving the system

$$\begin{cases} f_x = 2y - \dfrac{72{,}000}{x^2} = 0 \\ f_y = 2x - \dfrac{96{,}000}{y^2} = 0 \end{cases}.$$

The first equation gives $y = 36000/x^2$, which when substituted into the second equation yields

$$2x - 96{,}000\left(\frac{x^2}{36{,}000}\right)^2 = 0, \quad (36{,}000)^2 x - 48{,}000x^4 = 0$$

$$x(27{,}000 - x^3) = 0.$$

Solving this equation, we have $x = 0$, or $x = 30$. We reject the first root because $x \neq 0$. With $x = 30$, we find $y = 40$ and

$$f_{xx} = \frac{144{,}000}{x^3}, \; f_{xy} = 2, \text{ and } f_{yy} = \frac{192{,}000}{y^3}.$$

In particular, $f_{xx}(30,40) = 5.33$, $f_{xy} = (30,40) = 2$, and $f_{yy}(30,40) = 3$. So

$$D(30,40) = (5.33)(3) - 4 = 11.99 > 0$$

and since $f_{xx}(30,40) > 0$, we see that (30,40) gives a relative minimum. Physical considerations tell us that this is an absolute minimum. The minimal annual heating cost is

$$f(30,40) = 2(30)(40) + \frac{96{,}000}{40} + \frac{72{,}000}{30} = 7200, \text{ or } \$7{,}200.$$

31. False. Let $f(x,y) = xy$. Then $f_x(0,0) = 0$ and $f_y(0,0) = 0$. But $(0,0)$ does not afford a relative extremum of $(0,0)$. In fact, $f_{xx} = 0$, $f_{yy} = 0$, and $f_{xy} = 1$. Therefore, $D(x,y) = f_{xx}f_{yy} - f_{xy}^2 = -1$ and so $D(0,0) = -1$ which shows that (0, 0, 0) is a saddle point.

17.4 Problem Solving Tips

1. The method of Lagrange Multipliers allows us to find the critical points of the function $f(x, y)$ subject to the constraint $g(x, y) = 0$ (if the extrema exist). However, it does not tell us whether those critical points lead to a relative extrema. Instead we rely on the geometric or physical nature of the problem to classify these points.
2. To use the method of Lagrange Multipliers, first form the function $F(x, y, \lambda) = f(x, y) + \lambda g(x, y)$ and then solve the system of equations $F_x = 0$, $F_y = 0$, and $F_\lambda = 0$ for all values of x, y and λ. The solutions are candidates for the extrema of f.

17.4 CONCEPT QUESTIONS, page 1095

1. A constrained relative extremem of f is an extremum of f subject to a constraint of the form $g(x, y) = 0$.

EXERCISES 17.4, page 1095

1. We form the Lagrangian function $F(x, y, \lambda) = x^2 + 3y^2 + \lambda(x + y - 1)$. We solve the system

$$\begin{cases} F_x = 2x + \lambda = 0 \\ F_y = 6y + \lambda = 0 \\ F_\lambda = x + y - 1 = 0. \end{cases}$$

Solving the first and the second equations for x and y in terms of λ we obtain $x = -\frac{\lambda}{2}$ and $y = -\frac{\lambda}{6}$ which, upon substitution into the third equation, yields

$-\frac{\lambda}{2}-\frac{\lambda}{6}-1=0$ or $\lambda=-\frac{3}{2}$. Therefore, $x=\frac{3}{4}$ and $y=\frac{1}{4}$ which gives the point $(\frac{3}{4},\frac{1}{4})$ as the sole critical point of F. Therefore, $(\frac{3}{4},\frac{1}{4})=\frac{3}{4}$ is a minimum of F.

3. We form the Lagrangian function $F(x, y, \lambda) = 2x + 3y - x^2 - y^2 + \lambda(x + 2y - 9)$. We then solve the system

$$\begin{cases} F_x = 2-2x+\ \lambda=0 \\ F_y = 3-2y+2\lambda=0. \\ F_\lambda = x+2y\ -9=0 \end{cases}$$

Solving the first equation λ, we obtain $\lambda = 2x - 2$. Substituting into the second equation, we have $3 - 2y + 4x - 4 = 0$, or $4x - 2y - 1 = 0$. Adding this equation to the third equation in the system, we have $5x - 10 = 0$, or $x = 2$. Therefore, $y = 7/2$ and $f(2,\frac{7}{2})=-\frac{7}{4}$ is the maximum value of f.

5. Form the Lagrangian function $F(x, y, \lambda) = x^2 + 4y^2 + \lambda(xy - 1)$. We then solve the system

$$\begin{cases} F_x = 2x+\lambda y=0 \\ F_y = 8y+\lambda x=0\,. \\ F_\lambda = xy-1=0 \end{cases}$$

Multiplying the first and second equations by x and y, respectively, and subtracting the resulting equations, we obtain $2x^2 - 8y^2 = 0$, or $x = \pm 2y$. Substituting this into the third equation gives $2y^2-1=0$ or $y=\pm\frac{\sqrt{2}}{2}$. We conclude that $f(-\sqrt{2},-\frac{\sqrt{2}}{2})=f(\sqrt{2},\frac{\sqrt{2}}{2})=4$ is the minimum value of f.

7. We form the Lagrangian function

$$F(x, y, \lambda) = x + 5y - 2xy - x^2 - 2y^2 + \lambda(2x + y - 4).$$

Next, we solve the system

$$\begin{cases} F_x = 1-2y-2x+2\lambda=0 \\ F_y = 5-2x-4y+\lambda=0 \\ F_\lambda = 2x+y-4=0 \end{cases}$$

Solving the last two equations for x and y in terms of λ, we obtain

$$y=\tfrac{1}{3}(1+\lambda) \text{ and } x=\tfrac{1}{6}(11-\lambda)$$

which, upon substitution into the first equation, yields

$$1-\tfrac{2}{3}(1+\lambda)-\tfrac{1}{3}(11-\lambda)+2\lambda=0$$

$$\text{or } 1-\tfrac{2}{3}-\tfrac{2}{3}\lambda-\tfrac{11}{3}+\tfrac{\lambda}{3}+2\lambda=0$$

or $\lambda=2$. Therefore, $x = 3/2$ and $y = 1$. The maximum of f is

$$f(\tfrac{3}{2},1)=\tfrac{3}{2}+5-2(\tfrac{3}{2})-(\tfrac{3}{2})^2-2=-\tfrac{3}{4}.$$

9. Form the Lagrangian $F(x, y, \lambda) = xy^2 + \lambda(9x^2 + y^2 - 9)$. We then solve

$$\begin{cases} F_x = y^2 + 18\lambda x = 0 \\ F_y = 2xy + 2\lambda y = 0 \\ F_\lambda = 9x^2 + y^2 - 9 = 0. \end{cases}$$

The first equation gives $\lambda = -\dfrac{y^2}{18x}$. Substituting into the second gives

$$2xy+2y\left(-\frac{y^2}{18x}\right)=0, \text{ or } 18x^2y-y^3=y(18x^2-y^2)=0,$$

giving $y = 0$ or $y=\pm 3\sqrt{2}x$. If $y = 0$, then the third equation gives $9x^2 - 9 = 0$ or $x=\pm 1$. If $y=\pm 3\sqrt{3}/3$. Therefore, the points $(-1,0), (-\sqrt{3}/3,-\sqrt{6})$, $(-\sqrt{3}/3,\sqrt{6})$, $(\sqrt{3}/3,-\sqrt{6})$ and $(\sqrt{3}/3,\sqrt{6})$ give rise to extreme values of f subject to the given constraint. Evaluating $f(x,y)$ at each of these points, we see that $f(\sqrt{3}/3,-\sqrt{6})=(\sqrt{3}/3,\sqrt{6})=2\sqrt{3}$ is the maximum value of f.

11. We form the Lagrangian function $F(x, y, \lambda) = xy + \lambda(x^2 + y^2 - 16)$. To find the critical points of F, we solve the system

$$\begin{cases} F_x = y + 2\lambda x = 0 \\ F_y = x + 2\lambda y = 0 \\ F_\lambda = x^2 + y^2 - 16 = 0 \end{cases}$$

Solving the first equation for λ and substituting this value into the second equation yields $x-2\left(\dfrac{y}{2x}\right)y=0$, or $x^2 = y^2$. Substituting the last equation into the third equation in the system, yields $x^2 + x^2 - 16 = 0$, or $x^2 = 8$, that is, $x = \pm 2\sqrt{2}$. The corresponding values of y are $y = \pm 2\sqrt{2}$. Therefore the critical points of F are $(-2\sqrt{2},-2\sqrt{2})$, $(-2\sqrt{2},2\sqrt{2})$, $(2\sqrt{2},-2\sqrt{2})(2\sqrt{2}, 2\sqrt{2})$.

Evaluating f at each of these values, we find that $f(-2\sqrt{2}, 2\sqrt{2}) = -8$ and $f(2\sqrt{2}, -2\sqrt{2}) = -8$ are relative minimum values and $f(-2\sqrt{2}, -2\sqrt{2}) = 8$ and $f(2\sqrt{2}, 2\sqrt{2}) = 8$, are relative maximum values.

13. We form the Lagrangian function $F(x, y, \lambda) = xy^2 + \lambda(x^2 + y^2 - 1)$. Next, we solve the system

$$\begin{cases} F_x = \quad y^2 + 2x\lambda = 0 \\ F_y = 2xy + 2y\lambda = 0 \\ F_\lambda = x^2 + y^2 - 1 = 0 \end{cases}.$$

We find that $x = \pm\sqrt{3}/3$ and $y = \pm\sqrt{6}/3$ and $x = \pm 1$, $y = 0$. Evaluating f at each of the critical points $(-\frac{\sqrt{3}}{3}, -\frac{\sqrt{6}}{3}), (-\frac{\sqrt{3}}{3}, \frac{\sqrt{6}}{3})(\frac{\sqrt{3}}{3}, -\frac{\sqrt{6}}{3})(\frac{\sqrt{3}}{3}, \frac{\sqrt{6}}{3})$, (-1,0), and (1,0), we find that $f(-\frac{\sqrt{3}}{3}, -\frac{\sqrt{6}}{3}) = -\frac{2\sqrt{3}}{9}$ and $f(-\frac{\sqrt{3}}{3}, \frac{\sqrt{6}}{3}) = -\frac{2\sqrt{3}}{9}$ are relative minimum values and $f(\frac{\sqrt{3}}{3}, -\frac{\sqrt{6}}{3}) = \frac{2\sqrt{3}}{9}$ and $f(\frac{\sqrt{3}}{3}, \frac{\sqrt{6}}{3}) = \frac{2\sqrt{3}}{9}$ are relative maximum values.

15. Form the Lagrangian function $F(x, y, z, \lambda) = x^2 + y^2 + z^2 + \lambda(3x + 2y + z - 6)$. We solve the system

$$\begin{cases} F_x = 2x + 3\lambda = 0 \\ F_y = 2y + 2\lambda = 0 \\ F_z = 2x + \lambda = 0 \\ F_\lambda = 3x + 2y + z - 6 = 0. \end{cases}$$

The third equation give $\lambda = -2z$. Substituting into the first two equations gives

$$\begin{cases} 2x - 6z = 0 \\ 2y - 4z = 0. \end{cases}$$

So $x = 3z$ and $y = 2z$. Substituting into the third equation yields $9z + 4z + z - 6 = 0$, or $z = 3/7$. Therefore, $x = 9/7$ and $y = 6/7$. Therefore, $f(\frac{9}{7}, \frac{6}{7}, \frac{3}{7}) = \frac{18}{7}$ is the minimum value of F.

17. We want to maximize P subject to the constraint $x + y = 200$. The Lagrangian function is

$F(x, y, \lambda) = -0.2x^2 - 0.25y^2 - 0.2xy + 100x + 90y - 4000 + \lambda(x + y - 200)$.

Next, we solve

$$\begin{cases} F_x = -0.4x - 0.2y + 100 + \lambda = 0 \\ F_y = -0.5y - 0.2x + 90 + \lambda = 0 \\ F_\lambda = x + y - 200 = 0. \end{cases}$$

Subtracting the first equation from the second yields

$$0.2x - 0.3y - 10 = 0, \text{ or } 2x - 3y - 100 = 0.$$

Multiplying the third equation in the system by 2 and subtracting the resulting equation from the last equation, we find $-5y + 300 = 0$ or $y = 60$. So $x = 140$ and the company should make 140 finished and 60 unfinished units.

19. Suppose each of the sides made of pine board is x feet long and those of steel are y feet long. Then $xy = 800$. The cost is $C = 12x + 3y$ and is to be minimized subject to the condition $xy = 800$. We form the Lagrangian function

$$F(x, y, \lambda) = 12x + 3y + \lambda(xy - 800).$$

We solve the system

$$\begin{cases} F_x = 12 + \lambda y = 0 \\ F_y = 3 + \lambda x = 0 \\ F_\lambda = xy - 800 = 0. \end{cases}$$

Multiplying the first equation by x and the second equation by y and subtracting the resulting equations, we obtain $12x - 3y = 0$, or $y = 4x$. Substituting this into the third equation of the system, we obtain

$$4x^2 - 800 = 0, \text{ or } x = \pm\sqrt{200} = \pm 10\sqrt{2}.$$

Since x must be positive, we take $x = 10\sqrt{2}$. So $y = 40\sqrt{2}$. So the dimensions are approximately 14.14 ft by 56.56 ft.

21. We want to minimize the function $C(r,h)$ subject to the constraint $\pi r^2 h - 64 = 0$. We form the Lagrangian function $F(r, h, \lambda) = 8\pi rh + 6\pi r^2 - \lambda(\pi r^2 h - 64)$. Then we solve the system

$$\begin{cases} F_r = 8\pi h + 12\pi r - 2\lambda\pi rh = 0 \\ F_h = 8\pi r - \lambda r^2 = 0 \\ F_\lambda = \pi r^2 h - 64 = 0 \end{cases}$$

Solving the second equation for λ yields $\lambda = 8/r$, which when substituted into the first equation yields

$$8\pi h + 12\pi r - 2\pi rh(\tfrac{8}{r}) = 0$$
$$12\pi r = 8\pi h$$
$$h = \tfrac{3r}{2}.$$

Substituting this value of h into the third equation of the system, we find

$$3r^2(\tfrac{3r}{2}) = 64,\ \ r^3 = \frac{128}{3\pi},\ \text{or}\ r = \frac{4}{3}\sqrt[3]{\frac{18}{\pi}}\ \text{and}\ h = 2\sqrt[3]{\frac{18}{\pi}}.$$

23. Let the box have dimensions x' by y' by z'. Then $xyz = 4$. We want to minimize

$$C = 2xz + 2yz + \tfrac{3}{2}(2xy) = 2xz + 2yz + 3xy.$$

Form the Lagrangian function

$$F(x, y, z, \lambda) = 2xz + 2yz + 3xy + \lambda(xyz - 4).$$

Next, we solve the system

$$\begin{cases} F_x = 2z + 3y + \lambda yz = 0 \\ F_y = 2z + 3x + \lambda xz = 0 \\ F_z = 2x + 2y + \lambda xy = 0 \\ F_\lambda = xyz - 4 = 0. \end{cases}$$

Multiplying the first, second, and third equations by x, y, and z, respectively, we have

$$\begin{cases} 2xz + 3xy + \lambda xyz = 0 \\ 2yz + 3xy + \lambda xyz = 0 \\ 2xz + 2yz + \lambda xyz = 0. \end{cases}$$

The first two equations imply that $2z(x - y) = 0$. Since $z \neq 0$, we see that $x = y$. The second and third equations imply that $x(3y - 2z) = 0$ or $x = (3/2)y$. Substituting these values into the fourth equation in the system, we find

$y^2(\tfrac{3}{2}y) = 4$ or $y^3 = \tfrac{8}{3}$. Therefore, $y = \dfrac{2}{3^{1/3}} = \dfrac{2}{3}\sqrt[3]{9}$ and $x = \dfrac{2}{3}\sqrt[3]{9}$, and $z = \sqrt[3]{9}.$

So the dimensions are $\dfrac{2}{3}\sqrt[3]{9} \times \dfrac{2}{3}\sqrt[3]{9} \times \sqrt[3]{9}$.

25. We want to maximize $f(x,y) = 100x^{3/4}y^{1/4}$ subject to $100x + 200y = 200{,}000$. Form the Lagrangian function

$$F(x, y, \lambda) = 100x^{3/4}y^{1/4} + \lambda(100x + 200y - 200{,}000).$$

We solve the system

$$\begin{cases} F_x = 75x^{-1/4}y^{1/4} + 100\lambda = 0 \\ F_y = 25x^{3/4}y^{-3/4} + 200\lambda = 0 \\ F_\lambda = 100x + 200y - 200{,}000 = 0. \end{cases}$$

The first two equations imply that $150x^{-1/4}y^{1/4} - 25x^{3/4}y^{-3/4} = 0$ or upon multiplying by $x^{1/4}y^{3/4}$, $150y - 25x = 0$, which implies that $x = 6y$. Substituting this value of x into the third equation of the system, we have

$$600y + 200y - 200{,}000 = 0$$

giving $y = 250$ and therefore $x = 1500$. So to maximize production, he should spend 1500 units on labor and 250 units of capital.

27. False. See Example 1, Section 17.4..

17.5 Problem Solving Tips

1. To evaluate the double integral $\iint\limits_R f(x,y)\,dA = \int_a^b \left[\int_{g_1(x)}^{g_2(x)} f(x,y)\,dy \right] dx$, first evaluate the inside y-integral, treating x as a constant, and then evaluate the outside integral with respect to x.

To evaluate the double integral $\iint\limits_R f(x,y)\,dA = \int_a^b \left[\int_{h_1(y)}^{h_2(y)} f(x,y)\,dx \right] dy$, first evaluate the inside x-integral, treating y as a constant, and then evaluate the outside integral with respect to y.

17.5 CONCEPT QUESTIONS, page 1107

1. It gives the volume of the solid region bounded above by the graph of f and below by the region R.

3. $\iint\limits_R f(x,y)\,dA = \int_a^b \left[\int_{g_1(x)}^{g_2(x)} f(x,y)\,dy \right] dx$

5. The average value is $\dfrac{\iint_R f(x,y)\,dA}{\iint_R dA}$

EXERCISES 17.5, page 1107

1. $\int_1^2\int_0^1 (y+2x)\,dy\,dx = \int_1^2 \frac{1}{2}y^2+2xy\Big|_{y=0}^{y=1} dx = \int_1^2 (\frac{1}{2}+2x)\,dx = \frac{1}{2}x+x^2\Big|_1^2 = 5-\frac{3}{2}=\frac{7}{2}.$

3. $\int_{-1}^1\int_0^1 xy^2\,dy\,dx = \int_{-1}^1 \frac{1}{3}xy^3\Big|_0^1 dx = \int_{-1}^1 \frac{1}{3}x\,dx = \frac{x^2}{6}\Big|_{-1}^1 = \frac{1}{6}-(\frac{1}{6})=0.$

5. $\int_{-1}^2\int_1^{e^3} \frac{x}{y}\,dy\,dx = \int_{-1}^2 x\ln y\Big|_1^{e^3} dx = \int_{-1}^2 x\ln e^3\,dx = \int_{-1}^2 3x\,dx = \frac{3}{2}x^2\Big|_{-1}^2$

$= \frac{3}{2}(4)-\frac{3}{2}(1)=\frac{9}{2}.$

7. $\int_{-2}^0\int_0^1 4xe^{2x^2+y}\,dx\,dy = \int_{-2}^0 e^{2x^2+y}\Big|_{x=0}^{x=1} dy = \int_{-2}^0 (e^{2+y}-e^y)\,dy = (e^{2+y}-e^y)\Big|_{-2}^0$

$= [(e^2-1)-(e^0-e^{-2}) = e^2-2+e^{-2} = (e^2-1)(1-e^{-2}).$

9. $\int_0^1\int_1^e \ln y\,dy\,dx = \int_0^1 y\ln y - y\Big|_{y=1}^{y=e} dx = \int_0^1 dx = 1.$

11. $\int_0^1\int_0^x (x+2y)\,dy\,dx = \int_0^1 (xy+y^2)\Big|_{y=0}^{y=x} dx = \int_0^1 2x^2\,dx = \frac{2}{3}x^3\Big|_0^1 = \frac{2}{3}.$

13. $\int_1^3\int_0^{x+1} (2x+4y)\,dy\,dx = \int_1^3 2xy+2y^2\Big|_{y=0}^{y=x+1} dx = \int_1^3 [2x(x+1)+2(x+1)^2]\,dx$

$= \int_1^3 (4x^2+6x+2)\,dx = (\frac{4}{3}x^3+3x^2+2x)\Big|_1^3$

$= (36+27+6)-(\frac{4}{3}+3+2) = \frac{188}{3}.$

15. $\int_0^4\int_0^{\sqrt{y}} (x+y)\,dx\,dy = \int_0^4 \frac{1}{2}x^2+xy\Big|_{x=0}^{x=\sqrt{y}} dy = \int_0^4 (\frac{1}{2}y+y^{3/2})\,dy$

$= (\frac{1}{4}y^2+\frac{2}{5}y^{5/2})\Big|_0^4 = 4+\frac{64}{5}=\frac{84}{5}.$

17. $\int_0^2\int_0^{\sqrt{4-y^2}} y\,dx\,dy = \int_0^2 xy\Big|_0^{\sqrt{4-y^2}} dy = \int_0^2 y\sqrt{4-y^2}\,dy = -\frac{1}{2}(\frac{2}{3})(4-y^2)^{3/2}\Big|_0^2$

$= \frac{1}{3}(4^{3/2}) = \frac{8}{3}.$

19. $\int_0^1\int_0^x 2xe^y\,dy\,dx = \int_0^1 2xe^y\Big|_{y=0}^{y=x} dx = \int_0^1(2xe^x-2x)\,dx = 2(x-1)e^x - x^2\Big|_0^1 = (-1)+2 = 1.$

21. $\int_0^1\int_x^{\sqrt{x}} ye^x\,dy\,dx = -\int_0^1\int_x^{\sqrt{x}} ye^x\,dy\,dx = \int_0^1 -\frac{1}{2}y^2e^x\Big|_{y=\sqrt{x}}^{y=x} dx = -\frac{1}{2}\int_0^1(x^2e^x - xe^x)dx$

$= -\frac{1}{2}[x^2e^x\Big|_0^1 - 2\int_0^1 xe^x\,dx - \int_0^1 xe^x\,dx] = -\frac{1}{2}[x^2e^x\Big|_0^1 - 3\int_0^1 xe^x\,dx]$

$= -\frac{1}{2}[x^2e^x - 3xe^x + 3e^x]\Big|_0^1 = -\frac{1}{2}[e-3e+3e-3] = \frac{1}{2}(3-e).$

23. $\int_0^1\int_{2x}^2 e^{y^2}\,dy\,dx = \int_0^2\int_0^{y/2} e^{y^2}\,dx\,dy = \int_0^2 xe^{y^2}\Big|_{x=0}^{x=y/2} dy = \int_0^2 \frac{1}{2}ye^{y^2}\,dy$

$= \frac{1}{4}e^{y^2}\Big|_0^2 = \frac{1}{4}(e^4-1).$

25. $\int_0^2\int_{y/2}^1 ye^{x^3}\,dx\,dy = \int_0^1\int_0^{2x} ye^{x^3}\,dy\,dx = \int_0^1 \frac{1}{2}y^2e^{x^3}\Big|_{y=0}^{y=2x} dx = \int_0^1 2x^2e^{x^3}\,dx$

$= \frac{2}{3}e^{x^3}\Big|_0^1 = \frac{2}{3}(e-1).$

27. $V = \int_0^4\int_0^3(6-x)\,dy\,dx = \int_0^4(6-x)y\Big|_{y=0}^{y=3} dx = 3\int_0^4(6-x)\,dx$

$= 3(6x - \frac{1}{2}x^2)\Big|_0^4 = 3(24-8) = 48$, or 48 cu units.

29. $V = \int_0^2\int_0^{3-(3/2)z}(6-2y-3z)dy\,dz = \int_0^2 6y - y^2 - 3yz\Big|_{y=0}^{y=3-(3/2)z} dz$

$= \int_0^2[(6(3-\frac{3}{2}z) - (3-\frac{3}{2}z)^2 - 3(3-\frac{3}{2}z)z]dz$

$= -2(3-\frac{3}{2}z)^2 - \frac{2}{9}(3-\frac{3}{2}z)^3 - \frac{9}{2}z^2 + \frac{3}{2}z^3\Big|_0^2$

$= (-18+12)-(-18+6) = 6$, or 6 cu units.

31. $V = \int_0^1\int_0^{-2x+2}(4-x^2-y^2)dy\,dx = \int_0^1(4y - x^2y - \frac{1}{3}y^3)\Big|_{y=0}^{y=2(1-x)} dx$

$$= \int_0^1 [8(1-x) - 2x^2 + 2x^3 - \tfrac{8}{3}(1-x)^3]\,dx$$
$$= [(8x - 4x^2 - \tfrac{2}{3}x^3 + \tfrac{1}{2}x^4) + \tfrac{2}{3}(1-x)^4]\Big|_0^1$$
$$= (8 - 4 - \tfrac{2}{3} + \tfrac{1}{2}) - \tfrac{2}{3} = \tfrac{19}{6}, \text{ or } \tfrac{19}{6} \text{ cu units.}$$

33. $V = \int_0^2 \int_0^2 5e^{-x-y}\,dx\,dy = \int_0^2 -5e^{-x-y}\Big|_{x=0}^{x=2} dy = \int_0^2 -5(e^{-2-y} - e^{-y})dy$
$$= -5(-e^{-2-y} + e^{-y})\Big|_0^2 = -5(-e^{-4} + e^{-2}) + 5(-e^{-2} + 1) = 5(1 - 2e^{-2} + e^{-4}) \text{ cu units.}$$

35. $V = \int_0^2 \int_0^{2x} (2x + y)dy\,dx = \int_0^2 2xy + \tfrac{1}{2}y^2\Big|_0^{2x} dx = \int_0^2 (4x^2 + 2x^2)\,dx$
$$= \int_0^2 6x^2\,dx = 2x^3\Big|_0^2 = 16, \text{ or } 16 \text{ cu units.}$$

37. $V = \int_0^1 \int_0^{-x+1} e^{x+2y}\,dy\,dx = \int_0^1 \tfrac{1}{2}e^{x+2y}\Big|_{y=0}^{y=-x+1} dx = \tfrac{1}{2}\int_0^1 (e^{-x+2} - e^x)\,dx$
$$= \tfrac{1}{2}(-e^{-x+2} - e^x)\Big|_0^1 = \tfrac{1}{2}(-e - e + e^2 + 1) = \tfrac{1}{2}(e^2 - 2e + 1) = \tfrac{1}{2}(e-1)^2 \text{ cu units.}$$

39. $V = \int_0^4 \int_0^{\sqrt{x}} \frac{2y}{1+x^2}\,dy\,dx = \int_0^4 \frac{y^2}{1+x^2}\Big|_0^{\sqrt{x}} dx = \int_0^4 \frac{x}{1+x^2}\,dx$
$$= \tfrac{1}{2}\ln(1+x^2)\Big|_0^4 = \tfrac{1}{2}(\ln 17 - \ln 1) = \tfrac{1}{2}\ln 17 \text{ cu units.}$$

41. $V = \int_0^4 \int_0^{\sqrt{16-x^2}} x\,dy\,dx = \int_0^4 xy\Big|_{y=0}^{y=\sqrt{16-x^2}} dx = \int_0^4 x(16-x^2)^{1/2}\,dx$
$$= (-\tfrac{1}{2})(\tfrac{2}{3})(16-x^2)^{3/2}\Big|_0^4 = \tfrac{1}{3}(16)^{3/2} = \tfrac{64}{3}.$$

43. $A = \frac{1}{\frac{1}{2}}\int_0^1 \int_0^x (x + 2y)dy\,dx = 2\int_0^1 xy + y^2\Big|_0^x dx = 2\int_0^1 (x^2 + x^2)dx = 4\int_0^1 x^2\,dx$
$$= \frac{4x^3}{3}\Big|_0^1 = \frac{4}{3}.$$

45. The area of R is 1/2. The average value of f is

$$\frac{1}{1/2}\int_0^1\int_0^x e^{-x^2}\,dy\,dx = 2\int_0^1 e^{-x^2}y\Big|_{y=0}^{y=x} dx = 2\int_0^1 xe^{-x^2}\,dx = -e^{-x^2}\Big|_0^1 = -e^{-1}+1 = 1-\frac{1}{e}.$$

47. The area of the region is, by elementary geometry, $[4+\frac{1}{2}(2)(4)]$, or 8 sq units. Therefore, the required average value is

$$A = \tfrac{1}{8}\int_1^3\int_0^{2x} \ln x\,dy\,dx = \tfrac{1}{8}\int_1^3 (\ln x)y\Big|_0^{2x} dx = \tfrac{1}{4}\int_1^3 x\ln x\,dx$$
$$= \tfrac{1}{4}(\tfrac{x^2}{4})(2\ln x - 1)\Big|_1^3 \quad \text{(Integrating by parts)}$$
$$= \tfrac{9}{16}(2\ln 3 - 1) - \tfrac{1}{16}(-1) = \tfrac{1}{8}(9\ln 3 - 4).$$

49. The average population density inside R is $\dfrac{43{,}329}{20} \approx 2166$ people/sq mile.

51. The average weekly profit is

$$\frac{1}{(20)(20)}\int_{100}^{120}\int_{180}^{200}(-0.2x^2 - 0.25y^2 - 0.2xy + 100x + 90y - 4000)\,dx\,dy$$
$$= \frac{1}{400}\int_{100}^{120} -\tfrac{1}{15}x^3 - 0.25y^2x - 0.1x^2y + 50x^2 + 90xy - 4000x\Big|_{x=180}^{x=200} dy$$
$$= \frac{1}{400}\int_{100}^{120}(-144{,}533.33 - 5y^2 - 760y + 380{,}000 + 1800y - 80{,}000)\,dy$$
$$= \frac{1}{400}\int_{100}^{120}(155{,}466.67 - 5y^2 + 1040y)\,dy$$
$$= \frac{1}{400}(155{,}466.67y - \tfrac{5}{3}y^3 + 520y^2)\Big|_{100}^{120}$$
$$= \frac{1}{400}(3{,}109{,}333.40 - 1{,}213{,}333.30 + 2{,}288{,}000) \approx 10{,}460, \text{or } \$10{,}460/\text{wk}.$$

53. True. This result follows from the definition.

55. True. $\iint_R g(x,y)\,dA$ gives the volume of the solid bounded above by the surface $z = g(x,y)$. $\iint_R f(x,y)\,dA$ gives the volume of the solid bounded above by the surface $z = f(x,y)$. Therefore,

$$\iint_R g(x,y)\,dA - \iint_R f(x,y)\,dA = \iint_R [g(x,y) - f(x,y)]\,dA$$

gives the volume of the solid bounded above by $z = g(x,y)$ and below by $z = f(x,y)$.

CHAPTER 17 CONCEPT REVIEW, page 1111

1. xy; ordered pair; real number; $f(x,y)$
3. $z = f(x,y)$; f; surface
5. Fixed number; x
7. $\leq$; (a,b); $\leq$; domain
9. $g(x,y) = 0$; $f(x,y) + \lambda g(x,y)$; $F_x = 0$, $F_y = 0$, $F_\lambda = 0$; extrema
11. Iterated; $\int_3^5 \int_0^1 (2x + y^2)\,dx\,dy$

CHAPTER 17 REVIEW EXERCISES, page 1112

1. $f(0,1) = 0$; $f(1,0) = 0$; $f(1,1) = \dfrac{1}{1+1} = \dfrac{1}{2}$.
   $f(0,0)$ does not exist because the point (0,0) does not lie in the domain of *f*.

3. $h(1,1,0) = 1 + 1 = 2$; $h(-1,1,1) = -e - 1 = -(e + 1)$;
   $h(1,-1,1) = -e - 1 = -(e + 1)$.

5. $D = \{(x,y) \mid y \neq -x\}$

7. The domain of *f* is the set of all ordered triplets (*x*,*y*,*z*) of real numbers such that $z \geq 0$ and $x \neq 1$, $y \neq 1$, and $z \neq 1$.

9. $z = y - x^2$

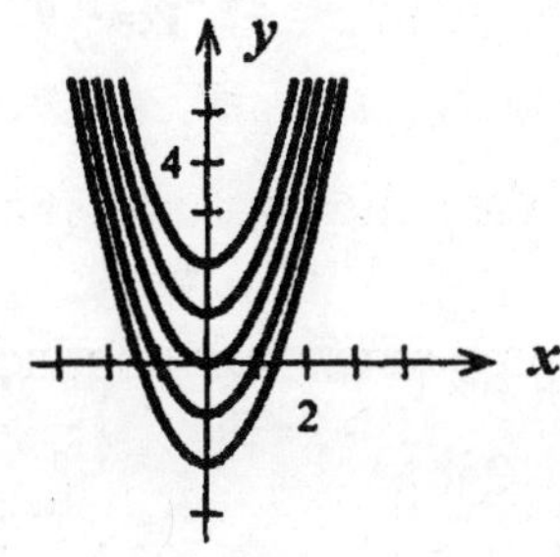

11. $z = e^{xy}$

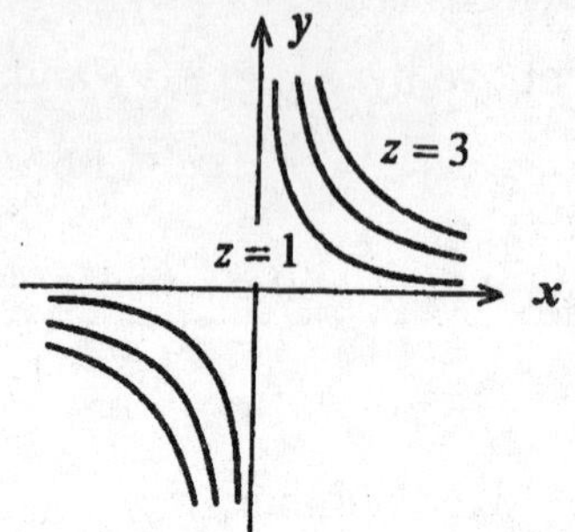

13. $f(x,y) = x\sqrt{y} + y\sqrt{x};\ f_x = \sqrt{y} + \dfrac{y}{2\sqrt{x}};\ f_y = \dfrac{x}{2\sqrt{y}} + \sqrt{x}$

15. $f(x,y) = \dfrac{x-y}{y+2x}.\ f_x = \dfrac{(y+2x)-(x-y)(2)}{(y+2x)^2} = \dfrac{3y}{(y+2x)^2}.$

$f_y = \dfrac{(y+2x)(-1)-(x-y)}{(y+2x)^2} = \dfrac{-3x}{(y+2x)^2}.$

17. $h(x,y) = (2xy+3y^2)^5; h_x = 10y(2xy+3y^2)^4;\ h_y = 10(x+3y)(2xy+3y^2)^4.$

19. $f(x,y) = (x^2+y^2)e^{x^2+y^2};$

$f_x = 2xe^{x^2+y^2} + (x^2+y^2)(2x)e^{x^2+y^2} = 2x(x^2+y^2+1)e^{x^2+y^2}.$

$f_y = 2ye^{x^2+y^2} + (x^2+y^2)(2y)e^{x^2+y^2} = 2y(x^2+y^2+1)e^{x^2+y^2}.$

21. $f(x,y) = \ln\left(1+\dfrac{x^2}{y^2}\right).\ f_x = \dfrac{\frac{2x}{y^2}}{1+\frac{x^2}{y^2}} = \dfrac{2x}{x^2+y^2};\ f_y = \dfrac{-\frac{2x^2}{y^3}}{1+\frac{x^2}{y^2}} = -\dfrac{2x^2}{y(x^2+y^2)}.$

23. $f(x,y) = x^4+2x^2y^2-y^4;\ f_x = 4x^3+4xy^2;\ f_y = 4x^2y-4y^3;$

$f_{xx} = 12x^2+4y^2,\ f_{xy} = 8xy = f_{yx},\ f_{yy} = 4x^2-12y^2.$

25. $g(x,y) = \dfrac{x}{x+y^2};\ g_x = \dfrac{(x+y^2)-x}{(x+y^2)^2} = \dfrac{y^2}{(x+y^2)^2},\ g_y = \dfrac{-2xy}{(x+y^2)^2}.$

Therefore, $g_{xx} = -2y^2(x+y^2)^{-3} = -\dfrac{2y^2}{(x+y^2)^3},$

$$g_{yy} = \frac{(x+y^2)^2(-2x)+2xy(2)(x+y^2)2y}{(x+y^2)^4} = \frac{2x(x^2+y^2)[-x-y^2+4y^2]}{(x+y^2)^4}$$

$$= \frac{2x(3y^2-x)}{(x+y^2)^3}.$$

and $$g_{xy} = \frac{(x+y^2)2y-y^2(2)(x+y^2)2y}{(x+y^2)^4} = \frac{2(x+y^2)[xy+y^3-2y^3]}{(x+y^2)^4}$$

$$= \frac{2y(x-y^2)}{(x+y^2)^3} = g_{yx}.$$

27. $h(s,t) = \ln\left(\frac{s}{t}\right)$. Write $h(s,t) = \ln s - \ln t$. Then $h_s = \frac{1}{s}$, $h_t = -\frac{1}{t}$.

Therefore, $h_{ss} = -\frac{1}{s^2}$, $h_{st} = h_{ts} = 0$, $h_{tt} = \frac{1}{t^2}$.

29. $f(x,y) = 2x^2 + y^2 - 8x - 6y + 4$; To find the critical points of f, we solve the system $\begin{cases} f_x = 4x - 8 = 0 \\ f_y = 2y - 6 = 0 \end{cases}$ obtaining $x = 2$ and $y = 3$. Therefore, the sole critical point of f is (2,3). Next, $f_{xx} = 4, f_{xy} = 0, f_{yy} = 2$. Therefore,

$$D = f_{xx}(2,3)f_{yy}(2,3) - f_{xy}(2,3)^2 = 8 > 0.$$

Since $f_{xx}(2,3) > 0$, we see that $f(2,3) = -13$ is a relative minimum.

31. $f(x,y) = x^3 - 3xy + y^2$. We solve the system of equations $\begin{cases} f_x = 3x^2 - 3y = 0 \\ f_y = -3x + 2y = 0 \end{cases}$ obtaining $x^2 - y = 0$, or $y = x^2$. Then $-3x + 2x^2 = 0$, and $x(2x - 3) = 0$, and $x = 0$, or $x = 3/2$ and $y = 0$, or $y = 9/4$. Therefore, the critical points are (0,0) and $(\frac{3}{2}, \frac{9}{4})$. Next, $f_{xx} = 6x, f_{xy} = -3$, and $f_{yy} = 2$ and $D(x,y) = 12x - 9 = 3(4x - 3)$. Therefore, $D(0,0) = -9$ so (0,0) is a saddle point. $D(\frac{3}{2}, \frac{9}{4}) = 3(6-3) = 9 > 0$, and $f_{xx}(\frac{3}{2}, \frac{9}{4}) > 0$ and therefore, $f(\frac{3}{2}, \frac{9}{4}) = \frac{27}{8} - \frac{81}{8} + \frac{81}{16} = -\frac{27}{16}$ is the relative minimum value.

33. $f(x,y) = f(x,y) = e^{2x^2+y^2}$. To find the critical points of f, we solve the system

$$\begin{cases} f_x = 4xe^{2x^2+y^2} = 0 \\ f_y = 2ye^{2x^2+y^2} = 0 \end{cases}$$

giving (0,0) as the only critical point of f. Next,

$$f_{xx} = 4(e^{2x^2+y^2} + 4x^2e^{2x^2+y^2}) = 4(1+4x^2)e^{2x^2+y^2}$$

$$f_{xy} = 8xye^{2x^2+y^2}$$

$$f_{yy} = 2(1+2y^2)e^{2x^2+y^2}.$$

Therefore, $D = f_{xx}(0,0)f_{yy}(0,0) - f_{xy}^2(0,0) = (4)(2) - 0 = 8 > 0$ and so (0,0) gives a relative minimum of f since $f_{xx}(0,0) > 0$. The minimum value of f is $f(0,0) = e^0 = 1$.

35. We form the Lagrangian function $F(x,y,\lambda) = -3x^2 - y^2 + 2xy + \lambda(2x + y - 4)$. Next, we solve the system

$$\begin{cases} F_x = 6x + 2y + 2\lambda = 0 \\ F_y = -2y + 2x + \lambda = 0. \\ F_\lambda = 2x + y - 4 = 0 \end{cases}$$

Multiplying the second equation by 2 and subtracting the resultant equation from the first equation yields $6y - 10x = 0$ so $y = 5x/3$. Substituting this value of y into the third equation of the system gives $2x + \frac{5}{3}x - 4 = 0$.
So $x = \frac{12}{11}$ and consequently $y = \frac{20}{11}$. So $(\frac{12}{11}, \frac{20}{11})$ gives the maximum value for f subject to the given constraint.

37. The Lagrangian function is $F(x,y,\lambda) = 2x - 3y + 1 + \lambda(2x^2 + 3y^2 - 125)$. Next, we solve the system of equations

$$\begin{cases} F_x = 2 + 4\lambda x = 0 \\ F_y = -3 + 6\lambda y = 0 \\ F_\lambda = 2x^2 + 3y^2 - 125 = 0. \end{cases}$$

Solving the first equation for x gives $x = -1/2\lambda$. The second equation gives $y = 1/2\lambda$. Substituting these values of x and y into the third equation gives

$$2\left(-\frac{1}{2\lambda}\right)^2 + 3\left(\frac{1}{2\lambda}\right)^2 - 125 = 0$$

$$\frac{1}{2\lambda^2} + \frac{3}{4\lambda^2} - 125 = 0$$

$$2+3-500\lambda^2=0, \text{ or } \lambda=\pm\frac{1}{10}.$$

Therefore, $x=\pm 5$ and $y=\pm 5$ and so the critical points of f are $(-5,5)$ and $(5,-5)$. Next, we compute

$$f(-5,5)=2(-5)-3(5)+1=-24.$$
$$f(5,-5)=2(5)-3(-5)+1=26.$$

So f has a maximum value of 26 at $(5,-5)$ and a minimum value of -24 at $(-5,5)$.

39. $$\int_{-1}^{2}\int_{2}^{4}(3x-2y)\,dx\,dy=\int_{-1}^{2}\tfrac{3}{2}x^2-2xy\Big|_{x=2}^{x=4}dy=\int_{-1}^{2}[(24-8y)-(6-4y)]\,dy$$
$$=\int_{-1}^{2}(18-4y)\,dy=(18y-2y^2)\Big|_{-1}^{2}=(36-8)-(-18-2)=48.$$

41. $$\int_0^1\int_{x^3}^{x^2}2x^2y\,dy\,dx=\int_0^1 x^2y^2\Big|_{y=x^3}^{y=x^2}dx=\int_0^1 x^2(x^4-x^6)\,dx$$
$$=\int_0^1(x^6-x^8)\,dx=\frac{x^7}{7}-\frac{x^9}{9}\Big|_0^1=\frac{1}{7}-\frac{1}{9}=\frac{2}{63}.$$

43. $$\int_0^2\int_0^1(4x^2+y^2)\,dy\,dx=\int_0^2 4x^2y+\tfrac{1}{3}y^3\Big|_{y=0}^{y=1}dx=\int_0^2(4x^2+\tfrac{1}{3})\,dx$$
$$=(\tfrac{4}{3}x^3+\tfrac{1}{3}x)\Big|_0^2=\tfrac{32}{3}+\tfrac{2}{3}=\tfrac{34}{3}.$$

45. The area of R is
$$\int_0^2\int_{x^2}^{2x}dy\,dx=\int_0^2 y\Big|_{y=x^2}^{y=2x}dx=\int_0^2(2x-x^2)\,dx=(x^2-\tfrac{1}{3}x^3)\Big|_0^2=\tfrac{4}{3}.$$
Then
$$AV=\frac{1}{4/3}\int_0^2\int_{x^2}^{2x}(xy+1)\,dy\,dx=\frac{3}{4}\int_0^2\frac{xy^2}{2}+y\Big|_{x^2}^{2x}dx$$
$$=\tfrac{3}{4}\int_0^2(-\tfrac{1}{2}x^5+2x^3-x^2+2x)\,dx=\tfrac{3}{4}(-\tfrac{1}{12}x^6+\tfrac{1}{2}x^4-\tfrac{1}{3}x^3+x^2)\Big|_0^2$$
$$=\tfrac{3}{4}(-\tfrac{16}{3}+8-\tfrac{8}{3}+4)=3.$$

47. $f(p,q)=900-9p-e^{0.4q}$; $g(p,q)=20{,}000-3000q-4p$.
We compute $\dfrac{\partial f}{\partial q}=-0.4e^{0.4q}$ and $\dfrac{\partial g}{\partial p}=-4$. Since $\dfrac{\partial f}{\partial q}<0$ and $\dfrac{\partial g}{\partial p}<0$ for all $p>0$ and $q>0$, we conclude that compact disc players and audio discs are complementary commodities.

49. Refer to the following diagram.

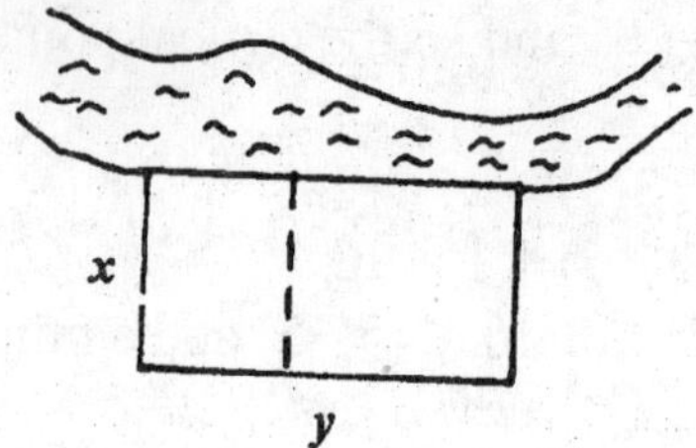

We want to minimize $C(x,y) = 3(2x) + 2(x) + 3y = 8x + 3y$ subject to $xy = 303{,}750$. The Lagrangian function is

$$F(x,y,\lambda) = 8x + 3y + \lambda(xy - 303{,}750).$$

Next, we solve the system

$$\begin{cases} F_x = 8 + \lambda y = 0 \\ F_y = 3 + \lambda x = 0 \\ F_\lambda = xy - 303{,}750 = 0 \end{cases}.$$

Solving the first equation for y gives $y = -8/\lambda$. The second equation gives $x = -3/\lambda$. Substituting this value into the third equation gives

$$\left(-\frac{3}{\lambda}\right)\left(-\frac{8}{\lambda}\right) = 303{,}750 \text{ or } \lambda^2 = \frac{24}{303{,}750} = \frac{4}{50{,}625},$$

or $\lambda = \pm\frac{2}{225}$. Therefore, $x = 337.5$ and $y = 900$ and so the required dimensions of the pasture are 337.5 yd by 900 yd.

CHAPTER 17 BEFORE MOVING ON, page 1114

1. We have the constraints $x \geq 0$, $y \geq 0$, $x \neq 1$ and $y \neq 2$. Therefore the domain of f is $D = \{(x, y) | x \geq 0,\ y \geq 0;\ x \neq 1 \text{ and } y \neq 2\}$.

2. $f_x = 2xy + ye^{xy}$, $f_y = x^2 + xe^{xy}$, $f_{xx} = 2y + y^2e^{xy}$, $f_{xy} = 2x + (1 + xy)e^{xy} = f_{yx}$
$f_{yy} = x \cdot xe^{xy} = x^2e^{xy}$

3.
$$\left.\begin{aligned} f_x &= 6x^2 + 6y = 6(x^2 - y^2) = 0 \\ f_y &= 6y^2 - 6x = 6(y^2 - x) = 0 \end{aligned}\right\} \text{ gives } y = x^2 \text{ and } x = y^2. \text{ Therefore,}$$

$x = x^4$, $x^4 - x = x(x^3 - 1) = 0$ giving $x = 0$ or 1. The critical points of f are (0,.0) and (1,1). $f_{xx} = 12x$, $f_{xy} = -6$, $f_{yy} = 12y$.

$D(x, y) = 144x^2 + 144y^2 - 36$; $D(0,0) = -36 < 0$; and so (0,0) does not yield a relative extrema. $D(1,1) = 252 > 0$ and $f_{xx}(1,1) = 12 > 0$, and so $f(1,1)$ yields a relative minimum. $f(1,1) = 2(1)^3 + 2(1)^3 - 6(1)(1) - 5 = -7$

4. $F(x, y, \lambda) = 3x^2 + 3y^2 + 1 + \lambda(x + y - 1)$
   $F_x = 6x + \lambda = 0$
   $F_y = 6y + \lambda = 0$
   $F_\lambda = x + y - 1 = 0$
   Gives $\lambda = -6x - 6y$ so $y = x$. Substituting into the third equation gives $2x = 1$ or $x = \frac{1}{2}$ and $y = \frac{1}{2}$. Therefore, $(\frac{1}{2}, \frac{1}{2}, \frac{5}{2})$ is the required minimum.

5. $$\iint_R (1 - xy)\, dA = \int_0^1 \int_{x^2}^x (1 - xy) dy\, dx$$
   $$= \int_0^1 [(y - \tfrac{1}{2}xy^2)\Big|_{x^2}^x] dx$$
   $$= \int_0^1 [x - \tfrac{1}{2}x^3 - x^2 + \tfrac{1}{2}x^5] dx = \tfrac{1}{2}x^2 - \tfrac{1}{8}x^4 - \tfrac{1}{3}x^3 + \tfrac{1}{12}x^6\Big|_0^1$$
   $$= \tfrac{1}{2} - \tfrac{1}{8} - \tfrac{1}{3} + \tfrac{1}{12} = \tfrac{1}{8}$$

each faculty member; and the development of a standardized evaluation form to rank faculty members.

The initial step of any planned change is to assess the present situation and to identify any deficiencies. First, the existing faculty evaluation system was reviewed. In that system, as explained before, each faculty member was expected to submit an annual activity report outlining his or her goals and professional accomplishments. Although there was no standardized form, the annual report was expected to include: progress toward the accomplishments of professional goals from previous years' annual report, teaching activities and evaluations, scholarship, university and community service, and projected goals for the upcoming year.

The department head then evaluated the annual report. The department head generally used the three areas of teaching, research, and service to establish rankings. The faculty received feedback regarding his or her performance via the next annual contract that reflected the new salary.

This system provided no formal two-way communication between the department head and individual faculty member. It is perceived that the increased communication between the individual faculty member and the department head would increase the ability of the faculty member to understand the goals of the department as well as understand the "value added" of each faculty member in relation to the merit-based pay decision process.

DESIGN AND IMPLEMENTATION OF THE NEW EVALUATION SYSTEM

The efforts at this step were turned on the creation and design of the evaluation instrument, and a process for conducting the faculty appraisals. The evaluation process that has been developed has two major components: the "annual agreement regarding duties and responsibilities" and the "annual faculty evaluation form." (See appendices 1 and 2)

The Annual Agreement is essentially a contract between the faculty member and the department head, which clearly sets forth the criteria that will be used to evaluate the performance of a faculty member and determine merit salary increases. The purpose of this contract is to recognize the diverse interests and abilities of each faculty member as well as to allow the department head to accurately assess whether or not various contracts with the faculty, in aggregate, meet the goals of the department.

The second document focuses on the accountability of each faculty member. In essence, the evaluation form would objectively evaluate the "value added" by each faculty member. This form also provides an exchange of mutual expectations between a faculty member and the department head.

A five level rating scale of excellent, exceeds expectations, meets all expectations, marginal, and unacceptable are to be used to evaluate each of these areas of responsibility. These ratings follow the management performance appraisal of a Fortune 500 company. The proposed instrument also requires a summary rating by both the participating faculty member as well as the department head in assessing the major areas of responsibilities. Similar to a typical performance appraisal in corporate America, a career development section is added to designate if a faculty member does intend to apply for promotion, sabbatical, or grants. The signature section assures that a conference and review has taken place between the faculty member and department head. It should be noted that the signatures are not intended to indicate agreement with the evaluation.

This system was implemented in the spring of 2003 and was used to evaluate twenty two faculty. In general, everybody has been receptive. The effects of this system and how it will help faculty to continuously improve will be discussed at a later paper. The effect of this system on overall continuous improvement of the programs will also be studied at a later time.

CONCLUSIONS AND OBSERVATIONS

The performance plan designed in this paper consists of two major components – The Annual Agreement Regarding Duties and Responsibilities and The Annual Faculty Evaluation. The Annual Agreement is essentially a contract, which clearly sets forth the criteria that will be used to evaluate the performance of a faculty member and determines merit salary increases. The Annual Faculty Evaluation focuses on the accountability of each faculty member in the areas of teaching, discovery, and engagement. This form will be used at the end of the evaluation period and rate the overall performance of the faculty member.

This process was used in the spring of 2003. Although it is a time consuming process, in general, all faculty members seem to be in favor of it. This process follows all other assessment process being done, for quality control, at the program level. The initial results show that a few minor changes need to be made to this document. One change is the period of evaluation. It was quickly observed that with the evaluation being done in March, it is better to have the evaluation period to extend to the end of February instead of the end of the calendar year. By the time the conferences between department head and faculty are done, over 30% of the new evaluation period has expired, not giving enough time to faculty to adjust. The overall effect of this process as well as future changes and observations will be reported in a follow up paper.

REFERENCES

[1] Caruth, Donald L. and Gail D. Handlogten (1997) *Staffing the Contemporary Organizations,* 2nd ed., Westport: Praeger Publishers, pp. 230-231

[2] Schermerhorn, Jr, John R., James G Hunt, and Richard N. Osborn (2000) *Organizational Behavior,* 7th ed., New York: John Wiley and Sons, Inc. p. 400.

[3] Society For Human Resource Management and Personnel Decisions International (2000) *"Performance Management Survey,"*http://www.shrm.org/surveys, Feb. 6, 2001.

[4] Lucas, John, J., Lori S. Feldman, and Philip H. Empey *"A Corporate Human Resource Approach to Evaluating University*

Faculty: Lessons in Development and Communication." Academy of Educational Leadership Journal, Vol. 5, Number 2, 2001.

APPENDIX 1

Department of Manufacturing Engineering Technologies & Supervision

ANNUAL AGREEMENT REGARDING DUTIES AND RESPONSIBILITIES

Name: ____________________ Rank: ____________________

The following is a statement of expectations based upon value added to the Department regarding your work activities for the upcoming year. The percentage to the right indicates the level of activities agreed upon and the weight that should be given each area in determining merit increases. The percentage of time spent must be at least the minimum for each activity as indicated below. With the Department Head, you will agree to the estimate of time spent for each of the activities listed below. Upon agreement, you and the Department Head will sign this document. A copy of this document will be provided to you. At your annual evaluation, the Department Head will review this document with you and discuss how effectively you have met your mutually agreed upon goals.

I. Teaching Activities: **Weight for Assessment: __________ (40% minimum)**

You have stated that you would like to contribute to the Department in the area of teaching. Your contribution will be evaluated by the courses taught, course/curriculum development, preparation and/or publication of instructional material, student evaluation, and special activities that have enhanced teaching effectiveness.

II. Publications and Scholarship activities: **Weight for Assessment: __________ (15% minimum)**

You have stated that you would like to contribute to the Department in the area of scholarship. Your contributions will be evaluated by articles in refereed journals, completed books, articles in non-refereed journals, manuscripts in progress and/or accepted for publication, book reviews, papers presented at technical and professional journals/societies, research grants and awards, and other scholarship that enhance the University's reputation through your scholarship.

III. Service to University Community, and Profession **Weight for Assessment: __________ (15% minimum)**

You have stated that you would like to contribute to the Department in the area of service to the University, Community, and the Profession. Your contributions will be evaluated by committee service at department, campus, and university levels, student advising, government, industry, or agency service, professional organization service, including dates and offices held, membership in professional societies, leadership role in workshops/conferences, and other activities that enhance the University's reputation through professional service.

The signatures below signify mutually agreed upon activities for the next assessment year.

Faculty Member: **Date:**

Department Head: **Date:**

APPENDIX 2

DEPARTMENT OF MANUFACTURING ENGINEERING TECHNOLOGIES & SUPERVIS ION
Purdue University Calumet

ANNUAL FACULTY EVALUATION FORM
Year: ________

Name of Faculty Member: ________________________________

Academic Rank: ________________________________

Directions:

The intent of this annual faculty evaluation form is to objectively assess the "value added" by each full-time faculty member to the Department.

This form is to be used as follows:

1. The faculty member is to complete appropriate sections and then forward the form to the Department Head by the second Friday of March. The evaluation period will be the "Calendar Year" previous to that March. Faculty members will submit their annual report, students' evaluation of faculty, annual agreement from prior year and other appropriate documents, which will be helpful in consideration of the performance.

2. The Department Head will review the submitted form and supporting documents. Additionally, the Department Head shall complete appropriate sections, and thereafter shall schedule a conference with the faculty member, at which time the completed form will be discussed. This conference should take place before mid-April.

3. At the conclusion of this conference, The faculty member will be asked to sign the form. The faculty member's signature signifies that he/she has discussed their performance with the Department Head and has also seen the Department Head's comments but not necessarily that he/she agrees with all of them. The Annual Agreement regarding Duties and Responsibilities will also be discussed and signed for the next assessment year. The faculty member may respond to the department head's evaluation in writing no later than 2 weeks from the date of his/her evaluation conference.

4. One copy of the evaluation form will be given to the faculty member and one form will be placed in the faculty member's personnel file.

Disclaimer: This document and process is used only for performance evaluation and career development. This evaluation instrument is not intended to be used as part of the tenure process.

Teaching Ability and Effectiveness
Potential subjects for assessment: Number and levels of Courses taught, student evaluation, contributions in Course/curriculum development, preparation and/or publication of instructional materials, and special activities that have enhanced teaching effectiveness. More activities could be initiated and included by faculty member.

Faculty Member's Assessment/Comments

Overall Rating For Teaching

________Excellent
________Exceeds Expectations
________Meets all Expectations
________Marginal
________Unacceptable
